Methods in Enzymology

Volume 203

MOLECULAR DESIGN AND MODELING: CONCEPTS AND APPLICATIONS

Part B

Antibodies and Antigens, Nucleic Acids, Polysaccharides, and Drugs

METHODS IN ENZYMOLOGY

EDITORS-IN-CHIEF

John N. Abelson   Melvin I. Simon

DIVISION OF BIOLOGY
CALIFORNIA INSTITUTE OF TECHNOLOGY
PASADENA, CALIFORNIA

FOUNDING EDITORS

Sidney P. Colowick and Nathan O. Kaplan

Methods in Enzymology

Volume 203

Molecular Design and Modeling: Concepts and Applications

Part B

Antibodies and Antigens, Nucleic Acids, Polysaccharides, and Drugs

EDITED BY

John J. Langone

MOLECULAR BIOLOGY BRANCH
DIVISION OF LIFE SCIENCES
CENTER FOR DEVICES AND RADIOLOGICAL HEALTH
FOOD AND DRUG ADMINISTRATION
ROCKVILLE, MARYLAND

ACADEMIC PRESS, INC.
Harcourt Brace Jovanovich, Publishers
San Diego New York Boston
London Sydney Tokyo Toronto

This book is printed on acid-free paper. ∞

Academic Press, Inc.
San Diego, California 92101

United Kingdom Edition published by
ACADEMIC PRESS LIMITED
24-28 Oval Road, London NW1 7DX

Library of Congress Catalog Card Number: 54-9110

ISBN 0-12-182104-8 (alk. paper)

PRINTED IN THE UNITED STATES OF AMERICA
91 92 93 94 9 8 7 6 5 4 3 2 1

Table of Contents

Section I. Antibodies and Antigens

A. Antibody Combining Site

B. Antibody–Antigen Interactions

C. Catalytic Antibodies and Vaccines

Section II. Nucleic Acids and Polysaccharides

Section III. Drugs

Section IV. Cross-Index to Prior Volumes

Contributors to Volume 203

Article numbers are in parentheses following the names of contributors.
Affiliations listed are current.

JEFFREY W. ALMOND (20), *Department of Microbiology, University of Reading, Reading RG1 5AQ, England*

JACOB ANGLISTER (10), *Department of Polymer Research, The Weizmann Institute of Science, Rehovot 76100, Israel*

ROBERT E. BIRD (4), *Molecular Oncology, Inc., Gaithersburg, Maryland, 20878*

NICOLAS BOISSET (13), *Laboratoire de Biochimie, Faculté de Pharmacie, Université François Rabelais, 37042 Tours Cedex, France*

MICHAEL B. BOLGER (2), *Department of Pharmaceutical Sciences, School of Pharmacy, University of Southern California, Los Angeles, California 90033*

JONATHAN J. BURBAUM (25), *Department of Biology, Massachusetts Institute of Technology, Cambridge, Massachusetts 02139*

ALAN D. CARDIN (28), *Department of Cancer Biology, Marion Merrill Dow, Inc., Cincinnati, Ohio 45215*

PIERRE-ALAIN CARRUPT (31), *School of Pharmacy, University of Lausanne, CH-1005 Lausanne, Switzerland*

JANET C. CHEETHAM (6, 9), *AMGEN, Thousand Oaks, California 91320*

DAVID A. DEMETER (28), *Department of Theoretical Chemistry, Marion Merrill Dow, Inc., Cincinnati, Ohio 45215*

CHRISTOPHER M. DOBSON (9), *Laboratory of Molecular Biophysics, University of Oxford, Oxford OX1 3QU, England*

DAVID J. EVANS (20), *Department of Microbiology, University of Reading, Reading RG1 5AQ, England*

GAIL G. FIESER (7), *Department of Molecular Biology, Research Institute of Scripps Clinic, La Jolla, California 92037*

DAVED H. FREMONT (7), *Department of Molecular Biology, Research Institute of Scripps Clinic, La Jolla, California 92037*

STEVEN A. FULLER (16), *ADI Diagnostics, Inc., Rexdale, Ontario M9W 4Z7, Canada*

ROBERT E. GRIEST (9), *Department of Biochemistry, University of Bath, Bath BA2 7AY, England*

T. M. GUND (32) *Department of Chemistry, Chemical Engineering, and Environmental Science, New Jersey Institute of Technology, Newark, New Jersey 07102*

DETLEF GÜSSOW (5), *Medical Research Council Collaborative Centre, London NW7 1AD, England*

MARIA GUTTINGER (19), *Pharma Research Technology, F. Hoffmann-La Roche, Ltd., CH-4002 Basel, Switzerland*

EDGAR HABER (3), *Massachusetts General Hospital, Harvard Medical School, Boston, Massachusetts 02114, and The Bristol-Myers Squibb Pharmaceutical Research Institute, Princeton, New Jersey 08543*

KENNETH W. HILL (18), *Departments of Chemistry and of Molecular Biology, Research Institute of Scripps Clinic, La Jolla, California 92037*

DONALD HILVERT (18), *Departments of Chemistry and of Molecular Biology, Research Institute of Scripps Clinic, La Jolla, California 92037*

MERLIN E. HOWDEN (15), *Department of Biological Sciences, Deakin University, Geelong, 3217 Victoria, Australia*

JAMES S. HUSTON (3),*Creative BioMolecules, Inc., Hopkinton, Massachusetts 01748*

SIRKKU JAAKKOLA (24), *Department of Medical Biochemistry, University of Turku, 20520 Turku, Finland*

RICHARD L. JACKSON (28), *Department of Research Sciences, Marion Merrill Dow, Inc., Cincinnati, Ohio 45215*

TERENCE C. JENKINS (22), *Cancer Research Campaign Biomolecular Structure Unit,*

The Institute of Cancer Research, Sutton, Surrey SM2 5NG, England

JOYCE E. JENTOFT (12), *Department of Biochemistry, School of Medicine, Case Western Reserve University, Cleveland, Ohio 44106*

SYD JOHNSON (4), *MedImmune Inc., Gaithersburg, Maryland 20874*

ELVIN A. KABAT (1), *Departments of Microbiology, Genetics and Development, and of Neurology, College of Physicians and Surgeons, Columbia University, New York, New York 10032*

LAWRENCE KAHAN (14), *Department of Physiological Chemistry, University of Wisconsin—Madison, Madison, Wisconsin 53706*

ANDERS KARLÉN (31), *School of Pharmacy, University of Lausanne, CH-1005 Lausanne, Switzerland*

JEAN N. LAMY (13), *Laboratoire de Biochimie, Faculté de Pharmacie, Université François Rabelais, 37042 Tours Cedex, France*

DAVID J. LIVINGSTONE (30), *Department of Medicinal Chemistry, SmithKline Beecham Research, Welwyn, Hertfordshire AL6 9AR, England*

EDWARD L. LOECHLER (23), *Department of Biology, Boston University, Boston, Massachusetts 02215*

ANDREW C. R. MARTIN (6), *Department of Biochemistry, University of Bath, Bath BA2 7AY, England*

YVONNE CONNOLLY MARTIN (29), *Computer Assisted Molecular Design Project, Pharmaceutical Products Division, Abbott Laboratories, Abbott Park, Illinois 60064*

JOHN MCCARTNEY (3), *Creative BioMolecules, Inc., Hopkinton, Massachusetts 01748*

MEREDITH MUDGETT-HUNTER (3), *Massachusetts General Hospital, Harvard Medical School, Boston, Massachusetts 02114, and The Bristol-Myers Squibb Pharmaceutical Research Institute, Princeton, New Jersey 08543*

FRED NAIDER (10), *Department of Chemistry, College of Staten Island, City University of New York, Staten Island, New York 10301*

STEPHEN NEIDLE (22), *Cancer Research Campaign Biomolecular Structure Unit, The Institute of Cancer Research, Sutton, Surrey SM2 5NG, England*

WILMA K. OLSON (21), *Department of Chemistry, Rutgers, The State University of New Jersey, New Brunswick, New Jersey 08903*

HERMANN OPPERMANN (3), *Creative BioMolecules, Inc., Hopkinton, Massachusetts 01748*

NORMAN R. PACE (26), *Department of Biology, Indiana University, Bloomington, Indiana 47405*

EDUARDO A. PADLAN (1), *Laboratory of Molecular Biology, National Institute of Diabetes and Digestive and Kidney Diseases, National Institutes of Health, Bethesda, Maryland 20892*

J. L. PELLEQUER (8), *Institut de Biologie Moléculaire et Cellulaire, CNRS, Strasbourg 67000, France*

JUHA PELTONEN (24), *Department of Medical Biochemistry, University of Turku, 20520 Turku, Finland*

SERGE PÉREZ (27), *Laboratoire de Physicochimie des Macromolécules, Institut National de la Recherche Agronomique, 44026 Nantes, France*

J. RICHARD L. PINK (19), *Pharma Research Technology, F. Hoffmann-La Roche, Ltd., CH-4002 Basel, Switzerland*

DANIEL P. RALEIGH (9), *Laboratory of Molecular Biophysics, University of Oxford, Oxford OX1 3QU, England*

M. RANCE (11), *Department of Molecular Biology, Research Institute of Scripps Clinic, La Jolla, California 92037*

CHRISTINA REDFIELD (9), *Laboratory of Molecular Biophysics, University of Oxford, Oxford OX1 3QU, England*

ANTHONY R. REES (6, 9), *Department of Biochemistry, University of Bath, Bath BA2 7AY, England*

JAMES M. RINI (7), *Department of Molecular Biology, Research Institute of Scripps Clinic, La Jolla, California 92037*

PAOLA ROMAGNOLI (19), *Immunology Laboratory, National Institute of Allergy and Infectious Diseases, Bethesda, Maryland 20892*

PAUL SCHIMMEL (25), *Department of Biology, Massachusetts Institute of Technology, Cambridge, Massachusetts 02139*

PETER G. SCHULTZ (17), *Department of Chemistry, University of California, Berkeley, Berkeley, California 94720*

GERHARD SEEMANN (5), *Research Laboratories of Behringwerke AG, D-3550 Marburg, Germany*

MARK A. SHERMAN (2), *Physical Biochemistry Section, Beckman Research Institute of The City of Hope, Duarte, California 91010*

KEVAN M. SHOKAT (17), *Department of Chemistry, University of California, Berkeley, Berkeley, California 94720*

FRANCESCO SINIGAGLIA (19), *Pharma Research Technology, F. Hoffmann-La Roche, Ltd., CH-4002 Basel, Switzerland*

C. E. SPIVAK (32), *Addiction Research Center, National Institute of Drug Abuse, Baltimore, Maryland 21224*

ENRICO A. STURA (7), *Department of Molecular Biology, Research Institute of Scripps Clinic, La Jolla, California 92037*

WAN-JR SYU (14), *Department of Microbiology, National Yang-Ming Medical College, Taipei, 11221 Taiwan, Republic of China*

MEI-SHENG TAI (3), *Creative BioMolecules, Inc., Hopkinton, Massachusetts 01748*

BELA TAKACS (19), *Pharma Research Technology, F. Hoffmann-La Roche, Ltd., CH-4002 Basel, Switzerland*

MIYOKO TAKAHASHI (16), *ADI Diagnostics, Inc., Rexdale, Ontario M9W 4Z7, Canada*

NABIL EL TAYAR (31), *School of Pharmacy, University of Lausanne, CH-1005 Lausanne, Switzerland*

BERNARD TESTA (31), *School of Pharmacy, University of Lausanne, CH-1005 Lausanne, Switzerland*

P. TSANG (11), *Department of Molecular Biology, Research Institute of Scripps Clinic, La Jolla, California 92037*

JOUNI UITTO (24), *Department of Dermatology, Jefferson Medical College, Philadelphia, Pennsylvania 19107*

M. H. V. VAN REGENMORTEL (8), *Institut de Biologie Moléculaire et Cellulaire, CNRS, Strasbourg 67000, France*

BRADLEY J. WALSH (15), *Centre for Immunology, St. Vincent's Hospital, Darlinghurst, 2010 New South Wales, Australia*

FREDERICK WARREN (3), *Creative BioMolecules, Inc., Hopkinton, Massachusetts 01748*

DAVID S. WAUGH (26), *Department of Biology, Massachusetts Institute of Technology, Cambridge, Massachusetts 02139*

HERSCHEL J. R. WEINTRAUB (28), *Department of Theoretical Chemistry, Marion Merrill Dow, Inc., Cincinnati, Ohio 45215*

E. WESTHOF (8), *Institut de Biologie Moléculaire et Cellulaire, CNRS, Strasbourg 67000, France*

IAN A. WILSON (7), *Department of Molecular Biology, Research Institute of Scripps Clinic, La Jolla, California 92037*

SCOTT WINSTON (16), *Univax Biologics, Inc., Rockville, Maryland 20852*

P. E. WRIGHT (11), *Department of Molecular Biology, Research Institute of Scripps Clinic, La Jolla, California 92037*

PEISEN ZHANG (21), *Department of Chemistry, Rutgers, The State University of New Jersey, New Brunswick, New Jersey 08903*

Preface

The construction of molecules based on a rational design that includes structural features representing binding and/or functional sites has evolved into a new era as a result of a greater understanding of how acceptors bind to ligands and the availability of powerful analytical techniques and methods of synthesis derived mainly from molecular biology. Powerful computer hardware and increasingly friendly software that can be used by experimentalists as well as by those with a more theoretical background have fostered an unparalleled level of research interest at a time when chemical and biological properties that not long ago were essentially the sole domain of naturally occurring products can now be incorporated into specifically designed macromolecules and low molecular weight drugs for use as biochemical reagents, biomedical materials, or in medical devices.

Volumes 202 and 203 of *Methods in Enzymology* include topics in which molecular modeling and applications of design principles and techniques are covered from both methodological and didactic approaches. Computer-based design and modeling, computational approaches, and instrumental methods for elucidating molecular mechanisms of protein folding and ligand–acceptor interactions are included as are genetic and chemical methods for the production of functional molecules, including antibodies and antigens, enzymes, receptors, nucleic acids and polysaccharides, and drugs.

The aim of these volumes is to present mainstream concepts and methodology and to give an up-to-date view of the subject matter. This is with the understanding that new developments appear constantly, and that it would not be prudent or desirable to attempt to include all available and often more specialized or limited techniques.

At the end of each of these volumes is a Cross-Index to Prior Volumes in which related chapters from previously published volumes of *Methods in Enzymology* are listed. Each list serves as a reference guide and allows us to avoid repeated coverage of topics presented in these earlier volumes.

Continued collaboration with John Abelson, Mel Simon, and our colleagues at Academic Press is gratefully acknowledged. A special note of thanks is due Dina Langone for expertly managing the many office tasks involved in assembling these volumes.

JOHN J. LANGONE

METHODS IN ENZYMOLOGY

Volume XXVII. Enzyme Structure (Part D)
Edited by C. H. W. Hirs and Serge N. Timasheff

Volume XXVIII. Complex Carbohydrates (Part B)
Edited by Victor Ginsburg

Volume XXIX. Nucleic Acids and Protein Synthesis (Part E)
Edited by Lawrence Grossman and Kivie Moldave

Volume XXX. Nucleic Acids and Protein Synthesis (Part F)
Edited by Kivie Moldave and Lawrence Grossman

Volume XXXI. Biomembranes (Part A)
Edited by Sidney Fleischer and Lester Packer

Volume XXXII. Biomembranes (Part B)
Edited by Sidney Fleischer and Lester Packer

Volume XXXIII. Cumulative Subject Index Volumes I–XXX
Edited by Martha G. Dennis and Edward A. Dennis

Volume XXXIV. Affinity Techniques (Enzyme Purification: Part B)
Edited by William B. Jakoby and Meir Wilchek

Volume XXXV. Lipids (Part B)
Edited by John M. Lowenstein

Volume XXXVI. Hormone Action (Part A: Steroid Hormones)
Edited by Bert W. O'Malley and Joel G. Hardman

Volume XXXVII. Hormone Action (Part B: Peptide Hormones)
Edited by Bert W. O'Malley and Joel G. Hardman

Volume XXXVIII. Hormone Action (Part C: Cyclic Nucleotides)
Edited by Joel G. Hardman and Bert W. O'Malley

Volume XXXIX. Hormone Action (Part D: Isolated Cells, Tissues, and Organ Systems)
Edited by Joel G. Hardman and Bert W. O'Malley

VOLUME XL. Hormone Action (Part E: Nuclear Structure and Function)
Edited by BERT W. O'MALLEY AND JOEL G. HARDMAN

VOLUME XLI. Carbohydrate Metabolism (Part B)
Edited by W. A. WOOD

VOLUME XLII. Carbohydrate Metabolism (Part C)
Edited by W. A. WOOD

VOLUME XLIII. Antibiotics
Edited by JOHN H. HASH

VOLUME XLIV. Immobilized Enzymes
Edited by KLAUS MOSBACH

VOLUME XLV. Proteolytic Enzymes (Part B)
Edited by LASZLO LORAND

VOLUME XLVI. Affinity Labeling
Edited by WILLIAM B. JAKOBY AND MEIR WILCHEK

VOLUME XLVII. Enzyme Structure (Part E)
Edited by C. H. W. HIRS AND SERGE N. TIMASHEFF

VOLUME XLVIII. Enzyme Structure (Part F)
Edited by C. H. W. HIRS AND SERGE N. TIMASHEFF

VOLUME XLIX. Enzyme Structure (Part G)
Edited by C. H. W. HIRS AND SERGE N. TIMASHEFF

VOLUME L. Complex Carbohydrates (Part C)
Edited by VICTOR GINSBURG

VOLUME LI. Purine and Pyrimidine Nucleotide Metabolism
Edited by PATRICIA A. HOFFEE AND MARY ELLEN JONES

VOLUME LII. Biomembranes (Part C: Biological Oxidations)
Edited by SIDNEY FLEISCHER AND LESTER PACKER

VOLUME LIII. Biomembranes (Part D: Biological Oxidations)
Edited by SIDNEY FLEISCHER AND LESTER PACKER

Volume 80. Proteolytic Enzymes (Part C)
Edited by Laszlo Lorand

Volume 81. Biomembranes (Part H: Visual Pigments and Purple Membranes, I)
Edited by Lester Packer

Volume 82. Structural and Contractile Proteins (Part A: Extracellular Matrix)
Edited by Leon W. Cunningham and Dixie W. Frederiksen

Volume 83. Complex Carbohydrates (Part D)
Edited by Victor Ginsburg

Volume 84. Immunochemical Techniques (Part D: Selected Immunoassays)
Edited by John J. Langone and Helen Van Vunakis

Volume 85. Structural and Contractile Proteins (Part B: The Contractile Apparatus and the Cytoskeleton)
Edited by Dixie W. Frederiksen and Leon W. Cunningham

Volume 86. Prostaglandins and Arachidonate Metabolites
Edited by William E. M. Lands and William L. Smith

Volume 87. Enzyme Kinetics and Mechanism (Part C: Intermediates, Stereochemistry, and Rate Studies)
Edited by Daniel L. Purich

Volume 88. Biomembranes (Part I: Visual Pigments and Purple Membranes, II)
Edited by Lester Packer

Volume 89. Carbohydrate Metabolism (Part D)
Edited by Willis A. Wood

Volume 90. Carbohydrate Metabolism (Part E)
Edited by Willis A. Wood

Volume 91. Enzyme Structure (Part I)
Edited by C. H. W. Hirs and Serge N. Timasheff

VOLUME 92. Immunochemical Techniques (Part E: Monoclonal Antibodies and General Immunoassay Methods)
Edited by JOHN J. LANGONE AND HELEN VAN VUNAKIS

VOLUME 93. Immunochemical Techniques (Part F: Conventional Antibodies, Fc Receptors, and Cytotoxicity)
Edited by JOHN J. LANGONE AND HELEN VAN VUNAKIS

VOLUME 94. Polyamines
Edited by HERBERT TABOR AND CELIA WHITE TABOR

VOLUME 95. Cumulative Subject Index Volumes 61–74, 76–80
Edited by EDWARD A. DENNIS AND MARTHA G. DENNIS

VOLUME 96. Biomembranes [Part J: Membrane Biogenesis: Assembly and Targeting (General Methods; Eukaryotes)]
Edited by SIDNEY FLEISCHER AND BECCA FLEISCHER

VOLUME 97. Biomembranes [Part K: Membrane Biogenesis: Assembly and Targeting (Prokaryotes, Mitochondria, and Chloroplasts)]
Edited by SIDNEY FLEISCHER AND BECCA FLEISCHER

VOLUME 98. Biomembranes [Part L: Membrane Biogenesis (Processing and Recycling)]
Edited by SIDNEY FLEISCHER AND BECCA FLEISCHER

VOLUME 99. Hormone Action (Part F: Protein Kinases)
Edited by JACKIE D. CORBIN AND JOEL G. HARDMAN

VOLUME 100. Recombinant DNA (Part B)
Edited by RAY WU, LAWRENCE GROSSMAN, AND KIVIE MOLDAVE

VOLUME 101. Recombinant DNA (Part C)
Edited by RAY WU, LAWRENCE GROSSMAN, AND KIVIE MOLDAVE

VOLUME 102. Hormone Action (Part G: Calmodulin and Calcium-Binding Proteins)
Edited by ANTHONY R. MEANS AND BERT W. O'MALLEY

VOLUME 103. Hormone Action (Part H: Neuroendocrine Peptides)
Edited by P. MICHAEL CONN

VOLUME 104. Enzyme Purification and Related Techniques (Part C)
Edited by WILLIAM B. JAKOBY

VOLUME 105. Oxygen Radicals in Biological Systems
Edited by LESTER PACKER

VOLUME 106. Posttranslational Modifications (Part A)
Edited by FINN WOLD AND KIVIE MOLDAVE

VOLUME 107. Posttranslational Modifications (Part B)
Edited by FINN WOLD AND KIVIE MOLDAVE

VOLUME 108. Immunochemical Techniques (Part G: Separation and Characterization of Lymphoid Cells)
Edited by GIOVANNI DI SABATO, JOHN J. LANGONE, AND HELEN VAN VUNAKIS

VOLUME 109. Hormone Action (Part I: Peptide Hormones)
Edited by LUTZ BIRNBAUMER AND BERT W. O'MALLEY

VOLUME 110. Steroids and Isoprenoids (Part A)
Edited by JOHN H. LAW AND HANS C. RILLING

VOLUME 111. Steroids and Isoprenoids (Part B)
Edited by JOHN H. LAW AND HANS C. RILLING

VOLUME 112. Drug and Enzyme Targeting (Part A)
Edited by KENNETH J. WIDDER AND RALPH GREEN

VOLUME 113. Glutamate, Glutamine, Glutathione, and Related Compounds
Edited by ALTON MEISTER

VOLUME 114. Diffraction Methods for Biological Macromolecules (Part A)
Edited by HAROLD W. WYCKOFF, C. H. W. HIRS, AND SERGE N. TIMASHEFF

VOLUME 115. Diffraction Methods for Biological Macromolecules (Part B)
Edited by HAROLD W. WYCKOFF, C. H. W. HIRS, AND SERGE N. TIMASHEFF

Volume 116. Immunochemical Techniques (Part H: Effectors and Mediators of Lymphoid Cell Functions)
Edited by Giovanni Di Sabato, John J. Langone, and Helen Van Vunakis

Volume 117. Enzyme Structure (Part J)
Edited by C. H. W. Hirs and Serge N. Timasheff

Volume 118. Plant Molecular Biology
Edited by Arthur Weissbach and Herbert Weissbach

Volume 119. Interferons (Part C)
Edited by Sidney Pestka

Volume 120. Cumulative Subject Index Volumes 81–94, 96–101

Volume 121. Immunochemical Techniques (Part I: Hybridoma Technology and Monoclonal Antibodies)
Edited by John J. Langone and Helen Van Vunakis

Volume 122. Vitamins and Coenzymes (Part G)
Edited by Frank Chytil and Donald B. McCormick

Volume 123. Vitamins and Coenzymes (Part H)
Edited by Frank Chytil and Donald B. McCormick

Volume 124. Hormone Action (Part J: Neuroendocrine Peptides)
Edited by P. Michael Conn

Volume 125. Biomembranes (Part M: Transport in Bacteria, Mitochondria, and Chloroplasts: General Approaches and Transport Systems)
Edited by Sidney Fleischer and Becca Fleischer

Volume 126. Biomembranes (Part N: Transport in Bacteria, Mitochondria, and Chloroplasts: Protonmotive Force)
Edited by Sidney Fleischer and Becca Fleischer

Volume 127. Biomembranes (Part O: Protons and Water: Structure and Translocation)
Edited by Lester Packer

VOLUME 140. Cumulative Subject Index Volumes 102–119, 121–134

VOLUME 141. Cellular Regulators (Part B: Calcium and Lipids)
Edited by P. MICHAEL CONN AND ANTHONY R. MEANS

VOLUME 142. Metabolism of Aromatic Amino Acids and Amines
Edited by SEYMOUR KAUFMAN

VOLUME 143. Sulfur and Sulfur Amino Acids
Edited by WILLIAM B. JAKOBY AND OWEN GRIFFITH

VOLUME 144. Structural and Contractile Proteins (Part D: Extracellular Matrix)
Edited by LEON W. CUNNINGHAM

VOLUME 145. Structural and Contractile Proteins (Part E: Extracellular Matrix)
Edited by LEON W. CUNNINGHAM

VOLUME 146. Peptide Growth Factors (Part A)
Edited by DAVID BARNES AND DAVID A. SIRBASKU

VOLUME 147. Peptide Growth Factors (Part B)
Edited by DAVID BARNES AND DAVID A. SIRBASKU

VOLUME 148. Plant Cell Membranes
Edited by LESTER PACKER AND ROLAND DOUCE

VOLUME 149. Drug and Enzyme Targeting (Part B)
Edited by RALPH GREEN AND KENNETH J. WIDDER

VOLUME 150. Immunochemical Techniques (Part K: *In Vitro* Models of B and T Cell Functions and Lymphoid Cell Receptors)
Edited by GIOVANNI DI SABATO

VOLUME 151. Molecular Genetics of Mammalian Cells
Edited by MICHAEL M. GOTTESMAN

VOLUME 152. Guide to Molecular Cloning Techniques
Edited by SHELBY L. BERGER AND ALAN R. KIMMEL

VOLUME 153. Recombinant DNA (Part D)
Edited by RAY WU AND LAWRENCE GROSSMAN

Volume 154. Recombinant DNA (Part E)
Edited by Ray Wu and Lawrence Grossman

Volume 155. Recombinant DNA (Part F)
Edited by Ray Wu

Volume 156. Biomembranes (Part P: ATP-Driven Pumps and Related Transport: The Na,K-Pump)
Edited by Sidney Fleischer and Becca Fleischer

Volume 157. Biomembranes (Part Q: ATP-Driven Pumps and Related Transport: Calcium, Proton, and Potassium Pumps)
Edited by Sidney Fleischer and Becca Fleischer

Volume 158. Metalloproteins (Part A)
Edited by James F. Riordan and Bert L. Vallee

Volume 159. Initiation and Termination of Cyclic Nucleotide Action
Edited by Jackie D. Corbin and Roger A. Johnson

Volume 160. Biomass (Part A: Cellulose and Hemicellulose)
Edited by Willis A. Wood and Scott T. Kellogg

Volume 161. Biomass (Part B: Lignin, Pectin, and Chitin)
Edited by Willis A. Wood and Scott T. Kellogg

Volume 162. Immunochemical Techniques (Part L: Chemotaxis and Inflammation)
Edited by Giovanni Di Sabato

Volume 163. Immunochemical Techniques (Part M: Chemotaxis and Inflammation)
Edited by Giovanni Di Sabato

Volume 164. Ribosomes
Edited by Harry F. Noller, Jr. and Kivie Moldave

Volume 165. Microbial Toxins: Tools for Enzymology
Edited by Sidney Harshman

Volume 166. Branched-Chain Amino Acids
Edited by Robert Harris and John R. Sokatch

VOLUME 167. Cyanobacteria
Edited by LESTER PACKER AND ALEXANDER N. GLAZER

VOLUME 168. Hormone Action (Part K: Neuroendocrine Peptides)
Edited by P. MICHAEL CONN

VOLUME 169. Platelets: Receptors, Adhesion, Secretion (Part A)
Edited by JACEK HAWIGER

VOLUME 170. Nucleosomes
Edited by PAUL M. WASSARMAN AND ROGER D. KORNBERG

VOLUME 171. Biomembranes (Part R: Transport Theory: Cells and Model Membranes)
Edited by SIDNEY FLEISCHER AND BECCA FLEISCHER

VOLUME 172. Biomembranes (Part S: Membrane Isolation and Characterization)
Edited by SIDNEY FLEISCHER AND BECCA FLEISCHER

VOLUME 173. Biomembranes [Part T: Cellular and Subcellular Transport: Eukaryotic (Nonepithelial) Cells]
Edited by SIDNEY FLEISCHER AND BECCA FLEISCHER

VOLUME 174. Biomembranes [Part U: Cellular and Subcellular Transport: Eukaryotic (Nonepithelial) Cells]
Edited by SIDNEY FLEISCHER AND BECCA FLEISCHER

VOLUME 175. Cumulative Subject Index Volumes 135–139, 141–167

VOLUME 176. Nuclear Magnetic Resonance (Part A: Spectral Techniques and Dynamics)
Edited by NORMAN J. OPPENHEIMER AND THOMAS L. JAMES

VOLUME 177. Nuclear Magnetic Resonance (Part B: Structure and Mechanism)
Edited by NORMAN J. OPPENHEIMER AND THOMAS L. JAMES

VOLUME 178. Antibodies, Antigens, and Molecular Mimicry
Edited by JOHN J. LANGONE

VOLUME 192. Biomembranes (Part W: Cellular and Subcellular Transport: Epithelial Cells)
Edited by SIDNEY FLEISCHER AND BECCA FLEISCHER

VOLUME 193. Mass Spectrometry
Edited by JAMES A. McCLOSKEY

VOLUME 194. Guide to Yeast Genetics and Molecular Biology
Edited by CHRISTINE GUTHRIE AND GERALD R. FINK

VOLUME 195. Adenylyl Cyclase, G Proteins, and Guanylyl Cyclase
Edited by ROGER A. JOHNSON AND JACKIE D. CORBIN

VOLUME 196. Molecular Motors and the Cytoskeleton
Edited by RICHARD B. VALLEE

VOLUME 197. Phospholipases
Edited by EDWARD A. DENNIS

VOLUME 198. Peptide Growth Factors (Part C)
Edited by DAVID BARNES, J. P. MATHER, AND GORDON H. SATO

VOLUME 199. Cumulative Subject Index Volumes 168–174, 176–194 (in preparation)

VOLUME 200. Protein Phosphorylation (Part A: Protein Kinases: Purification, Antibodies, Functional Analysis, Cloning, and Expression)
Edited by TONY HUNTER AND BARTHOLOMEW M. SEFTON

VOLUME 201. Protein Phosphorylation (Part B: Analysis of Protein Phosphorylation, Protein Kinase Inhibitors, and Protein Phosphatases)
Edited by TONY HUNTER AND BARTHOLOMEW M. SEFTON

VOLUME 202. Molecular Design and Modeling: Concepts and Applications (Part A: Proteins, Peptides, and Enzymes)
Edited by JOHN J. LANGONE

VOLUME 203. Molecular Design and Modeling: Concepts and Applications (Part B: Antibodies and Antigens, Nucleic Acids, Polysaccharides, and Drugs)
Edited by JOHN J. LANGONE

Section I

Antibodies and Antigens

A. Antibody Combining Site
Articles 1 through 5

B. Antibody–Antigen Interactions
Articles 6 through 16

C. Catalytic Antibodies and Vaccines
Articles 17 through 20

[1] Modeling of Antibody Combining Sites

By EDUARDO A. PADLAN and ELVIN A. KABAT

Introduction

Antibodies constitute an extremely large family of closely related serum proteins, termed immunoglobulins, produced in vertebrates by a class of lymphocytes, termed B lymphocytes, which are programmed to respond to contact (immunization) with foreign substances (antigens) or, under certain circumstances, to antigens of an individual's own tissues (autoantibodies). Antibodies may cause the elimination of antigens by phagocytosis, neutralization of toxins or viruses, lysis of tissue cells through the complement system, precipitation with the antigen used for immunization, or clumping of bacteria or red cells containing the antigen or the antigen adsorbed to inert particles (agglutination).

Antibodies may be produced to almost all classes of substances, e.g., proteins, polysaccharides, nucleic acids, and to more complex particles, e.g., pollens, infectious agents, viruses, and tissue cells. Unless there is some structural similarity between two antigens, one will generally not react with antibody to the other; when structural similarity does exist, reactions with both will occur (termed cross-reactions), the antibody reacting more strongly to the antigen used for immunization and to a lesser degree with the other. (For exceptional, poorly understood, anomalous cross-reactions, see Future Prospects and Problems.)

A second class of proteins, termed T cell receptors for antigen, is produced by a different class of lymphocytes, termed T lymphocytes. The T cell receptors for antigen do not circulate in the bloodstream but remain attached to the cells which synthesize them and, in conjunction with proteins of the major histocompatibility complex and cells of the macrophage system, can destroy tissue cells infected with viruses, tumor cells, etc. This is termed the cellular arm of the immune system.

Antibodies and T cell receptors for antigen show certain structural as well as certain genetic similarities, but appear to have arisen at different evolutionary periods and by different pathways. This chapter on modeling will focus exclusively on the antigen-binding sites of antibodies (the antibody combining sites).

The high degree of specificity of an antibody for the antigen used for immunization and the wide range of specificities that the immune system is capable of generating have been the subject of numerous investigations. The specificity of antibody:antigen interactions is accepted as being due to the complementarity of the antibody combining site structure and that of the antigenic determinant. The diversity of antigen-binding specificities is then due to variation in the combining site topography brought about by variation in primary and three-dimensional structure.

Recognition of the central importance of the antibody combining site in immune function has prompted intensive study of the primary and three-dimensional structures of antibodies. Considerable three-dimensional information has become available from X-ray crystallography,[1-9] although the number of antibody structures that will be elucidated by X-ray analysis can be only a very small fraction of the total number of different antibodies that higher organisms can produce. Other techniques are needed in the study of antibody combining sites and one that can make a significant contribution is modeling.

Here, we review the various modeling procedures that have been applied to antibodies, evaluate the success of these procedures in predicting combining site structures, and discuss potential improvements and problems.

Structural Background

Antibodies are multimers of a basic unit that consists of four polypeptide chains identical in pairs (Fig. 1[10]). There are two light (L) chains of about 220 amino acids and two heavy (H) chains of 450–575 amino acids. Both L and H chains are made up of regions of sequence homology (domains) of about 100–120 residues. There are two such domains in the L chain and four or five in the H chain. The N-terminal domains of both L and H chains are variable, i.e., they differ in sequence from antibody to antibody; the other domains are constant, i.e., they are the same in antibody chains of the same type except for single amino acid differences at a few positions. The variable domains of the L and H chains, V_L and V_H,

[1] R. J. Poljak, *Adv. Immunol.* **21,** 1 (1975).
[2] D. R. Davies, E. A. Padlan, and D. M. Segal, *Annu. Rev. Biochem.* **44,** 639 (1975).
[3] R. Huber, *Trends Biochem. Sci.* **1,** 87 (1976).
[4] E. A. Padlan, *Q. Rev. Biophys.* **10,** 35 (1977).
[5] L. M. Amzel and R. J. Poljak, *Annu. Rev. Biochem.* **48,** 961 (1979).
[6] D. R. Davies and H. M. Metzger, *Annu. Rev. Immunol.* **1,** 87 (1983).
[7] P. M. Alzari, M.-B. Lascombe, and R. J. Poljak, *Annu. Rev. Immunol.* **6,** 555 (1988).
[8] P. M. Colman, *Adv. Immunol.* **43,** 99 (1988).
[9] D. R. Davies, E. A. Padlan, and S. Sheriff, *Annu. Rev. Biochem.* **59,** 439 (1990).
[10] E. A. Kabat, T. T. Wu, M. Reid-Miller, H. M. Perry, and K. S. Gottesman, "Sequences of Proteins of Immunological Interest," 4th Ed. U.S. Department of Health and Human Services, Washington, D.C., 1987.

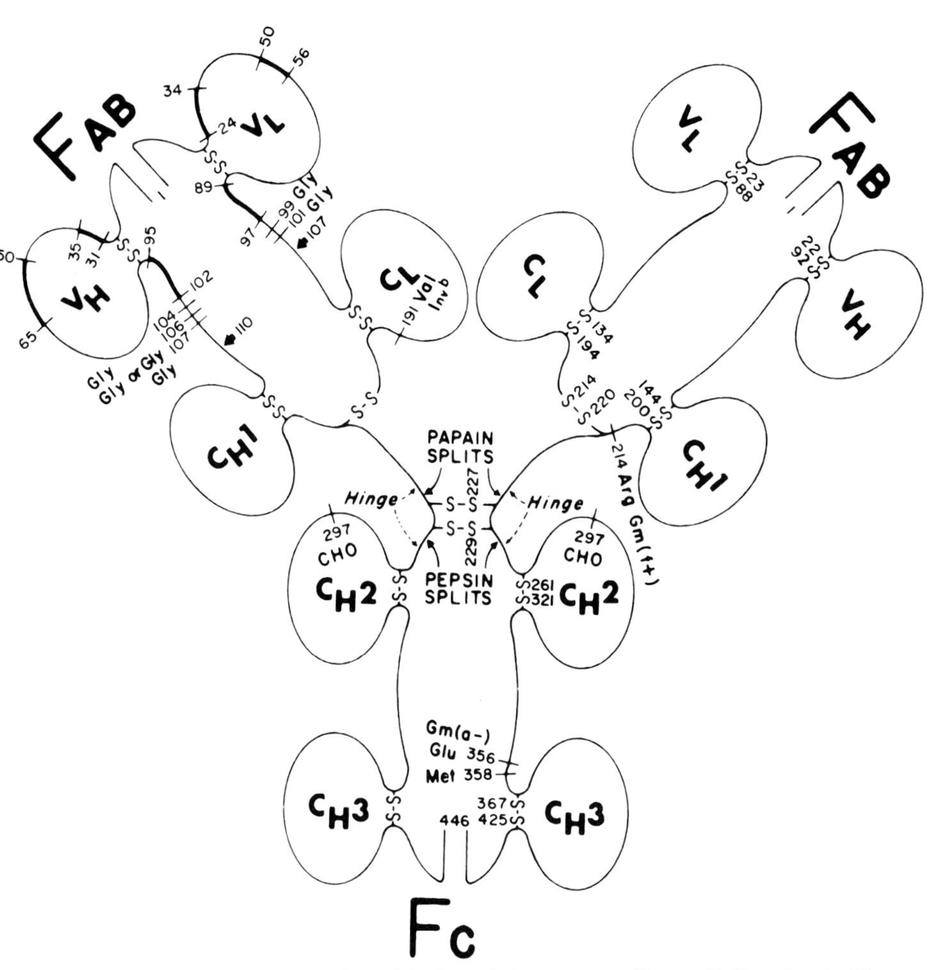

FIG. 1. Schematic representation of the four-chain structure of human IgG_1 molecule. The numbers on the right-hand side denote the actual residues of protein Eu [G. M. Edelman, B. A. Cunningham, W. E. Gall, P. D. Gottlieb, U. Rutishauser, and M. J. Waxdal, *Proc. Natl. Acad. Sci. U.S.A.* **63,** 78 (1969)]. The numbers of the Fab fragments on the left side are aligned for maximum homology: light and heavy chains are numbered according to T. T. Wu and E. A. Kabat, *J. Exp. Med.* **132,** 211 (1970), and E. A. Kabat and T. T. Wu, *Ann. N.Y. Acad. Sci.* **190,** 382 (1971). The heavy chains of Eu have residues 52A and 82A, B, C but lack the residues termed 100A–K and 35A, B. Thus, residue 110 (the end of the heavy-chain variable region) is 114 in the actual sequence. Hypervariable or complementarity-determining regions are shown by heavier lines. V_L and V_H denote the light- and heavy-chain variable regions; C_H1, C_H2, and C_H3 are domains of the constant region of the heavy chain; C_L is the constant region of the light chain. The hinge region, in which the two heavy chains are linked by disulfide bonds, is indicated approximately. The attachment of carbohydrate is at position 297. Arrows at positions 107 (in the light chain) and 110 (in the heavy chain) denote transition from variable to constant region. The sites of cleavage by papain and pepsin and the locations of various allotypic genetic factors [Gm(ft), Gm(a-), Inv^b] are indicated.[10]

display significant sequence similarity, as do the constant domains[10]; there is no obvious sequence similarity between variable and constant domains. The site on the antibody molecule that binds to antigen is formed by the association of V_L and V_H, the Fv module. The fragment formed by the association of the L chain and the two N-terminal domains of the H chain is called the Fab, or antigen-binding, fragment (Fig. 1). The antigen-binding properties of an antibody are determined entirely by the variable domains; indeed, a chimeric structure, in which the variable domains have been linked to the constant domains of the heterologous chains, was shown to display exactly the same ligand-binding characteristics.[11]

A comparison of the sequences of variable domains from a variety of immunoglobulins revealed the existence of regions of hypervariability, three each in the L and H chains.[12,13] These hypervariable regions were predicted by Wu and Kabat[12] to fold up three-dimensionally to form the walls of the antigen-binding site of the antibody (the antibody combining site) several years before any X-ray structures had been determined.

X-Ray crystallographic studies have allowed the direct visualization of the three-dimensional structure of complete antibodies,[14,15] although only at low resolution. Structures at high resolution, however, have become available for a number of antibody fragments, including several Fabs, which bear the antigen-binding sites. These studies have revealed the high degree of structural similarity of the homologous domains and, as would be expected, the similarity in the three-dimensional structure was found to parallel the similarity in amino acid sequence. Thus, the V_L and V_H domains were found to have very similar structures, as were the constant domains; constant and variable domains, on the other hand, share only a basic tertiary structure and appear to be rotational isomers.[16-18] The hypervariable segments were found to exist mainly as loops that are for the most part exposed and located at one end of the variable domains.

In spectacular confirmation of the prediction of Wu and Kabat,[12] the

[11] T. Simon and K. Rajewsky, *EMBO J.* **9,** 1051 (1990).

[12] T. T. Wu and E. A. Kabat, *J. Exp. Med.* **132,** 211 (1970).

[13] E. A. Kabat and T. T. Wu, *Ann. N.Y. Acad. Sci.* **190,** 382 (1971).

[14] E. W. Silverton, M. A. Navia, and D. R. Davies, *Proc. Natl. Acad. Sci. U.S.A.* **74,** 5140 (1977).

[15] S. S. Rajan, K. R. Ely, E. E. Abola, M. K. Wood, P. M. Colman, R. J. Athay, and A. B. Edmundson, *Mol. Immunol.* **20,** 797 (1983).

[16] R. J. Poljak, L. M. Amzel, H. P. Avey, B. L. Chen, R. P. Phizackerly, and F. Saul, *Proc. Natl. Acad. Sci. U.S.A.* **70,** 3305 (1973).

[17] M. Schiffer, R. L. Girling, K. R. Ely, and A. B. Edmundson, *Biochemistry* **12,** 4620 (1973).

[18] A. B. Edmundson, K. R. Ely, E. E. Abola, M. Schiffer, and N. Panagiotopoulos, *Biochemistry* **14,** 3953 (1975).

hypervariable regions were indeed found to form a continuous surface at the tip of the Fabs.[16,19] Furthermore, crystallographic studies of complexes of Fabs with specific ligands unequivocably established the essential identity of the hypervariable surface with the antibody combining site. It is the chemical nature of this surface that determines the particular specificity of an antibody and the affinity with which it binds to its specific ligand. Furthermore, the insertions and deletions that are frequently found in the hypervariable regions, together with the variation in the amino acid residues in these regions, result in the variability of the combining site topography, thus providing an obvious structural basis for the wide diversity of antigen-binding specificities. In view of the undisputed importance of the hypervariable regions in the binding interaction with the antigen, these regions are now also called complementarity-determining regions (CDRs).

The comparison of the three-dimensional structures of variable domains from different antibodies reveals that the nonhypervariable or framework regions of these domains are essentially superimposable, so that the structural variation is mainly confined to the hypervariable segments.[20,21] In addition to the close similarity in the tertiary structures of the homologous domains, it had also been found that the mode of association of the paired domains is essentially invariant.[6,22,23] Thus the antibody combining site can be viewed as being formed by a small number of segments of variable structure grafted onto a scaffolding of essentially invariant architecture.

Even the hypervariable regions have been found to display a high degree of structural similarity, so that canonical structures have been advanced for several of the CDRs. For example, it was found that hypervariable regions with the same number of residues, especially those with significant sequence similarity, have very similar backbone conformations.[4,20,21,24] Furthermore, Kabat *et al.*[25] noted that certain positions in the CDRs did not vary but were conserved and they suggested that those residues play a structural role. Indeed, structural comparisons of known CDR structures have shown that there is a small repertoire of main-chain conformations for at least five of the six CDRs and that the particular

[19] D. M. Segal, E. A. Padlan, G. H. Cohen, S. Rudikoff, M. Potter, and D. R. Davies, *Proc. Natl. Acad. Sci. U.S.A.* **71,** 4298 (1974).

[20] E. A. Padlan and D. R. Davies, *Proc. Natl. Acad. Sci. U.S.A.* **72,** 819 (1975).

[21] C. Chothia and A. M. Lesk, *J. Mol. Biol.* **196,** 901 (1987).

[22] J. Novotny and E. Haber, *Proc. Natl. Acad. Sci. U.S.A.* **82,** 4592 (1985).

[23] C. Chothia, J. Novotny, R. Bruccoleri, and M. Karplus, *J. Mol. Biol.* **186,** 651 (1985).

[24] P. de la Paz, B. J. Sutton, M. J. Darsley, and A. R. Rees, *EMBO J.* **5,** 415 (1986).

[25] E. A. Kabat, T. T. Wu, and H. Bilofsky, *J. Biol. Chem.* **252,** 6609 (1977).

conformation adopted is determined by a few key conserved residues, several of which are outside the CDRs.[21,26]

It is this high degree of structural similarity among the immunoglobulin domains that has been the stimulus for the modeling of antibody combining site structures.

Techniques for Modeling Antibody Combining Site Structures

All current attempts at modeling antibody combining sites assume the same framework structures for V_L and V_H, as well as the same quaternary association of these domains, as found in the crystal structures. With regard to the modeling of the CDRs, two approaches have been taken. One approach uses a known CDR or other loop structure as a template (homology or knowledge-based modeling), the other attempts to predict the CDR structures solely on the basis of energetic considerations (*ab initio* modeling). An automated procedure that combines both approaches has been developed.[27]

Homology Modeling of Antibody Combining Sites

The first structure-based modeling of an antibody combining site was done by Padlan *et al.*,[28] who built a model of the Fv of MOPC315. In constructing the model the following assumptions were made: (1) the framework structures of the V_L and the V_H domains of MOPC315 were essentially the same as those of the corresponding domains of the McPC603 Fab, the structure of which had been previously determined crystallographically[19]; (2) the quaternary association of the V_L and V_H domains of MOPC315 was the same as that found in McPC603; and (3) the CDRs of MOPC315 would have the same backbone conformations as those CDRs from other structures which had the same number of amino acid residues. Sequence similarity was taken into account in building loop structures for which no suitable starting model was available.

The same basic procedure was used in the modeling of the combining site structures of the rabbit BS-5 anti-polysaccharide antibody,[29] the inu-

[26] C. Chothia, A. M. Lesk, M. Levitt, A. G. Amit, R. A. Mariuzza, S. E. V. Phillips, and R. J. Poljak, *Science* **233,** 755 (1986).

[27] A. C. R. Martin, J. C. Cheetham, and A. R. Rees, *Proc. Natl. Acad. Sci. U.S.A.* **86,** 9268 (1989).

[28] E. A. Padlan, D. R. Davies, I. Pecht, D. Givol, and C. Wright, *Cold Spring Harbor Symp. Quant. Biol.* **41,** 627 (1976).

[29] D. R. Davies and E. A. Padlan, *in* "Antibodies in Human Diagnosis and Therapy" (E. Haber and R. M. Krause, eds.), p. 119. Raven, New York, 1977.

lin-binding EPC109 mouse myeloma protein,[30] and the W3129 and 19.1.2 mouse anti-$\alpha(1 \rightarrow 6)$-dextrans.[31] Very similar assumptions were used by Feldmann and co-workers[32,33] in their modeling of the Fv of the galactan-binding mouse immunoglobulin J539, by Smith-Gill *et al.*[34] in their modeling of the Fv of the lysozyme-specific HyHEL-10 antibody, and by de la Paz *et al.*,[24] who built models of three antibodies specific for an antigenic disulfide loop of lysozyme.

With the demonstration by Chothia and co-workers[21,26,35] that canonical structures exist for most of the hypervariable loops, CDR templates for modeling have been chosen on the basis of size as well as on the presence of the residues believed to be responsible for the different conformations. It should be pointed out that the hypervariable loops, as defined by Chothia and co-workers, overlap but do not entirely correspond to those defined by Kabat *et al.*[10] on the basis of sequence variation.

An automated approach to the homology modeling of antibody Fvs has been successfully implemented by Levitt and co-workers.[26,36,37] In this approach, the sequence for which a model is desired is divided into a number of segments, homologous regions from the known immunoglobulin structures are examined, and a search is made for segments that are most similar in size and sequence. The model is then pieced together using the most similar regions while inserting or deleting residues at positions chosen to maximize sequence similarity. Missing atoms and inserted residues are introduced using randomized coordinates; these are then properly incorporated into the model and breaks in the main chain rectified by limited energy minimization.

Ab Initio Modeling of Antibody Combining Sites

The first attempt to predict CDR structures *ab initio* was made by Stanford and Wu,[38] who constructed a model of the backbone structure of

[30] M. Potter, S. Rudikoff, E. A. Padlan, and M. Vrana, *in* "Antibodies in Human Diagnosis and Therapy" (E. Haber and R. M. Krause, eds.), p. 9. Raven, New York, 1977.

[31] E. A. Padlan and E. A. Kabat, *Proc. Natl. Acad. Sci. U.S.A.* **85,** 6885 (1988).

[32] R. J. Feldmann, M. Potter, and C. P. J. Glaudemans, *Mol. Immunol.* **18,** 683 (1981).

[33] C. R. Mainhart, M. Potter, and R. J. Feldmann, *Mol. Immunol.* **21,** 469 (1984).

[34] S. J. Smith-Gill, C. Mainhart, T. B. Lavoie, R. J. Feldmann, W. Drohan, and B. R. Brooks, *J. Mol. Biol.* **194,** 173 (1987).

[35] C. Chothia, A. M. Lesk, A. Tramontano, M. Levitt, S. J. Smith-Gill, G. Air, S. Sheriff, E. A. Padlan, D. Davies, W. R. Tulip, P. M. Colman, S. Spinelli, P. M. Alzari, and R. J. Poljak, *Nature (London)* **342,** 877 (1989).

[36] J. Anglister, M. W. Bond, T. Frey, D. Leahy, M. Levitt, H. M. McConnell, G. S. Rule, J. Tomasello, and M. Whittaker, *Biochemistry* **26,** 6058 (1987).

[37] R. Levy, O. Assulin, T. Scherf, M. Levitt, and J. Anglister, *Biochemistry* **28,** 7168 (1989).

[38] J. M. Stanford and T. T. Wu, *J. Theor. Biol.* **88,** 421 (1981).

the combining site of MOPC315 on the basis of amino acid sequence and steric considerations. These authors assumed the same framework structure for MOPC315 as that found in Fab NEW[39] and used the predictive method of Kabat and Wu[40] to construct models of the CDRs. Backbone dihedral angles for tripeptides were obtained from known protein structures, relying mostly on β-sheet proteins, and these angles were imposed on the segment being constructed based on its sequence. The peptide angles were allowed to vary from the initial values by as much as 30° in 5° intervals. The large number of resulting possible structures was reduced by imposing the conditions that the ends of the modeled CDRs should fit onto the assumed framework structure and that nonbonded atoms should not come within allowed minimum contact distances based on normally accepted van der Waals radii.

A related procedure was used by Bruccoleri *et al.*[41] to reconstruct the crystallographically determined hypervariable loops of McPC603 and HyHEL-5. These authors generated all the possible conformations for each loop by imposing peptide dihedral angles that are energetically accessible to each residue in the sequence, sampling torsional space at 30° intervals. Here also, the number of possible structures was limited by imposing the conditions that the ends of the loops should fit onto the framework and that bad steric contacts are avoided.

A different approach was taken by Fine *et al.*,[42] who modeled four of the CDRs of McPC603. These authors generated a large number of random conformations for the backbone of those CDRs, subjecting the conformations to molecular dynamics and energy minimization and then selecting those with lowest energy for further characterization of side-chain conformations. The random structures generated were required to fit onto the framework with correct geometry. In generating the conformations, random values were first assigned to each ϕ and ψ peptide angle along the backbone; then these angles were adjusted as minimally as possible in an iterative procedure to produce the desired fixed-end conditions. Side chains were then added to the lowest energy structures and energetically favorable conformations were obtained by varying side-chain torsional angles.

Combined Knowledge-Based and ab Initio Methods

The approach taken by Martin *et al.*[27] makes use of segments from all known structures (not restricted to antibody structures) as the database

[39] F. A. Saul, L. M. Amzel, and R. J. Poljak, *J. Biol. Chem.* **253,** 585 (1978).
[40] E. A. Kabat and T. T. Wu, *Proc. Natl. Acad. Sci. U.S.A.* **69,** 960 (1972).
[41] R. E. Bruccoleri, E. Haber, and J. Novotny, *Nature (London)* **335,** 564 (1988).
[42] R. M. Fine, H. Wang, P. S. Shenkin, D. L. Yarmush, and C. Levinthal, *Proteins: Struct. Funct. Genet.* **1,** 342 (1986).

from which to choose the most similar segment for the CDR loop of interest. A conformational search procedure[41,43] is then used to obtain the lowest energy conformations. Each conformation is then screened and the one with the smallest hydrophobic exposed area is selected. This algorithm has been implemented as a completely automated procedure.[27]

Evaluation of Predictive Methods

The assessment of the success of modeling algorithms is not straightforward, since crystal structures are not always available to allow the comparison of model with X-ray structure. Even when crystal structures are available, these are frequently of too-low accuracy themselves to permit meaningful comparisons. Indeed, most of the X-ray analyses of combining site structures had been done at medium resolution, i.e., between 2 and 3 Å. In these analyses, estimates of the error in atomic position in the most ordered regions, i.e., the most accurately determined parts of the molecule, run to about 0.5 Å; the error in the loop regions and for the side-chain atoms would be much larger. Thus, if a model agrees with the X-ray structure to within 1.5 Å, the agreement can be said to be within the limits of error. In fact, higher levels of disagreement could be tolerated since the CDR structures, being mostly loops, could be suffering from larger errors. Indeed, it is not unusual that a major rebuilding of external regions is required when a crystallographic analysis is extended to higher resolution. Thus, for example, when the crystal structure of J539 Fab was extended from a resolution of 2.6 Å[44] to 1.95,[45] large differences were noted in the CDRs. In one CDR, differences between corresponding C_α positions of as much as 8 Å were observed.

There are several cases where models for antibody combining sites can be compared with crystallographically determined structures. Crystal structures are available for the Fabs of McPC603,[46] J539,[44,45] D1.3,[47] NC41,[48] HyHEL-5,[49] HyHEL-10,[50] and Gloop2 (unpublished results referred to by Martin *et al.*[27]) for which models of the combining sites or of

[43] R. E. Bruccoleri and M. Karplus, *Biopolymers* **26,** 137 (1987).

[44] S. W. Suh, T. N. Bhat, M. A. Navia, G. H. Cohen, D. N. Rao, S. Rudikoff, and D. R. Davies, *Proteins: Struct. Funct. Genet.* **1,** 74 (1986).

[45] T. N. Bhat, E. A. Padlan, and D. R. Davies, in preparation.

[46] Y. Satow, G. H. Cohen, E. A. Padlan, and D. R. Davies, *J. Mol. Biol.* **190,** 593 (1986).

[47] A. G. Amit, R. A. Mariuzza, S. E. V. Phillips, and R. J. Poljak, *Science* **233,** 747 (1986).

[48] P. M. Colman, W. G. Laver, J. N. Varghese, A. T. Baker, P. A. Tulloch, G. M. Air, and R. G. Webster, *Nature (London)* **326,** 358 (1987).

[49] S. Sheriff, E. W. Silverton, E. A. Padlan, G. H. Cohen, S. J. Smith-Gill, B. C. Finzel, and D. R. Davies, *Proc. Natl. Acad. Sci. U.S.A.* **84,** 8075 (1987).

[50] E. A. Padlan, E. W. Silverton, S. Sheriff, G. H. Cohen, S. J. Smith-Gill, and D. R. Davies, *Proc. Natl. Acad. Sci. U.S.A.* **86,** 5938 (1989).

some of the CDRs had been constructed. When the models are compared to the crystal structures, some of the CDR structures are found to have been, by and large, correctly predicted, while in others large disagreements between model and X-ray structure are found.

For example, comparison of the crystal structure of the J539 Fv[44] with the model of Mainhart *et al.*[33] showed that the C_α root-mean-square deviations for individual CDRs ranged from 1.1 to 4.0 Å. When the modeled hypervariable loops of D1.3 were compared with the crystal structure,[47] the deviations of the model based on structural analysis ranged from 0.5 to 2.07 Å, and the deviations of the models built using conformational energy calculations ranged from 0.47 to 3.76 Å.[26] Comparing the model constructed for the Fv of HyHEL-10[34] with the X-ray structure,[50] deviations in C_α positions in the CDRs ranged from 0.44 to 4.27 Å.[51]

The use of canonical structures for CDRs in the modeling of combining site structures has been shown to be reasonably successful.[35] When models of five out of the six hypervariable loops of HyHEL-5, HyHEL-10, NC41, and NQ10 antibodies were compared to the crystal structures, the differences between the predicted and observed structures initially ranged from 0.4 to 7.2 Å in the C_α positions, with the larger errors being found in the predicted loops of HyHEL-5 and NC41. When better refined crystal structures became available for inclusion in the database and after the structures of the HyHEL-5 and NC41 hypervariable loops had been remodeled on the basis of these new data, the differences between predicted and observed ranged from 0.4 to 3.5 Å for the four models.[35]

The structures predicted *ab initio* for the hypervariable loops of McPC603 and HyHEL-5 by Bruccoleri *et al.*[41] matched the X-ray structures fairly well. The backbone differences between predicted and observed structures ranged from 0.7 to 2.6 Å (1.7 for all loops) for the six hypervariable loops of McPC603 and from 0.6 to 2.1 Å (1.4 for all loops) for HyHEL-5. The predictions of Fine *et al.*[42] of the hypervariable loop structures for McPC603 also yielded very reasonable results, with root-mean-square deviations between crystal and modeled structures of about 1.0 Å for backbone atoms.

Future Prospects and Problems

Possible Improvements

The use of hypervariable loops from known crystal structures as templates in knowledge-based modeling studies emphasizes the need for more,

[51] E. A. Padlan, unpublished results.

and especially more accurate, combining site structures. The relative ease with which Fab structures are now being determined crystallographically[52,53] could satisfy this need soon. The use of loops from proteins other than immunoglobulins[27] could fulfill this need even sooner.

Ab initio methods have already produced models that are in reasonable agreement with crystal structures. However, in view of the prohibitive computer time required, those methods have been restricted so far to only the shorter hypervariable segments. The advent of more powerful computers will remove this restriction. Eventually, the framework regions, which have been largely ignored, can also be included in the computations to obtain a more complete and more realistic situation.

In those cases where the structural consequences of only a single amino acid replacement is sought,[54,55] the magnitude of the problem is greatly reduced and more reliable results can be expected.

All models can benefit from further refinement using molecular dynamics.[27,56] This procedure is still somewhat limited, however, in that the relative weights given to the various potential terms used in the calculations are usually semiempirically assigned. Other limitations arise from the inadequate description of the electrical properties of the protein interior and of the protein environment.[57,58] Improvements along these lines should result in better prediction of structures.

Table I shows the combining site residues which have been found to be involved in the contact with the ligand in the antibody–ligand complexes of known structure. As further additions to Table I[58a–e] are made, each CDR will be further defined structurally. This information will be ex-

[52] M. Cygler, A. Boodhoo, J. S. Lee, and W. F. Anderson, *J. Biol. Chem.* **262,** 643 (1987).

[53] S. Sheriff, E. A. Padlan, G. H. Cohen, and D. R. Davies, *Acta Crystallogr., Sect. B* **46,** 418 (1990).

[54] M. E. Snow and L. M. Amzel, *Proteins: Struct. Funct. Genet.* **1,** 267 (1986).

[55] N. V. Chien, V. A. Roberts, A. M. Giusti, M. D. Scharff, and E. D. Getzoff, *Proc. Natl. Acad. Sci. U.S.A.* **86,** 5532 (1989).

[56] L. Holm, L. Laaksonen, M. Kaartinen, T. T. Teeri, and J. K. C. Knowles, *Protein Eng.* **3,** 403 (1990).

[57] M. K. Gilson and B. H. Honig, *Proteins: Struct. Funct. Genet.* **3,** 32 (1988).

[58] J. J. Wendoloski and J. B. Matthew, *Proteins: Struct. Funct. Genet.* **5,** 313 (1989).

[58a] L. M. Amzel, R. J. Poljak, F. Saul, J. M. Varga, and F. F. Richards, *Proc. Natl. Acad. Sci. U.S.A.* **71,** 1427 (1974).

[58b] E. A. Padlan, G. H. Cohen, and D. R. Davies, *Ann. Inst. Pasteur/Immunol.* **136C,** 271 (1985).

[58c] J. N. Herron, X.-M. He, M. L. Mason, E. W. Voss, Jr., and A. B. Edmundson, *Proteins: Struct. Funct. Genet.* **5,** 271 (1989).

[58d] R. L. Stanfield, T. M. Fieser, R. A. Lerner, and I. A. Wilson, *Science* **248,** 712 (1990).

[58e] W. R. Tulip, J. N. Varghese, R. G. Webster, G. M. Air, W. G. Laver, and P. M. Colman, *Cold Spring Harbor Symp. Quant. Biol.* **54,** 239 (1989).

TABLE I

COMBINING SITE RESIDUES IN CONTACT WITH THE LIGAND IN COMPLEXES OF KNOWN STRUCTURE[a]

Light-chain residues

	CDR1																FR2	CDR2							CDR3								
V_L	24	25	26	27	A	B	C	D	E	F	28	29	30	31	32	33	34	49	50	51	52	53	54	55	56	89	90	91	92	93	94	95	96	97
D1.3													His		Tyr			Tyr	Tyr									Phe	Trp	Ser				
HyHEL-5														Asn	Tyr				Asp									Trp	Gly	Arg		Pro		
HyHEL-10													Gly	Asn	Asn				Tyr			Gln						Ser	Asn				Tyr	
NEWM												Gly	Asn														Tyr			Ser	Leu			
McPC603																												Asp			Tyr	Pro	Leu	
4-4-20								His							Tyr		Arg											Ser					Trp	
B13I2											Asp		Asp		Tyr						Tyr							Gly			Val		Pro	
NC41																		Tyr	Trp		Ser	Thr		His	Ile				Tyr	Ser	Pro		Trp	

Heavy-chain residues

	FR1	CDR1							FR2	CDR2																			CDR3																			
V_H	30	31	32	33	34	35	35	A	B	47	50	51	52	A	B	C	53	54	55	56	57	58	59	60	61	62	63	64	65	95	96	97	98	99	100	A	B	C	D	E	F	G	H	I	J	K	101	102
D1.3	Thr	Gly	Tyr										Trp	Gly	Asp																Arg	Asp	Tyr	Arg														
HyHEL-5				Trp		Glu				Trp	Glu							Ser		Ser	Thr	Asn								Gly		Tyr																
HyHEL-10	Thr	Ser	Asp	Tyr							Tyr		Ser				Tyr	Ser		Ser		Tyr								Trp																		
NEWM										Trp			Tyr							Thr											Leu	Ile	Ala															
McPC603				Tyr									Arg																	Asn						Trp												
4-4-20				Trp																																												
B13I2		Arg		Ala								Ile	Ser	Ser			Gly		Ser	Tyr		Phe								Tyr				Pro	Phe													
NC41		Asn	Tyr										Asn				Asn														Glu	Asp	Asn	Phe		Ser	Leu											

[a] The contacts for D1.3 are from Amit *et al.*,[47] HyHEL-5 from Sheriff *et al.*,[49] HyHEL-10 from Padlan *et al.*,[50] NEWM from Amzel *et al.*,[58a] McPC603 from Padlan *et al.*,[58b] 4-4-20 from Herron *et al.*,[58c] B13I2 from Stanfield *et al.*,[58d] and NC41 from Tulip *et al.*[58e]

tremely useful in the prediction of which residues are probably exposed and available for interaction with the antigen.

Verification of the modeled binding site structures can be provided by techniques, like two-dimensional nuclear magnetic resonance (NMR), which examine the interaction with ligand more directly.[37,59,60] Modeling should probably rely more heavily on such techniques, since they can provide information on the involvement of side chains, the positions of which are the most uncertain in crystal structures and, consequently, the least reliably modeled.

The ultimate goal of modeling studies of combining site structures is the better understanding of ligand-binding properties. Indeed, the earliest modeling study, which involved the combining site of the dinitrophenol (DNP)-binding myeloma protein MOPC315,[28] resulted in a structure that could explain, at least qualitatively, most of the chemical data on the binding of a variety of ligands. Features of combining sites, which could explain observed ligand interactions, have been obtained in other studies also.[31,36,37] In one case, an improvement in ligand-binding affinity was accomplished by a single-residue mutation in an antibody against the antigenic loop of lysozyme, the change having been suggested by a modeled structure of the combining site.[61]

Conversely, the use of a series of structurally related ligands, or the systematic chemical modification of a ligand to alter its binding characteristics, provides information that is useful in assessing the validity of a predicted combining site structure. Indeed, it is desirable to broaden the concept of modeling to include all information which has a bearing on the three-dimensional structure of the binding site. This has proved especially useful with respect to modeling antibody combining sites to linear polymers, particularly to antibodies to linear carbohydrate polymers. This is illustrated in the following example.

The immunochemical probing of the fine structure of antibody combining sites was originally carried out by making use of the principle established by Landsteiner,[62] that a low-molecular-weight analog of an antigen, termed a hapten, could inhibit precipitation of an antibody by an antigen of related structure. All protein antigens used had been altered chemically by the introduction of a known chemical compound and the epitopes were all heterogeneous save for the grouping introduced. Thus,

[59] J. Anglister and B. Zilber, *Biochemistry* **29,** 921 (1990).

[60] J. Anglister, *Q. Rev. Biophys.* **23,** 175 (1990).

[61] S. Roberts, J. C. Cheetham, and A. R. Rees, *Nature (London)* **328,** 731 (1987).

[62] K. Landsteiner, "The Specificity of Serological Reactions," 2nd Ed. Harvard Univ. Press, Cambridge, Massachusetts, 1945.

although the technique was widely used, it could provide no evidence related to the structure of the native protein.

When monoclonal mouse and human myeloma proteins were discovered and found to have various antibody activities, it became possible to use the hapten inhibition technique and more recent modifications by radioimmunoassay and enzyme-linked immunosorbent assay (ELISA) to map the fine structures of these myeloma monoclonals and to characterize and define the antibody combining site itself. Moreover, it was demonstrated that $\alpha(1 \rightarrow 6)$-dextran was a primary antigen in humans and that two injections of 0.5 mg of this dextran 1 day apart would induce the formation of wheal and erythema skin sensitivity[63-66] and of precipitating antibodies, most of which were generally pauciclonal,[66] as recognized by isoelectric focusing, but others that were polyclonal hyperimmune antibodies. Using isomaltose oligosaccharides [$\alpha(1 \rightarrow 6)$-linked glucose] from the disaccharide to the nonasaccharide on a quantitative basis to inhibit precipitation of anti-$\alpha(1 \rightarrow 6)$-dextran by dextran served as a molecular ruler to probe the size of anti-$\alpha(1 \rightarrow 6)$-dextran combining sites. It was established with these pauciclonal human anti-dextrans that the upper limit for an anti-$\alpha(1 \rightarrow 6)$-dextran site would be complementary to a chain of six or seven $\alpha(1 \rightarrow 6)$-linked glucoses and that the lower limit could be complementary to about two $\alpha(1 \rightarrow 6)$-linked glucoses.

More insight into the three-dimensional site structure was gained when two monoclonal mouse anti-$\alpha(1 \rightarrow 6)$-dextrans were studied, W3129[66-68] and QUPC52.[66] W3219 had a site size complementary to five glucoses and QUPC52 to six glucoses; the K_a value of W3129 was 30 times higher for IM5 (isomaltopentaose) than was that of QUPC52. This was unusual in that antibodies with larger site sizes generally had higher K_a values. Measurements by equilibrium dialysis displacement of [^{3}H]IM5 by various isomaltose oligosaccharides showed that methyl-α-D-glucoside and IM2 (isomaltose) contributed 50% of the binding of the IM5 to the W3129 site but that these two sugars contributed less than 5% of the binding to the QUPC52 site. It thus appeared that the specificity of W3129 involved the terminal nonreducing glucose of the IM5 or of the dextran itself. Studies with a synthetic linear dextran of 200 $\alpha(1 \rightarrow 6)$-linked glucoses synthesized by Ruckel and Schuerch[69] showed that it did not precipitate with W3129

[63] E. A. Kabat and D. Berg, *Ann N.Y. Acad. Sci.* **55,** 471 (1952).
[64] E. A. Kabat and D. Berg, *J. Immunol.* **70,** 514 (1953).
[65] P. H. Maurer, *Proc. Soc. Exp. Biol. Med.* **83,** 879 (1953).
[66] J. Cisar, E. A. Kabat, M. M. Dorner, and J. Liao, *J. Exp. Med.* **142,** 435 (1975).
[67] M. Weigert, W. C. Raschke, D. Carson, and M. Cohn, *J. Exp. Med.* **139,** 137 (1974).
[68] J. Cisar, E. A. Kabat, J. Liao, and M. Potter, *J. Exp. Med.* **139,** 159 (1974).
[69] E. R. Ruckel and C. Schuerch, *Biopolymers* **5,** 515 (1967).

but inhibited the precipitation of W3129 by dextran. QUPC52, however, was actually precipitated by the linear dextran. These data indicated that the W3129 site was directed toward a chain of five $\alpha(1 \rightarrow 6)$-linked glucoses and specific for the nonreducing $\alpha(1 \rightarrow 6)$-linked glucose, whereas QUPC52 was specific for an internal chain of $\alpha(1 \rightarrow 6)$-linked glucoses not including the nonreducing end. These data were interpreted[66] as indicating that the terminal nonreducing glucose was held more firmly in the W3129 site, perhaps in a small cavity capable of more contacts, whereas in QUPC52, the internal chain of six $\alpha(1 \rightarrow 6)$-linked glucoses fitted into a groove-type site. Especially important was the finding that anti-idiotypic monoclonals to W3129 and to QUPC52 had distinctly different idiotypic specificities, a property which will be extremely important for recognizing cavity and groove-type sites.[70] Bennett and Glaudemans[71] confirmed the specificity of W3129 for the terminal nonreducing end and obtained the same K_a value for the linear dextran by fluorescence quenching as we had for the isomaltoheptitol by equilibrium dialysis, fluorescence quenching, and precipitin inhibition. Cloning and sequencing of W3129 by Borden and Kabat[72] made possible an attempt at molecular modeling. W3129 had a small cavity into which the nonreducing glucose would fit with the rest of the site being a groove, whereas 19.1.2, a groove-type antibody, showed only a groove.[31]

Additional immunochemical insight into site structure was obtained by the Glaudemans group.[73,74] They synthesized mono- and oligosaccharides of the isomaltose series in which fluorine or hydrogen replaced various hydroxyl groups. Fluorescence-quenching data were used to provide insight into the relative contribution to the K_a value of binding at various positions. Loss of or reduced binding by compounds with OH → H substitutions at that position was related to the bonding H being donated to or accepted from the protein. The fluorine-substituted compounds permit one to distinguish the two possibilities.[74] If both the OH → H and OH → F substitutions reduce or eliminate binding, one infers donation of the H of the OH, whereas if OH → F does not influence binding but OH → H reduces binding, this is considered as due to H-bond formation with the oxygen of the OH. If binding is not affected by either substitution, this suggests absence of H-bond formation. The binding data for these compounds for two cavity-type anti-$\alpha(1 \rightarrow 6)$-dextrans W3129 and 16.4.12E

[70] P. Borden and E. A. Kabat, *Proc. Natl. Acad. Sci. U.S.A.* **84,** 2240 (1987).

[71] L. G. Bennett and C. P. J. Glaudemans, *Carbohydr. Res.* **72,** 315 (1979).

[72] P. Borden and E. A. Kabat, *Proc. Natl. Acad. Sci. U.S.A.* **84,** 2440 (1987).

[73] C. P. J. Glaudemans, P. Kovac, and A. S. Rao, *Carbohydr. Res.* **190,** 267 (1989).

[74] E. M. Nashed, G. R. Perdomo, E. A. Padlan, P. Kovac, T. Matsuda, E. A. Kabat, and C. P. J. Glaudemans, *J. Biol. Chem.* **265,** 20699 (1990).

clearly showed that the major binding site in both antibodies is for the terminal nonreducing glucose and that all four hydroxyl groups in this glucose contribute to binding, thus defining a cavity-type site. Studies by Lemieux *et al.*[75] on a monoclonal antibody to human blood group I substance with various mono- and oligosaccharides gave similar findings indicative of the nonreducing galactose being in a small cavity. It would be of interest to introduce deuterium at nonexchangeable positions to evaluate the influence of steric hindrance. Similar studies on anti-$\alpha(1 \rightarrow 6)$-dextrans have defined groove-type sites.[66] It is highly significant that from an X-ray study of anti-single-stranded DNA, Smith *et al.*[76] stated, "based on Fourier analysis of X-ray diffraction patterns from crystalline Fab fragments of Bv01-01, the active site was a large irregular groove of the general type postulated by Kabat for linear polymers." It is evident that further studies of linear polymers and the correlation of fine structure mapping with three-dimensional structure by modeling as well as by X-ray crystallography have much to contribute to understanding antibody combining sites.

There could be some confusion as to the measurement of the site size, since one group[66] uses oligosaccharides and the other methyl glycosides of oligosaccharides.[71,73] Thus with W3129, the latter report the methyl glycoside of isomaltotetraose to be a subsite with the fourth sugar from the nonreducing end making a small but significant contribution to binding and defines site size as complementary to four sugars, whereas the other, studying binding of the free sugars, finds a similar small contribution ascribable to the fifth sugar; indeed, this difference is generally due to the small additional binding energy due to the OCH_2 linkage from the fourth to the fifth sugar.

This was clearly seen in the studies of Lemieux *et al.*[75] in which the best oligosaccharide inhibitor of precipitation of anti-blood group IMa was found to be β-D-Gal$(1 \rightarrow 4)\beta$-D-GalNAc$(1 \rightarrow 6)\beta$-D-Gal, with the binding energy of the reducing D-Gal for anti-IMa due exclusively to its OCH_2; changes in the rest of the reducing D-Gal had no effect[77] and the methyl glycoside of β-D-Gal$(1 \rightarrow 4)\beta$-D-GlcNAc-1β-OCH_3 was of comparable activity to the trisaccharide.[77]

[75] R. U. Lemieux, T. C. Wong, J. Liao, and E. A. Kabat, *Mol. Immunol.* **21,** 751 (1984).

[76] R. G. Smith, D. W. Ballard, P. R. Biler, P. E. Pace, A. L. M. Bothwell, J. N. Herron, A. B. Edmundson, and E. W. Voss, Jr., *J. Indian Inst. Sci.* **69,** 25 (1989).

[77] T. Feizi, E. A. Kabat, G. Vicari, B. Anderson, and W. L. Marsh, *J. Immunol.* **106,** 1578 (1971).

Potential Problems

In the original paper by Wu and Kabat in 1970,[12] they foresaw a very important problem in understanding antibody structure when, in describing hypervariable regions, they wrote that "it should be borne in mind that the contour of an antibody site having a given specificity or binding affinity for a given determinant could conceivably be formed by several kinds of amino acid sequence." This has now been confirmed in two studies. Andria *et al.*[78] found that different kinds of antibody combining sites were formed to a decapeptide of tobacco mosaic virus. Four different families of V_κ genes together with three J_κ minigenes, and two A/J antibodies of similar K_a and fine specificity, used very different V_L chains. H chains varied in lengths of CDR3 from 3 to 13 amino acids. Three V_H chains of the J558 family showed extensive differences in their CDRs.

A second instance studied by Nickerson *et al.*[79] has been found between a mouse monoclonal and a human monoclonal from an Epstein–Barr virus (EBV)-transformed cell line which synthesized antibodies to blood group A substance of essentially identical specificity and fine structure with all oligosaccharides tested. The human V_H belongs to the V_HIII subgroup and differs markedly from the mouse V_H with only 56% nucleotide identity between the two and 43% amino acid identity. In the CDRs the two sequences share only 2 of 5 amino acids in CDR1 and 2 of 17 in CDR2. However, there is also an additional shared moiety consisting of Tyr, –, Gly, Ser in CDR2 occurring at positions 52A, 54, and 55 in the human and positions 52, 53, and 54 in the mouse. CDR3 is nine amino acids in length in murine V_H and 12 in the human, with only the last residue, Tyr-102, in common. It is possible that the Gly and Gln at positions 94 and 95 in the mouse and positions 95 and 96 in the human are relevant to the shared binding specificities. The human light chain shows five amino acid differences in CDR1 from mouse anti-A, AC1001, and eight from A003 = 40/567; six and five in CDR2, and six and five in CDR3. One wonders how such differences in sequences of identical fine site specificity will influence the choice of canonical structures.

A major problem arises from the reported unusual cross-reactions for which no structural basis has yet been found. These have been detected exclusively by ELISA and ELISA inhibition. They are produced by a distinct population of B cells, $CD5^+$ (Leu 1) cells in humans[80] and Ly1 B

[78] M. L. Andria, S. Levy, and E. Benjamini, *J. Immunol.* **144,** 2614 (1990).

[79] K. G. Nickerson, H.-T. Chen, J. Larrick, and E. A. Kabat, in preparation.

[80] K. Hayakawa, R. R. Hardy, M. Honda, L. A. Herzenberg, A. D. Steinberg, and L. A. Herzenberg, *Proc. Natl. Acad. Sci. U.S.A.* **81,** 2494 (1984).

cells[81] in the mouse. These antibodies behave differently from the usual antibodies in that they show cross-reactions with a range of substances. Hybridoma heavy- and light-chain genes have been cloned and sequenced and the results indicate that polyreactive antibodies are encoded by germline genes.[82-85] For example, two mouse IgM_κ monoclonals, E7 and D23, showed strong polyreactivity with actin, tubulin, myosin, DNA, renin, α-fetoprotein, were negative with ovalbumin, but reacted with TNP-, PC-, ARS-, and OX-ovalbumins. Sequencing showed E7 to have one nucleotide substitution in the V_H region from germline gene $V_H1210.7$, which could have been due to V_HD joining, and it belongs to the 36–60 family.[84] D23 also had several substitutions from germline V_H101 in the leader and one at codon 94 with a (Ser-Ala) replacement, and belonged to the Q52 family. The V_κ chains of E7 and D23 differed from one another in length, CDR1 having 11 and 17 residues and CDR3 having 9 and 8 residues, respectively; only 3 of 7 residues in CDR2 were identical. E7 used $J_\kappa 2$ and D23 used $J_\kappa 5$. Several sequenced V_H chains were identical in V_H nucleotides to E7, with several others differing by one to five nucleotides. With D23, differences of one and two nucleotides from other sequenced V_H chains were seen. This wide variation in structure of polyreactive antibodies implies that it will not be easy to explain polyreactivity in structural terms.[84,85]

Conclusion

X-Ray crystallography and two-dimensional NMR are currently the only techniques that allow the direct visualization of antibody–antigen complexes and provide detailed three-dimensional information on the binding of antibody to antigen. Nevertheless, more and more model-building studies are producing structures that can then serve as the basis for definitive analyses of the interaction. It can be expected that, in the very near future, improvements in modeling algorithms and the utilization of data from all possible sources, especially immunochemical, will result in

[81] P. Casali, B. S. Prabhakar, and A. L. Notkins, *Int. Rev. Immunol.* **3,** 17 (1988).

[82] A. B. Hartman, C. P. Mallett, J. Srinivasappa, B. S. Prabhakar, A. L. Notkins, and S. J. Smith-Gill, *Mol. Immunol.* **26,** 359 (1989).

[83] I. Sanz, P. Casali, J. W. Thomas, A. L. Notkins, and J. D. Capra, *J. Immunol.* **142,** 4054 (1989).

[84] R. Baccala, T. V. Quang, M. Gilbert, T. Ternynck, and S. Avrameas, *Proc. Natl. Acad. Sci. U.S.A.* **86,** 4624 (1989).

[85] P. Casali and A. L. Notkins, *Immunol. Today* **10,** 364 (1989).

modeled structures of sufficient reliability to reduce the need for actual three-dimensional structure determination of antibody combining sites.

Acknowledgment

Work of the laboratories at Columbia University is supported by grants from the National Institute of Allergy and Infectious Diseases 1R01 AI-19042 and 1R01 AI-27508 and from the National Science Foundation DMB860078 and DMB-890-1840.

[2] Computer Modeling of Combining Site Structure of Anti-hapten Monoclonal Antibodies

By MICHAEL B. BOLGER and MARK A. SHERMAN

Introduction

Computer models of antibody combining sites can be extremely valuable tools in the study and understanding of small molecule–protein interactions. Rapidly expanding primary sequence data from molecular biology and polymerase chain reaction (PCR)[1] technology have provided a vast source of potential input data for the generation of three-dimensional models. In isolation, three-dimensional (3D) models and X-ray structures of antibody combining sites are of little value. Combined with biochemical binding data, quantitative structure affinity information, and the techniques of empirical force field theoretical calculations, the model or X-ray structure can provide useful information in the following fields.

Therapeutic Drug Monitoring by Radioimmunoassay. High affinity and high specificity are required for radioimmunoassay (RIA) of small drug molecules. Due to the complexity and expense, very few antibodies that are used in clinical assays have been rationally engineered for enhanced affinity and specificity. Antibody models may be used to guide the process of protein engineering to produce exquisite selectivity in anti-drug antibodies.[2]

Idiotype–Antiidiotype Interaction. Molecular mimicry by anti-idiotype

[1] W. D. Huse, L. Sastry, S. A. Iverson, A. S. Kang, M. Alting-Mees, D. R. Burton, S. J. Benkovic, and R. Lerner, *Science* **246,** 1275 (1989).

[2] S. Roberts, J. C. Cheetham, and A. R. Rees, *Nature (London)* **328,** 731 (1987).

antibodies has been demonstrated for a variety of drugs and peptides.[3-5] An example of combining site structure for the idiotype–antiidiotype complex has not yet been determined.[5a] The process of computer model building has the potential to add to our knowledge of the structural basis of antiidiotype recognition and perhaps to aid the process of designing therapeutic agents in autoimmune disease.

Pharmacophore Determination (3D Database Searching). Contemporary rational drug design strategies require a knowledge of the components and/or conformation of a drug molecule that is recognized by the drug receptor (i.e., the pharmacophoric region or active conformation). Such a description results in a concise definition of the minimal requirements for activity and allows the medicinal chemist to search a library of known structures for additional active conformations and new drug moieties. Currently, databases of three-dimensional drug structures are static and pharmacophores are determined from receptor-binding studies. A three-dimensional model of drug–antibody binding can provide additional pharmacophoric conformations that may not be apparent in receptor binding but can result in additional "lead compounds" from 3D database searching.

Mimotopes (Peptide Design). Advances in flexible multiple peptide synthesis technology allow for the preparation of novel peptides that bind with high affinity to monoclonal antibodies.[6] These peptides may have no relationship to the natural antigens in primary sequence but merely mimic the physicochemical interactions in an accurate fashion and result in very high binding affinity and specificity. Computer models of the interaction between antibodies and synthetic peptides might enhance the process of designing peptide mimotopes (*mim*ics of epi*topes*).

Catalytic Antibodies (Antibody Engineering). Polymerase chain reaction (PCR) technology has been applied to the production of Fab fragments of antibodies with potential catalytic properties. The large numbers of monoclonal antibodies and primary sequences that can be produced in this fashion are excellent candidates for analysis with programs designed to predict catalytic activity, such as Enzymix (A. Warshel, Department of Chemistry, USC, discussed below).

[3] D. S. Linthicum and M. B. Bolger, *BioEssays* **3(5),** 213 (1985).

[4] D. S. Linthicum, M. B. Bolger, P. H. Kussie, G. M. Albright, T. A. Linton, S. Combs, and D. Marcheti, *Clin. Chem.* **34(9),** 1676 (1988).

[5] J. J. Langone, ed., this series, Vol. **178.**

[5a] The three-dimensional structure of an idiotope–antiidiotope complex has been determined [G. A. Bentley, G. Boulot, M. M. Riottot, R. J. Poljak, *Nature (London)* **348**, 254 (1990)].

[6] H. M. Geysen, S. J. Rodda, T. J. Mason, G. Tribbick, and P. G. Schoofs, *J. Immunol. Methods* **102,** 259 (1987).

Homology Modeling

Qualitative Models and X-Ray Structures

Based on a physicochemical analysis of proteolytic fragments and the intact immunoglobulin G (IgG) molecule, Noelken *et al.*[7] first proposed a model of three relatively rigid fragments linked by a flexible hinge region. Since then much work has been accomplished in the detailed understanding of the three-dimensional structure of antibodies.[8,9] The first physical, and the first computer, model of immunoglobulin structure was constructed prior to the first X-ray crystallographic structure. Kabat and Wu calculated (ϕ,ψ) angles for all residues of a κ light chain based on amino acid sequence and the average (ϕ,ψ) angles of tripeptides from known X-ray structures.[10] Their model differs significantly from any of the subsequent X-ray structures but did account for alignment of Cys-23 and Cys-88 to produce the well-know intrasubunit disulfide bond of immunoglobulin domains. Subsequent modeling efforts have relied heavily on the results of X-ray crystallographic analysis and have benefited from the high degree of sequence and conformational similarity that exists in the immunoglobulin family [Table I[11-23]; all the structures except D1.3 and NC41 are available

[7] M. E. Noelken, C. A. Nelson, E. C. Buckley, and C. Tanford, *J. Biol. Chem.* **240,** 218 (1965).

[8] D. R. Davies, E. A. Padlan, and D. M. Segal, *Annu. Rev. Biochem* **44,** 639 (1975).

[9] D. R. Davies and E. A. Padlan, *Annu. Rev. Biochem.* **59,** 439 (1990).

[10] E. A. Kabat and T. T. Wu, *Proc. Natl. Acad. Sci. U.S.A.* **69(4),** 960 (1972).

[11] M. Schiffer, R. L. Girling, K. R. Ely, and A. B. Edmundson, *Biochemistry* **12,** 4620 (1973).

[12] R. J. Poljak, L. M. Amzel, B. L. Chen, R. P. Phizackerley, and F. Saul, *Proc. Natl. Acad. Sci. U.S.A.* **71,** 3440 (1974).

[13] O. Epp, E. Latham, M. Schiffer, R. Huber, and W. Palm, *Biochemistry* **14,** 4943 (1975).

[14] M. Marquart, J. Deisenhofer, and R. Huber, *J. Mol. Biol.* **141,** 369 (1980).

[15] W. Furey, B. C. Wang, C. S. Yoo, and M. Sax, *J. Mol. Biol.* **167,** 661 (1983).

[16] R. Satow, G. H. Cohen, E. A. Padlan, and D. R. Davies, *J. Mol. Biol.* **190,** 593 (1986).

[17] A. G. Amit, R. A. Mariuzza, S. E. V. Phillips, and R. J. Poljak, *Science* **233,** 585 (1986); Modified structure, G. A. Bentley, T. N. Bhat, G. Boulot, T. Fischmann, J. Navaza, R. J. Poljak, M. M. Riottot, and D. Tello, *Cold Spring Harbor Symp. Quant. Biol.* **54,** 239 (1989).

[18] S. W. Suh, T. N. Bhat, M. A. Navia, G. H. Cohen, D. N. Rao, S. Rudikoff, and D. R. Davies, *Proteins* **1,** 74 (1986).

[19] S. Sheriff, E. W. Silverton, E. A. Padlon, G. H. Cohen, S. J. Smith-Gill, B. C. Finzel, and D. R. Davies, *Proc. Natl. Acad. Sci. U.S.A.* **84,** 8075 (1987).

[20] E. A. Padlon, E. W. Silverton, S. Sheriff, G. H. Cohen, S. J. Smith-Gill, D. R. Davies, *Proc. Natl. Acad. Sci. U.S.A.* **86,** 5938 (1989).

[21] P. M. Colman, W. G. Laver, J. N. Varghese, A. T. Baker, P. A. Tulloch, G. M. Air, and R. G. Webster, *Nature (London)* **326,** 358 (1987).

[22] J. N. Herron, X. M. He, M. I. Mason, E. W. Voss, and A. B. Edmundson, *Proteins* **5,** 271 (1989).

[23] M. B. Lascombe, P. M. Alzari, G. Boulot, P. Saludjian, P. Tougard, C. Berek, S. Haba, E. M. Rosen, A. Nisonoff, and R. J. Poljak, *Proc. Natl. Acad. Sci. U.S.A.* **86,** 607 (1989).

TABLE I
IMMUNOGLOBULIN X-RAY STRUCTURES[a]

Name (PDB)	Type	Resolution (Å)	Haptens	Affinity (M^{-1})
Mcg[11] (1MCG)	γ Light chain	2.3	Enkephalin, DNP, etc.	10^4-10^6
NEW[12] (3Fab)	Human Fab	2.0	Vitamin K_1 OH	10^5
REI[13] (1REI)	κ Light chain	2.0	—	—
KOL[14] (2FB4)	IgG_1 λ-Fab	1.9	—	—
RHE[15] (2RHE)	Light-chain monomer	1.6	—	—
McPC603[16] (2MCP)	Mouse Fab	2.7	Phosphorylcholine	10^5
D1.3[17]	Mouse Fab	2.8	Lysozyme	10^9
J539[18] (1FBJ)	Mouse IgA Fab	2.7	β-(1→6)-D-Galactan	—
HyHEL-5[19] (2HFL)	Mouse Fab	2.5	Lysozyme	10^{10}
HyHEL-10[20] (2HFM)	Mouse Fab	3.0	Lysozyme	10^9
NC42[21]	Mouse Fab	3.0	Neuraminidase	—
4-4-20[22] (4Fab)	Mouse Fab	2.7	Fluorescein	10^{10}
F19.9[23] (1F19)	Mouse Fab	2.8	*p*-Azobenzene-As	—

[a] PDB, Brookhaven Protein Data Bank; As, arsonate.

from the Brookhaven Protein Data Bank (PDB) for use in model building]. The X-ray data have established the existence of a high degree of structural similarity in the framework region of immunoglobulins. The major job of model builders has been generation of hypervariable loops.

Loop Generation

Homology with X-Ray Structure

Several approaches have been used to model hypervariable loops. The first approach by Padlan *et al.* produced two physical models of MOPC315 [a dinitrophenol (DNP) binding mouse Fab] based on the structure of McPC603.[24] As with most current models the framework residues and intersubunit packing were not changed significantly from the X-ray base structure. Hypervariable loops with equal length to the base structure were maintained, and loops of different length were constructed to maximize structural stability and maintain the ϕ and ψ peptide angles within reasonable limits.[25] This model was able to explain qualitatively most of the data on the interaction of DNP in the combining site. A similar approach was used to make models for EPC109 [anti-β-(2 → 1)-fructosan], W3129 [anti-

[24] E. A. Padlan, D. R. Davies, I. Pecht, D. Givol, and C. Wright, *Cold Spring Harbor Symp. Quant. Biol.* **41,** 627 (1976).

[25] G. N. Ramachandran and V. Sasisekharan, *Adv. Protein Chem.* **23,** 283 (1968).

α-(1 → 6)-dextran], and GAR (anti-flavin).[26-28] We have previously constructed computer models of four mouse Fabs that bind with high affinity (K_d 0.7 nM) to the neuroleptic drug haloperidol.[29] Several restrictions applied to these antibodies improved the quality of the resulting models: (1) the allowed torsion angles on a given amino acid residue were dictated by the identity of its side chain; (2) certain residues within the hypervariable loop are involved in packing interactions with adjacent loop side chains, and therefore were restricted in their positioning[30]; (3) there exists a restricted set of tetrapeptides for which a tight β turn is energetically favorable[31]; and (4) extensive biochemical data on the binding affinity and quantitative structure–activity relationships for a series of haloperidol analogs were used to enhance the validity of the resultant models.[32]

De Novo Loop Generation

In contrast to the previous homology modeling approach, Wu and co-workers predicted the structure of MOPC315 and MOPC167 (anti-phosphorylcholine) based on the X-ray structure of NEW framework but used the *de novo* approach of adjusting (ϕ,ψ) angles to construct hypervariable loops.[33,34] The large number of possible conformations, 1000 to 2000, was reduced to about 5 by restricting the ends of the modeled loops to fit on the base framework structure. Their results have not been tested because the X-ray structures are not yet available. The CONGEN (Conformation Generation) search algorithm was used successfully to generate loops for McPC603 and HyHEL-5.[35] This program uses fixed end points in the framework region for the modeled loop and the Go and Sheraga chain-closure algorithm[36] to generate random conformations of closed loops that are evaluated by minimum energy calculations. The loop closures require computational times (Micro VAX II) of about 20 min for 7 amino acids (3.1 Å root-mean-square total to crystal structure of H2 in

[26] M. Vrana, S. Rudikoff, and M. Potter, *J. Immunol.* **122,** 1905 (1979).

[27] E. A. Padlan and E. A. Kabat, *Proc. Natl. Acad. Sci. U.S.A.* **85,** 6885 (1988).

[28] C. R. Kiefer, B. S. McGuire, E. F. Osserman, and F. A. Garver, *J. Immunol.* **131,** 1871 (1983).

[29] M. A. Sherman and M. B. Bolger, *J. Biol. Chem.* **263(9),** 4064 (1988).

[30] E. A. Kabat, T. T. Wu, and H. Bilofsky, *J. Biol. Chem.* **252,** 6609 (1977).

[31] P. Y. Chou and G. D. Fasman, *Adv. Enzymol.* **47,** 45 (1978).

[32] M. A. Sherman, D. S. Linthicum, and M. B. Bolger, *Mol. Pharmacol.* **29,** 589 (1986).

[33] J. M. Stanford and T. T. Wu, *J. Theor. Biol.* **88,** 421 (1981).

[34] S. E. Coutre, J. M. Stanford, J. G. Hovis, P. W. Stevens, and T. T. Wu, *J. Theor. Biol.* **92,** 417 (1981).

[35] R. E. Bruccoleri, E. Haber, and J. Novontny, *Nature (London)* **335,** 564 (1988).

[36] N. Go and H. A. Sheraga, *Macromolecules* **3,** 178 (1970).

HyHEL-5) and up to 7 days for 12 amino acids (3.0 Å root-mean-square total to crystal structure of L1 in McPC603).

Distance Matrix Searching

A derivative of the interatomic diagonal plot method of Phillips[37] has been implemented in the program FRODO by Jones and Thirup.[38] This procedure is useful in finding loops from a database of known protein structures that can be grafted onto the framework of a modeled antibody. This technique is not limited to known structures of other antibodies, because the database of known structures includes a variety of proteins rich in β sheets, loops, and tight turns. Individually, such distance matrix plots can be used to recognize domains and structural motifs.[39,40] In a searching algorithm, the framework C_α carbons and the number of residues to be included in the loop are specified as inputs. The goodness of fit of each fragment is judged by the sum $\Sigma(d_m - d_n)^2$, where d_m is an inter-C_α distance in one structure and d_n is the equivalent distance in the second structure, and the sum is taken over all matching atoms. The result is a set of loops that can be grafted onto the framework atoms to provide a backbone for building side chains appropriate to the sequence of interest. Such an algorithm has been nicely implemented in the BioSym, Inc., San Diego, CA, program "Insight."

Side-Chain Orientation

Once the backbone of the new structure has been developed, the next step is replacement of amino acids that appeared in the X-ray structure with amino acids from the sequence of the model antibody. Unfortunately, the automated routines found in some graphical modeling packages replace side chains in an arbitrary default conformation. This does not generate the best possible, nor the most correct, model. It has been shown that usually the maximum overlap procedure (MOP), i.e., following the conformation of the previous side chain when making a substitution, is the best procedure for the modeled mutation.[41] When the substitution involves replacing a small side chain with a much larger one, the best guess is to put the new side chain in its most frequently observed conformation.[42] In

[37] D. C. Phillips, *Biochem. Soc. Symp.* **31,** 11 (1970).

[38] T. A. Jones and S. Thirup, *EMBO J.* **5(4),** 819 (1986).

[39] M. G. Rossmann and A. Liljas, *J. Mol. Biol.* **85,** 177 (1974).

[40] F. M. Richards and C. E. Kundrot, *Proteins* **3,** 71 (1988).

[41] N. L. Summers, W. D. Carlson, and M. Karplus, *J. Mol. Biol.* **196,** 175 (1987).

[42] J. W. Ponder and F. M. Richards, *J. Mol. Biol.* **193,** 775 (1987).

addition, the rotamer selected can be chosen on the basis of the type of secondary structure in which the side chain is found.[43] Automated methods are currently being developed to aid the model builder in generating the best possible orientation for amino acid side chains that must be changed. One method, the coupled perturbation procedure (CPP), has been applied to calculating the conformation of a region of the Fab KOL based on the analogous region in the Fab NEW.[44] In the coupled perturbation procedure, minimum energy conformational searches are conducted not only for the side chain(s) being replaced but also for other side chains that are identified as belonging to the set of residues whose conformation is likely to be affected by the substitution. The coupled perturbation procedure applied to the multiple substitutions required to model a complete Fv region would be computationally intensive. Still, the results show favorable agreement with the X-ray crystallographic structures that were the targets of the modeling procedure.

Interface Packing

Nearly all currently available immunoglobulin X-ray structures strongly suggest that the Fab V_H and V_L domains pack against one another in much the same fashion.[45-47] Nearly 1800 Å^2 of protein surface area is buried between the two domains. Three-fourths of the interface residues are contributed by the framework regions and the remaining by the hypervariable (HV) loops. Of the 20 residues buried within the interface, 12 are absolutely or very strongly conserved in all known sequences. As expected, these residues form the central and lower regions of the interface. The eight residues that tend to vary somewhat in sequence identity are all located in the upper regions of the interface, adjacent to the HV loops. As such, these residues affect HV loop conformation and therefore need to be carefully evaluated in model-building studies.[29,48]

Two unrelated X-ray structures (LOC, NC41) have displayed alternative packing arrangements for V_H-V_L and V_L-V_L dimers which suggest that the two domains can slide past one another under appropriate conditions.[21,49] However, it has been suggested that the unusual packing ob-

[43] M. J. McGregor, S. A. Islam, and M. J. E. Sternberg, *J. Mol. Biol.* **198,** 295 (1987).

[44] M. E. Snow and L. M. Amzel, *Proteins* **1,** 267 (1986).

[45] C. Chothia, J. Novotny, R. Bruccoleri, and M. Karplus, *J. Mol. Biol.* **186,** 651 (1985).

[46] J. Novotny and E. Haber, *Proc. Natl. Acad. Sci. U.S.A.* **82,** 4592 (1985).

[47] C. Chothia and A. M. Lesk, *J. Mol. Biol.* **196,** 901 (1987).

[48] P. de la Paz, B. J. Sutton, M. J. Darsley, and A. R. Rees, *EMBO J.* **5,** 415 (1986).

[49] C. H. Chang, M. T. Short, F. A. Westholme, F. J. Stevens, B. C. Wang, W. Furey, A. Solomon, and M. Schiffer, *Biochemistry* **24,** 4890 (1985).

served in the LOC dimer is an artifact of the conditions used in crystallization (a second structure that matches the "consensus" arrangement has been solved for the same protein).[50] With the addition of several new Fab structures to the database, the unusual packing of NC41 is now viewed as being well within the range of natural variability.[23] Thus, it appears likely that sequence differences at the domain interface do affect the exact positioning of the heavy-chain complementarity-determining region (CDR) loops relative to the light chain CDRs.

Experimental evidence suggests that loop flexibility at the combining site may compensate for domain shifts of this sort. It was shown that upon grafting the heavy-chain CDR loops of a mouse antibody onto a human heavy-chain framework, full antigen binding activity was restored when the hybrid heavy chain was permitted to reassociate with the original mouse light chain, despite extensive sequence differences (56% identity) in the mouse vs human heavy-chain framework regions.[51] Thus, it appears that valid models can still be generated by grafting CDR loops onto a preexisting framework, provided that allowances are made for possible shifts in heavy- vs light-chain CDRs during antigen docking experiments. Additional Fab crystal structures should help define the key interface packing interactions responsible for the exact positioning of the heavy chain relative to the light chain.

Framework Overlap and Sequence Homology

Tertiary structure in framework regions of V_H and V_L domains is highly conserved despite sequence identity as low as 60%. Yet one cannot assume that framework residues have little or no effect on the conformation adopted by the attached CDR loops. Several studies have emphasized the importance of certain key framework residues in determining the overall architecture of the antigen combining site. For example, Chothia and Lesk[47] have suggested that the identity of the residue at position 71 of the heavy chain is largely responsible for the positioning of the β turn at the tip of H2. An arginine at this position tilts the adjacent H2 loop toward the center of the combining site, whereas an alanine in this position causes the H2 loop to tilt back such that proline-52a fills the cavity created by the small framework side chain.[52] Framework-induced changes in CDR con-

[50] M. Schiffer, C. Ainsworth, Z.-B. Xu, W. Carperos, K. Olsen, A. Solomon, F. J. Stevens, and C. H. Chang, *Biochemistry* **28,** 4066 (1989).

[51] P. T. Jones, P. H. Dear, J. Foote, M. S. Neuberger, and G. Winter, *Nature (London)* **321,** 522 (1986).

[52] A. Tramantano, C. Chothia, and A. M. Lesk, *J. Mol. Biol.* **215,** 175 (1990).

formation have also been observed indirectly through work on mutant antibodies. Novotny *et al.* examined various digitoxin-binding antibodies in which single mutations occur in framework regions adjacent to H3.[53] Based on computer-modeling studies, the authors suggested that the loss of a hydrogen bond between arginine-94 (heavy-chain framework) and aspartate-101 (H3) upon mutation of residue 94 to serine caused the specificity of the mutant antibody to change relative to the parent molecule. Apparently, H1 is subject to similar perturbations. In attempting to graft CDR loops from a rat anti-lymphocyte antibody onto a human framework, Reichmann *et al.* discovered that the hybrid antibody bound antigen poorly until serine-27 (heavy-chain framework) was changed to phenylalanine (as found in most antibodies) via site-directed mutagenesis.[54] As a result, side-chain packing against H1 was altered such that antigen-binding activity was restored. Therefore, given the key role played by certain framework residues in determining the conformation of adjacent hypervariable loops, it is essential that all CDR loops be modeled in the context of their surroundings rather than in isolation.

Possible Correlation of Loop Length and Type of Antigen (Protein vs Small Molecule)

Although sequence identity in the HV region is largely responsible for determining the fine specificities of a given antibody, X-ray data suggest that gross specificity (protein vs small molecule antigen) is largely determined by the lengths of the various CDR loops. Antibodies specific for small haptens often have concave combining sites, frequently found as a deep "pocket" or "groove." Modeling studies suggest that such pockets can often be formed by relatively long L1 and/or H3 loops. Antibodies that bind larger molecules, such as proteins, tend to have flat combining sites such that the surface of antibody–antigen contact is quite large.[17] Flat surfaces appear to result from relatively short L1 and H3 loops.[48] However, one must stress the term "relative" when describing loop lengths: a very short H3 loop (eight amino acids) in combination with other CDR loops of average size also will produce a pocket, as has been illustrated.[55] Nonetheless, given only the lengths of the CDR loops and limited sequence data for a given antibody, it is often possible to construct crude Fv region models that provide much insight into the overall architecture of the combining site.

[53] J. Novotny, R. E. Bruccoleri, and E. Haber, *Proteins* **7,** 93 (1990).

[54] L. Reichmann, M. Clark, H. Waldmann, and G. Winter, *Nature (London)* **332,** 323 (1988).

[55] L. Holm, L. Laasonen, M. Kaartinen, T. T. Teeri, and J. K. C. Knowles, *Protein Eng.* **3(5),** 403 (1990).

Empirical Force Field Calculations

Application of an empirical force field to optimization of the geometries of molecular models (molecular mechanics) has been an important technical advance over the last 10 years. The force field is a set of equations that describe the geometric properties of a molecule in terms of bond length, bond angle, dihedral angle, and nonbonded interactions. The force field is parameterized by filling a table with optimal values of the properties above for each type of bond, angle, and through space interaction expected to be found in the target molecule. Optimal values are obtained from X-ray crystallographic data and spectroscopic data (UV, IR) for small molecule examples. It has been shown that the atomic properties of small molecules are shared by larger macromolecules and that the results of such calculations have predictive validity. A few well-known programs that minimize the calculated empirical energy of a molecule include the following: QCFFPI,[56] MM2,[57] CHARMM,[58] SYBYL,[59] and AMBER.[60] The principal aim of energy minimization is to improve the stereochemistry of the model by relaxing strained bond lengths and angles, and relieving unfavorable contacts between nonbonded atoms.

The first attempt to use interactive computer graphics manually to construct hypervariable loops was conducted by Feldmann *et al.* for J539 based on the X-ray structure of McPC603.[61] Unfortunately, the model displayed loops containing residues with forbidden (ϕ,ψ) combinations. The J539 model was subsequently refined and subjected to energy minimization using the program CHARMM to produce a model that was significantly improved.[62] Subsequently, most immunoglobulin models have geometries that have been optimized by molecular mechanics. Several anti-lysozyme Fv regions have been modeled using molecular graphic splicing techniques and energy minimization.[48,63,64] The model that was developed for D1.3 has been confirmed by X-ray crystallography. When

[56] S. Lifson and A. Warshel, *J. Chem. Phys.* **49,** 5116 (1968).

[57] N. L. Allinger, Y. Yuh, *Quantum Chem. Program Exch.* **13,** 395 (1980).

[58] B. R. Brooks, R. E. Bruccoleri, B. D. Olafson, D. J. States, S. Swaminathan, and M. Karplus, *J. Comput Chem.* **4,** 187 (1983).

[59] *SYBYL Molecular Modeling System,* Tripos Associates Inc., St. Louis, MO (1985).

[60] U. C. Singh, P. K. Weiner, J. W. Caldwell, and P. A. Kollman, "*AMBER,* 3.0," Department of Pharmaceutical Chemistry, UCSF (1986).

[61] R. J. Feldman, M. Potter, and C. P. J. Glaudemans, *Mol. Immunol.* **18(8),** 683 (1981).

[62] C. R. Mainhart, M. Potter, and R. J. Feldmann, *Mol. Immunol.* **21,** 469 (1984).

[63] C. Chothia, A. M. Lesk, M. Levitt, A. G. Amit, R. A. Mariuzza, S. E. V. Phillips, and R. J. Poljak, *Science* **233,** 755 (1986).

[64] S. J. Smith-Gill, C. Mainhart, T. B. Lavoie, R. J. Feldmann, W. Drohan, and B. R. Brooks, *J. Mol. Biol.* **194,** 713 (1987).

the modeled hypervariable loops of D1.3 are compared to the X-ray root-mean-square, deviations range from 0.5 to 2.07 Å.[63] A similar comparison for the model of HyHEL-10 with the X-ray structure showed deviations in C_α positions from 0.44 to 4.27 Å.[64] Considering that the average resolution of X-ray crystallographic structures in Table I is 2.43 Å, the results of modeling even the most difficult portions of the immunoglobulin structure are quite good. In fact, a revised structure of D1.3 is even more in agreement with the predicted model.[17] In addition, modeling combined with protein engineering has resulted in enhance affinity for anti-lysozyme "loop" region peptide antibodies.[65]

Molecular Dynamics

Although many antibodies have a well-defined equilibrium geometry, their flexibility and structural fluctuations play an important role in their activity. Functional binding of antigens to the combining site sometimes requires conformational adjustments in both the ligands and the antibody. In a molecular dynamics simulation, one assigns initial positions and velocities to all the atoms in the system of interest, then solves the classical simultaneous equations of motion for the atoms with forces determined from a known potential (empirical force field) over a given time period.[66] The resulting atomic trajectories are analyzed to determine the motion of atoms with respect to time at a given temperature. An obvious advantage of a dynamic simulation compared to energy minimization is that the system is not constrained to the local minimum closest to the starting conformation and a more complete search of structural space can be achieved. Many programs associated with the force fields described can extend the calculation of minimum energy conformation to the dynamic properties of proteins.

Beyond molecular dynamics, one would like to be able to calculate the energetics of antigen binding and the activation barriers in catalytic antibodies. A new method has been developed to calculate relative changes in free energy between small molecules and proteins. It is based on a thermodynamic perturbation method with molecular dynamics being used to "mutate" the system from one state to another.[67-69] This type of calcula-

[65] S. Roberts, J. C. Cheetham, and A. R. Rees, *Nature (London)* **328,** 731 (1987).

[66] M. Karplus and J. A. McCammon, *Crit. Rev. Biochem.* **9,** 293 (1981).

[67] A. Warshel, *Proc. Natl. Acad. Sci. U.S.A.* **81,** 444 (1984).

[68] P. A. Bash, U. C. Singh, F. K. Brown, R. Langridge, and P. A. Kollman, *Science* **235,** 574 (1987).

[69] T. P. Lybrand, J. A. McCammon, and G. Wipff, *Proc. Natl. Acad. Sci. U.S.A.* **83,** 833 (1986).

tion evaluates a change in free energy ($\Delta\Delta G$) associated with relatively rare events such as spontaneous mutation of an amino acid, bond breaking in catalysis, or modification of the structure of a bound ligand. A direct simulation is impossible because the dynamic trajectories will never reach the activation barriers of such processes in a feasible simulation time. Consequently, several techniques have been developed to overcome this problem. One method that has been successful is the empirical valence bond (EVB) approach as implemented in the program Enzymix.[70] This program is a macromolecular simulation package designed to study the functions of proteins ranging from ligand binding to enzymatic free energy profiles. The EVB method is a simple way of including quantum mechanics into a free energy perturbation/molecular dynamics simulation. Using the EVB method it is possible to translate a postulated mechanism for enzymatic activity or hapten binding into a force field that can be used for calculation of the free energy profile of the postulated mechanism. The use of programs *in parallel* with experimental studies is a powerful way of determining reaction mechanisms or the molecular characteristics of small molecule–protein interactions.

Biochemical Validation by X-Ray Crystallography

The comparison of a predicted structure with a subsequently determined crystal structure is the most stringent and most desirable test of any modeling procedure. To date, five Fab structures (HyHEL-5, HyHEL-10, NC41, NQ10, and D1.3), previously modeled using the canonical structures approach, have been largely confirmed by subsequent crystallography data. Predicted vs observed main-chain conformations for all HV loops except H3 were in close agreement. All root-mean-square differences in atomic position were less than 1.3 Å. Upon superposition of predicted and observed framework residues, differences in the positions of CDR C_{α}s ranged from 0.4 to 3.5 Å. Thus, although all main-chain conformations were essentially correct, the relative positioning of the CDR loops relative to one another was not always correct. The differences are attributed to inability to predict correctly the affect of changes in inter- and intradomain side-chain packing interactions in the framework regions. Conformational changes induced by antigen binding were also cited as possible sources of error, since several of the observed structures consisted of antibody–antigen complexes. The results of these studies suggest that modeling is a valid substitute for X-ray structure determination when the model is to be used in a qualitative sense. Crystallography will continue to be the method

[70] A. Warshel, F. Sussman, and J. K. Hwang, *J. Mol. Biol.* **201,** 139 (1988).

of choice for studying combining site anatomy until the predictive algorithms are refined to the point of being able not only to reproduce the experimentally observed structure, but also to predict accurately the strength of binding interactions and kinetic constants of catalysis induced by the immunoglobulin.

Summary of Current Methods for Amino Acid Sequence Determination

Primary structure is the starting point for all modeling studies and we present a short summary of the most current methods. Because of major advances in the synthesis of oligonucleotides of predefined sequence, the determination of immunoglobulin variable region amino acid sequence via gene sequencing is now straightforward. Several approaches are currently being employed. Perhaps the most direct method involves sequencing heavy- and light-chain immunoglobulin mRNA isolated from hybridoma cells.[71-74] Oligonucleotides that specifically bind to conserved sequences near the variable–constant region junction of immunoglobulin mRNA are extended in the presence of dideoxynucleotides using reverse transcriptase to generate the sequencing ladder. Sequence ambiguities that commonly arise from failure of the enzyme to read through regions of strong secondary structure (a band across all four lanes) can often be resolved by incorporating terminal deoxynucleotidyltransferase into the sequencing protocol.[75] Fragments that terminated prematurely (lack of a 3′-dideoxynucleotide) continue to extend through random addition of deoxynucleotides, thus revealing the correct band. Others have chosen to avoid the problem altogether by end labeling the primer and omitting the dideoxynucleotides from the reaction[76-78]; full-length cDNA is then isolated and sequenced using a modified version of the base-specific chemical

[71] P. H. Hamlyn, G. G. Brownlee, C. C. Cheng, M. J. Gait, and C. Milstein, *Cell (Cambridge, Mass.)* **15,** 1067 (1978).

[72] M. Kaartinen, G. M. Griffiths, P. H. Hamlyn, A. F. Markham, K. Karjalainen, J. L. T. Pelkonen, O. Makela, and C. Milstein, *J. Immunol.* **130(2),** 937 (1983).

[73] G. M. Griffiths and C. Milstein, in "Hybridomas Technology in the Biosciences and Medicine" (T. A. Springer, ed.), p. 103. Plenum, New York, 1985.

[74] M. A. Sherman, R. J. Deans, and M. B. Bolger, *J. Biol. Chem.* **263(9),** 4059 (1988).

[75] D. C. DeBorde, C. W. Naeve, M. L. Herlocher, and H. F. Maassab, *Anal. Biochem.* **157,** 275 (1986).

[76] M. J. Schlomchik, D. A. Nemazee, V. L. Sato, J. V. Snick, D. A. Carson, and M. G. Weigert, *J. Exp. Med.* **164,** 407 (1986).

[77] A. K. Sood, H. L. Cheng, and H. Kohler, *J. Immunol. Methods* **95,** 227 (1986).

[78] D. J. Panka and M. N. Margolies, *J. Immunol.* **13,** 2385 (1987).

cleavage method.[79] This approach is especially useful for sequencing light-chain mRNA, which is often contaminated with transcripts arising from nonfunctional gene rearrangements.[77] The isolated full-length cDNA also can serve as a template for synthesis of double-stranded cDNA, which can then be cloned and sequenced using double-stranded plasmid sequencing.[80]

Variable region sequences are also easily determined using the polymerase chain reaction (PCR).[81] Pairs of mixed or degenerate oligonucleotide primers designed to hybridize to either end of practically any mouse heavy- or light-chain immunoglobulin mRNA have developed.[82,83] The primers also incorporate restriction sites that allow the cDNA to be force cloned into sequencing or expression vectors. This approach is especially useful when subsequent genetic manipulation is desirable (i.e., site-directed mutagenesis). As a result, the PCR method is currently the method of choice for V region sequence determination.

Modeling Antigen–Antibody Interactions

Several unique approaches to docking ligands to macromolecules have been developed. Developments in computer graphics have made the task of determination of bound geometry much easier, but the typical process still depends on chemical intuition.[84] Extensive quantitative structure activity and spectroscopic data are essential in validating the placement of hapten molecules in the combining site of antibodies. Displaying the chemical environment of a receptor site as a potential energy grid can help guide the chemist in modeling the most likely bound conformation.[85] When the structures of the ligand and protein are both well known, a strictly geometric approach (DOCK) can be used to find the possible modes of docking.[86] DOCK is a set of programs that automatically finds orientations of a small

[79] A. L. Maxam and W. Gilbert, this series, Vol. **65,** p. 499.

[80] T. Matsuda and E. A. Kabat, *J. Immunol.* **142(3),** 863 (1989).

[81] R. K. Saiki, S. Scharf, F. Faloona, K. B. Mullis, G. T. Horn, H. A. Ehrlich, and N. Arnheim, *Science* **230,** 1350 (1985).

[82] R. Orlandi, D. H. Gussow, P. T. Jones, and G. Winter, *Proc. Natl. Acad. Sci. U.S.A.* **86,** 3833 (1989).

[83] L. Sastry, M. Alting-Mees, W. D. Huse, J. M. Short, J. A. Sorge, B. N. Hay, K. D. Janda, S. J. Benkovic, and R. A. Lerner, *Proc. Natl. Acad. Sci. U.S.A.* **86,** 5728 (1989).

[84] J. M. Blaney, E. C. Jorgensen, M. L. Connolly, T. E. Ferrin, R. Langridge, S. J. Oatley, J. M. Burridge, and C. C. F. Blake, *J. Med. Chem.* **25,** 785 (1982).

[85] P. J. Goodford, *J. Med. Chem.* **28,** 849 (1985).

[86] I. D. Kuntz, J. M. Blaney, S. J. Oatley, R. Langridge, and T. E. Ferrin, *J. Mol. Biol.* **161,** 269 (1982).

molecule in a receptor site (process outlined below). The receptor site is represented as a cluster of overlapping spheres that characterize the shape of invaginations in the receptor site surface. It can be used in "single-ligand mode" to find multiple orientations of one hapten molecule in a site and score each orientation with a simple scoring routine that measures the geometric complementarity of the orientation for the site. It also can be used in "search mode" to orient many small molecules, score them, sort the molecules by the score of their best orientation, and write the highest scoring molecules to an output file. This technique has been used to discover a nonpeptide inhibitor of the HIV-1 protease.[87] A cluster of DOCK spheres was compared for steric complementarity with 10,000 molecules of the Cambridge Crystallographic Database and 1 molecule (haloperidol) was selected and found to inhibit protease activity with a K_i of about 100 μM. Although this compound is not a candidate for treatment of autoimmunodeficiency syndrome (AIDS) because of severe side effects at the doses required for inhibition of protease activity, it does provide an important target molecule for further drug discovery activity.

When docking of flexible ligands is required, this same approach has been successfully modified by allowing fragments of the complete ligand to be docked in unique orientations that tend to fill the receptor site, followed by reconnecting of the fragments and energy minimization to generate the best fit.[88] Finally, a sophisticated approach known as "simulated annealing" has been applied to docking of a flexible ligand to an antibody. Phosphorylcholine was rotated 180°, translated 5 Å away from the combining site, and twisted 90° about the two central torsion angles. The binding simulation yielded a final conformation with a low root-mean-square difference of 0.97 Å from the crystallographically observed conformation.[89]

Outline of Modeling Procedure Applied to Anti-Haloperidol Fab (MoAbA)

The anti-haloperidol mouse monoclonal Fv will be used as an example to illustrate the methods used in modeling the combining site of an anti-hapten antibody.

[87] R. L. DesJarlais, G. L. Seibel, I. D. Kuntz, P. S. Furth, J. C. Alvarez, P. R. Ortiz De Montellano, D. L. DeCamp, L. M. Babe, and C. S. Craik, *Proc. Natl. Acad. Sci. U.S.A.* **87,** 6644 (1990).

[88] R. L. DesJarlais, R. P. Sheridan, J. S. Dixon, I. D. Kuntz, and R. Venkataraghavan, *J. Med. Chem.* **29,** 2149 (1986).

[89] D. S. Goodsell and A. J. Olson, *Proteins* **8,** 195 (1990).

TABLE II

ALIGNMENT OF Fv REGIONS FOR MoAbA and Jy HEL-5 ANTI-LYSOZYME[a]

			Light Fv			
			| L1	|	|	
MoAbA	DILMTQSQKF	MSTSVGDRVS	VTCKASQNVG	NNVAWHQQKP	GQSPKALIYS	50
	** ***	** * * *	** ** *	* * ***	* *** **	
2HFL	DIVLTQSPAI	MSASPGEKVT	MTCSASSSV.	NYMYWYQQKS	GTSPKRWIYD	50
	L2 |			|	L3 |	
MoAbA	ASYRYSGVPD	RFTGSGSGTD	FTLTITNVQS	EDLAEYFCQQ	YNSYPYTFGG	100
	* ****	** ******	***	** *** ***	* ****	
2HFL	TSKLASGVPV	RFSGSGSGTS	YSLTISSMET	EDAAEYYCQQ	WGRNP.TFGG	99
MoAbA	GTKLEIK 107					
	*******					
2HFL	GTKLEIK 105					
			Heavy Fv			
				| H1 |	|	
MoAbA	QVQLQQSGPE	LVKPGASVRI	SCKASGYTFT	RYYIHWLKQR	PGQGPEWIGW	50
	******* *	* ****** *	*********	* * * ***	** * ****	
2HFL	ZVQLQQSGAE	LMKPGASVKI	SCKASGYTFS	DYWIEWVKQR	PGHGLEWIGE	50
	H2 |		71		|	
MoAbA	IYPGNVNTKY	NEKFKGKATL	TADKSSSTAY	LQLSSLTSED	SAVYFCAREG	100
	* ** * *	* ******	**********	** ******	* ** *	
2HFL	ILPGSGSTNY	HERFKGKATF	TADTSSSTAY	MQLNSLTSED	SGVYYCL . . .	97
	H3 |					
MoAbA	SYEYDEADYW	GQGTTLTVSS 120				
	** * *	**********				
2HFL	HGNYD.FDGW	GQGTTLTVSS 116				

[a] Amino acids are indicated in the standard single-letter amino acid code with the exception of Z, N-terminal proline (PCA). The sequence alignment of MoAbA and HyHEL-5 has an overall homology of 65%. Light-chain homology is 58% and heavy-chain homology is 71%. Average CDR homology is 31%. CDRs and residue 71 in the heavy chain are indicated above the alignment.

Sequence Alignment and Canonical Structure Determination

Tables II and III show the alignment of MoAbA with HyHEL-5 and provide information about canonical structures appropriate for modeling L1, L3, and H3 loops.

Selection of X-Ray Structure for Framework Region

Following the sequence alignment and determination of canonical structures, the choice of a framework X-ray structure had to be made. In the case of MoAbA, HyHEL-10 looked like a good candidate because of the identical canonical structures for L1, L2, and L3. It was apparent from a comparison of loop lengths that H3 would have to come from McPC603. However, the framework residue H71 became critical in selection of the appropriate X-ray structure. Only HyHEL-5 matched MoAbA in having an alanine at position H71. This meant that the orientation of H2 in those two antibodies would most likely be similar. If HyHEL-10 was used for the framework coordinates, then the large size of arginine at position H71 would create an inappropriate structure for H2 in MoAbA.

Selection of X-Ray Structures for Loop Modeling

Once the framework structure had been determined, the selection of X-ray sources for loop modeling became much easier. Replacements for L1, L3 and H3 of HyHEL-5 needed to be found. L1 and L3 from HyHel-10, and H3 from McPC603, were selected on the basis of sequence homology and appropriate length. By finding loops of exactly the right length and canonical structure in the Protein Data Bank (PDB), one avoids the necessity of *de novo* loop building as in our previous models.

Alignment of New Loops with Existing Loops

An interactive graphics workstation with MIDAS[90] software was employed for the initial loop alignments. Any text editor can be used to create small PDB files with just CDR loop coordinates. For L1, Cys-23 and Trp-35 were also taken from HyHEL-10 since these residues matched the corresponding framework residues in MoAbA exactly. Similarly for L2 taken from HyHel-10, residues Ile-48, Lys-49, and Gly-57 were also included for backbone matching. For H3 taken from McPC603, residues Ala-93, Arg-94, Trp-103, and Gly-104 were included. First, the complete HyHEL-5 PDB file was opened in MIDAS. Then each loop PDB file was

[90] R. Langridge, T. E. Ferrin, and C. Huang, *UCSF Computer Graphics Laboratory,* San Francisco, CA, 1990.

TABLE III

Canonical Structures of Importance in Modeling MoAbA

Canonical structure	Protein	26	27	28	29	30	31	a	b	c	d	e	f	32	2	25	33	71
L1																		
1	HyHEL-5	S	S	S	V	N	—	—	—	—	—	—	—	Y	I	A	M	Y
2	MoAbA	S	Q	N	V	G	N	—	—	—	—	—	—	N	I	A	V	F
2	HyHEL-10	S	Q	S	I	G	N	—	—	—	—	—	—	N	I	A	L	F

Canonical structure	Protein	50	51	52	48	64
L2						
1	MoAbA	S	A	S	I	G
1	HyHEL-10	Y	A	S	I	G
1	HyHEL-5	D	T	S	I	G

Canonical structure	Protein	91	92	93	94	95	96	90
L3								
1	MoAbA	Y	N	S	Y	P	Y	Q
1	HyHEL-10	S	H	S	W	P	Y	Q
3	HyHEL-5	W	G	R	N	P	—	Q

Canonical structure	Protein	26	27	28	29	30	31	32	34	94
H1										
1	MoAbA	G	Y	T	F	T	R	Y	I	R
1	HyHEL-5	G	Y	T	F	S	D	Y	I	R
1′	HyHEL-10	G	D	S	I	T	D	D	W	N

Canonical structure	Protein	52a	b	c	53	54	55	71
H2								
1	HyHEL-10	—	—	—	Y	S	G	R
2	MoAbA	P	—	—	G	N	V	A
2	HyHEL-5	P	—	—	G	S	G	A
3	J539	P	—	—	D	S	G	R
4	McPC603	N	K	G	N	K	Y	R

Canonical structure	Protein	95	96	97	98	99	100	a	b	c	101	102
H3												
?	MoAbA	E	G	S	Y	E	Y	D	E	A	D	Y
?	McPC603	N	Y	Y	G	S	T	W	Y	F	D	V
?	HyHEL-10	—	—	—	—	N	—	W	D	G	D	Y
?	HyHEL-5	H	G	N	Y	D		F	D	G		

opened as a separate model number. Figure 1 shows the results of loop alignments. Inclusion of framework residues from HyHEL-10 and McPC603 is important in the alignment procedure to establish proper overlap with the HyHEL-5 framework. The hydrophobic residue in position L29 of L1 is shown in Fig. 1a. When the PDB file for MoAbA was finished, the Ile of HyHEL-10 was easily swapped for the Val of MoAbA by simply removing the C_δ methyl group. L3 of HyHEL-10 and MoAbA is one residue longer than HyHEL-5 and falls into canonical structure 1 (see Table III). The characteristic position of Pro-95 can be seen in Fig. 1b. Although canonical structures have not been defined for H3, McPC603 and MoAbA have the same number of residues in H3 and have good homology in the framework region surrounding H3. Alignment of H3 from McPC603 and HyHEL-5 shows very little overall similarity, as seen

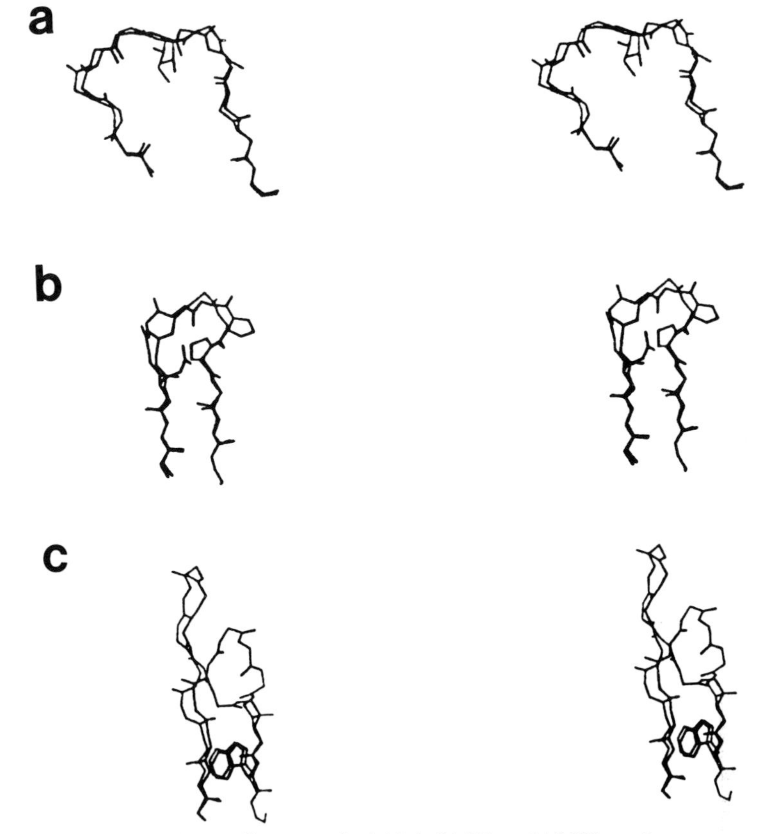

FIG. 1. Loop alignments for (a) L1, (b) L3, and (c) H3 overlaps.

in Fig. 1c. Yet framework residues Ala-93, Arg-94, and Trp-103 overlap nicely. Once aligned, the coordinates of each model can be copied to PDB files with the new overlapped coordinates using the MIDAS *Write* command. In order to identify the new PDB files as being models, MIDAS inserts a new line after every ATOM record label in the PDB file that reads "USER MIDAS 8001 8 8 8." This can be removed with a short Fortran, or C, routine to conserve disk space and make file editing easier. Again using a text editor, all the framework and loop atoms from HyHEL-5 were carefully removed and the newly aligned atoms from HyHEL-10 and McPC603 were copied into the new MoAbA PDB file.

Mutation of Amino Acids

Back in MIDAS, each residue in the light and heavy chain was checked for correspondence with the actual MoAbA sequence. The *Swapaa* routine in MIDAS can be used to mutate residues that need to be changed. Each replaced residue was oriented to have maximal overlap with the original residue side chain. Side-chain torsion angles were compared to the common values of Ponder and Richards.[42] In addition, the *Watch* routine in MIDAS was used visually to minimize bad contacts. When a bond rotation is started with MIDAS and *Watch* is active, atoms within 1.4 Å are connected with a dashed line (Fig. 2). In this manner bad contacts can be minimized quickly prior to force field energy minimization. Finally, a folded conformation of haloperidol was tentatively positioned in the combining site as previously described.[29]

Energy Minimization of Antibody Structure

Finally, CHARMM was used to minimize the total energy of the final model. Initially, CHARMM found 22 close contacts (< 1.4 Å), primarily in the newly modeled loop regions. This resulted in an initial force field energy of 3.496 million units. The van der Waals (VDW) term of the force field alone accounted for 3.495 million units. Following 1000 iterations of adopted basis Newton–Raphson energy minimization, the final energy was −7832 units. The VDW term had dropped to −664 units. A solvent-excluded molecular surface was calculated with the MS program and the final model displayed using MIDAS (Fig. 3). Haloperidol is shown in a tentative position in the combining site. Automated docking of this hapten molecule using DOCK is currently underway as described below for phosphorylcholine binding to McPC603.

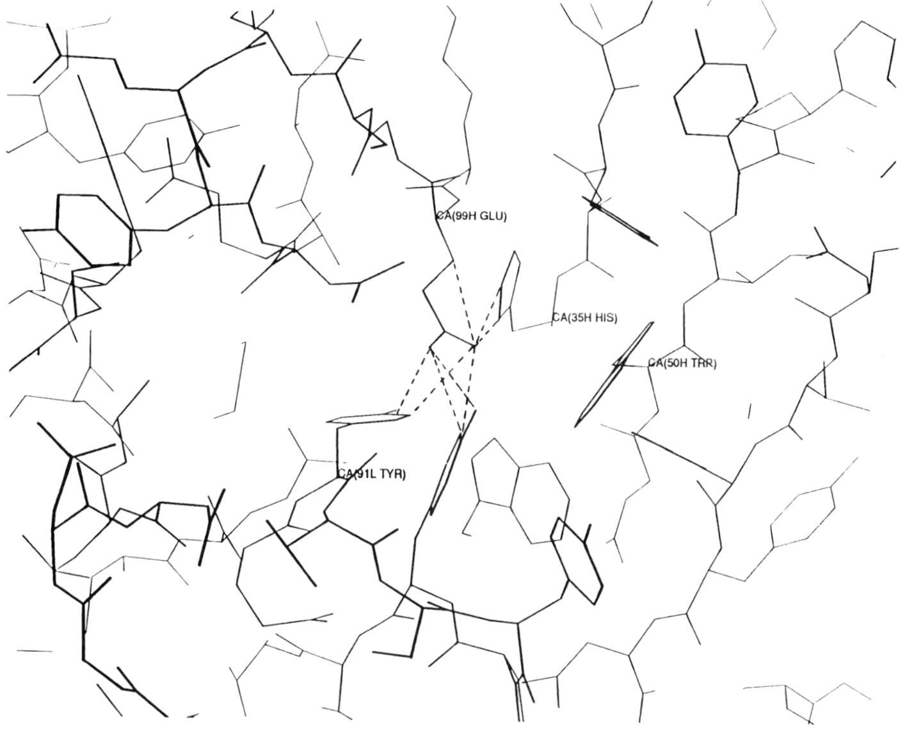

FIG. 2. Use of MIDAS to examine mutation of amino acids.

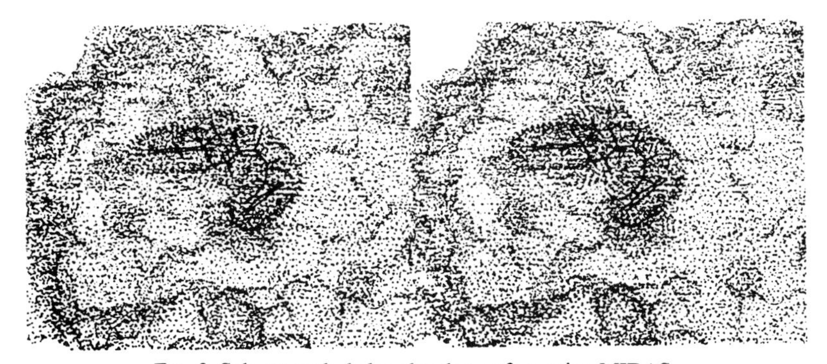

FIG. 3. Solvent-excluded molecular surface using MIDAS.

Application of DOCK to Orientation of Phosphorylcholine with McPC603 (2MCP)

Preparation of PDB File for DOCK Calculations

The Protein Data Bank file "2MCP" comes from Brookhaven as the complete Fab. The constant portions were removed with a text editor for ease of handling. This left light-chain residues Asp-1 to Lys-113 and heavy-chain residues Glu-1 to Ser-122. Phosphorylcholine is included in 2MCP as 11 HETATOMs, and was also removed and stored as a separate file for future use. DOCK prefers PDB files that are sequentially numbered as opposed to files with light- and heavy-chain subunits. A simple Fortran or C routine was used to renumber the PDB file to remove references to L and H subunits.

Generation of Spheres from Molecular Surface

The program SPHGEN comes with the DOCK package and generates spheres from the solvent-excluded molecular surface as calculated with the program MS by Connolly and co-workers.[84] A molecular surface should be calculated that is only as large as the region of interest for hapten binding. MIDAS comes with an "interior surface removal" (IRS) routine that selects atoms within a given radius of a probe file (phosphorylcholine). IRS was used to create a "site" file for input to the MS program, so that the molecular surface generated covered the combining site binding pocket. It is important to use the "-n" flag in the MS program to create a molecular surface with surface-normal values included. SPHGEN was then used to generate the spheres that define invaginations on the molecular surface. Each sphere touches the molecular surface at two points, i and j, and has its radius along the surface normal of point i. For the combining site, the sphere center is in the same direction of the surface normal of point i. One can think of SPHGEN as a pump attached to a molecular surface point. The pump inflates a balloon until it hits another surface point. If no contact is made, then the balloon is overfilled and bursts. In this way, only the binding pockets and grooves are filled with spheres for hapten alignment in the DOCK program. A clustering routine is also included in the DOCK package, so that the resulting set of spheres can be sorted into groups that define individual invaginations or binding domains. SPHGEN produces an output file to which the clustered spheres are written.

Automated Hapten Alignment with DOCK

The program DOCK is an automatic procedure for docking a small molecule into the negative image of a combining site generated by

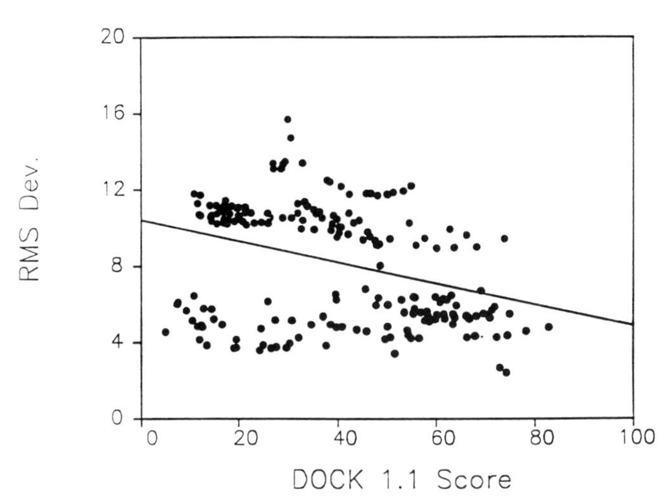

FIG. 4. DOCK score vs root-mean-square (RMS) differences.

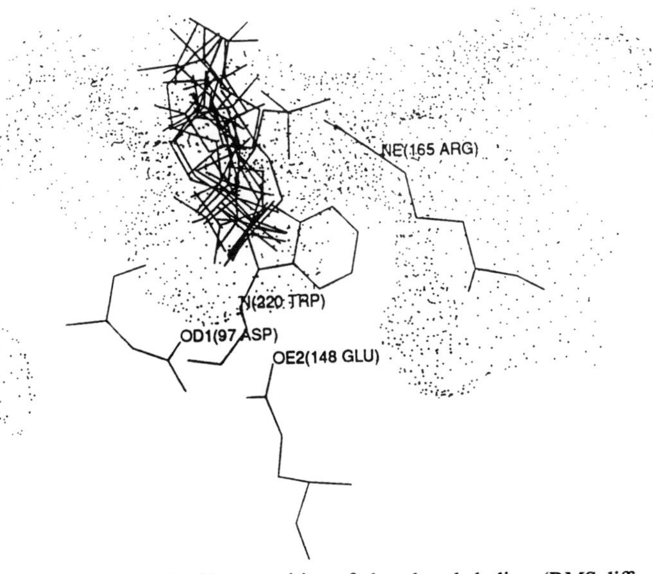

FIG. 5. Orientations closest to the X-ray position of phosphoryl choline. (RMS differences from X-ray position range from 2.4–3.8 Å.)

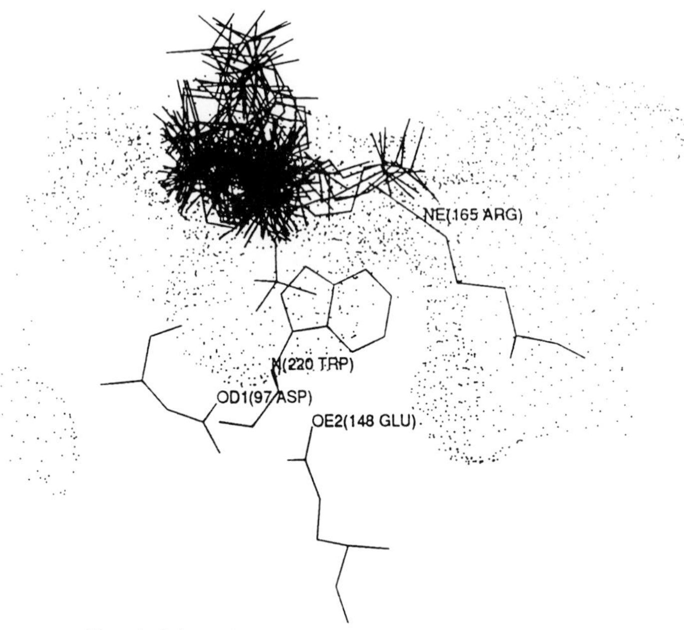

FIG. 6. Orientations near combining site pocket.

SPHGEN. The docking procedure is strictly based on geometric alignment of ligand atoms with combining site spheres. A DOCK score is generated for each unique alignment of the ligand in which four ligand center and combining site sphere center pairs are within a user-specified distance limit (default = 1.5 Å). Scoring is based on distances of ligand atoms to combining site atoms. Distances which are too close (default < 2.3 Å) are given negative (bad) scores. Distances which are too far (default > 5.0 Å) are given zero scores. Distances that are not too close but less than a maximal optimum distance (default 3.5 Å) are given a good score of 1.0. All other distances are rated by an exponential function (soft Lennard–Jones potential) which compares the observed distance to the maximal optimum distance (default 3.5 Å) and assigns a value between 0 and 1. The cumulative DOCK score for a given orientation is a sum of the individual scores for each atom. A DOCK run on McPC603 with phosphorylcholine as the hapten generated 221 orientations with cumulative DOCK scores from 5 to 83. Besides the DOCK score, we calculated the root-mean-square difference in position of the DOCK oriented hapten compared to the X-ray structure. Root-mean-square differences ranged from 2.38 to 15.66 Å. A plot of DOCK score vs root-mean-square difference is shown in Fig. 4. It shows that for a given DOCK score there is a broad range of root-mean-square differences observed for the hapten orientations. This type of plot

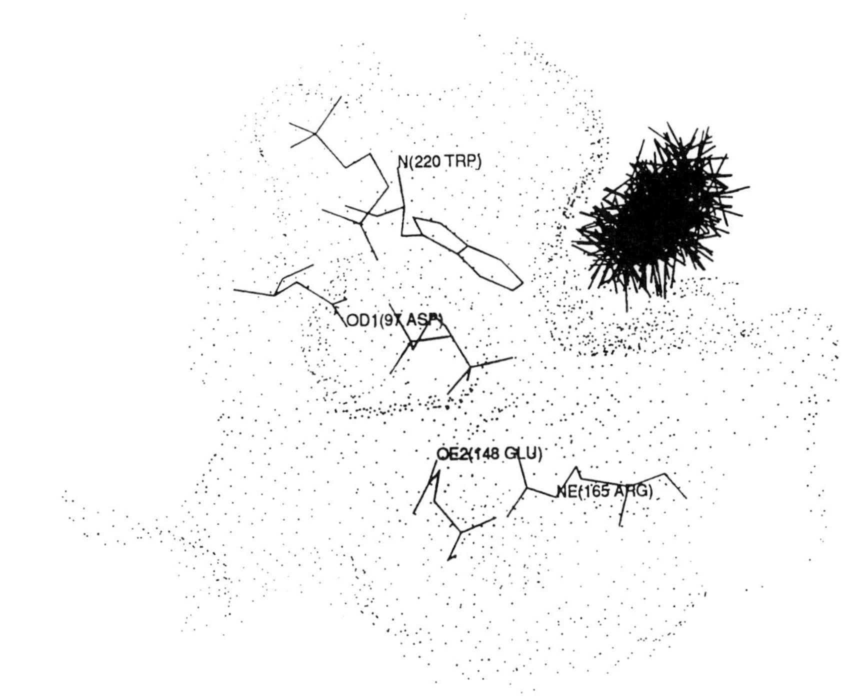

FIG. 7. Orientation binding on the side of the hypervariable loops.

can be used to cluster the orientations generated by DOCK into three or four families in space. Thirteen orientations with RMS differences from 2.4 to 3.8 Å and DOCK scores from 14 to 74 are shown in Fig. 5. All of these orientations have phosphorylcholine buried deep in the binding pocket with the combining site. Eighty-four orientations were found to lie near the combining site pocket, but close to the surface of the antibody with DOCK scores from 5 to 75 (Fig. 6). Finally, the rest of the orientations were observed to bind at a peripheral site on the side of the hypervariable loops (Fig. 7). The observation of a secondary binding site with DOCK scores ranging from 11 to 73 shows the value of an automated approach in orienting haptens. Using the automated procedure avoids user bias toward a classical orientation deep within the combining site pocket.

Acknowledgments

The authors acknowledge the support of the Northrup Corp. for providing molecular graphics computing facilities to the USC School of Pharmacy, for helpful discussions with I. D. Kuntz, T. Ferrin, B. Shoichet, and R. Langridge, and for the use of molecular graphics facilities at the UCSF Computer Graphics Laboratory during a sabbatical visit.

[3] Protein Engineering of Single-Chain Fv Analogs and Fusion Proteins

By JAMES S. HUSTON, MEREDITH MUDGETT-HUNTER, MEI-SHENG TAI, JOHN MCCARTNEY, FREDERICK WARREN, EDGAR HABER, and HERMANN OPPERMANN

Introduction

Proteolytic cleavage of IgA myeloma MOPC315 produced the first Fv fragment,[1] which was shown to be a noncovalent V_H-V_L heterodimer representing an intact antigen-binding site of M_r 25,000.[1a] Subsequent research indicated that other IgA and IgG antibodies examined were refractory to simple proteolytic release of Fv.[2] IgM antibodies were in general susceptible to proteolytic generation of Fv, but this has been of limited practical interest.[3] High affinity Fv fragments remained largely inaccessible to study until recombinant DNA methods were directed to this area.[4] We

[1] V_H, heavy-chain variable region; V_L, light-chain variable region; Fd, amino-terminal half of heavy chain (V_H-C_H1), which mates with light chain to form Fab; Fv, variable region fragment consisting of noncovalently associated V_H and V_L domains; sFv (single-chain Fv), biosynthetic Fv analog comprising both variable domains on a single polypeptide chain; CDR, complementarity-determining region; Fab, antigen binding fragment derived from IgG by papain cleavage; FB, fragment B (a 58-residue sequence) of staphylococcal protein A; FR, framework region; MLE, leader consisting of a modified *trp* LE sequence; MLE–sFv, the fusion protein consisting of the MLE leader followed by the sFv with an Asp-Pro peptide bond at their junction; PE40, 40-kDa cytotoxic portion of *Pseudomonas* exotoxin; PBSA, 0.15 *M* NaCl + 0.05 *M* potassium phosphate, pH 7.0, +0.03% NaN_3; SDS-PAGE, sodium dodecyl sulfate-polyacrylamide gel electrophoresis; GuHCl, guanidine hydrochloride; BSA, bovine serum albumin; TGFα, transforming growth factor α; PCR, polymerase chain reaction; Tac, designation for a monoclonal antibody directed against the interleukin-2 (IL-2) receptor p55 subunit, as opposed to *tac* promoter, a hybrid of the *trp* and *lac* promoters; TFR, transferrin receptor.

[1a] D. Inbar, J. Hochman, and D. Givol, *Proc. Natl. Acad. Sci. U.S.A.* **69,** 2659 (1972); J. Hochman, D. Inbar, and D. Givol, *Biochemistry* **12,** 1130 (1973); J. Hochman, M. Gavish, D. Inbar, and D. Givol, *Biochemistry* **15,** 2706 (1976).

[2] J. Sharon and D. Givol, *Biochemistry* **15,** 1591 (1976); J. Sen and S. Beychok, *Proteins: Struct. Funct. Genet.* **1,** 256 (1986).

[3] K. Kakimoto and K. Onoue, *J. Immunol.* **112,** 1373 (1974); L.-C. Lin and F. W. Putnam, *Proc. Natl. Acad. Sci. U.S.A.* **75,** 2649 (1978); M. Reth, T. Imanishi-Kari, and K. Rajewsky, *Eur. J. Immunol.* **9,** 1004 (1979); F. Goni and B. Frangione, *Proc. Natl. Acad. Sci. U.S.A.* **80,** 4837 (1983).

[4] L. Reichmann, J. Foote, and G. Winter, *J. Mol. Biol.* **203,** 825 (1988); A. Skerra and A. Plückthun, *Science* **240,** 1038 (1988); H. Field, G. T. Yarranton, and A. R. Rees, *in* "Vaccines 88" (H. Ginsberg, F. Brown, R. A. Lerner, and R. M. Channock, eds.), p. 29. Cold Spring Harbor Laboratory, Cold Spring Harbor, New York, 1988; A. Plückthun and A. Skerra, this series, Vol. 178, p. 497; H. Field, G. T. Yarranton, and A. R. Rees, *Protein Eng.* **3,** 641 (1990).

have studied the minimal antibody binding site through protein engineering of single-chain Fv analog and fusion proteins.[5,6] In this approach, the genes encoding V_H and V_L domains of a given monoclonal antibody are connected at the DNA level by an appropriate oligonucleotide, and on translation this gene forms a single polypeptide chain with a linker peptide bridging the two variable domains. This offers economy of design, wherein a single-chain Fv (sFv) gene encodes a single M_r 26,000 protein that forms the entire antibody combining site. Since the initial demonstration,[5] a considerable body of research on various single-chain Fv constructions has supported the broad applicability of this protein engineering concept.[6-22]

[5] J. S. Huston, D. Levinson, M. Mudgett-Hunter, M.-S. Tai, J. Novotny, M. N. Margolies, R. Ridge, R. E. Bruccoleri, E. Haber, R. Crea, and H. Oppermann, *Proc. Natl. Acad. Sci. U.S.A.* **85,** 5879 (1988).

[6] M.-S. Tai, M. Mudgett-Hunter, D. Levinson, G.-M. Wu, E. Haber, H. Oppermann, and J. S. Huston, *Biochemistry* **29,** 8024 (1990).

[7] D. Shealy, M. Nedelman, M.-S. Tai, J. S. Huston, H. Berger, J. Lister-James, and R. T. Dean, *J. Nucl. Med.* **31**(5) (Suppl.), 776 (1990).

[8] R. E. Bird, K. D. Hardman, J. W. Jacobson, S. Johnson, B. M. Kaufman, S.-M. Lee, T. Lee, S. H. Pope, G. S. Riordan, and M. Whitlow, *Science* **242,** 423 (1988).

[9] W. D. Bedzyk, K. M. Weidner, L. K. Denzin, L. S. Johnson, K. D. Hardman, M. W. Pantoliano, E. D. Ansel, and E. W. Voss, Jr., *J. Biol. Chem.* **265,** 18615 (1990).

[10] D. Colcher, R. Bird, M. Roselli, K. D. Hardman, S. Johnson, S. Pope, S. W. Dodd, M. W. Pantoliano, D. E. Milenic, and J. Schlom, *J. Natl. Cancer Inst.* **82,** 1191 (1990).

[11] B. L. Iverson, S. A. Iverson, V. A. Roberts, E. D. Getzoff, J. A. Tainer, S. J. Benkovic, and R. A. Lerner, *Science* **249,** 659 (1990).

[11a] R. A. Gibbs, B. A. Posner, D. R. Filpula, S. W. Dodd, M. A. J. Finkelman, T. K. Lee, M. Wroble, M. Whitlow, and S. J. Benkovic, *Proc. Natl. Acad. Sci. U.S.A.* **88,** 4001 (1991).

[12] R. Glockshuber, M. Malia, I. Pfitzinger, and A. Plückthun, *Biochemistry* **29,** 1362 (1990); A. Skerra, I. Pfitzinger, and A. Plückthun, *Bio/Technology* **9,** 273 (1991).

[13] J. H. Condra, V. V. Sardana, J. E. Tomassini, A. J. Schlabach, M.-E. Davies, D. W. Lineberger, L. Gotlib, D. J. Graham, and R. J. Colonno, *J. Biol. Chem.* **265,** 2292 (1990). For purposes of analyzing the length of each linker used in this work, we assumed that 113 residues were in the canonical V_L, their amino-terminal domain. If this V_L domain encompassed 116 residues (as implied by Condra *et al.*), the linker lengths for anti-HRV sFv proteins would be 3 residues shorter than the values noted in Table I (group B). Their 11-residue linker would then be only eight residues in length, and their addition of one or two (GGGGS) segments may have increased activity simply by increasing linker length.

[14] V. K. Chaudhary, C. Queen, R. P. Junghans, T. A. Waldmann, D. J. FitzGerald, and I. Pastan, *Nature (London)* **339,** 394 (1989).

[15] V. K. Chaudhary, J. K. Batra, M. G. Gallo, M. C. Willingham, D. J. FitzGerald, and I. Pastan, *Proc. Natl. Acad. Sci. U.S.A.* **87,** 1066 (1990).

[16] J. K. Batra, V. K. Chaudhary, D. FitzGerald, and I. Pastan, *Biochem. Biophys. Res. Commun.* **171,** 1 (1990).

[17] J. K. Batra, D. FitzGerald, M. Gately, V. K. Chaudhary, and I. Pastan, *J. Biol Chem.* **265,** 15198 (1990).

[18] R. J. Kreitman, V. K. Chaudhary, T. Waldmann, M. C. Willingham, D. J. FitzGerald, and I. Pastan, *Proc. Nat. Acad. Sci. U.S.A.* **87,** 8291 (1990).

The sFv is well suited to applications in immunotargeting, since a given sFv gene may be fused to a particular effector protein gene to yield a bifunctional sFv fusion protein.[6,14-22] Fusion of single-chain Fv and effector proteins obviates the need for chemical cross-linking in immunoconjugate preparations[22a] and simplifies fusion protein design relative to the multiple-chain species generated by transfectoma methodology.[22b] The biodistribution kinetics of sFv analogs were reported to be significantly faster than those of IgG or its Fab fragments,[7,10] which makes the sFv of particular interest for *in vivo* diagnostics and therapeutics. With a mass equal to one-half that of monovalent Fab and only one-sixth that of IgG (Fig. 1A[23-25]), the sFv exhibits rapid diffusion into the extravascular space, increased volume of distribution, and possibly improved renal elimination. The *in vivo* properties of sFv fusion proteins may likewise be expected to differ from those of conventional immunoconjugates, as suggested by observations that single-chain Fv immunotoxins exhibit greatly enhanced cytotoxicity over conventional cross-linked conjugates.[14-18] In the sections to follow, sFv analogs and fusion proteins will be discussed with a view to their design, production, and properties. Attention will be given to evi-

---

[19] J. McCartney, L. Lederman, E. Drier, G.-M. Wu, R. Batorsky, N. Cabral-Denison, J. S. Huston, and H. Oppermann, *ICSU Short Rep.* **10,** 114 (1990); manuscript in preparation (1991); the $(GGGGS)_3$ linker was also tested and found equivalent to $(ESGRS)(GGGGS)_2$.

[20] J. S. Huston, M. Mudgett-Hunter, G.-M. Wu, C. Yost, and H. Oppermann, unpublished results (1989).

[20a] E. Motez, C. Jaulin, F. Godeau, J. Choppin, J.-P. Levy, and P. Kourilsky, *Eur. J. Immunol.* **21,** 467 (1991). This applies $(GGIGS)(GGGGS)_2$ and related linkers to construction of a single-chain murine class I major transplantation antigen, expressed in COS-1 cells.

[21] V. K. Chaudhary, M. G. Gallo, D. J. FitzGerald, and I. Pastan, *Proc. Natl. Acad. Sci. U.S.A.* **87,** 9491 (1990).

[21a] J. K. Batra, D. J. FitzGerald, V. K. Chaudhary, and I. Pastan, *Mol. Cell. Biol.* **11,** 2200 (1991).

[22] J. McCafferty, A. D. Griffiths, G. Winter, and D. J. Chiswell, *Nature (London)* **348,** 552 (1990).

[22a] A. E. Frankel, ed., "Immunotoxins," 565 pp. Kluwer Academic Publishers, Boston, 1988.

[22b] S. L. Morrison, M. J. Johnson, L. A. Herzenberg, and V. T. Oi, *Proc. Natl. Acad. Sci. U.S.A.* **81,** 6851 (1984); M. S. Neuberger, G. T. Williams, and R. O. Fox, *Nature (London)* **312,** 604 (1984); S.-U. Shin and S. L. Morrison, this series, Vol. 178, p. 459; M. S. Runge, C. Bode, G. R. Matsueda, and E. Haber, *Proc. Natl. Acad. Sci. U.S.A.* **84,** 7659 (1987); T. W. Love, M. S. Runge, E. Haber, and T. Quertermous, this series, Vol. 178, p. 515.

[23] E. Haber and J. Novotny, *in* "Hybridoma Technology in the Biosciences and Medicine" (T. A. Springer, ed.), p. 57. Plenum, New York, 1985.

[24] D. M. Segal, E. A. Padlan, G. H. Cohen, S. Rudikoff, M. Potter, and D. R. Davies, *Proc. Natl. Acad. Sci. U.S.A.* **71,** 4298 (1974); Y. Satow, G. H. Cohen, E. A. Padlan, and D. R. Davies, *J. Mol. Biol.* **190,** 593 (1986).

[25] J. Novotny, R. E. Bruccoleri, J. Newell, D. Murphy, E. Haber, and M. Karplus, *J. Biol. Chem.* **258,** 14433 (1983).

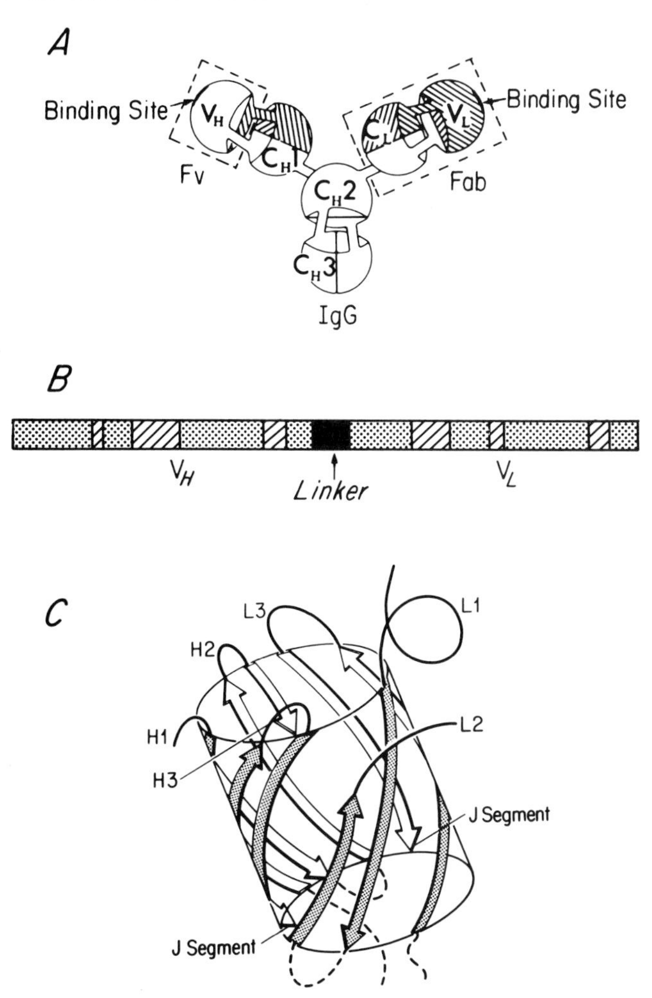

FIG. 1. Schematic representation of immunoglobulin G architecture, an sFv polypeptide chain, and the inner β barrel of the Fv region. (A) Schematic drawing of IgG indicating the Fab and Fv fragments, based on a synthesis of low-resolution X-ray structure data.[23] As discussed under linker chemistry, the switch region is emphasized by this low-resolution schematic diagram; the switch region in this drawing is a sticklike bridge that connects pairs of V and C domains. The apparent space between the Fv and constant regions is actually filled with side chains, according to the results of high-resolution crystal structures.[24] (B) Linear representation of an sFv in standard orientation, with the amino terminus at left and carboxyl terminus at right; the linker (black) bridges two variable regions, drawn to indicate alternating FRs (shaded) and CDRs (hatched lines). (C) Inner β barrel of the Fv region and sFv, indicating the geometry of β strands and CDR loops.[23,25] [Reproduced with permission from Plenum Press (Ref. 23) and *J. Biol. Chem.* (Ref. 25).]

dence supporting the fidelity of single-chain Fv binding sites, both as discrete molecules and as bifunctional fusion proteins.

General Design Considerations

In natural immunoglobulins, the antibody combining site is formed by complementarity-determining regions (CDRs) of the V_H and V_L variable domains within the Fv region. These variable domains associate by noncovalent interactions to constitute the Fv fragment, which is devoid of any interchain disulfide bonds. The CDRs form about one-quarter of the contacts at the interface between V_H and V_L,[26] and thus stability of Fv heterodimers may be expected to vary greatly and unpredictably, despite their highly conserved framework architecture.[27] The fusion of the V regions through a linker makes stability of the single-chain Fv independent of protein concentration and largely insensitive to the sometimes weak association between separate V_H and V_L domains. In contrast, Fv fragments dissociate into V_H and V_L domains at low protein concentrations,[12,28] resulting in disruption of the antigen-binding site; the dissociation constants reported for Fv formation range from 10^{-5} to 10^{-8} M.[12] Recombinant McPC603 Fv was shown to dissociate at micromolar concentrations, at which the McPC603 single-chain Fv was stable and exhibited essentially the same hapten-binding affinity as the parent antibody.[12,29]

In addition to noncovalent association between V_H and V_L domains, the combining site of an Fab fragment is stabilized by noncovalent constant region interactions (Fig. 1A) and usually by a disulfide bond that connects the C-terminal cysteinyl residue of the C_L to a cysteine in the C_H1 domain. In a sense, the artificial linker of an sFv substitutes for C_H1-C_L interactions that are present in an Fab fragment and helps to maintain integrity of the combining site. A fundamental question at the outset of this sFv project was whether the linker would have to mimic constant-region interactions that may orient V domains in some critical way to permit recovery of a native combining site. Experiments indicated full recovery of specific activity in an anti-digoxin sFv using an unstructured linker that primarily ensured proximity between pairs of V regions.[6] Thus, the linker need not specifically orient variable domains to generate a native antigen

[26] C. Chothia, J. Novotny, R. Bruccoleri, and M. Karplus, *J. Mol. Biol.* **186,** 651 (1985).

[27] J. Novotny and E. Haber, *Proc. Natl. Acad. Sci. U.S.A.* **82,** 4592 (1985).

[28] L. Riechmann, J. Foote, and G. Winter, *J. Mol. Biol.* **203,** 825 (1988).

[29] Domain interactions in heterodimeric McPC603 Fv were also shown to be stabilized by cross-linking with glutaraldehyde or by inserting another interchain disulfide between V_H and V_L; two alternative disulfide configurations were tried, L55 to H108 and L66 to H106, in which the cysteine mutations were at CDR residues, except L55 which flanked a CDR.[12] The general utility of these cross-linking methods remains to be demonstrated for Fv fragments, since introduction of chemical cross-links can produce heterogeneous products, while the design of innocuous CDR–CDR or FR–CDR disulfide bonds may be difficult for Fv fragments that lack a crystallographically determined three-dimensional structure.

binding site. Conversely, these results suggest that constant domains of an Fab region may primarily stabilize V region interactions rather than modulate interdomain contacts and binding site conformation. In fact, a comparison of the crystal structures for anti-lysozyme D1.3 Fv and Fab fragments showed that the interactions of the Fv region with lysozyme were almost identical whether or not constant regions were present.[30]

Polypeptide Chain Composition and Secondary Structure

The single-chain Fv consists of a single polypeptide chain with the sequence V_H–linker–V_L or V_L–linker–V_H, as opposed to the classical Fv heterodimer of V_H and V_L. About three-fourths of each variable-region polypeptide sequence is partitioned into four framework regions (FRs) that form a scaffold for the antigen binding site, which is made up of the remaining residues dispersed in three complementarity-determining regions (CDRs) that form loops connecting the FRs.[23] The sFv thus comprises eight FRs, six CDRs, and a linker segment, where the V_H sequence can be abbreviated as FR1-H1-FR2-H2-FR3-H3-FR4 and the V_L sequence as FR1-L1-FR2-L2-FR3-L3-FR4 (Fig. 1B).[23]

The predominant motif for secondary structure in immunoglobulin V regions is the twisted β sheet. A current interpretation of Fv architecture views the FRs as forming two concentric β barrels, with the CDR loops connecting antiparallel β strands of the inner barrel (Figs. 1C, 2A, and 2B).[25] The single-chain Fv thus incorporates the Fv as a single architectural unit and the connecting linker as a distinct structural element.[5] Although the linker is displayed only in a single conformation (Fig. 2), it is expected to be somewhat flexible in solution and should consequently occupy an ensemble of conformations.

The CDRs of a given murine monoclonal antibody may be grafted onto the FRs of human Fv regions in a process termed humanization or CDR replacement.[31] This methodology has been successfully applied to Fv fragments as well as to intact antibodies, and could possibly be extended to sFv proteins. Indications from initial studies are that humanized antibodies should offer minimal immunogenicity,[31] which may be still lower when sFv or Fv proteins are administered to patients. Antigenicity of different linkers remains to be analyzed, but their immunogenicity may be negligible given the low sFv molecular weight and paucity of side chains in linkers such as $(GGGGS)_3$. If residual immunogenicity were observed clinically, it

[30] T. N. Bhat, G. A. Bentley, T. O. Fischmann, G. Boulot, and R. J. Poljak, *Nature (London)* **347,** 483 (1990); A. G. Amit, R. A. Mariuzza, S. E. V. Phillips, and R. J. Poljak, *Science* **233,** 747 (1986).

[31] P. T. Jones, P. H. Dear, J. Foote, M. S. Neuberger, and G. Winter, *Nature (London)* **321,** 522 (1986); M. Verhoeyen, C. Milstein, and G. Winter, *Science* **239,** 1534 (1988); L. Riechmann, J. Foote, and G. Winter, *J. Mol. Biol.* **203,** 825 (1988); M. Bruggemann, G. Winter, H. Waldmann, and M. S. Neuberger, *J. Exp. Med.* **170,** 2153 (1989).

may be possible to vary linkers in a progressive way so that a different linker is used with each treatment course, for the same pair of humanized V domains.

Linker Chemistry

The general features of a viable linker are governed by the architecture and chemistry of Fv regions.[5] It is known that the sFv may be assembled in either domain order, as V_H–linker–V_L or V_L–linker–V_H (Table I), where the linker bridges the gap between C and N termini of the respective domains (Fig. 2).[5–22] For purposes of sFv design,[5] the C terminus of the amino-terminal V_H or V_L domain is considered to be the last residue of that sequence which is compactly folded, corresponding approximately to the end of the canonical V region sequence.[32] The amino-terminal V domain is thus defined to be free of switch region residues that link the V and C domains of a given H or L chain, which makes the linker sequence an architectural element in sFv structure that corresponds to bridging residues, regardless of their origin. In several examples, sFv linkers have incorporated residues from the switch region, even extending into the first constant domain (group B, Table I).[13,19]

The linker should be able to span the 3.5-nm distance between its points of fusion to the V domains without distortion of the native Fv conformation. Given the 0.38 nm distance between adjacent peptide bonds, a preferred linker should be at least about 10 residues in length. We have typically used a 15-residue linker, in order to avoid conformational strain from an overly short connection, while avoiding steric interference with the combining site from an excessively long peptide. A series of linkers, 5 to 25 residues in length, have been tested on the 26–10 sFv region of FB–sFv[26–10] fusion proteins. Using a microtiter plate assay, we detected digoxin-binding activity in all refolded species except that with a

[32] E. A. Kabat, T. T. Wu, M. Reid-Miller, H. M. Perry, and K. S. Gottesman, eds., "Sequences of Proteins of Immunological Interest," 4th Ed., 804 pp. U.S. Department of Health and Human Services, Washington, D.C., 1987.

FIG. 2. Computer-generated stereo views of a model sFv oriented as V_H–V_L with the $(GGGGS)_3$ linker,[12] based on the crystal structure of McPC603.[24] Ribbon diagrams display polypeptide backbone, emphasizing the β-barrel architecture of the sFv, sulfurs forming disulfide bonds (yellow spheres), and linker (orange ribbon); V_H framework is deep blue and V_L framework is light blue-green. (A) View looking into the binding site. (B) View looking at the side of the sFv showing the linker. (C) CPK (space-filling) representation of the same view as (B) showing the linker (pink), the V_H framework (turquoise), and CDRs (purple) on the right, and the V_L framework (light gray) and CDRs (red) on the left. Computations and molecular graphics presentations were made on a Silicon Graphics 4D-70 GTX (Silicon Graphics, Mountain View, CA) with Biosym Insight II software (Biosym Technologies, San Diego, CA) (modeling was contributed by Dr. Peter Keck, Creative Biomolecules, Hopkinton, MA).

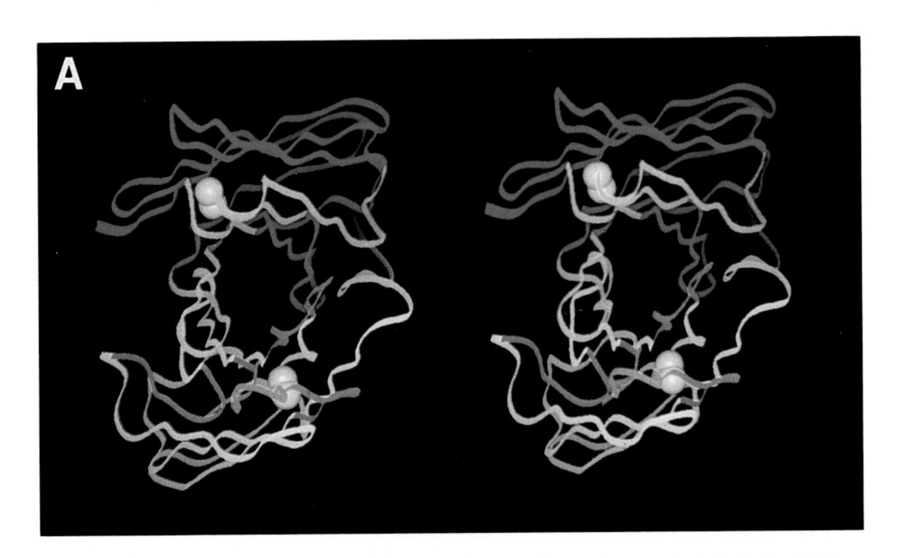

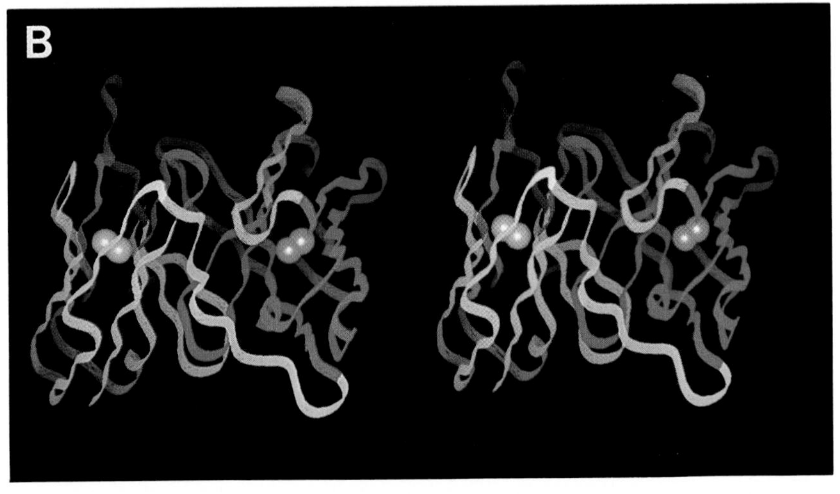

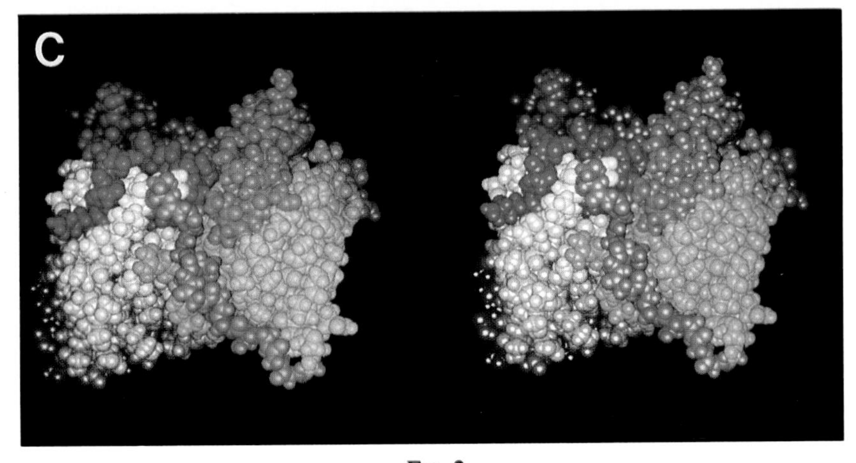

FIG. 2

TABLE I

SINGLE-CHAIN FV LINKER SEQUENCES

Identity/specificity	V order	Linker length[a]	Group	Linker sequence	Ref.
26–10 sFv, anti-digoxin	V_H-V_L	15	A	GGGGS GGGGS GGGGS	Huston *et al.* (1988)[5]
FB–sFv[26–10], anti-digoxin	V_H-V_L	15		GGGGS GGGGS GGGGS	Tai *et al.* (1990)[6]
	V_H-V_L	10–25		$(GGGGS)_n$, $n = 2–5$	Huston *et al.* (1989)[20]
	V_L-V_H	15		GGGGS GGGGS GGGGS	
McPC603 sFv, anti-phosphorylcholine	V_H-V_L	15		GGGGS GGGGS GGGGS	Glockshuber *et al.* (1990)[12]
Anti-Tac sFv–PE40, anti-IL-2 receptor	V_H-V_L	15		GGGGS GGGGS GGGGS	Chaudhary *et al.* (1989)[14] Kreitman *et al.* (1990)[18]
	V_L-V_H	15		GGGGS GGGGS GGGGS	Batra *et al.* (1990)[17]
Anti-TFR sFv-PE40, anti-transferrin receptor	V_H-V_L	15		GGGGS GGGGS GGGGS	Batra *et al.* (1991)[21a]
Diphtheria toxin-anti-Tac sFv	V_H-V_L	15		GGGGS GGGGS GGGGS	Chaudhary *et al.* (1990)[21]
Diptheria toxin-anti-TFR sFv	V_H-V_L	15		GGGGS GGGGS GGGGS	Batra *et al.* (1991)[21a]
D1.3 sFv–fd coat protein III, anti-lysozyme	V_H-V_L	15		GGGGS GGGGS GGGGS	McCafferty *et al.* (1990)[22]
Anti-HRV sFv, anti-human rhinovirus	V_L-V_H	11	B	$L^{114-116}$ I FPPSS EE	Condra *et al.* (1990)[13]
	V_L-V_H	14		$L^{114-116}$ I FPPSS GGGGS	
	V_L-V_H	19		$L^{114-116}$ I FPPSS GGGGS GGGGS	
FB–sFv[315] and 315 sFv, anti-DNP	V_H-V_L	15		ESGRS GGGGS GGGGS	McCartney *et al.* (1990)[19]
18-2-3/202 sFv, anti-fluorescein	V_L-V_H	14	C	EGKSS GSGSE SKST	Bird *et al.* (1988)[8]
4-4-20/202′ sFv, anti-fluorescein	V_L-V_H	15		EGKSS GSGSE SKSTQ	
Anti-Tac sFv–PE40, anti-IL-2-receptor	V_L-V_H	14		EGKSS GSGSE SKST	Batra *et al.* (1990)[17]
TGFα-sFv–PE40, anti-IL-2 receptor	V_L-V_H	14		EGKSS GSGSE SKST	Batra *et al.* (1990)[16]
OVB3 sFv–PE40, anti-tumor antigen	V_L-V_H	14		EGKSS GSGSE SKVD	Chaudhary *et al.* (1990)[15]
B6.2/212 sFv, anti-tumor antigen	V_L-V_H	14	D	GSTSG SGKSS EGKG	Colcher *et al.* (1990)[10]
4-4-20/212 sFv, anti-fluorescein	V_L-V_H	14		GSTSG SGKSS EGKG	Bedzyk *et al.* (1990)[9]
4-4-20/212 sFv, $V_L^{S89H,S91H,R34H,Y36L}$	V_L-V_H	14		GSTSG SGKSS EGKG	Iverson *et al.* (1990)[11]
7A4-1/212 sFv, catalytic esterase activity	V_L-V_H	14		GSTSG SGKSS EGKG	Gibbs *et al.* (1991)[11a]
3C2/59 sFv, anti-bovine growth hormone	V_L-V_H	18	E	KESGS VSSEQ LAQFR SLD	Bird *et al.* (1988)[8]
Anti-Tac sFv–PE40, anti-IL-2 receptor	V_L-V_H	16		ESGS VSSEE LA FR SLD	Batra *et al.* (1990)[17]

[a] Linker length is given in terms of the number of amino acid residues bridging V regions.

5-residue linker.[20] The linker sequence should also be hydrophilic to preclude intercalation within or between V domains during protein folding.

Based on these principles, *de novo* design yielded a 15-residue linker comprising 3 pentameric units of -$(Gly)_4$-Ser- (Fig. 2).[5,6] The serine residues confer extra hydrophilicity on the peptide backbone, which is otherwise free of side chains that might complicate domain refolding or association. The glycyl residues augment the range of accessible conformations,[33] which should reduce steric barriers to native interactions at the domain interface. The sFv molecules that were constructed with this linker are listed as group A in Table I. The molecules in group B combine part of this linker sequence with residues beyond the carboxyl terminus of the pertinent V domain, from the switch region or constant domain. Linkers noted in groups C and D also represent sequences based on *de novo* design, insofar as they show no significant homology to protein sequences in the PIR/NBRF database.[34] These linkers are significantly enriched in serine and glycine residues but are largely free of hydrophobic side chains. These sequences also contain charged residues which would enhance hydrophilicity, provided that they do not neutralize nearby V domain residues having the opposite charge.

Some workers have tried to build more rigid linkers derived from peptide bridges of proteins with known structures.[8,17] Bird *et al.* constructed an anti-bovine growth hormone sFv using residues 213 to 229 of human carbonic anhydrase I (carbonate dehydratase) as the basis for their linker.[8] This outer strand of carbonate dehydratase is partly extended and partly α helical, and in its natural conformation forms a 2.9 nm bridge. This natural sequence was modified by substitution of glycine for isoleucine-216 and addition of aspartate after leucine-229, making 18 residues that should easily span 3.5 nm given only slight flexibility in its structure. Batra *et al.* have compared this carbonate dehydratase bridge with other linkers in a series of sFv toxin fusions using the same V domains of the anti-Tac antibody (note groups A, C, and E, Table I).[17] They could not discern any significant differences in immunotoxin activity, but their sFv-PE40 proteins were not affinity purified, leaving the possibility that variable amounts of inactive protein could be present that would mask differences in affinity or specificity.

Native hapten-binding activity was found for MOPC315 sFv, whether its linker consisted of $(GGGGS)_3$ or $(ESGRS)(GGGGS)_2$, where the modified residues are from the switch region sequence at the C terminus of the V_H, bordering its pepsin cleavage site (group C, Table I).[19] However, using an sFv based on a monoclonal antibody to human rhinovirus, Condra *et*

[33] G. N. Ramachandran, C. M. Venkatachalam, and S. Krimm, *Biophys. J.* **6,** 849 (1966); G. N. Ramachandran, C. Ramakrishnan, and V. Sasisekharan, *J. Mol. Biol.* **7,** 95 (1963).

[34] PIR/NBRF Database. National Biomedical Research Foundation, Washington, D.C.; GenBank Database. IntelliGenetics, Inc., Mountain View, CA.

al. found a threefold increase in binding activity when one or two (GGGGS) segments were added to a linker comprising the switch region sequence of their light chain (group C, Table I).[13] Although data are limited, the literature surveyed in Table I suggests that some latitude exists in the design of linkers capable of producing functional sFv proteins. Linker fusion between V domains need not, in principle, compromise the fidelity of an sFv binding site. For example, the $(GGGGS)_3$ linker in several sFv constructions has been proven to yield sFv species that reproduce the binding affinity of the parent antibody.[6,12,19]

The observation that some sFv analogs or fusion proteins exhibit lower binding affinities than the parent antibody[8,10,13,14-18] may simply reflect a need for further purification of the sFv protein or additional refinement of antigen-binding assays. On the other hand, such behavior may require modification of sFv protein design, especially in view of the observation that a deletion of two framework residues at the N terminus of V_H led to a very large alteration in binding affinity and specificity in an anti-digoxin antibody.[35] Thus, changes at the amino termini of V domains may on occasion perturb a particular combining site, perhaps as a lowered affinity rather than the increased $K_{a,app}$ observed by Panka *et al.*[35] In a particular case, if an sFv were to exhibit a lower affinity for antigen than the parent Fab fragment, one could test for a possible N-terminal perturbation effect. For instance, given a V_L-V_H that was suspect, the V_H-V_L construction could be made and tested; if the initially observed perturbation were changed or eliminated in the alternate sFv species, then the effect could be traced to the initial sFv design. In the only published analysis of such sFv isomers, the two forms of anti-Tac sFv–PE40 showed minor differences in activity that were not considered significant.[17] These same concerns apply to instances where release of a leader from the amino terminus of an sFv causes changes in the actual V_H or V_L N-terminal sequence, or where an amino-terminal effector has been fused, perhaps without an appropriate hinge region at the junction.

Fusion Protein Architecture

The construction of single-chain Fv fusion proteins can involve two basic schemes, wherein an effector domain may be fused at the amino terminus of an sFv or at its carboxyl terminus (Fig. 3). Since constant domains are always attached to the C termini of V regions in the immuno-

[35] D. J. Panka, M. Mudgett-Hunter, D. R. Parks, L. L. Peterson, L. A. Herzenberg, E. Haber, and M. N. Margolies, *Proc. Natl. Acad. Sci. U.S.A.* **85,** 3080 (1988). This group found that deleting only two FR residues caused nearly a 50-fold increase of binding affinity as well as a significant change in specificity. Observation of this pronounced effect on hapten binding suggests the possibility that, in some situations, fusion of a linker segment between the pair of V domains of a particular Fv could have an effect on binding site properties in the resulting sFv analog.

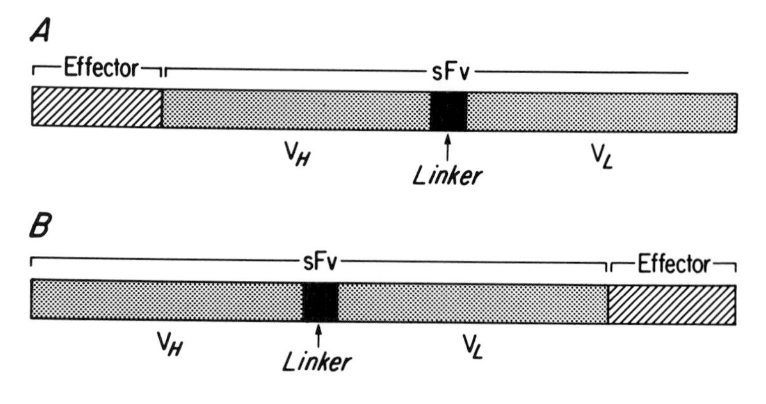

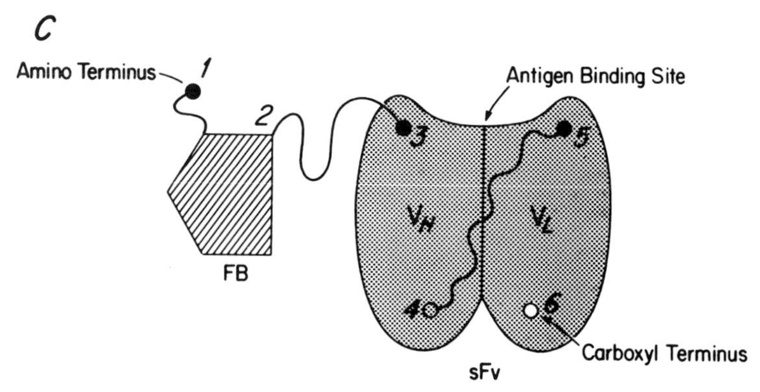

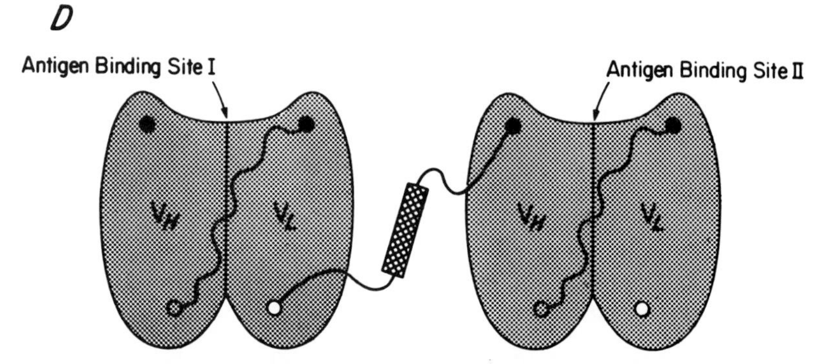

FIG. 3. Architecture of sFv fusion proteins. (A) Fusion protein with amino-terminal effector. (B) Fusion protein with carboxyl-terminal effector. (C) Schematic diagram of the folded bifunctional FB–sFv$^{26-10}$, drawn to emphasize the proximity of the sFv amino terminus to the antigen-binding site, based on analogy with known structures.[6] Integers on the diagram indicate the following aspects: 1, amino terminus of the FB–sFv$^{26-10}$ polypeptide; 2, residue FB-47 corresponding to the end of the FB core region[36]; 3, position of the amino terminus of the V_H and sFv, to which the FB sequence was fused; 4, carboxyl terminus of the V_H region, to which the 15-residue linker was fused[5]; 5, amino terminus of the V_L region, to which the linker was also fused[5]; 6, carboxyl terminus of the V_L and FB–sFv$^{26-10}$. (D) Schematic drawing of hypothetical bivalent sFv species (sites I and II are the same) or bispecific sFv species (sites I and II are different). A hypothetical spacer segment is drawn as a cross-hatched rectangle bridging the sFv pair. As drawn, linker is behind the V_H–V_L plane.

globulin superfamily, there is ample precedent for using this orientation to attach additional effector domains. Single-chain Fv immunotoxins have been studied that involved fusion of a 40-kDa core of *Pseudomonas* exotoxin to the carboxyl terminus of several sFv analogs,[14-18] resulting in noteworthy enhancements of cytotoxicity over chemically cross-linked IgG–PE40 conjugates.[14,15] In contrast, fusion of proteins to the amino termini of V domains is without precedent in the immunoglobulin superfamily and would be essential in construction of bivalent sFv fusions or complex multidomain fusion proteins (Fig. 3D).

The early development of V_H–V_L[5] or V_L–V_H[8] sFv analogs suggested that the V_L or V_H domain, respectively in each orientation, would tolerate amino-terminal fusion of the linker without significant impact on antigen binding site properties. While the linker fusion appears to be qualitatively similar to the amino-terminal fusion of an effector to an sFv, it is topologically and structurally quite different. Within a native sFv, the linker is constrained in its mobility by fusion at both of its ends (Fig. 2), but an amino-terminal effector domain is fused at only one end, thereby having the potential to sterically hinder the antigen binding site (Fig. 3).

Investigations have demonstrated that protein effector domains can be successfully fused to the amino terminus of the sFv.[6,16,21,21a] Fusion of the fragment B (FB)[36] of staphylococcal protein A to the amino terminus of 26–10 sFv produced an FB–sFv[26-10] fusion protein (Fig. 3C) that was virtually identical in its digoxin-binding properties to the free 26–10 sFv, Fab, and IgG.[6] Some of the success in obtaining native 26–10 binding properties may have been due to the presence of a natural hinge region between globular portions of the FB and 26–10 sFv.[36] The experiments of

[36] J. Deisenhofer, *Biochemistry* **20,** 2361 (1981). The FB, or fragment B, represents one of the Fc-binding domains of staphylococcal protein A, which has been used to make bifunctional fusion proteins with the 26–10 sFv.[6] Structural consideration of the FB domain was based on the X-ray crystallographic study of Deisenhofer. Electron density maps for the FB allowed identification of only 43 of its 58 residues (FB residues 5–47). The remaining FB residues (1–4 and 48–58) may represent disordered residues within the crystal, which could be an indication of their flexibility in solution. A very recent 2D-NMR analysis of FB fragment structure suggests that a third helical segment is present in solution. Helix III encompasses residues within the crystallographically disordered region (helix III corresponds to residues FB-41 through FB-54 in the fusion protein sequence given in Fig. 7) [H. Torigoe, I. Shimada, A. Saito, M. Sato, and Y. Arata, *Biochemistry* **29,** 8787 (1990)]. A helix-coil equilibrium may exist for this region which would be predominantly ordered for FB free in solution and largely disordered for FB associated with Fc [H. Torigoe, I. Shimada, M. Waelchli, A. Sato, M. Sato, and Y. Arata, *FEBS Lett.* **269,** 174 (1990)]. These new results are still consistent with the presence of a natural hinge region in the FB–sFv[26-10] forming a flexible linker between the FB core and sFv. In the random coil state of the helix III region, this linker segment would consist of 14 residues of the leader (FB-48 to FB-58 followed by Ser-Asp-Pro), which is connected to the amino terminus of V_H in the 26–10 sFv (Fig. 7A). In the helical state of FB-41 to FB-54, the next 7 residues preceding 26–10 V_H would presumably remain unstructured and provide a flexible linker [FB-55 to FB-58 followed by Ser-Asp-Pro preceding residue 2 at the start of V_H (Fig. 7A)].

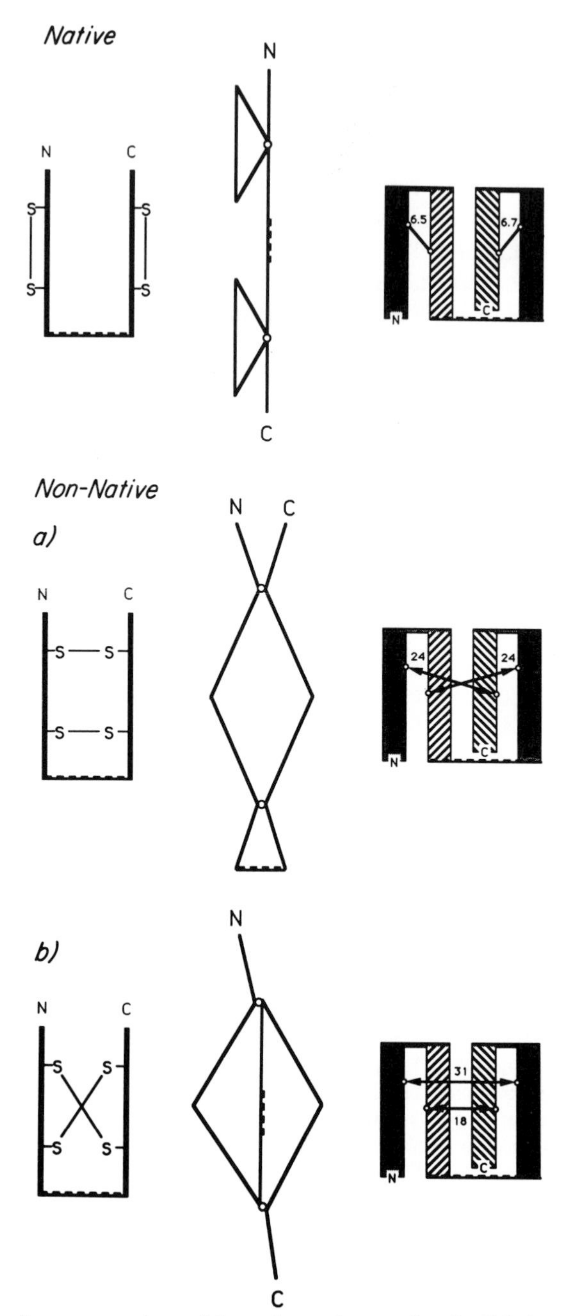

FIG. 4. Schematic representations of the native and nonnative disulfide-bonding patterns of fully oxidized sFv, based on the crystal structure of McPC603.[24] The crenulated line corresponds to the linker segment. The left-hand drawings show the linear connectivities of the four cysteinyl residues in forming three possible disulfide bond patterns for the sFv. The native form has one disulfide in each variable domain (vertical line), whereas the nonnative form has two disulfides between variable regions (a and b). The middle diagrams are drawn

Batra *et al.* further support this approach, in that a transforming growth factor α (TGFα)–anti-Tac sFv–PE40 fusion protein was made in which its two effectors and antigen-binding site were all shown to be functional.[16] The absence of a hinge region between the TGFα and sFv regions may have contributed to their triple fusion protein having about threefold lower cytotoxicity than the bifunctional sFv–PE40. In two other sFv immunotoxins, a diphtheria toxin fragment of 388 residues was fused to the amino terminus of anti-Tac sFv and anti-TFR sFv (Table I, group A linker).[21,21a] Active immunotoxins were obtained as in other cases[14–18] by dilution refolding. The cytotoxicity of the first fusion protein was comparable to that of anti-Tac sFv-PE40,[21] whereas the second fusion protein was less potent than anti-TFR sFv-PE40 against some target cell lines, but more potent against others.[21a] These results support the feasibility of more complicated constructions, such as a bivalent $(sFv)_2$ or a bispecific $sFv^{I}–sFv^{II}$, in which the binding sites would be the same or different, respectively (Fig. 3D). It should also be possible to make single-chain fusions with a mixture of different sFv and effector domains attached to each other.

Disulfide Bonding Patterns of sFv

Native sFv contains two disulfide bonds, one in each variable domain (Figs. 2A and B, and 4). Each disulfide connects β strands on the inner and outer barrels, with an inter-C_α distance of about 0.65 and 0.67 nm. The alternative distances between mispaired cysteines range from 1.8 to 3.1 nm and are clearly incompatible with the native sFv structure (Fig. 4). The mispaired forms are noted because of their relevance to the products of sFv protein folding. The four cysteine residues of the sFv can enter into three alternative intrachain disulfide-bonding patterns, assuming complete oxidation. The nonnative pairs of disulfides yield chain topologies that are very different from that of the native disulfide-bonded form (compare middle diagrams, Fig. 4). Although incorrect pairing does not strictly preclude reformation of some antigen-binding activity on renaturation, such forms cannot assume a native sFv conformation.

In general, the disulfide-bonded forms of many sFv analog or fusion proteins will display two or three separate bands on nonreduced sodium dodecyl sulfate-polyacrylamide gel electrophoresis (SDS-PAGE), although for some sFv proteins these forms comigrate. For example, the oxidized

linearly to scale, showing the size and disposition of disulfide-bonded loops in the three forms of oxidized sFv. The right-hand diagrams are drawn to convey the relative orientation of the sFv cysteinyl residues. They are part of either the inner β sheets (hatched walls) or outer β sheets (shaded walls). The numbers adjacent to black bars or arrows represent the distance in angstroms between α-carbons of these residues. The native distance is to be contrasted with the alternative nonnative pairings, which are clearly impossible in the context of a native sFv (also compare Fig. 2A and B).

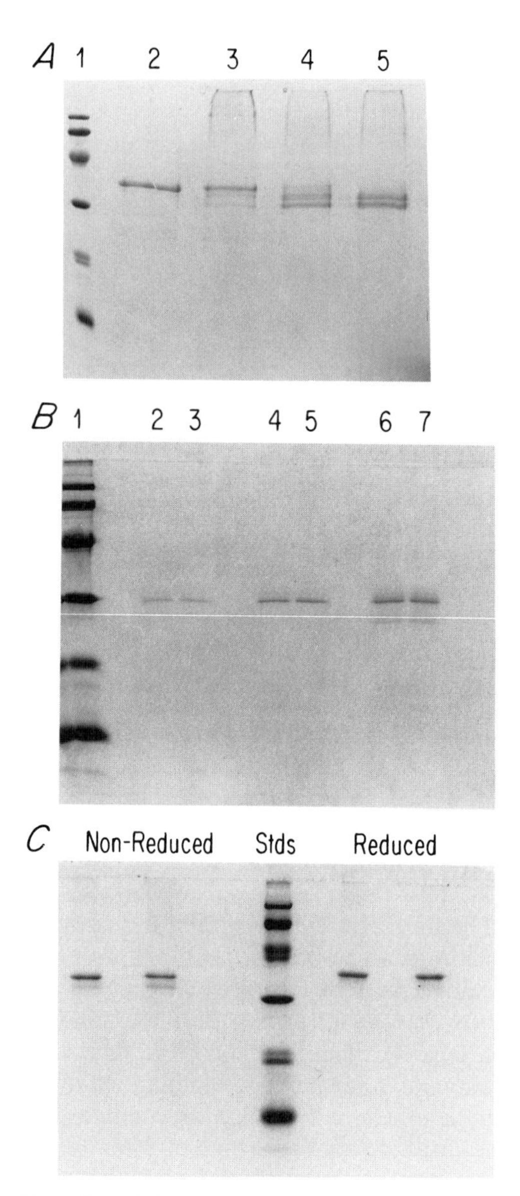

FIG. 5. SDS-PAGE profiles of mixtures of oxidized sFv or FB–sFv species. (A) Unreduced mixture of randomly oxidized FB–sFv^{315} analyzed by SDS-PAGE on 15% gels. Disulfide cross-linked species span a lower molecular weight range than the reduced chain at about 34K. Reduced and denatured FB–sFv^{315} (0.2 mg/ml) was oxidized overnight at 20° in 25 m*M* Tris, 10 m*M* EDTA, 6 *M* GuHCl, pH 9.0 in the presence of different concentrations of reduced and oxidized glutathione. Glutathione was removed by dialysis against 25 m*M* Tris, 10 m*M* EDTA, 8 *M* urea, pH 8.0, followed by renaturation of FB–sFv^{315} by dialysis against 166 m*M* H_3BO_3, 28 m*M* NaOH, 150 m*M* NaCl, 6 m*M* NaN_3, which is at pH 8.2. Samples were analyzed by 15% SDS-PAGE. Oxidized samples were electrophoresed in the absence of reducing agent. Numbered lanes contain the following samples: (1) Standards ($M_r \times 10^{-3}$), 94, 67, 43, 29, 20, 14; (2) FB–sFv^{315}, reduced; (3) FB–sFv^{315}, oxidized with

but unfractionated MOPC315 sFv protein shows up as a triplet of major bands (shown for FB–sFv^{315}, Fig. 5A). In this case, close examination of each main band reveals a doublet, which may be due to the presence of partial oxidation products; affinity-purified FB–sFv^{315} corresponds to the upper band in the middle doublet. The active form of 26–10 sFv is also distinguishable from its nonnative forms on SDS-PAGE after redox refolding; before affinity purification, gels of unreduced refolded 26–10 sFv and FB–sFv$^{26-10}$ exhibit only two bands, of which the dominant upper band corresponds to active protein (shown for 26–10 sFv in Fig. 5B and for FB–sFv$^{26-10}$ in Fig. 5C). The separation and observed molecular weight range of oxidized sFv proteins appears to result from anomalies associated with SDS binding to unreduced proteins.[37] The extent to which oxidized species are separable into distinct bands is characteristic for a given sFv protein, and its sequence dependence suggests that it results from anomalies on SDS-PAGE, possibly involving variations in the extent of detergent binding as well as differences in the hydrodynamic shape of protein–SDS complexes. Reduced 26–10 sFv proteins migrate as single bands, but use of the $(GGGGS)_3$ linker contributes to the appearance of an anomalously high molecular weight value, as the mean residue weight of the linker is very low. For example, the 26–10 sFv, with a known molecular weight of 26,354, appears from its migration on SDS-PAGE to have an M_r of 29,000, since the separation of proteins on SDS-PAGE is generally proportional to polypeptide chain length rather than mass.[37]

Practical Considerations

In principle, sFv proteins may be constructed to incorporate the Fv region of any monoclonal antibody, regardless of its class or antigen specificity. Departures from the parent V region sequences may also incorpo-

[37] R. Pitt-Rivers and F. S. A. Impiombato, *Biochem. J.* **109,** 825 (1968); J. A. Reynolds and C. Tanford, *Proc. Natl. Acad. Sci. U.S.A.* **66,** 1002 (1970); J. A. Reynolds and C. Tanford, *J. Biol. Chem.* **245,** 5161 (1970); W. W. Fish, J. A. Reynolds, and C. Tanford, *J. Biol. Chem.* **245,** 5166 (1970); K. Weber, J. R. Pringle, and M. Osborn, this series, Vol. 26, p. 3; Y. P. See and G. Jackowski, *in* "Protein Structure: A Practical Approach" (T. E. Creighton, ed.), p. 1. IRL Press, Oxford, 1989.

10 m*M* oxidized glutathione; (4) FB–sFv^{315}, oxidized with 1 m*M* oxidized and 0.1 m*M* reduced glutathione; (5) FB–sFv^{315}, oxidized with 0.1 m*M* oxidized and 1 m*M* reduced glutatione. (B) Unreduced mixture of 26–10 sFv oxidized forms made by (lanes 2 and 3) air oxidation during dilution refolding, (lanes 4 and 5) oxidation with glutathione in 3 *M* urea[6] during redox refolding, or (lanes 6 and 7) oxidation with glutathione in 6 *M* guanidine hydrochloride, during disulfide-restricted refolding. (C) Redox-refolded samples of oxidized FB–sFv$^{26-10}$ prior to affinity purification,[6] electrophoresed with or without reduction. Molecular weights calculated from animo acid sequence were the following: 26,354 for 26–10 sFv, 33,106 for FB–sFv$^{26-10}$, and 33,475 for FB–sFv^{315}.

rate discrete changes in CDRs to modify antigen affinity or specificity,[38] as well as wholesale alteration of framework regions to effect humanization.[31] In any event, an effective assay must be available for the parent antibody and its sFv analog. Fusion proteins such as sFv immunotoxins[14-18,21,21a] intrinsically provide an assay by their toxicity to target cells in culture. At the other extreme, low affinity binding sites such as those from IgM can challenge the limits of conventional assay procedures and make analysis of monovalent sFv a formidable problem. Even for high affinity antibodies, routine assays can unexpectedly perturb the sFv binding site, as exemplified in the later discussion of immunoprecipitation assays on anti-digoxin 26–10 sFv.

Molecular Construction of sFv Genes

The construction of a single-chain Fv typically is accomplished in three phases: (1) isolation of cDNA for the variable regions, (2) modification of the isolated V_H and V_L domains to permit their joining via a single-chain linker, and (3) expression of the single-chain Fv protein. The assembled sFv gene may then be progressively altered to modify sFv properties. In our first model sFv, the 26–10 sFv gene was made by synthetic DNA methods which permitted design of a DNA sequence that could facilitate future mutagenesis studies.[5,6] Such synthetic methods continue to be of value in circumstances where the V_H and V_L sequences are known, but the parent hybridoma or myeloma cell line is not available. *Escherichia coli* has provided the source of nearly all sFv proteins to date,[5-22] although other expression systems will certainly be used to generate sFv proteins in the future. Bacterial expression of antibody fragments and domains has become widely used,[4,5-22,39] as its cost is low, its production capability great, and its technical demands are limited compared to those of cultured mammalian cells. The known instability of hybridoma, mouse–human hybrid, or transformed human cell lines[40] can also be avoided by production of the corresponding sFv analogs in bacteria.

[38] S. Roberts, J. Cheetham, and A. R. Rees, *Nature (London)* **328,** 731 (1987). These researchers combined molecular modeling and site-directed mutagenesis with a *Xenopus* oocyte expression system to generate a series of mutant antibodies. They discovered a double mutant that showed an 8-fold increase in affinity and improved selectivity for lysozyme. Also discussed in Chapters [6] and [9], this volume.

[39] M. Better, C. P. Chang, R. R. Robinson, and A. H. Horwitz, *Science* **240,** 1041 (1988); M. Better and A. H. Horwitz, this series, Vol. 178, p. 476; A. Plückthun and A. Skerra, this series, Vol. 178, p. 497; A. Plückthun, *Bio/Technology* **9,** 545 (1991).

[40] V. R. Zurawski, Jr., P. H. Black, and E. Haber, *in* "Monoclonal Antibodies. Hybridomas: A New Dimension in Biological Analyses" (R. H. Kennett, T. J. McKearn, and K. B. Bechtol, eds.), p. 19. Plenum, New York, 1980.

cDNA Cloning or Synthesis of V Genes

The V_H and V_L genes for a given monoclonal antibody are most conveniently derived from the cDNA of its parent hybridoma cell line. Cloning of V_H and V_L from hybridoma cDNA has been facilitated by library construction kits using improved λ vectors such as Lambda ZAP (Stratagene, La Jolla, CA).[41] With only the amino acid sequences of the V domains, one can take a totally synthetic approach,[5,6] or alternatively a semisynthetic approach can be taken by appropriately modifying other available cDNA clones or sFv genes by site-directed mutagenesis. In two examples of sFv construction from our laboratory, the V genes for the 26–10 sFv were made by direct synthesis,[5,6] while the V genes of MOPC 315 were isolated from a cDNA library with specific oligonucleotide probes.[19] The synthetic 26–10 sFv gene was designed using preferred *E. coli* codons,[5] but we have since observed that mammalian cDNA sequences can also be expressed at high levels.

DNA Probes

Many alternative DNA probes have been used for V gene cloning from hybridoma cDNA libraries. Probes for constant regions have general utility provided that they match the class of the relevant heavy- or light-chain constant domain. Unrearranged genomic clones containing the J segments have even broader utility, but the extent of sequence homology and hybridization stringency may be unknown. Mixed pools of synthetic oligonucleotides based on the J regions of known amino acid sequence have been used.[42] If the parental myeloma fusion partner was transcribing an endogenous immunoglobulin gene, the authentic clones for the V genes of interest should be distinguished from the genes of endogenous origin by examining their DNA sequences in a Genbank homology search (Intelligenetics, Palo Alto, CA).[43]

Gene Isolation by Polymerase Chain Reaction Methodology

The cloning steps described above may be simplified by the use of polymerase chain reaction (PCR) technology. For example, immunoglobu-

[41] J. M. Short, J. M. Fernandez, J. A. Sorge, and W. D. Huse, *Nucleic Acids Res.* **16,** 7583 (1988).

[42] R. Orlandi, D. H. Gussow, P. T. Jones, and G. Winter, *Proc. Natl. Acad. Sci. U.S.A.* **86,** 3833 (1989); L. Sastry, M. Alting-Mees, W. D. Huse, J. M. Short, J. A. Sorge, B. N. Hay, K. D. Janda, S. J. Benkovic, and R. A. Lerner, *Proc. Natl. Acad. Sci. U.S.A.* **86,** 5728 (1989); E. S. Ward, D. Gussow, A. D. Griffiths, P. T. Jones, and G. Winter, *Nature (London)* **341,** 544 (1989).

[43] R. Strohal, G. Kroemer, G. Wick, and R. Kofler, *Nucleic Acids Res.* **15,** 2771 (1987).

lin cDNA can be transcribed from hybridoma cell mRNA by reverse transcriptase prior to amplification by Taq polymerase using specially designed primers.[44] Consensus PCR primer mixtures of general utility for direct cloning of V regions have been successful despite sequence divergence in signal peptides and in framework segments.[42,45] The primers should be designed to bind regions flanking the V_H and V_L genes, consisting of the secretion signal sequence and the beginning of the constant region.[45a] Signal sequences exhibit considerable variability, in contrast to the conservation shown by constant regions. Thus, a series of primers will be needed to cover the range of possible sequences. If consensus primers are designed to bind within the first or last framework, any mismatches may be incorporated into the amplified gene and may engender altered binding site properties that will be unacceptable. Relevant discussion is given in the last paragraph of the section on linker chemistry and in Ref. 35.)

PCR methodology is the current method of choice for constructing a single-chain Fv analog or fusion protein.[45] PCR was first exploited in this area to construct sFv immunotoxin genes,[15] where primers used for isolation of V genes also contained appropriate restriction sequences to speed sFv and fusion protein assembly. Extensions of the appropriate primers also encoded parts of the desired linker sequence, such that the PCR amplification products of V_H and V_L genes could be mixed to directly form the single-chain Fv gene. Once obtained, the sFv gene would be manipulated in the same manner as described for the 26–10 sFv gene. The application of PCR directly to human peripheral blood lymphocytes offers the opportunity to clone human V regions in bacteria.[42]

The construction of V region libraries by PCR has been coupled with bacterial secretion to produce V_H domains and active Fab fragments from combinatorial libraries[42]; the sFv should also be amenable to these approaches and may offer some advantages over other antibody fragments. Milstein has compared conventional methods for making monoclonal antibodies[46] to the combinatorial approach for generating diverse antibody-binding sites.[47] He states that successful development of these methods must incorporate a refinement process analogous to the successive responses to antigenic stimulation of the immune system.

[44] R. K. Saikai, D. H. Gelfand, S. Stoffel, S. J. Scharf, R. Higuchi, G. T. Horn, K. B. Mullis, and H. A. Erlich, *Science* **293,** 487 (1988).

[45] Y. L. Chiang, R. Sheng-Dong, M. A. Brow, and J. W. Larrick, *BioTechniques* **7,** 360 (1989); W. D. Huse, L. Sastry, S. A. Iverson, A. S. Kang, M. Alting-Mees, D. R. Burton, S. J. Benkovic, and R. A. Lerner, *Science* **246,** 1275 (1989); G. T. Davis, W. D. Bedzyk, E. W. Voss, and T. W. Jacobs, *Bio/Technology* **9,** 165 (1991).

[45a] S. T. Jones and M. M. Bendig, *Bio/Technology* **9,** 88 (1991); and erratta, *ibid.,* **9,** 579 (1991).

[46] G. Kohler, and C. Milstein, *Nature (London)* **245,** 495 (1975).

[47] C. Milstein, *Proc. R. Soc. London B* **239,** 1 (1990).

Refinement of antibody binding sites now appears possible, including provision for addressing Milstein's concerns, by using filamentous bacteriophage that allow the expression of peptides or polypeptides on their surface. These methods have permitted the construction of phage antibodies that express functional sFv on their surface,[22] as well as epitope libraries that can be searched for peptides that bind to particular combining sites.[48] McCafferty *et al.*[22] fused the gene encoding anti-lysozyme D1.3 sFv (Table I, group A linker) to gene III of fd bacteriophage, which after transfection into *E. coli* resulted in a bifunctional fusion protein comprising the D1.3 sFv attached to the amino terminus of the gene III coat protein. Since the D1.3 binding site of the sFv fusion protein retained its specific activity, these phage antibodies were specifically enriched from a one-million-fold excess of the parental fdCAT1 bacteriophage by two cycles of lysozyme affinity chromatography. Subsequent transfection into bacteria allowed amplification of the phage bearing D1.3 sFv on their surface, from which the sFv gene could be recovered. With appropriate affinity isolation steps, this sFv–phage methodology offers the opportunity to generate mutants of a given sFv with desired changes in specificity and affinity, as well as provide for a refinement process in successive cycles of modification. These new phage technologies show great promise, and eventually such methods should augment conventional hybridoma methods and site-directed mutagenesis[38] for the generation of high-affinity antibody binding sites.

General Design of sFv Genes

We have standardized the placement of restriction sites in sFv genes to facilitate the exchange of individual V_H and V_L linker elements, or leaders in sFv constructions. The selection of particular restriction sites can be governed by the choice of stereotypical sequences that may be fused to different sFv genes. In mammalian secretion, secretion signal peptides are cleaved from the N termini of secreted proteins by signal peptidases.[49] The same applies to bacterial secretion, as discussed below, while production of sFv proteins by intracellular accumulation of inclusion bodies has generally involved fusion protein expression.[5–11,13–21,50] In such cases, a restriction site for gene fusion and a corresponding peptide cleavage site are placed at the N terminus of either V_H or V_L. We have frequently chosen a cleavage site susceptible to mild acid for release of the leader, which in the

[48] S. F. Parmley and G. P. Smith, *Gene* **73,** 305 (1988); J. K. Scott and G. P. Smith, *Science* **249,** 386 (1990); J. J. Devlin, L. C. Panganiban, and P. E. Devlin, *Science* **249,** 404 (1990); S. E. Cwirla, E. A. Peters, R. W. Barrett, and W. J. Dower, *Proc. Natl. Acad. Sci. U.S.A.* **87,** 6378 (1990).

[49] M. S. Briggs and L. M. Gierasch, *Adv. Protein Chem.* **38,** 109 (1986).

[50] A. Mitraki and J. King, *Bio/Technology* **7,** 690 (1989); C. H. Schein, *Bio/Technology* **7,** 1141 (1989).

case of the 26–10 anti-digoxin sFv entailed the placement of an Asp-Pro bond at the leader–V_H junction.[5,6] This dipeptide can overlap with a *Bam*HI restriction site that serves well as an adapter to mate sFv genes with genes for leader sequences. In our general scheme, a *Sac*I site serves as an adapter at the C-terminal end of V_H. A large number of V_H regions end in the sequence -Val-Ser-Ser-, which is compatible with the codons for a *Sac*I site (G-AGC-TCT), to which the single-chain linker may be attached. The C-terminal -Gly-Ser- of the $(\text{-GGGGS-})_3$ linker can be encoded by GGA-TCC to generate a *Bam*HI site, which is useful provided that the same site was not chosen for the beginning of V_H. Alternatively, an *Xho*I site can be placed at this end of the linker by including another serine to make a -Gly-Ser-Ser- sequence that can be encoded by -GGC-TCG-AGC/T-, which contains the *Xho*I site (CTCGAG). For sFv genes encoding V_H–linker–V_L, we typically place a *Pst*I or *Hin*dIII site at the end of V_L following the new stop codon, which forms a standard site for ligation to expression vectors. If any of these restriction sites occur elsewhere in the cDNA, one should remove them by silent base changes using site-directed mutagenesis. Similar designs can be used to develop a standard architecture for V_L–V_H constructions.

Protein Production

Expression Strategies

At this juncture, the strategies used to produce sFv proteins have primarily relied on bacterial expression. Among these various methods, fusion protein expression has provided a reliable source of sFv proteins, but direct expression and secretion appear to offer great promise for more efficient methods to produce sFv proteins. Direct expression can potentially produce the sFv protein without a leader but in need of refolding, whereas secretion can ideally produce native sFv protein for subsequent isolation. Although there can be problems of intracellular breakdown in both of these approaches, active McPC603 sFv has been prepared in 0.2 to 0.3 mg/liter yield by secretion from *E. coli.*[12] Several groups have also achieved secretion or direct expression of heterodimeric Fv fragments from *E. coli.*[4] Furthermore, as an example of eukaryotic expression, Fv fragments have been secreted from myeloma cells at levels of 8 mg/liter.[28] Ultimately, the development of optimized secretion systems may be expected to provide vehicles for routine sFv production.

The expression of fusion proteins in *E. coli* as insoluble inclusion bodies provides a basis for large-scale production of sFv proteins that may otherwise be unstable due to intracellular degredation. In the FB–sFv^{26-10} example to be discussed, the final yield of active bifunctional fusion protein was 110 mg/liter of fermentation.[6] This case typifies amino-terminal fusion partners that do not interfere with the antigen binding,[6,19] which may simplify screening for sFv fusion protein during purification or mutant selection. Fusion protein derived from inclusion bodies must be purified and refolded *in vitro* to recover antigen binding activity, as was done for the 26–10 sFv analog and FB fusion protein.[5,6] The FB fragment from staphylococcal protein A has provided several advantages as a leader fused to the 26–10 sFv: it facilitated high-level expression of the fusion protein without compromising the antigen binding site in the native sFv; it bound to IgG-Sepharose during affinity purification; it improved the solubility of the 26–10 sFv. The leader can typically be removed from the expressed fusion protein by taking advantage of an appropriate cleavage strategy. In the case of 26–10 sFv, mild acid hydrolysis was used to cleave a labile Asp-Pro peptide bond between the leader and sFv, yielding proline at the sFv amino terminus.[5] In other situations, leader cleavage can rely on chemical or enzymatic hydrolysis at specifically engineered sites, such as CNBr cleavage of a unique methionine, hydroxylamine cleavage of the peptide bond between Asn-Gly, and limited proteolytic digestion at specific cleavage sites, such as those recognized by factor Xa, enterokinase, V8 protease, or pepsin.

Direct expression of intracellular Fv proteins, which yields the desired sFv without a leader attached, has been achieved for both single-chain Fv analogs[50a] and sFv fusion proteins.[6,14–22] This approach avoids the steps needed for leader removal, but the isolation of inclusion bodies must be followed by refolding and purification. Several sFv analogs[50a] and fusion proteins [14–18,21,21a] have been directly expressed by cloning the assembled gene behind a T7 promoter and transforming into the *E. Coli* host strain BL21(DE3).[50b] This strain carries the T7 RNA polymerase gene under the control of the inducable *lac* UV5 promoter. After induction with IPTG, T7-specific transcripts increasingly predominate over cellular RNA due to the very high elongation rate of T7 RNA polymerase. This leads to high levels of expression which usually result in formation of inclusion bodies by the sFv protein of interest.

Refolding of sFv Proteins

The denaturation transitions of Fab fragments from polyclonal antibodies are known to cover a broad range of denaturant.[51] Monoclonal antibody Fab fragments are similarly dispersed in general, but specific Fab

[50a] S. M. Andrew, P. Perez, P. J. Nicholls, A. J. T. George, and D. M. Segal, unpublished results (1990); D. Jin, M.-S. Tai, J. S. Huston, and H. Oppermann, unpublished results (1991).

[50b] F. W. Studier and B. A. Moffat, *J. Mol. Biol.* **189,** 113 (1986); F. W. Studier, A. H. Rosenburg, J. J. Dunn, and J. W. Dubendorff, this series, Vol. 185, p. 60.

[51] C. E. Buckley III, P. L. Whitney, and C. Tanford, *Proc. Natl. Acad. Sci. U.S.A.* **50,** 827 (1963); M. E. Noelken and C. Tanford, *J. Biol. Chem.* **239,** 1828 (1964).

fragments or component domains exhibit relatively sharp denaturation transitions over a limited range of denaturant.[52] Thus sFv proteins can be expected to differ similarly,[9] covering a broad range of stabilities and denaturation properties, which appear to be paralleled by their preferences for distinct refolding procedures. Initially, there was some question about whether the V_H and V_L domains would be able to refold properly when connected to each other in an sFv, instead of to their respective constant domains in the parent antibody. We found that this was not an issue, since both the 26-10 and MOPC 315 sFv and FB-sFv proteins recovered their specific antigen-binding activity after refolding.[5,6,19] The refolding protocols that we have used consist of dilution refolding, redox refolding, and disulfide-restricted refolding (Fig. 6).

Dilution refolding is the method of protein folding used successfully in our original 26-10 sFv experiments (Fig. 6A),[5] which relies on the early observation that fully reduced and denatured antibody fragments can refold on removal of denaturant and reducing agent with recovery of specific binding activity.[53] Our procedure started with ion exchange-purified 26-10 sFv in 6 *M* urea + 2.5 m*M* Tris-HCl + 1 m*M* EDTA + 0.1 *M* 2-mercaptoethanol, at pH 8, which was diluted 100-fold into 0.01 *M* sodium acetate, pH 5.5, followed by dialysis against the pH 5.5 buffer with changes over 3 days to effect reoxidation of the disulfides. Ouabain-Sepharose affinity chromatography of 26-10 sFv produced a 12.8% yield of active protein. This method takes advantage of the protein folding process to enhance correct disulfide pairing. Refolding was conducted at sFv concentrations below 10 μg/ml to minimize aggregation. Poor refolding conditions often show the formation of aggregated protein, manifested as visible precipitation or turbidity, or on nonreducing SDS-PAGE as high-molecular-weight material that is disaggregated only if reduced prior to electrophoresis. Such aggregation may necessitate an alternative refolding protocol.

Redox refolding utilizes a glutathione redox couple to catalyze disulfide interchange[54] as the protein refolds into its native state (Fig. 6B).[6] As

[52] E. S. Rowe and C. Tanford, *Biochemistry* **12,** 4822 (1973); E. S. Rowe, *Biochemistry* **15,** 905 (1976); E. S. Rowe, Ph.D Dissertation, Duke University, 1971; J. Buchner and R. Rudolph, *Bio/Technology* **9,** 157 (1991).

[53] E. Haber, *Proc. Natl. Acad. Sci. U.S.A.* **52,** 1099 (1964); P. L. Whitney and C. Tanford, *Proc. Natl. Acad. Sci. U.S.A.* **53,** 524 (1965). The Haber renaturation procedure utilized dialysis to remove denaturant and reducing agent, since straight dilution resulted in precipitation of the dissociated chains of rabbit anti-ribonuclease Fab fragments. In contrast, the Whitney and Tanford procedure for refolding of rabbit anti-DNP Fab fragments relied on dilution followed by dialysis to effect complete renaturation and reoxidation. The differences in refolding protocols for specific sFv analog or fusion proteins are thus reminiscent of features observed in these first experiments on the renaturation of fully reduced and denatured Fab fragments of polyclonal antibodies.

[54] P. Saxena and D. B. Wetlaufer, *Biochemistry* **9,** 5015 (1971).

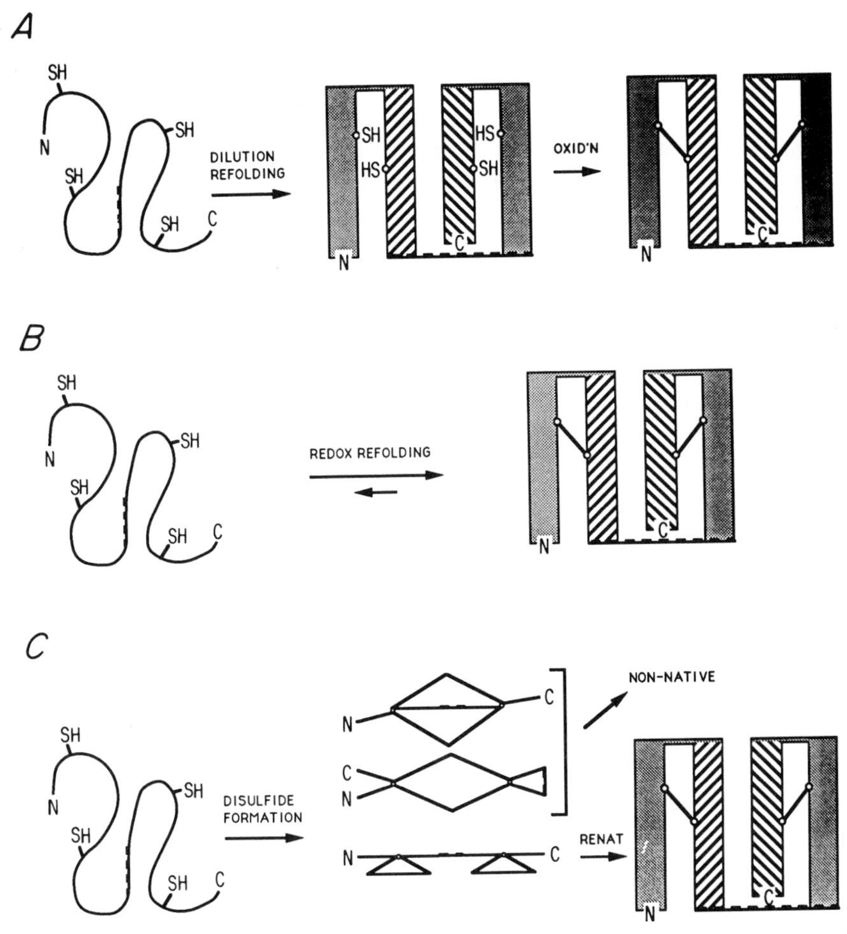

FIG. 6. Refolding schemes for sFv proteins. The left-hand figures represent the fully reduced and denatured sFv polypeptide, with the crenulated line indicating the linker segment. As noted in the legend to Fig. 4, the native disulfide-bonding pattern is depicted by the right-hand figures, while the disposition of intermediate forms is noted in the middle diagrams. (A) Dilution refolding involves a two-stage refolding process, with renaturation of the reduced protein followed by air oxidation. (B) The redox-refolding process does not involve isolated intermediates, since refolding and reoxidation occur concurrently in solution. (C) Disulfide-restricted refolding leads to three potential forms of the fully oxidized monomeric protein, shown as intermediate species (note Fig. 4), of which only the bottom species can yield a native sFv.

conducted in our experiments, the 26–10 sFv protein was diluted from a fully reduced state in 6 *M* urea into 3 *M* urea + glutathione redox couple (1 m*M* oxidized/0.1 m*M* reduced) + 25 m*M* Tris-HCl + 10 m*M* EDTA, pH 8, to yield a final concentration of approximately 0.1 mg/ml. The 26–10 sFv unfolding transition begins around 3 *M* urea, and consequently the refolding buffer represents near-native solvent conditions for the 26–10 sFv. Under these conditions, the protein can presumably reform approximations to the V domain structures, wherein rapid disulfide interchange can occur until a stable equilibrium is attained. After incubation at room temperature for 16 hr, the material was dialyzed first against urea buffer lacking glutathione, and then against 0.01 *M* sodium acetate + 0.25 *M* urea, pH 5.5, with a refolding yield of 23%. Similar experiments with the FB–sFv$^{26-10}$ fusion protein gave 46% refolding yields, with 110 mg of active protein obtained per liter of fermented cells. The higher protein concentrations of this redox-refolding procedure and sometimes better yields of active product can combine to produce a more manageable purification scheme than that associated with the dilution procedure, which generates very large sample volumes when procedures are applied to large-scale production.

Disulfide-restricted refolding offers still another route to obtaining active sFv, which involves initial formation of intrachain disulfides in the fully denatured sFv (Fig. 6C).[19] This capitalizes on the favored reversibility of antibody refolding when disulfides are kept intact, which has been known since the early studies on antibody refolding of Tanford and co-workers.[51] Disulfide cross-links should restrict the initial refolding pathways available to the molecule, as well as tether residues adjacent to cysteinyl residues that are close in the native state. For chains with the correct disulfide pairing, the recovery of a native structure should be favored, while those chains with incorrect disulfide pairs must necessarily produce nonnative species on removal of denaturant. This procedure was originally applied to MOPC315 Fv by Givol and co-workers,[1a] as no other procedure was found to regenerate active Fv. McCartney *et al.*[19] have found that the sFv based on MOPC315 also shows an obligatory requirement for the restricted disulfide-refolding procedure. Although this refolding method may give a lower yield than other procedures, it may be able to tolerate higher protein concentrations during refolding.

Secretion of sFv

Proteins secreted into the periplasmic space of Gram-negative bacteria or into the culture medium from a variety of cell types appear to refold properly with formation of the correct disulfide bonds.[12,39] In the majority of cases, the signal peptide sequence is removed by a signal peptidase to generate a product with its natural amino terminus. Even though most heterologous secretion systems currently give considerably lower yields than intracellular expression, the rapidity of obtaining correctly folded and active sFv proteins can be of decisive value for protein engineering. Plückthun and co-workers have reported the expression of McPC603 sFv and Fv fragment into the periplasm of *E. coli,* with yields of 0.2 to

0.3 mg/liter for sFv and 0.5 mg/liter for Fv.[12] They used the *ompA* signal sequence to direct secretion of the sFv, and previously secreted the Fv using a dicistronic operon with the *ompA* signal directing secretion of V_H and *phoA* signal sequence directing secretion of V_L. Better *et al.* have succeeded in secreting active Fab fragments from *E. coli* into the culture medium.[39] Their expression vector consisted of the *Salmonella araB* promoter for expression of a dicistronic message comprising the heavy and light chain gene fragments fused to the *pelB* signal sequence from *Erwinia carotovora.* Although titers of the secreted Fab were only 0.5–1 mg/liter in shake flask cultures, optimization of fermentation led to Fab titers as high as 561 mg/liter.[55]

More recently, the synthesis and secretion of a mouse monoclonal antibody against *Pseudomonas aeruginosa* was achieved in cells from fall armyworm, *Spodoptera frugiperda,* which used the baculovirus polyhedrin promoter.[56] Quantitation by immunoassay demonstrated the synthesis of 25–30 μg of antibody per milliliter of medium containing 5×10^5 cells. Monoclonal antibodies have also been produced in mature regenerated transgenic tobacco plants.[57] Functional antibody accumulated to 1.3% of total leaf protein in plants expressing full-length cDNAs containing signal sequences. In all likelihood, sFv production should be amenable to nonbacterial methods such as these.

Specific Examples: Anti-Digoxin 26–10 sFv Analog and FB–sFv$^{26-10}$ Fusion Protein

This section presents results on the anti-digoxin 26–10 sFv and its fusion with an amino-terminal effector domain, the fragment B of staphylococcal protein A.[6] The FB–sFv$^{26-10}$ fusion protein has been expressed as inclusion bodies in *E. coli;* the sequence of the FB–sFv$^{26-10}$ gene and fusion protein as well as a diagram of the expression plasmid are shown in Fig. 7.[6,58] Fusion of the modified trp LE (MLE)[5] and FB[6] leaders to the 26–10 sFv at the DNA level resulted in high-level expression of fusion proteins which formed birefringent inclusion bodies.[5,6] Designs of plasmids and constructions have been described for both the 26–10 MLE–sFv$^{26-10}$ and the FB–sFv$^{26-10}$ fusion protein.[5,6] As noted previously, the FB–sFv$^{26-10}$ was designed with a potentially flexible 14-residue spacer or hinge region between the structured core of the FB and first V_H residue

[55] M. Better, J. Weickmann, and Y.-L. Lin, *ICSU Short Rep.* **10,** 105 (1990).

[56] J. zu Putlitz, W. L. Kubansek, M. Duchene, M. Marget, B.-U. von Specht, and H. Domdey, *Bio/Technology* **8,** 651 (1990).

[57] A. Hiatt, R. Cafferkey, and K. Bowdish, *Nature (London)* **342,** 76 (1989).

[58] J. Brosius and A. Holy, *Proc. Natl. Acad. Sci. U.S.A.* **81,** 6929 (1984); H. A. DeBoer, L. J. Comstock, and M. Vasser, *Proc. Natl. Acad. Sci. U.S.A.* **80,** 21 (1983).

A

```
                                          FB-10                                      FB-20                                      FB-30
Ala Asp Asn Lys Phe Asn Lys Glu Gln Gln Asn Ala Phe Tyr Glu Ile Leu His Leu Pro Asn Leu Asn Glu Glu Gln Arg Asn Gly Phe
GCT GAC AAC AAA TTC AAC AAG GAA CAG CAG AAC GCG TTC TAC GAG ATC TTG CAC CTG CCG AAC CTG AAC GAA GAG CAG CGT AAC GGC TTC
                                          FB-40                                      FB-50                                      FB-60
Ile Gln Ser Leu Lys Asp Asp Pro Ser Gln Ser Ala Asn Leu Leu Ala Glu Ala Lys Lys Leu Asn Asp Ala Gln Ala Pro Lys Ser Asp
ATC CAA AGC TTG AAA GAC GAC CCG TCT CAG AGC GCT AAC CTG CTG GCA GAG GCC AAG AAA CTG AAC GAC GCT CAG GCG CCG AAG AGT GAT

                                             10                                          20                                          30
Pro Glu Val Gln Leu Gln Gln Ser Gly Pro Glu Leu Val Lys Pro Gly Ala Ser Val Arg Met Ser Cys Lys Ser Ser Gly Tyr Ile Phe
CCC GAA GTT CAA CTG CAA CAG TCT GGT CCT GAA TTG GTT AAA CCT GGC GCC TCT GTG CGC ATG TCC TGC AAA TCC TCT GGG TAC ATT TTC
                                             40                                          50                                          60
Thr Asp Phe Tyr Met Asn Trp Val Arg Gln Ser His Gly Lys Ser Leu Asp Tyr Ile Gly Tyr Ile Ser Pro Tyr Ser Gly Val Thr Gly
ACC GAC TTC TAC ATG AAT TGG GTT CGC CAG TCT CAT GGT AAG TCT CTA GAC TAC ATC GGG TAC ATT TCC CCA TAC TCT GGG GTT ACC GGC
                                             70                                          80                                          90
Tyr Asn Gln Lys Phe Lys Gly Lys Ala Thr Leu Thr Val Asp Lys Ser Ser Ser Thr Ala Tyr Met Glu Leu Arg Ser Leu Thr Ser Glu
TAC AAC CAG AAG TTT AAA GGT AAG GCG ACC CTT ACT GTC GAC AAA TCT TCC TCA ACT GCT TAC ATG GAG CTG CGT TCT TTG ACC TCT GAG
                                            100                                         110                                         120
Asp Ser Ala Val Tyr Tyr Cys Ala Gly Ser Ser Gly Asn Lys Trp Ala Met Asp Tyr Trp Gly His Gly Ala Ser Val Thr Val Ser Ser
GAC TCC GCG GTA TAC TAT TGC GCG GGC TCC TCT GGT AAC AAA TGG GCC ATG GAT TAT TGG GGT CAT GGT GCT AGC GTT ACT GTG AGC TCT
                                            130                                         140                                         150
Gly Gly Gly Gly Ser Gly Gly Gly Gly Ser Gly Gly Gly Gly Ser Asp Val Val Met Thr Gln Thr Pro Leu Ser Leu Pro Val Ser Leu
GGT GGC GGT GGC TCG GGC GGT GGT GGG TCG GGT GGC GGC GGA TCT GAC GTC GTA ATG ACC CAG ACT CCG CTG TCT CTG CCG GTT TCT CTG
                                            160                                         170                                         180
Gly Asp Gln Ala Ser Ile Ser Cys Arg Ser Ser Gln Ser Leu Val His Ser Asn Gly Asn Thr Tyr Leu Asn Trp Tyr Leu Gln Lys Ala
GGT GAC CAG GCT TCT ATT TCT TGC CGC TCT TCC CAG TCT CTG GTC CAT TCT AAT GGT AAC ACT TAC CTG AAC TGG TAC CTG CAA AAG GCT
                                            190                                         200                                         210
Gly Gln Ser Pro Lys Leu Leu Ile Tyr Lys Val Ser Asn Arg Phe Ser Gly Val Pro Asp Arg Phe Ser Gly Ser Gly Ser Gly Thr Asp
GGT CAG TCT CCG AAG CTT CTG ATC TAC AAA GTC TCT AAC CGC TTC TCT GGT GTC CCG GAT CGT TTC TCT GGT TCT GGT TCT GGT ACT GAC
                                            220                                         230                                         240
Phe Thr Leu Lys Ile Ser Arg Val Glu Ala Glu Asp Leu Gly Ile Tyr Phe Cys Ser Gln Thr Thr His Val Pro Pro Thr Phe Gly Gly
TTC ACC CTG AAG ATC TCT CGT GTC GAG GCC GAA GAC CTG GGT ATC TAC TTC TGC TCT CAG ACT ACT CAT GTA CCG CCG ACT TTT GGT GGT

Gly Thr Lys Leu Glu Ile Lys Arg *oc
GGC ACC AAG CTC GAG ATT AAA CGT TAA CTGCAG
```

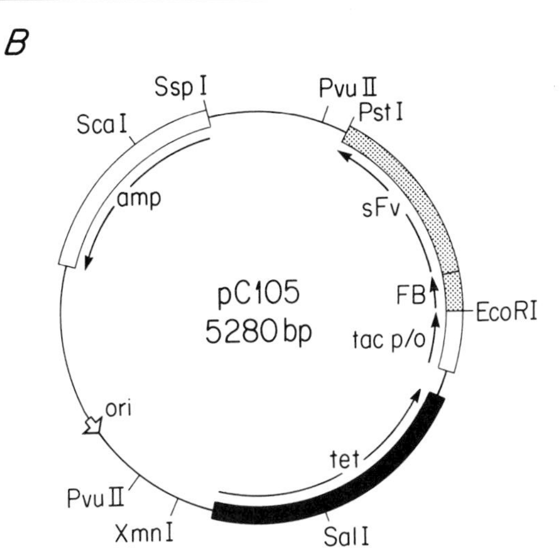

FIG. 7. Gene and protein sequences of the FB–sFv$^{26-10}$ and the expression plasmid as used for its production.[6] (A) Sequence of FB–sFv$^{26-10}$ fusion protein and corresponding gene. Upper sequence is that of FB (FB-1 to FB-58) followed by -Ser-Asp. The lower sequence is the 26–10 sFv (residues 1–248), in which the 15 residues bridging the V regions are underlined and complementarity-determining regions (CDRs) are shaded. The actual gene began with an *Eco*RI site followed by the initiation codon (ATG); the first amino acid (*N*-formylmethionine) of the translation product was removed in the bacterial host. (B) Circular map of the FB–sFv$^{26-10}$ expression plasmid. The FB gene was fused to a gene encoding the lower sequence for the 26–10 sFv protein. The genes were derived from synthetic DNA and present numerous 6-b restriction sites to facilitate site-directed mutagenesis. The entire FB–sFv gene was inserted between the *Eco*RI and *Pst*I sites of the pKK223-3 expression vector[58] and this plasmid was used to transform *E. coli* strain RB791 (*lacI^q*). The fusion protein was expressed under control of the *tac* promoter by addition of isopropyl-β-D-thiogalactoside (IPTG). [Reproduced with permission from *Biochemistry* **29,** 8024 (1990), American Chemical Society.]

(Glu-2) of 26–10 sFv.[36] Assay results are discussed below that verified the recovery of authentic 26–10 digoxin-binding properties in the fusion protein, which confirmed that access to the binding site was not obstructed by the amino-terminal fragment B.

Protein Refolding and Purification Schemes

Purification protocols will be outlined in this section for several 26–10 single-chain Fv species. These procedures all share a common starting point, in that the recombinant protein initially was derived from *E. coli* inclusion bodies.[6] The cell paste for small-scale preparation was suspended in 25 m*M* Tris-HCl, pH 8, and 10 m*M* EDTA, treated with 0.1 mg/ml lysozyme overnight, sonicated at a high setting for three 5-min periods in the cold, and spun in a preparative centrifuge at 11,300 *g* at 4° for 30 min. The pellet was then washed with a buffer containing 3 *M* urea, 25 m*M*

Tris-HCl, pH 8, and 10 mM EDTA; after recentrifugation, the protein pellet served as the starting material for each of the following protocols, or it was stored at $-20°$ until use. For large-scale preparation of inclusion bodies, our fermentation facility concentrates cells by ultrafiltration and then lyses the cells with a laboratory homogenizer (Model 15MR; APV Gaulin, Inc., Everett, MA). The inclusion bodies are then collected by centrifugation.

Method A: Purification of 26–10 sFv from MLE–sFv$^{26-10}$ Fusion Protein[5]

1. Solubilization of fusion protein in acid cleavage buffer [6 M guanidine hydrochloride (GuHCl), 10% (v/v) acetic acid, pH 2.5]: Dissolve MLE–sFv$^{26-10}$ inclusion bodies in 6.7 M GuHCl, and then dissolve solid GuHCl equal to the weight of the inclusion body pellet; 6 M GuHCl is 50.2% GuHCl by weight, and this compensates for water in the pellet. Add glacial acetic acid to 10% of total volume and adjust to pH 2.5 with concentrated HCl.

2. Cleavage of the unique Asp-Pro bond at the junction of leader and 26–10 sFv: Incubate the solution from step A-1 above at 37° for 96 hr. Stop the reaction by adding 9 vol of cold ethanol, storing at $-20°$ for several hours, followed by centrifugation to yield a pellet of precipitated 26–10 sFv and uncleaved fusion protein. The cleavage mixture may also be stored as a pellet at $-20°$ until needed.

3. Transfer of cleaved protein into urea buffer for ion-exchange chromatography: Dissolve the precipitated cleavage mixture in 6 M GuHCl, 0.2 M Tris-HCl, pH 8.2, with addition of solid GuHCl equal to the pellet weight; adjust pH if needed before adding 2-mercaptoethanol to 0.1 M. Dialyze this solution into starting buffer for ion-exchange chromatography consisting of 6 M urea, 2.5 mM Tris-HCl, pH 7.5, 1 mM EDTA, 5 mM dithiothreitol.

4. Removal of uncleaved fusion protein from 26–10 sFv: Chromatograph the solution on a column of DE-52 (Whatman, Clifton, NJ) equilibrated in starting buffer (step A-3). SDS-PAGE is used to monitor purity of the flowthrough protein, and leading fractions that contain pure 26–10 sFv are pooled.

5. Renaturation of 26–10 sFv by the dilution refolding procedure: Dilute the DE-52 pool of 26–10 sFv at least 100-fold into 0.01 M sodium acetate, pH 5.5, to a concentration below 10 μg/ml, and dialyze at 4°. The two disulfides reform by air oxidation of four cysteinyl residues in the refolded 26–10 sFv.

6. Affinity purification of the anti-digoxin 26–10 sFv from the mixture of refolded species: The sample from step A-5 is loaded onto a column containing ouabain-amine-Sepharose 4B, and column is washed succes-

sively with 0.01 *M* sodium acetate, pH 5.5, followed by two column volumes of 1 *M* NaCl, 0.01 *M* sodium acetate, pH 5.5, and then 0.01 *M* sodium acetate once again to remove salt. Finally, the active protein is displaced from the resin by 20 m*M* ouabain in 0.01 *M* sodium acetate, pH 5.5. Absorbance measurements at 280 nm indicate which column fractions contain active protein. However, the spectra of the protein and ouabain overlap, and consequently ouabain must be removed by exhaustive dialysis against 0.01 *M* sodium acetate, pH 5.5, 0.25 *M* urea, in order to quantitate the protein yield accurately. If the protein is to be kept at 4° for long periods, sodium azide is added to a concentration of 0.03%.

Method B: Preparation of 26–10 sFv by Redox Refolding[6]

1. Renaturation of 26–10 sFv: Starting with the DE-52 pool of 26–10 sFv in 6 *M* urea buffer (step A-4), dilute to a final concentration below 0.1 mg/ml by dropwise addition to 3 *M* urea, 25 m*M* Tris-HCl, pH 8, 1 m*M* oxidized glutathione, 0.1 m*M* reduced glutathione, and incubate at room temperature for 16 hr.
2. Removal of redox couple from 26–10 sFv: Dialyze in the cold against 3 *M* urea, 25 m*M* Tris-HCl, pH 8, 10 m*M* EDTA.
3. Removal of denaturant from 26–10 sFv: Dialyze in the cold for 2 days against 0.01 *M* sodium acetate, pH 5.5, 0.25 *M* urea.
4. Affinity isolation of 26–10 sFv: Follow step A-6 for ouabain-amine-Sepharose affinity chromatography, adding 0.25 *M* urea to all buffers.

Method C: Purification of FB–sFv$^{26-10}$ Fusion Protein (Fig. 8)[6]

1. Solubilization and dialysis of inclusion bodies: Dissolve in 6 *M* GuHCl, 0.2 *M* Tris-HCl, pH 8.2, 0.1 *M* 2-mercaptoethanol, and incubate at room temperature for 1.5 hr. Dialyze the denatured and reduced inclusion body protein into 6 *M* urea, 2.5 m*M* Tris-HCl, pH 8, 1 m*M* EDTA, 5 m*M* dithiothreitol.
2. Purification of FB–sFv$^{26-10}$ in denaturant: Chromatograph the inclusion body protein on a column of Whatman DE-52. Pool the flowthrough fractions that are free of contaminants according to SDS-PAGE analysis.
3. Renaturation of FB–sFv$^{26-10}$ by redox refolding procedure: Follow step B-1.
4. Removal of denaturant and redox couple: The 3 *M* urea buffer and glutathione redox couple are removed in one step by dialysis for 2 days against 50 m*M* potassium phosphate, pH 7, 0.25 *M* urea.
5. Affinity isolation of FB–sFv$^{26-10}$: Follow step A-6, except that all column buffers contain PBSA (0.15 *M* NaCl, 0.05 *M* potassium phosphate, pH 7, 0.03% NaN_3) instead of acetate buffer, and the final dialysis to remove ouabain is against PBSA.

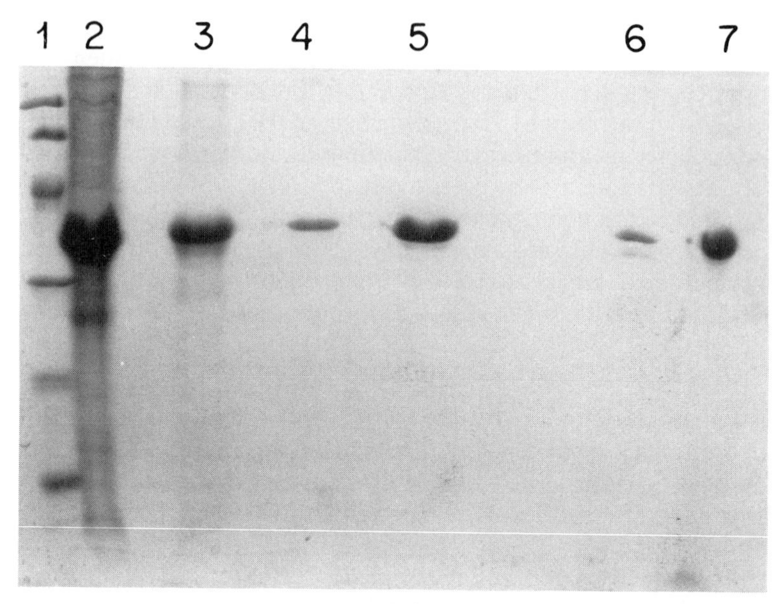

FIG. 8. SDS-PAGE analysis of FB–sFv$^{26-10}$ during purification.[6] SDS-PAGE analysis of FB–sFv$^{26-10}$ at progressive stages of purification. Samples in numbered lanes of the Coomassie Blue-stained 15% gel were the following: (1) Standards ($M_r \times 10^{-3}$), 14.4, 20.1, 29, 43, 67, 94; (2) unpurified inclusion bodies; (3) DE-52-purified FB–sFv$^{26-10}$; (4) refolded FB–sFv$^{26-10}$ mixture (reduced); (5) affinity-purified FB–sFv$^{26-10}$ (reduced); (6) refolded FB–sFv$^{26-10}$ mixture (unreduced); (7) affinity-purified FB–sFv$^{26-10}$ (unreduced). As discussed for 26–10 sFv,[5] the FB–sFv$^{26-10}$ migrated with a higher apparent molecular weight than its calculated molecular weight of 33,106; this is at least partially due to the low residue weight of glycine and serine in the linker sequence. [Reprinted with permission from *Biochemistry* **29**, 8024 (1990), American Chemical Society.]

Method D: Purification of 26–10 sFv from FB–sFv$^{26-10}$ or FB–FB–sFv$^{26-10}$

Figure 10 and the discussion under Improved Methods of 26–10 sFv Production detail this approach further, which parallels other applications of such FB fusion protein expression.[58a]

1. Cleavage of fusion protein: Since crude inclusion bodies will not dissolve in acetic acid, the purified, or at least refolded, fusion protein offers the starting material for this procedure. The purified and refolded fusion protein, prepared according to steps C-1 to C-5, is the starting material for this procedure. The active fusion protein in PBSA is mixed with 0.11 vol of glacial acetic acid to make a 10% solution, and the solution is incubated at 37° for 96 hr. The acetic acid conditions are favored if the

[58a] E. Samuelsson, H. Wadensten, M. Harmanis, T. Moks, and M. Uhlén, *Bio/Technology* **9**, 363 (1991).

fusion protein contains the natural FB sequence, because it avoids unwanted cleavage products. Cleavage in 10% acetic acid results in acid hydrolysis of the Asp-Pro bond at leader–sFv junctions; the fragment B has two acid-labile peptide bonds between residues FB-36 to FB-38 (Asp-Asp-Pro), which are not cleaved in 10% acetic acid but are partially hydrolyzed in 6 *M* GuHCl, 10% acetic acid, pH 2.5. The cleavage mixture is ethanol precipitated as noted under step A-2, and following centrifugation the pelleted material can be stored at $-20°$ until use.

2. Renaturation of proteins in cleavage mixture: The pellet is dissolved in 6 *M* GuHCl, 25 m*M* Tris-HCl, pH 8, 10 m*M* EDTA, dialyzed into 6 *M* urea, 25 m*M* Tris-HCl, pH 8, 10 m*M* EDTA. The dialysate is diluted to 3 *M* urea, 25 m*M* Tris-HCl, pH 8, 10 m*M* EDTA, followed by dialysis against the 3 *M* urea buffer in the cold for 2 days. Alternatively, renaturation can be performed starting with the fully reduced random coil (step C-1 followed by steps B-1 to B-4).

3. Removal of uncleaved fusion protein: The remaining intact fusion protein is removed from the dialyzed cleavage mixture by affinity chromatography on human IgG-Sepharose in 3 *M* urea, 25 m*M* Tris-HCl, pH 8, 10 m*M* EDTA. Under these conditions aggregation is negligible, but the FB or FB–FB effector domains bind to the immobilized immunoglobulin. The flowthrough fractions contain the purified sFv. The column can be successfully regenerated after elution with 0.2 *M* glycine, pH 2.9, followed by extensive washing with PBSA.

4. Final refolding and affinity isolation of 26–10 sFv: Follow steps B-3 and B-4, to prepare affinity-purified 26–10 sFv.

Analysis of sFv Antigen-Binding Properties

Full characterization of any antibody combining site requires that both its affinity and specificity for antigen be determined. Several model antibodies have been used in the careful analysis of sFv combining sites.[5,6,8,9,12] Ideally, the measurement of binding affinity should use a thermodynamically rigorous approach such as equilibrium dialysis or ultrafiltration.[59] In the absence of such methods, agreement between two distinct methods is desirable to reinforce the veracity of data. Antigen binding to high-affinity antibody binding sites frequently involves very fast rates of binding and slow rates of dissociation. This makes measurement of their association constants amenable to routine immunoassay procedures. However, our experience suggests that sFv analogs and fusion proteins can require some-

[59] J. A. Sophianopoulos, S. J. Durham, A. J. Sophianopoulos, H. L. Ragsdale, and W. P. Cropper, *Arch. Biochem. Biophys.* **187**, 132 (1978); W. MacMahon, J. Stallings, and D. Sgoutas, *Clin. Chem.* **28**, 240 (1982); J. A. Sophianopoulos, A. J. Sophianopoulos, and W. MacMahon, *Arch. Biochem. Biophys.* **223**, 350 (1983); A. J. Sophianopoulos and J. A. Sophianopoulos, this series, Vol. 117, p. 354.

what more care than the parent antibody to determine binding properties accurately, because the absence of all constant domains increases the chance for perturbation of V region conformation.

We have observed that sFv solubility and optimal storage conditions may differ substantially from those of the parent antibody or Fab. These differences may in part derive from evolutionary pressures having operated to improve the solubility and stability of the whole immunoglobulin rather than of the Fv alone. The constant domains of H and L chains appear to mitigate the solution properties of the variable domains, since the parent 26–10 IgG or Fab are very stable and soluble at neutrality, while the pure 26–10 sFv protein loses its activity and aggregates in PBSA at pH 7.3. The pure 26–10 sFv can be maintained as a stable active protein in 0.01 *M* sodium acetate + 0.25 *M* urea, pH 5.5. It is possible to modulate the 26–10 sFv solution properties by attaching ancillary domains, since fusion proteins typically exhibit some average of the solution properties shown by their component proteins. The considerable solubility of the FB fragment improved the 26–10 sFv solution properties when they were combined to form FB–sFv^{26-10} fusion protein, which was highly soluble and stable in PBSA at neutrality.[6] Radioimmunoassays have indicated that the binding activity of 26–10 sFv is stable in PBSA (pH 7.3) at concentrations below 10^{-9} *M* only if it has been diluted into buffer containing 0.1% (w/v) gelatin or 1% (w/v) horse serum.

The absence of constant domains in sFv proteins may sometimes result in an sFv having little or no cross-reactivity with antisera to nonspecific mouse IgG or Fab fragments. However, the 26–10 sFv was found to cross-react well with one of several commercially available goat anti-mouse Fab (GaMFab) antisera.[5] In contrast, MOPC 315 sFv assays were improved by using antiserum raised specifically against the parent IgA.[19] An interim approach has been to use competition assays for the survey of activity in sFv proteins, but this does not readily permit determination of the antigen-binding affinity in an impure mixture, as would direct binding analysis.[5,6] Perturbation of the sFv binding site can also result from interaction of sFv with the precipitating antibody, which is described below for the 26–10 sFv at low pH.

Immunoprecipitation Assay of Digoxin-Binding Affinity

For reasons of solubility described above, the 26–10 sFv was assayed at pH 5.5 in our early studies, using a standard immunoprecipitation procedure.[5] According to this method, GaMFab antibody was coincubated overnight with the 26–10 antibody species and [^{3}H]digoxin prior to addition of rabbit anti-goat IgG antibody to effect precipitation. We have since ascertained that GaMFab binding to the 26–10 sFv perturbs its interaction with digoxin at acid or neutral pH, while GaMFab binding to the other

26–10 species at pH 5.5 induced a perturbation that was pH dependent; a significant reduction in $K_{a,app}$ was observed at pH 5.5 for FB–sFv$^{26-10}$ and the controls, compared to that obtained at pH 7.3 (in Table II,[60] compare A columns for method 1 at pH 5.5 and pH 7.3). By first incubating the 26–10 species with digoxin overnight, and then separately adding GaM-Fab for immunoprecipitation, this pH-related perturbation was eliminated for the FB–sFv$^{26-10}$ and the control 26–10 species (in Table II, compare B to A columns for method 1), yielding $K_{a,app}$ values under both pH conditions of $2-3 \times 10^9\ M^{-1}$. Overnight incubation with digoxin prior to GaMFab addition conferred protection from GaMFab perturbation at pH 5.5 on all 26–10 combining sites except for that of the 26–10 sFv (Table II, method 1, B columns).

Under neutral buffer conditions, the order of addition in the immunoprecipitation assay appears to have had little affect on the observed $K_{a,app}$ values. In all cases, the $K_{a,app}$ increased by no more than about a factor of two. Interestingly, at pH 7.3 the 26–10 sFv exhibited a value for $K_{a,app}$ of $5-7 \times 10^8\ M^{-1}$ that was still significantly lower than the corresponding values for the FB–sFv$^{26-10}$, 26–10 Fab, and IgG (Table II); even when GaMFab was added after ligand, a reduction in the $K_{a,app}$ for 26–10 sFv was observed under neutral conditions.

In order to test the hypothesis that the observed pH 5.5 perturbation was caused by antibody binding to determinants on or near the V domains, the $K_{a,app}$ was also determined in another series of double antibody-precipitation assays that used second antibodies directed to epitopes that were likely to be distant from the variable domains (Table II, method 2). Although this assay could not be applied to 26–10 sFv, the $K_{a,app}$ for the other 26–10 species remained between 1.3×10^9 and $3.0 \times 10^9\ M^{-1}$, regardless of pH, even when the rabbit anti-mouse IgG antibody or rabbit anti-FB antibody was coincubated with the 26–10 species and [^{3}H]digoxin (data not shown).

Ultrafiltration Assay of Digoxin-Binding Affinity

When ultrafiltration was used to separate bound from free hapten, the perturbation associated with the double antibody-precipitation method disappeared. Ultrafiltration is physicochemically equivalent to equilibrium dialysis and in this case provided an unambiguous measure of antigen–antibody association, since the assay involved no precipitating antibody.[6,59] At both pH 5.5 and 7.3, the 26–10 sFv and FB–sFv$^{26-10}$ were shown to exhibit the same high affinity for digoxin as the parent 26–10 IgG and Fab, within expeimental error ($K_{a,app} = 2.4 \times 10^9\ M^{-1}$) (Table III).

[60] G. A. McPherson, "Kinetic, EBDA, LIGAND, Lowry. A Collection of Radioligand Binding Analysis Programs." Elsevier BIOSOFT, Cambridge, 1985.

TABLE II
MEASUREMENT OF 26–10 BINDING SITE AFFINITIES BY IMMUNOPRECIPITATION[a]

26–10 species	Digoxin affinity ($K_{a,app} \times 10^{-9}\ M^{-1}$)					
	Method 1[b]				Method 2[c]	
	pH 5.5		pH 7.3		pH 5.5	pH 7.3
	A	B	A	B		
sFv	0.4 ± 0.05	1.2 ± 0.08	0.5 ± 0.2	0.7 ± 0.1	nd[d]	nd[d]
FB–sFv	0.4 ± 0.05	2.7 ± 0.2	1.3 ± 0.2	2.9 ± 0.4	1.3 ± 0.2	1.3 ± 0.1
Fab	0.4 ± 0.05	3.0 ± 0.6	2.1 ± 0.3	2.8 ± 0.5	3.0 ± 0.5	2.2 ± 0.3
IgG	0.5 ± 0.07	1.8 ± 0.1	1.5 ± 0.2	2.0 ± 0.4	2.0 ± 0.3	1.6 ± 0.2

[a] Affinity measurements of 26–10 binding species were determined at both pH 5.5 (0.01 *M* sodium acetate, 0.25 *M* urea with 0.1% gelatin or 1% BSA) and pH 7.3 (PBSA, 1% horse serum). In all methods, the 26–10 binding species ($2-9 \times 10^{-10}\ M$) were incubated with varying concentrations of [^{3}H]digoxin (New England Nuclear, Boston, MA) ranging from 1×10^{-10} to $1 \times 10^{-7}\ M$, overnight at 4°. Precipitation was effected as described in footnote *b* or *c*. Precipitates were collected on filters as described below, placed in 5.0 ml scintillation fluid (Ultima Gold; Packard Instruments, Meriden, CT) and counted on a model 1500 Tri-Carb liquid scintillation analyzer (Packard Instruments, Downers Grove, IL). In each instance, $K_{a,app}$ was calculated from the equilibrium binding data using the binding analysis program LIGAND.[60]

[b] Goat anti-mouse Fab serum (GaMFab) was used as primary precipitating antibody in either of two ways: for column A, GaMFab was incubated overnight at 4° with the 26–10 binding species and [^{3}H]digoxin; for column B, GaMFab was added following the overnight incubation of the 26–10 species with [^{3}H]digoxin. Quantitative precipitation was effected by the addition of affinity-purified rabbit anti-goat IgG (10 μl; ICN ImmunoBiologicals, Lisle, IL) and protein A-Sepharose (25 μl of a 10% slurry in PBSA; Sigma, St. Louis, MO). Following a 2-hr incubation on ice, the precipitate was collected on glass fiber filters (Schleicher and Schuell, Keene, NH) on a Millipore filtration manifold. The filters were washed with 10 ml of the appropriate buffer (without added carrier protein) and counted as described above (footnote *a*).

[c] Primary precipitating antibodies were as follows: rabbit anti-mouse IgG was used for 26–10 Fab and 26–10 IgG; rabbit anti-FB was used for FB–sFv$^{26-10}$. In all cases the primary precipitating antibody was incubated with the 26–10 binding species and [^{3}H]digoxin overnight at 4°. The resulting precipitate was collected on filters and counted as described above in footnotes *a* and *b*.

[d] nd, Not determined. $K_{a,app}$ was not determined for sFv since there are no determinants present on sFv to which rabbit anti-mouse IgG or rabbit anti-FB showed binding.

In addition to the low pH effect on the assay, protein concentration was also a complicating factor in early assays of the 26–10 sFv, in which concentrations of $10^{-8}\ M$ or higher resulted in observations of low values of digoxin-binding affinity for the 26–10 sFv and Fab.[5] This problem was avoided by using subnanomolar protein concentrations that were less than or equal to the $K_{d,app}$ of the 26–10 combining site ($5 \times 10^{-10}\ M$), in both

TABLE III
MEASUREMENTS OF 26–10 BINDING SITE AFFINITIES BY ULTRAFILTRATION[a]

26–10 species	Digoxin affinity[b] ($K_{a,app} \times 10^{-9}\ M^{-1}$)	
	pH 5.5	pH 7.3
sFv	2.6 ± 0.3	2.4 ± 0.2
FB–sFv	2.6 ± 0.3	2.5 ± 0.3
Fab	2.4 ± 0.2	2.2 ± 0.3
IgG	2.5 ± 0.4	2.4 ± 0.5

[a]Affinity measurements were conducted under the same conditions as noted for radioimmunoassays (Table II, footnote *a*), with the exception that 26–10 sFv at pH 7.3 was in PBSA containing 0.1% gelatin or 1% horse serum (IgG-free horse serum; GIBCO Laboratories, Grand Island, NY).

[b]Aliquots (0.2 ml) of each 26–10 binding species and [^{3}H]digoxin were incubated overnight at 4° in a total volume of 1.0 ml. Aliquots (0.9 ml) were then placed in Centricon 10 microconcentrators (Amicon, Danvers, MA) centrifuged at 1150 *g* for 30 min at 4° and aliquots (0.3 ml) of the ultrafiltrate were collected and counted as described (Table II, footnote *a*). [Reprinted with permission from *Biochemistry* ***29,*** 8024 (1990), American Chemical Society.]

the double antibody-precipitation and ultrafiltration studies (Tables II and III).

Filter-Binding Assay of Digoxin-Binding Affinity

A recently developed assay has allowed protein concentration to be decreased still further for some digoxin-binding species, by separating unbound [^{3}H]digoxin by filtration on glass fiber filters (Schleicher and Schuell, Keene, NH), which adsorbed the 26–10 antibody and its complexes with digoxin.[61] Using this approach, an investigation of the 26–10 IgG at very low protein concentrations ($10^{-11}\ M$) yielded a value of $K_{a,app}$ for digoxin binding of $2.5 \times 10^{10}\ M^{-1}$.[61] Although 26–10 Fab and sFv could not be measured in this assay because they failed to adsorb to the

[61] J. F. Schildbach, D. J. Panka, D. R. Parks, G. C. Jager, J. Novotny, L. A. Herzenberg, M. Mudgett-Hunter, R. E. Bruccoleri, E. Haber, and M. N. Margolies, *J. Biol. Chem.* **266,** 4640 (1991).

filters, measurements were conducted on the FB–sFv$^{26-10}$ which was retained by the filters.[62] Under the same conditions of buffer and protein concentration used for 26–10 IgG, the $K_{a,app}$ of $2.0 \times 10^{10}\ M^{-1}$ for digoxin binding to the FB–sFv$^{26-10}$ was the same as that for 26–10 IgG, within experimental error. The filter assay thus yields binding affinities that are an order of magnitude greater than those obtained at higher concentrations by ultrafiltration or radioimmunoassay. The basis for this difference is under investigation. However, the salient feature of the glass filter assay is its internal consistency, such that the digoxin binding sites of both FB–sFv$^{26-10}$ and 26–10 IgG exhibit the same binding characteristics under the same conditions of assay, which is likewise true for the other methods used.[5,6]

Measurement of Specificity Profiles

Specificity analysis of the 26–10 binding site was facilitated by the availability of a family of digoxin analogs that have discrete differences in structure.[6] Comparison of the relative binding affinities of several such cardiac glycosides provided a readily accessible measure of combining site specificity. Each 26–10 species was immobilized on a microtiter plate with goat anti-mouse Fab anti-serum and the binding of ^{125}I-labeled digoxin was measured in the presence of each digoxin analog at several concentrations spanning its inhibition curve. Such specificity profiles are usually tabulated by dividing the 50% inhibition concentration value for each competitive glycoside by the 50% inhibition concentration value obtained for unlabeled digoxin. These normalized values represent indices of specificity, and the data from several experiments have been averaged in Table IV, which is considered to indicate recovery of native 26–10 binding specificity by the 26–10 sFv and FB–sFv$^{26-10}$. A given specificity index is interpreted to be the same as the control 26–10 Fab or IgG, within experimental error, if the values correspond to the control indices within approximately a factor of two.

Analysis of the 26–10 sFv,[6] FB–sFv$^{26-10}$,[6] FB–sFv^{315},[19] and the McPC 603 sFv,[12] all of which used the same [(-Gly)$_4$-Ser-)]$_3$ linker, has proven the feasibility of making single-chain Fv proteins that faithfully reproduce the binding properties of the parent antibody. The earlier discussion on linker chemistry suggested some possible sources of interference with recovery of a native single-chain Fv combining site. In fact, some linker sequences described in the literature may perturb the combining site, as evidenced by values of antigen-binding affinity that were lower than those of the parent antibody.[8,13–18]

[62] M. Mudgett-Hunter, unpublished results (1990).

TABLE IV
SPECIFICITY ANALYSIS OF 26–10 DIGOXIN-BINDING SITES[a]

Cardiac glycoside	sFv	FB–sFv	Fab	IgG
Digoxigenin	1.3	1.2	0.8	0.9
Digitoxin	2.4	2.1	1.3	1.1
Digitoxigenin	1.8	2.1	1.1	1.0
Acetylstrophanthidin	2.8	1.7	2.6	1.1
Gitoxin	12	13	11	8.9
Ouabain	34	40	39	25

[a] Results are expressed as normalized concentration of inhibitor giving 50% inhibition of ^{125}I-labeled digoxin binding. Relative affinities for each cardiac glycoside were calculated by dividing the concentration of each at 50% inhibition by the concentration of digoxin that gave 50% inhibition for each type of 26–10 species. [Reprinted with permission from *Biochemistry* **29,** 8024 (1990), American Chemical Society.]

Bifunctionality of sFv Fusion Proteins

The binding affinity of the FB–sFv$^{26-10}$ fusion protein for human ^{125}I-labeled IgG was determined to be $(3.2 \pm 0.5) \times 10^7 \ M^{-1}$, which is nearly the same as the $K_{a,app}$ of $1.2 \times 10^7 \ M^{-1}$ measured for recombinant FB. FB specificity within the fusion protein was also measured, using an assay in which we monitored the inhibition of radioligand binding to FB–sFv$^{26-10}$. FB–sFv$^{26-10}$ was bound through its sFv site with digoxin–bovine serum albumin (BSA) on the microtiter plate. The relative specificity profiles for the FB moeity were generated from measurements of the amount of human ^{125}I-labeled IgG bound in the presence of a range of concentrations of each inhibitory species (Fig. 9). The known specificity of protein A was reproduced by the FB site of the fusion protein, as indicated by the relative affinities of the inhibitors tested. These results on the Fc affinity and specificity of the FB domain within the fusion protein are consistent with its full recovery of native properties.[63] Furthermore, bifunctionality of the fusion protein was confirmed by this assay, since the FB–sFv$^{26-10}$ necessarily had a functional antigen-combining site for it to have bound digoxin–BSA on the plate. Subsequently added human ^{125}I-labeled IgG could only bind to the FB moiety of truly bifunctional species on the plate. The percentage of material with dual activity was determined to be at least 77%.[6]

[63] J. J. Langone, *Adv. Immunol.* **32,** 157 (1982).

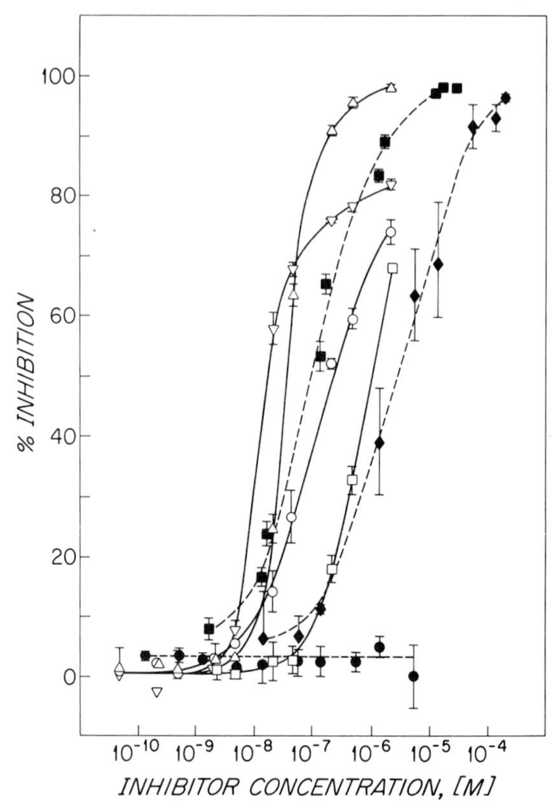

FIG. 9. FB specificity profiles for FB–sFv[26-10].[6] (△), Human IgG; (▽), rabbit IgG; (□), murine IgG_1; (○), murine IgG_{2a}; (◆), FB; (■), protein A; (●), BSA control. Association between human ^{125}I-labeled IgG and FB–sFv[26-10] was inhibited by immunoglobulins, free FB, and protein A, with BSA as a nonspecific control. Standard deviations are noted as error bars. [Reproduced with permission from *Biochemistry* **29**, 8024 (1990), American Chemical Society.]

Improved Methods of 26–10 sFv Production

Purification of the 26–10 sFv from FB–sFv or FB–FB–sFv Fusion Proteins

The published method for production of 26–10 sFv from the MLE–sFv fusion protein[5] can be difficult to apply to other antibodies, since acid cleavage of the starting fusion protein yields a mixture of sFv and MLE–sFv that cannot be separated by a general procedure. In order to have a more general procedure, the following methods were developed to take advantage of the Fc-binding capacity of the FB leader to effect a nearly quantitative removal of fusion protein from the reaction mixture. Using

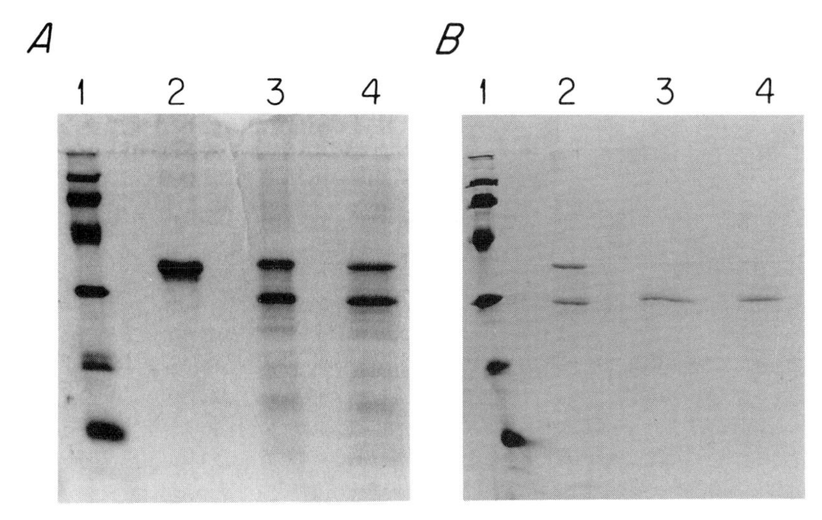

FIG. 10. SDS-PAGE of FB–sFv$^{26-10}$ acid cleavage mixture before and after fractionation on human IgG-Sepharose. (A) Lane 2 is fusion protein before cleavage; lanes 3 and 4 are after cleavage in 10% acetic acid, PBSA. (B) Lane 2 is the sample before affinity chromatography; lanes 3 and 4 are the pooled protein after affinity chromatography.

the example of 26–10 sFv released from FB–sFv$^{26-10}$ by cleavage in 10% acetic acid, the major species after the acid hydrolysis are 26–10 sFv and FB–sFv$^{26-10}$, as shown in an SDS-PAGE analysis of this cleavage mixture in Fig. 10A (note Method D in the section, Protein Refolding and Purification Schemes); the FB leader appears to partition into ethanol during precipitation, and therefore is not present in the precipitate analyzed in Fig. 10A. The intact fusion protein was removed by passage of the cleavage mixture through a human IgG-Sepharose affinity resin (Fig. 10B). We have found that removal of FB–sFv$^{26-10}$ from the mixture of refolded cleavage products, prior to isolation of active 26–10 species by ouabain-amine-Sepharose chromatography, is best accomplished by keeping the cleavage mixture in 3 *M* urea, 25 m*M* Tris-HCl, pH 8, 10 m*M* EDTA during the IgG-Sepharose chromatography; under these conditions the FB fusion protein binds to the affinity resin while the urea prevents problems of aggregation and precipitation that complicate this separation in PBSA or acetate buffer prior to affinity purification. Once it has been affinity purified, the 26–10 sFv was found to be monomeric according to sedimentation equilibrium analysis of protein in 0.01 *M* sodium acetate, pH 5.5, 0.25 *M* urea, with and without 0.15 *M* NaCl.[64] Unlike the 26–10 sFv, the separation of MOPC315 sFv from uncleaved FB fusion protein was best

[64] W. F. Stafford III and J. S. Huston, unpublished results (1990).

done under normal aqueous buffer conditions. These examples suggest a range of conditions that could be applicable to other systems.

As indicated in the protocol for this method of sFv purification, a single or double FB leader can be used, and the FB–FB leader sometime offers more quantitative adsorption to the affinity resin. We have found that the natural FB sequence can be cleaved at two bonds following aspartyl residues at positions FB-36 and FB-37 (Fig. 7), but these cleavages are negligible if the leader sequence has been removed by cleavage in 10% acetic acid, without 6 *M* GuHCl. The acid lability of these particular peptide bonds can be eliminated by conversion of Asp to Glu at both positions.

Secretion of 26–10 sFv

In light of the successful secretion of active McPC603 sFv from *E. coli*,[12] there is great interest in this method for the production of sFv proteins. Studies described in this section detail the application of secretion methods to 26–10 sFv production.

The gene for the 26–10 sFv was fused downstream from synthetic DNA coding for the signal sequence from the pectate lyase (*pelB*) gene of *Erwinia carotovora*. The *Bam*HI restriction site used for this construction added an aspartyl residue to the amino terminus of the sFv. The gene was placed under the control of the *tac* promoter in a vector similar to the expression vector used for intracellular expression of the FB–sFv^{26-10} fusion protein (Fig. 7B). *Escherichia coli* RB791 containing the *lacI*^{q} gene was transformed with this plasmid and assays confirmed the presence of active 26–10 sFv in periplasmic extracts.

Initial expression studies were performed in shaker flask cultures and small-scale (2 liter) fermenters at 30°. Low expression levels and slow growth rates at lower temperatures have been shown to increase the secretion efficiency of some heterologous proteins in *E. coli*.[12,65] Since the *pelB* signal sequence had previously been shown to secrete an active Fab fragment into the culture medium,[39,55] the culture medium and the periplasmic extract were both assayed for 26–10 sFv protein. Although digoxin-binding activity was not detected in the culture medium, expression of the 26–10 sFv in the periplasmic extract reached levels of 375 μg/liter of fermentation, which is comparable to that obtained for secretion of McPC603 sFv.[12] The secreted material was bound to a ouabain-Sepharose affinity column, and retained protein eluted with 6 *M* GuHCl was subjected to SDS-PAGE. The resulting gel was electrophoretically transferred to ProBlott (Applied Biosystems, Foster City, CA) and the band in the

[65] H. Ikemura, H. Takagi, and M. Inouye, *J. Biol. Chem.* **262,** 7859 (1987); N. T. Keen and S. Tamaki, *J. Bacteriol.* **168,** 595 (1986).

vicinity of M_r 29,000 was subjected to automated protein sequencing with a pulse-liquid sequencer (model 477A; Applied Biosystems). The resulting sequence confirmed that the secreted product was 26–10 sFv that was correctly processed by the *E. coli* signal peptidase. This suggests that active 26–10 binding sites were formed during secretion, although the 26–10 sFv may have been in an aggregated form that required elution with denaturant instead of 20 m*M* ouabain.[65a]

We also examined the secretion of a fusion protein made with the FB domain of protein A fused to the carboxyl terminus of the 26–10 sFv, making 26–10 sFv–FB, representing an isomer of the previously studied FB–sFv$^{26-10}$. The $(GGGGS)_3$ linker sequence was inserted both between the V_H and V_L domains and between the sFv and FB. This 26–10 sFv–FB molecule was isolated from the periplasmic extract in a single step by IgG-Sepharose affinity chromatography. This affinity-purified protein also exhibited digoxin-binding activity, indicating that the 26–10 sFv–FB was bifunctional, as observed for the FB–sFv$^{26-10}$ fusion protein.[6]

Comments

Single-chain Fv analogs and fusion proteins have been shown to offer a novel approach for protein engineering of the minimal antibody combining site. Several constructions of the 26-kDa single-chain Fv have yielded functionally native antibody combining sites with binding site properties that are equivalent to those of the parent antibody.[6,12,19] Furthermore, the fusion of effector domains to either chain terminus of the sFv appears to be practical without perturbation of the antigen combining site.[6,14–19] As detailed understanding of antigen–antibody interactions improves, based on their characterization by X-ray crystallography,[26,29,66,67] the availability of single-chain Fv species as single-gene products should be of considerable value for testing new structural insights through protein engineering. By the application of mutagenesis methods, single-chain Fv analogs and fusion proteins can be expected to be useful vehicles for studying modulated binding properties[38] and altered solution behavior.[68,69] The first reports on their biological applications clearly indicate that these biosynthetic antibody binding site proteins will be of potentially unique value for *in vivo* imaging with sFv analogs,[7,10] for cancer treatment with single-chain Fv

[65a] M. S. Tai and F. Warren, unpublished results (1991).

[66] G. A. Bentley, G. Boulot, M. M. Riottot, and R. J. Poljak, *Nature (London)* **348**, 254 (1990).

[67] D. R. Davies, E. A. Padlan, and S. Sheriff, *Annu. Rev. Biochem.* **59**, 439 (1990).

[68] T. F. Kumosinski and S. N. Timasheff, *J. Am. Chem. Soc.* **88**, 5635 (1966).

[69] J. S. Huston, I. Björk, and C. Tanford, *Biochemistry* **11**, 4256 (1972).

immunotoxins,[14-18] and for other biomedical applications still to be explored.

Acknowledgments

We thank Clare Corbett, Peter Keck, Walter F. Stafford III, and Brenda Giust for their assistance in the conduct of this research and in the preparation of this chapter. Supported in part by the National Institutes of Health, Grants CA 39870, CA 51880, and HL 19259.

[4] Construction of Single-Chain Fv Derivatives of Monoclonal Antibodies and Their Production in *Escherichia coli*

By SYD JOHNSON and ROBERT E. BIRD

A single-chain Fv (scFv) molecule consists of the variable domains of an antibody tethered together by a designed protein linker such that the antigen combining site is regenerated in a single protein.[1,2] Either the V_L or V_H can be used as amino-terminal domain of an scFv. Because of several unique properties such as small size, ease of engineering, and stability even at low concentrations, scFvs may eventually be useful in the diagnosis and/or therapy of diseases such as cancer, where target antigens are often expressed preferentially on the surface of cells. Applications under development include the use of genetic fusions of scFvs to potent toxins[3] and the use of radiolabeled scFv molecules to image tumors that express antigens recognized by the scFv.[4] The generality of the scFv technology is based on the observation that the basic structure of the individual immunoglobulin domains is conserved in all antibodies.[5] It is then assumed that this framework structure is sufficiently stable that, so long as a linker or other

[1] R. E. Bird, K. D. Hardman, J. W. Jacobson, S. Johnson, B. M. Kaufman, S.-M. Lee, T. Lee, S. H. Pope, G. S. Riordan, and M. Whitlow, *Science* **242,** 423 (1988).

[2] J. S. Huston, D. Levinson, M. Mudgett-Hunter, M.-S. Tai, J. Novotny, M. N. Margolies, R. J. Ridge, R. E. Bruccoleri, E. Haber, R. Crea, and H. Oppermann, *Proc. Natl. Acad. Sci. U.S.A.* **85,** 5879 (1988).

[3] V. K. Chaudhary, C. Queen, R. P. Junghans, T. A. Waldmann, D. J. FitzGerald, and I. Pastan, *Nature (London)* **339,** 394 (1989).

[4] D. Colcher, R. Bird, M. Roselli, K. D. Hardman, S. Johnson, S. Pope, S. W. Dodd, M. W. Pantoliano, D. E. Milenic, and J. Schlom, *J. Natl. Cancer Inst.* **82,** 1191 (1990).

[5] E. A. Padlan, *Q. Rev. Biophys.* **10,** 35 (1977).

adjoining region does not interfere with the domain structure and the ability of the V_L and V_H to associate, the individual domains will fold and associate to form a molecule that mimicks the antigen combining region of an antibody heterodimer.

In addition to the applications described above, since the scFv is a stable replica of an Fv, it provides a useful research tool for testing modifications to the Fv, including complementarity-determining region (CDR) grafting and the study of antibody–antigen interactions. One important reason for this is the simplicity of the system; once the scFv gene is constructed, the gene can be modified and the resulting mutant scFv protein is easy to produce and subsequently test. Several different expression/production systems for scFv molecules have been developed, including secretion of active scFvs using *Escherichia coli*,[6] *Bacillus subtilis*,[7] yeast,[8] and mammalian cells,[9] and the refolding of insoluble, *E. coli*-produced protein.[1,2] We believe that the refolding procedures that have been used offer the most reproducible system for producing large quantities of any scFv protein and are adaptable to almost any laboratory for the production of active protein. The secretion of active scFv protein will be extremely useful as a tool to develop genetic screens and selections for mutagenized or combinatorial libraries of scFvs. For example, active scFvs have been displayed on the surface of filamentous bacteriophage using the fusion phage technology developed by Smith.[10] Eventually secretion of active scFvs will replace refolding as the method of choice for production. However, we will restrict the content of this chapter to the development of scFvs by refolding the protein that has been produced in *E. coli*.

X-Ray crystallographic determination of the three-dimensional structures of antibody variable domains has demonstrated that they fold into a nine-strand β sheet with the residues that contact the antigen being in general confined to loops generated at the turns between strands of the β structure. These residues were initially referred to as hypervariable regions, based on comparative sequence data,[11] and are now referred to as complementarity-determining regions (CDRs) based on their function.[12] Each

[6] R. Glockshuber, M. Malia, I. Pfitzinger, and A. Plückthun, *Biochemistry* **29,** 1362 (1990).

[7] M. Pantoliano, P. Alexander, S. Dodd, P. Bryan, M. Rollence, J. Wood, and S. Fahnestock, *J. Cell. Biochem.* **13**A, 91 (1989).

[8] G. T. Davis, W. D. Bedzyk, E. W. Voss, and T. W. Jacobs, *Bio/Technolgy* **9,** 165 (1990).

[9] S. Johnson, unpublished results (1990).

[10] G. Smith, *Science* **228,** 1315 (1985).

[11] E. A. Kabat, T. T. Wu, M. Reid-Miller, H. M. Perry, and K. S. Gottesman, "Sequences of Proteins of Immunological Interest," 4th Ed., U.S. Department of Health and Human Services, Washington, D.C., 1987.

[12] T. T. Wu and E. A. Kabat, *J. Exp. Med.* **132,** 211 (1970).

TABLE I
LINKERS USED TO CONNECT V_L AND V_H SEQUENCES

Position			Antibody
V_H-linker-V_L			
H113	Linker	L1	
...S	-GGGGSGGGGSGGGGS	-.D...	26-10[a]
...S	-GGGGSGGGGSGGGGS	-Q...	Anti-tac[b]
...S	-SGGGGSGGGGSGGGGSQ	-...	Anti-tac[c]
V_L-linker-V_H			
L105	Linker	H1 H2	
...L	-KESGSVSSEQLAQFRSLD	-V...	Anti-BGH[d]
...K	-EGKSSGSGSESKST	-Q...	4-4-20[d]
L107	Linker	H1	
...K	-GSTSGSGKSSEGKG	-Q...	B6.2[e]

[a] J. S. Huston, D. Levinson, M. Mudgett-Hunter, M.-S. Tai, J. Novotny, M. N. Margolies, R. J. Ridge, R. E. Bruccoleri, E. Haber, R. Crea, and H. Opperman, *Proc. Natl. Acad. Sci. U.S.A.* **85,** 423 (1988).

[b] V. K. Chaudhary, C. Queen, R. P. Junghans, T. A. Waldman, D. J. Fitz-Gerald, and I. Pastan, *Nature (London)* **339,** 394 (1989).

[c] J. K. Batra, D. FitzGerald, M. Gately, V. K. Chaudhary, and I. Pastan, *J. Biol. Chem.* **265,** 15198 (1990).

[d] BGH, bovine growth hormone. R. E. Bird, K. D. Hardman, J. W. Jacobson, S. Johnson, B. M. Kaufman, S.-M. Lee, T. Lee, S. H. Pope, G. S. Riordan, and M. Whitlow. *Science* **242,** 423 (1988).

[e] D. Colcher, R. Bird, M. Roselli, K. D. Hardman, S. Johnson, S. Pope, S. W. Dodd, M. W. Pantoliano, D. E. Milenic, and J. Schlom, *J. Natl. Cancer Inst.* **82,** 1191 (1990).

variable domain contains three CDRs; thus an Fv or scFv has six. The amino terminus of the heavy or light chain is in relatively close proximity to CDR1, while the respective carboxyl terminus is at the other end of the structure.

When the first scFvs were designed, the three-dimensional structure of the myeloma protein MCPC603 was used to model potential linkers.[13,14]

[13] D. M. Segal, E. A. Padlan, G. H. Cohen, S. Rudikoff, M. Potter, and D. R. Davies, *Proc. Natl. Acad. Sci. U.S.A.* **71,** 4298 (1974).

[14] Y. Satow, G. H. Cohen, E. A. Padlan, and D. R. Davies, *J. Mol. Biol.* **190,** 593 (1986).

This structure was chosen with the assumption that the framework regions of most other antibodies would be as similar to this protein in structure as are those in the protein structure database. At a minimum, a structural model is required to determine the spatial distance that must be spanned be the peptide linker, usually 20–40 Å. Peptides of 12–25 amino acids have been used for this purpose (Table I). Linkers were first chosen from the structures of proteins contained in the Brookhaven Protein Structure database and examined by computer graphics to determine whether they would link the carboxyl terminus of the V_L to the amino terminus of the V_H without interfering with the structure of the Fv.[1] Subsequently, linkers were designed to span the distance between the carboxyl terminus of one domain to the amino terminus of the other.[1,4] It appears that linkers must be (1) sufficiently long to span the distance between the V_L and V_H in the Fv structure, (2) sufficiently flexible to allow the association of the V_L and V_H, and (3) relatively hydrophilic, being on the water-accessible surface of the molecule.

Both the V_L and V_H have been used successfully as the amino-terminal domain in an scFv molecule.[1,2] This reflects the approximate symmetry of the variable region structure. Several carefully designed linkers have been designed to join the V_L to the V_H of an anti-fluorescein scFv, resulting in a molecule with an affinity constant within a factor of two of Fab fragments of the parent antibody.[15] A simple $(G_4S)_3$ linker has been used successfully both to join the V_H to the V_L[2] and the V_L to the V_H.[6] Linker sequences that have been successfully used in an scFv for one antibody are good starting points for the design of scFv molecules based on other antibodies. For example, we have successfully used the same linkers to construct both anti-fluorescein[1,15] and anti-tumor antigen scFvs,[4] and the $(G_4S)_3$ linker developed by Huston *et al.*[2] has been used in other scFvs.[3,6]

Cloning of V_H and V_L Segments of Rearranged Immunoglobulin Genes

In order to construct an scFv gene it is necessary to clone the two variable regions that will be used to construct the gene. The variable region cDNAs can be cloned by conventional methodology using specific primers for the first-strand synthesis[16,17] (Fig. 1A) or they can be amplified from first-strand cDNA by polymerase chain reaction (PCR) to facilitate the

[15] M. W. Pantoliano, unpublished (1989).

[16] S. Levy, E. Mendel, and S. Kon, *Gene* **54,** 167 (1987).

[17] W. D. Bedzyk, L. S. Johnson, G. S. Riordan, and E. W. Voss, Jr., *J. Biol. Chem.* **264,** 1565 (1989).

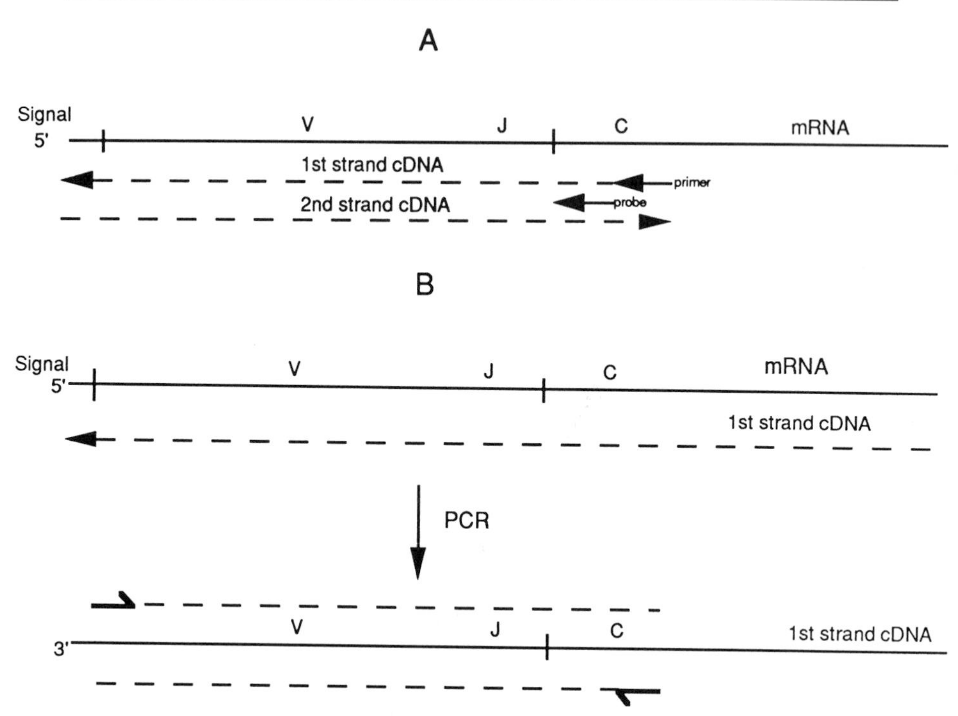

FIG. 1. Methods for obtaining cDNA copies of immunoglobulin variable region segments: (A) cDNA cloning using specific primers; (B) use of the polymerase chain reaction (PCR) with at least one degenerate, mixed, or consensus primer.

cloning (Fig. 1B).[18,19] Polymerase chain reaction can also be used to assemble an scFv gene from the variable cDNAs.

Direct cDNA Cloning

cDNA copies of the V_H and V_L of the target antibody are generated as follows. The first-strand cDNA reaction is carried out using a phosphorylated oligonucleotide complementary to a segment of the mRNA coding for the constant region of the particular heavy- or light-chain isotype.[16,17] The primer thus anneals to a segment of the mRNA adjacent to the variable region. Second-strand cDNA synthesis is carried out using RNase H and *E.*

[18] R. Orlandi, D. H. Gussow, P. T. Jones, and G. Winter, *Proc. Natl. Acad. Sci. U.S.A.* **86,** 3833 (1989).

[19] L. Sastry, M. Alting-Mees, W. D. Huse, J. M. Short, J. A. Sorge, B. N. Hay, K. D. Janda, S. J. Benkovic, and R. A. Lerner, *Proc. Natl. Acad. Sci. U.S.A.* **86,** 5728 (1989).

coli DNA polymerase I, as described by Gubler and Hoffman,[20] followed by T4 DNA polymerase to assure that blunt ends are produced.

The ds-cDNA is ligated into pUC18 (or M13mp18) that has been digested with *Sma*I and treated with alkaline phosphatase. The ligation is used to transform *E. coli* DH5α by the method of Hanahan.[21] Colony hybridization is used to identify transformants carrying the desired cDNA segment. The probe for the hybridization is a second segment of the C-region sequence lying between the first-strand cDNA primer and the V region. Probes and primers corresponding to the J segments of the variable regions can also be used. Because the 5′ end of the first-strand cDNA is fixed, a nested set of sequences can be chosen for sequencing by picking positive clones of approximately 450 bp (full length), 300 bp, and 150bp, subcloning into M13mp18 and mp19 and sequencing. After the sequence is determined in this way, full-length clones are sequenced entirely on both strands to assure that this agrees with the composite sequence.

Amplification of First-Strand cDNA by Polymerase Chain Reaction

By taking advantage of the fact that immunoglobulin variable region genes are flanked by conserved sequences at the 3′ end and by semiconserved sequences at the 5′ end, it is possible to design consensus primers to selectively amplify the variable regions of rearranged immunoglobulin genes by the use of the polymerase chain reaction (PCR).[18,19] This approach has been used effectively to clone the cDNAs for many MAbs and is the basis of the generation of combinatorial libraries in order to generate new specificities.[22] It should be noted that such consensus primers may not allow for somatic changes at the 3′ end of the primer sequences and may not correspond to the entire germline repertoire of variable region genes. Thus, it is desirable to obtain N-terminal sequence of the heavy and light chains. With that caveat, the technique is incredibly powerful and the vast majority of rearranged variable region segments should be able to be cloned in this manner. Additionally, one should be aware of the possible presence of extraneous light- and/or heavy-chain mRNAs expressed by various versions of common fusion partners.[23] If several cDNAs have been cloned that agree in sequence, it may not be necessary to sequence the amino termini of the two protein chains of the monoclonal. However, if it is necessary to sequence these termini to confirm cDNA sequence, a good

[20] U. Gubler, and B. J. Hoffman, *Gene* **25,** 263 (1983).

[21] D. Hanahan, *J. Mol. Biol.* **166,** 557 (1983).

[22] W. D. Huse, L. Sastry, S. A. Iverson, A. S. Kang, M. Alting-Mees, D. R. Burton, S. J. Benkovic, and R. A. Lerner, *Science* **246,** 1275 (1989).

[23] S. Cabilly and A. D. Riggs, *Gene* **40,** 157 (1985).

method for chain separation has been provided.[16] We have found that the purification of poly(A)$^+$ RNA, using oligo(dT) cellulose in a format sold as the "Fast Track" kit (Invitrogen, San Diego, CA), is rapid and efficient. First-strand cDNA is made from 1 μg (less can certainly be used) with random primers by using a kit from Boehringer Mannheim (Indianapolis, IN) in a reaction volume of 10 μl at 37° for 1 hr. One microliter of the cDNA reaction is used directly in a PCR using 2.5 units Taq DNA polymerase with 0.5 μM each of the two primers in a reaction buffer consisting of 10 mM Tris-HCl (pH 8.3), 50 mM KCl, 1.5 mM $MgCl_2$, 0.1 mg/ml gelatin, 0.2 mM dNTPs. We generally use 30 cycles of 1 min at 94°, 1 min at 55°, and 1 min at 72°. The annealing temperature may need to be adjusted according to the product(s) observed.

Assembly and Expression of scFv Proteins

scFv genes can be efficiently assembled in either of two ways: (1) site-directed mutagenesis of the ends of the variable region segments and assembly with a linker segment generated from oligonucleotides, or (2) direct assembly of the scFv gene using PCR.[24] To assemble the gene by PCR, the linker segment is generated by overlap in two of the primers used to amplify the two variable regions. This overlap is illustrated in Fig. 2. Hence, when the first two PCR fractions are mixed, one of the two possible cross-annealing products will provide 3 OH groups that will be extended by the polymerase to give a full-length scFv gene with the linker sequence joining the V_L and V_H. This strategy, and the PCR products produced using the primers described below to generate an scFv gene from a human MAb, are shown in Fig. 2. In this example the primer sequences are as follow:

1. 5′ V_L
   AACCGTCGACGGATATCGTGATCACCCAGTCTCCGTCC
2. 3′ V_L + 5′ end of the linker
   CGGAAGATTTACCAGAACCAGAGGTGGACCCTTTTATTTCAAGCTTGGTCCCCC
3. 5′ V_H + 3′ end of the linker
   GGTTCTGGTAAATCTTCCGAAGGTAAAGGTCTCCTGCAGCTGCAGGAGTCCGGC
4. 3′ V_H + stop codons
   CGCAGATCTTTATGAGCTCACAGAGACCAGGGTGCC

There is a 19-base overlap generated in the linker region by the primers at the 3′ end of the V_L and 5′ end of the V_H. This set of oligonucleotides could be used with first-strand cDNA or with isolated cDNA clones. In this case we had sequence data on cDNAs of the V_L and V_H regions. One could

[24] R. Higuchi, *in* "PCR Protocols" (M. A. Innis, D. H. Gelfand, J. J. Sninskey, and T. J. Whitepp, eds.), p. 177. Academic Press, San Diego, California, 1990.

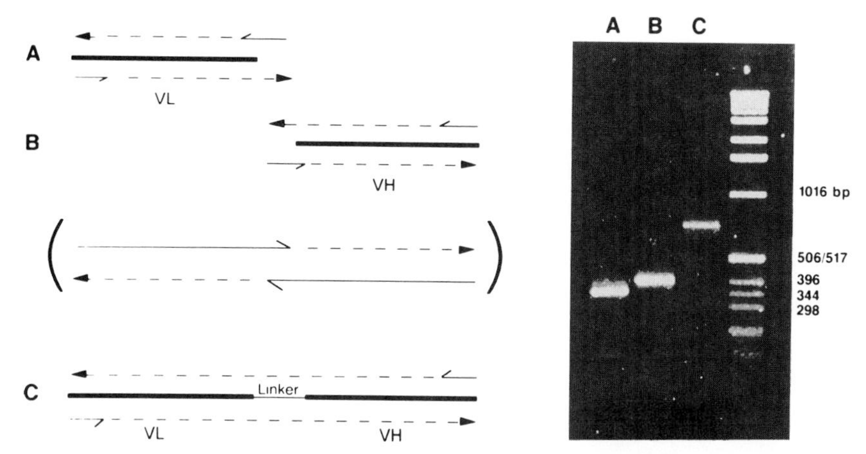

FIG. 2. Assembly of scFv genes using "overlapping" PCR. scFv genes can be assembled from cDNA clones (or directly from first strand cDNA) using the polymerase chain reaction. The individual V_H (lane A) and V_L (lane B) segments are amplified using primers that introduce overlapping segments of the linker region at the 3′ end of the V_L segment and the 5′ end of the V_H. These PCR fragments are then purified, annealed, and amplified using only primers at the 5′ end of the V_L and the 3′ end of the V_H (lane C).

design consensus primers with the linker and terminator segments for a rapid technique, given the cautions described above with regard to the possible presence of extraneous immunoglobulin mRNA in the hybridoma cells.

The segment coding for the linker can be designed to accommodate the majority of variable region segments. The V_L of human and mouse sequences can be modified to generate a *Hin*dIII site at the 3′ end and the V_H to generate a *Pst*I or *Pvu*II site at the 5′ end.

For efficient expression in *E. coli*, we have fused scFv genes to the *E. coli ompA* signal sequence.[4] This has resulted in consistent levels of expression after induction of expression using a strong inducible promoter, such as λ pL or pR, or the T7 promoter/polymerase system, and the signal sequence is efficiently removed. The structure of the resulting gene and expression plasmid is shown in Fig. 3. In the case in which the gene is placed under the control of the hybrid OL/PR promoter, expression is induced by temperature shift from 30 to 42°. scFvs are expressed at 5 to 15% of total cell protein after 15 to 30 min and levels do not increase much thereafter. Unprocessed material may accumulate after this point, however. The scFv protein expressed in this manner is insoluble, but does not generally appear as inclusion bodies. Soluble scFvs have been secreted onto the *E. coli* periplasm using the *lac* promoter and *ompA* signal.[6]

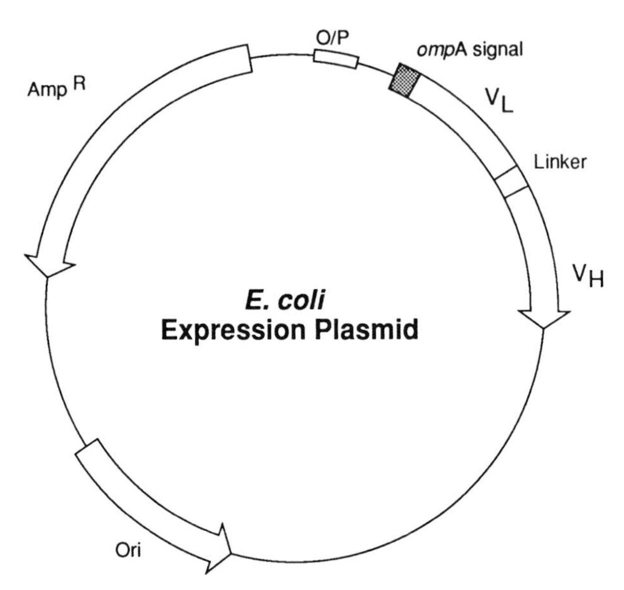

FIG. 3. Structure of generic plasmid for the expression of an scFv gene as a fusion to the *E. coli ompA* signal sequence. Any inducible promoter may be used.

Insoluble protein is recovered after expression and cell lysis. The cell pellet is resuspended in 10 ml/g of 50 m*M* Tris-HCl, pH 8.0, 1 m*M* EDTA, 0.1 m*M* phenylmethylsulfonyl fluoride (PMSF) and the cells are disrupted mechanically using either a Manton Gaulin (Gaulin Corp., Everett, MA) apparatus or a French pressure cell. The insoluble protein is recovered by differential centrifugation. After two washes in the lysis buffer, the protein is dissolved in 6 *M* guanidine-HCl, 50 m*M* Tris-HCl, pH 8.0, 50 m*M* KCl, 0.1 m*M* PMSF at a concentration of about 10 mg/ml. Renaturation is achieved by a 10- to 200-fold dilution into the same buffer without guanidine. The denatured protein should be added with gentle mixing and then left undisturbed for the renaturation to occur. After 12 hr the protein is concentrated and the buffer is exchanged by tangential flow ultrafiltration using a Pellicon (Millipore, Bedford, MA) apparatus with an M_r 10,000 cutoff membrane. This process can be accelerated by first filtering the solution to remove any components that became insoluble during the renaturation process.

Purification of Active scFv Protein

The renatured scFv can be purified to near homogeneity by cation-exchange chromatography. We have utilized a high-performance liquid chromatography (HPLC) system using either aspartic acid or carboxymethyl functional groups. Most of the improperly folded scFv remains insoluble and that which remains in solution after refolding differs in its charge from active scFv protein and either does not bind to the column or elutes earlier than the latter species in a salt gradient. This may vary from antibody to antibody. It is therefore advisable to have a reliable and rapid assay for the binding activity of the scFv to antigen.

Antigen affinity chromatography can also be used to purify scFv proteins where the antigen is known and can be obtained in sufficient quantity to prepare an affinity resin for chromatography. This methodology has been used to purify scFvs that bind bovine growth hormone (BGH),[1] fluorescein,[1] digoxin,[2] and phosphorylcholine.[6] In the case of the anti-BGH scFv, affinity chromatography on BGH bound to Sepharose yielded protein that was pure and had 90% bioactivity.[1] For the antigens that can be obtained and immobilized, this methodology offers a fast way to purify active scFv proteins.

In designing assays to monitor the purification or to determine the affinity of an scFv molecule, keep in mind that reagents that are generally used to detect antibodies do so by binding to the constant domains and should not be expected to cross-react with the variable domains and thus with the scFv. This includes not only bacterial antibody-binding proteins but also polyclonal sera to general classes of or species-specific immunoglobulin. Only antisera to the starting Mab, Fab, or scFv itself will contain enough variable region-specific antibody to be used to detect the scFv in direct binding formats. The lack of binding of these reagents to the scFv relative to its parent MAb and Fab fragments allows for simple competitive assays to be designed. In such experiments it is advisable to use Fab fragments as the competitor and thus avoid any effects related to the bivalent nature of the MAb (i.e., avidity vs affinity). In general, antibodies in which there is a significant difference between avidity and affinity (more than 10-fold) may not be good choices for scFvs unless their uniqueness or some other overriding factor precludes the selection of another antibody.

The absence of constant region segments is also an important consideration in modifying the protein by iodination or conjugation to reporter molecules such as biotin or enzymes. Since the scFv molecule contains only the binding domain, it is quite likely that any modification that either perturbs the structure of the domain or modifies residues within the binding pocket will decrease or abolish the binding activity of the molecule. Thus, if one is going to use such methods to modify the scFv for subsequent detection after it has bound antigen, a second method should be used to not only measure general immunoreactivity, but also to estimate retention of affinity in the immunoreactive fraction. This could be either a comparison to metabolically labeled scFv as a "gold standard" or competition assay using the Fab.

For monoclonal antibodies specific for small ligands is may be possible to use a change in the fluorescence of the ligand or the antibody as a measure of binding. We and others have used such assays to measure the binding of scFvs to fluorescein,[1] digoxin,[2] and phosphorylcholine,[6] respectively.

Conclusion

Genetic and protein engineering applied to immunoglobulin molecules has produced novel new ways of generating and preserving specificities. This area was the subject of a review.[25] The ability to amplify the rearranged variable region segments from single lymphocytes and methods for the stimulation of B cells *in vitro* make more realistic the generation or preservation of specificities from species other than mouse. The scFv molecule preserves the specificity of an antibody in a much smaller form that is simple to generate and manipulate. The combination of these new methods with each other, such combinatorial libraries of Fabs secreted from *E. coli*,[19] and combinatorial or mutagenized libraries of scFvs expressed on the surface of bacteriophage,[26] are extremely powerful methods of screening or selecting and preserving specificities.

Finally, the incorporation of an scFv into a genetic fusion with other functional polypeptides, such as metal-binding domains, modified toxin domains, catalytic domains, or other scFvs, has the potential to generate novel and useful functional combinations.[27-29]

[25] G. Winter and C. Milstein, *Nature (London)* **349**, 293 (1991).

[26] C. A. K. Borrebaeck, L. Danielsson, and S. A. Moller, *Proc. Natl. Acad. Sci. U.S.A.* **85**, 3995 (1988).

[27] J. Larrick, L. Danielsson, C. A. Brenner, E. Wallace, M. Abrhamson, K. Fry, and C. Borrebaeck, *Bio/Technology* **7**, 934 (1989).

[28] J. McCafferty, A. D. Griffiths, G. Winter, and D. J. Chiswell, *Nature (London)* **348**, 552 (1990).

[29] V. A. Roberts, B. L. Iverson, S. A. Iverson, S. J. Benkovic, R. A. Lerner, E. D. Getzoff, and J. A. Tainer, *Proc. Natl. Acad. Sci. U.S.A.* **87**, 6654 (1990).

[5] Humanization of Monoclonal Antibodies

By DETLEF GÜSSOW and GERHARD SEEMANN

Since Köhler and Milstein's breakthrough in hybridoma technology,[1] monoclonal antibodies (MAbs) have become ever more important tools in all fields of biology and medicine. A large panel of diagnostic as well as therapeutic MAbs have been developed that are of potential value for *in vivo* diagnosis and therapy in humans.[2,3] Virtually all MAbs are of rodent origin and exposure of humans to such heterologous MAbs can severely limit their use because of allergic reactions. Immunocompetent individuals may develop human anti-mouse immunoglobulin (Ig) antibodies (HAMA).[4,5] Human MAbs would obviously eliminate such difficulties but the technology to produce them will require considerable refinement before it becomes generally applicable.[6] However, standard molecular biology techniques can offer a way out of the dilemma. Working on the level of the immunoglobulin genes, we are now able to recombine variable-region genes of the heavy or light chains (V_H or V_L) with constant-region genes of any desired isotype, and, more importantly, we also have the choice between constant-region genes of different species.[7,8] Shuffling the variable domains between different antibody genes is now an established technique for creating so-called chimeric antibodies where the variable domain is of murine origin and the constant region is human. These chimeric MAbs usually resemble their parental mouse MAb in specificity as well as affinity for the antigen.[9,10] Chimeric MAbs are already more suitable for therapeutic use in humans. However, the mouse variable region still carries a host of immunogenic epitopes and it is likely that chimeric MAbs are still immu-

[1] G. Köhler and C. Milstein, *Nature (London)* **256,** 495 (1975).
[2] J. L. Murray and M. W. Unger, *Crit. Rev. Oncol./Hematol.* **8**, 227 (1988).
[3] H.-H. Sedlacek, G. Schulz, A. Steinsträsser, L. Kuhlmann, A. Schwarz, L. Seidel, G. Seemann, H.-P. Kraemer, and K. Bosslet, *in* "Contributions to Oncology" (S. Eckhardt, J. H. Holzner and G. A. Nagel, eds.), Vol. 32. Karger, Basel, 1988.
[4] R. A. Miller, A. R. Oseroff, P. T. Stratte, and R. Levy, *Blood* **62**(5), 988 (1983).
[5] M. J. P. G. Van Kroonenburgh and E. K. J. Pauwels, *Nucl. Med. Commun.* **9,** 919 (1988).
[6] J. E. Boyd, K. James, and D. B. L. McClelland, *Trends Biotechnol.* **2**(3), 70 (1984).
[7] V. T. Oi, S. L. Morrison, L. A. Herzenberg, and P. Berg, *Proc. Natl. Acad. Sci. U.S.A.* **80,** 825 (1983).
[8] S. L. Morrison and V. T. Oi, *Adv. Immunol.* **44,** 65 (1989).
[9] G. L. Boulianne, N. Hozumi, and M. J. Shulman, *Nature (London)* **312,** 643 (1984).
[10] S. L. Morrison, M. J. Johnson, L. A. Herzenberg, and V. T. Oi, *Proc. Natl. Acad. Sci. U.S.A.* **81,** 6851 (1984).

nogenic as it has been shown in a mouse model system with chimeric MAbs containing murine V and human constant domains.[11]

Jones *et al.*[12] and Riechmann *et al.*[13] have taken another more subtle approach, in order to make a more human form of a rodent MAb. Antigen-binding sites, composed of the three CDRs (complementarity determining regions) of the heavy chain and the three CDRs of the light chain, can be taken from a rodent MAb and inserted directly into the framework of a human antibody, thus transplanting only the CDRs rather than the entire variable domain of a rodent antibody. This is achieved by the use of oligonucleotides to replace the CDRs of an appropriate human immunoglobulin gene with the CDRs from a rodent MAb with a desired specificity. The CDRs or hypervariable regions are unlikely to carry any species-specific characteristics, and therefore such a "reshaped" human antibody should be indistinguishable from genuine human immunoglobulins. Although anti-idiotypic antibodies against a reshaped or "humanized" MAb can be generated *in vivo,* the occurrence of such antibodies does not seem to be a problem in therapy (H. Waldmann, personal communication, 1989).

The general applicability of the antibody-reshaping technique relies on two assumptions: (1) the antigen-binding site is fashioned by the CDRs as defined by Kabat *et al.*,[14] no other parts of the variable region taking part in binding the antigen; (2) the frameworks of the variable domains serve as a scaffold to support the CDRs in a specific way that facilitates antigen binding. Subsequently it is of great importance to retain the interactions between the donor (rodent) CDRs and the acceptor (human) framework as closely as possible to the CDR–framework interactions of the original MAb.

However, the affinity of the first fully reshaped antibody, CAMPATH1, was nearly 40-fold lower compared to the original rat MAb.[13] Close inspection of the sequences of the hypervariable loops in the human and rat antibodies, in particular at their junctions with the framework regions of the V_H domain, showed the possible origin of the flaw. A comparison of the amino acid sequence in the flanking region of CDR1 (residues 31–35, Kabat *et al.*[14]) showed that the original rat antibody has a phenylalanine at

[11] M. Brüggemann, G. Winter, H. Waldmann, and M. S. Neuberger, *J. Exp. Med.* **170,** 2153 (1989).

[12] P. T. Jones, P. H. Dear, J. Foote, M. S. Neuberger, and G. Winter, *Nature (London)* **321,** 522 (1986).

[13] L. Riechmann, M. Clark, H. Waldmann, and G. Winter, *Nature (London)* **332,** 323 (1988).

[14] E. A. Kabat, T. T. Wu, M. Reid-Miller, H. M. Perry, and K. S. Gottesman, "Sequences of Proteins of Immunological Interest," 4th Ed., U.S. Department of Health and Human Services, US Washington, D.C., 1987.

position 27 (phenylalanine and tyrosine are the most common amino acids at this position). Existing structural data of the human myeloma protein KOL show that Phe-27 packs against residues 32 and 34 to support CDR1.[15] The NEW framework (the human acceptor for the CDR transplants of the heavy chain)[13] has a serine at position 27 and thus fails to support the CDR in the same way as in the original rat antibody. A subsequent engineered mutation, serine to phenylalanine at position 27, restored the binding affinity of the humanized CAMPATH1 antibody close to the original affinity. This example demonstrates that amino acids in the framework, in particular those that are located close to the CDRs, must be considered. We now routinely compare the sequences of the mouse hybridoma and the human target variable sequences with each other and their respective family consensus sequences.

Here we describe the humanization of the murine monoclonal antibody BW431/26,[16] which has binding specificity for carcinoembryonic antigen (CEA) and is presently used as a murine MAb for immunoscintigraphy of CEA-producing tumors such as colorectal, breast, and lung carcinomas.[17]

Cloning of Immunoglobulin V-Region Genes

A prerequisite for the humanization of MAb is knowledge of the nucleotide sequences of the V-region genes. To date the most elegant way to clone and sequence Ig V-region genes has been described by Orlandi and colleagues.[18] They used the polymerase chain reaction (PCR)[19] to amplify specifically Ig V-region genes and cloned the amplified DNA fragments into vectors that allow easy sequencing and expression. To do this they identified conserved regions at each end of the nucleotide sequences encoding V domains of mouse Ig heavy (V_H) and light (V_L) chains by comparing the frequencies of the most common nucleotides in V_H and V_L gene sequences.[14] Since oligonucleotide primers used in PCR do not need to match their target sequence exactly[20] they were able to design oligonu-

[15] M. Marquart, J. Deisenhofer, and R. Huber, *J. Mol. Biol.* **141,** 369 (1980).

[16] K. Bosslet, A. Steinsträsser, A. Schwarz, H. P. Harthus, G. Lüben, L. Kuhlmann, and H. H. Sedlacek, *Eur. J. Nucl. Med.* **14,** 523 (1988).

[17] R. P. Baum, A. Hertel, M. Lorenz, A. Schwarz, A. Encke, and G. Hör, *Nucl. Med. Commun.* **10,** 345 (1989).

[18] R. Orlandi, D. H. Güssow, P. T. Jones, and G. Winter, *Proc. Natl. Acad. Sci. U.S.A.* **86,** 3833 (1989).

[19] R. K. Saiki, S. Scharf, F. Faloona, K. B. Mullis, G. T. Horn, H. A. Erlich, and N. Arnheim, *Science* **230,** 1350 (1985).

[20] C. C. Lee, X. Wu, R. A. Gibbs, R. G. Cook, D. M. Muzny, and C. T. Caskey, *Science* **239,** 1288 (1988).

cleotides that map to the 5′ and 3′ regions of V_H and V_L genes and contain restriction sites for subsequent forced cloning into suitable vectors (Fig. 1 and Ref. 19).

The amplified V_H and V_L genes cover the V gene exons only. For expression in a mammalian system it is necessary to bring them into genomic configuration. The V_H and V_L genes are force cloned into KS^+ vectors (pBluescript II KS^+; Stratagene, La Jolla, CA) containing the 5′ region of immunoglobulin heavy and light chain genes, respectively (including promoter, signal exon, and irrelevant V gene exon) as a unique *Hin*dIII/*Bam*HI restriction fragment (Fig. 2). The endogenous V gene exons contain the same or compatible restriction sites as the amplification primers in the appropriate positions and are exchanged for the amplified V genes of the hybridoma by forced cloning (see the section, Cloning of V Genes). These vectors can be used for nucleotide sequence determination of the specific V genes and for forced cloning of the amplified V_H and V_L cassettes into expression vectors that contain Ig constant-region genes, using the *Hin*dIII and *Bam*HI restriction sites (Fig. 3).

Mouse V_H Forward:

5′ TGAGGAGACGGTGACCGTGGTCCCTTGGCC 3′
Bst E II

Mouse V_H Backward:

5′ AGGT(C/G)(C/A)A(G/A)CTGCAG(G/C)AGTC(T/A)GG 3′
Pst I

Mouse V_K Forward:

5′ GTTAGATCTCCAGCTTGGTCCC 3′
Bgl II

Mouse V_K Backward:

5′ GACATTCAGCTGACCCAGTCTCCA 3′
Pvu II

FIG. 1. Oligonucleotide primer for the amplification of V_H and V_K genes. The mouse V_H backward primer is a mixture of 32 primers. The locations of the restriction sites are indicated.

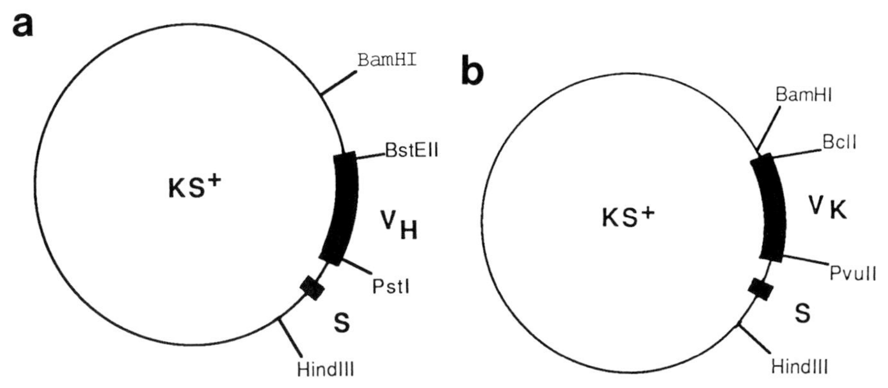

FIG. 2. Restriction maps and intron/exon organization of the V gene inserts of the KS^+ plasmids containing a V_H (a) or a V_L (b) gene. The inserts are flanked by *Bam*HI and *Hin*dIII restriction sites. Exons are represented by black bars. S, Signal exon; V_H/V_K, V gene exons. The restriction sites for cloning of the amplified V genes are indicated. The *Bcl*I restriction site in the V_K exon is compatible with *Bgl*II.

Preparation of mRNA

Poly(A)$^+$ mRNA is prepared following one of the standard procedures. An example is as follows:

Centrifuge hybridoma cells (825 *g*; room temperature; 15 min)
Discard supernatant
Resuspend in 100 ml cold phosphate-buffered saline (PBS) (0°)
Centrifuge (275 *g*; room temperature; 10 min)
Discard supernatant
Resuspend in 100 ml cold PBS
Centrifuge as above
Discard supernatant
Add 22 ml lysis buffer to the pellet, vortex for 5 sec, incubate on ice (5 min)
Centrifuge nuclei (1630 *g*; 4°; 15 min)
Transfer supernatant to new tube and add 25 ml 2× proteinase K, buffer
Incubate at 50° for 60 min
Add 0.1 vol 5 *M* NaCl and let cool down to room temperature
Apply the solution to an oligo(dT)-Sepharose column (1-ml) column volume equilibrated with binding buffer)
Wash with 10 ml binding buffer and 15 ml washing buffer
Elute poly(A)$^+$ mRNA with 7.5 ml elution buffer and collect three fractions of 2.5 ml
Add 0.1 vol of 3 *M* sodium acetate and 2.5 vol ethanol to each fraction
Store at −20°

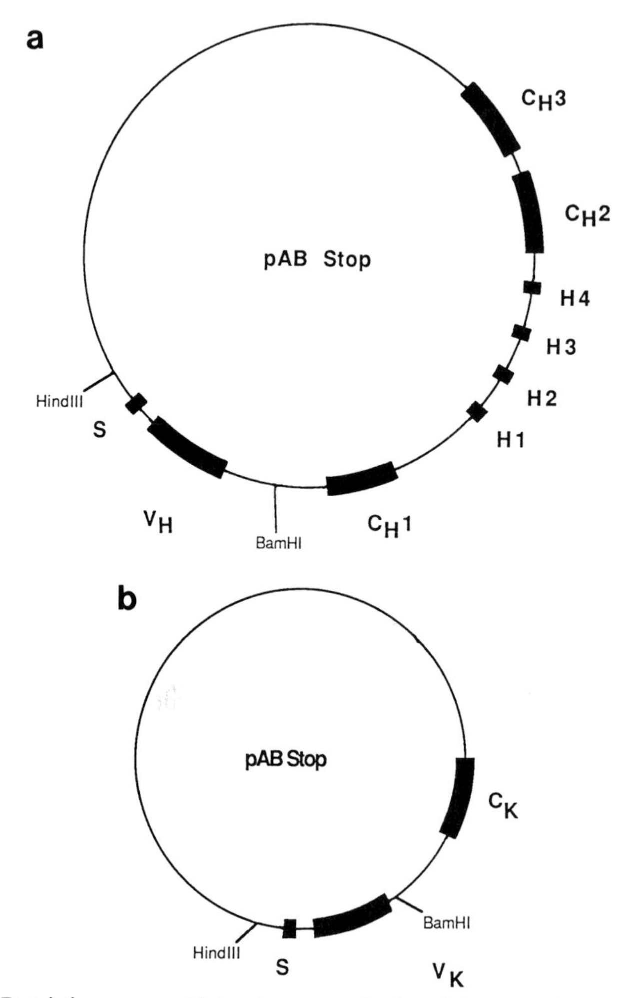

FIG. 3. Restriction maps and intron/exon organization of the expression vectors containing the human (a) heavy (IgG_3) and (b) light chain (κ) constant-region genes. The intrinsic V genes are exchanged for the hybridoma V genes using the *Hin*dIII and *Bam*HI restriction sites. Exons are represented by black bars. S, Signal exon; V_H/V_K, V gene exons; C_H1, C_H2, C_H3, and C_K, constant-domain exons; H1, H2, H3, and H4, hinge region exons.

The average yield is approximately 40 μg of poly(A)$^+$ mRNA per 1×10^8 hybridoma cells.

Lysis buffer: 0.14 *M* NaCl, 2.0 m*M* $MgCl_2$, 10.0 m*M* Tris, pH 8.6, 0.5% Nonidet P-40 (NP-40)

Proteinase K buffer (2×): 200 m*M* Tris, pH 7.4, 300 m*M* NaCl, 25 m*M* ethylenediaminetetraacetic acid (EDTA), 2% sodium dodecyl sulfate (SDS), 2 mg/ml proteinase K

Binding buffer: 0.5 *M* NaCl, 10.0 m*M* Tris, pH 7.8, 5.0 m*M* EDTA, 0.5% SDS

Washing buffer: 0.15 *M* NaCl, 10.0 m*M* Tris, pH 7.8, 5.0 m*M* EDTA, 0.5% SDS

Elution buffer: No NaCl, 10.0 m*M* Tris, pH 7.8, 5.0 m*M* EDTA, 0.5% SDS

Amplification and Cloning of V Genes

Amplification of the V genes can be performed either by synthesis of first-strand cDNA followed by polymerase chain reaction (PCR) or by PCR directly from the poly(A)$^+$ mRNA.[21]

First-Strand cDNA Synthesis. Mix in the following order:

Reverse transcriptase buffer (5×)	10.0 μl
Dithiothreitol (DTT) (0.1 *M*)	2.5 μl
Human placental ribonuclease inhibitor (50 U/ul)	3.0 μl
dNTPs (10 m*M*)	5.0 μl
Primer (10 p*M*/μl)	5.0 μl
Poly(A)$^+$ mRNA (5–10 μg)	
ddH_2O	to 47.0 μl

Mix gently and centrifuge for a few seconds, then add

Reverse transcriptase (25 U/μl)	3.0 μl

Incubate at 42° for 1 hr.

PCR Reaction from cDNA

H_2O:	29.5 μl
dNTPs (2.5 m*M*):	5.0 μl
PCR buffer (10×):	5.0 μl
Oligonucleotide primers (5 p*M*/μl):	5.0 μl
Taq polymerase (The Perkin-Elmer Corp., Emmeryville, CA)	0.5 μl
(2.5 U/μl)[21a]	0.5 μl

At this point the reaction mix can be treated for 5 min with UV light (254 or 300 nm) to avoid contamination (for example, with previously amplified DNA).

[21] W. T. Tse and B. G. Forget, *Gene* **88**, 293 (1990).

[21a] In case of the V_H primers the Taq polymerase was added to the reaction mixture after the first denaturation and annealing steps to avoid primer dimer formation.

Add 5.0 μl of first-strand cDNA synthesis reaction and seal the surface of the reaction mixture with a drop of paraffin oil. Amplify for 30 to 50 cycles. The standard programs we use are the following: 94°, 1 min; 52° (for V_K, V_λ) or 57° (for V_H), 1 min; 72° for 2 min.

Analyze 5 μl of the reaction mixture on a 2% agarose gel.

PCR Reaction from mRNA. For PCR for mRNA resuspend 1–5 μg of poly(A)$^+$ mRNA in 50 μl PCR reaction mix and proceed as described above for the PCR from cDNA.

If the amplification reactions are unsuccessful under these conditions, the addition of 5.0 μl Perfect Match (Stratagene) and a modification of the annealing and extention conditions as well as $MgCl_2$ concentrations may improve the results.

Cloning of V Genes

If the amplification is successful, the remaining 45 μl of the reaction mixture is purified with Geneclean (BIO 101, Inc., La Jolla, CA). The purified amplification products (in 20 μl doubly distilled H_2O) are treated as follows: 10 μl is double digested with *Pst*I/*Bst*EII for the V_H gene or *Pvu*II/*Bgl*II for the V_L gene. The digested samples are analysed on a 2% TAE agarose gel with 3 μl of undigested amplification product as control. The residual amplification product is stored at −20° as the template for the synthesis of additional material, if required.

The vectors for cloning the amplified V genes have been constructed by isolation of the *Hin*dIII/*Bam*HI inserts of the M13 VHPCR and M13 VKPCR vectors described in Orlandi *et al.*[18] into KS$^+$ plasmids. Two internal *Pvu*II sites were removed from the KS$^+$ plasmid backbone prior to the construction of the KS$^+$ VKPCR vector (C. Weber, unpublished data, 1990).

These KS$^+$ vectors (KS$^+$ VHPCR and KS$^+$ VKPCR) are digested with *Pst*I and *Bst*EII (V_H) or *Pvu*II and *Bcl*I (V_L), respectively, and the vector fragment is purified on a 0.8% agarose gel.

Note: The KS$^+$ VKPCR vector was prepared in *Escherichia coli* JM110 (ATCC No. 47013) to avoid dam methylation of the *Bcl*I site.

From the resulting colonies DNA minipreparations are performed and the size of the inserts is analyzed by a *Hin*dIII/*Bam*HI digestion. Clones containing inserts of the correct size (approximately 800 bp for V_H and 650 bp for V_L) are sequenced[22] using primers based 3′ of the CDR3 regions of the V genes (Fig. 4).

Since hybridoma cells may produce more than one heavy or light chain and can contain additional nonfunctional mRNAs it is advisable to clone the isolated V genes into appropriate expression vectors containing Ig C_H

[22] F. Sanger, S. Nicklen, and A. R. Coulson, *Proc. Natl. Acad. Sci. U.S.A.* **74,** 5463 (1977).

VK/VL: 5′ GGA TCC AAC TGA GGA AGC 3′

VH: 5′ TGT CCC TAG TCC TTC ATG ACC T 3′

FIG. 4. Oligonucleotides for sequencing of the cloned V genes.

and C_L genes and express them in eukaryotic cells (see section entitled "Expression of Humanized MAb"). The resulting chimeric Ig molecules contain the complete mouse V domains and should have the identical antigen-binding properties as the original mouse MAb, if the right pair of V_H and V_L genes has been amplified and cloned.

Reverse transcriptase buffer (5×): 250 m*M* Tris, pH 8.3, 30 m*M* $MgCl_2$, 500 m*M* NaCl
PCR buffer (10×): 100 m*M* Tris, pH 8.3, 500 m*M* KCl, 15 m*M* $MgCl_2$, 0.1% (w/v) gelatin
TAE buffer (50×): 2 *M* Tris base, 1 *M* acetic acid, 50 m*M* EDTA
The solutions are autoclaved before use, distributed in 0.5-ml aliquots, and stored at −20°.

Design of Oligonucleotides for Mutagenesis

After determination of the nucleotide sequences of the hybridoma V genes (Fig. 5) they are translated into the corresponding amino acid sequences and compared to the consensus sequences of the murine Ig heavy and light chain V gene families to identify their families of origin (Fig. 6).

Comparison of the translated BW431/26 V_H sequence with the mouse VH consensus sequences marks it as a member of family IA. Likewise BW431/26 V_L is identified as a member of family VI of the κ light chains.

For designing the reshaped V regions, it is necessary to identify amino acid residues that are not conserved or only semiconserved in the framework regions of the human and mouse antibodies in question. Framework regions that are not conserved between human and mouse antibodies, but which might help to shape the CDRs or which might interact directly with the antigen, have to be identified. Unusual amino acid replacements are changes from hydrophilic to hydrophobic, large to small, or change in electrostatic charge. The following amino acid exchanges are considered as conserved replacements and are excluded from the computer inspection: F - Y, L - M, V - I, Q - H, D - E, R - K, W - R, N - D, G - A, S - T, Q - D/E, H - R/N, I - L/M, L - F/V, and V - L/M. The framework and CDR regions are defined by Kabat *et al.* as shown in Fig. 7.

The decision whether framework exchanges must be introduced in addition to the CDR grafting is then made by inspection of the location of the identified amino acid residues in the acceptor V_H or V_K three-dimensional structure.

```
       10                   30                   50
L  Q  E  S  G  P  D  L  V  K  P  S  Q  S  L  S  L  T  C  T
ctgcaggagtcaggacctgacctggtgaaaccttctcagtcactttcactcacctgcact

       70                   90                  110
V  T  G  Y  S  I  T  S  G  Y  S  W  H  W  I  R  Q  F  P  G
gtcactggctactccatcaccAGTGGTTATAGCTGGCACtggatccggcagtttccagga

      130                  150                  170
N  K  L  E  W  M  G  Y  I  Q  Y  S  G  I  T  N  Y  N  P  S
aacaaactggaatggatgggcTACATACAGTACAGTGGTATCACTAACTACAACCCCTCT

      190                  210                  230
L  K  S  R  I  S  I  T  R  D  T  S  K  N  Q  F  F  L  Q  L
CTCAAAAGTcgaatctctatcactcgagacacatccaagaaccagttcttcctgcagttg

      250                  270                  290
N  S  V  T  T  E  D  T  A  T  Y  Y  C  A  R  E  D  Y  D  Y
aattcagtgactactgaggacacagccacatattactgtgcaagaGAAGACTATGATTAC

      310                  330                  350
H  W  Y  F  D  V  W  G  A  G  T  T  V  T  V  S  S
CACTGGTACTTCGATGTCtggggcgcagggaccacggtcaccgtctcctca
```

b

```
       10                   30                   50
Q  L  T  Q  S  P  A  I  M  S  A  S  L  G  E  E  I  T  L  T
cagctgacccagtctccagcaatcatgtctgcatctctaggggaggagatcaccctaacc

       70                   90                  110
C  S  T  S  S  S  V  S  Y  M  H  W  Y  Q  Q  K  S  G  T  S
tgcAGTACCAGCTCGAGTGTAAGTTACATGCACtggtaccagcagaagtcaggcacttct

      130                  150                  170
P  K  L  L  I  Y  S  T  S  N  L  A  S  G  V  P  S  R  F  S
cccaaactcttgatttatAGCACATCCAACCTGGCTTCTggagtcccttctcgcttcagt

      190                  210                  230
G  S  G  S  G  T  F  Y  S  L  T  I  S  S  V  E  A  E  D  A
ggcagtgggtctgggacctttattctctcacaatcagcagtgtggaggctgaagatgct

      250                  270                  290
A  D  Y  Y  C  H  Q  W  S  S  Y  P  T  F  G  G  G  T  K  L
gccgattattactgcCATCAGTGGAGTAGTTATCCCACGttcggaggggggaccaagctg

E  I
gagatca
```

FIG. 5. Nucleotide sequences of the murine BW431/26 (a) V_H and (b) V_K genes. The sequences start with the *Pst*I/*Pvu*II restriction sites and continue to the end of the V exon. CDR sequences are typed in capital letters. The amino acid sequences are given in the single-letter code on top of the first base of the triplets.

A three-step comparison of the framework amino acid residues consists of (1) the donor (mouse) antibody V region with its own V region family consensus sequence, (2) the acceptor (human) antibody V region with its own V region family, and (3) the mouse and human families, which should allow the differences to be dissected in a systematic way, identifying those differences peculiar to the individual antibodies or their entire families.

Heavy Chain V_H

Step 1. The sequence of the (mouse) donor is compared with the consensus sequence for its own family of V_H genes (Fig. 8).

Step 2. The sequence of the (human) acceptor is compared with the consensus sequence for its own family of V_H genes (human V_H II; HuVH.CON2 in Fig. 9).

Step 3. The consensus sequences of the donor (mouse) and acceptor (human) V_H families are compared (Fig. 10).

Light Chain V_K

Step 1. The sequence of the (mouse) donor is compared with the consensus sequence of its own V_K family (MOVK VI) (Fig. 11).

Step 2. The sequence of the (human) acceptor is compared with the consensus of its own V_K family. The human REI V_K sequence is based on Ref. 13. REI is a member of the human V_K family I (HuVKCON SUB1 in Fig. 12).

Step 3. The consensus sequences of the acceptor (human) and donor (mouse) V_K families are compared (Fig. 13).

All unusual amino acid replacements that occur outside the CDRs, as defined by Kabat *et al.*,[14] are marked and the positions at which replacements occur are inspected by computer graphics (Evans & Sutherland Corp., Salt Lake City, UT; Frodo[23]) using as a basis the X-ray structures of NEW[24] and REI.[25] All positions in question are highlighted with a 200% van der Waals radius, to see whether the amino acid residue at a certain position is close to, or actually makes contact with, any of the amino acids composing the CDRs. In such a case an amino acid replacement in the framework of the human V_H or V_K region must be considered.

[23] T. A. Jones, *in* "Computational Crystallography" (D. Sayre, ed.), p. 303. Oxford Univ. Press (Clarendon), London and New York, 1982.

[24] R. J. Poljak, L. M. Amzel, H. P. Avey, B. L. Chen, R. P. Phizackerley, and F. Saul, *Proc. Natl. Acad. Sci. U.S.A.* **70,** 3305 (1973).

[25] O. Epp, P. Colman, H. Fehlhammer, W. Bode, M. Schiffer and R. Huber, and W. Palm, *Eur. J. Biochem.* **45,** 513 (1974).

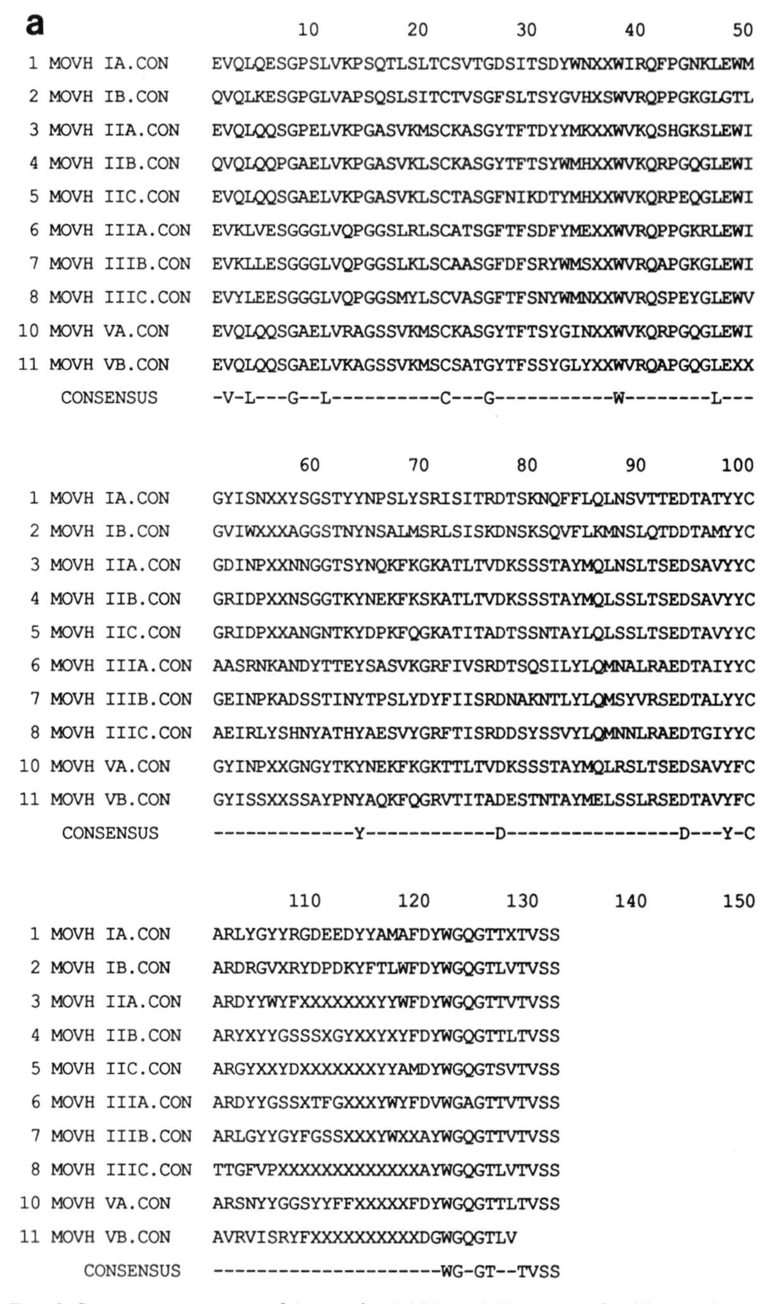

a
10 20 30 40 50
1 MOVH IA.CON EVQLQESGPSLVKPSQTLSLTCSVTGDSITSDYWNXXWIRQFPGNKLEWM
2 MOVH IB.CON QVQLKESGPGLVAPSQSLSITCTVSGFSLTSYGVHXSWVRQPPGKGLGTL
3 MOVH IIA.CON EVQLQQSGPELVKPGASVKMSCKASGYTFTDYYMKXXWVKQSHGKSLEWI
4 MOVH IIB.CON QVQLQQPGAELVKPGASVKLSCKASGYTFTSYWMHXXWVKQRPGQGLEWI
5 MOVH IIC.CON EVQLQQSGAELVKPGASVKLSCTASGFNIKDTYMHXXWVKQRPEQGLEWI
6 MOVH IIIA.CON EVKLVESGGGLVQPGGSLRLSCATSGFTFSDFYMEXXWVRQPPGKRLEWI
7 MOVH IIIB.CON EVKLLESGGGLVQPGGSLKLSCAASGFDFSRYWMSXXWVRQAPGKGLEWI
8 MOVH IIIC.CON EVYLEESGGGLVQPGGSMYLSCVASGFTFSNYWMNXXWVRQSPEYGLEWV
10 MOVH VA.CON EVQLQQSGAELVRAGSSVKMSCKASGYTFTSYGINXXWVKQRPGQGLEWI
11 MOVH VB.CON EVQLQQSGAELVKAGSSVKMSCSATGYTFSSYGLYXXWVRQAPGQGLEXX
CONSENSUS -V-L---G--L----------C---G-----------W--------L---
60 70 80 90 100
1 MOVH IA.CON GYISNXXYSGSTYYNPSLYSRISITRDTSKNQFFLQLNSVTTEDTATYYC
2 MOVH IB.CON GVIWXXXAGGSTNYNSALMSRLSISKDNSKSQVFLKMNSLQTDDTAMYYC
3 MOVH IIA.CON GDINPXXNNGGTSYNQKFKGKATLTVDKSSSTAYMQLNSLTSEDSAVYYC
4 MOVH IIB.CON GRIDPXXNSGGTKYNEKFKSKATLTVDKSSSTAYMQLSSLTSEDSAVYYC
5 MOVH IIC.CON GRIDPXXANGNTKYDPKFQGKATITADTSSNTAYLQLSSLTSEDTAVYYC
6 MOVH IIIA.CON AASRNKANDYTTEYSASVKGRFIVSRDTSQSILYLQMNALRAEDTAIYYC
7 MOVH IIIB.CON GEINPKADSSTINYTPSLYDYFIISRDNAKNTLYLQMSYVRSEDTALYYC
8 MOVH IIIC.CON AEIRLYSHNYATHYAESVYGRFTISRDDSYSSVYLQMNNLRAEDTGIYYC
10 MOVH VA.CON GYINPXXGNGYTKYNEKFKGKTTLTVDKSSSTAYMQLRSLTSEDSAVYFC
11 MOVH VB.CON GYISSXXSSAYPNYAQKFQGRVTITADESTNTAYMELSSLRSEDTAVYFC
CONSENSUS -------------Y------------D----------------D---Y-C
110 120 130 140 150
1 MOVH IA.CON ARLYGYYRGDEEDYYAMAFDYWGQGTTXTVSS
2 MOVH IB.CON ARDRGVXRYDPDKYFTLWFDYWGQGTLVTVSS
3 MOVH IIA.CON ARDYYWYFXXXXXXXYYWFDYWGQGTTVTVSS
4 MOVH IIB.CON ARYXYYGSSSXGYXXYXYFDYWGQGTTLTVSS
5 MOVH IIC.CON ARGYXXYDXXXXXXXYYAMDYWGQGTSVTVSS
6 MOVH IIIA.CON ARDYYGSSXTFGXXXYWYFDVWGAGTTVTVSS
7 MOVH IIIB.CON ARLGYYGYFGSSXXXYWXXAYWGQGTTVTVSS
8 MOVH IIIC.CON TTGFVPXXXXXXXXXXXXXXAYWGQGTLVTVSS
10 MOVH VA.CON ARSNYYGGSYYFFXXXXXFDYWGQGTTLTVSS
11 MOVH VB.CON AVRVISRYFXXXXXXXXXXDGWGQGTLV
CONSENSUS ---------------------WG-GT--TVSS

FIG. 6. Consensus sequences of the murine (a) V_H and (b) V_K gene families are listed in the single-letter code. The amino acid positions that are conserved among all V_H/V_K gene families are outlined in the bottom lines.

b

```
                          10        20        30        40        50
1  MOVK I.CON     DIVMTQSPSSLAVSAGEKVTXSCTASESLYSSKHKVHYLAWYQKKPEQS
2  MOVK II.CON    DVVMTQTPLSLPVSLGDQASISCRSSQSLVHSXNGNTYLNWYLQKPGG
3  MOVK III.CON   DIVLTQSPASLAVSLGQRATISCRASESVDXXXYGNSFMHWYQQKPGQPP
4  MOVK IV.CON    EXVLTQSPAIMAASPGEKVTMTCXASSSXXXXXVSSSYLHWYQQKPGASP
5  MOVK V.CON     DIQMTQSPSSLSASLGDRVTITCRASQXXXXXXDISNYLNWYQQKPGGTP
6  MOVK VI.CON    QIVLTQSPAIMSASPGEKVTMTCSASSXXXXXXXSVSYMHWYQQKSGTSP
7  MOVK.MBR       DIQLTQSPPSLTVSVGERVTISCKSNQNLLWSGNRRYCLGWHQWKPGQTP
   CONSENSUS      ----TQ-P-----S-G------C-----------------W---K----P

                          60        70        80        90       100
1  MOVK I.CON     KLLIYGASNRYIGVPDRFTGSGSGTDFTLTISSVQVEDLTHYYCAQFYSY
2  MOVK II.CON    KLLIYKVSNRFSGVPDRFSGSGSGTDFTLKISRVEAEDLGVYYCFQGTHV
3  MOVK III.CON   KLLIYAASNLESGVPARFSGSGSGTDFTLNIHPVEEDDAATYYCQQSNED
4  MOVK IV.CON    KLXIYXTSNLASGVPARFSGSGSGTSYSLTISSXEAEDDATYYCQQWSGY
5  MOVK V.CON     KLLIYYASRLHSGVPSRFSGSGSGTDYSLTISSLEZEDIATYFCQQGNSL
6  MOVK VI.CON    KRWIYDTSKLASGVPARFSGSGSGTSYSLTISSMEAEDAATYYCQQWSSN
7  MOVK.MBR      TPLITWTSDRFSGVPDRFIGSGSVTDFTLTISSVQAEDVAVYFCQQHLDL
   CONSENSUS      ---I---S----GVP-RF-GSGS-T---L-I------D---Y-C-Q----

                         110       120       130       140       150
1  MOVK I.CON     PXXXXXXXLTFGAGTKLELYRX
2  MOVK II.CON    PXXXXXXXYTFGGGTKLEIKRA
3  MOVK III.CON   PXXXXXXXXTFGGGTKLEIKRA
4  MOVK IV.CON    PFXXXXXXXTFGXGTKLEIKRX
5  MOVK V.CON     PXXXXXXXRTFGGGTKLEIXKRA
6  MOVK VI.CON    PPMXXXXXLTFGAGTKLELKRX
7  MOVK.MBR       PXXXXXXXYTFGGGTKLEI
   CONSENSUS      P-------TFG-GTKLE----A
```

FIG. 6b.

	V_H	V_K
FR 1	1-30	1-23
CDR1	31-35	24-34
FR2	36-49	35-49
CDR2	50-65	50-56
FR3	66-94	57-88
CDR3	95-102	89-97
FR4	103-113	98-107

FIG. 7. Definition of CDR regions according to Kabat *et al.*[14]

```
                      10        20         30   35ab   40
MOVH IA       EVQLQESGPSLVKPSQTLSLTCSVTGDSITSDYWNXXWIRQFPGNKLEWM
BW431/26VH    QVQLQESGPDLVKPSQSLSLTCTVTGYSITSGYSWH-WIRQFPGNKLEWM
              *        *      *     *   *

                52abc     60        70         80 abc     90
MOVH IA       GYISNXXYSGSTYYNPSLYSRISITRDTSKNQFFLQLNSVTTEDTATYYC
BW431/26VH    GYI----YSGITNYNPSLKSRISITRDTSKNQFFLQLNSVTTEDTATYYC


                    100abcdefghijk       110
MOVH IA       ARLYGYYRGDEEDYYAMAFDYWGQGTTXTVSS
BW431/26VH    AR---------EDYDYHWFDVWGAGTTVTVSS
                                     *
```

FIG. 8. Comparison of BW431/26 V_H sequence with the mouse V_H IA (MOVH IA) consensus sequence. Relevant framework differences are marked with an asterisk. The differences are as follows: EQ1, SD10, TS17, ST23, DY27, QA105. X represents amino acids that are variable in the consensus sequence. These are not considered as exchanges. Dashes in the BW431/26 sequence are introduced to accommodate for the variable length of the CDRs.

In the case of the BW431/26 antibody, for V_H the exchange of a phenylalanine for a serine at position 27, and that of an isoleucine for a phenylalanine at position 29 in the NEW V_H gene have been considered as additional framework exchanges. With the exchange of the amino acid at position 27 from serine to phenylalanine we followed Riechmann's strategy for the humanization of the CAMPATH 1 antibody.[13] Phe-27 is present in most human V_H domains and may play an important role in positioning CDR1. Since Phe-27 is changed to serine in the NEW V_H domain it seems necessary to introduce Phe-27 into the humanized V_H.

Phe-29 of NEW in the immediate vicinity of CDR1 contains an aro-

```
                      10        20         30   35ab   40
HuVH.NEW      QVQLQESGPGLVRPSQTLSLTCTVSGSTFSDYYSTXXWVRQPPGRGLEWI
HuVH.CONR2    XVTLRESGPXLVKPTETLTLTCTVSGFSLSTXGMXVGWIRQPPGKXLEWL
                * *                       * * *

                52abc     60        70           82abc     90
HuVH.NEW      GYVFYHGTSDTDTPLRSXXXRVTMLVDTSKNQFSLRLSSVTAADTAVYYC
HuVH.CON2     ARINXXXWDDDKYYSTSLRSRLTISYDTSKNQVVLXXXXXDPXDTATYYC
               * *   **  *****     *  **      **      **   *

                   100abcdefghijk      110
HuVH.NEW      ARNLIAGCIXXXXXXXXXXXDVWGQGSLVTVSS
HuVH.CON2     ARRXPRXXXGDXGXYXXAFDVWGQGTTVTVSS
                * **                     *
```

FIG. 9. Comparison of the NEW V_H sequence with the human V_H II (HuVH.CON2) family consensus sequence. The NEW V_H sequence is based on Ref. 13. Framework differences are marked with an asterisk. The differences are as follows: QT3, QR5, SF27, FL29, DT31, YR50, FN52, TW53, SD54, DK57, TY58, PY59, LS60, RT61, VL67, LS70, VY71, FV78, SV79, TD83, AP84, VT89, NR95, IP97, AR98, LT109.

```
                      10        20        30   35ab   40
HuVH II       -VTLRESGPXLVYPTETLTLTCTVSGFSLSTXGMXVGWIRQPPGKXLEWL
MOVH IA       EVQLQESGPSLVKPSQTLSLTCSVTGDSITSDYWNXXWIRQFPGNKLEWM
                * *       *  *          *              *  *

                52abc      60        70        80 abc       90
HuVH II       ARINXXXWDDDKYYSTSLRSRLTISYDTSKNQVVLXXXXXDPXDTATYYC
MOVH IA       GYISNXXYSGSTYYNPSLYSRISITRDTSKNQFFLQLNSVTTEDTATYYC
                                       *      **      ***

                   100abcdefghijk       110
HuVH II       ARRXPRXXXGDXGXYXXAFDVWGQGTTVTVSS
MOVH IA       ARLYGYYRGDEEDYYAMAFDYWGQGTTXTVSS
                * **   ** *       *
```

FIG. 10. Comparison of the acceptor (human V_H II; HuVH II) and the donor mouse V_H IA; MOVH IA) consensus sequences. Relevant differences in the framework regions are marked with an asterisk. Differences between donor and acceptor consensus sequences are as follows: DEL E1, TQ3, RQ5, YK13, EQ16, FD27, PF40, KN43, YR71, VF78, VF79, XQ81, XS82C, XV83, DT84, PT85, XE86, RL95, XY96 (X = T, C, V, P, Q, H, R, L, N), PG97, RY98, XY99 (X = P, Q, V, R, M, T, G), DE100C, XE100D (X = Y, L, M, G, V), GD100E, XA100H (X = N, S, D, Y), XM100I (X = S, D, G), VY102.

```
                      10        20     27abcdef 30        40
MOVK VI       QIVLTQSPAIMSASPGEKVTMTCSASSXXXXXXXSVSYMHWYQQKSGTSP
BW431/26VK    --------AILSASPGEKVTMTCRASSXXXXXXXSVSYMHWYQQKPGSSP
                                                           *

                  50        60        70        80        90
MOVK VI       KRWIYDTSKLASGVPARFSGSGSGTSYSLTISSMEAEDAATYYCQQWSSN
BW431/26VK    KPWIYATSNLASGVPARFSGSGSGTSYSLTIIRVEAEDAATYYCQQWSSN
               *                              **

             95abcdef  100   106a
MOVK VI       PPMXXXXLTFGAGTKLELXKRX
BW431/26VK    PXXXXXXLTFGAGTKLEIX---
```

FIG. 11. Comparison of the 431/26 V_K sequence with the mouse V_K VI (MOVK VI) family consensus sequence. The relevant differences in the framework regions are marked with an asterisk. The differences are as follows: SP40, RP46, SI76, SR77.

```
                      10        20     27abcdef 30        40
HuVKCON SUB1  DIQMTQSPSSLSASVGDRVTITCRASQSVXXSXDISSYLNWYQQKPGKAP
HuVK REI      DIQMTQSPSSLSASVGDRVTITCQASQ------DISDYLNWYQQKPGKAP

                  50        60        70        80        90
HuVKCON SUB1  KLLIYXASSLESGVPSRFSGSGSGTDFTLTISSLQPEDFATYYCQQYNSL
HuVK REI      KLLIYEASNLQAGVPSRFSGSGSGTDFTFTISSLQPEDIATYYCQQYQSL
                                                     *

             95abcdef  100   106a
HuVKCON SUB1  PXXYDXXYTFGQGTKVEIXKRT
HuVK REI      P------YTFGQGTKVEI-KR-
```

FIG. 12. Comparison of the human acceptor sequence REI with the consensus sequence of the human V_K I family (HuVKCON SUB1). The relevant framework difference is marked with an asterisk; the difference is FI83.

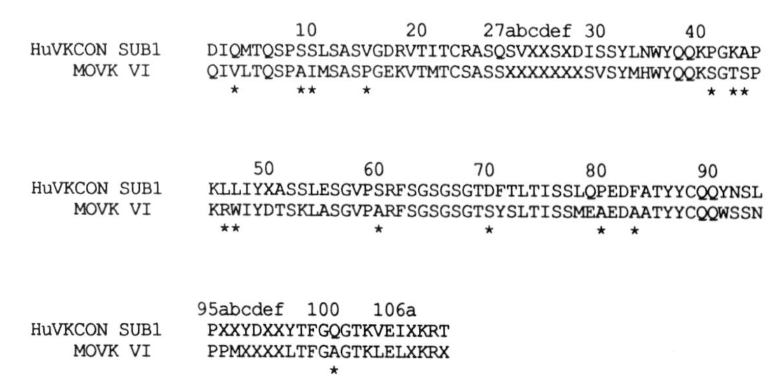

FIG. 13. Comparison of the consensus sequences of the acceptor (HuVKCON SUB1) and donor (MOVK VI) V_K families. Relevant differences in the framework regions are marked with an asterisk. The differences are as follows: QV3, SA9, SI10, VP15, PS41, KT43, AS44, LR47, LW48, SA60, DS70, PA80, FA83, QA100.

matic ring and is larger than Ile-29 of BW431/26. Therefore it might interact with CDR1 side chains in a different way than does Ile-29, so we decided to use Ile-29 in the humanized MAb. For the light chain no additional exchanges were necessary.

On the basis of these data the oligonucleotides for the mutagenesis of the human V genes were designed. The oligonucleotides span the CDR coding region, including additional exchanges in the adjacent frameworks, and at their 5′ and 3′ regions they contain 12 bp that is complementary to the flanking framework regions of the human V genes. The CDR1 oligonucleotide for the light chain had to be placed differently (Fig. 14).

For the VK CDR1 oligonucleotide the overlap with the frameworks at its 3′ and 5′ ends had to be shortened to 9 b. The full-length (12-b overlap) oligonucleotide exhibited cross-hybridization with framework three sequences and this gave rise to misincorporation during mutagenesis.

Mutagenesis of Human V Genes

The mouse CDRs are placed into the human frameworks by oligonucleotide-directed mutagenesis. The mutagenesis of the human V genes can be performed with any mutagenesis system available. We use the gapped duplex (gd) DNA method and the pMa/c phasmid vectors described by Stanssens and colleagues.[26] The gapped duplex method routinely yields a

[26] P. Stanssens, C. Opsomer, Y. M. McKeown, W. Kramer, M. Zabeau, and H.-J. Fritz, *Nucleic Acids Res.* **17**, 4441 (1989).

VK - CDR1

5′ CTG GTA CCA **GTG CAT GTA ACT TAC ACT CGA GCT GGT ACT** ACA GGT GAT 3′

VK - CDR2

5′ GCT TGG CAC ACC **AGA AGC CAG GTT GGA TGT GCT** GTA GAT CAG CAG 3′

VK - CDR3

5′ CCC TTG GCC GAA **CGT GGG ATA ACT ACT CCA CTG ATG** GCA GTA GTA GGT 3′

VH - CDR1

5′ CTG TCT CAC CCA **GTG CCA GCT ATA ACC ACT** GCT **GAT** GGT **GAA** GCC AGA CAC GGT 3′

VH - CDR2

5′ CAT TGT CAC TCT **ACT TTT GAG AGA GGG GTT GTA GTT AGT GAT ACC ACT GTA CTG TAT GTA** TCC AAT CCA CTC 3′

VH - CDR3

5′ GCC TTG ACC CCA **GAC ATC GAA GTA CCA GTG GTA ATC ATA GTC TTC** TCT TGC ACA ATA 3′

FIG. 14. Oligonucleotides for the CDR exchange mutagenesis. The bases originated from the BW431/26 sequence and the codon for Phe-27 in V_H are printed in bold letters.

larger percentage of recombinant clones with all three CDRs mounted in a single step (5–20%), as compared with classical M13 mutagenesis.[13] The template for mutagenesis of the V_H gene is the human myeloma protein NEW and for the V_L gene it is human myeloma protein REI, or any CDR-grafted version of these genes. We used the humanized versions containing the CDR regions of the anti-lysozyme MAb D1.3[27] (Fig. 15a and b).

[27] M. Verhoeyen, C. Milstein, and G. Winter, *Science* **239,** 1534 (1988).

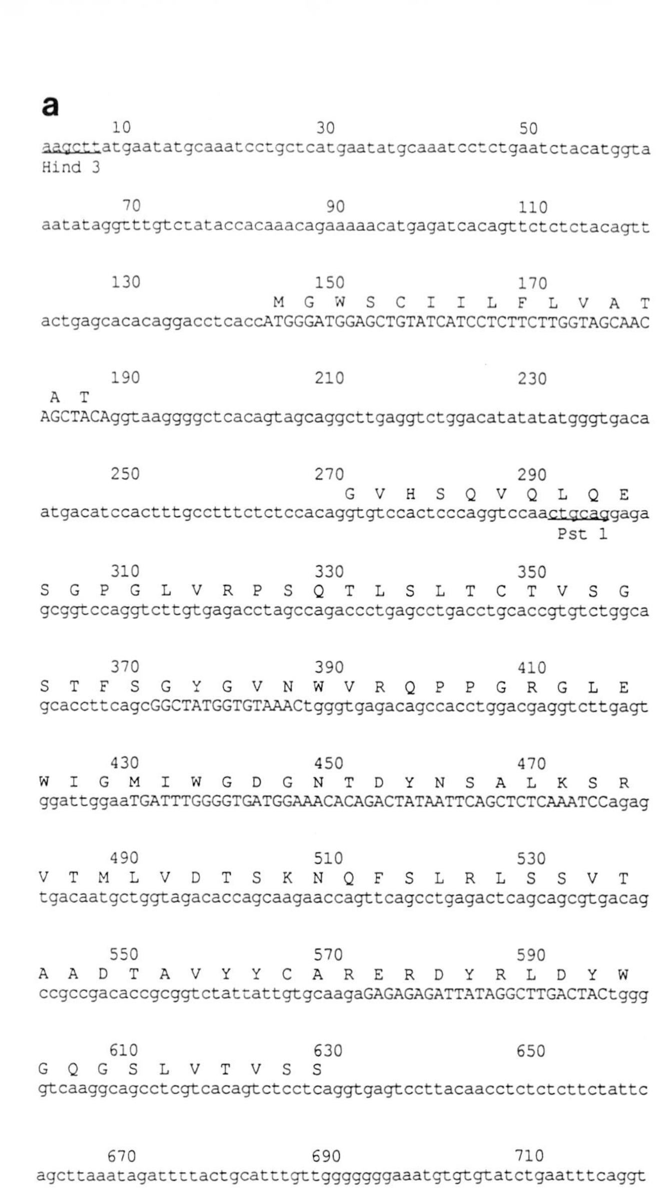

FIG. 15. Nucleotide sequences of the (a) acceptor V_H (anti-Lys-24) and (b) V_K gene (anti-Lys-18) segments of the pMc vectors used for CDR exchange mutagenesis. The amino acid sequences are printed in single-letter code on top of the central base of a triplet. The CDR regions are given in capital letters.

b

```
         10                                  30                                  50
aagcttatgaatatgcaaatcctctgaatctacatggtaaatataggtttgtctatacca
Hind 3

         70                                  90                                 110
caaacagaaaaacatgagatcacagttctctctacagttactgagcacacaggacctcac

        130                                 150                                 170
  M  G  W  S  C  I  I  L  F  L  V  A  T  A  T
cATGGGATGGAGCTGTATCATCCTCTTCTTGGTAGCAACAGCTACAggtaaggggctcac

        190                                 210                                 230
agtagcaggcttgaggtctggacatatatatgggtgacaatgacatccactttgcctttc

        250                                 270                                 290
    G  V  H  S  D  I  Q  M  T  Q  S  P  S  S  L  S  A
tctccacaggtgtccactccgacatccagatgacccagagcccaagcagcctgagcgcca

        310                                 330                                 350
S  V  G  D  R  V  T  I  T  C  R  A  S  G  N  I  H  N  Y  L
gcgtgggtgacagagtgaccatcacctgtAGAGCCAGCGGTAACATCCACAACTACCTGG

        370                                 390                                 410
A  W  Y  Q  Q  K  P  G  K  A  P  K  L  L  I  Y  Y  T  T  T
CTtggtaccagcagaagccaggtaaggctccaaagctgctgatctacTACACCACCACCC

        430                                 450                                 470
L  A  D  G  V  P  S  R  F  S  G  S  G  S  G  T  D  F  T  F
TGGCTGACggtgtgccaagcagattcagcggtagcggtagcggtaccgacttcaccttca

        490                                 510                                 530
T  I  S  S  L  Q  P  E  D  I  A  T  Y  Y  C  Q  H  F  W  S
ccatcagcagcctccagccagaggacatcgccacctactactgcCAGCACTTCTGGAGCA

        550                                 570                                 590
T  P  R  T  F  G  Q  G  T  K  V  E  I  K  R
CCCCAAGGACGttcggccaagggaccaaggtggaaatcaaacgtgagtagaatttaaact

        610
ttgcttcctcagttggatcc
              BamH1
```

FIG. 15b.

Preparation of Template and Gapped Duplex DNA

The human V genes were cloned into the *Hin*dIII and *Bam*HI sites of modified pMc and pMa phasmids. In order to use the mutagenic oligonucleotides as designed, the polylinkers of pMa/c had to be inverted.

For the preparation of template DNA and gapped duplex DNA we follow the protocol described in Ref. 26. The single-stranded DNA (ssDNA) is prepared from the pMc version of the vector. The double-stranded DNA (dsDNA) is prepared from the pMa version using the restriction enzymes *Xba*I and *Eco*RI to release the V gene insert from the vector.

Mutagenesis

The gdDNA is used as template for the site-specific mutagenesis to exchange the CDR regions. Usually we try to exchange all three CDRs in one round of mutagenesis. A typical experiment is as follows.

Generation of gdDNA. ssDNA (0.5 p*M*: 770 ng for V_L and 805 ng for V_H) is mixed with 0.1 p*M* dsDNA (*Eco*RI/*Xba*I cut) + 5 μl 1.5 *M* KCl in 100 m*M* Tris, pH 7.5, + H_2O to 40 μl. Incubate at 100° for 4 min and then at 65° for 10 min in water baths. Use 8 μl for gel analysis.

Annealing and Polymerase Reaction. To 8 μl of gdDNA, add 3 μl of each mutagenic oligonucleotide (10 μ*M*/ml, kinased). Incubate at 68° for 5 min and at room temperature for 15 min. Add the following:

Fill-in buffer (10×): 500 m*M* NaCl, 100 m*M* Tris-HCl, pH 7.5, 100 m*M* $MgCl_2$, 10 m*M* DTT, 10 m*M* dNTPs; 4 μl
T4 ligase (2U), 2 μl
Klenow polymerase (2.5 U), 0.5 μl
H_2O to 40 μl

Incubate at room temperature for 45 min.

Selection and Screening for Mutants. Transform 5 μl of the reaction mixture into WK6mutS[28] bacteria and plate aliquots on 1× YT[29] (yeast Tryptone)-Amp plates (200 μg/ml ampicillin) and YT-chloramphenicol plates (30 μg/ml chloramphenicol) to determine the efficiency of transformation. The remaining transformation mixture is used to inoculate 10 ml LB[29] (Luria-Bertani) medium (200 μg/ml ampicillin). Bacteria are grown overnight to allow segregation of strands. Plasmid DNA is isolated[29] and used to transform WK6 bacteria[28] selecting for Amp resistance (200 μg/ml ampicillin).

The resulting colonies are screened with the CDR3 oligonucleotides described in Fig. 4 using standard colony hybridization methods.[29] From the colonies hybridizing with the probe, 1.5-ml cultures are grown (YT, 200 μg/ml ampicillin), DNA is prepared and the nucleotide sequence determined using the V_H and V_L gene sequencing oligonucleotides (Fig. 4). Five to 20% of the clones hybridizing with the CDR3 mutagenic oligonucleotide have all three CDRs placed properly.

For antibody BW431/26 the sequences of the humanized V gene versions were confirmed (Fig. 16).

[28] R. Zell and H.-J. Fritz, *EMBO J.* **6,** 1809 (1987).

[29] T. Maniatis, E. F. Fritsch, and J. Sambrook, "Molecular Cloning: A Laboratory Manual." Cold Spring Harbor Laboratory, Cold Spring Harbor, New York, 1982.

BW431/26VHhum

```
        10                  30                  50
Q  V  Q  L  Q  E  S  G  P  G  L  V  R  P  S  Q  T  L  S  L
caggtccaactgcaggagagcggtccaggtcttgtgagacctagccagaccctgagcctg

        70                  90                 110
T  C  T  V  S  G  F  T  I  S  S  G  Y  S  W  H  W  V  R  Q
acctgcaccgtgtctggcTTCaccATCagcAGTGGTTATAGCTGGCACtgggtgagacag

       130                 150                 170
P  P  G  R  G  L  E  W  I  G  Y  I  Q  Y  S  G  I  T  N  Y
ccacctggacgaggtcttgagtggattggaTACATACAGTACAGTGGTATCACTAACTAC

       190                 210                 230
N  P  S  L  K  S  R  V  T  M  L  V  D  T  S  K  N  Q  F  S
AACCCCTCTCTCAAAAGTagagtgacaatgctggtagacaccagcaagaaccagttcagc

       250                 270                 290
L  R  L  S  S  V  T  A  A  D  T  A  V  Y  Y  C  A  R  E  D
ctgagactcagcagcgtgacagccgccgacaccgcggtctattattgtgcaagaGAAGAC

       310                 330                 350
Y  D  Y  H  W  Y  F  D  V  W  G  Q  G  S  L  V  T  V  T  V
TATGATTACCACTGGTACTTCGATGTCtggggtcaaggcagcctcgtcacagtcacagtc
```

BW431/26VKhum

```
        10                  30                  50
G  V  H  S  D  I  Q  M  T  Q  S  P  S  S  L  S  A  S  V  G
ggtgtccactccgacatccagatgacccagagcccaagcagcctgagcgccagcgtgggt

        70                  90                 110
D  R  V  T  I  T  C  S  T  S  S  S  V  S  Y  M  H  W  Y  Q
gacagagtgaccatcacctgtAGTACCAGCTCGAGTGTAAGTTACATGCACtggtaccag

       130                 150                 170
Q  K  P  G  K  A  P  K  L  L  I  Y  S  T  S  N  L  A  S  G
cagaagccaggtaaggctccaaagctgctgatctacAGCACATCCAACCTGGCTTCTggt

       190                 210                 230
V  P  S  R  F  S  G  S  G  S  G  T  D  F  T  F  T  I  S  S
gtgccaagcagattcagcggtagcggtagcggtaccgacttcaccttcaccatcagcagc

       250                 270                 290
L  Q  P  E  D  I  A  T  Y  Y  C  H  Q  W  S  S  Y  P  T  F
ctccagccagaggacatcgccacctactactgcCATCAGTGGAGTAGTTATCCCACGttc

       310                 330
G  Q  G  T  K  V  E  I  K  R
ggccaagggaccaaggtggaaatcaaacgt
```

FIG. 16. Nucleotide sequences of the humanized V_H (top) and V_K (bottom) of MAb BW431/26. The sequences start with the codon for the first amino acids of the mature heavy and light chains. The human framework sequences are printed in lower-case letters while the murine CDR sequences and the murine framework sequences that have been transferred to the human V genes are printed in capital letters. The amino acids are given in the single-letter code and are printed on top of the first base of the codon.

Expression of Humanized MAb

The humanized V genes were cloned into separate eukaryotic expression vectors that contain Ig constant-region genes for the heavy and light chain, respectively (Fig. 3). The resulting V_H and V_L expression vectors were subsequently transfected into BHK cells (ATCC CCL10) using the calcium phosphate precipitation technique with methotrexate and G418 as selection markers.[30] We were able to isolate clones that produce up to 15 μg/ml antibody product in serum-free roller bottle cultures without prior amplification.

Characterization of Humanized MAb

The humanized BW431/26 binds to purified CEA and to CEA on tissue sections. Competition experiments with the original mouse MAb reveal that the affinity of the reshaped antibodies, although it has not yet been determined exactly, closely resembles that of the mouse MAb, which is approximately 1×10^{-10} liter/mol.

Anti-idiotypic antibodies (anti-id) against the murine BW431/26 MAb were tested for their binding to the humanized version. Surprisingly, 17 out of 18 anti-id tested bind the humanized version as well as the mouse MAb; binding of one anti-id is less efficient but still detectable. This finding suggests a close immunological relationship between the variable domains of the mouse and the humanized BW431/26 (K. Bosslet *et al.*, unpublished data, 1991).

A comparison of the framework amino acid sequences of the donor (the murine MAb BW431/26) V_H and V_L with the acceptor V_H (human NEW) and V_L (human REI) shows a high degree of sequence similarity (70% for V_H and 68% for V_K). The homologies are clustered in the regions close to the CDRs, especially in the light chain domains. Even in positions where the amino acid sequences differ from each other the amino acid exchanges are often conservative. This may have contributed to the successful humanization of the BW431/26 antibody. The crystal structures of the REI and NEW antibodies are known and we use their respective V regions as general acceptors for the CDRs of any murine MAb. Computer modeling is used for refinement to identify potential framework amino acids in the mouse antibody that might interact with the CDRs or directly with antigen and these amino acids are transferred to the human frameworks along with the CDRs. Using the reshaping scheme described here we estimate the success rate for humanizing any particular monoclonal antibody to be approximately 80%.

[30] M. Wirth, J. Bode, G. Zettlmeissl, and H. Hauser, *Gene* **73,** 419 (1988).

Queen *et al.*[31] have reshaped an antibody by selecting human acceptor V domains from the Kabat database to match the framework sequences of the murine donor V domains as closely as possible. This alternative reshaping strategy raises the possibility that any human antibody framework can be used as acceptor for the murine CDRs. Humanization of murine MAbs is a technique that is now widely applied to MAbs of potential use in diagnosis and therapy in humans. Whether one of the reshaping schemes will offer any particular advantage in successful humanization of MAbs will become clearer as the number of reshaped antibodies increases.

Acknowledgments

Initial work was done by D. G. at the MRC/LMB, CB2 2QH Cambridge, England. We thank Dr. Greg Winter and P. T. Jones for their advice and help in establishing the humanization technique in our laboratory, and for their technical help with the computer analyses, Mrs. K. Müller and R. Bier for excellent technical assistance, Dr. K. Bosslet for helpful discussion, and Dr. H.-H. Sedlacek for support.

[31] C. Queen, W. P. Schneider, H. E. Selick, P. W. Payne, N. F. Landolfi, J. F. Duncan, N. M. Avdalovitc, M. Levitt, R. P. Junghans, and T. A. Waldmann, *Proc. Natl. Acad. Sci. U.S.A.* **86,** 10029 (1989).

[6] Molecular Modeling of Antibody Combining Sites

By ANDREW C. R. MARTIN, JANET C. CHEETHAM, and ANTHONY R. REES

Introduction

Both the variable and constant domains of an antibody Fab consist of two twisted antiparallel β sheets which form a β-sandwich structure. The constant regions have three- and four-stranded β sheets arranged in a Greek key-like motif,[1] while variable regions have a further two short β strands producing a five-stranded β sheet.

The two β sheets of the variable domain are inclined at 30° to one another,[2] with a conserved disulfide bridge in each domain linking the two β sheets. Lesk and Chothia[3] have shown the relative orientation of the two

[1] J. S. Richardson, *Adv. Protein Chem.* **34,** 168 (1981).

[2] C. Chothia and J. Janin, *Proc. Natl. Acad. Sci. U.S.A.* **78,** 4146 (1981).

[3] A. M. Lesk and C. Chothia, *J. Mol. Biol.* **160,** 325 (1982).

β sheets to vary by up to 18° in variable domains and up to 10° in constant domains. This characteristic "immunoglobulin fold"[4-6] is also seen in a number of molecules of related and unrelated function,[7] one of the most recent examples being a bacterial protein, PapD, which mediates assembly of pili in *Escherichia coli.*[8]

The V_L and V_H domains interact via the five-stranded β sheets to form a nine-stranded β barrel of about 8.4 Å radius, with the strands at the domain interface inclined at approximately 50° to one another.[9] The loops linking the β strands form the CDRs with the domain pairing bringing these loops from both V_L and V_H into close proximity. The CDRs themselves form some 25% of the V_L/V_H domain interface.[10]

The six hypervariable loops [complementarity determining regions (CDRs)] that are supported on the β-barrel framework form the antigen combining site (ACS). While their sequence is hypervariable in comparison with the rest of the immunoglobulin structure,[11] some of the loops show a relatively high degree of both sequence and structural conservation. In particular, CDR-L2 and CDR-H1 are highly conserved in conformation.[12] Analysis of conserved key residues has led Chothia and co-workers to define canonical ensembles into which the CDRs may be grouped.[13,14]

Requirement for Modeling

Since the first X-ray crystallographic structure determinations, sequences of immunoglobulin light and heavy chain variable regions have accumulated at an ever increasing rate. X-Ray structures, however, are accruing at a relatively slow pace with no more than three or four immunoglobulin structures being published each year and fewer than that being deposited in the Brookhaven Protein Databank.[15] Protein crystallography

[4] R. J. Poljak, L. M. Amzel, H. P. Avey, B. L. Chen, R. P. Phizackerley, and F. Saul, *Proc. Natl. Acad. Sci. U.S.A.* **70,** 3305 (1973).

[5] M. Schiffer, R. L. Girling, K. R. Ely, and A. B. Edmunds, *Biochemistry* **12,** 4620 (1973).

[6] C. Chothia, J. Novotný, R. E. Bruccoleri, and M. Karplus, *J. Mol. Biol.* **186,** 651 (1985).

[7] A. F. Williams and A. N. Barclay, *Annu. Rev. Immunol.* **6,** 381 (1988).

[8] A. Holmgren and C.-I. Bränden, *Nature (London)* **342,** 248 (1989).

[9] J. Novotný, R. E. Bruccoleri, J. Newell, D. Murphy, E. Haber, and M. Karplus, *J. Biol. Chem.* **258,** 14433 (1983).

[10] P. M. Colman, *Adv. Immunol.* **43,** 99 (1988).

[11] T. T. Wu and E. A. Kabat, *J. Exp. Med.* **132,** 211 (1970).

[12] P. de la Paz, B. J. Sutton, M. J. Darsley, and A. R. Rees, *EMBO J.* **5,** 415 (1986).

[13] C. Chothia and A. M. Lesk, *J. Mol. Biol.* **196,** 901 (1987).

[14] C. Chothia, A. M. Lesk, A. Tramontano, M. Levitt, S. J. Smith-Gill, G. Air, S. Sheriff, E. A. Padlan, D. Davies, W. R. Tulip, P. M. Colman, S. Spinelli, P. M. Alzari, and R. J. Poljak, *Nature (London)* **342,** 877 (1989).

[15] F. C. Bernstein, T. F. Koetzle, G. J. B. Williams, E. F. Meyer, M. D. Brice, J. R. Rodgers, O. Kennard, T. Shimanouchi, and M. Tasumi, *J. Mol. Biol.* **112,** 535 (1977).

is limited by two major factors: the time required to collect, process, and refine an X-ray data set and the intractability of certain proteins to crystallographic analysis (for example, certain proteins prove impossible to crystallize, or do not diffract to acceptable resolutions once crystallized). In addition to the technical complexity of crystallography itself, the supporting biochemistry can be difficult. The protein may be difficult to purify, available in limited quantities, or unstable.

Currently, the only experimental alternative to X-ray techniques is nuclear magnetic resonance (NMR).[16,17] However, at the moment, the technique is applicable only to the solution of relatively small proteins (<20 kDa).[18] Thus, with the experimental limitations and the extreme importance of antibodies in therapy and research, there have been many attempts to develop methods by which to model antibody combining sites and thus circumvent experimental procedures. The availability of accurate and reliable modeling procedures would also allow the prediction of the effects of site-directed mutagenesis (SDM) experiments and allow the intelligent application of SDM as well as larger modifications to the combining site (CDR replacement, introduction of catalytic activity, and metal binding sites) and, eventually, tailoring of combining sites to new antigens.

Antibody Modeling—Approaches to Date

Various workers have attempted to model antibody combining sites using, in the main, knowledge-based approaches.[12,19–24] More recently, attempts have been made to use *ab initio* methods[25–27] where the conformational space available to a loop is saturated, followed by a screening procedure. Generally this involves selecting the lowest energy conformation calculated using an empirically derived potential function.[28,29]

Modeling of antibody combining sites may be divided into three classes:

1. modeling panels of mutant antibodies given a structure (determined either by crystallography or by molecular modeling) for the parent antibody

[16] R. M. Cooke and I. D. Campbell, *BioEssays* **8,** 52 (1988).

[17] K. Wüthrich, "NMR of Proteins and Nucleic Acids." Wiley, New York, 1986.

[18] G. M. Clore and A. M. Gronenborn, *Protein Eng.* **1,** 275 (1987).

[19] M. J. Darsley, P. de la Paz, D. C. Phillips, A. R. Rees, and B. J. Sutton, *in* "Investigation and Exploitation of Antibody Combining Sites" Methodological Surveys in Biochemistry and Analysis Vol. 15 (E. Reid, G. M. W. Cook, and D. J. Morré, eds.), Plenum, New York, 1985.

[20] C. R. Mainhart, M. Potter, and R. J. Feldmann, *Mol. Immunol.* **21,** 469 (1984).

[21] M. E. Snow and L. M. Amzel, *Proteins: Struct. Funct. Genet.* **1,** 276 (1986).

2. modeling insertions, deletions, and CDR replacements given a structure for the parent antibody
3. the larger problem of modeling an unknown antibody structure from its amino acid sequence alone. This involves two stages—building the framework region and building the CDRs

Modeling Single-Site Mutations

This involves the replacement of one or more amino acid residues by others. In modeling such replacements, it is necessary to assess the possibility of backbone conformational changes in addition to side chain placement. The work of Sibanda and Thornton,[30] Thornton *et al.*,[31] and of Greer[32] indicates that changes in loops tend to be accommodated locally and one can thus be reasonably confident that the overall structure of the antibody will not be modified. Support for this assumption also comes from X-ray structures of mutant hemoglobins,[33] T4 lysozyme[34] and crambins,[35] the isomorphous crystallization of mutant proteins,[36] and solution data for phage λ-repressor mutants.[37,38] If the residue replacements are conservative (especially if they are surface residues not making inter-CDR interactions), it may only be necessary to consider the placement of the side chain, but if nonhomologous changes are being made, changes to

[22] E. A. Padlan, D. R. Davies, I. Pecht, D. Givol, and C. Wright, *Cold Spring Harbor Symp. Quant. Biol.* **41,** 627 (1976).
[23] S. J. Smith-Gill, C. R. Mainhart, T. B. Lavoie, R. J. Feldmann, W. Drohan, and B. R. Brooks, *J. Mol. Biol.* **194,** 713 (1987).
[24] E. A. Padlan and E. A. Kabat, *Proc. Natl. Acad. Sci.* **85,** 6885 (1988).
[25] R. M. Fine, H. Wang, P. S. Shenkin, D. L. Yarmush, and C. Levinthal, *Proteins: Struct. Funct. Genet.* **1,** 342 (1986).
[26] R. E. Bruccoleri, E. Haber, and J. Novotný, *Nature (London)* **335,** 564 (1988).
[27] J. Moult and M. N. G. James, *Proteins: Struct. Funct. Genet.* **1,** 146 (1986).
[28] B. Brooks, R. E. Bruccoleri, B. D. Olafson, D. J. States, S. Swaminathan, and M. Karplus, *J. Comput. Chem.* **4,** 187 (1983).
[29] J. Åqvist, W. F. van Gunsteren, M. Leifonmark, and O. Tapia, *J. Mol. Biol.* **183,** 461 (1985).
[30] B. L. Sibanda and J. M. Thornton, *Nature (London)* **316,** 170 (1985).
[31] J. M. Thornton, B. L. Sibanda, M. S. Edwards, and D. J. Barlow, *BioEssays* **8,** 63 (1988).
[32] J. Greer, *J. Mol. Biol.* **153,** 1027 (1981).
[33] G. Fermi and M. Perutz, *in* "Atlas of Molecular Structures in Biology" (D. C. Phillips and F. M. Richards, eds.). Oxford Univ. Press (Clarendon) London and New York, 1981.
[34] M. G. Grütter, R. B. Hawkes, and B. W. Matthews, *Nature (London)* **277,** 667 (1979).
[35] W. A. Hendrickson and M. M. Teeter, *Nature (London)* **290,** 107 (1981).
[36] M. Knossow, R. S. Daniels, A. R. Douglas, J. J. Skehel, and D. C. Wiley, *Nature (London)* **311,** 678 (1984).
[37] M. H. Hecht, H. C. M. Nelson, and R. T. Sauer, *Proc. Natl. Acad. Sci.* **80,** 2676 (1983).
[38] M. A. Weiss, M. Karplus, D. J. Patel, and R. T. Sauer, *J. Biomol. Struct. Dynamics* **1,** 151 (1983).

backbone conformation are likely. For example, changing a surface hydrophilic residue to a hydrophobic one may result in a refolding of the loop to bury the hydrophobic group away from the solvent.

A number of approaches have been used to model single-site mutations. One of the simplest is the maximum overlap procedure (MOP).[12,20,23,38a,39-41] Amino acid replacements are made such that the new residue is constructed with as many atoms as possible in identical positions to the atoms in the parent structure; that is, wherever possible, the parent dihedrals are inherited by the mutated residue. The entire structure is then subjected to energy minimization to obtain a prediction for the mutant protein.

Sidechain replacement by MOP is readily achieved using the REPLACE and REFI commands of the interactive molecular graphics program FRODO.[42] REFI uses an optimization technique to fit a template side chain conformation to the atoms present in the parent structure. This "molten atom" or "local change" approach[43] moves only one atom at a time such that each movement improves the immediate environment of the atom by decreasing the differences from ideality of bond lengths, bond angles, and dihedral angles near this atom. Small displacements are calculated and applied to each atom in turn. The process is thus iterative, as moving one atom will affect the environment of another. Typically, up to 50 cycles are required to achieve convergence to less than 0.01 Å. Similar procedures may be implemented using HYDRA[44] or QUANTA (Polygen, Waltham, MA).

However, when using MOP, it is possible for the structure to be trapped in a local energy minimum resulting in an incorrect model. The minimum perturbation procedure (MPP)[45] helps to overcome this problem. After making the appropriate residue replacement(s), a conformational search is performed about the dihedral angles of the new side chain. Low-energy conformations are identified and subjected to restrained energy minimization.[28] Restraints are applied such that conformational changes are assumed to be local to the area of the residue replacement(s).

[38a] The terms MOP (maximum overlap procedure), MPP (minimum perturbation procedure), and CPP (coupled perturbation procedure) were introduced by Snow and Amzel.[21]

[39] R. J. Feldmann, M. Potter, and C. P. J. Glaudemans, *Mol. Immunol.* **18,** 683 (1981).

[40] M. K. Swenson, A. W. Burgess, and H. A. Scheraga, *in* "Frontiers in Physicochemical Biology" (B. Pulman, ed.), p. 115. Academic Press, New York, 1978.

[41] P. Warme, F. A. Momany, S. V. Rumball, R. W. Tuttle, and H. A. Scheraga, *Biochemistry* **13,** 768 (1974).

[42] T. A. Jones, this series, Vol. 115, p. 157.

[43] J. Hermans and J. E. McQueen, *Acta Crystallogr. Sect. A* **30,** 730 (1974).

[44] R. E. Hubard, *Proc. Comput.-Aided Mol. Des. Conf.* p. 99 (1984).

[45] H. L. Shih, J. Brady, and M. Karplus, *Proc. Natl. Acad. Sci.* **82,** 1697 (1985).

Simple approaches such as MOP and MPP can prove extremely successful, especially when the replacement is conservative and the residue makes few interactions with neighboring residues. Karplus's group[45] cites an example where they have modeled a Gly → Asp mutation in influenza virus hemagglutinin by MPP. They accurately positioned the aspartate side chain in the region indicated by a difference density peak from crystallographic techniques.[36]

The coupled perturbation procedure (CPP)[21] extends the MPP approach by searching the conformational space not only of the replaced residue, but also of a "dependency set." The dependency set for each replaced residue constitutes those residues whose conformations are likely to be affected by the replacement, i.e., those which are capable of making interactions with any conformation of the parent or replacement residue. As in MPP, the structure is then subjected to restrained energy minimization. By using more comprehensive conformational search procedures such as CONGEN,[46] CPP can search side chain conformations throughout the whole structure, effectively extending the dependency set to all residues in the molecule. However, the computer time required for such a search is likely to become prohibitive.

All three methods (MOP, MPP, CPP), however, are likely to prove inadequate if a drastic change in the nature of a residue is made, since none accounts for the possibility of backbone conformational change.

Modeling Insertions, Deletions, and CDR Replacements

As with modeling single-site mutations, such changes are likely to be accommodated locally within the CDR. However, the magnitude of these local changes may be large and it is thus always necessary to consider conformational changes in the backbone of the loop. CDR replacement is thus an extension of the insertion/deletion problem as it is obviously necessary to consider the whole loop conformation. However, in contrast to modeling the entire combining site, the environment in which the loop is constructed is largely defined by the presence of the other five loops of the combining site.

To date, most modeling in this category has relied on the presence of CDRs of the required length in the database of known antibody structures. When these are not present, the closest available CDR has been modified manually using computer graphics.[12,23,24] When loops of the correct length are available, such methods are likely to be quite successful. However, when manual insertions or deletions need to be made, the results are highly

[46] R. E. Bruccoleri and M. Karplus, *Biopolymers* **26,** 137 (1987).

dependent on user interaction and their reproducibility is thus low. Alternatively, it has been possible to employ complete conformational searches,[27,46] not only of the side chains, but also of the backbone in the vicinity of the replacement. Such methods are extremely computer intensive and again require user interaction (see Conformational Search Methods, below).

Modeling the Framework

When modeling a completely unknown antibody from its sequence alone, it is first necessary to build the framework. This is relatively straightforward as it is highly conserved,[47] although differences do occur in packing of the V_L and V_H domains with respect to one another. A higher degree of variability is seen around the takeoff region of CDR-H3 and an analysis of this effect is currently under way (J. Pedersen and A. R. Rees, unpublished, 1991).

A single known high-resolution antibody structure with high sequence homology to the structure being modeled may be used as a starting point for maximum overlap (MOP)-type correction of sequence differences.[12] Alternatively, light and heavy chains may be modeled from separate starting structures.[14] V_L and V_H domains are selected separately on the basis of sequence homology. If the two domains are chosen from different antibodies composed of V_L1/V_H1 and V_L2/V_H2, respectively, the composite is formed by least squares fitting V_L1 onto V_L2 before removing V_H1 and V_L2 to leave the V_L1/V_H2 composite which inherits the V_H/V_L packing of V_L1/V_H1.

Problems have been identified with poorly defined framework regions influencing the construction of CDRs.[48] For example, the N terminus of the heavy chain packs against CDR-H3 and in the crystal structure of the antibody HyHEL-5 the first two residues of V_H are placed differently from the equivalent residues in Gloop2 and other antibody crystal structures. The temperature factors of the HyHEL-5 atoms were examined and those higher than the mean plus three standard deviations ($\bar{x} + 3\sigma$) noted. This identified five residues with temperature factors falling outside the normal distribution: PCA-1H, Val-2H, Gln-3H, Arg-40H, and Asp-102H. This suggests these residues are either extremely mobile, or difficult to place in the crystal structure. The analysis was repeated examining just main chain atoms (N, $C\alpha$, C, O) and, again, the N-terminal residues of V_H were identified: PCA-1H, Val-2H, and Gln-3H. Thus, these residues seem to be poorly defined in the crystal structure, with the side chains of the first two

[47] P. M. Alzari, M.-B. Lascombe, and R. J. Poljak, *Annu. Rev. Immunol.* **6,** 555 (1987).

[48] A. C. R. Martin, D.Phil. Thesis University of Oxford, Oxford (1990).

residues and the main chain of the third being particularly poor.[49] There is, however, no evidence for any other conformation being more correct.

Main-chain temperature factors should thus be examined to identify poorly defined regions of the structure and such regions should be replaced with consensus conformations from other known crystal structures with lower temperature factors in these regions. An automated framework-building procedure has now been developed (J. Pedersen and A. R. Rees, unpublished, 1991).

Modeling the CDRs of an Entire Antibody Fv

Modeling the CDRs themselves is a more difficult task. By their very nature, these loops are hypervariable, showing, in some cases, extreme variability in sequence, length, and conformation.

Most approaches to the problem so far have used the available CDRs from antibody crystal structures as starting models for those in the antibody to be modeled.[12,13,20,23,24,39,50]

MOP-Based Methods. Procedures based on the MOP method for single-residue replacements have been used independently by at least two groups.[12,20,23] Their approaches differ in the way in which they select loops as starting points. The groups of Feldmann and Smith-Gill[20,23] choose a single antibody on which to model all the hypervariable loops, selecting loops from other antibodies only when the required loop length is not present in that starting antibody structure. In contrast, the group of Rees[12] has used a single, high-resolution framework chosen by sequence homology. Individual loops are selected from a database of the available antibody crystal structures in the Brookhaven Protein Databank,[15] first on length and, if two or more loops of equally suitable length exist, one loop is selected on sequence homology. These loops are then attached to the framework model.

The selected loop is fitted onto the model framework using interactive computer graphics and the sequence of the loop is then corrected using the MOP protocol. In both methods, if no loop of the required length is available, the loop of closest length to the unknown is used and insertions or deletions are made using interactive molecular graphics while attempting to maintain the overall shape of the loop and intraloop hydrogen bonding.

When this procedure has been repeated for each of the six loops and the

[49] S. Sheriff, personal communication.

[50] C. Chothia, A. M. Lesk, M. Levitt, A. G. Amit, R. A. Mariuzza, S. E. V. Phillips, and R. J. Poljak, *Science* **233**, 755 (1986).

side chains of the framework region have been resequenced using MOP, the whole modeled Fv is subjected to energy minimization, either restrained[23] or unrestrained,[12,51] using the GROMOS[29] or CHARMM[28] potential.

Key Residue Method. The approach devised by Chothia and coworkers[13,14,50] also exploits the antibody structure database. Loop conformation is predicted on the basis of the presence of critical or key residues. These are residues which affect loop packing (e.g., bulky residues such as Trp, Tyr, or Phe), can form hydrogen bonds or salt bridges (e.g., Ser, Thr, Asn, Gln, Asp, Glu, Arg, or Lys), or are able to adopt unusual conformations (e.g., Gly or Pro). As used by Chothia *et al.* the method is completely manual although it is a prime candidate for automation.

All Protein Database Method. The above methods are limited by the restricted size of the antibody crystal structure database, which contains only around a dozen structures. When the database is restricted to the known immunoglobulin crystal structures, making insertions and deletions in CDRs when conformations of the required lengths are not available among the known structures is one of the most unreliable parts of the procedures. Such changes are generally made by hand, using interactive computer graphics.[12,23,24] Clearly, such a procedure is highly subjective and its repeatability is low. The site at which insertions or deletions are made is critical in determining the resulting conformation.[48] When more crystal structures become available, the probability of the database containing a CDR of identical length and high sequence homology to that being modeled and of identifying all key residues (as defined by Chothia) will increase. At the present time, however, the number of antibody structures in the Brookhaven Protein Databank is increasing only at the rate of one or two per year.

One way of expanding the database of loop structures is to examine loops from all known protein crystal structures rather than restricting the database to antibodies. The use of nonhomologous structures in modeling protein loops was proposed by Jones and Thirup.[52] Sutcliffe *et al.*[53] have developed a method for modeling loops (implemented in the program COMPOSER) which utilizes a database of all high-resolution crystal structures. They first define a structurally conserved framework region (SCR) by structural comparison of all homologous known structures (here the antibody structures). Loop regions not defined within this SCR are then constructed. If the loop is a tight turn, it is built *ab initio.* Otherwise, geometri-

[51] A. R. Rees and P. de la Paz, *Trends Biochem. Sci.* **11,** 144 (1986).

[52] T. A. Jones and S. Thirup, *EMBO J.* **5,** 819 (1986).

[53] M. J. Sutcliffe, I. Haneef, D. Carney, and T. L. Blundell, *Protein Eng.* **1,** 377 (1987).

cal constraints for the adjoining section of the SCR are determined and loops of appropriate length which satisfy the same constraints in the framework to which they originally were attached are selected from the database. Selection from this group is performed on the basis of structurally determining residues (SDRs). When attaching a loop to the SCR, a sliding weighting scale is used to match the SCR and framework onto which the chosen loop was originally attached.

From analysis of residue replacements in homologous proteins, they also provide a rule for each possible side-chain replacement in sheet, helix, or loop (i.e., a total of 1200 rules).[54] The side-chain χ torsion angles within the SCRs of globins and immunoglobulins were calculated. Their distributions were analyzed and the most probable torsion angles were calculated with their standard deviations. Correlations between side-chain atom positions where substitutions have occurred were examined and these form the basis of the rules. Side chains are built by matching the parent side-chain conformation against the correlated parent positions for the required replacement. If all correlated conformations score equally, the side chain is built up in the most probable conformation. A similar set of rules has been suggested by Karplus's group,[55] although they do not split the rules into the three groups for sheet, helix, and loop. They state that for structurally and functionally similar proteins, there is a high probability that side-chain orientation of the parent can be transferred to the mutant side chain and that the basic rule for this transfer is that maximal occupancy of atom positions in the two structures is maintained. By implication this supports the use of MOP in side-chain replacement. For mutated residues, the transferability is in the range 60–89% for γ atoms and 56–84% for δ atoms. This compares with transferabilities of 60–97% (γ) and 50–90% (δ) for identical residues. Higher degrees of homology between proteins result in higher degrees of transferability, although highly accessible residues appear to have more random orientations and this is important in surface loops.

Stanford and Wu Method. Stanford and Wu[56] proposed a method for generating backbone conformation of the CDRs by analysis of backbone torsion angles for tripeptides observed in other β-sheet proteins. Since several torsion angles are generally possible for each amino acid, 1000–2000 conformations were generated for each CDR. However, this number is reduced by the requirement to link the two takeoff points of the framework onto which the loops must be fitted. The presence of the framework

[54] M. J. Sutcliffe, F. R. F. Hayes, and T. L. Blundell, *Protein Eng.* **1,** 385 (1987).

[55] N. L. Summers, W. D. Carlson, and M. Karplus, *J. Mol. Biol.* **196,** 175 (1987).

[56] J. M. Stanford and T. T. Wu, *J. Theor. Biol.* **88,** 421 (1981).

also applies van der Waals constraints on the possible conformations of the loops. An iterative method is applied to adjust the torsion angles to fit the end-points of the loops onto the framework and the conformation with the best fit onto the framework is selected.

Random Search Method. Levinthal's group has suggested a method which combines molecular dynamics and generation of random sets of conformations for the CDRs.[25] These are screened to eliminate structures with van der Waals overlap either within the loop or with the rest of the molecule. Energy minimization, or molecular dynamics followed by energy minimization, is then applied to the system. By generating a large number of random structures, they attempt to saturate conformational space, selecting possible structures for the loop to be modeled by refining a set of the randomly generated structures and selecting loops of low energy. The results obtained by applying this method to modeling the anti-phosphorylcholine antibody McPC603[57] are mixed. The method works well for the shortest loops studied (CDR-H1, five residues; CDR-L2, seven residues) especially in the presence of the other CDRs from the crystal structure. When constructions were made in the absence of the other CDRs the prediction for CDR-H1 was still good, but the prediction for CDR-L2 in the region where it interacts with the other CDRs was poor. For these shorter loops, they find that the length of the loop to be modeled is the primary factor in determining the structure of the loop. For the longer loops studied (CDR-L3, 9 residues; CDR-H3, 11 residues) multiple energy minima were found, some significantly lower than the energy of predicted loops showing minimal root mean square (RMS) deviation from the crystal structure. No actual RMS deviations are cited as the nature of the method means that clusters of similar possible conformations are selected. At the time of publication of these results, the computer power available to them was insufficient to tackle the two longest loops (CDR-L1 and CDR-H2, 17 and 19 residues, respectively) in McPC603.

Conformational Search Methods. An alternative method to saturate conformational space is to generate all possible loop conformations by a tree search.[27,46] A large number of alternative loop conformations is generated by rotation about the dihedral angles, ϕ and ψ, of the peptide main chain together with side-chain χ dihedral angles. One of the major problems with this type of approach is that the size of the tree search grows exponentially as the number of degrees of freedom included in the search increases. Thus, the conformational search procedure becomes impractical for longer loops. Bruccoleri *et al.*[26] have employed methods such as "real

[57] S. Rudikoff, Y. Satow, E. A. Padlan, D. R. Davies, and M. Potter, *Mol. Immunol.* **18,** 705 (1981).

space renormalization"[58-60] to reduce the search time for longer loops. Here, loops are built from both points at which they leave the framework simultaneously. The bottom residues are constructed first and the conformations generated are written into a conformations file or "CG" file. Around 10 low-energy conformations for this pair of residues are read back from the CG file and used on which to build the next pair of residues. The 10 lowest energy conformations of these four residues are, in turn, used to construct the next pair of residues until the size of the loop remaining for construction is small enough for a full conformational search. Alternatively, rather than constructing one residue at a time, fragments of two or three residues may be built. Numerous construction protocols thus become possible in order to construct a long loop. In Bruccoleri and Karplus's procedure, CONGEN,[46] final "chain closure" of the loop over three residues is performed analytically using a modification[61] of Gō and Scheraga's chain closure algorithm.[62] Additionally, cycles of CHARMM energy minimization[28] can be incorporated within the conformational search; this frees the resulting model from being restricted to the step size (generally 30°) used for the conformational search, thus permitting structures which might otherwise be rejected on the basis of van der Waals overlap.

The ϕ and ψ angles for an amino acid are combined into a single degree of freedom and the amount of conformational space explored is restricted by looking up ϕ/ψ combinations in a tabulated form of the Ramachandran plot[63] and examining only those combinations with energies below a specified cutoff. Three such Ramachandran maps are used: one for glycine, one for proline and one for alanine, which is taken as being representative of all the other amino acids. Three residues of the section being constructed are generated using the analytical chain closure algorithm of Gō and Scheraga[62] (as modified by Bruccoleri and Karplus[61]). Thus, for a five-residue section, two residues are constructed by full conformational search while three are constructed by chain closure. Residues constructed by full conformational search will be termed *backbone* while those constructed by chain closure will be termed *chain.*

In order to improve the efficiency of the algorithm, after each combination of *backbone* torsion angles is generated, the distance between the endpoints is checked to ensure that it is possible to span the remaining gap with the number of residues left to construct (when these are in all-trans

[58] H. A. Scheraga, *Biopolymers* **22,** 1 (1983).
[59] M. R. Pincus and R. D. Klausner, *Proc. Natl. Acad. Sci.* **79,** 3413 (1982).
[60] M. R. Pincus, R. D. Klausner, and H. A. Scheraga, *Proc. Natl. Acad. Sci.* **79,** 5107 (1982).
[61] R. E. Bruccoleri and M. Karplus, *Macromolecules* **18,** 2767 (1985).
[62] N. Gō and H. A. Scheraga, *Macromolecules* **3,** 178 (1970).
[63] G. N. Ramachandran, C. Ramakrishnan, and V. Sasisekharan, *J. Mol. Biol.* **7,** 95 (1963).

configuration with the bond angles stretched by 5° from the optimum). Only if this condition is met are further *backbone* residues constructed followed by the *chain* residues and, finally, the side chains. In addition, conformations are rejected with van der Waals energy above a specified cutoff (typically 20 kcal/mol for *backbone,* 100 kcal/mol for *chain,* and 20 kcal/mol for side chains).

Two major problems exist with the conformational search algorithms apart from the extreme expense in CPU time. First, although the method is very good at reproducing crystal structures, including quite precise side-chain orientations[26]—typically, RMS deviations below 2 Å may be achieved for each loop, including side-chains, a large number of low-energy structures is generated from which only one must be selected. Generally the conformation observed in the crystal structure does not represent the lowest potential energy structure. Bruccoleri *et al.*[26] have suggested the use of solvent accessibility as a filter. If the bottom two energy structures differ in energy by less than 2 kcal (three times the Boltzmann factor kT, indicating both conformations are well populated at room temperature) then the conformation with the smaller solvent accessible surface[64] is chosen.[26] However, this is frequently still unable to select the best conformations generated by CONGEN. Second, where real space renormalization is employed, the order in which the segments are constructed affects the outcome. In a real modeling situation any such decisions about how a loop is put together will be relatively arbitrary.

Results of Antibody Database Methods

So far, all these methods have had only limited success; where crystal structures have become available after modeling has been performed, quite large deviations between predicted and observed structures have been obtained, especially in side-chain conformations.[50,65] In a comparison of the modeled structure[20] of J539 with the crystal structure,[65] RMS deviations of between 1.1 Å (CDR-H1 and CDR-L2) and 4.0 Å (CDR-H3) are seen for C_αs and between 2.0 Å (CDR-L1) and 6.5 Å (CDR-H3) for all atoms. Although it is most likely that these differences result from inadequacies in the modeling, the authors cannot rule out the possibility that crystal packing leads to distortions in loop conformation. Chothia's model of the antibody D1.3[50] is better, with RMS deviations of 0.50–0.97 Å (backbone—N, C_α, C, C_β) for five of the six loops, CDR-H1 having an RMS of 2.07 Å. (All atom RMS deviations have not been published; a

[64] B. K. Lee and F. M. Richards, *J. Mol. Biol.* **55,** 379 (1971).

[65] S. W. Suh, T. N. Bhat, M. A. Navia, G. H. Cohen, D. N. Rao, S. Rudikoff, and D. R. Davies, *Proteins: Struct. Funct. Genet.* **1,** 74 (1986).

higher resolution map for D1.3, the map used in the comparison is at 2.8-Å resolution,[66] is required to make a more critical evaluation of the accuracy of side-chain predictions. However, examination of Fig. 1 in Chothia *et al.*[50] suggests that the side-chain predictions in some of the loops may be poor as the C_β positions inferred from peptide plane orientations appear to be badly misplaced. However, it should be noted that, at 2.8-Å resolution, the exact orientation of peptide planes outside regions of defined secondary structure may not be well defined in the electron density map.

In general, when the key residues defined by Chothia are present and loops of identical length are available in the database, the predictions are good, a least on backbone, as indicated above. However, for loops of lengths not represented in the database or lacking the critical residues required for this type of analysis, the predictions tend to be poor. In addition, relatively unconserved residues can have a profound effect on loop packing.[67] A variable residue at the base of CDR-H2 in Gloop2 (an antibody against the loop region of lysozyme), when changed from Glu to Ser, resulted in abolition of binding. Roberts *et al.*[67] argued that, since this residue is relatively buried in the model of the antibody,[12] it is most likely that the Glu50H → Ser mutation disrupts the packing of CDRs in the ACS, thus drastically altering the topology of the combining site surface and hence abolishing antigen binding. Chothia's approach would not identify this residue as being critical in loop packing. The recent crystal structure of Gloop2[68,69] supports this prediction, as Glu-50H is not accessible on the surface of the ACS and is involved in inter-CDR contacts.

A Combined Algorithm

Clearly, none of the methods presented above is wholly satisfactory, although all show a limited amount of success. The ideal modeling procedure may thus involve a combination of the best features of a number of these approaches. Database methods can be used to reduce the search times of conformational search procedures for long loops, thus eliminating the need for real-space renormalization.[60] A number of conformations can be extracted from a database, either by using methods such as that of Sutcliffe *et al.*[53] (see All Protein Database Method, above) or, alternatively, by using distance constraints within the loops themselves defined from the known antibody structures. A conformational search program such as

[66] A. G. Amit, R. A. Mariuzza, S. E. V. Phillips, and R. J. Poljak, *Science* **233,** 747 (1986).
[67] S. Roberts, J. C. Cheetham, and A. R. Rees, *Nature (London)* **328,** 731 (1987).
[68] P. D. Jeffrey, R. E. Griest, G. L. Taylor, and A. R. Rees, manuscript in preparation (1991).
[69] P. D. Jeffrey, D.Phil. Thesis University of Oxford, Oxford (1989).

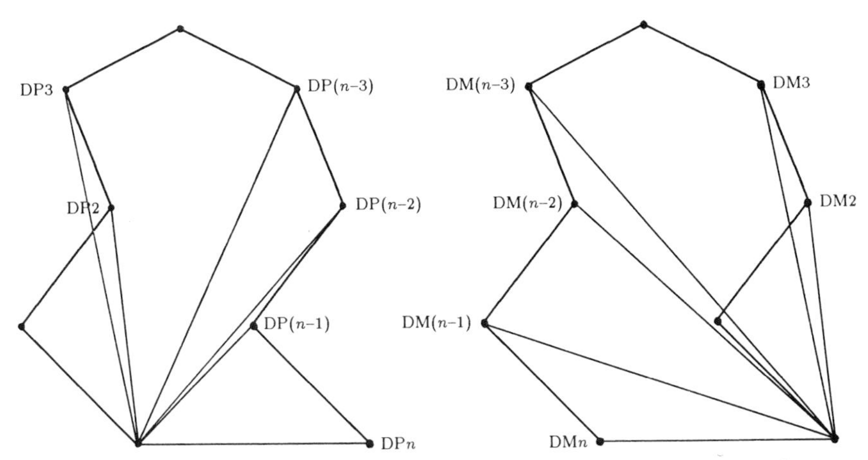

FIG. 1. C_α distances are measured from the N and C termini of the loops of known antibody and Bence–Jones protein structures as shown in the diagram. These distances are then used to search the database of all known protein structures for loops of similar conformation.

CONGEN[46] may then be used to select, from this set of database loops, the best backbone conformations for the bases of the loops, while reconstructing only the top of the loops and the side chains by conformational searching. By removing the requirement for manual insertions or deletions from the knowledge-based methods and the requirement for real space renormalization from the *ab initio* method, a protocol results which is minimally dependent on user choice.

Such a method has been developed[48,70] and is described below.

Database Searching

In order to identify protein loops of the same general shape as loops seen in antibodies and, most importantly, of the required length for the antibody being modeled, sets of inter-C_α distance constraints within the loops are used as shown in Fig. 1. These distance constraints are derived from an analysis of known antibody crystal structures. For the light-chain CDRs, both antibody and Bence–Jones structures were used, while antibody structures were used for the heavy chain CDRs. The range used to search the database is the mean $\pm$ 3.5σ (σ = standard deviation). 3σ covers virtually 100% of a normal distribution; since the sample size is small, 3.5σ was chosen in order not to exclude values falling just outside the current distribution and to prevent the search from being overrestrictive.

[70] A. C. R. Martin, J. C. Cheetham, and A. R. Rees, *Proc. Natl. Acad. Sci.* **86,** 9268 (1989).

A database has been created containing inter-C_α distances for every pair of residues in the range $C_{\alpha_i} \rightarrow C_{\alpha_{i+n}}$ $(-20 \leqslant n \leqslant +20)$ within each protein. The database is searched by indexing a column from the database and performing a binary search on the sorted column.

Loop Processing

Having identified a set of loops matching the distance constraints applied at both the N and C termini, redundancies resulting from updated entries in the protein databank and from crystal structures containing multiple copies of a protein fragment are removed. The loops are extracted from the protein databank and are fitted onto the framework (the known crystal structure in the case of single-loop replacements, or a modeled framework where a complete antibody is being modeled). In order to orientate the loops correctly, they are overlapped onto the original loops present on the framework. Overlapping is performed in four stages:

1. The N terminus of the database loop (N_D) is moved onto the N terminus of the framework loop being replaced (N_F).
2. The C terminus of the database loop (C_D) is rotated onto the vector between N_F and the C terminus of the framework loop (C_F).
3. The database loop is translated along the vector between N_F and C_F such that the distance $(N_F - N_D) = (C_F - C_D)$.
4. The database loop is rotated about the $N_D - C_D$ vector such that the plane formed by the center of gravity of its backbone atoms, N_D and C_D, is coplanar with that formed by the center of gravity of the framework loop backbone atoms, N_F and C_F.

This procedure relies on the assumption that the loop takeoff angles defined by the planes through the backbone center of gravity, and the N and C terminal takeoff points, do not vary greatly between different antibodies with loops of different lengths. This has indeed been shown to be the case.[48]

The sequences of the database loops are then corrected to match the loop being modeled by placing side chains using template conformations which are least squares fitted onto the backbone atoms (N, C_α, C, O) of the residue being replaced. Since the side chains are later repositioned using the conformational search program, CONGEN, the actual positioning of side-chain atoms with the exception of the C_β is not critical. CONGEN treats the C_β as a backbone atom, as its position is defined by the orientation of the backbone atoms, N, C_α, C, and O. Thus a side-chain conformational search which first rotates the χ_1 torsion angle $(C_\alpha - C_\beta)$ is unable to position the C_β. The position of the C_β is taken from the parent amino acid except when this is a Gly in which case it is taken from the template conformation. More sophisticated side-chain replacement techniques have been proposed[54,55]; however, since only the C_β position is important here, they were not considered worthwhile.

As with the majority of molecular mechanics-based programs, CONGEN reduces the computational burden by treating most hydrogen atoms implicitly by using "extended" (or "united") atoms[71] (i.e., parameters for the "heavy" atom to which one or more hydrogens are attached are modified to account for the presence of the hydrogens). Only those hydrogens potentially involved in hydrogen bonding are treated explicitly.[71a] Since coordinates for hydrogens are not provided by X-ray crystallographic techniques, such "polar" hydrogens are added to the heavy atoms in standard orientations.

The atom order is corrected to a standard order and, having specified which residues will be constructed by conformational searching, a CONGEN conformations file is generated for further processing by CONGEN.

In some cases, the number of loops extracted from the database is too large to process—the computer time required would become excessive. In such cases, the number of database fragments may be reduced to a more manageable number by using a structurally determining residue filter, using the algorithm of Sutcliffe.[72] Each residue in each database fragment is assigned as being structurally determining if it causes the next C_α to be moved relative to the position of that C_α in any of the other database loops and the structurally determining residues are scored against the sequence being modeled using the Dayhoff mutation matrix.[73] (For details, see Sutcliffe[72])

Conformational Searching

Conformational searching is performed using the program CONGEN,[46] which saturates the conformational space available to a peptide fragment by rotation about the backbone ϕ and ψ torsion angles and the sidechain χ angles.

Side chains are constructed using the ITERATIVE algorithm,[46] which starts with an energetically acceptable position for all the side chains. All possible conformations for the first side chain are then regenerated and the conformation with the lowest energy is selected and its energy recorded.

[71] J. A. McCammon, P. G. Wolynes, and M. Karplus, *Biochemistry* **18,** 927 (1979).

[71a] J. A. McCammon and S. C. Harvey, *in* "Dynamics of Proteins and Nucleic Acids," p. 182. Cambridge Univ. Press, Cambridge, 1987.

[72] M. J. Sutcliffe, Ph.D. Thesis University of London Birkbeck College, London (1988).

[73] W. C. Barker and M. O. Dayhoff, *in* "Atlas of Protein Sequence and Structure" (M. O. Dayhoff, ed.), Vol. 5. National Biomedical Research Foundation, Silver Spring, Maryland, 1972.

The process is repeated for each side chain in turn and, when all side chains have been searched, the process starts again from the first side chain until the energy does not change, or an iteration limit is reached. This procedure generates only one energetically feasible side-chain conformation per backbone conformation and has been shown to be the most effective[46] in generating an accurate conformation in a minimum amount of computer time.

In practice, it is rarely practical to construct more than five residues (occasionally six or seven if the conformational flexibility of the region is restricted). As discussed previously, longer regions may be constructed by "real space renormalization"[60] CONGEN treats the extracted set of database loops which have been processed into the form of a CONGEN conformations file as it would the base residues of a peptide being constructed by real space renormalization (see Conformational Search Methods, above).

A five-residue fragment in the middle of each database loop is reconstructed using a 30° search of the torsion angles of the outer two *backbone* residues and the Gō and Scheraga chain closure on the middle three *chain* residues. If this fails to find more than 100 conformations, the torsional search is reduced to 15° or, if necessary, 5°. If the search still fails, additional *backbone* residues are searched until conformations are generated.

It has been shown[48] that when building a complete combining site, the constructions are best performed in the absence of the other loops. If a loop is built between two other previously built loops, cumulative errors can cause the current loop to be built badly. In the case of short loops, however, where the FILTER algorithm (see below) cannot be used, energy is the only criterion on which the screening is performed and these loops must thus be built last in the presence of the other loops.

Energy Screening

CONGEN is built around the CHARMM[28] energy minimization and dynamics package. The work of Bruccoleri in using CONGEN has selected a conformation from those generated by selecting the lowest energy conformation as evaluated using the CHARMM potential *in vacuo.* The *in vacuo* CHARMM potential, however, does not appear to rank the conformations very well—conformations of low RMS deviation compared with the crystal structure frequently do not fall among the lowest energy structures, as shown in Fig. 2. An examination of the low-energy conformations of CDR-H2 of HyHEL-5 using molecular graphics shows the low energy to result from an optimization of the van der Waals packing. In solvent, this attractive effect would be counterbalanced by an attraction of the loop

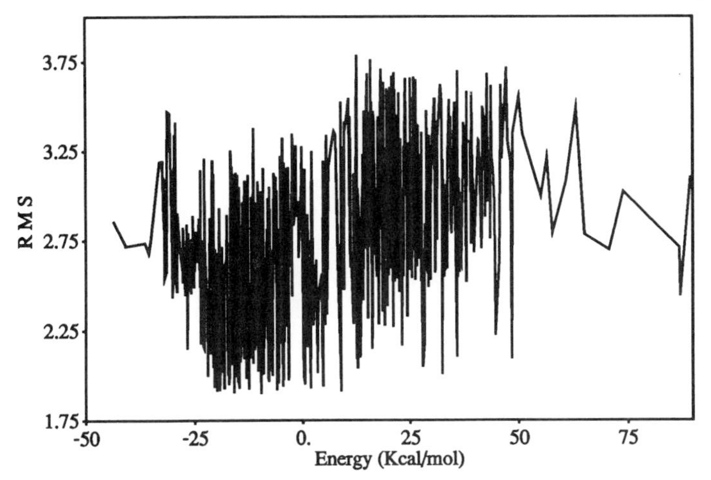

FIG. 2. Root mean square (RMS) deviation from the crystal structure is plotted against energy for each conformation generated for CDR-H2 of HyHEL-5. The energies are calculated using the CHARMM potential *in vacuo*. It is clear that those conformations which rank low in energy are not those of most similar structure to that observed in the crystal structure.

toward the solvent. This is analagous to the situation seen in a dynamics simulation of T1 ribonuclease,[74] where *in vacuo* the active site is seen to collapse, while in the presence of solvent, the native conformation is retained.

Use of a static random water box is not practical since some conformations, by chance, have favorable interactions with the solvent while others do not. In order to overcome this problem, it would be necessary to simulate the waters dynamically and computer time becomes unacceptably large when it is necessary to screen many hundreds of conformations.

As an alternative, in order to prevent the optimization of van der Waals packing which occurs *in vacuo,* the attractive $(-B/r_{ij}^6)$ part of the 12-6 Lennard-Jones potential is removed. By doing this, the concept of a van der Waals radius is lost, since all atoms repel each other to some extent, whatever the distance between them [even though this repulsive effect drops very rapidly with distance (A/r_{ij}^{12})]. If the atoms were being simulated dynamically (whether by molecular dynamics, or simply by movements in attempting to minimize the potential energy), this would have to be taken into account. Since atoms are not being allowed to move, but the potential is simply being used to calculate comparative energies, this step is not necessary. In addition, the presence of solvent around the loops will lead to a high dielectric constant (around 50), reducing the importance of electro-

[74] A. D. Mackerell, Jr., L. Nilsson, and R. Rigler, *Biochemistry* **27,** 4547 (1988).

statics. Thus the electrostatic potential is also removed. Similar "solvent-modified" potentials have been used in molecular dynamics of L-arabinose-binding protein,[75] in simulations of carbohydrates[76] and in analysis of misfolded proteins.[77] In the latter case, Novotný *et al.* omit the attractive part of the Lennard–Jones potential for all sulfur atoms and carbon atoms other than those in carbonyl groups, where either of the atoms in the pair being considered has a solvent accessibility greater than 0.1 Å^2. Since the loops of interest in antibody combining sites are on the protein surface and thus the majority of their atoms are solvent exposed, removing the attractive part of the potential for *all* atoms is effective and saves the computation time required to calculate the solvent accessibility. Novotný *et al.* calculate the solvent-modified electrostatic energy by multiplying the atomic charges by a factor dependent on the ratio of the distance of the atom from the center of the closest surface atom. The resulting "dielectric screening factor," first described by Northrup *et al.*[78] decreases the effective charge of an atom linearly from 1.0 in the center of the protein to 0.3 on the surface. For our purposes of modeling surface loops, we have found that removal of the electrostatic interaction altogether results in a greatly improved simulation and the added complexity of accounting more accurately for the center of the protein is unnecessary. This solvent-modified potential has been implemented in a modified version of GROMOS.

The rankings obtained by this solvent-modified potential are subsequently better than those from the *in vacuo* CHARMM potential, with low RMS conformations falling among the low-energy conformations (Fig. 3).

Filtering

In general, however, the lowest energy conformation is still not the lowest RMS conformation observed amongst the bottom five energies. Selection from this group is performed using one of two filtering procedures described below.

Solvent Accessibility. The solvent-modified potential used to rank the conformations considers only the enthalpic term of the free energy—entropy is ignored. In considering the local folding of small fragments of the structure of a protein, the major entropic component is that relating to the hydrophobic effect[79]—the tendency for hydrophobic atoms to pack away from the solvent. By calculating the solvent-accessible surface area of

[75] B. Mao, M. R. Pear, J. A. McCammon, and F. A. Quicho, *J. Biol. Chem.* **257,** 1131 (1982).
[76] K. Bock, M. Meldal, D. R. Bundle, T. Iversen, P. J. Garegg, T. Norberg, A. A. Lindberg, and S. B. Svenson, *Carbohydr. Res.* **130,** 23 (1984).
[77] J. Novotný, A. A. Rashin, and R. E. Bruccoleri, *Proteins: Struct. Funct. Genet.* **4,** 19 (1988).
[78] S. H. Northrup, M. R. Pear, J. D. Morgan, J. A. McCammon, and M. Karplus, *J. Mol. Biol.* **153,** 1087 (1981).
[79] C. Tanford, "The Hydrophobic Effect," 2nd Ed. Wiley, New York, 1980.

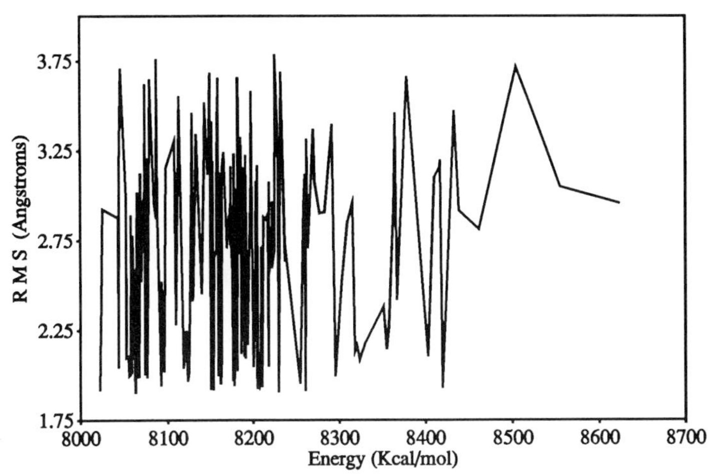

FIG. 3. RMS deviation from the crystal structure is plotted against energy for each conformation generated for CDR-H2 of HyHEL-5. The energies are calculated using the solvent-modified GROMOS potential described in the text. The ranking of conformations is much improved, with two of the lowest RMS conformations appearing in the bottom five energies.

hydrophobic atoms, that conformation with the lowest exposure of hydrophobics to the solvent can be selected. The atoms defined as hydrophobic for this purpose are listed in Table I. This list was compiled by testing a number of possible combinations to find the optimum set and represents all side-chain carbons (from C_α out), excluding those in charged, delocalized groups or adjacent to hydrophilic groups.

However, screening by solvent accessibility is oversensitive to positioning of atoms which are not part of the protein fragment being modeled.

TABLE I

ATOMS DEFINED AS HYDROPHOBIC FOR THE PURPOSE OF SCREENING CONFORMATIONS ON THE BASIS OF THE SOLVENT ACCESSIBILITY OF HYDROPHOBIC ATOMS

Ala-C_α,C_β	Met-C_α,C_β,C_γ,S_δ,C_ϵ
Cys-C_α,C_β,S_γ	Asn-C_α,C_β
Asp-C_α,C_β	Pro-C_α,C_β,C_γ,C_δ
Glu-C_α,C_β,C_γ	Glu-C_α,C_β,C_γ
Phe-C_α,C_β,C_γ,$C_{\delta1}$,$C_{\delta2}$,$C_{\epsilon1}$,$C_{\epsilon2}$,C_ζ	Arg-C_α,C_β,C_γ
His-C_α,C_β,C_γ,$C_{\delta2}$,$C_{\epsilon1}$	Thr-C_α,C_γ
Ile-C_α,C_β,$C_{\gamma1}$,$C_{\gamma2}$,$C_{\delta1}$	Val-C_α,C_β,C_γ,$C_{\delta1}$,$C_{\delta2}$
Lys-C_α,C_β,C_γ,C_δ	Trp-C_α,C_β,C_γ,$C_{\delta1}$,$C_{\delta2}$,$C_{\epsilon2}$,$C_{\epsilon3}$,$C_{\zeta2}$,$C_{\zeta3}$,$C_{\eta2}$
Leu-C_α,C_β,C_γ,$C_{\delta1}$,$C_{\delta2}$	Tyr-C_α,C_β,C_γ,$C_{\delta1}$,$C_{\delta2}$,$C_{\epsilon1}$,$C_{\epsilon2}$

Thus, while it performs well in testing the modeling procedure by reconstructing individual CDRs of a known crystal structure with the other CDRs present[70] and should perform equally well in modeling CDR replacements or mutations, it performs poorly when framework residues have also been modeled and the other CDRs are absent.

In addition, solvent accessibility cannot be used as a screen when loops are being constructed in a combining site with no prior knowledge of the structure of the other loops, since the solvent accessibility is meaningless with no other loops present. These problems led to the development of an alternative filtering method.

Structurally Determining Residues—FILTER. As an alternative to solvent accessibility, a structurally determining residue algorithm has been developed. This is applicable only to loops of six residues or longer, as it requires information from the database search (see later). For each residue of the loop in turn, the database loops are searched for the required residue type and the ϕ and ψ torsion angles are recorded. If the required residue type is not identified in any of the database loops, then the most similar residue types are identified from the Dayhoff mutation matrix and the confidence for the prediction at this residue is reduced by 5%. This procedure is repeated until residues are identified among the database loops.

The ϕ and ψ angles for each of the conformations being screened are then calculated and scored against the torsion angles observed in the database. The terminal residues start with a confidence of 50% since one of the two torsion angles is undefined. Two scoring schemes are implemented. In the first scheme, torsion angles are scored as $1/\theta$, where θ is the difference between model and database torsion angles (in radians) and θ is less than a specified cutoff. This value is then multiplied by the confidence. The second scheme simply scores a 1 for a residue if it has ϕ/ψ angles within the cutoff of any ϕ/ψ angles observed in the database conformations and thus represents the distribution of scores across the loops. Both scoring schemes are examined. If the first scoring scheme is unable to distinguish between the model conformations, the distribution of the scores is also examined—conformations which have their score distributed across the amino acids are chosen in preference to ones which score highly on only a few residues. In addition, if the conformation scoring highest in scheme 1 does not have a high distribution, this selection is rejected and a new angular cutoff is used.

The angular cutoff is initially chosen at 15°. If this is unable to select a conformation, the cutoff is increased to 30° and if this is still unable to distinguish between the models, the cutoff is further increased to 45°. This protocol is based on the ability of the different angular cutoffs to distinguish correctly between the low-energy conformations. Figure 4 shows the ability of different angular cutoffs to select the correct conformation. If the algorithm still fails to select a unique conformation, it is a fair indication that all the low-energy conformations are extremely similar and the lowest energy conformation may be selected.

Variations to the Procedure

For loops of six or seven residues, only one or two residues will be constructed from the database. In these cases, several hundred loops are extracted from the database and the conformational space available to the backbone of a loop appears to be well saturated in the current database of crystal structures. Conformational search of the backbone is thus unnecessary and is used only to construct the side chains. Conformational searching of the side chains gives better fits to the crystal structure than does use of the MOP protocol.

For loops of five residues or shorter, the limited number of distance constraints means that an enormous number of conformations (> 10,000) is selected from the database. Processing this number of loops becomes impractical and thus CONGEN is used alone in these cases. Because the database is not searched, the FILTER algorithm cannot be used. In these cases, it has been shown that the lowest energy structure is acceptable.

While the combined algorithm performs very well, it is extremely computer intensive. Chothia's approach of defining canonical ensembles

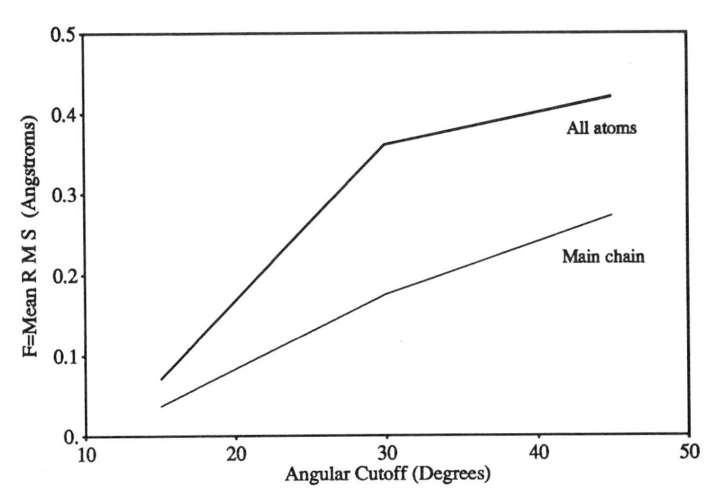

FIG. 4. Ability of the FILTER algorithm to distinguish correctly from the low-energy conformations of Gloop2 and HyHEL-5 CDRs at different angular cutoffs. For each cutoff, $F = \sum(R_{sel} - R_{min})/n$, the mean difference in RMS deviation between the selected conformation and the conformation with the minimum RMS deviation, is plotted. Cases where FILTER is unable to select a conformation are excluded, as are those where consideration of the distribution selects against the highest scoring conformation.

into which the CDRs fit requires very little computer time, but is limited to those CDRs which fit the currently defined set of canonicals. Where these structures do exist, the method works well (especially on backbone conformation) and the two approaches may successfully be combined. Thus a model may be produced by constructing those loops which fit canonicals using Chothia's approach, the remaining loops being built by the combined algorithm.

Summary

The overall protocol may be summarized as follows:

1. Construct the framework on the basis of sequence homology, rejecting sections with unusually high temperature factors and selecting them from other structures.
2. For loops of five residues or shorter, construct the loop using CONGEN. Go to step 8.
3. For loops of six residues or longer, search the distance database for backbone conformations.
4. Overlap the database loops onto the framework.
5. Correct the sequences.
6. Add explicit hydrogens.
7. For loops of eight or more residues, delete the midsection of the loop and reconstruct with CONGEN.
8. Reconstruct side chains with CONGEN.
9. Calculate the energy for each conformation with the solvent-modified potential.
10. Screen the bottom five energy conformations with the FILTER algorithm.

The procedures described for modeling by the combined algorithm (independent of both CHARMM and GROMOS) will be made available in an integrated modeling package from Oxford Molecular Limited (Oxford, England).

Gloop2—A Case Study

Gloop2[80] is a monoclonal antibody raised against the loop peptide region of lysozyme and selected to cross-react with the native protein. This antibody has been the target of numerous site-directed mutagenesis experiments[67,81,82] (S. Roberts, *et al.* unpublished, 1991) and modeling exercises[12,48,70] (C. Chothia *et al.*, unpublished). The crystal structure has be-

[80] M. J. Darsley and A. R. Rees, *EMBO J.* **4,** 393 (1985).
[81] S. Roberts, D.Phil. Thesis University of Oxford, Oxford (1988).
[82] K. Hilyard, D.Phil. Thesis University of Oxford, Oxford (1991).

TABLE II
SEQUENCE OF THE V_L REGION OF GLOOP2[a]

LFR1	D I Q M T Q S P S S L S A S L G E R V S L T C
L1	R A S Q E I S G Y L S
LFR2	W L Q Q K P D G T I K R L I Y
L2	A A S T L D S
LFR3	G V P K R F S G R R S G S D Y S L T I S S L E S E D F A D Y Y C
L3	L Q Y L S Y P L T
LFR4	F G A G T K L E
HFR1	Q V Q L Q Q S G T E L A R P G A S V R L S C K A S G Y T F T
H1	T F G I T
HFR2	W V K Q R T G Q G L E W I G
H2	E I F P G N S K T Y (Y A E R F K G)[b]
HFR3	K A T L T A D K S S T T A Y M Q L S S L T S E D S A V Y F C A R
H3	E I R Y
HFR4	W G Q G T T L T V

[a] Light-chain CDRs are named L1, L2, and L3; heavy-chain CDRs are H1, H2, and H3; the light-chain framework regions are named LFR1, LFR2, LFR3, and LFR4 while those from the heavy chain are named HFR1, HFR2, HFR3, and HFR4.

[b] The section of this CDR enclosed in parentheses was considered as framework for the modeling exercise.

come available,[68,69] allowing some of the predictions from the modeling to be tested.

The Gloop2 structure has been modeled using the combined algorithm described above and will be used as a case study to describe the approach used. The Gloop2 V_L sequence is shown in Table II. The protocol used to model each loop is shown in Table III and will be described in turn. The loops are all built onto an "empty" combining site with the exception of the two short CDRs, H1 and H3, which are built last in the presence of the other four loops. The notation used in Table III to describe the constructions encloses the region built by CONGEN in square brackets and the region used for chain closure in parentheses. Residues not thus enclosed come from database structures. In all cases, the side chains of all residues are constructed by CONGEN.

Framework

The V_L and V_H domains of the framework for the Gloop2 model were built separately from REI[83] and HyHEL-5,[84] respectively. The β strands of

[83] O. Epp, P. Colman, H. Fehlhammer, W. Bode, M. Schiffer, R. Huber, and W. Palm, *Eur. J. Biochem.* **45,** 513 (1974).

[84] S. Sheriff, E. W. Silverton, E. A. Padlan, G. H. Cohen, S. J. Smith-Gill, B. C. Finzel, and D. R. Davies, *Proc. Natl. Acad. Sci.* **84,** 8075 (1987).

TABLE III
CONSTRUCTION OF GLOOP2 USING THE COMBINED ALGORITHM

Loop	Sequence	Number of conformations	CPU time (hr)	Global fit RMS Deviation			
				$P2_1$		$P1$	
				All (Å)	C_α (Å)	All (Å)	C_α (Å)
L1	RAS[Q(EIS)G]YLS	1368	59 + 25	1.90	0.92	2.09	0.86
L2	AASTLDS	157	4 + 1	1.25	0.68	1.10	0.66
L3	LQ[Y(LSY)P]LT	721	24 + 14	2.08	0.76	2.00	0.75
H1	[T(FGI)T]	375	18 + 8	2.06	0.79	2.04	1.03
H2	EI[F(PGN)S]KTY	3066	123 + 57	2.51	1.35	2.23	1.20
H3	[R(EIR)Y]	144	7 + 3	2.17	1.00	1.76	0.76
Overall			235 + 108	2.11	1.23	1.96	1.25

[a] For each of the six CDRs of Gloop2, the construction protocol is indicated together with the number of conformations generated, the computer time required for the conformational search and the energy screening (on a VAX 3200), and global RMS deviations compared with the $P2_1$ and $P1$ crystal structures. These RMS deviations were calculated by fitting the structures on the framework regions and calculating the deviation over the CDRs on all atoms and C_αs. The all-atom RMS deviations drop between 0.02 and 0.55 Å while the C_α deviations drop between 0.03 and 0.44 Å when a local RMS fit is calculated.

REI were fitted to the β strands of the HyHEL-5 V_L, using the following residue ranges (numbered consecutively):

REI	HyHEL-5
34–39	33–38
43–47	42–46
84–90	83–89
100–104	98–102

The V_L domain of HyHEL-5 was then deleted to leave a hybrid of REI V_L and HyHEL-5 V_H. Temperature factors of the main-chain atoms (N, C_α, C, O) of the REI and HyHEL-5 domains were examined as described above. The N-terminal residues of HyHEL-5 V_H were identified as having high-temperature factors (greater than the mean plus three standard deviations): PCA-1H, Val-2H, and Gln-3H. These residues were replaced with the equivalent conformation from J539.[65] Finally, side-chain replacements were performed with the REPLACE and REFI functions of the interactive molecular graphics program FRODO.[42]

CDR-L1

CDR-L1 is 11 residues long. The middle five residues were constructed using CONGEN, the remaining residues coming from database loops. The

FILTER algorithm (Table IV) is unable to distinguish between the conformations at any angular cutoff, suggesting that all the low-energy conformations are very similar. Thus the lowest energy conformation was selected.

CDR-L2

CDR-L2 is seven residues long. It has been shown[48] that, for loops of this length, the main-chain conformational space is sufficiently well saturated by the available database of protein crystal structures. Thus only the side chains were built using the conformational search; the conformations were ranked with the solvent-modified potential and the FILTER algorithm was used to select a final conformation (Table IV). At the 15° cutoff, conformation 9 scores highest, but this score is achieved on a single residue (distribution = 1); similarly at 30°, conformation 9 has the highest score, but ranks lowest on distribution; at 45°, conformation 93 ranks highest on both score and distribution and is thus the conformation selected.

CDR-L3

CDR-L3 is nine residues long. The middle five residues were constructed using CONGEN, the remaining residues coming from database loops. Conformation 82 was selected using FILTER at 45°. The proline at the seventh position within the loop is correctly predicted in the unusual cis conformation.

CDR-H1

CDR-H1 is only five residues long. The number of available distance constraints is thus too small to perform a database search (tens of thousands of conformations are extracted from the database) and the whole loop was constructed using CONGEN. The absence of the database search means the FILTER algorithm cannot be used and, since energy is thus the only available criterion on which to select a conformation, it is important that the other loops are present in order to calculate the energy. Thus CDR-H1 (and CDR-H3, which is also short) was constructed after, and in the presence of, CDR-L1, CDR-L2, CDR-L3, and CDR-H2. A repeat construction was performed in the absence of the other loops and, as expected, a worse conformation was selected (RMS: 3.21 Å for all atoms, 1.42 Å for N, C_α, C) compared with the conformation selected in the presence of the other loops (RMS: 2.18 Å for all atoms, 1.24 Å for N, C_α, C).

TABLE IV
LOW-ENERGY CONFORMATIONS FOR THE SIX CDRs OF GLOOP2 CONSTRUCTED ONTO THE EMPTY COMBINING SITE[a]

CDR	Conformation	Energy (kcal/mol)	RMS (Å)		FILTER score			Identifier
			All	B/B	15°	30°	45°	
L1	247*	5975.5	2.94	2.04	1.36(3)	2.60(6)	2.56(6)	1REI-A-0024
	319	5975.7	2.97	2.11	1.36(3)	2.22(5)	2.62(7)	1REI-A-0024
	320	5978.2	2.96	2.09	1.36(3)	2.51(6)	2.62(7)	1REI-A-0024
	241	5985.9	2.64	1.88	1.36(3)	2.60(6)	2.56(6)	1REI-A-0024
	253	5986.0	2.99	2.11	1.36(3)	2.60(6)	2.56(6)	1REI-A-0024
L2	43	5273.0	4.22	2.56	1.92(2)	2.97(2)	9.48(4)	1PYP-0-0109
	93*	5297.9	1.51	1.08	0.27(1)	9.25(5)	26.23(7)	2RHE-0-0051
	11	5300.6	1.86	1.33	3.52(5)	5.81(5)	21.58(7)	1FB4-L-0049
	109	5358.7	2.26	1.33	2.22(3)	5.08(4)	7.81(5)	2YHX-0-0383
	9	5366.7	2.06	1.33	6.64(1)	12.45(2)	26.06(6)	1CN1-A-0143
L3	98	5816.3	3.67	1.95	1.61(3)	2.82(4)	3.40(4)	1REI-A-0089
	84	5816.7	3.68	1.88	1.61(3)	2.82(4)	3.40(4)	1REI-A-0089
	107	5816.9	3.23	2.04	1.61(3)	2.82(4)	3.71(5)	1REI-A-0089
	82*	5818.4	2.82	1.99	1.61(3)	2.82(4)	3.87(6)	1REI-A-0089
	96	5818.9	3.70	1.94	1.61(3)	2.82(4)	3.40(4)	1REI-A-0089
H1	14*	6228.4	2.18	1.24	—	—	—	—
	12	6233.0	2.30	1.40	—	—	—	—
	16	6238.7	2.21	1.54	—	—	—	—
	4	6239.4	2.37	1.40	—	—	—	—
	11	6239.8	2.28	1.53	—	—	—	—
H2	240	5559.3	3.81	1.87	0.63(2)	2.28(5)	2.72(5)	1FB4-H-0050
	463*	5559.7	3.00	1.06	2.20(5)	3.42(6)	4.51(7)	1FBJ-H-0050
	1657	5566.7	2.50	1.10	0.96(3)	2.71(5)	4.10(6)	2HFL-H-0050
	471	5567.9	3.04	1.07	1.41(4)	2.84(6)	4.26(7)	1FBJ-H-0050
	722	5568.8	3.78	1.63	0.43(1)	1.47(3)	2.66(5)	1IG2-H-0050
H3	128*	7495.3	2.36	1.45	—	—	—	—
	125	7191.4	2.38	1.59	—	—	—	—
	41	7495.9	4.01	3.03	—	—	—	—
	120	7500.6	2.26	1.52	—	—	—	—
	138	7509.7	2.85	1.49	—	—	—	—

[a] FILTER scores show the normal score with the distribution (i.e., the number of residues over which the score is achieved) in parentheses. RMS deviations are calculated against the $P2_1$ crystal structure by least squares fitting over the framework and are quoted over the regions constructed. The identifier refers to the parent database loop from which the conformation was built. The first four characters are the PDB code, followed by the chain identifier (0 if no identifier) and the residue number of the first residue of the loop.

CDR-H2

CDR-H2, as defined by Wu and Kabat,[11] is 17 residues long (sequence: EIFPGNSKTYYAERFKG). This contrasts with the Chothia and Lesk definition[13] of just four residues at the top of the loop (sequence: PGNS). The latter definition would ignore residues such as Glu-50H, known to affect antigen binding (see above). The Wu and Kabat definition seems excessive since it includes, on the C-terminal side, a complete strand of β sheet and part of the loop at the back of the β barrel. The definition used here, and recommended for all future modeling exercises using the combined algorithm, consists of the Kabat and Wu loop, but with the C-terminal residue defined as the strand partner of the N-terminal residue of the loop. For Gloop2, CDR-H2 is thus defined as 10 residues (sequence: EIFPGNSKTY). Conformation 463 was selected using a FILTER cutoff of 15°. The all-atom RMS deviation is poor (3.00 Å), owing to rotations of the Phe at position 3 in the loop and Tyr at position 10 by approximately 120° when compared with the $P2_1$ crystal structure. Gloop2 has been solved in two separate crystal forms, $P2_1$ and $P1$.[68,69] When compared with the $P1$ structure, the side chains are placed almost perfectly and the all-atom RMS drops to 2.23 Å (global fit).

This concerted side-chain motion between crystal forms illustrates the effects of crystallization conditions on surface side-chain placement. Even though surface side chains may show relatively low temperature factors as a result of crystal packing interactions, their mobility in solution may be high. In the Gloop2 $P1$ structure, the mean side-chain temperature factor for the Fv domain is 13.46 ($\sigma = 8.20$) while the side chains of these two residues of H2 show mean temperature factors of 5.56 ($\sigma = 0.68$) for the Phe at position 3 and 710 ($\sigma = 1.73$) for the Tyr at position 10, suggesting they are relatively immobile.

CDR-H3

CDR-H3 is only four residues long. As was the case with CDR-H1, this short length precludes the use of the database. It was actually necessary to construct five residues rather than four, using CONGEN, as it was not possible to construct four residues by conformational search. Thus an extra residue was built at the N terminus of the loop. Once again, the FILTER algorithm could not be used and conformation 128 was selected on energy.

Summary

Each of the six CDRs of Gloop2 is shown with the modeled structure in Fig. 5.

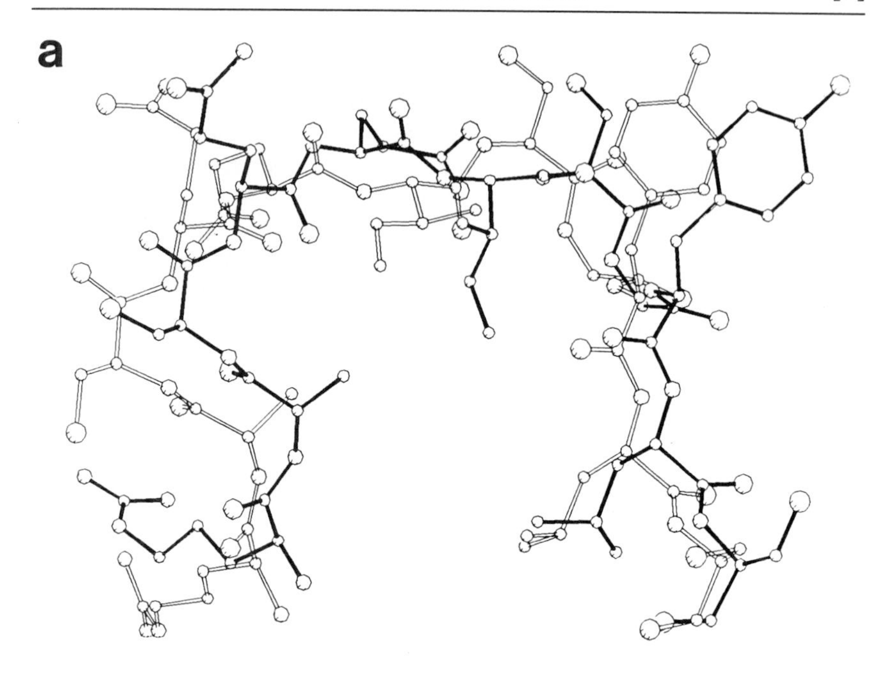

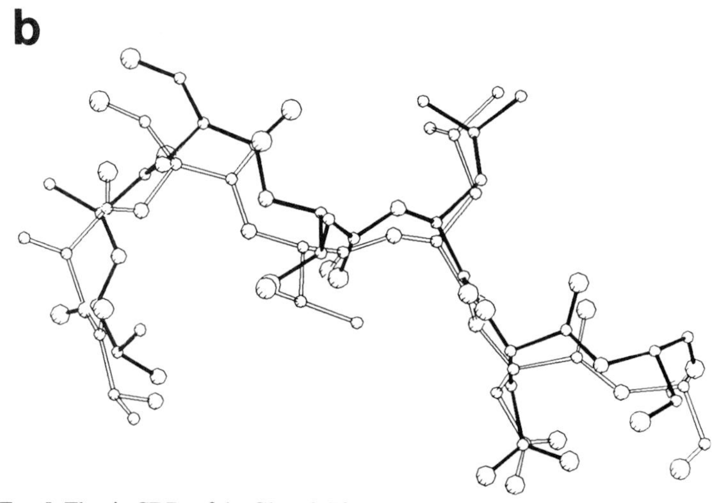

FIG. 5. The six CDRs of the Gloop2 $P2_1$ crystal structure are shown in filled lines with the modeled loops generated by the combined algorithm shown in open lines. All the fits are global (i.e., the framework regions are fitted rather than a local fit being performed on the CDRs themselves). (a) CDR-L1; (b) CDR-L2; (c) CDR-L3; (d) CDR-H1; (e) CDR-H2; (f) CDR-H3.

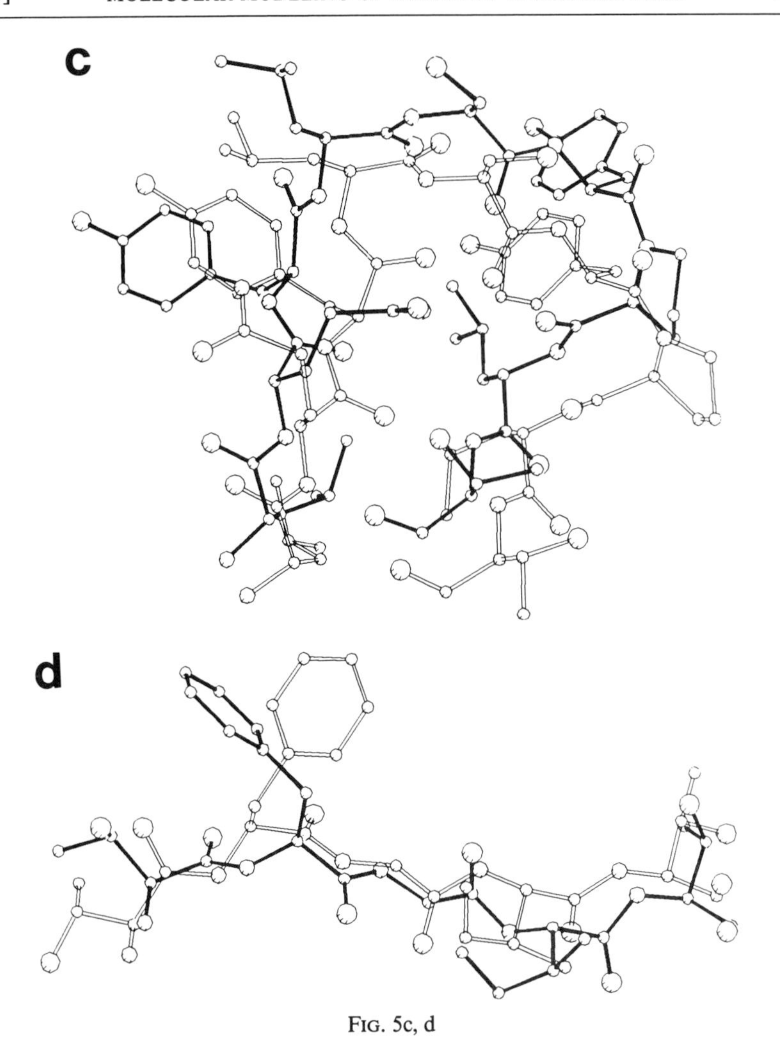

FIG. 5c, d

Overall, the results obtained using the combined algorithm are similar in accuracy to those achieved using the canonical method of Chothia *et al.* However, the canonical method is limited to those loops where the key residues identified by Chothia are present. With the number of antibody structures currently available, it is not possible to classify CDR-H3 into canonical ensembles. Additionally, a small percentage of examples in the remaining CDRs do not match the current canonical classifications and the protein engineer may well wish to mutate the key residues, precluding the use of Chothia's method for modeling the resulting conformation.

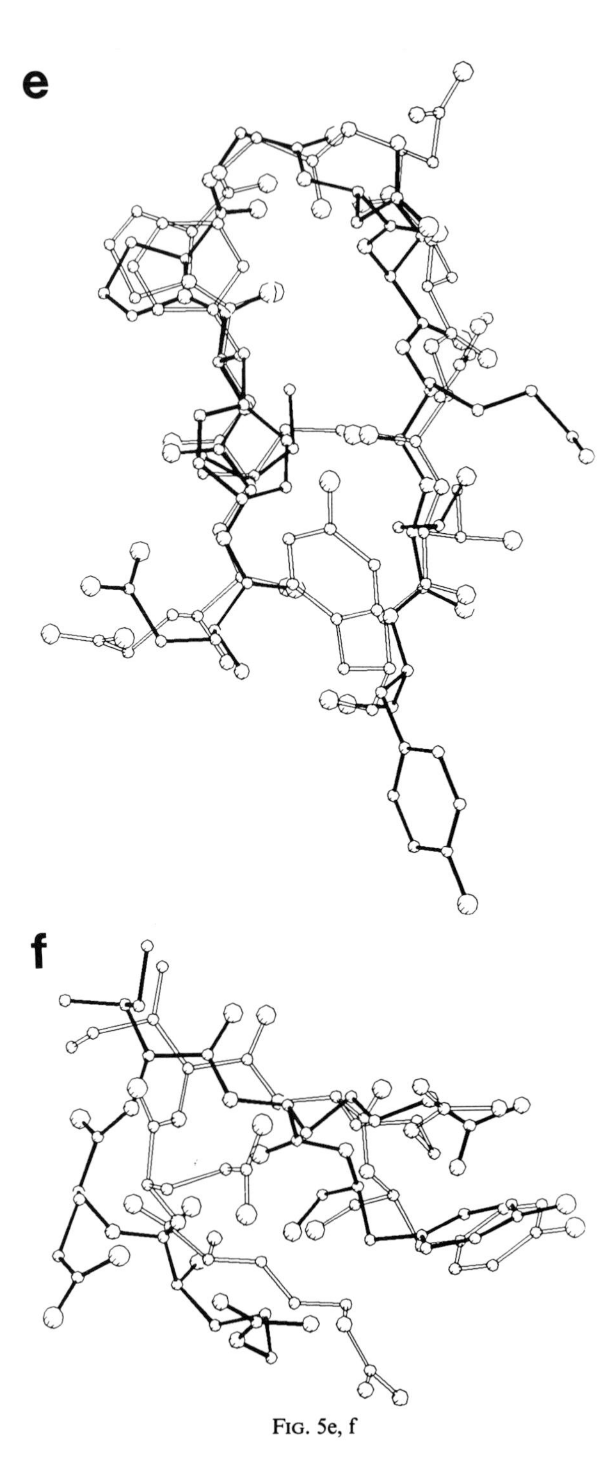

FIG. 5e, f

Thus the best approach appears to be to use Chothia's method (at least to model the backbone conformation) when the loop to be modeled is represented in the database of canonical structures. Any other loops, either unrepresented among the known canonicals (including CDR-H3), or where mutations have been made to the key residues, may then be modeled by the combined algorithm presented here.

Acknowledgments

The authors would like to thank Celltech for providing a V53000 on which the computations were performed, Phil Jeffrey and Steven Sheriff for providing coordinates of HyHEL-5 and Gloop2, and Bob Bruccoleri for providing CONGEN and its source code. We also thank SERC, UK, for financial support.

[7] X-Ray Crystallographic Analysis of Free and Antigen-Complexed Fab Fragments to Investigate Structural Basis of Immune Recognition

By IAN A. WILSON, JAMES M. RINI, DAVED H. FREMONT, GAIL G. FIESER, and ENRICO A. STURA

Introduction

The development of hybridoma technology[1] made it possible to produce unlimited numbers of antibodies of known specificity against any given antigen. This opportunity led to an upsurge of interest in obtaining detailed structural information concerning antibody–antigen recognition and has driven the present investigations of the structural basis of immune recognition. The structures of many monoclonal antibodies have now been solved both as free Fabs[1a] and as Fab–antigen complexes. These structure determinations, combined with four crystal structures of Fabs derived from multiple myeloma antibodies, have markedly increased our knowledge of antibody–antigen interactions.

The rapid increase in the number of free and antigen-bound antibody

[1] G. Kohler and C. Milstein, *Nature (London)* **256,** 495 (1975).

[1a] Throughout this review, the term Fab will frequently be used as the general terminology for Fab or Fab′ fragments of antibodies. In specific instances, where appropriate, Fab′ as distinct from Fab will be used.

structures being determined is also a reflection of improved methodologies for antibody production, crystallization, and Fab structure determination. Indeed, the present success rates in crystallizing Fabs and Fab–antigen complexes, in combination with improvements in methodologies for Fab structure determinations, have resulted in the structure solutions of many Fabs and Fab–antigen complexes in several different laboratories.

The single most significant advance in the structure determination of Fabs and Fab complexes has been the use of molecular replacement (MR). The main obstacle to the use of MR was thought to be the conformational flexibility of the Fab molecule. The variable and constant domains that constitute the intact Fab are connected by a flexible hinge or elbow. As a result, there are significant differences in the relative disposition of the variable and constant domains among different Fabs. In fact, it was only recently that Cygler and colleagues[2–4] showed that the rotation and translation problem could be solved by MR using the variable and constant domains as separate probes. This method has now been used successfully for many Fab structure determinations. In addition, rapid refinement methods, such as simulated annealing, have made it possible to take these solutions and proceed to a refined structure in a comparatively short time.[5–9] These methods, although of general applicability in protein crystallography, have now been tailored explicitly to the solution of antibody structures.

In this chapter, we will discuss the methods available for solving the three-dimensional structures of free and antigen-complexed antibody fragments by X-ray crystallography, focusing mainly on experiences within our own laboratory. The main aspects that will be addressed are antibody production and purification, crystallization, molecular replacement, and refinement. An analysis of Fab and Fab complexes solved by X-ray crystallography in the past 3 years will follow and show the impressive increase in our knowledge of the structural details of antibody–antigen interactions.

[2] M. Cygler and W. F. Anderson, *Acta Crystallogr., Sect. A* **44,** 38 (1988).
[3] M. Cygler and W. F. Anderson, *Acta Crystallogr., Sect. A* **44,** 300 (1988).
[4] M. Cygler, A. Boodhoo, J. S. Lee, and W. F. Anderson, *J. Biol. Chem.* **262,** 643 (1987).
[5] A. T. Brünger, *Acta Crystallogr., Sect. A* **46,** 46 (1990).
[6] A. T. Brünger, D. J. Leahy, and R. O. Fox, *J. Mol. Biol.* in press (1991).
[7] A. T. Brünger, J. Kurijan, and M. Karplus, *Science* **235,** 458 (1987).
[8] A. T. Brünger, "X-PLOR Manual, Version 1.5." Yale Univ., New Haven, Connecticut, 1988.
[9] A. T. Brünger, *Acta Crystallogr., Sect. A* **45,** 42 (1989).

Antibody Production

In the past decade, the overwhelming majority of antibodies for research use have been produced by hybridoma technology.[1] The isolation of immunoglobulins of known specificity from mouse ascites fluid has provided most of the antibodies used for crystallographic studies. Other cell culture systems, such as tissue culture suspensions and hollow fibers, have been used to produce large quantities of immunoglobulin required for crystallographic studies. Recombinant DNA technology has now led to cloned antibody fragments, such as Fabs[10] and Fvs,[11,12] which can be expressed in either *Escherichia coli* or mammalian cells. These methods of protein production have the distinct advantage that Fabs or Fvs of greater purity can be obtained without the difficulties associated with the isolation of antibody from ascites fluid and the subsequent cleavage to produce Fab or Fab′ fragments. The structure of such an Fv has been reported,[13] as has the V_L dimer of a recombinant Fv fragment of McPC603.[14]

Purification and Production of Fab and Fab′ Fragments

Structural studies of antibodies have concentrated mainly on Fab and Fab′ fragments because of the problems encountered in crystallizing intact immunoglobulins. These difficulties are presumably due to flexibility in the intact immunoglobulin, as observed in the crystal structure of the intact Kol antibody, where the Fc domain was found to be disordered.[15] The smaller size of the Fabs has the additional advantage that the effort required to solve these structures is reduced. Smaller fragments such as Fvs, which would reduce the crystallographic problem even further, have been more difficult to produce by proteolytic cleavage. The use of bioengineering to produce Fvs obviates such a problem.

Most of the Fabs studied to date have been made from murine monoclonal antibodies produced in ascitic tumors. The procedure by which these monoclonal antibodies are isolated, cleaved, and purified in our laboratory basically follows classical methods.[16] However, close communi-

[10] M. Better, C. P. Chang, R. R. Robinson, and A. H. Horwitz, *Science* **240,** 1038 (1988).

[11] A. Skerra and A. Plückthun, *Science* **240,** 1038 (1988).

[12] L. Riechmann, J. Foote, and G. Winter, *J. Mol. Biol.* **203,** 825 (1988).

[13] T. N. Bhat, G. A. Bentley, T. O. Fischmann, G. Boulot, and R. J. Poljak, *Nature (London)* **347,** 483 (1990).

[14] R. Glockshuber, B. Steipe, R. Huber, and A. Plückthun, *J. Mol. Biol.* **213,** 613 (1990).

[15] M. Matsushima, M. Marquart, T. A. Jones, P. M. Colman, K. Bartels, R. Huber, and W. Palm, *J. Mol. Biol.* **121,** 441 (1978).

[16] P. Parham, *J. Immunol.* **131,** 2895 (1983).

cation between the biochemist and the crystallographer significantly enhances the likelihood of obtaining good quality Fab crystals.

Purification and Cleavage of Immunoglobulin. At present IgG is most commonly purified from mouse ascites fluid. Ammonium sulfate is added to the ascites fluid to a final concentration of 20% to precipitate extraneous proteins and other unwanted components. After centrifugation, the supernatant is brought to a final concentration of 50% saturated ammonium sulfate. The precipitate is then resuspended in phosphate-buffered saline and dialyzed overnight against the same buffer at pH 7. Mouse IgG_1 is typically cleaved with pepsin to yield $F(ab')_2$, while IgG_{2a} and IgG_{2b} are usually cleaved by papain to yield Fab. In practice both proteases are tried at several concentrations and several digestion times and the best combination of conditions is selected on the basis of the analysis of the products by sodium dodecyl sulfate polyacrylamide gel electrophoresis (SDS-PAGE). Reduction of the $F(ab')_2$ fragments is done with ~10 m*M* cysteine at 25°, followed by acetylation with iodoacetamide in the absence of light. Methods for Fab and Fab′ production are presented elsewhere in more detail in several publications that describe the different susceptibility of mouse, rat, and rabbit antibodies to proteolytic cleavage.[16-18]

Purification of Fab and Fab′ Fragments. The cleavage and reduction of immunoglobulins does not usually yield a homogeneous product. Uncleaved immunoglobulin, overreduction, and multiple cleavages can produce extremely heterogeneous protein samples. The digested material is typically purified by size-exclusion chromatography in 0.1 *M* sodium acetate, pH 5.5. The fractions recovered are analyzed by absorbance measurements at 280 nm[19] and checked for purity by SDS-PAGE and isoelectric focusing gel electrophoresis (IEF). The fractions containing relatively pure Fab or Fab′ are pooled and concentrated. A preliminary crystallization screening is often performed at this stage regardless of how heterogeneous the sample may still be. Ion-exchange chromatography on Mono S or Mono Q columns (Pharmacia-LKB, Piscataway, NJ) is then performed to further purify the Fabs. Samples are eluted using a salt gradient, such as 0–1 *M* ammonium sulfate or sodium chloride, in buffers ranging from pH 5.0 to 8.5. This separation method yields distinct fractions that are again analyzed by SDS-PAGE and IEF. These fractions are then pooled, concentrated, and analyzed for purity and reactivity.

We have also used affinity columns to try to purify the Fab. However, harsh conditions are often necessary to remove the Fab from the affinity

[17] J. Rousseaux, G. Biserte, and H. Bazin, *Mol. Immunol.* **17,** 469 (1980).

[18] R. R. Porter, *J. Immunol.* **75,** 119 (1959).

[19] R. Arnon, this series, Vol. 19, p. 226.

column. Chromatofocusing is also useful for Fab purification, but we have found that different preparations show substantial variation in solubility, possibly due to the retention of varying amounts of ampholines. However, chromatofocusing has been successfully used by others.[20,21]

Crystallization of Fabs and Fab–Antigen Complexes

As previously stated, Fab and Fab′ fragments produced by proteolytic cleavage of immunoglobulins are in most cases microheterogeneous. In some cases, additional heterogeneity may result from glycosylation of the Fab domain. The degree of microheterogeneity has been found to greatly influence the ability of Fabs to crystallize. Each of these Fab components often has a different potential for crystallization. The use of seeding techniques has been very successful at this stage of the crystallization investigations and has led to the production of many large, single crystals of Fabs and Fab–antigen complexes.[22,23] For instance, Fab protein fractions that after purification fail to nucleate spontaneously can often be successfully seeded from other fractions of the same preparation which have nucleated. Currently, with preparations that do not yield crystals spontaneously in any of the purified fractions, we are trying cross-seeding with any of the 30 Fab–antigen complex crystals and 20 native Fabs that have been crystallized so far in our laboratory. To date, this approach has been successful in cross-seeding one anti-peptide Fab with an Fab from a different antibody raised against the same peptide.[22–24]

Screening for Crystallization Conditions. Crystallization of Fabs is carried out using microvapor diffusion methods and other standard crystallization techniques (reviewed by McPherson[25]). In our laboratory, we routinely use multiwell sitting drop vapor diffusion plates in a constant-temperature environment.[22–24,26] The initial search for crystallization conditions is now standardized and serves to characterize the solubility of each Fab with different precipitants (footprint). The conditions scanned in this

[20] D. R. Rose, M. Cygler, R. J. To, M. Przybylska, B. Sinnott, and D. R. Bundle, *J. Mol. Biol.* **215,** 489 (1990).

[21] A. L. Gibson, J. N. Herron, X.-M. He, V. A. Patrick, M. L. Matson, J.-N. Lin, D. M. Kranz, E. W. Voss, Jr., and A. B. Edmundson, *Proteins: Struct. Funct. Genet.* **3,** 155 (1988).

[22] E. A. Stura and I. A. Wilson, *Methods* **1,** 38 (1990).

[23] E. A. Stura and I. A. Wilson, *J. Cryst. Growth* **110,** 270 (1991).

[24] E. A. Stura and I. A. Wilson, *in* "Protein Crystallization. A Practical Approach" (A. Ducruix and R. Geige, eds.), in press. IRL Press, London and Washington, D.C., 1991.

[25] A. McPherson, "Preparation and Analysis of Protein Crystals." Wiley, New York, 1982.

[26] E. A. Stura, D. L. Johnson, J. Inglese, J. M. Smith, S. J. Benkovic, and I. A. Wilson, *J. Biol. Chem.* **264,** 9703 (1989).

standardized screen include ammonium sulfate, sodium and potassium phosphate, sodium citrate, and polyethylene glycols of M_r 600, 4000, and 10,000. Only the pH values 5.5, 7.0, and 8.5 are tested at this stage. The simplicity of the footprint enables a rapid screening of all the different combinations of purified Fab fractions with all the available antigens. In fact, for any given Fab, the various fractions obtained in the Fab purification steps are tested separately in order to decide which is the most promising fraction for crystallization. By this we mean conditions that yield microcrystalline precipitates or polycrystalline growth and certain types of granular precipitates. The same footprint screen is also used for Fab–antigen complex crystallization. Control trays are set up in parallel for the free antigen and the free Fab to ascertain whether the Fab or the antigen crystallizes under similar conditions to those of the complex.

From our analysis of many Fabs, we have found that the solubility of these molecules varies widely, and that crystallization occurs over a wide range of conditions. An exception to this occurs for several anti-hemagglutinin peptide Fabs that, when complexed with peptide antigen, form crystals or microcrystals under similar conditions (15–24% polyethylene glycol 10,000, 0.2 *M* imidazole malate, pH 8.5).[22–24,27,28]

Streak Seeding and Cross-Seeding. Conditions that produce small crystals or polycrystals of Fabs can often be found quickly. Seeding has been extremely effective in utilizing this crystalline material as a source of seeds to search for better conditions under which large, single Fab and Fab–complex crystals can be obtained. Streak seeding is the method of choice in our laboratory for this search because of its simplicity.[22–26] Briefly, the probe to be used for streak seeding is made by mounting an animal whisker onto the end of a small, thick-walled capillary with wax. This whisker is used to touch an existing crystal or microcrystalline precipitate and dislodge seeds from it. Some seeds remain attached to the whisker and are introduced into a preequilibrated protein-precipitant drop. Since the seeds are deposited in a straight line, it is easy to determine whether subsequent crystalline growth has resulted from the seeding procedure. The use of this method has also been valuable for finding conditions that support and enhance single crystal growth but which may not permit spontaneous nucleation. A more extensive discussion of this method and its various applications can be found in Stura and Wilson.[22–24]

Improving Crystal Quality. The use of streak seeding in the optimization of the crystallization conditions also provides information on the

[27] U. Schulze-Gahmen, J. M. Rini, J. H. Arevalo, E. A. Stura, J. H. Kenten, and I. A. Wilson, *J. Biol. Chem.* **263,** 17100 (1988).

[28] E. A. Stura, G. G. Fieser, and I. A. Wilson, in preparation (1991).

degree of supersaturation necessary for the implementation of more conventional seeding methods, such as macroseeding and microseeding. In fact, the use of macroseeding and microseeding techniques has been essential in the production of X-ray-quality crystals of Fab–antigen complexes of three different anti-hemagglutinin peptide Fabs.[22,23,27,28]

Other factors, such as the use of additives, have been important both in the initial crystallization of Fabs and in increasing the size and quality of existing crystals or polycrystals. For example, very small amounts of high-molecular-weight polyethylene glycols and ethanol have been added to the precipitant, ammonium sulfate, in the crystallization of an anti-progesterone Fab′ DB3.[29,30] Methylpentanediol (1%) was found to be essential for the nucleation of crystals of the native anti-peptide Fab′ B13I2.[31] Small organic molecules, divalent ions, and diamines, together with various anions such as acetate, formate, malate, phosphate, and others, have also been effective in the improvement of Fab crystal size and quality.[29–31]

Analysis of Crystals. SDS-PAGE can be used to determine whether crystals of a putative complex contain both the antibody and the antigen. However, this technique is not applicable when the antigen is small, such as for a peptide or hapten. In this instance, native-PAGE[32,33] can be used to evaluate whether the crystal is a complex or only the free Fab. A PhastGel analysis (8–25%, w/v, Pharmacia, Piscataway, NJ) is carried out on the solutions of native and complexed Fab and on dissolved crystals of the putative complex. The bound ligand usually changes the mobility of the Fab and can be observed as a shift in the position and distribution of the bands in the native gel (see Fig. 1).

Results of Crystallization of Fabs and Fab–Antigen Complexes. Using the combination of the methods described above, we have achieved a high success rate in the crystallization of Fabs and Fab–antigen complexes[22–24,26–31] In some cases, we have obtained Fab crystals in more than one form, as for the anti-HIV-1 peptide Fab 50.1,[34–36] where we have obtained

[29] E. A. Stura, A. Feinstein, and I. A. Wilson, *J. Mol. Biol.* **193,** 229 (1987).

[30] E. A. Stura, J. H. Arevalo, A. Feinstein, R. B. Heap, M. J. Taussig, and I. A. Wilson, *Immunology* **62,** 511 (1987).

[31] E. A. Stura, R. L. Stanfield, T. M. Fieser, R. S. Balderas, L. R. Smith, R. A. Lerner, and I. A. Wilson, *J. Biol. Chem.* **264,** 15721 (1989).

[32] A. T. Andrews, *in* "Electrophoresis. Theory, Techniques and Biochemical and Clinical Applications" (A. R. Peacocke and W. F. Harrington, eds.), p. 63. Oxford Univ. Press (Clarendon), Oxford, 1981.

[33] B. D. Hames, *in* "Gel Electrophoresis. A Practical Approach" (B. D. Hames and D. Rickwood, eds.), p. 1. IRL Press, London, 1981.

[34] K. Javaherian, A. J. Langlois, C. McDanal, K. L. Ross, L. I. Echler, C. L. Jellis, A. T. Profy, J. R. Rusche, D. P. Bolognesi, S. D. Putney, and T. J. Matthews, *Proc. Natl. Acad. Sci. U.S.A.* **86,** 6768 (1989).

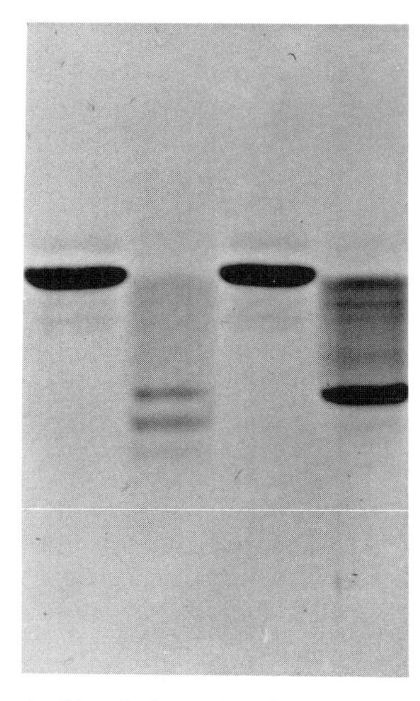

FIG. 1. Native polyacrylamide gel electrophoresis (8–25%) at pH 8.8 comparing the mobility of free and peptide bound Fabs with crystals of the putative complex. Samples are native Fab 17/9 (lane A), washed and dissolved crystals of the complex of Fab 17/9 with peptide D30 (sequence YDVPDYASL; lane B), and Fab 17/9 and peptide D30 in solution mixed in a 1:8 (Fab:peptide) molar ratio (lane C). Consistent with the two additional net negative charges of the D30 peptide, the mobility of the Fab–peptide complex is increased with respect to the free Fab. The double band observed for the Fab–peptide complex might indicate that the protein is modified during the crystallization procedure.

native crystals is the space groups, $P1$, $P2_1$, $P2_12_12_1$, and $I222$ or $I2_12_12_1$. Initial crystals or crystalline material suitable for seeding are obtained for most Fabs within the first footprint analysis as described above, or in subsequent experiments where the search is extended to scan a wider pH range with finer precipitant concentrations steps.

Native Fabs and their respective Fab–antigen complexes often show

[35] S. D. Putney, T. J. Matthews, W. G. Robey, D. L. Lynn, M. Robert-Guroff, W. T. Mueller, A. J. Langlois, J. Ghrayeb, S. R. Petteway, Jr., K. J. Weinhold, P. J. Fischinger, F. Wong-Staal, R. C. Gallo, and D. P. Bolognesi, *Science* **234,** 1392 (1986).

[36] S. Matsushita, M. Robert-Guroff, J. R. Rusche, A. Koito, T. Hattori, H. Hoshino, K. Javaherian, K. Takatsuki, and S. D. Putney, *J. Virol.* **62,** 2107 (1988).

marked differences in their solubility even with relatively small ligands. For example, the peptide complexes of five different anti-hemagglutinin peptide Fabs show differential solubility in polyethylene glycol (PEG) relative to the free Fab. Fabs from clones 26/9, 21/8, 126D4, and 12CA5 generally are more soluble in PEG as Fab–peptide complexes compared to the free Fab. The situation is reversed for the 17/9 crystallization, which requires only 7–12% polyethylene glycol to crystallize the Fab–peptide complex while 30–39% PEG 600 is required for crystallization of the free Fab.[24,27]

In conclusion, the low yields of Fab obtained from proteolytic cleavage of immunoglobulins and the subsequent difficulties in the purification steps due to the heterogeneity of the cleaved Fabs seem to be the major obstacles that prevent a higher success rate in producing X-ray-quality single crystals. Indeed, Fab crystals do not often diffract to particularly high resolution and most structure determinations reported so far are in the 2.5- to 3.2-Å resolution range. It is unclear whether, in those cases, the heterogeneity of the Fab sample is the problem or if the structural flexibility of the Fab molecule itself contributes to poor crystal quality.

Molecular Replacement

Structure determinations by molecular replacement can be much less labor intensive and time consuming than those using heavy atom methods. Consequently, attempts to solve Fab structures have focused on this approach. As might be expected, successful application of this technique depends critically on the structural similarity between the available molecular replacement models and the unknown structure. In fact, relatively subtle structural differences between the molecular replacement model and the unknown structure can result in uninterpretable rotation and translation functions. Nevertheless, various Fabs and Fab–antigen complexes have now been solved by molecular replacement even though Fabs show considerable variation in their relative domain dispositions and loop structures. Among these are several liganded Fabs,[4,37–39] three Fab–hapten complexes,[6,40,41] a Fab–carbohydrate complex,[20] a Fab–peptide com-

[37] J. Vitali, W. W. Young, V. B. Schatz, S. E. Sobottka, and R. H. Kretsinger, *J. Mol. Biol.* **198,** 351 (1987).

[38] L. Prasad, M. Vandonselaar, J. S. Lee, and L. T. J. Delbaere, *J. Biol. Chem.* **263,** 2571 (1988).

[39] R. L. Stanfield, T. M. Fieser, R. A. Lerner, and I. A. Wilson, *Science* **248,** 712 (1990).

[40] M.-B. Lascombe, P. M. Alzari, G. Boulot, P. Saludjian, P. Tougard, C. Berek, S. Haba, E. M. Rosen, A. Nisonoff, and R. J. Poljak, *Proc. Natl. Acad. Sci. U.S.A.* **86,** 607 (1989).

[41] J. N. Herron, X. He, M. L. Mason, E. W. Voss, and A. B. Edmundson, *Proteins: Struct. Funct. Genet.* **5,** 271 (1989).

plex,[39] two Fab–lysozyme complexes[42–44] and an Fv structure both unliganded and in complex with lysozyme.[13] In addition, a combination of multiple isomorphous replacement and molecular replacement phases was used in the solution of an anti-phenyl arsenate Fab,[45] two Fab–neuraminidase complexes,[46,47] and an idiotope Fab–anti-idiotope Fab complex.[48]

Despite these successes, phase determination by this technique has not been altogether a routine procedure, as discussed above. However, the increased database of Fab structures combined with the development of new techniques, such as intensity-based domain refinement of molecular replacement models,[5,49] show that molecular replacement is the most facile method of determining the initial set of phases for the Fab structure determination.

Given the success in obtaining crystals of Fab–antigen complexes and of solving Fab structures by molecular replacement, Fabs are now being used to obtain crystalline complexes of molecules of significant biological interest when attempts to obtain crystals of the molecule by itself have failed. For example, both the Sendai virus HN (hemagglutinin-neuramindase) protein[50] and the rat CD4 molecule[51] failed to crystallize alone but formed crystals when complexed with Fabs. If the antigen in the complex is relatively small, molecular replacement phases calculated from the Fab portion alone may result in an electron density map in which the antigen can be traced without the use of heavy atom phase information. In cases where the molecular replacement phases do not produce an interpretable electron density map, they can be combined with phase constraints from a heavy atom or anomalous scatterer to obtain a solution to the structure.

[42] S. Sheriff, E. W. Silverton, E. A. Padlan, G. H. Cohen, S. J. Smith-Gill, B. C. Finzel, and D. R. Davies, *Proc. Natl. Acad. Sci. U.S.A.* **84,** 8075 (1987).

[43] S. Sheriff, E. A. Padlan, G. H. Cohen, and D. R. Davies, *Acta Crystallogr., Sect. B* **46,** 418 (1990).

[44] E. A. Padlan, E. W. Silverton, S. Sheriff, G. H. Cohen, S. J. Smith-Gill, and D. R. Davies, *Proc. Natl. Acad. Sci. U.S.A.* **86,** 5938 (1989).

[45] D. R. Rose, R. K. Strong, M. N. Margolies, M. L. Gefter, and G. A. Petsko, *Proc. Natl. Acad. Sci. U.S.A.* **87,** 338 (1990).

[46] P. M. Colman, W. G. Laver, J. N. Varghese, A. T. Baker, P. A. Tulloch, G. M. Air, and R. G. Webster, *Nature (London)* **326,** 358 (1987).

[47] P. M. Colman, W. R. Tulip, J. N. Varghese, P. A. Tulloch, A. T. Baker, W. G. Laver, G. M. Air, and R. G. Webster, *Philos. Trans. R. Soc. London B* **323,** 511 (1989).

[48] G. A. Bentley, G. Boulot, M.-M. Riottot, and R. J. Poljak, *Nature (London)* **348,** 254 (1990).

[49] T. O. Yeates and J. M. Rini, *Acta Crystallogr., Sect. A* **46,** 352 (1990).

[50] W. G. Laver, S. D. Thompson, K. G. Murti, and A. Portner, *Virology* **171,** 291 (1989).

[51] S. J. Davis, R. L. Brady, A. N. Barclay, K. Harlos, G. G. Dodson, and A. F. Williams, *J. Mol. Biol.* **213,** 7 (1990).

The Model. Analysis of the available structures shows that, although the overall protein fold is highly conserved, Fabs display considerable structural variation. Most prominent are the differences in the elbow angles relating the constant and variable domains, which range from approximately 130 to 180°. Within each variable and constant domain structural variation is also found. Significant variation in the pseudo twofold axes relating the heavy and light chains of the variable and constant domains is observed (up to 12° around and 3 Å along the rotation axis). Prior to running the rotation function, sequence similarity is often a good indicator of the likely usefulness of a given model. However, this is not always the case. In fact, as will be discussed below, rotation function results using the individual V_L, V_H, C_H1, and C_L subdomains suggest that structural differences within subdomains themselves are probably a determining factor in the success of a given model. Therefore, we have relied on rotation function analyses to determine which model gives the best results, and based on this information assemble the most suitable "hybrid" model for the translation search.

Rotation Searches. Our current approach to solving the rotation problem is to perform the rotation search with each of the available Fab models in turn. Although the variable and constant domains have traditionally been treated as separate models, due to the large variation in elbow angles, we now routinely use intact Fabs as well. The intact Fabs currently available span a wide range of elbow angles. In general, we have found that a comparison of all models usually leads to a consensus on the correct orientation for either the variable or constant domain, if not both. In those cases where only one domain can be oriented with confidence, the use of intact Fabs as models has allowed us to place limits on the orientation of the second domain.

Analysis of the rotation function results have been facilitated by superimposing the variable and constant domains from each Fab model onto the corresponding domains of an arbitrarily chosen intact Fab reference model. Since the relative domain dispositions reflect that of an intact Fab, the rotation required to orient the variable domain model in the unknown cell will be similar to that required to orient the constant domain. The rotation angles will differ by an amount that depends primarily on the difference in elbow angle between the reference model and the unknown structure. Furthermore, the elbow angle of the reference model can be oriented so that this difference is manifested in only one rotation function angle.[43] When intact Fabs are used, we superimpose all models on conserved residues of the variable domain.

All rotation function analyses performed in our laboratory have been done using the Crowther fast rotation function[52] as implemented in the

MERLOT package.[53] The results of a typical set of rotation function searches for an anti-peptide Fab′ B13A2, solved in our laboratory, are shown graphically in Fig. 2. In this case, the variable domain models give much cleaner solutions than those found for the constant domains. We have found this to be true in other instances. Since the sequence similarities among constant domains are much higher than those for the variable domains, this is somewhat surprising. These results presumably reflect either underlying structural differences among constant domains, or a greater likelihood that the constant domain or parts of it are somewhat disordered in these crystals.

Figure 2 also shows that only those intact Fab models with elbow angles close to that of the unknown structure confirm the variable domain solution. Even Fabs whose variable domains alone give reasonable solutions yield uninterpretable results with intact Fab models when the elbow angle differs significantly from that of the unknown. Because of this correlation, very tight limits on the possible orientation of the constant domain are established. In cases where only one domain can be oriented with confidence, this result is particularly helpful. As will be shown below, the new methods for intensity-based domain refinement are ideally suited to optimizing such intact Fab models. Both the orientation and relative translational disposition of the domains within the model can then be refined prior to translation function analysis.

Another example of a molecular replacement analysis is illustrated by the solution of the B13I2 Fab′–peptide complex. As in the previous example, the results showed that the variable domain search models gave much cleaner results than those obtained with the constant domain models. Again, only those intact Fab search models with elbow angles close to that of the unknown yielded interpretable results. Rotation function results with the separate V_L, V_H, C_L, and C_H1 subdomains, which represent only one-quarter of the Fab structure, are not unlike those found for Fab HED10 by Cygler and Anderson.[2] Although these searches were much noisier, solutions can be found for these subdomains that are close to, if not at the top of, the peak list for a given rotation function search. More importantly, the results show that a given subdomain may serve as a good model for determining the position of either the variable or constant domain when used alone, but when in association with the other subdo-

[52] R. A. Crowther, *in* "The Molecular Replacement Method" (M. G. Rossmann, ed.), p. 173. Gordon & Breach, New York, 1972.

[53] P. D. M. Fitzgerald, *J. Appl. Crystallogr.* **21,** 273 (1988).

[53a] F. C. Bernstein, T. F. Koetzle, G. J. B. William, E. F. Meyer, M. D. Brice, J. R. Rodgers, O. Kennard, T. Shimanouchi, and M. Tasumi, *J. Mol. Biol.* **112,** 535 (1977).

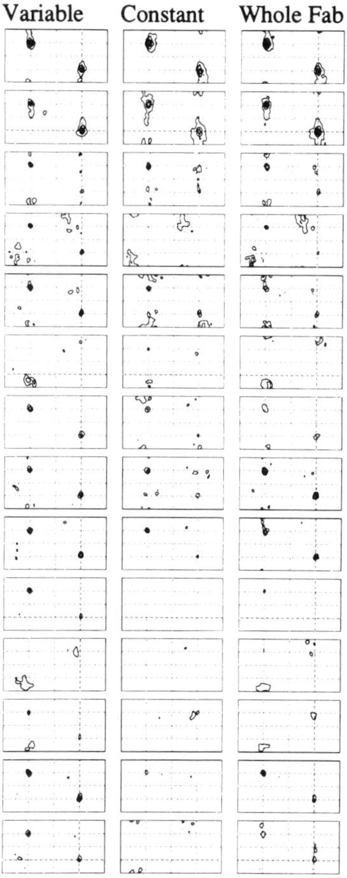

Variable Constant Whole Fab

Antibody	Seq L	Seq H	Elbow
B13A2_n1	100%	100%	145°
B13A2_n2	100%	100%	146°
B13I2_n1	83%	82%	155°
B13I2_p	83%	82%	157°
F17/9_n1	80%	78%	161°
DB3_n	85%	72%	179°
pdbFFC	84%	73%	175°
pdbHFM	82%	72%	146°
pdbHFL	79%	72%	163°
pdbFDL	74%	72%	170°
pdbF19	77%	63%	179°
pdbFBJ	78%	53%	144°
pdbMCP	82%	48%	133°
pdbFB4	42%	65%	167°

FIG. 2. Contour plots representing the results of the cross-rotation function analysis of Fab′ B13A2 (space group $P2_1$, $a = 42.2$ Å, $b = 123.8$ Å, $c = 100.0$ Å, $\beta = 96.8°$, with two molecules per asymmetric unit). Coordinates of B13A2, B13I2, F17/9, and DB3 are all from our laboratory. Structures preceded by "pdb" are the most recently deposited coordinates from the Brookhaven Protein Databank.[53a] The Crowther fast rotation functions[52] were calculated with the program suite MERLOT[53] using 4- to 8-Å data, a Patterson cutoff radius of 23 Å, and a uniform B factor of 25 Å^2. The orientation of the probe models was chosen such that both molecules in the asymmetric unit could be visualized in the same β section (α runs from 0 to 180°, γ from 0 to 360°, and $\beta = 90°$). Variable (V_L–V_H) and constant (C_L–C_H1) domains were independently superimposed onto the partially refined coordinates of the B13A2 Fab′ molecule 1. The intact Fab probes used were superimposed on their variable domains alone. The sequence identity is based on amino acid differences of the light and heavy chains between B13A2 and the probe Fabs. The elbow angle is defined between the pseudo-twofold axes of symmetry of the variable and constant domain pairs. The translation function for B13A2 was solved with the intact coordinates of Fab HFM (HyHEL-10[44]) and was primarily facilitated by the similarity of the elbow angles between the two structures; "n" refers to native and "p" to peptide–fab complex.

main yields uninterpretable results. Again, by comparing the results from the entire series of models, a consensus for the correct solution can be obtained. The individual domains giving the highest correlations might also be used, in conjunction with intensity-based domain refinement, to assemble superior models for the translation function analysis.

Refining Rotational Orientation. Since a successful solution to the translation function is very dependent on an accurately oriented model, it is desirable to optimize the orientation of the model before proceeding to the translation search. The Lattman rotation function,[54] provided for in the MERLOT package, and the program BRUTE[55] have been widely used for this purpose. However, procedures for performing intensity-based domain refinement[5,49] have been developed, and these techniques will likely become the method of choice for optimizing the orientation of the molecular replacement model (see below).

Translation Searches. Although we have successfully used the Crowther and Blow[56] translation function (as implemented in the MERLOT package), the translation searches based on the correlation between observed and calculated intensities seem to give cleaner results. The BRUTE and X-PLOR[8] correlation searches are similar and both allow for the inclusion of a fragment whose position in the unit cell is known. This option is particularly useful when there is more than one molecule in the asymmetric unit. This facility has also been utilized in determining the relative position of the variable and constant domains along the unique axis in polar space groups.

We routinely use data in the 8.0- to 4.0-Å resolution range and find that the best results are obtained with at least a 0.5-Å grid. With a moderately sized orthorhombic cell, the X-PLOR search can take 10–20 hr of central processing unit (CPU) time on a CONVEX C2 computer. In contrast, a program that calculates the pseudocorrelation between observed and calculated intensities defined by Harada *et al.*[57] (as implemented by Dr. D. Filman in our laboratory), executes in approximately 30 min on a similar problem. The results are comparable with those obtained using either X-PLOR or BRUTE.

Intensity-Based Domain Refinement. With multidomain proteins such as the Fabs, differences in the relative domain dispositions between the model and unknown structure make it impossible to accurately orient the intact model with respect to all domains. In an attempt to address this problem, Yeates and Rini[49] have developed a procedure (INTREF) that

[54] E. E. Lattman and W. E. Love, *Acta Crystallogr., Sect. B* **26,** 1854 (1970).
[55] M. Fujinaga and R. Read, *J. Appl. Crystallogr.* **20,** 517 (1987).
[56] R. A. Crowther and D. M. Blow, *Acta Crystallogr., Sect. A* **44,** 554 (1967).
[57] Y. Harada, A. Lifchitz, and J. Berthou, *Acta Crystallogr., Sect. A* **37,** 398 (1981).

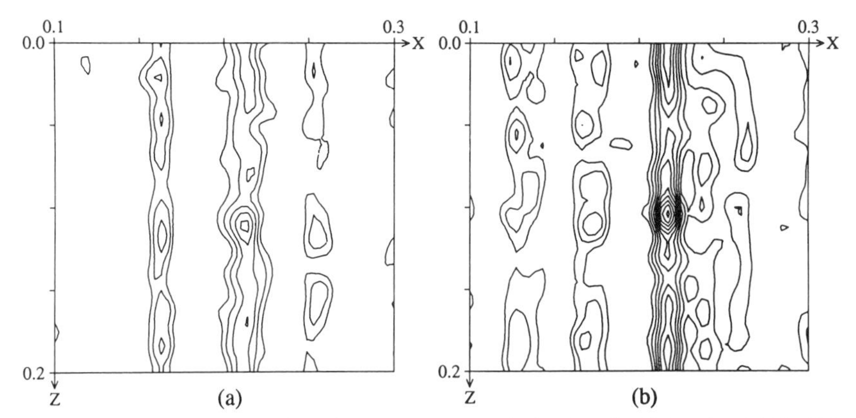

FIG. 3. Partial y section of the full-symmetry translation function[57] for the Fab′ B13I2–peptide complex (taken from Ref. 49). (a) Kol variable domain model oriented by the fast rotation function. (b) Variable domain model obtained after refining the orientation and relative dispositions of the V_H, V_L, C_H1, and C_L subdomains in INTREF.

can not only refine the orientation of individual domains but also their relative translational disposition as well. The model is refined at the rotation function stage and serves to produce models for translation function analysis, which are more complete and structurally more accurate. The power of this method is illustrated in the solution of a Fab–peptide complex of an antibody (50.1; see Refs. 34–36) raised against a peptide from gp120 of the human immunodeficiency virus type 1 (HIV-1) virus (see below). Brünger[5] has taken a similar approach and employed such rigid body refinement to screen the top peaks in the rotation function. He has used this method in the solution of both an anti-digoxin Fab complex[58] and an anti-dinitrophenyl spin-label Fab–hapten complex.[6]

Using INTREF refinement we initially treat the variable and constant domains as individual "groups" so that interactions between domains are not included in the minimization. Therefore, the orientation of each domain is refined independently. Subsequent refinement defining two "parts" for each group is then performed. In this way the rotations and relative translations between the light and heavy chains are refined. The results of the optimization are reflected in the improved signal-to-noise ratios and the accuracy of the translation function solutions obtained with INTREF-refined models. Such results are graphically illustrated by the work of Yeates and Rini,[49] using the solved structure of the Fab′ B13I2–peptide complex[39] as a test case. Figure 3, taken from their paper, shows a

[58] A. T. Brünger, *Acta Crystallogr. Sect. A* **47**, 195 (1991).

partial y section of the translation function map containing the correct solution for the variable domain model before and after INTREF refinement. Prior to refinement, the highest peak in the map (which is the correct peak), was 6.9σ over the mean and 0.9σ greater than the next highest peak. After refinement, the peak was 10.4σ over the mean and 3.0σ greater than the next highest peak. Such improvement in the signal-to-noise ratio allows the correct solution to the translation function to be identified with much greater confidence. The corresponding results for the constant domain search are shown in Fig. 4. Although an interpretable translation function solution could not be found with the unrefined constant domain model (Fig. 4a shows the y section where peak is expected), after INTREF refinement the highest peak in the map is now the correct solution (Fig. 4b). The peak height was 5.1σ over the mean and 0.5σ greater than the next highest peak. To illustrate the gains that can be made by assembling more complete models with INTREF, the relative orientation and translational disposition of the variable and constant domains were refined. As shown in Fig. 5, the resulting model gave a very clean solution in the translation search with a peak height 14.7σ over the mean and 4.9σ over the next highest peak.

The 50.1–peptide complex solved in our laboratory further illustrates the power of this technique. The molecule crystallizes in space group $P2_1$ with two molecules in the asymmetric unit. Although clean rotation function solutions were obtained for both variable domains, interpretable re-

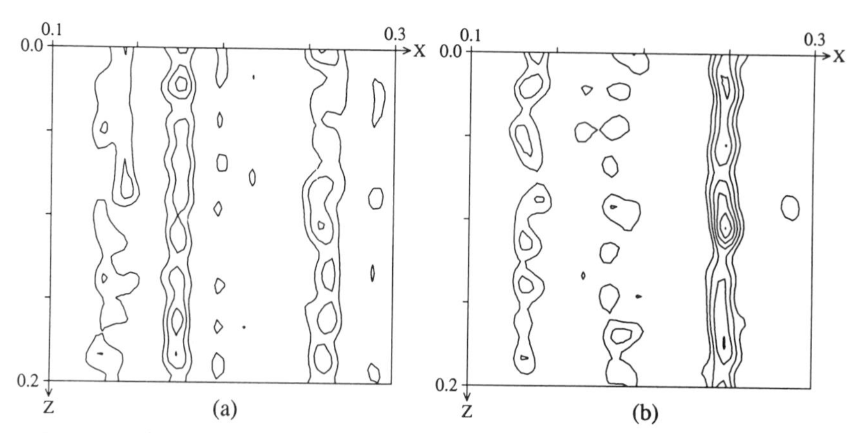

FIG. 4. Partial y section of the full-symmetry translation function[57] for the Fab′ B13I2–peptide complex (taken from Ref. 49). (a) Fab 17/9 constant domain oriented by the fast rotation function. (b) Constant domain model obtained after refining the orientation and relative dispositions of the V_H, V_L, C_H1, and C_L subdomains in INTREF.

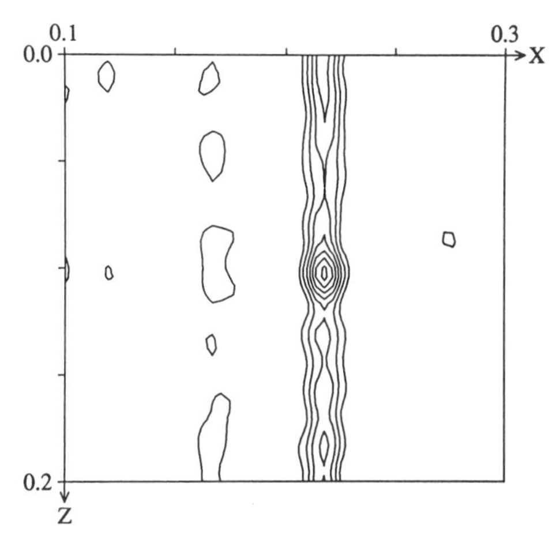

FIG. 5. Partial y section of the full-symmetry translation function[57] for the Fab′ B13I2–peptide complex using the complete INTREF-refined model (taken from Ref. 49).

sults for the constant domains were not. However, one of the intact Fab models (17/9) gave peaks in the rotation function, which confirmed those obtained with the variable domain searches. The signal-to-noise ratio for intact molecule 1 increased significantly over that found with the variable domain alone, while that for molecule 2 decreased. This observation suggested that the elbow angle for molecule 1 was quite similar to that of Fab 17/9, but that for molecule 2 it differed somewhat. In an attempt to correct these errors and thereby produce more accurate models for the translation function analysis, the intact oriented Fabs were then refined with INTREF. The variable and constant domains of each Fab model were initially treated as independent groups so as to improve their rotational orientation only. Once convergence had been achieved, each molecule was then defined as a single group with two parts to refine the relative disposition between the variable and constant domains. Quite remarkably, the procedure resulted in a reorientation of the variable and constant domains of 1.9 and 4.9°, respectively, for molecule 1 and 3.7 and 12.5° for molecule 2. The large correction of over 12° in the constant domain of molecule 2 is almost entirely a result of the difference in elbow angle between Fab17/9 and molecule 2 in the 50.1 Fab–peptide crystal. This difference explains the decrease in signal-to-noise ratio observed for molecule 2 using the intact Fab model. The INTREF-refined intact models were then used in the X-PLOR translation function, resulting in very clean solutions for the translational position of each molecule.

Refinement

Rigid Body Domain Refinement. Prior to manual rebuilding and atomic refinement of the structure, the model is typically subjected to rigid body domain refinement. This is a crucial step, since small errors in the absolute position and orientation of the model or its composite domains can lead to a structure that can be refined but which requires an unnecessary amount of manual intervention.[59] Although the program CORELS[60] has been widely used for that purpose, we now routinely use X-PLOR. In all cases low-resolution data (usually 10–6 Å) are used at the outset to increase the radius of convergence of the method. Data to 5.0 Å resolution are then added, followed by a final run using data in the 8.0- to 4.0-Å resolution range. In the first stage, only the variable and constant domains are treated as rigid bodies. Once convergence has been reached, the individual V_L, V_H, C_L, and C_H1 subdomains are refined independently. The R factors at this stage are typically between 0.45 and 0.41 for data in the 8.0- to 4.0-Å resolution range.

Atomic Refinement. Our current refinement strategy involves automated refinement with X-PLOR and manual rebuilding from omit maps using FRODO.[61] Although the refinement of each of the Fab structures solved in our laboratory has been approached slightly differently, the same steps have generally been involved. In the first step, those residues that are not conserved between the model and the unknown are truncated to alanines. Amino acid insertions in the model relative to the unknown structure are deleted. The resulting model is refined with the high-temperature molecular dynamics refinement protocol[7,62] of X-PLOR and then used to calculate phases for a 2Fo–Fc electron density map. The side chains are not included in the model appear as positive density in the map and allow the correct side chain to be built into the structure. Once completed, the model is then refined with the X-PLOR positional refinement protocol. The R factor at this stage is typically around 0.25 for data in the 8.0- to 3.0-Å range. The resulting model is used to calculate a series of omit maps in which approximately 10% of the residues from the phasing model is sequentially deleted. To reduce the effects of residual model bias, 40 steps of X-PLOR positional refinement are then performed on the omit model. This is conveniently achieved by using the constraint interaction and selection cards in X-PLOR to eliminate all terms involving the omit-

[59] Z. S. Derewenda, *Acta Crystallogr., Sect. A* **45,** 227 (1989).

[60] J. L. Sussman, S. R. Holbrook, G. M. Church, and S.-H. Kim, *Acta Crystallogr., Sect. A* **33,** 800 (1985).

[61] T. A. Jones, *J. Appl. Crystallogr.* **11,** 268 (1978).

[62] A. T. Brünger and A. Krukowski, *Acta Crystallogr., Sect. A* **46,** 585 (1990).

ted residues. The phases from the resulting omit model are used to calculate either Fo – Fc or 2Fo – Fc electron density maps for manual rebuilding with FRODO. For the Fab–peptide complexes solved in our laboratory, we have found that the electron density for the peptide ligand is very weak in these initial maps. Interpretable density is not obtained until after the first cycle of manual rebuilding. Special care must be taken with the side chains in the vicinity of the ligand-binding site. With an anti-progesterone Fab′ DB3, for example, a tryptophan side chain in the native structure occupies the site where progesterone is bound in the complex (Arevalo, Stura, Taussig and Wilson unpublished). After the complete structure has been manually rebuilt, a second round of molecular dynamics and manual rebuilding from omit maps is typically performed. At this stage, solvent molecules can be added and regions of the structure still showing poor geometry are rebuilt again relying exclusively on omit maps.

Based on our own experience and that of others,[2–4,6,13,20,37–44] we feel that molecular replacement is now the method of choice for solving Fab structures. The rapidly growing database of Fab structures and the new methodologies for intensity-based domain refinement have largely eliminated the difficulties associated with the solution of the rotation and translation problems. In addition, the new refinement techniques, incorporating molecular dynamics, have helped to reduce the time required to refine the structure. Indeed, once data have been collected, the rate-determining step is now often the manual rebuilding of the structure.

Structures of Fabs and Fab–Antigen Complexes

Early studies with Fabs obtained from the myeloma proteins, Fab′ NEW,[63] Kol,[64] McPC603[65] and J539,[66] and the two Fab complexes, Fab′ NEW with vitamin K_1OH,[67] and McPC603 with phosphocholine,[68] showed that the antibody combining site had a concave shape into which the antigen fits (see Refs. 69 and 70). Crystallographic structures of several

[63] F. A. Saul, L. M. Amzel, and R. J. Poljak, *J. Biol. Chem.* **253,** 585 (1978).

[64] M. Marquart, J. Deisenhofer, R. Huber, and W. Palm, *J. Mol. Biol.* **141,** 369 (1980).

[65] Y. Satow, G. H. Cohen, E. A. Padlan, and D. R. Davies, *J. Mol. Biol.* **190,** 593 (1986).

[66] S. W. Suh, T. N. Bhat, M. A. Navia, G. H. Cohen, D. N. Rao, S. Rudikoff, and D. R. Davies, *Proteins: Struct. Funct. Genet.* **1,** 74 (1986).

[67] L. M. Amzel, R. J. Poljak, F. Saul, J. M. Varga, and F. F. Richards, *Proc. Natl. Acad. Sci. U.S.A.* **71,** 1427 (1974).

[68] D. M. Segal, E. A. Padlan, G. H. Cohen, S. Rudikoff, M. Potter, and D. R. Davies, *Proc. Natl. Acad. Sci. U.S.A.* **71,** 4298 (1974).

[69] L. M. Amzel and R. J. Poljak, *Annu. Rev. Biochem.* **48,** 961 (1979).

[70] D. R. Davis and H. Metzger, *Annu. Rev. Immunol.* **1,** 87 (1983).

mouse monoclonal Fab fragments of known specificity have now been determined. These structure determinations again show small ligands are bound in a slot or a pocket in the antibody combining site. Examples are Fab 4-4-20 with fluorescein,[41] Fab′ B13I2 with a 19-mer peptide corresponding to residues 69–87 of myohemerythrin,[39] and Fab NQ10/12.5 with 2-phenyloxazolone.[71] However, our view of antibody combining sites dramatically changed with the determination of five Fab–protein complexes; three of these are Fab complexes with lysozyme, D1.3,[72]HyHEL-5,[42] and HyHEL-10,[44] and two with neuraminidase, NC41 and NC10.[46,47] The determination of these Fab–protein structures showed that the contact area between the antibody and antigen is more extensive, as expected, but the interacting surfaces were more undulating and less concave than the pockets or grooves seen previously (see Refs. 73–76). Indeed, for HyHEL-10[44] the antibody actually protrudes into the active site of lysozyme and in that sense the antigen actually envelopes the antibody.

Some representative Fab–antigen complexes are shown in Fig. 6. It is clear that there is a significant difference in the shape of the antibody combining site, depending on whether a large ligand, such as a protein, or a small ligand, such as a peptide or hapten, is bound (Figs. 6 and 7). The solvent-accessible surface buried on the antibody by the antigen varies from 161 $Å^2$ for McPC603 with phosphocholine to 886 $Å^2$ for NC41 with neuraminidase.[76] The antigens, whether large or small, interact with four, five, or six of the hypervariable loops through specific hydrogen bonds, salt bridges, or van der Waals interactions. There is frequently a high proportion of aromatic residues on the antibody that interact with the antigen (see Refs. 77 and 78). There appears to be no correlation at present between the elbow angle and the state of ligation of the Fab. Other detailed analyses of Fab–antigen interactions are inappropriate here but can be found in various excellent reviews (see Refs. 73–76).

In addition to the complexes reported above, other free Fabs have been determined. These include an autoimmune anti-poly(dT) Fab HED10,[4] an anti-neuraminidase Fab S10/1,[46] an anti-histidine-containing protein Fab

[71] P. M. Alzari, S. Spinelli, R. A. Mariuzza, G. Boulot, R. J. Poljak, J. M. Jarvis, and C. Milstein, *EMBO J.* **9,** 3807 (1990).

[72] A. G. Amit, R. A. Mariuzza, S. E. V. Phillips, and R. J. Poljak, *Science* **233,** 747 (1986).

[73] P. M. Colman, *Adv. Immunol.* **43,** 99 (1988).

[74] P. M. Alzari, M.-B. Lascombe, and R. J. Poljak, *Annu. Rev. Immunol.* **6,** 555 (1988).

[75] D. R. Davies, S. Sheriff, and E. A. Padlan, *J. Biol. Chem.* **263,** 10541 (1988).

[76] D. R. Davies, E. A. Padlan, and S. Sheriff, *Annu. Rev. Biochem.* **59,** 439 (1990).

[76a] M. L. Connolly, *J. Appl. Crystallogr.* **16,** 548 (1983).

[76b] D. A. Case and M. Karplus, *J. Mol. Biol.* **132,** 343 (1979).

[77] E. A. Padlan, *Proteins: Struct. Funct. Genet.* **7,** 112 (1990).

[78] I. S. Mian, A. R. Bradwell, and A. J. Olson, *J. Mol. Biol.* **216,** 133 (1991).

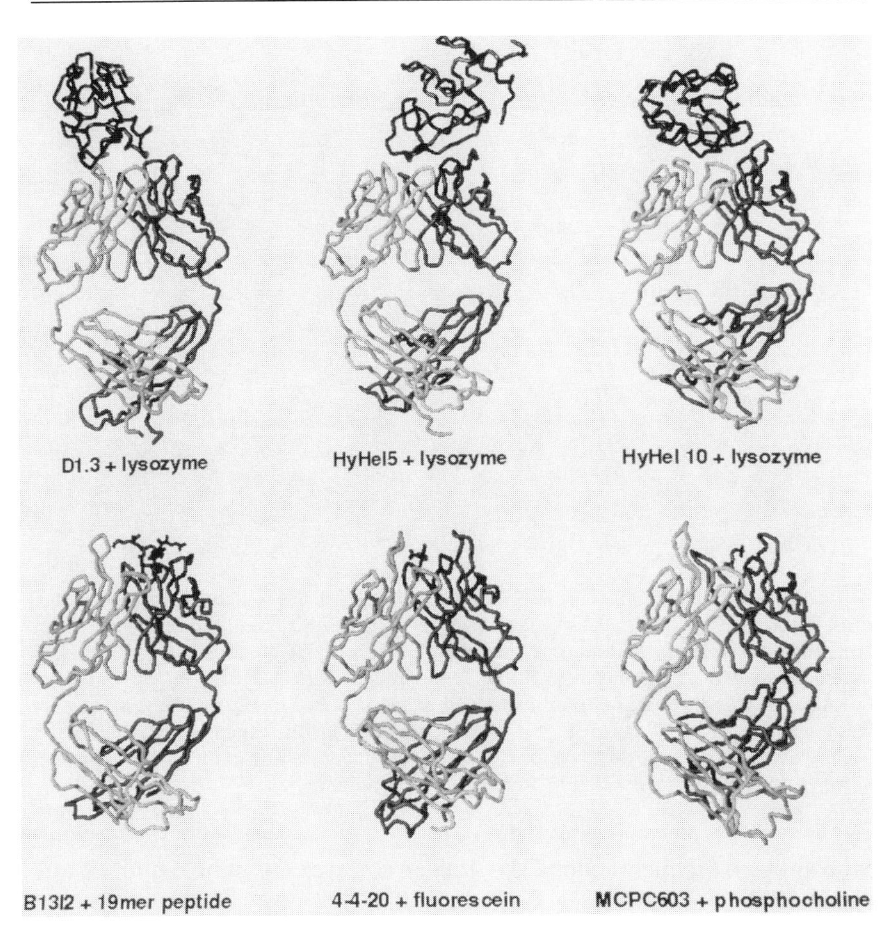

FIG. 6. The structures of antibody–antigen complexes. The C_α traces of six Fab–antigen complexes are depicted. The heavy chain of the Fab is displayed in dark gray, the light chain is pale gray, and the antigen is colored in black. The coordinates for the HyHEL-5,[42] HyHEl-10,[44] 4-4-20,[41] and McPC603[65] complexes were taken from the Brookhaven Protein Databank. The coordinates for the B13I2 peptide complex are from our laboratory[39] and the D1.3 lysozyme coordinates were provided by Drs. S. E. V. Phillips and R. J. Poljak. The figures were calculated using the program MCS[76a] and displayed on a Sun-4 Sparc workstation.

JEL42,[38] an anti-*p*-azobenzenearsenate Fab R19.9,[40] and an anti-arsenate Fab 36-71.[45] Many other Fabs have been solved but not yet published.

Until recently, we have been unable to study Fab structures of known specificities both in their free and antigen-bound states. The problem has been the difficulty in obtaining crystals of both the Fab and its complex

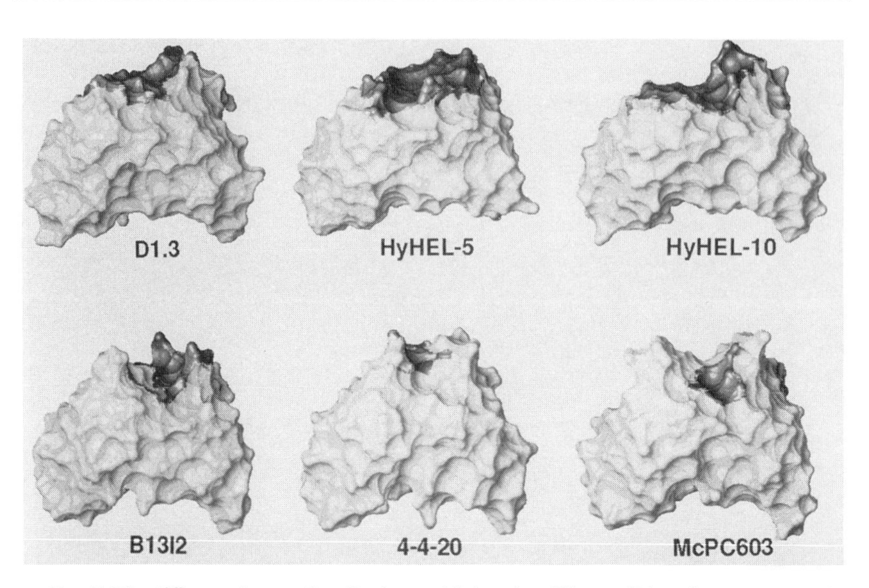

FIG. 7. The different shapes of antibody combining sites. These solid surfaces represent the variable domains of the six Fabs shown in Fig. 6. These Fabs were solved in complex with antigen but for clarity the antigen is not shown. The variable domains are in equivalent orientations to the Fabs shown in Fig. 6. The antibody surface buried by the antigen is highlighted in dark gray. The coordinates used are as described in Fig. 6. The surfaces were calculated with the program MCS,[76a] using standard atomic radii[76b] and a probe radius of 3.5 Å. The use of such a large probe radius emphasizes the major features of the antibody surface, giving a smoother representation of the shape of the combining site.

with antigen. Frequently, the Fab–antigen complex crystallizes under completely different conditions than the native Fab even if the antigen is relatively small (see for example, Refs. 22, 23, 27, 28, and 31). However, this is not always the case, as isomorphous crystals have been obtained of the free and bound Fabs of an anti-progesterone Fab′ DB3,[29,30] an anti-digoxin Fab,[79] and an anti-2-phenyloxazolone Fab NQ10/12.5.[72] Whether to cocrystallize or diffuse in the ligand to the native Fab is a question frequently posed. Cocrystallization allows for the accommodation of changes in either the antibody or the antigen. An interesting example with the anti-phenyloxazolone Fab showed that diffusion of the hapten into the native Fab crystals was probably unsuccessful due to the intermolecular contacts in the crystals. We had a similar experience with Fab 17/9, where the variable and constant domains from symmetry-related Fabs form head-to-tail interactions in the native Fab crystal lattice that would prevent

[79] D. R. Rose, B. A. Seaton, G. A. Petsko, J. Novotny, M. N. Margolies, E. Locke, and E. Haber, *J. Mol. Biol.* **165,** 203 (1983).

peptide binding. Isomorphous crystals of the NQ10/12.5 complex were in fact made by cocrystallization.[72] On the other hand, under nearly identical conditions, the BV04-01 native Fab crystallizes in a triclinic space group, whereas crystals of the complex with bound trinucleotide are monoclinic (A. B. Edmundson, personal communication).

These structure determinations of free and bound Fabs have now begun to answer the question of whether the antibody changes conformation on antigen binding. The structures that have provided these data include an anti-MHR peptide Fab′ B13I2,[39] an anti-lysozyme Fab D1.3 and its Fv,[13] an anti-single-stranded DNA Fab BV04-01 (A. B. Edmundson, personal communication, 1991), and an anti-2-phenyloxazolone Fab NQ10/12.5.[72] The D1.3 and the BV04-01 structures both show significant variation in the relative disposition of V_H and V_L between free and bound Fabs. In addition, small but significant changes (~ 1–2 Å) are seen in the B13I2 Fab′ hypervariable loops on peptide binding.[39] These changes are similar in magnitude to those seen in the antigens, lysozyme and neuraminidase, when they bind to their respective Fabs (see Refs. 73, 75, and 76). Other free and bound Fab structures solved include an anti-hemagglutinin peptide Fab 17/9 (J. M. Rini, U. Schulze-Gahmen and I. A. Wilson, unpublished, 1991), an anti-progesterone Fab′ DB3 with three different steroids (J. H. Arevalo, E. A. Stura, M. J. Taussig and I. A. Wilson, unpublished, 1991), an anti-dinitrophenyl spin-label Fab ANO2,[6] and an anti-*O*-polysaccharide Fab′ Se-155.4.[20]

The present situation is that we now have antibody–antigen structures with bound peptides, proteins, carbohydrate, nucleic acid, and a variety of haptens. Other important antibody complexes that are being determined include the idiotype–anti-idiotype complexes. The major question addressed by these studies is whether the anti-idiotype antibody mimicks the original antigen. The Fab D1.3–Fab E225 complex shows the idiotype–anti-idiotype interactions to be extensive but there does not appear to be any correlation between the E225 paratope and HEL epitope.[48,80] Crystals of another anti-idiotypic antibody against M315 have been obtained with Fv and Fab′ fragments of M315.[81] Interestingly, Colman and colleagues have previously reported that two different antibodies can recognize essentially the same epitope on neuraminidase but in a completely different way.[82]

[80] G. A. Bentley, T. N. Bhat, G. Boulot, T. Fischmann, J. Navaza, R. J. Poljak, M.-M. Riottot, and D. Tello, *Cold Spring Harbor Symp. Quant. Biol.* **54,** 239 (1989).

[81] Y. Matsuura, K. Inaka, M. Kusunoki, Y. Katsube, N. Sakato, and H. Fujio, *Biochim. Biophys. Acta* **916,** 524 (1987).

[82] W. R. Tulip, J. N. Varghese, R. G. Webster, G. M. Air, W. G. Laver, and P. M. Colman, *Cold Spring Harbor Symp. Quant. Biol.* **54,** 257 (1989).

Conclusions

Unquestionably, the use of X-ray crystallography to investigate the interaction of antibodies with antigens has been highly successful. In the future, the number of solved free and bound Fab structures each year will increase dramatically as it is now possible to solve such structures, by the methods discussed here, in a matter of months rather than the years taken previously. Such progress heralds the structure determinations of a myriad of Fabs with a multitude of specificities. Hence, one can look forward to a more thorough investigation of the enormous antibody repertoire of the immune system. Such studies will be invaluable for the future design and use of antibodies in biology, biotechnology, chemistry, and medicine.

Acknowledgments

We would like to thank Dr. Robyn Stanfield and Chris Hassig for help with Figs. 6 and 7 and Dr. Ursula Schulze-Gahmen for Fig. 1. Warren Densley is thanked for excellent technical assistance. We are also grateful to Dr. Al Profy and RepliGen for help and collaboration with the 50.1 antibody and Drs. Michael Taussig and Arnold Feinstein for the DB3 antibody. The work presented here was supported in part by National Institutes of Health Grants AI-23498, GM-38794, and GM-38419 (to I.A.W.). J.M.R. is a Centennial Fellow of the MRC (Canada). D.H.F. is a graduate student in the Department of Chemistry at the University of California, San Diego. This is Publication #6651-MB from the Research Institute of Scripps Clinic.

[8] Predicting Location of Continuous Epitopes in Proteins from Their Primary Structures

By J. L. PELLEQUER, E. WESTHOF, and M. H. V. VAN REGENMORTEL

Introduction

The antigenic determinants, or epitopes, of a protein correspond to those parts of the molecule that are recognized by the binding sites, or paratopes, of certain immunoglobulin molecules. When such specific binding is observed experimentally, the particular immunoglobulin becomes known as an antibody specific for the protein. In the same way that the antibody specificity of an immunoglobulin becomes established only after its complementary antigen has been identified, the epitope nature of a cluster of amino acids in a protein can be established only by means of an immunoglobulin molecule. Epitopes are thus relational entities that can be defined only in a functional sense (i.e., in an immunoassay) by the binding of complementary paratopes.[1]

It has been suggested that the structure of epitopes can be elucidated only by X-ray crystallography of antigen–antibody complexes.[2] At present, five epitopes have been analyzed by this method, three of lysozyme[3–5] and two of influenza neuraminidase.[6,7] In all five cases, a large area of the protein surface comprising 15–22 amino acid residues was identified as being "in contact" with the antibody paratope. As discussed elsewhere[8] the stereochemical information obtained from such studies is entirely structural and the contribution of individual atomic interactions to the overall functional binding remains a matter of interpretation of how to relate structure to function. Molecular modeling and energy calculations suggest that only a small number of amino acids (as few as three to five residues) in the so-called "contact epitope" contribute to the binding energy of interaction.[9] In the present chapter, epitopes are considered as functional epitopes, i.e., portions or fragments of a protein that are able to bind to antibody in an immunoassay and not as "contact epitopes" or "energetic epitopes" identifiable only in structural studies.

At present the most common way of classifying epitopes consists in distinguishing continuous and discontinuous epitopes.[10] The label continuous epitope is given to any short linear peptide fragment of the antigen that is able to bind to antibodies raised against the intact protein. In most cases these antibodies cross-react only weakly with any linear peptide fragment of the antigen. It is usually assumed that this weak cross-reactivity is due to the fact that the peptide fragment represents only a portion of a more complex epitope and is not able to assume the correct conformation present on the folded protein.

A more extreme viewpoint has been advocated, according to which all

[1] M. H. V. Van Regenmortel, *Philos. Trans. R. Soc. London, Ser. B* **323,** 451 (1989).

[2] W. G. Laver, G. M. Air, R. G. Webster, and S. J. Smith-Gill, *Cell (Cambridge, Mass.)* **61,** 553 (1990).

[3] A. G. Amit, R. A. Mariuzza, S. E. V. Phillips, and R. J. Poljak, *Science* **233,** 747 (1986).

[4] S. Sheriff, E. W. Silverton, E. A. Padlan, G. H. Cohen, S. J. Smith-Gill, B. C. Finzel, and D. R. Davies, *Proc. Natl. Acad. Sci. U.S.A.* **84,** 8075 (1987).

[5] E. A. Padlan, E. W. Silverton, S. Sheriff, G. H. Cohen, S. J. Smith-Gill, and D. R. Davies, *Proc. Natl. Acad. Sci. U.S.A.* **86,** 5938 (1989).

[6] P. M. Colman, W. G. Laver, J. N. Varghese, A. T. Baker, P. A. Tulloch, G. M. Air, and R. G. Webster, *Nature (London)* **326,** 358 (1987).

[7] W. R. Tulip, J. N. Varghese, R. G. Webster, G. M. Air, W. G. Laver, and P. M. Colman, *Cold Spring Harbor Symp. Quant. Biol.* **54,** 257 (1990).

[8] M. H. V. Van Regenmortel, *Immunol. Today* **10,** 266 (1989).

[9] J. Novotny, R. E. Bruccoleri, and F. A. Saul, *Biochemistry* **28,** 4735 (1989).

[10] M. Z. Atassi and J. A. Smith, *Immunochemistry* **15,** 609 (1978).

continuous epitopes represent "unfoldons," i.e., unfolded regions of the antigen that cross-react only with antibodies specific for the denatured protein.[2] Such antibodies may be present in antisera raised against the protein because some of the antigen molecules used for immunization were denatured, or they could be obtained by immunization with peptide fragments. In this latter case, any cross-reactivity observed with the parent protein could be ascribed to the presence of denatured protein molecules in the preparation used in the immunoassay. Although it is true that in many solid-phase assays in which the protein antigen is adsorbed to plastic, the antigen is likely to be at least partly denatured, it seems unjustified to assume that all reported cases of cross-reactivity between proteins and peptides are due to antibodies specific only for the denatured form of the antigen.[11-13] It cannot be ruled out that a continuous epitope may possess distinctive conformational features resembling those of the cognate structure in the parent protein or that an appropriate conformation could arise in the peptide by a process of induced fit occurring during binding to the antibody paratope.

The second type of epitopes, known as discontinuous or assembled topographic epitopes,[14] is believed to correspond to the vast majority of epitopes found in proteins.[15] Discontinuous epitopes consist of a group of residues that are not contiguous in the sequence but are brought together by the folding of the polypeptide chain or by juxtaposition of two separate peptide chains. Most antibodies to discontinuous epitopes bind to the protein antigen only if the protein is intact and its conformation is preserved. When the protein is fragmented into a number of peptides, it is to be expected that the relative positions in space of the residues making up a discontinuous epitope will not be preserved. This explains why the majority of monoclonal antibodies (MAbs) raised against intact proteins, being specific for discontinuous epitopes, do not react with any peptide fragment derived from the parent protein antigen.

However, it is also true that a significant number of MAbs (typically of the order of 10%) raised against intact proteins are able to react with the immunogen as well as with certain peptide fragments of the protein. It is this fraction of the total immune response against a protein that forms the

[11] M. H. V. Van Regenmortel, *Trends Biochem. Sci.* **12,** 237 (1987).

[12] R. Arnon, "Synthetic Vaccines I and II." CRC Press, Boca Raton, Florida 1987.

[13] M. H. V. Van Regenmortel, J. P. Briand, S. Muller, and S. Plaue, "Synthetic Peptides as Antigens." Elsevier, Amsterdam, New York, and Oxford, 1988.

[14] D. C. Benjamin, J. A. Berzofsky, I. J. East, F. R. N. Gurd, C. Hannum, S. J. Leach, E. Margoliash, J. G. Michael, A. Miller, E. M. Prager, M. Reichlin, E. E. Sercarz, S. J. Smith-Gill, P. E. Todd, and A. C. Wilson, *Annu. Rev. Immunol.* **2,** 67 (1984).

[15] D. J. Barlow, M. S. Edwards, and J. M. Thornton, *Nature (London)* **322,** 747 (1986).

basis of the observed cross-reactions between peptides and proteins[11] and that will be discussed in the present chapter. Such cross-reactive antibodies that can also be obtained by immunization with synthetic peptides have many useful applications in biological studies.[16-18] Anti-peptide antibodies may react either with the cognate, native protein or with the denatured protein and both types of interaction may be useful in certain studies. For instance, anti-peptide antibodies used for isolating and characterizing gene products need not necessarily recognize the native protein, since the test antigen is likely to be denatured in many of the immunoassays in current use.

Nature of Prediction Algorithms

General Basis of Predictions

Many attempts have been made to predict the position of continuous epitopes in proteins from certain features of their primary structure. This is particularly relevant since for the majority of proteins known today, the only available structural information concerns their amino acid sequence usually deduced from the nucleotide sequence of the corresponding gene.

Parameters such as hydrophilicity, accessibility, and mobility of short segments of polypeptide chains have been correlated with the location of continuous epitopes in a few well-characterized proteins. This has led to a search for empirical rules that would allow the position of continuous epitopes to be predicted from certain features of the protein sequence. All prediction calculations are based on propensity scales for each of the 20 amino acids. These scales describe the tendency of each residue to be associated with properties such as surface accessibility, hydrophilicity, or segmental atomic mobility. Scales of hydrophilicity and hydrophobicity are mostly derived from the study of partition coefficients of amino acids in two noninteracting isotropic phases. Scales of accessibility and mobility are based on the study of proteins of known three-dimensional structure. Accessibility scales are constructed by measuring the accessible surface of all the residues in a number of proteins while mobility scales are derived from atomic temperature factors (*B* values) obtained by refinement of X-ray structures. Scales of secondary structure are based on the prediction

[16] R. A. Lerner, *Adv. Immunol.* **36,** 1 (1984).
[17] G. J. Walter, *J. Immunol. Methods* **88,** 149 (1986).
[18] S. Modrow and H. Wolf, *in* "Immunochemistry of Viruses II. The Basis for Serodiagnosis and Vaccines" (M. H. V. Van Regenmortel and A. R. Neurath, eds.), p. 83. Elsevier, Amsterdam, 1990.

of turns and loops obtained from statistical analysis of proteins of known structure.

Each scale consists of 20 values assigned to each of the 20 amino acid residues on the basis of their relative propensity to possess the property described by the scale. On a hydrophilicity scale, for instance, the most hydrophilic residues will possess the highest values. When different prediction methods are compared, it is important to normalize the scales.

When prediction profiles are constructed, the scale value for each amino acid is not assigned to the corresponding position in the sequence. Instead, a so-called window is used that allows a certain segment of the protein to be analyzed independently of the remainder of the protein. In principle, the size of the window should be adapted to the type of structural property that is examined. For example, a window of 19 residues can be used for detecting transmembrane helices,[19] while a window of 5 residues is suitable for localizing loops in proteins. When analyzing antigenic regions, a window of seven residues is used by most authors.

The corresponding value of the scale is introduced for each of the seven residues and the arithmetical mean of the seven values is assigned to the center of the window, i.e., to the fourth residue. Each mean value will be placed on a graph where the abscissa represents the protein sequence and the ordinate the average propensity of the amino acid according to a particular scale. In a hydrophilicity profile, high values will give rise to peaks corresponding to hydrophilic regions, whereas valleys will correspond to hydrophobic regions of the protein (Fig. 1).

In order to eliminate excessive fluctuations in the profiles, it is customary to use a smoothing procedure. There are two main smoothing methods. In the first, the curve is modified by eliminating indentations, while in the second method, the calculated values are replaced by other values, which might be integers, for example.

Among the first type of smoothing methods, the classical approach consists of assigning a certain weight to each position in the window by which the values of the propensity scale corresponding to the given amino acid will be multiplied before averaging. Karplus and Schulz[20] have used the following sequence of arithmetical weights on a window of seven residues: 0.25/0.5/0.75/1/0.75/0.5/0.25. With such a sequence the center of the window has more weight than its extremities. Van Regenmortel and Daney de Marcillac[21] have used instead a Gaussian smoothing curve, leading to the following sequence of residue weights in a window of seven

[19] J. Kyte and R. F. Doolittle, *J. Mol. Biol.* **157,** 105 (1982).

[20] P. A. Karplus and G. E. Schulz, *Naturwissenschaften* **72,** 212 (1985).

[21] M. H. V. Van Regenmortel and G. Daney de Marcillac, *Immunol. Lett.* **17,** 95 (1988).

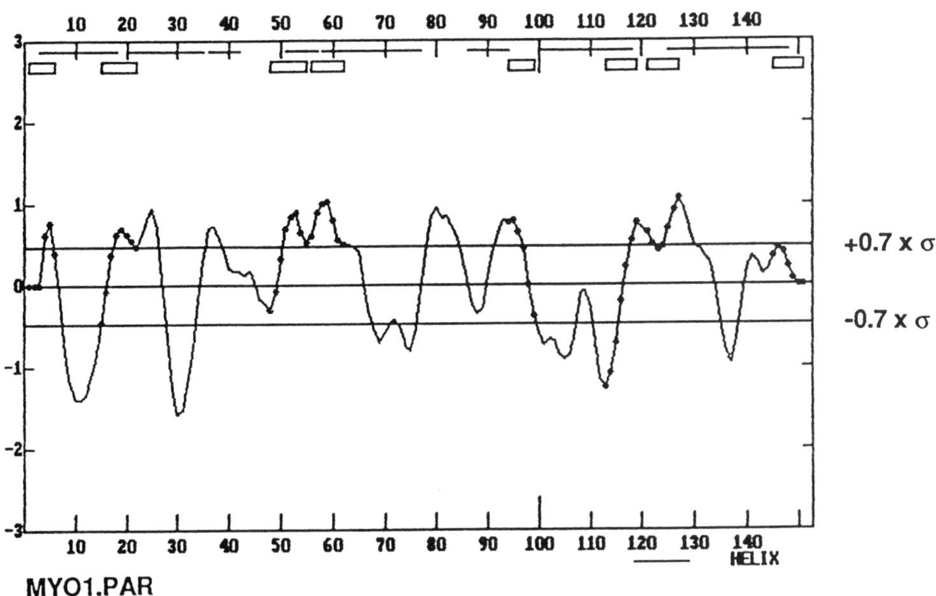

FIG. 1. Hydrophilicity profile of myoglobin constructed with the scale of Parker *et al.*[39] This graph uses a scale normalized between +3 and −3. The two lines on each side of the mean correspond to ±0.7 × standard deviation. Such an interval includes 50% of the amino acids of the protein. Rectangles at the top of the curves correspond to the known protein epitopes; the circles drawn on the curve correspond to the same residues. The secondary structure pattern, if known, is shown above the rectangles. A line corresponds to a helix (the only pattern on this curve). In the program a dashed line corresponds to a sheet and a dotted line to turns.

residues (σ is equal to 2 and the weights are 0.05/0.11/0.19/0.22/0.19/0.11/0.05).

A second type of smoothing, used by Jameson and Wolf,[22] is based on the smoothing procedure used in University of Wisconsin Genetics Computer Group (UWGCG) programs.[23] Clearly, the use of very large windows will tend to smooth the resulting curves while small windows will give rise to jagged curves. In order to compare the profiles obtained by different methods, it is essential to normalize the various scales. However, normalization can lead to arbitrary values for some amino acids. For instance, in the normalization procedure used for obtaining Table III (see below), the original values of each scale are set between +3 and −3 [new value = (old value × 3)/maximum old value]. Such a normalization procedure is

[22] B. A. Jameson and H. Wolf, *CABIOS* **4,** 181 (1988).

[23] J. Devereux, P. Haeberli, and O. Smithies, *Nucleic Acids Res.* **12,** 387 (1984).

meaningful only if the original scale is evenly distributed around zero. If this is not the case, the mean of the scale is subtracted from the original values.

Prediction of Secondary Structure

The secondary structure elements of proteins are constituted either of regular and repeated motifs linked by hydrogen bonds (e.g., helices and sheets) or of irregular motifs with side-chain and backbone atoms exposed to solvent (e.g., turns and loops). On an average over all known proteins, approximately 50% of residues are in turns and loops and 50% in regular and repeated motifs (25% in helices and 25% in sheets). Since the knowledge of the secondary structure of a protein is very useful for predicting antigenicity, a number of algorithms for predicting secondary structure have been applied to the prediction of epitopes.

The core of proteins usually contains a combination of helices and sheets, which are hydrophobic. Since the core is mostly devoid of water molecules, the formation of intramolecular hydrogen bonds is favored. In contrast, turns and loops are situated on the surface of the protein in contact with solvent atoms. Thus, turns and loops are accessible and hydrophilic, two characteristics of antigenic regions.[24] Besides, loops are flexible, as judged from the blurred electron-density maps often observed around those regions. Generally, not much antigenicity is found in helices and sheets of the core, except on their extremities, which are more flexible and more hydrophilic than the central hydrophobic core.

Two main algorithms for predicting two-dimensional structure exist, namely those proposed by Chou and Fasman[25] and Garnier *et al.*[26] Tests made to assess the success rate of those two prediction algorithms show that at most 55–70% of two-dimensional elements are successfully predicted.[27-29] Discrimination between helices and sheets is often imperfect and the ends of these motifs are not well defined and tend to run over turns and loops. The prediction of turns and loops that sets limits to regular secondary structure elements is particularly important for predicting continuous epitopes smaller than about 10 residues. On the other hand, since a short linear peptide is unable to assume a helical conformation, it would be found to be nonantigenic in an immunological assay, although

[24] G. D. Rose, L. M. Gierasch, and J. A. Smith, *Adv. Protein Chem.* **37,** 1 (1985).
[25] P. Y. Chou and G. D. Fasman, *Adv. Enzymol.* **47,** 45 (1978).
[26] J. Garnier, D. J. Osguthorpe, and B. Robson, *J. Mol. Biol.* **120,** 97 (1978).
[27] W. Kabsch and C. Sander, *FEBS Lett.* **155,** 179 (1983).
[28] G. D. Fasman, ed., "Prediction of Protein Structure and the Principles of Protein Conformation." Plenum, New York and London, 1989.
[29] M. J. Rooman and S. J. Wodak, *Nature (London)* **335,** 45 (1988).

the entire corresponding helical motif could be an antigenic site in the intact protein.[30,31] Epitope mapping with short or long peptides may thus identify antigenic regions in different parts of the protein.

Review of Propensity Scales

Hydrophilicity Scale of Hopp and Woods

Hopp and Woods[32] were the first to base predictions on the observed link between the location of antigenicity at the surface of proteins and the hydrophilic character of these regions, i.e., the fact that they show a high degree of exposure to solvent. The authors also suggested that charged amino acids are likely to be antigenic because such residues are mainly located at the surface of proteins. Once a link between hydrophilicity and antigenicity was established, the authors constructed a scale for the hydrophilicity of the 20 amino acids. For this purpose, the solvent parameters assigned to amino acids by Levitt[33] were used. The four charged residues (Asp, Glu, Lys, and Arg) were given the maximum value of 3.0 because this seemed to improve antigenicity prediction within the set of 12 proteins studied. In order to eliminate all wrong predictions, the authors modified some of the scale values of Levitt.[33] The values for Asp and Glu were raised from 2.5 to 3.0 and that for Pro was increased from −1.4 to 0. Although the peaks in profiles constructed with this scale correspond to hydrophilic regions, the authors remarked that not all the peaks corresponded to epitopes and that all known epitopes of a protein were not located in the most hydrophilic regions. In order to evaluate the quality of predictions, the highest peaks were counted and labeled as correct epitopes (C) or as wrongly placed (W). The ratio $C/(C + W)$ was taken as the percentage of correct prediction. In the case of some proteins of known antigenic structure (myoglobin, lysozyme, cytochrome *c*, etc.) this approach led to a level of correct prediction of 100% using a window of six residues.

Hydropathy Scale of Kyte and Doolittle

Kyte and Doolittle[19] tried to devise a scale that would make it possible to predict which residues are located on the outside of a protein. This scale, which was called a hydropathy scale, took into account two types of

[30] Z. Al Moudallal, J. P. Briand, and M. H. V. Van Regenmortel, *EMBO J.* **4,** 1231 (1985).
[31] M. H. V. Van Regenmortel, *in* "The Plant Viruses Volume 2" (M. H. V. Van Regenmortel and H. F. Conrat, eds.), p. 79. Plenum, New York, 1986.
[32] T. P. Hopp and K. R. Woods, *Proc. Natl. Acad. Sci. U.S.A.* **78,** 3824 (1981).
[33] M. Levitt, *Biochemistry* **17,** 4277 (1976).

correlated interactions, i.e., the access of hydrophilic side chains to the aqueous solvent and the minimization of the contact between hydrophobic side chains and water. Furthermore, since it had been shown by Janin and Chothia[34] that both the hydrophobicity and large size of certain residues were linked to their tendency to be buried, Kyte and Doolittle[19] incorporated in their scale the solvent accessibility values of amino acids determined by Chothia.[35] They measured amino acid partition coefficients between water and a noninteracting isotropic phase and calculated the free energy change. These values, together with the values of fraction of side chains 100% buried and 95% buried,[35] were used to construct the scale. Most hydropathy values are the mean of these three scales (for instance, residues Ile, Leu, Cys, Met) although others are rather arbitrary (Arg, Lys, Pro, Tyr). For instance, the value for Arg was the lowest and was set at -4.5 because the highest value in the scale was $+4.5$. One of the limitations of the Kyte and Doolittle[19] scale is that it does not consider polar residues that are buried in the protein and are stabilized by salt bridges or hydrogen bonds.[35]

Acrophilicity Scale of Hopp

Hopp[36] developed another method for predicting surface residues of proteins using a so-called acrophilicity scale based on the relative degree of exposure of amino acids in 49 proteins of known structure. This author used stereo paired α-carbon plots and identified the amino acids located in each protein protrusion. The scale was obtained by measuring the distance between the center of the protein and each amino acid in protruding regions. Residues with the highest acrophilicity values are Gly, Pro, Asn, and Ser, which are also the residues most frequently found in β turns.

Flexibility Scale of Karplus and Schulz

In view of the observed link between antigenicity and segmental mobility,[37,38] Karplus and Schulz[20] developed a method for predicting mobility of protein segments on the basis of the known temperature *B* factors of the α-carbons of 31 proteins of known structure. Amino acids could be sepa-

[34] J. Janin and C. Chothia, *J. Mol. Biol.* **143,** 95 (1980).

[35] C. Chothia, *J. Mol. Biol.* **105,** 1 (1976).

[36] T. P. Hopp, *Synth. Antigens, Ann. Sclavo* **2,** 47 (1984).

[37] E. Westhof, D. Altschuh, D. Moras, A. C. Bloomer, A. Mondragon, A. Klug, and M. H. V. Van Regenmortel, *Nature (London)* **311,** 123 (1984).

[38] J. A. Tainer, E. D. Getzoff, H. Alexander, R. A. Houghten, A. J. Olson, R. A. Lerner, and W. A. Hendrickson, *Nature (London)* **312,** 127 (1984).

rated into 2 classes comprising 10 flexible and 10 rigid residues. The rigid residues (Ala, Leu, His, Val, Tyr, Ile, Phe, Cys, Trp, and Met) possessed average *B* values lower than 1. The authors derived three scales for the amino acids called BNORM0, BNORM1, and BNORM2 corresponding to different degrees of rigidity in the neighboring residues. BNORM0 is the scale that applies when none of the neighboring residues is rigid; BNORM1 is the scale applying when one neighboring residue is rigid while BNORM2 applies when the two neighbors are rigid. As a result, the prediction takes into account the flexible nature of a stretch of residues and not only the propensity of single residues.

Hydrophilicity Scale of Parker, Guo, and Hodges

Parker *et al.*[39] constructed another hydrophilicity scale based on peptide retention times during high-performance liquid chromatography (HPLC) on a reversed-phase column. The scale consisted of retention times normalized from $+10$ to -10.

When the retention time of peptides containing charged residues was measured, it was found that the addition of a charge at the C or N terminus drastically affected the retention time. The scale, therefore, incorporated values for increased hydrophilicity brought about by the presence of charged groups at the C and N termini of proteins.

Hydrophilicity profiles were constructed using a window of seven residues. Any residues with a profile value greater than 25% above the average hydrophilicity value were defined as surface sites (the 25% value was calculated as 25% of the difference between the maximum value in the plot and the average hydrophilicity value). Parker *et al.*[39] also produced composite surface profiles by superimposing profile plots corresponding to hydrophilicity, accessibility,[40] and flexibility[20] parameters. The most significant peaks were selected by scaling the surface sites (defined by profile values greater than 25% above the average parameter value) from 0 to 100, where the maximum surface site value was set equal to 100 and the 25% surface site value was equal to 0. The maximum value of each residue in any of the superimposed plots was used to produce the composite profile value. A 50% cutoff line in the composite profile was taken as threshold for the prediction of antigenic residues. A program for generating composite profiles is available on diskettes for the IBM-PC or Apple Macintosh computer from the Biochemistry Department at the University of Alberta (Edmonton, Canada).[39]

[39] J. M. D. Parker, D. Guo, and R. S. Hodges, *Biochemistry* **25,** 5425 (1986).

[40] J. Janin, *Nature (London)* **277,** 491 (1979).

Antigenicity Scale of Welling et al.

Welling *et al.*[41] constructed an antigenicity scale on the basis of statistical analysis of the 606 amino acids found in 69 continuous epitopes of 20 well-studied proteins. Each amino acid was characterized by its frequency of appearance in antigenic regions and the scale values were obtained by dividing this frequency by the frequency of each amino acid in all the proteins. The database used by Welling *et al.*[41] is restricted since it considered, for instance, only one of the epitopes of influenza virus hemagglutinin and two epitopes of hepatitis B virus surface antigen. Another limitation stems from the fact that all residues in the continuous epitopes were given equal importance. Because systematic replacement studies (in which each residue of an epitope is replaced by the 19 possible other residues) have established that some positions within an antigenic peptide can be occupied by any of the 20 residues without affecting binding activity,[42] the inclusion of such indifferent residues in the calculations must clearly reduce the significance of any observed correlation.

Antigenic Index of Jameson and Wolf

A new method was proposed by Jameson and Wolf[22] that combines different profiles, each one representing a different aspect of protein architecture. The chosen parameters were hydrophilicity, accessibility, flexibility, and secondary structure (using two scales). After giving to each parameter a certain weight, the sum of the four curves was computed and gave rise to the so-called "antigenic index." The weights were chosen so that 40% of the antigenic index was derived from the secondary structure component. The accessibility and flexibility parameters, which according to these authors do not correlate well with antigenicity, were each given a weight of 15%. The remaining 30% was allocated to the inverted hydrophobicity. The choice of weights is somewhat arbitrary and it is possible that a systematic search might improve the prediction success of the antigenic index.

The five scales used for computing the index are the inverted hydrophobicity of Kyte and Doolittle,[19] the accessibility of Janin *et al.*,[43] the flexibility of Karplus and Schulz,[20] and finally the two secondary structure predictions of Chou and Fasman[25] and Garnier *et al.*[26] Each profile was constructed with a window of seven residues. The accessibility profile was

[41] G. W. Welling, W. J. Weijer, R. Van Der Zee, and S. Welling-Wester, *FEBS Lett.* **118,** 215 (1985).

[42] H. M. Geysen, T. J. Mason, and S. J. Rodda, *J. Mol. Recog.* **1,** 32 (1988).

[43] J. Janin, S. Wodak, M. Levitt, and B. Maigret, *J. Mol. Biol.* **125,** 357 (1978).

obtained using the formula of Emini *et al.*[44]: $S_n = (\Pi\delta_{n+4-i})(0.37)^{-6}$, where S_n is the surface probability, δ_n is the fractional surface probability values of Janin *et al.*,[43] and i varies from 1 to 6. The final curve was smoothed according to the UWGCG method. The curves obtained by this procedure display a strong bias in favor of positive values (i.e., peaks), which raises the problem of the selection of significant peaks. In the case of myoglobin, for example, the largest peak was found to lie outside known epitopes.

Comparison Between Prediction Efficacies of Different Scales

Criteria of Prediction Efficacy

Several approaches can be used to measure the efficacy of antigenicity prediction obtained with different scales. The simplest method is to compare the position of the peaks in the profile with the position of known epitopes in the protein. Hopp and Woods,[32] for instance, took into account the largest peak in the profile as well as the valleys located in known epitopes. If the largest peak was located in an epitope and no valley coincided with an epitope, the prediction was taken to be 100% correct. This method suffers from the limitations that it considers only one correctly predicted epitope in a protein and that the highest hydrophilic peaks are not necessarily located in epitopes. In their analysis, Hopp and Woods[32] allowed for an overlap of ± 2 residues in the localization of the highest peak since each value does not in fact correspond to a single residue but to an average.

Hopp[45] normalized several of the scales and compared the level of correct prediction achieved by hydrophilicity, accessibility, inverted hydrophobicity, antigenicity, and secondary structure profiles. In general, hydrophilicity peaks were found to correspond to loops and turns of the polypeptide chain, although in some cases they corresponded to the extremity of β strands and to exposed helices. In the secondary structure profiles, the peaks and valleys occurred in the same places as in hydrophilicity plots. For instance, the Chou and Fasman[25] helix prediction profile and the Garnier *et al.*[26] β-strand prediction profile generated peaks in regions of maximum hydrophobicity. This agreement is due to the fact that the centers of the largest helices and β strands usually correspond to the tightly packed hydrophobic cores of proteins. By inverting the secondary structure scales, plots are obtained that show peaks at the same position as in hydrophilicity plots.[45]

[44] E. A. Emini, J. V. Hugues, D. S. Perlow, and J. Boger, *J. Virol.* **55,** 836 (1985).
[45] T. P. Hopp, *J. Immunol. Methods* **88,** 1 (1986).

Other criteria have also been used. For instance, Parker *et al.*[39] took into account only peaks higher than a certain threshold, whereas Getzoff *et al.*[46] used a statistical method based on the random draw of certain values (Monte Carlo method). In the latter method the peaks to be tested were arranged in decreasing order of size and the peptides found to be reactive in the pepscan technique[47] were scored by giving the value 1 to their N-terminal residue and the value 0 to all other residues of the protein. One thousand random draws were then carried out, choosing from the total number of residues of the protein, and the total number of positive draws was added up. Using a window of seven residues, it was found that none of the three prediction scales that were tested predicted better than random.[46]

The predictive value of eight different scales has also been compared by Van Regenmortel and Daney de Marcillac,[21] who used as a criterion of success the number of residues correctly predicted to be antigenic in four well-studied proteins. It was found that none of the methods achieved a high level of correct prediction. A χ^2 statistical analysis showed that the segmental mobility scale of Karplus and Schulz[20] and the hydrophilicity scale of Parker *et al.*[39] were slightly more successful than the others.

In the analysis described below we used a 2×2 contingency table analysis for testing the homogeneity of a population. In this case, the test was used to evaluate the capacity of each scale to distinguish antigenic regions (i.e., a stretch of amino acids) from nonantigenic regions.

The 2×2 contingency table has four entries (see Table I): one entry for the number of amino acids in epitopes above the chosen threshold (A); one for the number of amino acids in epitopes but below the threshold (B); one for the number of amino acids outside epitopes above the threshold (C); and finally one for the number of amino acids outside epitopes below the threshold (D). The threshold for A and C is $+ 0.7 \times$ standard error (25% of the upper population) while the threshold for B and D is $-0.7 \times$ standard error (25% of the lower population).

The homogeneity of a prediction is the difference between the number of amino acids well predicted and the number of amino acids wrongly predicted. Thus, predictions giving about the same number of amino acids inside and outside epitopes would be homogeneous and would give low values of χ^2. On the other hand, when there is an imbalance in the number of well-predicted and wrongly predicted amino acids, the prediction is heterogeneous, with a high value of χ^2. However, while a homogeneous prediction is always of bad quality, a heterogeneous prediction is not

[46] E. D. Getzoff, J. A. Tainer, R. A. Lerner, and H. M. Geysen, *Adv. Immunol.* **43,** 1 (1988).

[47] H. M. Geysen, J. A. Tainer, S. J. Rodda, T. J. Mason, H. Alexander, E. D. Getzoff, and R. A. Lerner, *Science* **235,** 1184 (1987).

TABLE I
CONTINGENCY TABLE (2 × 2) USED TO CALCULATE χ^2 VALUES

Number of amino acids	Number of amino acids	
	In epitopes	Outside epitopes
Above threshold $+0.7\sigma$	A[a]	C
Below threshold -0.7σ	B	D

[a] A and D correspond to a correct prediction (peak within an epitope and valley outside an epitope). B and C correspond to a wrong prediction (peak outside an epitope and valley within an epitope). The threshold is set to $\pm 0.7 \times$ standard error, which corresponds to 50% of the population. The χ^2 is calculated as follows:

$$\chi^2 = (AD - BC)^2(A + B + C + D)/(A + B)(C + D) \times (A + C)(B + D)$$

Values above 3.84 are significant (one degree of freedom). A negative sign is used to characterize a high χ^2 corresponding to a wrong prediction; this sign has no mathematical meaning. In the case of myoglobin analyzed by the method of Parker *et al.*[39] (see Fig. 1), the values are $A = 26$, $B = 4$, $C = 18$, and $D = 35$.

$$\chi^2 = (910 - 72)^2(83)/(30 \times 53 \times 44 \times 39) = 21.36.$$

necessarily of good quality. Indeed, the direction in the homogeneity is not apparent. From Table I, one can notice that a high χ^2 value can be obtained in either of the following situations: the product (AD) is larger than (BC) or (BC) is larger than (AD). The former case is obtained for a good prediction and the latter one for a bad prediction. Instances of high χ^2 values associated with a bad prediction of antigenicity are marked with a negative sign. Values of χ^2 below 3.84 are not statistically significant at the level of 0.05.

We have also used two other criteria, one simple and one more complex. The simple one consists of counting only the amino acids above a threshold (entries A and C) and calculating the ratio. Since usually the aim of predicting continuous epitopes is to obtain guidance for the synthesis of peptides, it is natural to consider only the amino acids falling within peaks. In this case, a 100% correct prediction would yield a profile in which all the peaks belong to known epitopes.

TABLE II
CONTINGENCY TABLE (3 × 2) USED TO CALCULATE χ^2 VALUES

Number of amino acids	Number of amino acids	
	In epitopes	Outside epitopes
Above threshold $+0.7\sigma$	A[a]	C
Between $+0.7\sigma$ and -0.7σ	E	F
Below threshold -0.7σ	B	D

[a] The thresholds are the same as in Table I. E and F correspond to doubtful predictions in the central area of the profile. The χ^2 is calculated as follows:

$$\chi^2 = (N-1)(A+B+C+D+E+F)$$

where $N = M_1/(A+E+B) + M_2/(C+F+D)$; $M_1 = A/(A+C) + E/(E+F) + B/(B+D)$; $M_2 = C/(A+C) + F/(E+F) + D/(B+D)$. In the case of myoglobin analyzed by the method of Parker *et al.*,[39] the values are $A = 26$, $B = 4$, $C = 18$, $D = 35$, $E = 24$, $F = 40$, $M_1 = 24.77$, $M_2 = 63.77$, $N = 1.14$, and $\chi^2 = 20.58$.

We also used a 3 × 2 contingency table (see Table II) for testing the possibility of the independence of the two variables ("epitope" and "propensity threshold"). Independence would imply that the number of correctly and wrongly predicted amino acids is the same within random sampling error for each scale.

In this case, Table II has three lines and two columns and the degree of freedom is $(3-1)(2-1) = 2$. The threshold value for significance is 5.99.

The calculation was made according to the simplified form given by Skory.[48]

Comparison of 22 Scales Applied to 11 Proteins

Description of Computer Program: PREDITOP. Plots displaying regions of a protein with antigenic potential were generated with an in-house program written in Turbo-Pascal 5.5 for IBM PC-compatible computers (see Fig. 1). The program is user friendly and its use does not require computer knowledge. To obtain a prediction plot, a number of questions must be answered concerning the width and the center of the window,

[48] J. Skory, *Biometrics* **8,** 380 (1952).

whether or not one wishes to smooth out the curves, etc. The program works with several files: a file containing the sequence of the protein, a dictionary file containing the propensities for a particular characteristic of each amino acid, a file containing the known epitopes, and finally a file containing the secondary structure elements in case the three-dimensional structure of the protein analyzed has been determined. File names are composed of a fixed part corresponding to its function and of a keyword denoting the protein being analyzed. Thus, the file name DATAMYO corresponds to the sequence file (DATA) of myoglobin (MYO). The various scales have the author's name attached to them (if not longer than eight characters). All files are formatted (American Standard Code for Information Interchange, ASCII) and can thus be modified with a text editor. Further, since the scales have been normalized between −3 and +3 it is possible to superimpose several prediction plots. The user can also create a scale corresponding to some physicochemical or structural parameter (e.g., the charge of the amino acid) via the text editor as long as the order of the amino acid is respected. Answering by default is systematically implemented except for the protein keyword, which the user is free to choose. The program can be obtained from the authors on $5\frac{1}{4}$-in. IBM-compatible diskettes.

Comparative Value of Different Scales. The validity of the 22 scales listed in Table III for predicting antigenicity has been analyzed using 11 well-studied proteins containing a total of 66 identified epitopes. The 22 scales are grouped as follows: 9 scales of inverted hydrophobicity (DOOLITTL, HEIJNE, MANAVALA, PRILS, ROSE, SWEET, TOTLS, GES, ZIMMERMA); 2 scales of hydrophilicity (HOPP, PARKER); 4 scales of accessibility (JANIN, CHOTHIA, CHOTHIA8, ACROPHIL); 2 scales of flexibility (KARPLUS, RAGONE); 1 scale of antigenicity (WELLING); 3 scales of secondary structure (CHOUF3, GARNIER3, LEVITT), and the particular scale EMINI. The latter scale is not normalized between +3 and −3 because the authors,[44] using the original values of Janin *et al.*,[43] based their calculation on multiplication of the amino acids inside a window of 6 residues.

The epitopes of 9 of the 11 proteins have been extensively studied: cholera toxin (CHO), cytochrome *c* (CYT), histone H2A (H2A), hepatitis B virus (HBV), leghemoglobin (LEG), lysozyme (LYS), myohemerythrin (MHR), myoglobin (MYO), and tobacco mosaic virus (TMV). The antigenic structures of two of the proteins, bovine serum albumin (BSA) and dengue virus envelope (DEN), are less extensively known. All epitopes that were considered are listed in Table IV. Their size ranged from 6 to 17 residues.

TABLE III

NORMALIZED SCALES FOR 20 COMMON AMINO ACIDS USED TO CONSTRUCT ANTIGENICITY PREDICTION PROFILES

Scale[a]	Arg	Asp	Glu	Lys	Ser	Asn	Gln	Gly	Pro	Thr	Ala	His	Cys	Met	Val	Ile	Leu	Tyr	Phe	Trp	Ref.
DOOLITTL	3.00	2.30	2.30	2.60	0.50	2.30	2.30	0.30	1.10	0.50	−1.20	2.10	−1.70	−1.30	−2.80	−3.00	−2.50	0.90	−1.90	0.60	*19*
HEIJNE	3.00	1.82	1.37	0.86	0.04	0.46	0.31	−0.40	0.59	−0.13	−0.70	0.62	−0.85	−1.21	−1.00	−1.15	−1.12	−0.07	−1.42	−1.00	*b*
MANAVALA	1.22	2.14	1.04	1.60	1.74	1.53	1.17	0.47	1.59	1.25	−0.10	0.75	−1.85	−1.60	−3.00	−2.95	−2.14	−0.57	−1.19	−1.11	*c*
PRILS	−0.12	2.63	1.49	2.05	1.13	0.95	1.69	0.83	1.27	1.45	0.55	0.39	−1.61	−2.14	−2.26	−2.16	−3.00	−1.34	−2.14	0.29	*49*
ROSE	1.10	1.50	1.50	3.00	0.80	1.30	1.50	−0.10	1.10	0.20	−0.40	−1.00	−3.00	−2.10	−2.30	−2.50	−2.10	−0.70	−2.50	−2.10	*d*
SWEET	0.92	2.04	1.90	1.04	0.86	1.43	1.42	1.04	0.76	0.43	0.62	1.00	−0.26	−1.59	−1.42	−1.95	−1.90	−2.60	−3.00	−0.78	*e*
TOTLS	0.40	2.99	1.33	2.58	1.34	2.10	1.27	1.11	2.08	1.32	0.42	−0.60	−1.56	−2.47	−2.41	−2.74	−3.00	−1.65	−2.75	0.22	*49*
GES	3.00	2.14	1.87	2.03	−0.54	0.94	0.74	−0.65	−0.32	−0.70	−0.81	0.44	−0.92	−1.30	−1.08	−1.22	−1.14	−0.18	−1.39	−0.89	*f*
ZIMMERMA	0.74	1.06	1.04	−0.54	1.89	1.98	2.13	1.96	−2.38	1.22	0.74	0.29	−0.34	−0.20	−0.86	−3.00	−2.08	−2.83	−2.46	1.61	*g*
HOPP	3.00	3.00	3.00	3.00	0.30	0.20	0.20	0.00	0.00	−0.40	−0.50	−0.50	−1.00	−1.30	−1.50	−1.80	−1.80	−2.30	−2.50	−3.40	*32*
PARKER	0.87	2.46	1.86	1.28	1.50	1.64	1.37	1.28	0.30	1.15	0.03	0.30	0.11	−1.41	−1.27	−2.45	−2.78	−0.78	−2.78	−3.00	*39*
JANIN	2.10	0.30	0.60	3.00	−0.80	0.10	0.60	−1.70	−0.30	−0.60	−1.70	−0.80	−3.00	−1.90	−2.30	−2.60	−2.10	−0.10	−2.10	−1.70	*43*
CHOTHIA	3.00	1.60	1.30	2.80	0.90	1.90	2.40	−0.60	1.30	0.80	−0.80	1.40	−2.00	−1.00	−2.40	−3.00	−1.50	1.60	−2.00	0.40	*35*
CHOTHIA8	−1.75	0.95	−0.11	−0.68	1.84	0.58	−0.32	2.98	0.78	1.04	2.11	−0.54	1.18	−0.66	0.35	−0.59	−0.56	−1.81	−1.80	−3.00	*h*
ACROPHIL	0.30	2.10	0.50	1.40	1.80	2.30	−0.20	3.00	2.60	−0.10	−0.50	−0.40	−2.60	−1.80	−1.70	−2.50	−2.50	−2.00	−2.70	−3.00	*36*
KARPLUS	−0.20	−0.30	1.20	1.10	3.00	1.70	2.90	2.30	0.20	0.60	−0.10	−1.60	−2.10	−2.50	−1.60	−1.10	−2.00	−2.10	−2.90	−3.00	*20*
RAGONE	−0.88	1.39	0.28	0.05	1.92	0.96	0.17	2.57	1.12	1.22	1.76	−0.13	0.55	−1.08	−0.56	−1.50	−1.18	−1.70	−1.96	−3.00	*i*
WELLING	0.80	0.90	−0.30	2.10	0.10	−0.30	0.20	−1.30	−0.10	−0.10	1.30	3.00	−0.70	−3.00	0.20	−2.20	1.00	0.40	−0.90	−0.70	*41*

CHOUF3	−0.21	2.47	−1.32	0.09	2.31	3.00	−0.06	3.00	2.78	−0.16	−1.74	−0.21	1.04	−2.06	−2.59	−2.75	−2.11	0.78	−2.06	−0.16	*25*
GARNIER3	1.01	1.50	−2.35	0.46	1.26	2.05	0.17	2.79	1.75	0.12	−2.50	−0.17	2.15	−2.40	−3.00	−2.30	−2.80	1.40	−0.91	1.75	*26*
LEVITT	−0.22	1.43	0.11	0.02	1.15	1.02	0.08	2.15	3.00	0.27	−0.56	−0.85	−0.44	−1.69	−1.50	−1.38	−1.16	0.30	−1.13	−0.60	*33*
EMINI	0.95	0.81	0.84	0.97	0.65	0.78	0.84	0.48	0.75	0.70	0.49	0.66	0.26	0.48	0.36	0.34	0.40	0.76	0.42	0.51	*44*

[a] The name of each scale usually corresponds to the first eight letters of the author's name, however, ACROPHIL refers to the accessibility scale of Hopp[36] and GES to the scale of Engelman *et al.*[f] The scales are grouped as follows: Inverted hydrophobicity scales: PRILS and TOTLS (scales of Cornette *et al.*[49]), DOOLITTL (Kyte and Doolittle[19]), GES (Engelman *et al.*[f]), HEIJNE (Von Heijne[b]), MANAVALA (Manavalan and Ponnuswamy[c]), ROSE (Rose *et al.*[d]), SWEET (Sweet and Eisenberg,[e]), and ZIMMERMA (Zimmerman *et al.*[g]); hydrophilicity scales: HOPP (Hopp and Woods[32]) and PARKER (Parker *et al.*[39]); accessibility scales: CHOTHIA,[35] CHOTHIA8,[h] JANIN (Janin *et al.*[43]), ACROPHIL (Hopp[36]), and EMINI (Emini *et al.*[44]); flexibility scales: KARPLUS (Karplus and Schultz[20]) and RAGONE (Ragone *et al.*[i]); antigenicity scale: WELLING (Welling *et al.*[41]); scales for tunrns: CHOUF3 (Chou and Fasman[25]), GARNIER3 (Garnier *et al.*[26]), and LEVITT (Levitt[33]). The published original values of the scales have been normalized between +3 and −3 by the program PREDITOP, except in the case of the EMINI scale, which was left unchanged (see text).

[b] G. Von Heijne, *Eur. J. Biochem.* **116,** 419 (1981).

[c] P. Manavalan and P. K. Ponnuswamy, *Nature (London)* **275,** 673 (1978).

[d] G. D. Rose, A. R. Geselowitz, G. J. Lesser, R. H. Lee, and M. H. Zehfus, *Science* **229,** 834 (1985).

[e] R. M. Sweet and D. J. Eisenberg, *Mol. Biol.* **171,** 479 (1983).

[f] D. M. Engelman, T. A. Steitz, and A. Goldman, *Annu. Rev. Biophys. Biophys. Chem.* **15,** 321 (1986).

[g] J. M. Zimmerman, N. Eliezer, and R. J. Simha, *Theor. Biol.* **21,** 170 (1968).

[h] C. Chothia, *Annu. Rev. Biochem.* **53,** 537 (1984).

[i] R. Ragone, F. Facchiano, A. Facchiano, A. M. Facchiano, and G. Colonna, *Protein Eng.* **2,** 497 (1989).

TABLE IV
CONTINUOUS EPITOPES OF 11 PROTEINS USED FOR EVALUATING DIFFERENT PREDICTION SCALES IN THE PRESENT STUDY[a]

Source	Epitope	Ref.
Cytochrome (CYT)	13–25	*b*
	42–50	*c*
	61–73	*d*
Leghemoglobin (LEG)	15–23	*e*
	52–59	*e*
	92–98	*e*
	107–116	*e*
	132–142	*e*
Lysozyme (LYS)	18–27	*3*
	38–54	*f*
	64–80	*g*
	116–121	*3*
Myoglobin (MYO)	1–6	*h*
	15–22	*i*
	48–55	*j*
	56–62	*r*
	94–99	*i*
	113–119	*i*
	121–127	*h*
	145–151	*i*
Myohemerythrin (MHR)	4–9	*k*
	16–21	*k*
	37–46	*38*
	54–58	*k*
	63–72	*k*
	80–85	*k*
	90–95	*k*
	110–115	*k*
Tobacco mosaic virus (TMV)	1–10	*l*
	19–32	*30*
	34–39	*m*
	55–61	*l*
	62–68	*n*
	76–88	*30*
	103–112	*o*
	134–146	*30*
	149–158	*m*
Histone 2A (H2A)	1–15	*p*
	5–18	*p*
	28–42	*p*
	44–61	*p*
	56–70	*p*
	65–85	*p*
	85–100	*p*
	90–105	*p*
Bovine serum albumin (BSA)	137–146	*q*
	168–179	*q*
	308–314	*q*
	328–337	q
	359–362	*q*
	526–535	*q*
	553–556	*q*
	559–565	*o*
Dengue polyprotein virus (DEN)	198–211	*r*
	259–272	*r*
	298–311	*r*
	319–334	*r*
Hepatitis B virus (HBV)	2–16	*s*
	22–35	*s*
	69–79	*t*
	95–109	*s*
	125–137	*u*
	139–147	*v*
Cholera toxin (CHO)	50–64	*w*
	69–85	*w*
	83–97	*w*

[a] Overlapping continuous epitopes were joined end to end, so as not to count residues twice.
[b] G. Corradin, M. A. Juillerat, C. Vita, and H. D. Engers, *Mol. Immunol.* **20,** 763 (1983).
[c] M. Z. Atassi, *Mol. Immunol.* **18,** 1021 (1981).
[d] R. Jemmerson and Y. Paterson. *BioTechniques* **4,** 18 (1986).
[e] J. G. R. Hurrell, J. A. Smith, and S. J. Leach, *Immunochemistry* **15,** 297 (1978).
[f] Y. Takagaki, A. Hirayama, H. Fujio, and T. Amano, *Biochemistry* **19,** 2498 (1980).

The criteria used for comparing the scales are χ^2 values and the ratio of correctly predicted over wrongly predicted residues.

The χ^2 values calculated for each method on the 11 tested proteins are listed in Table V. It is immediately apparent that most χ^2 values for lysozyme and myoglobin are abnormally high. This is due to the fact that these two proteins, which have been extensively studied, were used to construct and refine most of the prediction scales. If one ignores the data for these two proteins, it can be seen that χ^2 values higher than the threshold value for statistical significance, i.e., 3.84, are not very numerous (only 34 χ^2 values out of 198). When the χ^2 values for each scale obtained with the different proteins (except LYS and MYO) were averaged, only the scales CHOTHIA8 and LEVITT gave values slightly higher than the threshold, i.e., 4.06 and 3.86 respectively. However, these average values are strongly influenced by a few large χ^2 values and do not represent a general trend in all the proteins. In general, there are only between one and two good predictions (above 3.84) per scale per protein. The homogeneity of these data indicates that none of the prediction methods achieves a high level of correct prediction and that none is strikingly better than the others.

[g] E. Teicher, E. Maron, and R. Arnon, *Immunochemistry* **10,** 265 (1973).
[h] H. E. Schmitz, H. Atassi, and Z. Atassi, *Immunol. Commun.* **12,** 161 (1983).
[i] M. Z. Atassi, *Immunochemistry* **12,** 423 (1975).
[j] S. J. Rodda, H. M. Geysen, T. J. Mason, and P. G. Schoofs, *Mol. Immunol.* **23,** 603 (1986).
[k] E. D. Getzoff, H. M. Geysen, S. J. Rodda, H. Alexander, J. A. Tainer, and R. A. Lerner, *Science* **235,** 1191 (1987).
[l] D. Altschuch, D. Hartmann, J. Reinbolt, and M. H. V. Van Regenmortel, *Mol. Immunol.* **20,** 271 (1983).
[m] D. Altschuh, Z. Al Moudallal, J. P. Briand, and M. H. V. Van Regenmortel, *Mol. Immunol.* **22,** 329 (1985).
[n] R. C. de L. Milton and M. H. V. Van Regenmortel, *Mol. Immunol.* **16,** 179 (1979).
[o] P. R. Morrow, D. M. Rennick, C. Y. Leung, and E. Benjamini, *Mol. Immunol.* **21,** 301 (1984).
[p] S. Muller, S. Plaue, M. Couppez, and M. H. V. Van Regenmortel, *Mol. Immunol.* **23,** 593 (1986).
[q] M. Z. Atassi, *Eur. J. Biochem.* **145,** 1 (1984).
[r] L. J. Markoff, M. Bray, C.-J. Lai, R. M. Chanock, K. Eckels, P. Summers, M. K. Gentry, R. A. Houghten, and R. A. Lerner, *in* "Vaccines 88," (H. Ginsberg, F. Brown, R. A. Lerner, and R. M. Chanock, eds.), p. 161. Cold Spring Harbor Laboratory, Cold Spring Harbor, New York, 1988.
[s] R. A. Lerner, R. Green, H. Alexander, F. T. Liu, J. G. Sutcliffe, and T. M. Shinnick, *Proc. Natl. Acad. Sci. U.S.A.* **78,** 3403 (1981).
[t] A. R. Neurath, S. B. H. Kent, and N. Srick, *Virus Res.* **1,** 321 (1984).
[u] J. L. Gerin, H. Alexander, J. Wai-Kuo Shih, R. H. Purcell, G. Dapolito, R. Engle, N. Green, J. G. Sutcliffe, T. M. Shinick, and R. A. Lerner, *Proc. Natl. Acad. Sci. U.S.A.* **80,** 2365 (1983).
[v] P. K. Bhatnagar, E. Papas, H. E. Blum, D. R. Milich, D. Nitecki, M. J. Karels, and G. N. Vyas, *Proc. Natl. Acad. Sci. U.S.A.* **79,** 4400 (1982).
[w] C. O. Jacob, M. Sela, and R. Arnon, *Proc. Natl. Acad. Sci. U.S.A.* **80,** 7611 (1983).

TABLE V

COMPARATIVE VALUES OF DIFFERENT SCALES APPLIED TO 11 PROTEINS DETERMINED FROM χ^2 VALUES FROM 2 × 2 CONTINGENCY TABLE[a,b]

Scales	BSA	CHO	CYT	DEN	H2A	HBV	LEG	LYS	MYO	MHR	TMV
DOOLITTL	0.68	0.06	2.25	1.07	0.03	0.01	3.77	21.47	12.72	1.59	0.42
HEIJNE	3.20	0.28	0.50	0.36	0.78	0.00	**6.09**[c]	16.25	7.12	3.63	2.67
MANAVALA	0.00	0.21	0.32	0.01	0.40	0.02	1.32	26.45	16.61	**3.86**	**4.63**
PRILS	0.11	1.32	0.28	**7.33**	2.54	0.25	0.60	16.31	18.65	3.19	1.63
ROSE	1.12	2.61	1.73	3.14	0.18	0.41	0.10	18.03	17.37	2.54	0.00
SWEET	0.32	3.67	0.95	0.08	0.24	0.46	0.18	7.52	18.62	1.36	0.14
TOTLS	0.00	0.83	0.28	**5.04**	1.30	0.00	1.65	21.22	17.45	2.33	1.06
GES	**4.68**	2.61	0.23	0.00	**5.40**	0.14	1.10	6.68	8.59	0.23	**7.74**
ZIMMERMA	2.11	0.08	**3.95**	0.07	1.38	**6.13**	0.00	3.30	8.99	0.41	0.27
HOPP	1.32	1.83	1.74	1.68	3.25	0.75	0.00	2.89	10.70	0.01	0.53
PARKER	0.05	0.16	1.24	0.70	0.45	0.26	0.33	27.34	21.36	3.06	0.02
JANIN	**4.49**	3.72	−4.21	2.30	0.87	3.20	0.01	3.25	3.94	1.76	2.29
CHOTHIA	1.11	0.10	0.27	1.88	0.00	0.06	3.21	23.16	8.40	3.61	0.00
CHOTHIAS	3.28	0.05	0.01	0.00	**10.48**	3.30	0.00	4.38	6.68	**5.91**	**13.58**
ACROPHIL	0.00	0.40	1.91	0.10	0.50	0.08	**3.89**	26.56	12.67	**4.74**	**4.74**
KARPLUS	−8.65	0.02	0.05	0.02	**8.40**	1.18	1.37	20.03	2.50	**4.47**	0.52
RAGONE	0.96	0.94	0.51	0.34	**6.71**	0.51	0.10	14.70	11.65	3.68	**12.21**
WELLING	**10.96**	**16.37**	1.90	**6.15**	**9.11**	0.07	−11.97	−6.85	1.83	1.66	0.40
CHOUF3	1.08	0.16	1.94	0.02	1.76	0.16	**7.88**	42.10	3.76	0.37	**5.00**
GARNIER3	1.35	0.19	0.68	1.87	0.31	1.35	**7.12**	22.91	0.35	0.00	3.19
LEVITT	0.08	0.54	0.13	1.86	0.35	0.47	**10.76**	42.97	4.18	**9.43**	**11.14**
EMINI	0.72	**4.16**	3.18	**5.32**	0.09	2.34	**4.59**	−14.56	4.58	3.31	0.01

[a] See Table I.

[b] Values of χ^2 below 3.84 are not statistically significant. A high χ^2 corresponds to a heterogeneous prediction, while a low χ^2 corresponds to a homogeneous prediction. Zero means that χ^2 was $< 10^{-3}$.

[c] Values in boldface are statistically significant.

The results in Table V were confirmed by another method of analysis illustrated in Table VI, which calculates how many residues were correctly or wrongly predicted to be antigenic. Two of the proteins, i.e. BSA and DEN, have not been extensively studied as far as their antigenic structure is concerned, and only 11–12% of their residues has been implicated in known epitopes (Table VI). This is no doubt the reason why the ratio of correctly over wrongly predicted residues is particularly low in these two cases (0.14 and 0.18). If these two proteins are ignored, the average ratio for all proteins is 1.1, corresponding to 53% correct predictions (Table VI).

The various parameters used for comparing the efficacy of the different prediction scales are summarized in Table VII. The first column of Table VII gives the average 3 × 2 χ^2 values for all proteins except BSA and DEN.

It can be seen that almost all values are above the χ^2 threshold of significance (5.99). This means that results obtained with each scale are not random. Thus, for instance, a hydrophobicity scale makes significant distinction between the contribution of hydrophobic and hydrophilic residues to epitopes.

From the ratios of correctly over wrongly predicted residues (Table VII), it appears that only two scales, LEVITT and EMINI, correctly identified 60% of the antigenic residues (ratios of 1.56 and 1.47, respectively).

The various hydrophobicity and hydrophilicity scales gave similar results, leading to 51–57% correct predictions (Table VII). In this case, the best values were obtained with the methods of Parker *et al.*[39] and Cornette *et al.*[49] (scale TOTLS). In contrast, the accessibility scales gave only 46–52% correct predictions, whereas the scales that predict turns gave a slightly higher level of correct prediction (53–61%). Krchnak *et al.*[50] also showed that peptide sequences that are predicted to possess a β-turn conformation tend to induce antibodies that are able to cross-react with the parent protein.

The lowest level of correct prediction was obtained with the scale WELLING, a finding that agrees with the rating of different methods reported previously.[21] In this earlier study only eight scales and four proteins had been studied, two of which were myoglobin and lysozyme. In that case only the scales of Parker *et al.*[39] and Karplus and Schulz[20] were found to give slightly better results than the others. In the present, more extensive comparisons, five scales gave marginally better results than the others: those of Emini *et al.*,[44] Levitt,[33] Chou and Fasman,[25] Parker *et al.*,[39] and Cornette *et al.*[49]

Conclusions

The results presented in Tables V–VII confirm earlier studies of the comparative value of different antigenicity prediction methods[21,46] and demonstrate that none of the prediction algorithms in current use gives a high level of correct prediction. This may seem surprising in view of the wide popularity of some of these prediction methods. The high scores of successful predictions noted by some authors[45,51] in the past probably result from the use of unreliable criteria for assessing the validity of prediction. Since the 2×2 χ^2 values (Table V) obtained with the different scales

[49] J. L. Cornette, K. B. Cease, H. Margalit, J. L. Spouge, J. A. Berzofsky, and C. De Lisi, *J. Mol. Biol.* **195,** 659 (1987).

[50] V. Krchnak, O. Mach, and A. Maly, this series, Vol. 178, p. 586.

[51] W. J. Weijer, G. W. Welling, and S. Welling-Wester, "Vaccines 86." (F. Brown, R. M. Chanock, and R. A. Lerner, eds.), p. 71. Cold Spring Harbor Laboratory, Cold Spring Harbor, New York, 1986.

TABLE VI

COMPARATIVE VALUE OF DIFFERENT PREDICTION SCALES ANALYZED ON 11 PROTEINS[a]

Scales	BSA 64/582 = 11%		CHO 46/103 = 45%		CYT 35/104 = 34%		DEN 58/494 = 12%		H2A 91/129 = 70%		HBV 77/171 = 45%		LEG 45/153 = 30%		LYS 50/129 = 39%		MYO 56/153 = 37%		MHR 55/118 = 47%		TMV 90/158 = 57%	
	A	*C*	*A*	*C*	*A*	*C*	*A*	*C*	*A*	*C*	*A*	*C*	*A*	*C*	*A*	*C*	*A*	*C*	*A*	*C*	*A*	*C*
DOOLITTL	27	144	15	14	11	21	14	114	24	14	24	22	11	20	31	14	24	27	17	17	23	22
HEIJNE	27	118	9	10	5	15	9	103	29	14	24	10	15	15	22	10	15	24	10	14	14	16
MANAVALA	23	120	14	13	11	18	7	101	24	15	24	25	11	25	29	6	22	19	21	15	33	14
PRILS	20	114	18	9	13	19	19	102	22	14	24	27	9	29	26	5	25	18	19	16	28	16
ROSE	21	116	14	6	6	19	13	83	24	11	19	24	8	23	25	15	18	10	16	11	20	15
SWEET	28	120	19	9	13	18	16	112	24	14	29	25	10	28	25	9	27	22	14	11	21	21
TOTLS	21	118	17	10	13	19	17	109	25	12	27	26	13	24	26	5	25	13	17	13	27	19
GES	29	140	15	10	6	22	15	90	29	12	22	11	13	18	16	11	19	21	12	18	14	22
ZIMMERMA	15	108	14	15	16	12	15	115	18	15	30	11	12	26	26	16	18	16	13	15	30	19
HOPP	29	138	13	8	5	21	16	94	26	14	24	16	10	19	20	16	24	23	11	16	17	14
PARKER	27	143	18	13	14	18	16	116	25	14	28	21	7	25	25	6	26	18	19	13	28	17

(Header spanning all protein columns: Percentage of antigenic amino acids)

JANIN	19	93	13	5	3	17	13	58	17	9	17	5	4	11	9	9	12	13	4	12	8	15
CHOTHIA	27	142	14	13	11	21	13	115	26	15	23	26	11	23	31	10	23	24	17	13	24	22
CHOTHIA8	23	147	11	14	15	16	25	130	17	12	18	30	7	40	19	25	17	23	19	8	32	16
ACROPHIL	20	112	12	11	10	21	10	104	24	13	18	25	15	16	20	6	18	19	18	10	25	13
KARPLUS	6	128	13	12	11	15	8	111	16	15	24	22	13	20	22	12	12	18	14	10	27	14
RAGONE	25	141	16	12	14	17	25	133	23	17	20	30	9	34	23	18	20	22	19	8	31	13
WELLING	31	146	18	10	8	26	21	99	29	12	10	23	5	37	1	23	23	31	18	15	21	22
CHOUF3	20	115	13	8	16	10	11	107	16	11	14	23	15	12	30	7	11	15	13	10	23	12
GARNIER3	19	98	12	9	13	14	7	102	12	8	19	27	14	12	28	13	6	15	11	10	20	12
LEVITT	18	115	12	10	17	19	12	116	20	13	16	25	15	10	28	15	15	14	18	5	28	7
EMINI	14	68	16	3	9	9	6	66	13	9	17	8	13	15	19	6	15	20	11	8	25	17
Total[b]	489	2684	316	224	240	387	308	2280	483	283	471	462	240	482	501	257	415	425	331	268	519	358
Ratio[c]	0.18		1.41		0.62		0.14		1.71		1.02		0.50		1.95		0.98		1.24		1.45	

[a] For each protein, the first column corresponds to correctly predicted amino acids (*A*, Table I), the second to wrongly predicted amino acids (*C*, Table I). The total number of correctly and wrongly predicted residues for the 11 proteins was 4313 and 8110, respectively. The ratio of correctly predicted over wrongly predicted residues is thus 0.53, corresponding to 35% correct predictions. If the values for the less studied proteins, BSA and DEN, are subtracted, the ratio becomes 1.1 and the percentage correct predictions is 53%.

[b] Total number of correctly and wrongly predicted amino acids by various methods for each protein.

[c] The ratio corresponds to the ratio of the two totals in *b*.

TABLE VII

COMPARATIVE VALUE OF DIFFERENT PREDICTION SCALES APPLIED TO NINE PROTEINS[a]

Scales	Average $3 \times 2\ \chi^2$ [b]	σ of χ^2 [c]	Correctly predicted[d]	Wrongly %	Ratio[e]	Correct prediction[b]
DOOLITTLE	6.96	7.31	180	171	1.05	51
HEIJNE	11.39	9.25	143	128	1.11	53
MANAVALA	8.70	10.54	189	150	1.26	51
PRILS	8.88	9.50	184	153	1.20	55
ROSE	6.48	6.67	150	134	1.11	53
SWEET	8.88	8.55	182	157	1.15	54
TOTLS	8.69	10.28	190	141	1.34	57
GES	10.14	8.12	149	145	1.02	51
ZIMMERMA	6.15	6.35	177	145	1.22	55
HOPP	7.98	6.36	150	147	1.02	51
PARKER	8.92	9.57	190	145	1.31	57
JANIN	5.30	2.37	87	96	0.90	48
CHOTHIA	9.40	6.89	180	167	1.07	52
CHOTHIA8	6.23	8.96	155	184	0.84	46
ACROPHIL	5.98	5.25	142	124	1.14	53
KARPLUS	5.87	5.57	152	138	1.10	52
RAGONE	8.52	6.10	175	181	0.96	49
WELLING	9.77	6.21	133	236	0.56	36
CHOUF3	8.69	11.76	151	111	1.36	58
GARNIER3	6.31	6.37	135	120	1.12	53
LEVITT	11.21	12.46	169	108	1.56	61
EMINI	5.99	4.65	138	95	1.45	60

[a] All proteins of Table VI except BSA and DEN.

[b] The χ^2 column corresponds to the mean χ^2 for each protein obtained from 3×2 contingency table.

[c] The σ column corresponds to the standard deviation of the χ^2.

[d] The columns correctly predicted and wrongly predicted correspond, respectively, to the number of correctly predicted and wrongly predicted amino acids above the cutoff level ($+0.7 \times$ standard error).

[e] Ratio of correctly predicted/wrongly predicted amino acids.

[f] Ratio [correctly predicted/(correctly predicted + wrongly predicted)] expressed as percentage correct prediction.

are very variable, it seems best to rely on the percentage of correctly predicted residues obtained with each method (Table VII). According to this criterion, the widely used hydrophilicity scale of Parker *et al.*,[39] the scales of Cornette *et al.*,[49] the turn scales of Chou and Fasman[25] and

Levitt,[33] and the method of Emini *et al.*[44] give slightly better results than the other scales.

These globally mediocre results reflect the fact that no single scale contains enough information to allow a transformation from primary structure data to a tertiary structure entity (the epitope). Clearly, available prediction methods based on a unidimensional analysis do not cope satisfactorily with the three-dimensional reality of antigenic sites. The weak correlation between amino acid type and secondary structure element or three-dimensional motif makes matters worse. This is highlighted by the fact that proteins with only 20–30% overall sequence similarity can possess identical tertiary foldings.[52] Furthermore, identical short peptide sequences in unrelated proteins can have different conformation.[53] Thus, even a scale derived from a set of antigenically well-known proteins will give no guarantee of success on an unknown protein. Since antigenic sites defined as continuous epitopes are strongly dependent on interactions with other parts of the overall tertiary fold, antigenicity predictions are plagued by the same problems encountered with secondary structure prediction algorithms.[29]

It seems likely that more successful prediction methods will have to include both a combination of scales and information related to the three-dimensional structure of the proteins. In the absence of any clear-cut superiority for any of the prediction methods in current use (Table VII), there is obviously a need for improving existing algorithms and prediction scales. Such improvement is likely to be possible only if the level of correct prediction achieved with any proposed method is submitted to quantitative evaluation and compared with other methods. The approach used in the present chapter should help to achieve that goal.

[52] D. J. Neidhart, G. L. Kenyon, J. A. Gerlt, and G. A. Petsko, *Nature (London)* **347**, 692 (1990).

[53] I. A. Wilson, D. H. Haft, E. D. Getzoff, J. A. Tainer, R. A. Lerner, and S. Brenner, *Proc. Natl. Acad. Sci. U.S.A.* **82,** 5255 (1985).

[9] Use of Two-Dimensional ^1H Nuclear Magnetic Resonance to Study High-Affinity Antibody–Peptide Interactions

By JANET C. CHEETHAM, CHRISTINA REDFIELD, ROBERT E. GRIEST, DANIEL P. RALEIGH, CHRISTOPHER M. DOBSON, and ANTHONY R. REES

Introduction

The mammalian immune system invokes a complex assembly of molecules that have evolved over millions of years. A key player in this defense mechanism is the antibody, showing a vast repertoire of specificities ($> 10^8$), for antigens that range in size from small-molecule haptens to large glycoproteins. The genetic origins of this diversity are now well understood,[1] but our knowledge of the structural parameters that control antibody specificity is still rudimentary. To improve our understanding of this special problem in molecular recognition, considerable effort has been focused on structural studies of antibodies and their antigen-bound complexes.

One of the most intriguing aspects of antibody recognition is the ability of some antibodies raised against peptide fragments of proteins to cross-react with the intact native molecules.[2-4] This phenomenon not only has considerable significance from the point of view of protein structure and folding, but has also generated much interest in the possibility of using small peptides in a pharmacological role as synthetic vaccines.[5,6] To address the question of what determines whether a given peptide is able to act as a good mimic of a protein epitope, it is clearly important to understand the way in which antibodies interact with both protein and peptide antigens. At the structural level, X-ray diffraction studies have revealed much about the nature of the complex between an antibody and a protein antigen.[7] The dearth of similar information relating to the binding of peptide antigens from this source, however, reflects the difficulty of ob-

[1] S. Tonegawa, *Nature (London)* **302,** 575 (1983).
[2] M. Sela, B. Schechter, I. Schechter, and F. Borek, *Cold Spring Harbor Symp. Quant. Biol.* **32,** 537 (1967).
[3] R. A. Lerner, *Nature (London)* **299,** 592 (1982).
[4] H. J. Dyson, R. A. Lerner, and P. E. Wright, *Annu. Rev. Biophys.* **17,** 305 (1988).
[5] R. Porter and J. Whelan, in "Synthetic Peptides as Antigens: CIBA Foundation Symposium," p. 119. John Wiley & Sons, New York, 1986.
[6] R. Arnon, *Trends Biochem. Sci.* **11,** 512 (1986).
[7] D. Davies, E. A. Padlan, and S. Sheriff, *Annu. Rev. Biochem.* **59,** 439 (1990).

taining good crystals of such complexes, probably due to the rather flexible nature of peptides, perhaps even when bound to an antibody. Only one crystallographic structure of an antipeptide antibody complexed with its peptide antigen has been reported to date.[8]

Nuclear magnetic resonance (NMR) spectroscopy has evolved into a powerful tool for the determination of the three-dimensional structure of small proteins in solution[9] of molecular weights up to about 15,000. Antibody–protein and antibody–peptide complexes are too large for present NMR methods to be used to obtain structural information at anywhere near the resolution of X-ray crystallography. However, at somewhat lower resolution, NMR can provide a good deal of valuable information, both static and dynamic, on the nature of the interactions within these complexes, particularly in systems that involve peptide[10,11] and small protein[12] antigens. Antibodies bind to these antigens with a wide range of different affinities. Elsewhere in this volume,[13] proton (^{1}H) NMR methods are described that can be used to study antibody–peptide interactions in systems where the binding affinity is low enough for rapid exchange to occur between the bound and free ligand. The binding constants for these antibodies are typically 10^5 to 10^6, and observation of the bound resonances of the peptide is made via magnetization transfer and transferred nuclear Overhauser (NOE) experiments.[14] For antibodies that exhibit tighter antigen binding, the NMR spectrum of the bound peptide can be studied directly. This can be achieved by employing isotope-edited NMR techniques[15,16] or, as we shall describe in this chapter, through the use of ^{1}H NMR methods. Such studies can reveal not only the identity of those residues of the antigen with which the antibody interacts (and in some instances the specific involvement of individual groups within the combining site[17]) but also information at a more detailed level on the differing strengths of this interaction across the antigen sequence.

[8] R. L. Stanfield, T. M. Fieser, R. A. Lerner, and I. A. Wilson, *Science* **248,** 712 (1990).

[9] K. Wüthrich, *Science* **243,** 45 (1989).

[10] W. Ito, M. Nishimura, N. Sakato, H. Fujio, and Y. Arata, *J. Biochem.* **102,** 643 (1987).

[11] J. Anglister, C. Jacob, O. Assulin, G. Ast, R. Pinker, and R. Arnon, *Biochemistry* **27,** 717 (1988).

[12] Y. Paterson, S. W. Englander, and H. Roder, *Science* **249,** 755 (1990).

[13] J. Anglister and F. Naider, this volume [10].

[14] G. M. Clore and A. M. Gronenborn, *J. Magn. Reson.* **48,** 402 (1982).

[15] P. Tsang, T. M. Fieser, J. M. Ostresh, R. A. Lerner, and P. E. Wright, *Peptide Res.* **1,** 87 (1988).

[16] P. Tsang, M. Rance, and P. E. Wright, this volume [11].

[17] K. Kato, C. Matsunaga, Y. Nishimura, M. Waelchli, M. Kainosho, and Y. Arata, *J. Biochem.* **105,** 867 (1989).

Overview of the Method

The antibody molecule, and its peptide antigen, fall at opposite ends of a broad molecular weight band, ranging from ~ 150,000 at the top end to ~ 1000 at the bottom. Resonances in the antibody are typically broad, with large linewidths (>20 Hz), while those in the peptide antigen are narrow (linewidths <5 Hz). Our approach exploits this difference in linewidth to distinguish, in the peptide–antibody complex, peptide residues that do not interact strongly with the antibody combining site (and thus retain some degree of independent mobility) from those that are immobilized on binding. The method is applicable to systems where the binding affinity of the antibody for peptide is in the range to allow slow exchange (on the NMR time scale) between bound and free ligand ($K_A \sim 10^{-6}$ M and above). Where resonances that correspond to the *bound* peptide are observed in the NMR spectrum, these can be monitored *directly*. The proton NMR experiment best suited to this approach is two dimensional (2D) ^{1}H *J*-correlated spectroscopy[18,19] (COSY). Dynamic filtering is an intrinsic feature of the 2D COSY experiment, since the cross-peak amplitude decays rapidly as linewidths approach the coupling constant, *J*.[20] As a consequence it is possible to observe selectively those residues in the peptide–antibody complex that give narrow resonances against a high background of resonances from protons in the macromolecule. Furthermore, for the COSY experiment changes in intensities of resonances can be correlated with the mobilities of the residues from which they derive, in a relatively straightforward manner. For example, from simulations of the effect of increased linewidth of the ^{1}H resonances in the peptide on the intensity of the corresponding COSY cross-peaks, the dynamic properties of the bound antigen can be investigated. The ideas embraced here in the interpretation of linewidth differences among resonances of the peptide as a reflection of differential residue mobilities, are analogous to those that have been employed in ^{15}N studies of similar peptide–antibody complexes.[15,16]

An outline of the general protocol is given in Fig. 1. The essential steps involved in the procedure are as follow:

1. complete sequence-specific assignment of all the ^{1}H resonances in the free peptide antigen
2. titration of the peptide with the antibody Fab fragment

[18] W. P. Aue, E. Bartholdi, and R. R. Ernst, *J. Chem. Phys.* **64,** 2229 (1976).
[19] A. Bax and R. Freeman, *J. Magn. Reson.* **61,** 306 (1981).
[20] M. Weiss, J. L. Eliason, and D. J. States, *Proc. Natl. Acad. Sci. U.S.A.* **81,** 6019 (1984).

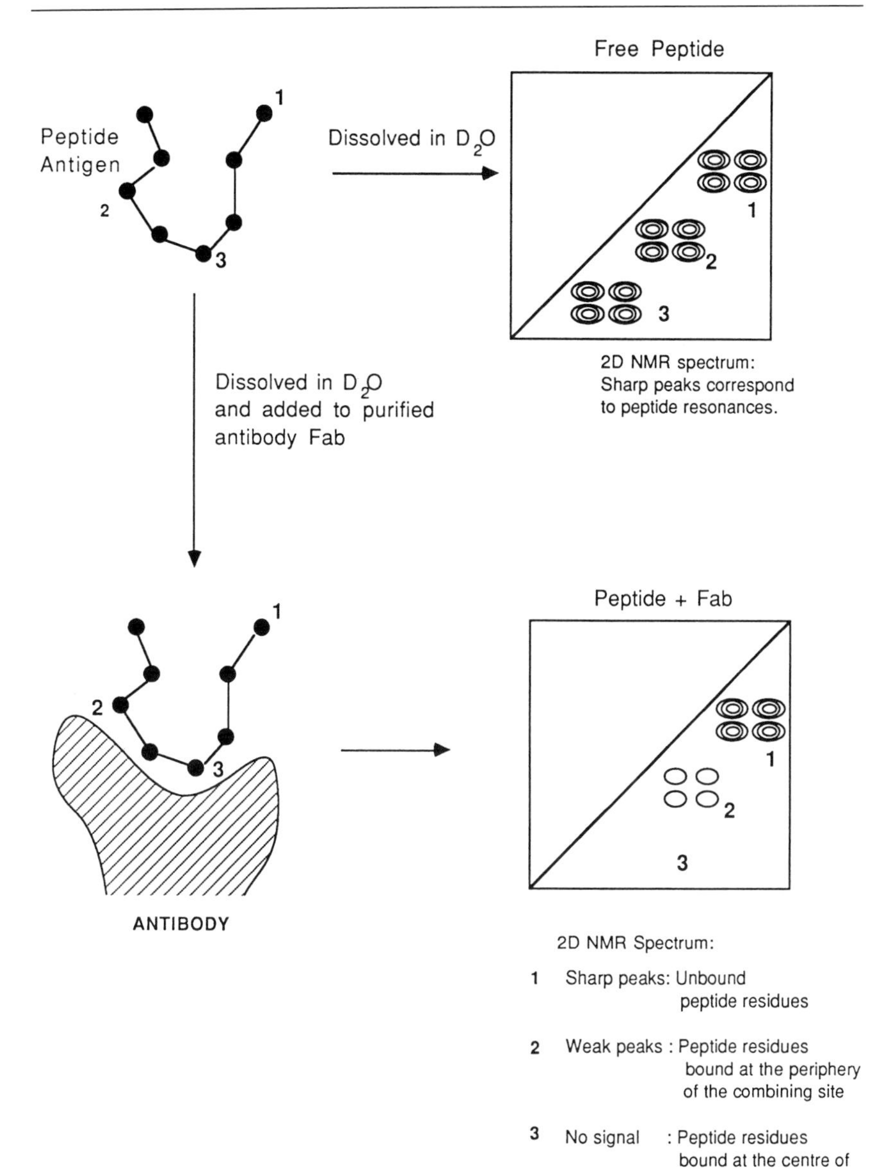

FIG. 1. Overall scheme for the study of high-affinity antibody–peptide complexes by 2D ¹H COSY NMR. In the 2D spectra, cross-peaks on one side of the diagonal only have been illustrated.

3. acquisition of 2D COSY [or double-quantum filtered (DQF) COSY[21]] data sets for the free peptide, free Fab, and bound peptide
4. assignment of the bound peptide spectrum, and analysis of the intensities of the COSY cross-peaks, through spectral simulations

We have explored a number of different systems in our laboratory, but for the purposes of discussion in the following sections we will refer to some specific details taken from experiments carried out to study the interaction of an anti-peptide antibody Gloop1 with a 28-residue peptide antigen known as the "loop" peptide.[22-24]

Generating Antibody and Peptide Samples for NMR

The total quantities of material that will be needed to carry out this type of analysis will be typically in the region of 100–200 mg of the antibody Fab, and 15–20 mg of peptide, depending on the molecular weight of the peptide.

Production of Antibody Fab from IgG

Each NMR experiment requires a relatively large amount of Fab (25–35 mg) to give a 1 m*M* sample of the protein. Tissue culture methods such as the hollow fiber bioreactor system[25] are potentially useful in acquiring sufficient antibody for experiments to be carried out. For example, in the case of Gloop1 we were able to obtain several grams of IgG with a typical yield of 1 mg ml^{-1} of antibody in the culture supernatant.

The complete antibody IgG is not well suited to these solution studies of peptide–antibody interactions since it tends to aggregate readily at the high protein concentrations needed to observe the 2D spectrum of the bound peptide. Therefore the smaller Fab fragments of the antibody are used. These can be prepared from IgG by proteolytic digestion.[26] First the IgG is purified from the cell culture supernatant and the Fab is then generated by enzymatic cleavage of the IgG. The Fab is purified from the

[21] The double-quantum filtered (DQF) COSY experiment [Neuhaus *et al., Eur. J. Biochem.* **151,** 257 (1985) and Rance *et al., Biochem. Biophys. Res. Commun.* **117,** 479 (1983)] is a modified form of the conventional COSY experiment. The multiple-quantum filtered experiment has the advantage of selectively suppressing the strong singlet diagonal peak in the COSY spectrum, which is dispersive and may otherwise mask nearby cross-peaks.

[22] J. C. Cheetham, D. P. Raleigh, R. Griest, C. Redfield, C. M. Dobson, and A. R. Rees, *PNAS,* in press (1991).

[23] M. Darsley and A. R. Rees, *EMBO J.* **4,** 383 (1985).

[24] E. Maron, C. Shiozawa, R. Arnon, and M. Sela, *Biochemistry* **10,** 763 (1971).

[25] P. Knight, *Biotechnology* **7,** 459 (1989).

[26] P. Parham, *J. Immunol.* **131,** 2895 (1983).

reaction mixture and concentrated into a buffer solution. It is important that these samples are stored at low temperature (4°) and not frozen or freeze–dried; Fab once cleaved from the IgG undergoes denaturation on freezing.

Below we give a sample protocol that was used to prepare Gloop1 antibody Fab. This might be appropriate for any other IgG of subclass IgG2b.

1. The cell culture supernatant was precipitated with 50% ammonium sulfate.
2. The sample was centrifuged to separate the precipitate, and then the solid material was resuspended in minimum volume, dialyzed against 10 m*M* Tris at pH 8.0, and the sample then run on a 250-ml DEAE column using a salt gradient (0.0 to 0.2 *M* NaCl); total volume 500 ml.
3. The pooled IgG fractions were digested with a 1 : 100 ratio of papain to protein (pH 7.08, 10 m*M* cysteine) over 4 hr, at 25°; the reaction was then stopped with iodoacetamide.
4. The digested mixture was concentrated using an Amicon (Danvers, MA) filter, and applied to an S200 Sephadex gel-filtration column (700 mm × 2.4 cm) to separate the Fab from partially digested IgG.
5. Column fractions containing Fab were concentrated on an Amicon filter.
6. In a step specific to the preparation of the Gloop1 Fab, designed to remove contaminating, small-molecule ligands, the Fab was then mixed with a molar equivalent of its protein antigen, hen egg lysozyme (HEL), and the sample run through a TSK3000 gel-filtration column (pH 5.5, phosphate buffer) to clean further the Fab, and to separate it from the HEL. The Fab was then concentrated using an Amicon filter. The HEL used was obtained from Sigma (St. Louis, MO) at 95% purity (purified three times by recrystallization).

Fab Sample Preparation for NMR

The antibody–peptide binding studies, designed to study nonexchangeable protons, are best carried out in a deuterated medium, because the dynamic range in D_2O is better than in H_2O; the peptide concentration in a sample with the antibody at close to the 1 : 1 complex is extremely low (~ 1 m*M*), due to the limitation imposed by the intrinsic solubility of the Fab. NMR samples of the Fab must therefore be exchanged from protonated into deuterated solvent. This can be achieved using an Amicon filter (e.g., Centricon 30) to give a final protein concentration (as measured by UV absorption at 280 nm) of ~ 1.2 m*M* in 0.5 ml D_2O. Different Fabs will have varying solvent requirements (buffers, salts, etc), and these should be chosen in each case to optimize both the solubility of the protein and its stability in solution over the time course of the experiments. In addition, either nonprotonated, e.g., phosphate, or deuterated buffers should be used. Spectra in H_2O can be acquired to observe amide (NH) protons, but the sensitivity of the experiment is reduced.

Impurity Problems in the Fab

At an early stage in our studies with Gloop1 we noted that the Fab molecule produced by proteolytic digestion of the IgG was often contaminated by small molecule species, presumably bound within the combining site. The presence of the impurities in our Fab sample was not evident in the 2D spectra until peptide antigen was added, displacing them from the combining site; their resonances were narrow, and the corresponding cross-peaks in the 2D spectrum intense. It was therefore essential that these molecules should be eliminated prior to the addition of any peptide ligand, since their presence would interfere with the interpretation of the bound peptide spectrum. This was accomplished by introducing an additional step into the purification procedure for the Fab (step 6 in the protocol given above), in which the antibody is prebound to the protein antigen—in this instance lysozyme. The small molecules are then displaced from the combining site by binding to the protein antigen and the cleaned Fab is then dissociated from the column, under acid conditions, into the NMR sample buffer. Clearly a similar procedure could be used with most other antibody–antigen pairs.

Gloop1 is an anti-peptide antibody and the antigen-binding site, in common with many other antibodies, has a number of exposed hydrophobic groups,[27] making the interior surface of this region somewhat "sticky." Therefore the presence of bound fragments such as small peptide molecules at this site, after cleavage of the IgG, is not too surprising. Bjorkmann *et al.*[28] similarly noted in their crystallographic study of the HLA-A2 molecule, purified by papain digestion of plasma cells, unexpected density at the recognition site that may have originated from a mixture of small peptide fragments. These observations would suggest that the prebinding of antigen as a final step in the purification of a particular Fab may prove to be an important part of the preparative scheme.

Preparation of the Peptide Antigen

The peptide antigen is required in sufficient quantities to allow both the complete assignment of its ^1H NMR spectrum and titration experiments

[27] R. Griest and A. R. Rees, manuscript in preparation. (1991).

[28] P. J. Bjorkman, M. A. Saper, B. Samroui, W. S. Bennett, J. L. Strominger, and D. C. Wiley, *Nature (London)* **329,** 506 (1987).

with the antibody Fab. In most instances the easiest route to generate supplies of the peptide will be through automated peptide synthesis,[29] rather than, for example, enzymatic cleavage of a protein source. In particular the synthetic route has the additional advantage of allowing amino acid substitutions to be readily introduced into the antigen sequence, and so enables the factors that control the specificity of the peptide–antibody interaction to be readily probed in parallel with studies of the binding of the native, immunogenic peptide. The effects of any changes introduced into the synthetic peptide sequence, however, should always be monitored by determining the relative antibody-binding affinities of the mutant peptides.

Peptide Purity

For the NMR studies it is important that the purity of the peptide material used is high (>90%). This can normally be achieved using reversed-phase high-performance liquid chromatography (HPLC). The purity of a given sample should then be established by two methods. First, by the use of analytical reversed-phase HPLC, and second using one- or two-dimensional ^{1}H NMR spectroscopy. The latter technique offers the most sensitive test, and can be particularly valuable in showing up any differences between one peptide preparation and another. For example, if a sample contains, say, 90% of one peptide form and 10% of a different species, two sets of resonance lines would be seen, with relative intensities proportional to their concentrations in the mixture. After HPLC, the pH of the fractions containing the peptide should be adjusted to that of the NMR experiment with NaOH/HCl, and the sample lyophilized and stored in the freezer at $-20°$.

Sequence-Specific Assignments in the Peptide Antigen

Before studying the peptide in its antibody-bound complex, all of the ^{1}H NMR resonances and cross-peaks in the 2D spectrum originating from the side-chain and main-chain protons of all the residues of the unbound antigen must be fully assigned.

Sample Requirements

For the purposes of assignment two different types of peptide sample are usually required. The first is a sample with amide protons exchanged for deuterons, which can be prepared simply by dissolving the lyophilized peptide material in D_2O (99.96% isotopic purity), at pH 4.0. The labile protons are then allowed to exchange overnight. In practice the exchange is

[29] E. Atherton and R. C. Sheppard, "Solid phase peptide synthesis." IRL Press, Oxford, 1989.

usually complete in less than an hour, but this will depend on the degree of ordered structure within the molecule, which may give some amide NH groups protection from the solvent medium.[30] Second, a sample of peptide is prepared in which all the amide hydrogens are present. This can be done by dissolving lyophilized peptide in 0.5 ml of a solution containing 90% H_2O/10% D_2O; ideally the final pH of the solution should be close to the pH minimum for hydrogen exchange. The spectrum of a sample prepared in this way should contain resonances from all the NH protons.

Method of Assignment

A comprehensive description of the general strategy used to obtain sequence-specific assignments from NMR data is given by Wüthrich.[31] It should be feasible to perform most of the NMR experiments described below, in a routine manner, on all the commercial high field (500 and 600 MHz) NMR solution spectrometers currently available. General guidelines to acquiring these data can be obtained from practical textbooks on NMR methods,[32-34] and more specific details extracted from current journals describing NMR studies of similar molecules.

The individual spin systems in the peptide can, in most instances, be identified by using a combination of COSY, DQF COSY, and single and double relayed coherence transfer (RELAY) or total correlation spectroscopy (TOCSY) experiments.[35-38] Sequence specific assignments can then be made using interresidue proton-distance information [e.g., dαN $(i, i+1)$ connectivities], obtained from nuclear Overhauser enhancement spectroscopy (NOESY)[39-41] experiments. Many peptides appear to be relatively unstructured, flexible molecules, in which there is rapid conforma-

[30] G. Wagner and K. Wüthrich, this series, Vol. 131, p. 307.

[31] K. Wüthrich, "NMR of Proteins and Nucleic Acids." John Wiley & Sons, New York, 1986.

[32] M. L. Martin, J.-L. Delpuech, and G. J. Martin, "Practical NMR Spectroscopy." Heyden and Son, London, 1980.

[33] J. K. M. Sanders and B. K. Hunter, "Modern NMR Spectroscopy." Oxford University Press, Oxford, 1987.

[34] A. Derome, "Modern NMR Techniques for Chemistry Research." Pergamon, Oxford, 1987.

[35] G. Eich, G. Bodenhausen, and R. R. Ernst, *J. Am. Chem. Soc.* **104,** 3731 (1982).

[36] A. Bax and G. P. Drobny, *J. Magn. Reson.* **61,** 306 (1985).

[37] L. Braunschweiler and R. R. Ernst, *J. Magn. Reson.* **53,** 521 (1983).

[38] A. Bax and D. G. Davis, *J. Magn. Reson.* **65,** 355 (1985).

[39] J. Jeener, B. H. Meier, P. Bachmann, and R. R. Ernst, *J. Chem. Phys.* **71,** 4546 (1979).

[40] Anil Kumar, R. R. Ernst, and K. Wüthrich, *Biochem. Biophys. Res. Commun.* **95,** 1 (1980).

[41] S. Macura, Y. Huang, D. Suter, and R. R. Ernst, *J. Magn. Reson.* **43,** 259 (1981).

tional averaging in solution.[42,43] In such cases the conventional NOESY experiment may not give observable NOE cross-peaks, even with long mixing times (> 1 sec). A rotating frame NOESY (ROESY) experiment[42] can then be tried as an alternative to obtain the necessary interresidue connectivities. It should also be noted that in some cases the NH—C_αH cross-peaks of residues may appear to be absent from the normal COSY spectrum recorded in water, disappearing when the C_αH resonance is saturated along with the solvent during data collection. This problem can be overcome either by either collecting spectra at different temperatures, since the H_2O resonance shifts significantly with temperature, or by using experiments that incorporate a pre-TOCSY pulse sequence.[43]

Once the sequential assignment of the peptide is complete, it will then normally be necessary to redetermine the chemical shift positions at the exact pH of the antibody-binding experiment (usually ~ pH 5–8). Since the chemical shift values of nonamide protons are relatively insensitive to pH change, over the range pH 4 to 8, this can be done in most instances simply by extrapolating the data between the different pH values. Exceptions to this will be those resonances that derive from residues whose pK_a values fall within this range, and resonances of peptides that exhibit significant structural changes as a function of pH. All chemical shift values of resonances in the peptide should be referenced to a standard such as dioxane or tetramethyl silane (TMS).

NMR Titration of the Antibody Fab with Peptide

With a full set of assignments for the ^{1}H resonances of the unbound peptide, the next step is to confirm that in the presence of Fab, the exchange rate between the bound and the free forms of the peptide antigen is slow on the NMR time scale. Where the measured relative affinity binding constant of the system is high (between antibody and peptide), one would expect this to be the case.

The ratio of peptide to Fab should be varied, in equal increments from well below 1 : 1 at the start of the experiment, through the fully bound complex (1 : 1), to an end point where the ligand is present in a significant excess. In practical terms the titration can be carried out by placing the Fab sample in an NMR tube, and then adding the peptide in small-volume (e.g., 5 μl) aliquots via a Hamilton syringe. The exact volume of each aliquot will depend on how concentrated the peptide can be prepared in a

[42] A. A. Bothner-By, R. L. Stephens, J. Lee, C. D. Warren, and R. W. Jeanloz, *J. Am. Chem. Soc.* **106,** 811 (1984).

[43] G. Otting and K. Wüthrich, *J. Magn. Reson.* **75,** 546 (1987).

solution of D_2O. However, it is important to keep the overall volume change in the NMR sample to a minimum, to avoid the need to make large adjustments to the shim settings of the spectrometer during the titration experiment. The pH of the peptide solution should be identical to that of the Fab sample to which it is added. For every point in the titration (including that in the absence of added peptide), a DQF COSY spectrum is recorded under comparable experimental conditions (temperature and pH), and using the same spectral parameters. As the peptide concentrations in the early stages of the titration are very low (< 1 m*M*), a period of about 12 hr is normally required to collect a typical 2D data set for each point.

If the exchange between the bound and free forms of the peptide is slow, then the following pattern of results can be expected. At below 1 : 1 a set of cross-peaks corresponding only to bound peptide resonances, or resonances from free Fab or the Fab in the complex, is observed. When the peptide is present in excess, an additional set of cross-peaks will be observed, deriving from resonances of the residues in the unbound antigen. The chemical shift positions and linewidths of both sets of cross-peaks should be independent of the amount of added ligand. Figure 2 provides a good illustration of the type of changes that one can expect.

In cases where the titration results suggest the system is not in slow exchange (and thus in either the fast or intermediate exchange range), the procedure we describe in this section will not be applicable. For these antibodies the methods such as those described in the accompanying chapters of this volume by Anglister and others should be used. Line-broadening effects that are due to intermediate exchange can be identified by analyzing the temperature and frequency dependence of the spectrum.

Probing the Antibody:Peptide Complex by 2D ^{1}H NMR

In a general description of the approach we have taken in studying peptide–antibody complexes by ^{1}H NMR, given at the beginning of this chapter, we indicated that the COSY experiment was a particularly appropriate 2D NMR method by which to locate highly mobile residues within the peptide–antibody complex. The reasons for this can be illustrated by considering the NMR properties of the antibody and peptide molecules individually, and then together in the bound complex.

Choice of the 2D COSY Experiment

The antibody Fab fragment has a molecular weight of ~ 50,000–60,000. In a standard 1D ^{1}H NMR spectrum of the protein, a large envelope of broad, unresolved resonances is observed (Fig. 3a). The broad

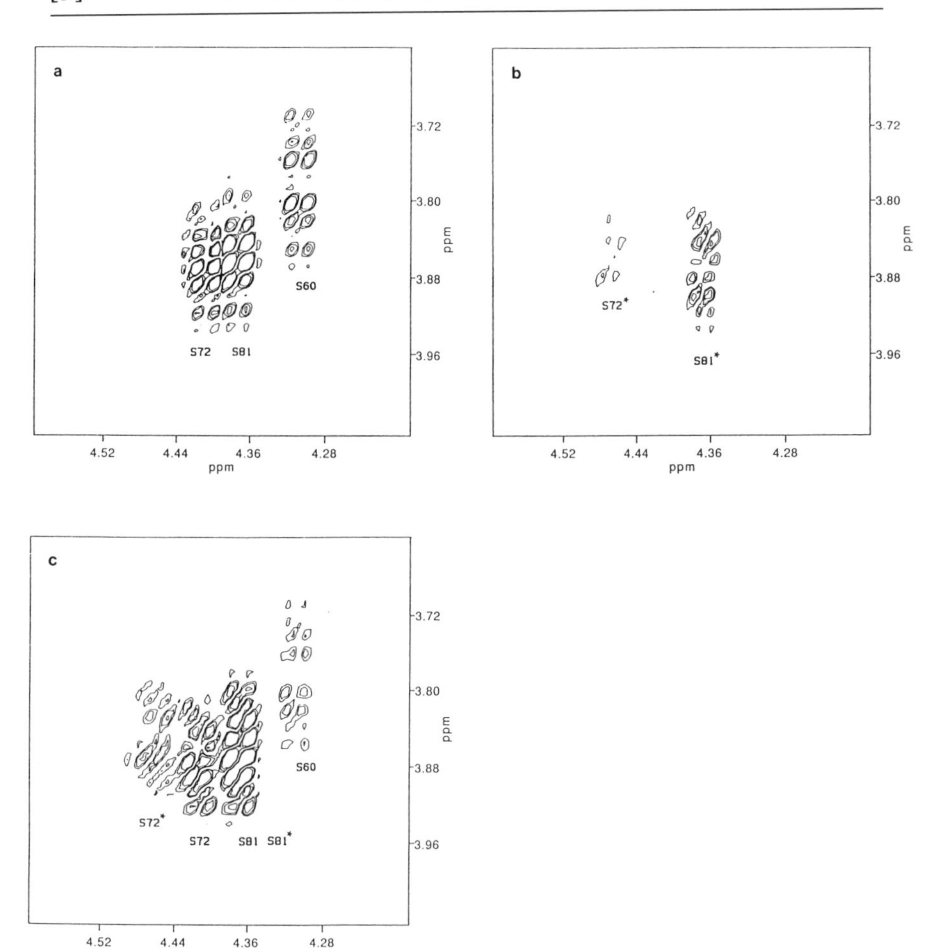

FIG. 2. Titration of the loop peptide with Gloop1 Fab by 2D ¹H NMR. Sections taken from the DQF COSY spectra of samples with ligand:antibody ratios of 0 : 1 (a), 0.8 : 1 (b), and 1.8 : 1 (c), corresponding to free peptide, fully bound antigen, and a mixture of the bound and free peptides, respectively. All spectra were recorded at 500 MHz, pH 7.1, and at 27°.

spectral lines are characteristic of the long correlation times associated with slow molecular tumbling. By contrast, the ¹H resonances in the spectrum of a smaller, peptide molecule (Fig. 3b) have much narrower linewidths (typically <5 Hz). If we compare the 2D DQF COSY spectrum from the same two molecules, the effect of this difference in linewidths is even more dramatic (Fig. 3c and d). In the case of the Fab, only a small number of cross-peaks can be observed (Fig. 3c); for the few residues that do give narrow resonances, we can deduce that these regions must have substantial

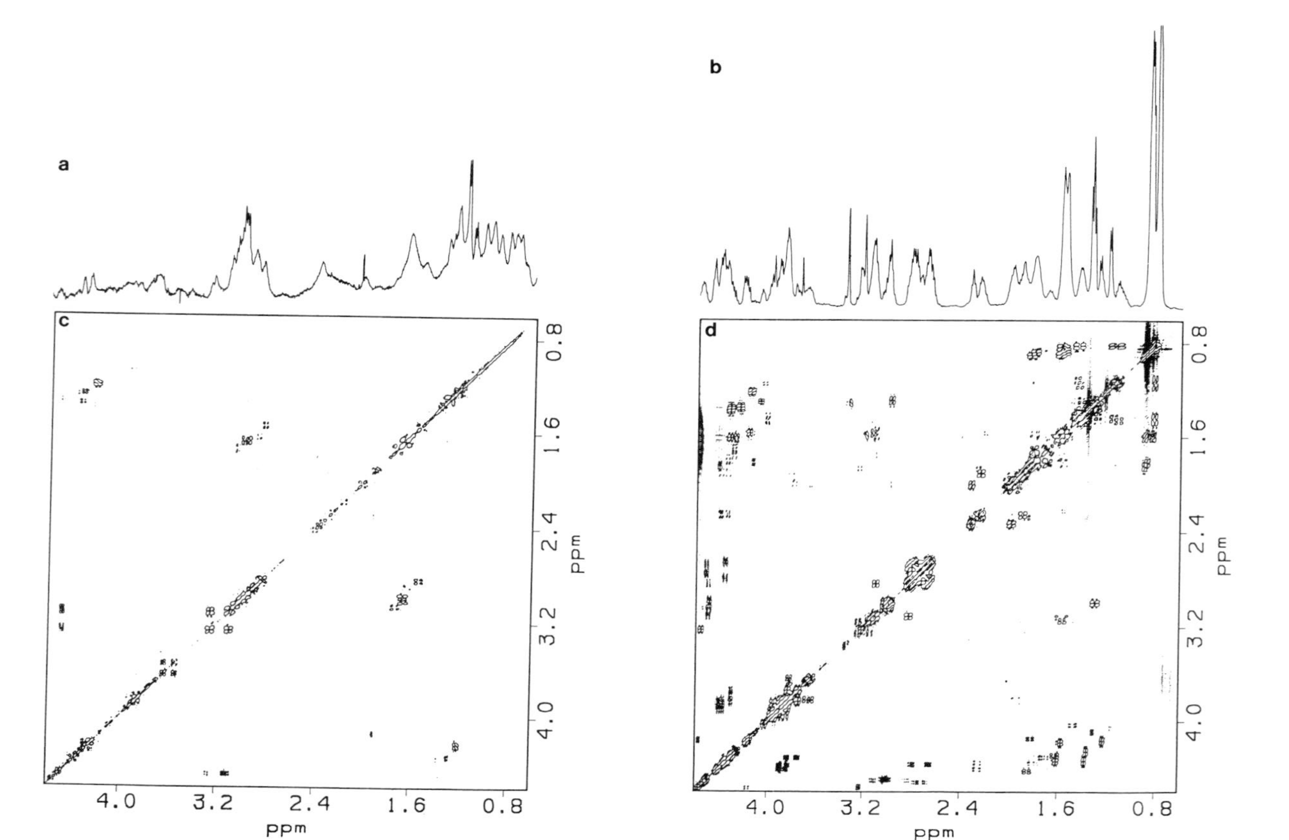

FIG. 3. ^{1}H NMR spectra (500 MHz) of Gloop1 Fab and the loop peptide antigen. The 1D spectra (a and b) are shown directly above those from the corresponding 2D DQF COSY experiments (c and d). Only the aliphatic regions of the spectra are displayed; both 2D spectra have been contoured at identical levels, the 1D spectra of the Fab and peptide are drawn to different scales. The sample concentrations were 1.2 m*M* (Fab) and 1.1 m*M* (peptide). All data were collected at 27°.

independent mobility relative to the rest of the protein structure[44–46] (e.g., residues at the ragged N and C termini of the Fab, produced at the stage of enzymatic digestion of the IgG). The corresponding spectrum from the peptide (Fig. 3d) shows a large number of cross-peaks, as would be expected for a molecule in which the residues have a high degree of motional freedom.

What are the ramifications of these observations for the case where the peptide is not free in solution, but bound to an antibody combining site? When a peptide binds to an antibody, the mobility of its residues will be reduced. This loss of conformational freedom will derive in part from the general increase in correlation time for the antibody-bound peptide molecule, but for many residues as a result of specific interactions between the peptide and the antibody combining site. Two different scenarios can be envisaged. First, all the peptide residues may be strongly immobilized on binding. The linewidths of the bound peptide resonances will then be comparable to those of the free Fab, and the 2D COSY spectrum of the Fab–peptide complex essentially identical to that of the free Fab. Alternatively, the peptide may interact with the combining site to differing degrees along the linear sequence, with certain residues retaining some degree of independent mobility in the bound complex, while others are tightly bound. In this case, the COSY spectrum of the Fab–peptide complex will contain a number of additional cross-peaks, arising from the peptide residues that remain highly mobile when bound to the antibody (Fig. 4). Furthermore, since the intensity of the COSY cross-peak is reduced as the linewidths of the resonances from which it derives increase, an analysis of the peptide cross-peak intensities for these residues in the bound and free spectra will indicate the reduction in their motional freedom on binding to the antibody.

Experimental Methods

Three separate 2D data sets must be collected, under comparable conditions (pH, temperature, concentration, etc.). These are the unbound peptide antigen, the free Fab, and the bound peptide antigen (1 : 1 complex with the Fab). A set of experimental parameters, taken from our own

[44] H. R. Wilson, R. J. P. Williams, J. A. Littlechild, and H. C. Watson, *Eur. J. Biochem.* **170,** 529 (1988).

[45] R. E. Oswald, M. J. Bogusky, M. Bamberger, R. A. G. Smith, and C. M. Dobson, *Nature (London)* **337,** 579 (1989).

[46] R. N. Perham, H. W. Duckworth, and G. C. K. Roberts, *Nature (London)* **292,** 474 (1981).

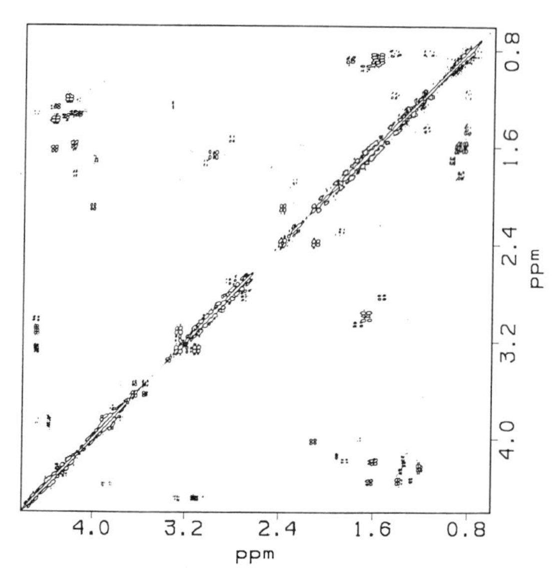

FIG. 4. Aliphatic region of a 500-MHz 2D ^{1}H COSY spectrum of the bound loop peptide. The peptide:antibody ratio in the sample was 0.9 : 1.

study of the binding of the loop peptide antigen to Gloop1 Fab, are given below.

1. The sample concentrations were 1.2 m*M* (Fab) and 1.1 m*M* (peptide); all NMR data were recorded at 27°, pH 7.
2. DQF COSY spectra were recorded in D_2O on a Bruker AM-500, using 1024 t1 (time) increments with 32 scans per increment and a sweep width of 12.5 ppm. Residual water was saturated between scans and phase-sensitive data were collected using the method of time proportional phase incrementation (TPPI).[47]
3. The NMR data were processed with resolution enhancement by a phase-shifted sine bell[48,49] (6°) in both dimensions, using FTNMR (Hare Research, Inc., Woodinville, WA).[50] After zero filling once in t2, and twice in t1, the resolution in the final spectrum was 3.0 Hz/point.

Several important practical details in this scheme should be emphasized. First, the data set from the unbound peptide provides a negative control for

[47] A. G. Redfield and S. D. Kunz, *J. Magn. Reson.* **19,** 250 (1975).

[48] M. Gueron, *J. Magn. Reson.* **30,** 515 (1978).

[49] G. K. Wagner, K. Wüthrich, and H. Tschesche, *Eur. J. Biochem.* **86,** 67 (1978).

[50] FTNMR is a software package for NMR data processing, available from Hare Research, Inc.

the antigen, indicating the appearance of the spectrum if there was no interaction between the peptide and the antibody Fab. It is therefore important that the peptide concentration in this sample should be identical to that in the peptide–Fab sample corresponding to the fully bound complex (close to 1 : 1). Second, the internal consistency between the free Fab, unbound peptide, and peptide–Fab complex data sets should be maximized. This can be accomplished by recording the three data sets sequentially, on the same spectrometer and using identical spectral parameters. For the same reason, the peptide–Fab complex should also be generated "*in situ*" from the Fab sample used to record the 2D data set for the free antibody. Peptide antigen, concentrated in D_2O at the same pH as the Fab, is added to the Fab, in the NMR tube, using a Hamilton syringe. As in the titration experiment, the volume of solution added at this point should be kept to a minimum (e.g., 10–50 μl for a 0.5-ml sample). After thorough mixing of the two solutions, the NMR tube is returned to the spectrometer and equilibrated for 30 min, and the third data set recorded. The molar ratio of peptide:Fab that gives a sample corresponding to the fully bound (1 : 1) complex, with no detectable free ligand, should be determined from the titration experiment. This value could also be estimated from the measured binding affinity, to a first approximation, although a slight excess of the antibody is needed to ensure that no residual unbound peptide is present. Third, the experimental variables such as temperature and sample conditions (pH, buffer composition) will be primarily controlled by the binding characteristics of each peptide:antibody system under study.

Assignment of Resonances in the Bound Peptide

Residues of the peptide whose resonances remain narrow in the bound complex must retain a significant degree of independent mobility within the antibody combining site. By analogy, therefore, these residues cannot be involved in strong interactions with the Fab. As a consequence, the changes in the chemical shift positions of their resonances between the free and bound states will be typically small (e.g., 0.0–0.1 ppm). Assignments in the bound peptide can therefore be made on the basis of the resonance positions in the unbound molecule. Similarity of the patterns of cross-peak intensities between the free and bound spectra should also be taken into account when making the assignments. The procedure is illustrated for several of the residues in the bound loop peptide in Fig. 5. Resonances from 14 of the 28 residues could be assigned in this way. These assignments can be confirmed via additional experiments such as single and double RELAY or TOCSY.[35–38]

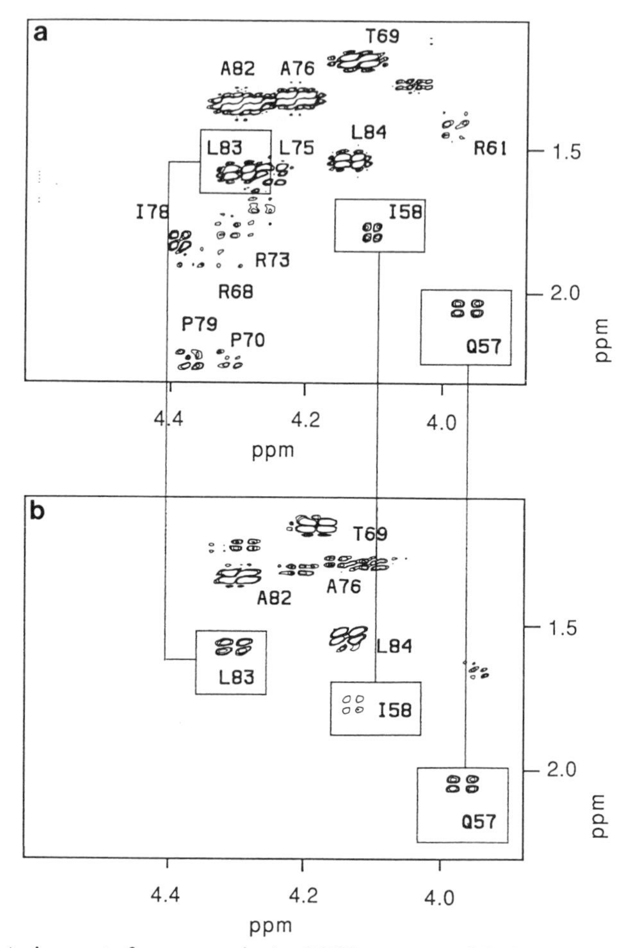

FIG. 5. Assignment of resonances in the COSY spectrum of the bound peptide. Chemical shift positions of the cross-peaks in the free peptide spectrum (a) can be used to identify the cross-peaks of the corresponding resonances in the bound peptide spectrum (b). The boxed peaks highlight specific examples of the correlation between the chemical shift positions in the two different spectra.

Where residues do not give observable cross-peaks in the bound spectrum, no assignments can be made for the corresponding protons. However, from the absence of their cross-peaks one can conclude that these peptide residues are interacting, either directly (epitope) or indirectly (e.g., through conformational change), with the antibody in the bound complex.

Cross-Peak Intensity Changes in the Bound Spectrum

The intensity of a cross-peak in the COSY spectrum of a molecule is a function of both linewidth and J coupling. The natural linewidth is largely determined by the mobility of the molecule, with linewidths increasing as the correlation time increases. For a particular proton pair, the value of the spin–spin coupling constant (J) is related to the intervening torsion angle between them (i.e., to the molecular conformation).[51] When the magnitude of the linewidth exceeds J, the cross-peak intensity is very weak. Within any residue, changes in either the resonance linewidths, and/or the magnitude of the J coupling constant, may bring about differences in the intensity and pattern of the COSY cross-peak(s) deriving from a particular pair of protons. It is therefore important to consider the influence of both these parameters on the appearance of the cross-peaks. This can be accomplished through the use of spectral simulation methods.

Simulation of Intensity Changes in COSY Cross-Peaks with Increasing Linewidth for Resonances in the Bound Peptide

To simulate the effect of increasing linewidth on the intensity of cross-peaks in the COSY spectrum of the peptide, the 2D data set collected from the unbound antigen can be reprocessed with a number of different simulated line broadenings. Each row of the data matrix is multiplied by an exponential function, with a line-broadening factor, prior to resolution enhancement in both the t1 and t2 dimensions in the normal manner. The resulting series of spectra can then be contoured at identical levels and compared to the spectrum of the bound peptide. In this way, for peptide residues that give rise to COSY cross-peaks in both the free and bound states, the observed reduction in cross-peak intensity in the COSY spectrum on binding to antibody can be correlated with an apparent increase in linewidth. A comparison of observed and simulated cross-peak intensities, at different applied line broadenings, taken from data for the loop peptide and the loop–Fab Gloop1 complex, is given in Fig. 6. In the case of those residues that give no cross-peaks in the bound spectrum, only a minimum estimate of the degree of broadening that would result in the disappearance of the COSY peaks can be made.

It should be noted that for any given peptide, the intensities of COSY cross-peaks corresponding to the different residues, and even to the different individual ^{1}H couplings within them, can be expected to show dramatically differing dependencies on linewidth. If we take the loop peptide as an

[51] M. Karplus, *J. Phys. Chem.* **30,** 11 (1959).

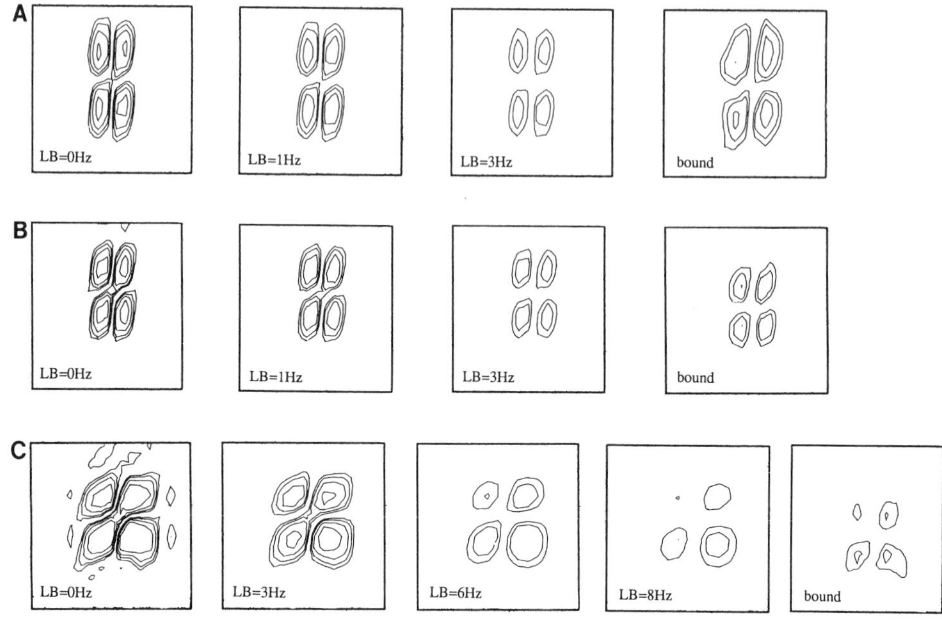

FIG. 6. The simulated effect of varying linewidth increases on the intensity of the COSY cross-peaks from the $\alpha\beta$ protons of Ile-58 (A), Ala-82 (B), and the $\beta\beta$ protons of Ser-60 (C). Artificial line broadenings of between 1 and 8 Hz were applied to the 2D DQF COSY data set collected from a 1.2 m*M* sample of the free loop peptide, using FTNMR.[50] The cross-peak intensities observed in the spectrum of the bound peptide are shown for comparison.

example, for the $\alpha\beta$ cross-peaks from residues such as R68, C64, and N74, an increase in linewidth of only 4 Hz is sufficient to cause the disappearance of the COSY cross-peaks, while for some, e.g., A82, L83, and L84, an applied line broadening of >10 Hz is required. For cross-peaks that disappear at a simulated line broadening of 3 Hz, the side-chain groups from which they derive could be either tightly or weakly bound within the combining site, and in both instances might give no observable peaks in the bound peptide spectrum. Thus, for a given peptide, residues that fall into the latter category, where a significant change in mobility is required to eliminate their cross-peaks from the 2D spectrum, obviously represent the best reporter groups, since their cross-peaks will completely disappear only if the residues are strongly interacting in the bound complex.

Changes in the Appearance of COSY Cross-Peaks as a Function of J

If we look at the dependence of cross-peak intensities on *J* value, two distinct groups of protons can be identified:

Group 1: proton pairs for which the value of the active coupling constant is fixed, as a result of fixed geometry or rapid averaging (e.g., within the aromatic ring protons of Trp, or to terminal methyl groups in the side chains of residues such as Ile). For these proton resonances, if the cross-peak intensity is strong in the spectrum of the free peptide, the absence of the corresponding peaks in the bound spectrum can be attributed wholly to an increase in linewidth.

Group 2: protons for which the coupling constant varies as a function of the side-chain conformation. In these circumstances, the observed loss in intensity seen for some of the cross-peaks in the COSY spectrum of the bound peptide may result, at least in part, from specific conformational changes within the peptide residues on binding to the antibody, as well as from an overall reduction in the degree of mobility of the residue from which the peaks derive.

In this latter case, for a given cross-peak, a transition of the residue side chain from an averaged to a particular, fixed rotamer conformation (or vice versa) would be expected to produce a significant change in J, resulting in a change of the cross-peak pattern and intensity. To evaluate the magnitude of this type of change we simulated the variation in the cross-peak intensity as a function of J for a number of different spin systems.[52] If we examine the $\alpha\beta$ cross-peak of residues such as Ser, Asn, Asp, and Cys, as an example, for a peptide in which the natural linewidths are of the order of ~ 3 Hz, simulations show that the transition of the side chain from an averaged conformation to one of the fixed χ_1 rotamers would give a change in cross-peak intensity of ~ 50% (Fig. 7). For the majority of residues within a peptide antigen, therefore, changes in the coupling constant for a particular pair of protons, on binding to the antibody, would not be expected to cause the complete disappearance of the associated cross-peak from the COSY spectrum.

Where the residue gives cross-peaks in both the free and bound peptide spectra, we can take this analysis one stage further. The relative intensities of the components of the corresponding cross-peaks will change on going from one preferred side-chain conformation to the other, as will the magnitude of the individual peak intensities (Fig. 7). Therefore, if no change is observed in the pattern or relative intensities of the cross-peak components between the bound and free peptide spectra, one can conclude that it is unlikely that the conformation of the residue side chain is altered significantly as a result of binding to the antibody. Conversely, if the cross-peak

[52] The simulation of cross-peak shape and intensity changes, with variations in J value (spin–spin coupling constant), was performed using SIMULATION, implemented on a SUN workstation. The program, written by Dr. C. Redfield, is available from the author on request.

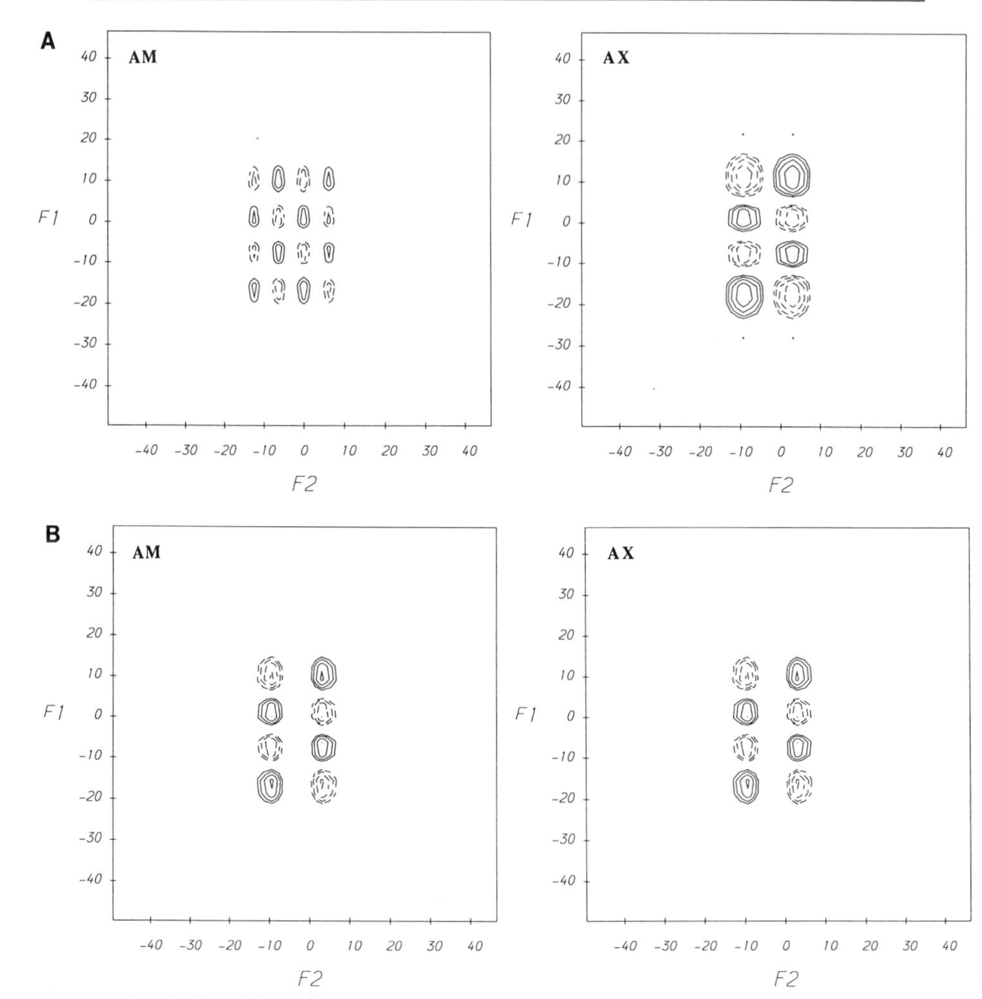

FIG. 7. Simulation of the COSY cross-peaks for an AMX (e.g., Ser $\alpha\beta$ protons, where A, M, and X denote the α, β, and β' spins, respectively) spin system (e.g., Ser $\alpha\beta$ protons), showing the pattern and relative intensities of the peaks obtained for (A) $J_{AM} = 3.4$ Hz, $J_{AX} = 12.4$ Hz, $J_{MX} = 16.0$ Hz and (B) $J_{AM} = 6.6$ Hz, $J_{AX} = 6.6$ Hz, $J_{MX} = 16.0$ Hz. Each pair of peaks is contoured at the same level. All peaks were simulated with 512 real points in the t1 domain and 1024 real points in the t2 domain, and using a resonance linewidth of 3.0 Hz.

pattern does alter between the two, the calculated degree of line broadening will not represent the true linewidth increase. The simulation of cross-peak behavior, when the number of spins involved in the system is small and the individual coupling constants between the protons in the different rotamers are well defined, is relatively straightforward. The complexity of these calculations increases rapidly, however, as both the number of spins that must be incorporated becomes larger, and as a consequence the estimation of the individual coupling constants for a given proton pair is more difficult.

Summary

In general, we can conclude that, while in some cases changes in the coupling constants within a given residue may contribute to the observed changes in the intensities of the COSY cross-peaks in the bound peptide spectrum, for many their absence from the spectrum will also reflect an increase in the linewidth of the corresponding resonances from which the peak derives. Moreover, since many peptides undergo rapid conformational averaging in solution,[53,54] changes in the J values are likely to correspond to the localization of the amino acid side chains in specific antibody:peptide interactions. From the analysis of the COSY spectrum of the bound peptide, as described above, we can therefore extract information pertaining both to the dynamic behavior of residues in the peptide when bound within the antibody combining site, and to the degree of interaction of individual residues in the peptide with those of the antibody in the bound complex.

A Practical Illustration of the Method

Table I gives the chemical shift changes, and the estimated degree of line broadening of the ^{1}H COSY cross-peaks, observed when the loop peptide antigen binds to Gloop1. The data, and their interpretation as we describe below, serve to exemplify the type of picture that one can build up of the molecular interactions in an antibody–peptide complex, using the general protocol and methods that we have described. The characteristics of this system are as follow:

1. The antibody was raised against a peptide immunogen and is specific for the loop determinant of hen egg lysozyme (HEL).[23]
2. The native sequence of the loop, corresponding to residues 57 to 84 in HEL, is ^{57}QINSRWWCNDGRTPGSRNLCNIPCSALL^{84}. In the peptide used for the NMR studies, for ease of synthesis the cysteine at position 76 (underlined) was replaced by an Alanine.[55]

[53] R. L. Baldwin, *Trends Biochem. Sci.* **11,** 6 (1986).

[54] H. J. Dyson, K. J. Cross, R. A. Houghten, I. A. Wilson, P. E. Wright, and R. A. Lerner, *Nature (London)* **318,** 480 (1985).

[55] The Ala-substituted peptide was shown using radioimmunoinhibition assays to bind to the antibody with an affinity comparable to that of the native peptide.

TABLE I

CHEMICAL SHIFT CHANGES AND ESTIMATED DEGREE OF LINE BROADENING OF THE ^{1}H COSY CROSS-PEAKS WHEN LOOP ANTIGEN BINDS TO GLOOP1

Peptide residue	On binding of the peptide to antibody: Chemical shift change (ppm)			Linewidth increase (Hz)		Residue accessibility in the protein (Å^2)
	α	β	Others	$\alpha\beta$	Side chain	
Q57	0.01	0.01	0.03	1–3	1–3	0.0
I58	0.02	0.00		1–3	1–3	0.0
			0.00 (δCH_3)			
			0.02 (γCH_2)			
			0.03 ($\gamma CH_2'$)			
N59				>5	>5	0.0
S60		0.00	0.00 (β')	>6	9–10	0.0
R61				>3	>10	2.0
W62				>8	>10	2.0
W63				>3	>10	0.0
C64				>5	>5	0.0
N65				>5	>5	2.7
D66				>6	>6	0.2
G67						
R68				>3	>3	6.0
T69		0.03	0.10 (γCH_3)	>5	6–8	0.0
P70				>5	>5	8.2
G71						
S72	0.05	0.08	0.04 (β')	5–6	5–6	0.3
R73				>3	1–3	11.0
			0.00 (δCH_2)			
			0.00 ($\delta CH_2'$)			
N74				>6	>6	0.0
L75			0.08 (γCH_2)	>6	6–8	1.5
			0.10 (δCH_3)			
A76	0.07	0.04		6–8	6–8	0.0
N77				>6	>6	7.8
I78				>6	>10	1.0
P79			0.01 0.00 (β')	>5	6–8	3.5
C80	0.02	0.03		5–6	5–6	0.0
S81	0.01	0.01		1–3	1–3	3.0
A82	0.00	0.00		1–3	1–3	0.0
L83	0.00	0.01	0.01 (γCH_2)	1–3	1–3	0.0
			0.01 (δCH_3)			
L84	0.01	0.01	0.01 (γCH_2)	1–3	1–3	0.0
			0.01 (δCH_3)			

3. The antibody binds two forms of the loop peptide, in which the disulfide between Cys-64 and Cys-80 is either reduced or oxidized, with comparable affinities ($K_A \sim 10^8\ M$).[23,56]

4. The antibody is strongly cross-reactive with the native protein (HEL), binding with only a 10-fold lower affinity. Limited regions on the protein epitope have been identified from serological studies.

5. A high-resolution crystal structure is available for the protein antigen.[57]

Chemical Shift Changes for ^{1}H Resonances of the Peptide on Binding to Gloop1

Resonances from both the main chain and side chains of residues at the N and C termini of the peptide (Q57–I58 and S60; P79–L84) have chemical shift positions that are virtually identical to those in the free ligand (shift ≤ 0.03 ppm). A second group, clustered between residues 69 and 76, show somewhat larger changes in chemical shift (0.03 to 0.1 ppm). For 14 residues of the peptide, no cross-peaks are seen in the bound spectrum.[22] In these cases the magnitude of the chemical shift changes that accompany the binding of the peptide to antibody cannot therefore be determined by ^{1}H spectroscopy.

Distribution of Linewidth of Resonances of the Bound Peptide along the Amino Acid Sequence

From the estimated degree of broadening of the bound peptide cross-peaks (Table I), three separate groups of residues can be identified:

1. The regions with the narrowest resonances in the bound complex, that is those comprising the most mobile regions, are primarily from residues of the N- and C-terminal segments of the peptide. We therefore concluded that these residues, Q57–I58 and C80–L84, do not interact strongly with the antibody in the bound complex.

2. In the intervening region (N59–P79) a somewhat reduced level of mobility is retained by a number of residues in the bound peptide, e.g., S72 and A76, while others show a very substantial degree of immobilization.

3. The largest measured increases in linewidth are found for four residues that do not correspond to a single continuous region of the peptide sequence. Three (R61, W62, and W63) lie adjacent to one another toward the N terminus of the peptide, but the fourth, I78, is some 15 residues

[56] S. Roberts, J. C. Cheetham, and A. R. Rees, *Nature (London)* **328,** 731 (1987).

[57] H. H. G. Handoll, D. Phil. thesis, University of Oxford, 1985.

further down the peptide chain. The degree of line broadening (> 10 Hz) suggests that these residues are strongly immobilized in the bound complex.

4. For all the residues where cross-peaks could be observed in both the free and bound COSY spectra, the pattern of the peaks in the two data sets were essentially identical, suggesting that for these residues changes in the coupling constants were not contributing to the observed reductions in cross-peak intensities.

A Comparison of the NMR Results with the Crystal Structure of the Native Protein Antigen

The conformation of the corresponding sequence in the loop peptide in the structure of hen egg lysozyme (with which the antibody strongly cross-reacts) has already been determined by X-ray crystallography.[57] If we examine the mobility of the residues in the bound peptide, as indicated by the estimated linewidths of their ^1H resonances, then we note that the majority of residues that are associated with high mobility in the bound peptide are located in ordered regions of the crystal structure of the protein, with low thermal factors, and low accessibility (Table I). Conversely, three of the four residues identified as having low mobility in the peptide complex are surface residues, with high accessibility (Table I). R61, W62, W63, and I78. These all show very large increases in linewidth in the bound peptide and, although not contiguous in the primary sequence, are grouped together in the tertiary structure of the protein, forming a hydrophobic projection on the surface of the molecule (Fig. 8). A similar grouping of these four hydrophobic residues in both the bound peptide and native protein molecules may therefore account for the high cross-reactivity shown by the Gloop1 antibody.[22]

Concluding Remarks

In this chapter we have demonstrated how the 2D COSY experiment can be used to determine the degree of mobility of residues in a peptide antigen bound within the combining site of a high-affinity antibody. This dynamic behavior can be correlated with the nature of the peptide interactions with the antibody in the complex. The technique is able to differentiate between tightly bound and weakly associated residues of the peptide in the antibody complex. Moreover, we have shown from the results of our studies of the Gloop1 antibody/loop peptide system that the areas of comparable mobility, or immobility, need not comprise residues from a contiguous region in the primary sequence. The method does not provide a

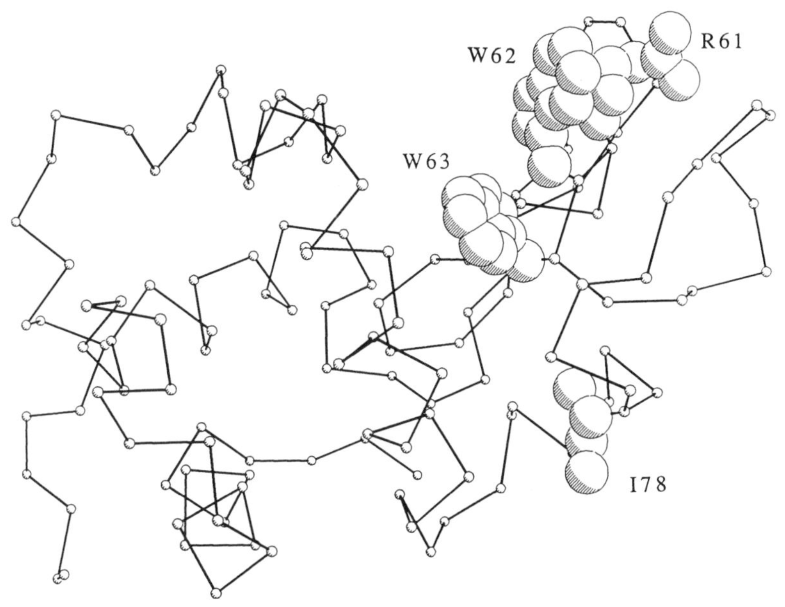

FIG. 8. Conformation of the main chain of hen egg lysozyme.[57] The four residues (R61, W62, W63, I78) that constitute a hydrophobic ridge on the protein surface, and which are observed to be strongly immobilized in the peptide on binding to the Gloop1 antibody, are indicated.

structure for the bound peptide, but the dynamic picture it provides of the antigen in its bound complex is consistent with the current description of antibody–peptide complexes obtained from X-ray crystallographic studies. Furthermore, the ^{1}H NMR studies offer an entirely complementary partner to such work. Residues that are located in very flexible regions of a polypeptide chain, such as those in parts of the peptide antigen that may not be tightly bound within the antibody complex, are often difficult to study using X-ray methods, giving disordered or absent regions in the electron density. High-resolution NMR techniques, however, are particularly well suited as a probe of these more mobile residues within a large protein structure.

In a wider context, the 2D ^{1}H NMR approach that we have described has a clear and important role to play as one member of a group of NMR-based approaches to the study of peptide antigens and their antibody-bound complexes in solution.[13,16] With the rapidly expanding use of molecular biology techniques to reduce the molecular weight of the immunoglobulin fragments in these systems to more tractable levels for NMR studies, and to incorporate mutations and isotopic labels directly into the protein molecule,[58-60] these methods should increasingly provide both new insight into the conformational and dynamic features of antigen specificity, and give a viable alternative to X-ray crystallographic methods, which ultimately rely on the availability of suitable crystals for analysis. In realizing their full potential, one would predict that over the next few years a rapid growth will be seen in the structural database relating to antibody–peptide interactions, and a concurrent improvement in our general level of understanding of the molecular basis for one of the natural recognition processes.

[58] L. Riechmann, J. Foote, and G. Winter, *J. Mol. Biol.* **203,** 825 (1988).
[59] A. Skerra and A. Pluckthun, *Science* **240,** 1038 (1988).
[60] H. Field, G. T. Yarrington, and A. R. Rees, *Protein Eng.* **3,** 641 (1989).

[10] Nuclear Magnetic Resonance for Studying Peptide–Antibody Complexes by Transferred Nuclear Overhauser Effect Difference Spectroscopy

By JACOB ANGLISTER and FRED NAIDER

Introduction

Short synthetic peptides have been useful as probes for the antigenic structure of proteins.[1] Immunization with a flexible synthetic peptide that is homologous in amino acid sequence with a segment of a protein elicits the production of a spectrum of antibodies. These polyclonal antibodies may recognize different conformations of the peptide and only a fraction of these antibodies may bind the native protein. We have initiated a detailed analysis of the interaction of an immunizing peptide with monoclonal antibodies that either do or do not recognize the native protein. Studies of such complexes should provide insights into antibody function and selectivity. The principal aims of our research are (1) to study the molecular basis of specific antibody–peptide interaction and to understand how the diversity in the amino acid sequence of the antibody affects combining site structure, antibody affinity, and specificity; and (2) to characterize the different antigenic structures of peptide antigens, and to relate peptide conformation to the production of antibodies cross-reactive with the parent protein.

[1] H. J. Dyson, R. A. Lerner, and P. E. Wright, *Annu. Rev. Biophys. Chem.* **17,** 305 (1988).

In order to learn about the molecular basis for the cross-reactivity of antibodies with proteins and peptides derived from them it is necessary to determine the three-dimensional structure of peptide–antibody complexes. Although crystal structures are available for several Fab fragments, only limited success has been achieved on Fab–peptide or Fv–peptide complexes.[2] In principle, nuclear magnetic resonance (NMR) spectroscopy can be used to study the three-dimensional structure of antibody–peptide complexes in solution. Complete structure determination by this technique, however, is currently limited to small proteins of up to 150 amino acid residues.[3-5] Studies on larger proteins are hindered by the loss of spectral resolution, which stems from resonance broadening due to enhanced relaxation pathways and overlap of individual resonances from the increased number of protons. Consequently very few NMR studies of large proteins have been reported. This chapter describes an approach to studying ligand–protein interaction using two-dimensional (2D) transferred nuclear Overhauser effect (NOE) difference spectroscopy. The interaction of antibodies with peptide antigens provides a specific case study.

Methodology

The antibody molecule has a molecular weight of about 150,000 and contains more than 6000 nonexchangeable protons. To obtain the simplest proton NMR spectrum possible, it is necessary to work with the smallest fragment of the antibody that retains its affinity for the antigen. This fragment is the antibody Fv, which contains about 220 amino acids and has a molecular weight of approximately 25,000.[6] The Fv fragment was used by Dwek and co-workers in their pioneering NMR studies of the antibody MOPC315 complexed with a dinitrophenyl hapten.[7,8] Unfortunately the Fv can be obtained by proteolytic cleavage only for a very limited number of antibodies. Therefore, until recently, only the larger Fab fragment was readily available for NMR and crystallographic studies. In addition to the variable domains, the Fab contains one constant domain from each chain. The molecular weight of the Fab is about 50,000. It

[2] R. L. Stanfield, T. M. Fieser, R. A. Lerner, and I. A. Wilson, *Science* **242,** 712 (1990).

[3] K. Wüthrich, "NMR of Proteins and Nucleic Acids." Wiley, New York, 1986.

[4] K. Wüthrich, *Acc. Chem. Res.* **22,** 36 (1989).

[5] D. A. Torchia, S. W. Sparks, and A. Bax, *Biochemistry* **28,** 5509 (1989).

[6] J. Hochman, D. Inbar, and D. Givol, *Biochemistry* **12,** 1130 (1973).

[7] S. K. Dower and R. A. Dwek, *in* "Biological Applications of Magnetic Resonance" (R. G. Shulman, ed.), p. 271. Academic Press, New York, 1979.

[8] R. A. Dwek, J. C. A. Knott, D. Marsh, A. C. McLaughlin, E. M. Press, N. C. Price, and A. I. White, *Eur. J. Biochem.* **53,** 25 (1975).

contains about 440 amino acids and more than 2000 nonexchangeable protons. Genetic engineering techniques have been used to express the Fv in myeloma cells and *Escherichia coli*,[9-11] and it is expected that future NMR studies will take advantage of the increased resolution expected from studies on Fv–antigen complexes (see Wright *et al.*[12] for a preliminary NMR study on an Fv fragment).

Deuterium Labeling of Antibody

The ^{1}H NMR spectrum of the Fab is very poorly resolved even when measured on a 500- or 600-MHz spectrometer. In order to simplify the spectrum and assign resonances of the Fab to specific amino acid types the antibody is labeled with deuterated amino acids.[13] This biosynthetic labeling is accomplished by feeding hybridoma cells producing the monoclonal antibody a controlled diet containing selected deuterated amino acids. In studies of antibody–antigen interactions it is beneficial to concentrate on simplifying the spectrum of the aromatic protons for three reasons: (1) that region of the spectrum is *a priori* less crowded; (2) all aromatic hydrogens can be very efficiently labeled; and (3) aromatic amino acids have a major role in antibody–antigen interactions, with solvent exclusion contributing significantly to the binding energy.[14]

Two-Dimensional Transferred Nuclear Overhauser Effect Difference Spectroscopy

Three-dimensional structure determination by NMR is based on the observation of magnetization transfer between neighboring protons due to the NOE.[15] The magnitude of the NOE is dependent on the reciprocal of the sixth power of the distance between interacting nuclei and is observed up to a distance of approximately 5 Å. A two-dimensional (2D) NOE experiment (NOESY) contains the information about magnetization transfer due to the NOE between all pairs of neighboring protons in the molecule.[3] Despite the power of 2D NOE spectroscopy this technique has not been applied to the study of the structure of large proteins because of

[9] A. Skerra and A. Plueckthun, *Science* **240,** 108 (1988).

[10] L. Riechmann, J. Foote, and G. Winter, *J. Mol. Biol.* **203,** 825 (1988).

[11] R. E. Bird, K. D. Hardman, J. W. Jacobson, S. Johnson, B. M. Kaufman, S.-M. Lee, T. Lee, S. H. Pope, G. S. Riordan, and M. Whitlow, *Science* **242,** 423 (1988).

[12] P. E. Wright, H. J. Dyson, R. A. Lerner, L. Riechmann, and P. Tsang, *Biochem. Pharmacol.* **40,** 83 (1990).

[13] J. Anglister, T. Frey, and H. M. McConnell, *Biochemistry* **23,** 1138 (1984).

[14] E. A. Padlan, *Proteins: Funct. Struct. Genet.* **7,** 112 (1990).

[15] J. H. Noggle and R. E. Schirmer, "The Nuclear Overhauser Effect: Chemical Applications." Academic Press, New York, 1971.

the poor spectral resolution found in 2D NMR spectra of these macromolecules. Specific deuteration of aromatic amino acids significantly improves the resolution in the NOESY spectrum; however, cross-peak overlap still prevents resonance identification.[16]

In order to discern protein–ligand interactions and the conformation of the bound ligand using the NOESY spectrum of the complex, one must discriminate between cross-peaks due to intermolecular interactions and the numerous cross-peaks due to intramolecular interactions in the complexed protein. Most pertinent is the fact that it is almost impossible to identify the bound ligand resonances, since they are at least as broad as the protein resonances and often considerably broader due to a fast ligand off-rate. Therefore the bound ligand resonances are unresolved from the background. To circumvent this problem we take advantage of a phenomenon known as transferred nuclear Overhauser effect spectroscopy (TRNOE).[17,18] This phenomenon is observed when the ligand molecules exchange rapidly between the bound and the free state. Under conditions where the ligand off-rate is fast relative to the T_1 relaxation time of both the Fab and the ligand protons, and to the mixing time used in the NOESY experiment, the spectrum of the Fab in the presence of excess ligand (excess spectrum) contains TRNOE cross-peaks. These cross-peaks reflect magnetization transfer that occurs between protons in the bound state. Transfer can be between Fab protons and protons of the bound ligand and/or between different protons of the bound ligand. This transferred magnetization is observed via the free ligand due to the relatively rapid exchange between the bound and free ligand. These inter- and intramolecular TRNOE cross-peaks are accompanied by numerous cross-peaks due to intramolecular interactions between Fab protons. Since there is essentially no free ligand, none of the TRNOE cross-peaks appears in the NOESY spectrum of the Fab saturated with ligand at a 1 : 1 molar ratio (saturated spectrum). However, the numerous cross-peaks due to intramolecular interactions in the saturated Fab are the same as those in the NOESY spectrum of the Fab in the presence of a ligand excess. By subtracting the saturated spectrum from the excess spectrum, one obtains a difference spectrum in which only TRNOE cross-peaks are observed.[19] This TRNOE difference spectrum is usually well resolved. Its resolution may be further improved by specific deuteration, which simultaneously allows the assignment of cross-peaks to a specific type of amino acid.

[16] J. Anglister, Q. Rev. Biophys. **23,** 173 (1990).

[17] P. Balaram, A. A. Bothner-By, and E. Breslow, *J. Am. Chem. Soc.* **94,** 4017 (1972).

[18] J. P. Albrand, B. Birdsall, J. Feeney, G. C. K. Roberts, and A. S. V. Burgen, *Int. J. Biol. Macromol.* **1,** 37 (1979).

[19] J. Anglister, R. Levy, and T. Scherf, *Biochemistry* **28,** 3360 (1989).

It should be noted that intramolecular NOE in small molecules has a positive sign, with intensity decreasing as a function of molecular weight. For molecules having a molecular weight of about 1500 the NOE is vanishingly small or zero. For larger molecules it becomes negative, and the intensity increases to a maximum of -1 with increasing molecular weight. Therefore, in the presence of a binding protein, the sign of transferred NOE cross-peaks due to intramolecular interactions in the bound ligand usually will be negative. The exact dependence of the TRNOE intensity as a function of the binding constant, ligand off-rate, molar ratio between ligand and protein, averaging of the resonances of bound and free ligand, and the cross-relaxation rate in the free ligand is thoroughly described by Clore and Gronenborn.[20,21]

Assignment of TRNOE Cross-Peaks

Cross-peaks appearing in the TRNOE difference spectrum arise from the following: (1) exchange between bound and free ligand, (2) magnetization transfer between antibody protons and free ligand protons via the bound state, (3) intramolecular magnetization transfer within the bound ligand via exchange with the free. Therefore at least one of the two frequency values of a TRNOE cross-peak should be of a free ligand proton, while the corresponding second resonance is (1) the same proton of the bound ligand, (2) an antibody proton or (3) another proton of the free ligand.

The assignment of cross-peaks to their respective free ligand protons is based on the 2D homonuclear correlated spectrum (COSY) of the ligand.[19] If changes in chemical shift of the ligand protons on binding are large relative to the ligand off-rate, averaging of bound and free ligand proton resonances is prevented and the COSY spectrum of the ligand can be used. However, for fast off-rates, resulting in averaging of the resonances of the bound and free ligand, one should measure the COSY spectrum of the Fab solution in the presence of the same ligand excess used for the NOESY measurements. Contributions of the Fab to the COSY spectrum mostly vanish due to the linewidth of the Fab protons, and only the contributions of the much smaller ligand are observed. When the resonance of a ligand proton interacting with the antibody falls in a region where a few ligand resonances overlap, unambiguous assignment requires repetition of the experiment with a specifically deuterated ligand.

[20] G. M. Clore and A. M. Gronenborn, *J. Magn. Reson.* **48,** 402 (1982).
[21] G. M. Clore and A. M. Gronenborn, *J. Magn. Reson.* **53,** 423 (1983).

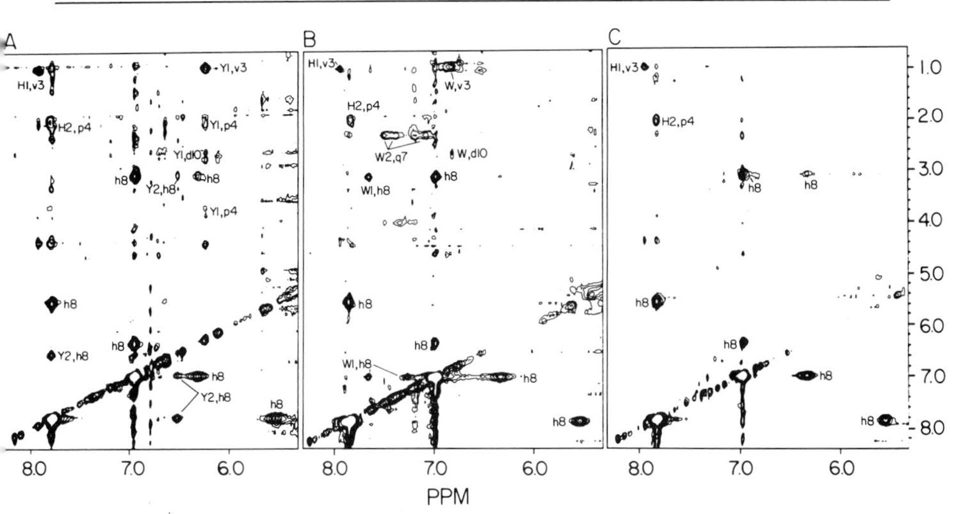

FIG. 1. Two-dimensional difference spectra between the NOESY spectrum of TE32 Fab with a fourfold excess of the peptide CTP3 (VEVPGSQHIDSQKKA) and the NOESY spectrum of the peptide-saturated Fab, showing interactions of specific types of antibody aromatic residues with peptide residues. Assignment to antibody residues is marked by capital letters and arbitrary numbers, while assignment to peptide residues is marked by lower-case letters and their location in the sequence. (A) Interactions of antibody tyrosine and histidine residues with peptide residues. Antibody phenylalanine and tryptophen residues are perdeuterated, while tyrosine residues are deuterated at 2,6-phenyl positions. (B) Interactions of antibody tryptophan and histidine residues with peptide residues. Antibody phenylalanine and tyrosine residues are perdeuterated. (C) Interactions of antibody phenylalanine and histidine residues with peptide residues. Antibody tyrosine and tryptophan residues are perdeuterated.

Our studies concentrate on the interactions between aromatic residues of the antibody and the amino acids of a peptide of cholera toxin (CTP3). CTP3 (VEVPGSQHIDSQKKA)[21a] contains only one aromatic residue, histidine, and the resonances of the free peptide histidine can be easily identified. When interaction between a nonaromatic peptide proton and an aromatic proton is observed, and the chemical shifts of the aromatic protons differ from those of the peptide histidine imidazole protons, it is assigned to antibody aromatic residue. The assignment to a specific type of amino acid is based on specific deuteration of the aromatic amino acids of the antibody.[19] Figure 1 shows the assignments of the TRNOE cross-peaks

[21a] Single-letter abbreviations used for amino acids: A, alanine; R, arginine; N, asparagine; D, aspartic acid; C, cysteine; Q, glutamine; E, glutamic acid; G, glycine; H, histidine; I, isoleucine; L, leucine; K, lysine; M, methionine; F, phenylalanine; P, proline; S, serine; T, threonine; W, tryptophan; Y, tyrosine; V, valine.

to the aromatic residues of the TE32 anti-CTP3 antibody based on the measurements of the 2D TRNOE difference spectrum for three preparations of Fab differing in the labeling of antibody aromatic residues.

When one of the two chemical shifts of a cross-peak is of a free ligand proton and the difference between the two is less than about 2 ppm, this cross-peak may be due to exchange. Assignment of such cross-peaks is facilitated by repeating the measurements at different temperatures. At lower temperature the Fab resonances become broader, whereas those of the bound peptide become narrower due to a slower off-rate. Such narrowing can be easily detected if the off-rate is on the order of or faster than the reciprocal of the intrinsic T_2 of the protons in the complex. When both chemical shifts of a cross-peak in the TRNOE difference spectrum are the same as those of the free ligand protons and the fine structure of the two resonances involved is observed, this cross-peak is assigned to intramolecular interaction in the bound ligand.

Experiments Involving Slow Ligand Off-Rate

When the ligand off-rate is slow relative to the mixing period used in the NOESY experiment and relative to the spin–lattice relaxation time of the Fab and peptide protons, the TRNOE becomes much weaker or nonobservable. We encountered this problem while studying the interactions between the CTP3 peptide and the TE34 antibody.[22] In order to be able to use TRNOE difference spectroscopy we searched for minor modifications of the peptide that would decrease the binding constants and enhance its off-rate. We found that conversion of the C-terminal carboxyl into an amide decreases the binding constant of the peptide to TE34 by two orders of magnitude while increasing its off-rate similarly. Indeed, TRNOE difference spectra with excellent signal-to-noise ratio were obtained using the modified peptide. To verify that this modification did not alter the interactions of the antibody with the unmodified portion of the peptide we repeated the experiments with a truncated peptide, acetyl-IDSQKKA, corresponding to residues 9 to 15 of CTP3. This peptide analog lacks one residue (His-8) from the epitope found to be recognized by TE34; however, it contains an intact C-terminal carboxyl. The binding constant of this peptide analog to TE34 is only one order of magnitude less than that for CTP3 binding to the antibody. Our experiments show that both CTP3 with an amide at the carboxyl terminus and the truncated analog exhibit very similar cross-peaks. Thus, we conclude that minor modification of the epitope recognized by the antibody does not significantly affect antibody interaction with the remainder of the epitope.[22] The use of modified peptides that overlap the antigenic epitope permits study of the interac-

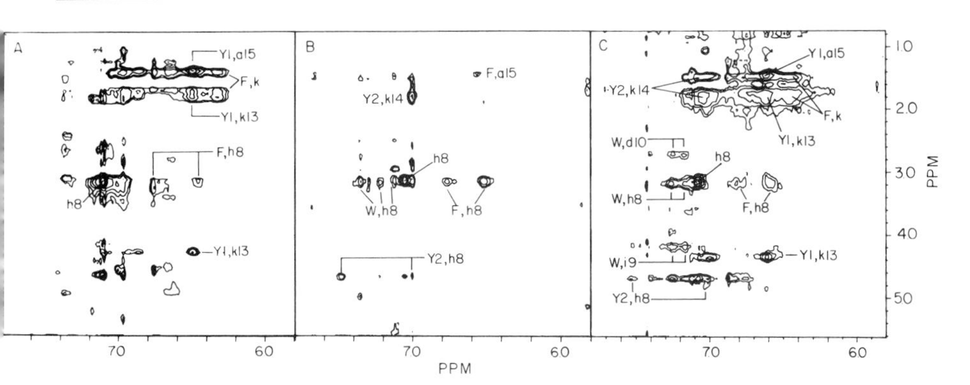

FIG. 2. Two-dimensional TRNOE difference spectra obtained after specific chain labeling. (A) Interactions of the light-chain aromatic residues of TE34 Fab with the NCA peptide (Ac-VEVPGSQHIDSQKKA-NH_2). (B) Interactions of heavy-chain aromatic residues with the NCA peptide. (C) Interactions of aromatic residues from both chains with the NCA peptide. Assigned antibody residues are marked by capital letters and arbitrary numbers; peptide protons are marked by small letters and numbers denoting the location in sequence of the residues to which the protons belong.

tions between the antibody and the whole peptide. The strategy that we developed can be used to extend the applicability of TRNOE difference spectroscopy to the study of other protein–ligand complexes involving slow ligand off-rates.

Chain Labeling

Antibody light and heavy chains can be separated and later recombined to reconstitute the native conformation of the antibody.[23,24] This procedure can be used to label each of the chains and subsequently assign resonances to a specific chain.[25] For example, two differently labeled Fab samples (one deuterated, the other not deuterated) are prepared, the interchain disulfide bond is reduced and alkylated, and the chains separated by ion-exchange chromatography under denaturing conditions. The deuterium-labeled heavy-chain fragment is recombined with the unlabeled light chain and vice versa. The native conformation is then recovered by dialyzing away the denaturant. Figure 2 shows the TRNOE difference spectra obtained for reconstituted Fab in which the chains were specifically labeled.[26] Figure 2A

[22] J. Anglister and B. Zilber, *Biochemistry* **29,** 921 (1990).
[23] F. Franek and R. S. Nezlin, *Folia Microbiol.* **8,** 128 (1963).
[24] H. Metzger and M. Mannik, *J. Exp. Med.* **120,** 765 (1964).
[25] J. Anglister, T. Frey, and H. M. McConnell, *Nature (London)* **315,** 65 (1985).
[26] B. Zilber, T. Scherf, M. Levitt, and J. Anglister, *Biochemistry* **29,** 10032 (1990).

shows the interactions of the aromatic protons of the light chain and Fig. 2B shows the interactions of the aromatic protons of the heavy chain. Figure 2C was obtained for native unlabeled Fab and shows the interactions of aromatic protons from both chains. Figure 2C correspond very well to the sum of the spectra presented in Fig. 2A and B.

Assignment of the Antibody Interactions to Specific Residues

The assignment of TRNOE cross-peaks allows us to define interactions between various types of amino acids. Given the present state of the art it is impossible to make a complete structure determination of a Fab fragment using NMR spectroscopy. Thus, we are confronted with the problem of extending a given NMR assignment from a type of amino acid to a given residue in the Fab. This problem has been approached using computer model building and conformational energy calculations to generate a starting model for the NMR studies. It was first used by Dwek and his coworkers to study the combining site of an anti-DNP antibody.[7,8] The computerized model building is based on principles introduced by Padlan *et al.*[27] This procedure takes advantage of the homology in the three dimensional structure between various antibodies, which is especially high in the conserved regions that form the immunoglobulin fold. The main obstacle is the modeling of the hypervariable loops, and different approaches to this problem have been developed.[28]

Using the calculated models, we can judge what amino acids in the combining site are exposed to the solvent and, therefore, have the potential to interact with the antigen. By combining the NMR information, the amino acid sequence data of the six CDRs and the preliminary model, most ligand–protein interactions can be assigned to specific antibody residues. Additional information about interresidue interactions in the antibody combining site can facilitate the assignment of the interactions to the specific residues. Such information can be obtained by calculating the difference between the excess spectrum and the NOESY spectrum of the Fab itself. This difference spectrum contains cross-peaks due to interactions with the antigen and those due to intra-Fab interactions of antibody protons that change their chemical shift on antigen binding.

[27] E. A. Padlan, D. R. Davies, I. Pecht, D. Givol, and C. Wright, *Cold Spring Harbor Symp. Quant. Biol.* **41,** 627 (1976).

[28] C. Chothia, A. M. Lesk, A. Tramontano, M. Levitt, S. J. Smith-Gill, G. Air, S. Sheriff, E. A. Padlan, D. Davies, W. R. Tulip, P. M. Coleman, S. Spinelli, P. M. Alzari, and R. J. Poljak, *Nature (London)* **342,** 877 (1989).

Derivation of Constraints on Proton–Proton Distances

The 2D TRNOE difference spectrum contains cross-peaks due to all the short-range interactions (proton–proton distance <4 Å) between the antibody and the peptide. Measurements of the intensity of intramolecular magnetization transfer between a pair of neighboring protons after a short irradiation period of one of them or after a short mixing time in a NOESY experiment are used to evaluate the distance between the protons. Such measurements have been carried out to obtain constraints on proton–proton distances that were subsequently used to calculate the three-dimensional structure of small proteins.[3] Clore and Gronenborn extended the application of such measurements to transferred NOE in protein–ligand complexes.[20,21] Their analysis distinguishes three regions of chemical exchange. In the case of fast exchange between bound and free ligand, when the ligand off-rate is faster then 10 times the spin–lattice relaxation time of its protons and when the resonances of the bound and free ligand are averaged, the magnitude of the transferred NOE ($N_{H,A}$) between an antibody proton, A, and a peptide ligand proton, H, is given by $N_{H,A} \approx (1 - a)\sigma_{H,A}\tau_m$, where $\sigma_{H,A}$ is the cross-relaxation rate between an antibody proton and neighboring bound peptide proton, a is the mole fraction of the free peptide, and τ_m is the mixing time in the NOESY experiment or the irradiation time in a 1D NOE experiment. This simple relationship applies to both 1D and 2D TRNOE experiments. The distance ratios between different pairs of protons in the complex can be determined from $\sigma_{H,A}$, which is inversely proportional to the sixth power of the distance between the protons.

When the chemical shifts of the bound and free peptide resonances are not averaged but the off-rate is still faster than T_1 and $\sigma_{H,A}$, and the peptide is present in large molar excess, the TRNOE intensity between an antibody proton and a free peptide proton is still approximated by the above equation. In the case of medium off-rate the efficiency of the TRNOE will depend on the off-rate[20,21] and the above equation for $N_{H,A}$ can serve only as an upper limit. If the exchange between bound and free ligand is slow on the cross-relaxation and spin–lattice relaxation scale the transferred NOE intensity is vanishingly small.

To translate cross-peak intensities into restraints on proton–proton distances one needs a calibration usually obtained by measuring the intensity of a cross-peak between a pair of protons separated by a known and fixed distance. To obtain such calibration we measured the difference spectrum between the NOESY spectrum of TE34 Fab in the presence of 10-fold peptide excess and the NOESY spectrum of the uncomplexed Fab (without spectrum). In addition to cross-peaks due to transferred NOE this type of difference spectrum shows cross-peaks due to intra-Fab interactions of protons experiencing changes in chemical shift on antigen binding. When all tryptophen and phenylalanine residues of the antibody were deuterated while tyrosine residues were unlabeled the difference spectrum measured for the Fab contains very well-resolved cross-peaks due to intra-residue interactions between the ϵ_1 and δ_1 and between the ϵ_2 and δ_2 tyrosine protons. These cross-peaks are not observed when the C_δ protons of the antibody tyrosine residues are deuterated. The intensity of the resolved cross-peaks is then compared to the intensity of the exchange cross-peak of $C_{\delta 2}H$ of the peptide histidine. Since the exchange cross-peak appears in all the TRNOE difference spectra, and all spectra are measured at the same mixing time and with the same molar ratio of peptide, this cross-peak can serve as an internal standard for calibrating the intensities in all TRNOE difference spectra. Through this internal standard one gets the ratio between cross-peak intensities in the TRNOE difference spectrum and the cross-peak due to intratyrosine interactions and subsequently the ratio between the distances of the corresponding pairs of protons. This analysis indicates that all the cross-peaks due to intermolecular interactions observed in the TRNOE difference spectra are due to strong interactions between protons that are less than 3.5 Å apart.

Experimental Procedures

Sample Preparation

Hybridoma cells producing antibodies are grown as monolayers in 700 ml T flasks on RPMI 1640 medium[29] supplemented with only 2% fetal calf serum to increase the efficiency of the labeling. Supernatant is collected every 2–3 days and 50 ml of fresh medium is supplied to each flask. Antibody production is dependent on the specific cell line and varies between 20 and 60 mg/liter. Biosynthetic specific deuteration of the antibody is accomplished by feeding the hybridoma cells producing the antibodies medium containing the selected deuterated amino acids according to the RPMI recipe. Antibodies are purified utilizing protein A-Sepharose CL-4B column chromatography.[30,31] Antibody solution is concentrated by an M_r 25K cutoff collodion bag (Schleicher & Schuell, Dassel, Germany) to

[29] G. E. Moore, R. E. Gerne, and H. A. Franklin, *J. Am. Med. Assoc.* **199,** 519 (1967). GIBCO Laboratories, Grand Island, New York, Technical Manual, 1982.

[30] G. Otting, H. Widmer, G. Wagner, and K. Wüthrich, *J. Magn. Reson.* **66,** 187 (1986).

[31] V. T. Oi and L. A. Herzenberg, *in* "Selected Methods in Cellular Immunology" (B. B. Mishell and S. M. Shiigi, eds.), p. 351. Freeman, San Francisco California, 1980.

a concentration of 30 mg/ml. Fab is obtained by papain digestion[32] and purified by applying the reaction mixture to a Sephadex G-100 column linked in series to a protein A column (50 ml protein A-Sepharose CL-4B/200 mg of antibody). The columns are preequilibrated and developed with 50 m*M* Tris-HCl, pH 9, containing 0.15 *M* sodium chloride and 0.05% (w/v) sodium azide. The large peak of Fab is preceded by a much smaller peak of $(Fab')_2$. The Fc binds to the protein A and is eluted with citrate buffer, pH 4.5. The Fab is concentrated to 3 m*M* with a collodion bag and the aqueous solution is dialyzed against four changes of 10 m*M* phosphate-buffered dueterium oxide, pH 7.1, containing 0.05% sodium azide. A typical NMR sample contains 500 μl of 3 m*M* Fab. Antibodies are never lyophilized or frozen.

NMR Measurements

NOESY spectra are measured on a 500-MHz spectrometer in the phase-sensitive mode using a proton-selective probe. A mixing time of 100 msec, which was found to be optimal for obtaining maximum intensities for the strong cross-peaks observed, is used. The carrier frequency is set on the HDO line, and a spectral width of 6000 Hz is used. The HDO line is presaturated, using minimal power, for 3 sec before the first 90° pulse is applied. Spectra are Fourier transformed in both dimensions after application of a squared cosine window function. A considerable reduction in t_1 ridges is achieved by using the method of Otting *et al.*,[30] already incorporated in the standard 2D Fourier transformation in Bruker's software. Spectra are not symmetrized. The two spectra used in the difference calculation are measured consecutively, Fab concentration is between 2 and 3.5 m*M* in 0.01 *M* phosphate-buffered D_2O, pD 7.15, and the temperature is 42°. Spectra are recorded with 256 values of t_1; for each value, 64 or 80 scans are collected that are preceded by 4 dummy scans. Measurements of the TRNOE difference spectra for 60- and 100-msec mixing times exclude the possibility that the cross-peaks observed in the TRNOE difference spectra at 100 msec are due to spin diffusion.

Difference Spectra Calculations

The phases of the initial measurements (first t_1 value) of both 2D experiments are matched to obtain a 1D difference with minimal distortion to resonances and baseline. This is accomplished first by visual comparison of the two spectra and then by fine numerical phase correction of

[32] R. R. Porter, *Biochem. J.* **73,** 119 (1959).

one to match the other. The phase corrections thus obtained are applied to the preliminary 2D Fourier transformation. From the NOESY spectrum of the Fab with excess peptide, we select a row crossing a free peptide resonance. The same row is selected from the NOESY spectrum of the peptide-saturated Fab, and the two are subtracted. The phase of the resulting 1D difference spectrum is numerically corrected to obtain a pure Lorentzian for the resonance of the peptide proton. This phase correction is now used to reprocess the two NOESY spectra, and the difference between these rows is reexamined to verify that the resonance of the free peptide retains a pure Lorentzian shape. This procedure ensures the matching of phases between the two 2D spectra in the region containing cross-peaks between nonaromatic protons. The rows and columns crossing the diagonal at the resonances of the histidine imidazole protons of the peptide require additional slight phase correction. Baseline distortions due to t_2 ridges are corrected for by subtracting from the 2D difference spectrum the value for a column with no signal at the intersection with the diagonal of the 2D spectrum. Further baseline correction of individual rows is carried out where necessary by fitting a baseline with a fourth-order polynomial.

Conclusions and Perspectives

The methodology developed in our laboratory allows one to obtain detailed information concerning the interactions between residues in a peptide antigen and residues in the combining site of a monoclonal antibody. Transferred NOE difference spectroscopy, in conjunction with specific deuteration of the antibody, provides parameters that give information about distances between the above residues. Unfortunately, the number of NOE cross-peaks are not sufficient to calculate an unequivocal structure for the bound antigen. To circumvent this problem we apply models for the antibody combining site that are predicted based on structures for CDRs determined by X-ray crystallography. These models help us to verify assignment of the peptide–antibody interactions and can be used along with NOE distance constraints, energy minimization, and molecular dynamics calculations to provide a 3D model of the structure of the bound peptide antigen. Moreover, comparison of NMR data with calculated models allows one to examine the validity of a given model and to exclude "bad models" that result from choosing improper segments from the known three-dimensional structures of other antibodies. It is important to emphasize that the above approaches are warranted by the severe overlap problem that is confronted in studies of the interaction of peptides with large proteins. This problem has been encountered by others and elegant procedures involving isotope editing[33,34] and 2D difference spectroscopy with deuterated ligands have been developed.[35] In all cases to date, information forthcoming from such studies requires comparison with crystal structures to rationalize interactions that are apparent in the NMR spectra. The combined application of 2D TRNOE difference spectroscopy and isotope editing procedures holds great promise for unraveling the mysteries of peptide–protein interactions. At present, despite some obvious shortcomings, these are the only approaches to study the conformation of peptides bound to antibodies, enzymes, or receptors in solution.

Acknowledgments

We are most grateful to Michael Levitt, Rina Levy, Tali Scherf, and Barbara Zilber for their important contributions to the work described in this chapter. This work was supported by the United States–Israel Binational Science Foundation. F. N. is a visiting professor on a Fulbright International Exchange Program.

[33] S. W. Fesik, J. R. Luly, J. W. Erickson, and C. Abad-Zapatero, *Biochemistry* **27**, 8297 (1988).
[34] P. Tsang, T. M. Feiser, J. M. Ostresh, R. A. Houghten, R. A. Lerner, and P. E. Wright, *Fronteirs NMR Mol. Boil.*, 63 (1990).
[35] S. W. Fesik and E. R. P. Zuiderweg, *J. Am. Chem. Soc.* **111**,5013, (1989).

[11] Isotope-Edited Nuclear Magnetic Resonance Studies of Fab–Peptide Complexes

By P. TSANG, M. RANCE, and P. E. WRIGHT

Introduction

Through the development of isotope-edited techniques,[1-3] the application of nuclear magnetic resonance (NMR) spectroscopy to studies of higher molecular weight systems such as Fab–antigen or Fab′–antigen complexes has become feasible.[4-7] The general concept and principles of this method will be described briefly in this chapter, and illustrated with results obtained in our laboratory.

[1] R. H. Griffey and A. G. Redfield, *Q. Rev. Biophys.* **19,** 51 (1987).
[2] A. Bax, S. W. Sparks, and D. A. Torchia, this series, Vol. 176, p. 134.
[3] G. Wagner, this series, Vol. 176, p. 93.
[4] P. Tsang, T. M. Fieser, J. M. Ostresh, R. A. Lerner, and P. E. Wright, *Pept. Res.* **1,** 87 (1988).

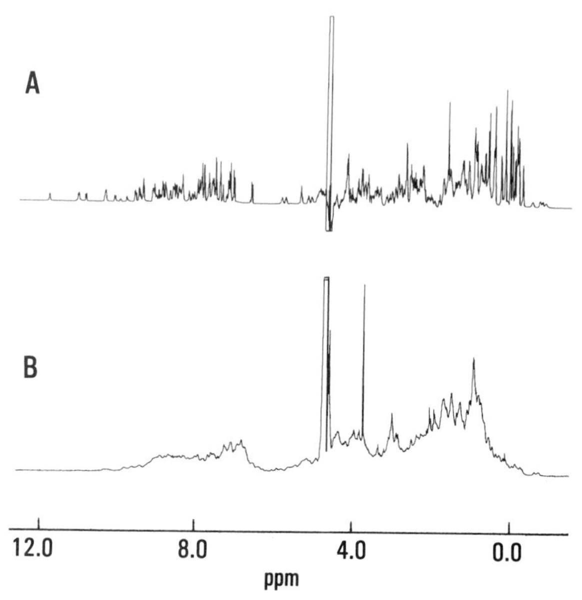

FIG. 1. ^1H NMR spectra (500 MHz, 308 k) recorded from 90% H_2O/10% D_2O samples of 3 m*M* BPTI (A) and a 0.5 m*M* Fab′–peptide complex (B). Both spectra were acquired using a Hahn-echo pulse sequence along with solvent presaturation and a homo-spoil pulse.

Detailed structural studies by solution-state ^1H NMR spectroscopy have been successfully conducted on proteins up to ~20 kDa.[8] While in principle these methods may be applied to systems the size of the Fab–peptide complexes discussed here (~56 kDa), significant physical limitations exist that frustrate these attempts, thus necessitating a different approach and technique such as isotope editing. The importance of spectral editing for studies of systems greater than 20 kDa is emphasized in Fig. 1, which shows proton one-dimensional (1D) spectra obtained from a small protein, Basic Pancreatic Trypsin Inhibitor (BPTI) (Fig. 1A) and a Fab′–peptide complex (Fig. 1B). One major effect of increasing the molecular weight from 6500 (BPTI) to 56,000 (Fab′–peptide complex) is a dramatic

[5] P. Tsang, T. M. Fieser, J. M. Ostresh, R. A. Houghten, R. A. Lerner, and P. E. Wright, *in* "Frontiers of NMR in Molecular Biology" (D. Live, I. Armitage, and D. Patel, eds.), p. 63. Alan R. Liss, New York, 1990.

[6] P. E. Wright, H. J. Dyson, R. A. Lerner, L. Riechmann, and P. Tsang, *Biochem. Pharmacol* **40,** 83 (1990).

[7] P. Tsang, M. Rance, T. M. Fieser, R. A. Lerner, J. M. Ostresh, R. A. Houghten, and P. E. Wright, manuscript in preparation.

[8] K. Wüthrich, this series, Vol. 177, p. 125.

decrease in spectral resolution as a result of the greater density of resonances (the complex contains more than 5000 protons) from the higher molecular weight system. In addition, there is a reduction in sensitivity that results primarily from increased T_2 relaxation rates and, consequently, broader resonances. This has meant a degradation in the quality of obtainable experimental data.[9] Thus, the 56-kDa protein complex represents a formidable system for study by NMR due to the *overabundance* of information and inherent decreases in experimental sensitivity.

NMR studies of high-molecular-weight systems stand to benefit considerably from spectral editing techniques. Isotope editing allows one to monitor selectively not only a chosen component of a biomolecular complex, e.g., the peptide in a peptide–receptor complex, but also individual residues, at the level of atomic resolution. An example of the effects of isotope-editing techniques applied to a Fab′–peptide complex in which the peptide is ^{15}N-labeled at two residues[4] is given in Fig. 2. The full proton spectrum of this complex is shown, along with the amide proton–aromatic proton region of an ^{15}N isotope-edited spectrum recorded from the same sample. The upper, isotope-edited spectrum contains resonances from the amide protons of the labeled residues of the peptide. The observed resonances arise from the free and bound forms of the peptide in its complex with the antibody Fab′.[4]

A crucial parameter in NMR studies of complexes in general is the rate of exchange of ligand between its free and bound states. The peptide exchange rate of peptide–anti-peptide antibody complexes is generally slow relative to the chemical shift time scale.[10,11] In the case of the complexes studied in this laboratory, the association constant[12] is $\sim 5 \times 10^6$ M^{-1}, while the exchange rate has been estimated[7] to be less than ~ 40 sec^{-1}. The slow exchange behavior of the peptide is illustrated by the NMR titration spectra[6] shown in Fig. 3. The isotope-edited spectra in Fig. 3 were recorded during a titration of a Fab′ sample with the peptide MHKDFLE-KIGGL-NH_2,[12a] ^{15}N-labeled at the two glycine residues. The sequence of this peptide corresponds to residues 79–87 of the protein myohemer-

[9] D Neuhaus, G. Wagner, M. Vasak, J. H. R. Kägi, and K. Wüthrich, *Eur. J. Biochem.* **151,** 257 (1985).

[10] J. Kaplan and G. Fraenkel, "NMR of Chemically Exchanging Systems." Academic Press, New York, 1980.

[11] J. Sandström, "Dynamic NMR Spectroscopy." Academic Press, New York, 1982.

[12] T. M. Fieser, unpublished results.

[12a] Single-letter abbreviations for amino acids: A, alanine; R, arginine; N, asparagine; D, aspartic acid; C, cysteine; Q, glutamine; E, glutamic acid; G, glycine; H, histidine; I, isoleucine; L, leucine; K, lysine; M, methionine; F, phenylalanine; P, proline; S, serine; T, threonine; W, tryptophan; Y, tyrosine; V, valine.

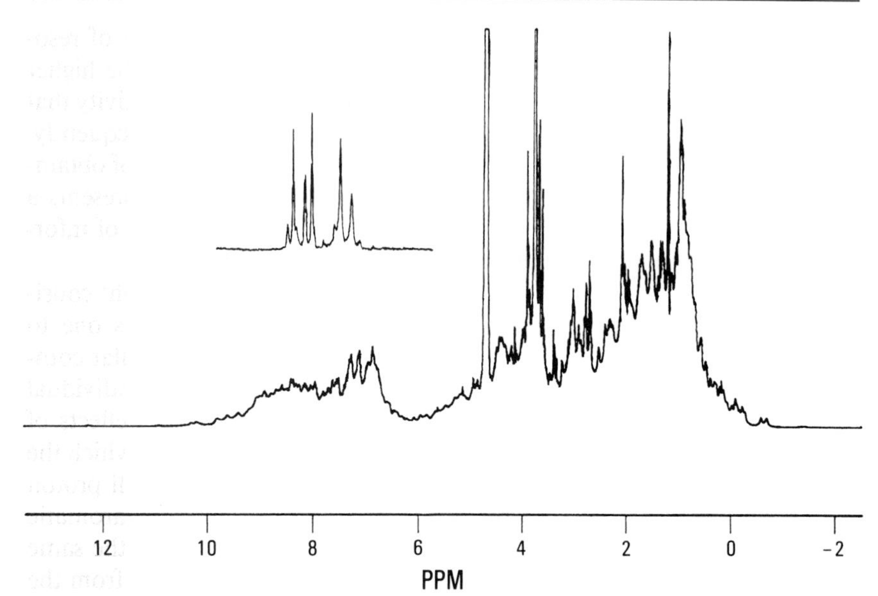

FIG. 2. One-dimensional proton spectra acquired from a complex of Fab′ and peptide (of sequence MHKDFLEKIGGL, corresponding to residues 76–87 of myohemerythrin) labeled with ^{15}N at G85 and G86. The full spectrum (lower) of the complex was acquired using a Hahn-spin echo version of a one-pulse sequence. The ^{15}N-edited spectrum (upper), corresponding to the amide-aromatic region only, was acquired using an HMQC sequence with ^{15}N decoupling during acquisition (see Fig. 5A). Both spectra were recorded from Fab′ sample of approximately 0.5 m*M* in a pH 5, 0.1 *M* sodium deuteroacetate buffer at 308 K. The sample contains ~ 2 : 1 peptide : Fab. Reproduced from *Pept. Res.* **1,** 87 (1988) with permission from Eaton Publishing.

ythrin.[13] The monoclonal antibody used in these experiments was raised against a peptide corresponding to residues 69–87 of myohemerythrin.[4] Two sets of resonances were detected in these experiments. At equimolar or lower peptide concentrations, a relatively broad set of resonances corresponding to the bound form of the peptide are visible. A second group of significantly sharper resonances appears above equimolar concentrations and is due to the excess free peptide. Similar results are obtained from titrations using identical peptides but labeled with ^{15}N at different residues.[7] For all the cases examined, the peptide exchange was determined to be slow on the NMR chemical shift time scale.

NMR studies of other antibody complexes have been reported in the literature.[14–22] In several of these cases, the chemical exchange rate of the various antigens was fast relative to the longitudinal relaxation (T_1) rates of

[13] S. Sheriff, W. A. Hendrickson, and J. L. Smith, *J. Mol. Biol.* **197,** 273 (1990).
[14] A. M. Goetz and J. H. Richards, *Biochemistry* **17,** 1773 (1978).

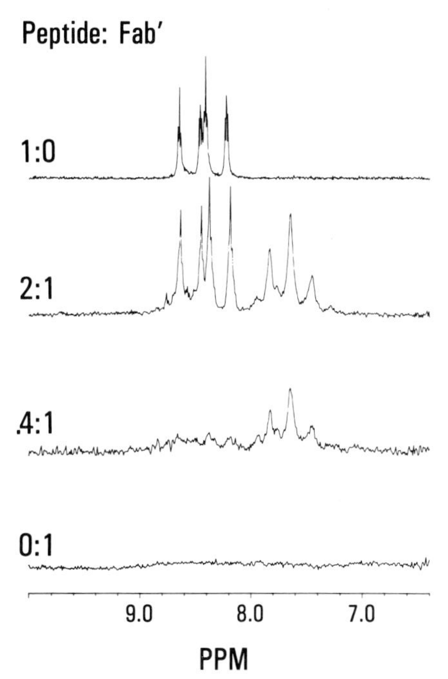

FIG. 3. One-dimensional ^{15}N isotope-edited spectra acquired at different points during a titration of Fab′ with peptide labeled with ^{15}N at G85 and G86. The approximate ratios of peptide : Fab′ are indicated at the left of each spectrum. The spectra shown were acquired at 500 MHz and 308 K using a version of the heteronuclear multiple quantum coherence (HMQC) sequence (Fig. 5A) with a jump-return pulse for solvent suppression. Reproduced from *Biochem. Pharmacol.* **40,** 83 (1990), courtesy of Pergamon Press, Oxford.

the antigen and Fab protons of interest.[20-22] An important point to be made is that isotope-editing represents one of the few *direct* methods of monitoring and investigating the bound component of these complexes regardless of what the actual exchange rates relative to these T_1 rates are. Methods utilizing the transferred NOE, such as transferred NOE difference

[15] S. K. Dower and R. A. Dwek, *in* "Biological Applications of Magnetic Resonance" (R. G. Shulman, ed.), p. 271. Academic Press, New York, 1979 and references cited therein.

[16] W. Ito, M. Nishimura, N. Sakato, H. Fujio, and Y. Arata, *J. Biochem. (Tokyo)* **102,** 643 (1987).

[17] J. Anglister, T. Frey, and H. M. McConnell, *Biochemistry* **23,** 1138 (1984).

[18] J. Anglister, T. Frey, and H. M. McConnell, *Biochemistry* **23,** 5372 (1984).

[19] J. Anglister, C. Jacob, O. Assulin, G. Ast, R. Pinker, and R. Arnon, *Biochemistry* **27,** 717 (1988).

[20] J. Anglister, R. Levy, and T. Scherf, *Biochemistry* **28,** 3360 (1989).

[21] J. Anglister and B. Zilber, *Biochemistry* **29,** 921 (1990).

[22] J. A. Glasel and P. N. Borer, *Biochem. Biophys. Res. Commun* **141,** 1267 (1986).

spectroscopy, have been employed frequently to study a range of high-molecular-weight systems, including other Fab complexes.[21,22] These methods do not allow the direct observation of resonances due to the bound ligand (peptide antigen). A potential shortcoming of the method is that the observed transferred NOEs can provide physically meaningful structural information only in cases where there is a single binding site, i.e., no weak and nonspecific binding occurs, and where there is a single bound conformation. The application of the transferred NOE technique has been limited to systems for which the exchange rate of the ligand on and off the receptor is faster than the longitudinal relaxation rates of the ligand and receptor protons and the mixing time of the NOE experiment.[23] Modification of the peptide antigen to reduce the binding constant has been proposed to allow study of complexes whose antigen-exchange rates are slow with respect to these time scales.[21] However, there is a significant risk that such modification might perturb important antigen–antibody interactions.

Isotope editing has been used successfully in studies of a great number of biological macromolecules and macromolecular complexes.[4–7,24–31] These studies have utilized both uniform and specific isotopic labeling, accomplished using either synthetic or biosynthetic means. Some general advantages of the isotope-editing techniques include spectral simplification and aid in resonance assignments. A direct benefit of editing is the ability to acquire highly specific information about the structure and interactions of a designated region of the molecule. In the absence of editing, such information is otherwise difficult to obtain due to spectral overlap problems. In the specific case of anti-peptide antibodies, isotope labeling of the peptide antigen allows studies of the conformational properties of the Fab′-bound peptide, along with detailed studies of the nature of the interactions between the peptide and the antibody combining site. In this chapter, we focus on the approach involving selective labeling of the

[23] J. Anglister, *Q. Rev. Biophys.* **23,** 175 (1990).
[24] R. H. Griffey, M. A. Jarema, S. Kunz, P. R. Rosevear, and A. G. Redfield, *J. Am. Chem. Soc.* **107,** 711 (1985).
[25] S. Fesik, *Biochemistry* **27,** 8297 (1988).
[26] G. M. Clore, P. C. Driscoll, P. T. Wingfield, and A. M. Gronenborn, *Biochemistry* **29,** 7387 (1990).
[27] D. A. Torchia, S. W. Sparks, and A. Bax, *Biochemistry* **28,** 5509 (1989).
[28] R. H. Griffey, C. D. Poulter, Z. Yamaizumi, S. Nishimura, and B. L. Hawkins, *J. Am. Chem. Soc.* **105,** 143 (1983).
[29] L. P. McIntosh, F. W. Dahlquist, and A. G. Redfield, *J. Biomol. Struct. Dyn.* **5,** 21 (1987).
[30] M. A. Weiss, A. G. Redfield, and R. H. Griffey, *Proc. Natl. Acad. Sci. U.S.A.* **83,** 1325 (1986).
[31] S. Roy, M. Z. Papastavros, V. Sanchez, and A. G. Redfield, *Biochemistry* **23,** 4395 (1984).

peptide. However, an alternative approach, which relies on isotope labeling of the "receptor," has also been used to examine a lysozyme–Fv complex.[6] Studies utilizing the latter approach could yield additional and complementary information from that obtainable by labeling of the antigen alone.

Some of the inherent disadvantages of isotope-editing methods include the potential expense and difficulty of obtaining the appropriate synthetically or biosynthetically labeled molecules necessary to achieve a given experimental goal. Most of the common amino acids may be purchased commercially at reasonable cost with ^{15}N labels at the α-amino group. Some multiply labeled amino acids, such as $^{13}C/^{15}N$, $^{13}C/^{2}H$, and $^{15}N/^{2}H$, are also obtainable occasionally without custom synthesis. Also, most amino acids may be selectively deuterated at the α, β, and certain side-chain positions by fairly straightforward and inexpensive methods.[32] With these deuteration schemes it is possible to modify ^{13}C or ^{15}N isotopically enriched amino acids to produce $^{13}C/^{2}H$, $^{15}N/^{2}H$, or other multiply labeled amino acid compounds. These multiply labeled amino acids are particularly useful for NOE studies of high-molecular-weight systems.[33] Thus, despite the issues of cost and availability, it remains feasible to buy or to produce many singly and multiply labeled amino acids.

Isotope-Edited Experiments

Isotope-editing experiments provide an extremely powerful means for simplifying an otherwise complex, proton NMR spectrum. In its most general usage, isotope editing refers to any technique that exploits the presence of a heteronucleus such as ^{13}C, ^{15}N, or ^{31}P to reduce the number of proton resonances observed in an NMR spectrum. While it is possible in principle to record NMR spectra directly for the heteronuclei, the sensitivity of such measurements is poor relative to proton observation.[2,3] Many experiments have been proposed for isotope editing; for the most part, these techniques rely on one of two basic schemes to achieve spectral editing. The conceptually simplest class of techniques is based on difference decoupling. The second, broader class of experiments is based on the perturbation of proton single-quantum coherences during evolution under heteronuclear scalar couplings. Both classes of techniques are dependent on the presence of heteronuclear scalar couplings, and are effective as long as the proton resonance linewidth is less than the magnitude of the coupling constant. Fortunately, $^{15}N-^{1}H$ and $^{13}C-^{1}H$ scalar couplings are normally greater than ~90 and ~125 Hz, respectively, for one bond coup-

[32] D. M. LeMaster, this series, Vol. 177, p. 23, and references cited therein.

[33] P. Tsang, P. E. Wright, and M. Rance, *J. Am. Chem. Soc.* **112,** 8183 (1990).

lings, so that the isotope-editing techniques are applicable to a wide range of biomolecules and complexes. Modifications to incorporate isotope editing can be made in virtually all high-resolution NMR experiments; several excellent reviews have appeared on isotope-editing techniques and their applications to biomolecules.[1-3,29,34,35] The present discussion will be restricted to a few techniques that are particularly important in studies of high-molecular-weight (by solution NMR standards) systems such as antibody Fab fragments. The principal factor limiting the applicability of isotope-editing techniques in studies of antibody Fab fragments is the short spin–spin (T_2) relaxation times characteristic for such slowly tumbling molecules.

In addition to the classification of techniques based on the method employed for isotope editing, NMR experiments can be distinguished according to the type of information obtained. Three of the most useful categories of experiments in applications to biomolecules in general are (1) resolved experiments, in which individual resonances are observed and chemical shift and linewidth information is obtained; (2) correlated experiments, in which coherence transfer among scalar-coupled nuclei is observed, thereby establishing through-bond connectivities; and (3) nuclear Overhauser effect experiments, in which information regarding internuclear distances is obtained. Isotope-edited versions of all of these experiments are applicable to studies of antibody Fab fragments.

Isotope editing is achieved in difference decoupling experiments by acquiring two spectra, one of which has been obtained with on-resonance or broadband heteronuclear decoupling and the other of which has been obtained without decoupling. By taking the difference of these two spectra those proton resonances unaffected by the heteronuclear decoupling will be eliminated, while those that are significantly affected will show up as antiphase triplet resonances with a spacing of $J/2$ between components of the triplet; J is the relevant heteronuclear coupling constant. This class of experiments was introduced by Llinás *et al.*[36] and was referred to as FINDS (Fourier internuclear difference spectroscopy); the pulse sequence is shown in Fig. 4a. Griffey, Redfield, and co-workers have employed this technique, which they refer to as an INDOR (internuclear double resonance) experiment, in numerous studies to observe imino ^{1}H resonances in spectra of medium-sized biomolecules such as tRNA[28,31,37] and ^{13}C-bound proton resonances in phage λ repressor.[30] By selective decoupling and acquisition

[34] L. P. McIntosh and F. W. Dahlquist, *Q. Rev. Biophys.* **23,** 1 (1990), and references cited therein.

[35] G. Otting and K. Wüthrich, *Q. Rev. Biophys.* **23,** 39 (1990).

[36] M. Llinás, W. J. Horsley, and M. P. Klein, *J. Am. Chem. Soc.* **98,** 7554 (1976).

[37] R. H. Griffey, D. Davis, Z. Yamaizumi, S. Nishimura, A. Bax, B. Hawkins, and C. D. Poulter, *J. Biol. Chem.* **260,** 9734 (1985).

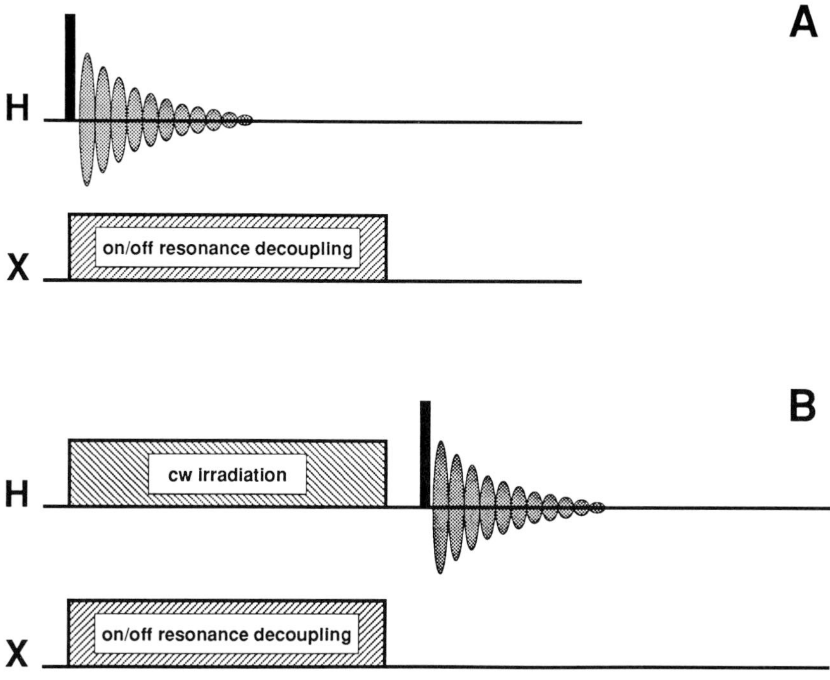

FIG. 4. Pulse sequences for difference decoupling techniques. (A) Scheme for FINDS or INDOR experiment. The heteronuclear (X) decoupling is applied during proton (H) signal acquisition, with the frequency alternately shifted on and off resonance so that effective decoupling is achieved only on alternate scans; the FIDs (free induction decay) are accordingly added or subtracted from computer memory. (B) One-dimensional isotope-directed NOE experiment. The proton saturation is applied at the decoupled resonance frequency of the isotope-labeled proton of interest, at a power level low enough so that the undecoupled resonance frequencies of this proton are not significantly affected. The heteronuclear decoupling is applied on and off resonance, and the resulting FIDs are added and subtracted, respectively.

of a set of difference decoupled spectra as a function of the heteronuclear decoupling frequency, correlations can be obtained between the proton and heteronuclear resonance frequencies. Extensions of the FINDS or INDOR experiment have also been devised to obtain NOE data.[24,30] By applying proton irradiation at a given frequency while simultaneously applying on/off resonance decoupling at a heteronuclear resonance frequency, NOEs from a proton attached to the heteronucleus can be selectively observed; this pulse scheme is shown in Fig. 4b. Analogous techniques for observing NOEs to protons scalar-coupled to specific heteronuclei also exist.[30]

The second class of isotope-editing experiments referred to above also relies on the principle of double resonance, in this case a perturbation of the evolution of proton single-quantum coherences under the heteronuclear scalar coupling. Many experiments have been proposed that fall into this general category; however, most are simply variants of three basic techniques. These three methods may be loosely described as spin-echo difference (SED), heteronuclear multiple quantum coherence (HMQC), and heteronuclear single-quantum coherence (HSQC) experiments. The spin-echo difference technique is applicable in those experiments designed for spectral simplification through isotope editing, while the HMQC and HSQC experiments are usually employed when ^{1}H-heteronuclear correlations are desired in addition to possible spectral editing.

The spin-echo difference experiment was first introduced by Emshwiller *et al.*[38] The basic pulse sequence is shown in Fig. 5A; this experiment is very simple in concept and has been described in detail elsewhere.[38,39] A Hahn spin–echo sequence is applied to the protons, causing the proton magnetization to be refocused at a time 2τ, assuming τ^{-1} is significantly larger than the homonuclear scalar coupling. If a 180° pulse on the heteronuclei is applied simultaneously with the proton refocusing pulse, the transverse magnetization of those protons directly coupled to a heteronucleus will still be refocused but inverted relative to that for protons not coupled to heteronuclei; it is assumed that $\tau = 1/2J$, where J is the heteronuclear scalar coupling constant. By taking the difference of signals acquired with 0 and 180° pulses applied to the heteronuclei, only the resonances for protons coupled to heteronuclei will be observed. This provides an extremely powerful yet simple means for simplifying a complex proton NMR spectrum. In applications of the spin–echo difference technique to large molecules, two practical considerations must be taken into account. The first is that spin–spin relaxation of the protons of interest will reduce the sensitivity of the experiment due to irreversible loss of magnetization during the 2τ spin–echo period. Taking this relaxation process into account, the optimum value of τ will be given by the expression, $\tau_{opt} = [\tan^{-1}(\pi J T_2)]/(\pi J)$; thus, as T_2 decreases τ should be set to a value somewhat shorter than $1/2J$. The signal intensity will be scaled according to $\exp(-2\tau_{opt}/T_2)\sin^2(\pi J \tau_{opt})$. The second practical consideration is that it is often necessary to perform the experiments on molecules dissolved in H_2O, and therefore some scheme must be employed to suppress the large H_2O resonance, as discussed below.

The spin–echo difference sequence can be appended to virtually any experiment, thus allowing the possibility of isotope editing the resulting spectra.[1,35] Three of the most important sequences in studies of large

[38] M. Emshwiller, E. L. Hahn, and D. Kaplan, *Phys. Rev.* **118,** 414 (1960).

[39] R. Freeman, T. H. Mareci, and G. A. Morris, *J. Magn. Reson.* **42,** 341 (1981).

biomolecules are shown in Fig. 5. The sequence in Fig. 5B is useful for obtaining one-dimensional (1D) NOE spectra[1,40]; in addition to acting as an isotope filter, the spin–echo difference sequence has the effect of selectively inverting the magnetization of protons coupled to a heteronucleus, and the experiment is fundamentally equivalent to the well-known transient NOE experiment.[41,42] The sequence shown in Fig. 5C is commonly referred to as an ω_2-filtered NOESY (two-dimensional nuclear Overhauser enhancement spectroscopy) experiment[35]; using this sequence a two-dimensional (2D) $^1H-^1H$ NOESY spectrum can be obtained that contains only resonances in the ω_2 dimension for protons coupled to heteronuclei. If the spin–echo difference sequence were appended to the beginning of the experiment, an ω_1-filtered NOESY spectrum would be obtained. Similarly, a pulse sequence for an ω_2-filtered TOCSY (total correlation spectroscopy) experiment is known in Fig. 5D; using this sequence, an isotope-edited 2D $^1H-^1H$ TOCSY spectrum can be recorded. In the TOCSY experiment,[43,44] magnetization is coherently transferred among scalar-coupled ^1H nuclei, which allows these nuclei to be identified as belonging to the same amino acid residue. The TOCSY experiment is greatly preferred to other coherence transfer methods in studies of large biomolecules due to improved sensitivity in the presence of large linewidths. In such applications it is necessary to use relatively short isotropic mixing periods to minimize the loss of magnetization through relaxation processes; for the same reason, it is also preferrable to employ *z*-filters[45,46] rather than spin-locking pulses[44] to eliminate undesirable signals. The use of short mixing times will effectively limit the observation of long-range coherence transfer.

By modifying the spin–echo difference sequence it is possible to obtain 2D data that correlate the protons with heteronuclei. The simplest method, shown in Fig. 6A and generally referred to as the HMQC or "forbidden echo" experiment, is to insert a variable time period t_1 between the two heteronuclear 90° pulses, keeping the proton 180° pulse centered in the t_1 period.[47–49] The first 90° heteronuclear pulse converts some proton single-quantum coherence to heteronuclear multiple-quantum coherence; the

[40] M. Rance, P. E. Wright, B. A. Messerle, and L. D. Field, *J. Am. Chem. Soc.* **109,** 1591 (1987).

[41] I. Solomon, *Phys. Rev.* **99,** 559 (1955).

[42] S. L. Gordon and K. Wüthrich, *J. Am. Chem. Soc.* **100,** 7094 (1978).

[43] L. Braunschweiler and R. R. Ernst, *J. Magn. Reson.* **53,** 521 (1983).

[44] A. Bax and D. G. Davis, *J. Magn. Reson.* **65,** 355 (1985).

[45] M. Rance, *J. Magn. Reson.* **74,** 557 (1987).

[46] R. Bazzo and I. D. Campbell, *J. Magn. Reson.* **76,** 358 (1988).

[47] L. Müller, *J. Am. Chem. Soc.* **101,** 4481 (1979).

[48] A. Bax, R. H. Griffey, and B. L. Hawkins, *J. Magn. Reson.* **55,** 301 (1983).

[49] M. R. Bendall, D. T. Pegg, and D. M. Doddrell, *J. Magn. Reson.* **52,** 81 (1983).

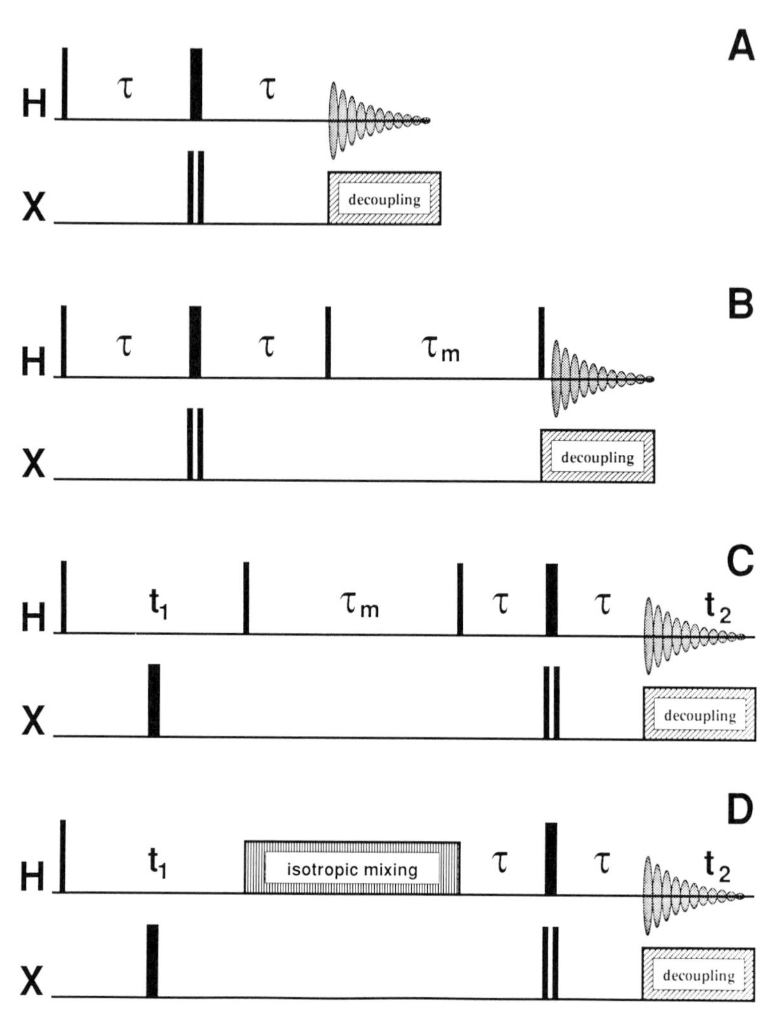

FIG. 5. Pulse sequences for isotope-filtered, ^{1}H NMR spectra. The thin bars represent 90° pulses and the thick bars 180° pulses; recommended phase-cycling procedures can be found in the literature (see text). The delay τ is set to $\sim 1/2J$ (see text), where J is the heteronuclear coupling constant. (A) Spin-echo difference sequence for obtaining 1D, isotope-filtered proton spectra. Phase cycling results in the two heteronuclear 90° pulses combining on alternate scans to give 0 or 180° net rotations, and the FIDs consequently are alternately added and subtracted from computer memory. Heteronuclear decoupling can be applied during data acquisition. (B) One-dimensional, isotope-edited (or isotope-directed) NOE experiment; τ_m is the NOE mixing period. (C) ω_2-filtered, 2D ^{1}H–^{1}H NOESY experiment. The 180° pulse in the middle of the t_1 period eliminates the heteronuclear couplings in the ω_1 dimension. (D) ω_2-filtered, 2D ^{1}H–^{1}H TOCSY experiment. During the isotropic mixing period some multipulse sequence such as DIPSI-2 [A. J. Shaka, C. J. Lee, and A. Pines, *J. Magn. Reson.* **77,** 274 (1988)] is applied to induce coherence transfer among the scalar-coupled protons.

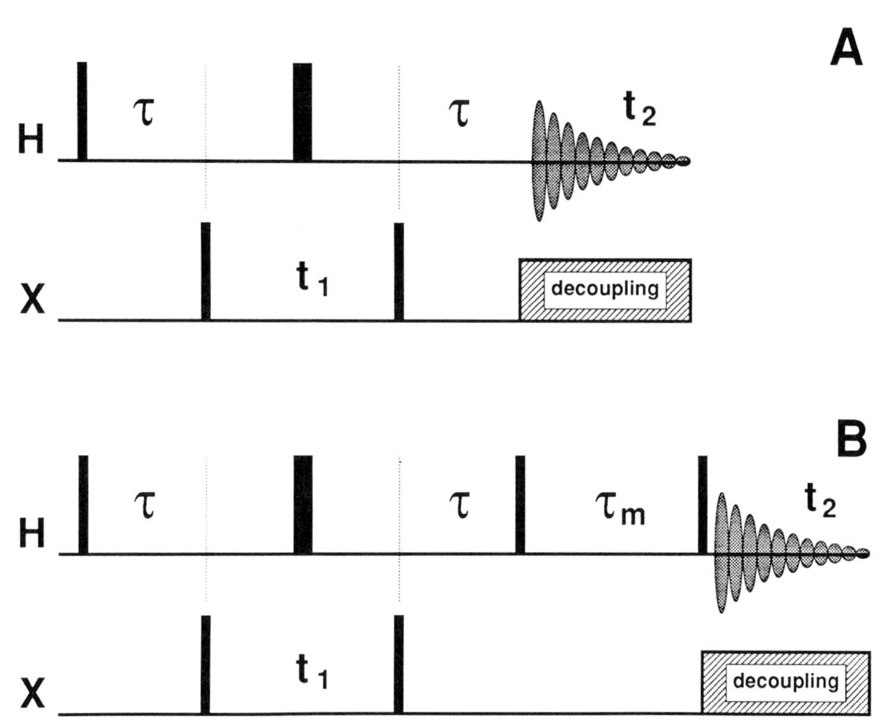

FIG. 6. Two-dimensional, heteronuclear correlation experiments. (A) HMQC experiment and (B) HMQC-NOESY experiment. Heteronuclear decoupling can be applied during data acquisition if desired. Recommended phase-cycling procedures can be found in the literature (see text).

latter evolves under the influence of the heteronuclear chemical shifts (the proton contribution is refocused by the 180° pulse) and possibly some $^1H-^1H$ scalar couplings before it is reconverted to proton single-quantum coherence for detection. The resulting 2D map will display heteronuclear chemical shifts in the ω_1 dimension and the correlated proton shifts in the ω_2 dimension. Alternative methods for obtaining proton–heteronuclear correlation spectra have been proposed, such as the HSQC or Overbodenhausen experiment[50] and the Hahndor experiment.[51] One advantage of these techniques over the HMQC experiment is the elimination of multiplet structure in the ω_1 (heteronuclear shift) dimension, which results in improved sensitivity; however, in studies of high-molecular-weight systems this advantage is usually minimal due to the inherently large resonance linewidths. Another potential sensitivity advantage of the HSQC or Hahn-

[50] G. Bodenhausen and D. J. Ruben, *Chem. Phys. Lett.* **69,** 185 (1980).
[51] A. G. Redfield, *Chem. Phys. Lett.* **96,** 537 (1983).

dor techniques for $^1H-^{15}N$ correlation experiments arises from the fact that the spin–spin relaxation rate of ^{15}N single quantum coherence is normally smaller than the rate for $^1H-^{15}N$ multiple quantum coherence in the long correlation time limit.[52-54] The degree of improvement in relaxation rates will be dependent on the distribution or protons in the neighborhood of the scalar-coupled $^1H-^{15}N$ pair; in fact, if the neighboring protons are replaced by deuterons, the relaxation rates in the HMQC experiment will be more favorable. In $^1H-^{13}C$ correlation experiments, the HMQC method is usually more advantageous than the HSQC or Hahndor techniques in terms of relaxation rates due to the larger heteronuclear dipolar interaction.[54] A disadvantage of the latter experiments is that they require more radio frequency pulses than the HMQC method, making them more susceptible to pulse errors and more difficult to modify for solvent suppression.

As with the spin–echo difference sequence, the HMQC sequence can be appended to almost any conventional experiment to obtain an isotope-edited, proton–heteronuclear correlation spectrum. One of the most important examples is shown in Fig. 6B, where an HMQC sequence is followed by a NOESY mixing period; the resulting 2D spectrum will correlate proton and heteronuclear resonances as well as NOEs to protons spatially close to the heteronuclear-coupled protons.[1,40]

Solvent Suppression

The simplest and one of the most effective procedures for suppressing the H_2O signal in NMR studies of small proteins is to presaturate the solvent nuclei.[55,56] However, this becomes problematic in experiments on larger molecules due to undesirable saturation transfer to the proton nuclei of interest,[57] which reduces the sensitivity of the measurement. Saturation transfer occurs via chemical exchange processes, direct cross-relaxation with solvent nuclei, and spin diffusion effects. Thus, when possible, it is usually advisable to employ alternate methods for solvent suppression. For studies of molecules in D_2O, presaturation at low power levels to suppress the residual HDO signal generally works well.

[52] T. J. Norwood, J. Boyd, and I. D. Campbell, *FEBS Lett.* **255,** 369 (1989).
[53] T. J. Norwood, J. Boyd, J. E. Heritage, N. Soffe, and I. D. Campbell, *J. Magn. Reson.* **87,** 488 (1990).
[54] A. Bax, M. Ikura, L. E. Kay, D. A. Torchia, and R. Tschudin, *J. Magn. Reson.* **86,** 304 (1990).
[55] J. Schaefer, *J. Magn. Reson.* **6,** 670 (1972).
[56] D. I. Hoult, *J. Magn. Reson.* **21,** 337 (1976).
[57] J. D. Stoesz, A. G. Redfield, and D. Malinowski, *FEBS Lett.* **91,** 320 (1978).

To avoid a loss of magnetization to saturation transfer effects, frequency-selective pulse schemes for solvent suppression are commonly employed. Such schemes require that the nuclei of interest be excited with the maximum efficiency possible while the solvent resonance is minimally excited. Many frequency-selective pulse schemes have been proposed for solvent suppression[58,59]; the most appropriate method will depend largely on the exact details of the experiment. For the difference decoupling experiments shown in Fig. 4, it is possible to replace the proton 90° pulse with a "jump-return" composite pulse[60] ($90^\circ_x - \tau - 90^\circ_{-x}$); this is one of the simplest, yet most effective, schemes for solvent suppression. In practical applications, it is necessary to adjust the relative phase and amplitude of the two pulses to compensate for instrumental imperfections and phase transients. The spin–echo difference sequence shown in Fig. 5A can be modified in several ways to incorporate frequency-selective pulses. One method is to replace both the 90° and 180° proton pulses with jump-return composite pulses,[31,61] so that the magnetization is always left along the $+z$ axis (i.e., the direction of the equilibrium magnetization). One drawback to this scheme is that the use of two frequency-selective pulses narrows the spectral regions over which efficient excitation is achieved. It is possible to replace only the proton 90° pulse with a $90^\circ_x - \tau - 90^\circ_x$ composite pulse, which inverts the solvent resonance, and rely on the nonselective proton 180° pulse to return the solvent magnetization to the $+z$ axis; it is highly undesirable to leave the solvent magnetization inverted, since radiation damping will subsequently cause a large transverse magnetization to appear, which will swamp the signals of interest.

As the number of proton pulses in a sequence increases, solvent suppression usually becomes more difficult. In such cases, it is sometimes necessary to resort to presaturation of the solvent resonance; in ω_2-filtered experiments, where the magnetization of interest normally originates from nonexchangable protons, presaturation causes a minimal loss of sensitivity. In NOE experiments, for mixing times longer than 40–50 msec the solvent magnetization has returned to the $+z$ axis due to the effect of radiation damping, and subsequent pulses can be replaced by composite, frequency-selective pulses.

Some of the techniques that are commonly employed to improve solvent suppression are not advisable in studies of large biomolecules. Undesirable, transverse magnetization is often eliminated through the use

[58] P. J. Hore, this series, Vol. 176, p. 64.

[59] A. G. Redfield, this series, Vol. 49, p. 253.

[60] M. Guéron and P. J. Plateau, *J. Am. Chem. Soc.* **104,** 7310 (1982).

[61] R. H. Griffey, A. G. Redfield, R. E. Loomis, and F. W. Dahlquist, *Biochemistry* **24,** 817 (1985).

of homogeneity-spoiling pulsed-field gradients; however, this normally requires that the magnetization of interest be transferred to the longitudinal axis during the "homo-spoil" pulse and the subsequent recovery period. During this period of time, which can typically be 10–30 msec, significant cross-relaxation of the longitudinal magnetization can occur for slowly tumbling macromolecules; thus, homo-spoil pulses are recommended only for NOE experiments with sufficiently long mixing times. Spin-locking pulses, which are sometimes used to dephase undesirable magnetization, should be avoided due to excessive relaxation during the pulse. Shaped pulses are also problematic due to the typical length of such pulses.

Postacquisition data-processing procedures can be extremely effective in eliminating troublesome dispersive components of large, residual solvent signals. Postacquisition methods have been described that selectively eliminate undesirable signals in both the frequency and time domains.[62,63] These techniques are particularly useful for observing resonances close to the solvent frequency in isotope-edited spectra and for eliminating t_2 ridges in 2D spectra.

Experimental Considerations

Sample Considerations

In general, NMR experiments require the use of the most highly concentrated samples obtainable of a given molecule to optimize sensitivity. The isotope-edited NMR studies described here and in the literature[4-6] are performed on samples that range in concentration from approximately 0.2–1.0 m*M* in Fab′ concentration. This range is dictated by the particular solubility of the Fab′. In terms of protein quantity, this translates to 5 to 27 mg of Fab′/NMR sample. The NMR samples are prepared in a relatively "low" pH buffer, e.g., sodium deuteroacetate, pH 5, in order to improve the observation of amide protons bound to the ^{15}N-labeled residues by reducing their exchange rates with solvent protons.[64]

The peptides employed in these studies are synthesized by solid-phase methods[65] using *t*-Boc (tert-butyloxycarbonyl) derivatives of the labeled amino acids. The biosynthetic incorporation of stable isotopes into the antibody instead can be accomplished by appropriate growth of the hybridoma cells on labeled media.[6,23] General principles of biosynthetic labeling

[62] P. Tsang, P. E. Wright, and M. Rance, *J. Magn. Reson.* **88,** 210 (1990), and references cited therein.

[63] P. Plateau, C. Dumas, and M. Guéron, *J. Magn. Reson.* **54,** 46 (1983).

[64] K. Wüthrich and G. Wagner, *J. Mol. Biol.* **130,** 1 (1979).

[65] R. Houghten, *Proc. Natl. Acad. U.S.A.* **82,** 5131 (1985).

of antibodies and proteins with stable isotopes such as ^{13}C, ^{15}N, and ^{2}H are discussed and reviewed elsewhere.[23,34,66] The amount of pure, labeled peptide typically utilized in titrations of Fab′ with peptide is on the order of a milligram or less.

NMR Hardware

For experimental implementation of the pulse sequences in Figs. 4–6, there are a number of spectrometer hardware requirements. The first of these is the ability of apply radio frequency pulses simultaneously to protons and heteronuclei, with the receiver reference frequency set to that of the proton transmitter. This is known on some commercial spectrometers as operation in the "reverse" mode. A second prerequisite involves the use of an appropriate "reverse" or "inverse" detection probe, so designated because of the relative orientations of the proton (inner) and heteronucleus (outer) coils. Finally, an additional low-power broad-band amplifier is necessary for heteronuclear decoupling via WALTZ-16, GARP-1, or other such low-power multipulse decoupling scheme.[67,68]

Aspects of Interactions and Conformation of Bound Peptide

Using the techniques outlined above, it is possible to assign many of the resonances of nuclei in the bound peptide in a Fab–peptide or Fab′–peptide complex. Resonances corresponding to individual nuclei such as the α-proton or α-carbon of a given residue may be assigned using specifically ^{13}C-labeled peptide. Similarly, the assignments for a given amide nitrogen and amide proton may be determined through the use of peptide labeled specifically with ^{15}N at the desired residue. Much of the structural information to be derived from NMR studies of proteins depends on the assignment of resonances of main-chain atoms of a peptide and protein and the observation of scalar and dipolar interactions among these nuclei.[69] Hence, the assignment of resonances of the α and amide protons of the large protein or bound peptide will facilitate NMR experiments aimed eventually at the determination of the folding of the main chain of the molecule. Deuterium substitution can serve as a further assignment tool for protons in both the protein and the peptide.[23,32]

With the availability of specific resonance assignments, it becomes feasible to examine with a high degree of specificity the local interactions

[66] D. M. LeMaster, *Q. Rev. Biophys.* **23,** 133 (1990), and references cited therein.

[67] A. J. Shaka, J. Keeler, T. Frenkiel, and R. Freeman, *J. Magn. Reson.* **52,** 335 (1983).

[68] A. J. Shaka, P. B. Barker, and R. Freeman, *J. Magn. Res.* **64,** 547 (1985).

[69] K. Wüthrich, "NMR of Proteins and Nucleic Acids." Wiley, New York, 1986.

between a particular residue of the bound peptide and residues within the antibody combining site. Examples of NMR parameters potentially capable of providing information on such interactions include chemical shift, temperature coefficients, NOE, and coupling constants. The effects of hydrogen bonding and local electronic effects from nearby aromatic side chains may manifest themselves as observed resonance frequency shifts. The relative direction and size of a frequency shift from a "standard" (or "random coil") frequency[70] of a particular nucleus may be used to characterize environmental changes of that nucleus upon binding. The relative solvent accessibility of exchangeable protons such as backbone or side-chain amide and amino protons is often reflected in their rates of exchange with solvent deuterons and in the temperature coefficients of the amide or amino proton resonances.[71]

On the basis of the changes observed for these several NMR parameters, it has been possible to detect changes in the peptide environment on binding.[4] Significant effects on the chemical shift and linewidths of several of the amide ^{1}H and ^{15}N resonances have been observed on binding of peptide antigen to the antibody.[5] The relative broadening of resonances on binding to the Fab is greatest for residues within the immunologically determined epitope,[72,73] which suggests that the residues of the peptide that interact the most strongly with the antibody are also those that are the most immobilized, as judged from NMR.[7]

The dynamic differences observed among the bound peptide residues can be further examined using a combined approach of relaxation measurements and isotope-editing techniques. Although applications to antibody–antigen complexes have not yet been reported, relaxation measurements have been used successfully to determine motional parameters of other proteins and peptides.[26,74,75] Through the measurements of relaxation parameters such as T_1, T_2, and heteronuclear NOE, characterization of the overall motion of a protein or peptide and identification of local motions at specific residues becomes possible. Thus, NMR provides new opportunities to investigate the mobility or dynamics of residues of a peptide bound within the combining site of an antibody.

NOE information directly pertaining to individual amino acid protons of an antibody–peptide complex may also be obtained using isotope-

[70] A. Bundi and K. Wüthrich, *Biopolymers* **18,** 285 (1979).

[71] M. Llinás and M. P. Klein, *J. Am. Chem. Soc.* **97,** 4731 (1975).

[72] T. M. Fieser, J. A. Tainer, H. M. Geysen, R. A. Houghten, and R. A. Lerner, *Proc. Natl. Acad. Sci. U.S.A.* **84,** 8586 (1987).

[73] J. K. Scott and G. P. Smith, *Science* **249,** 386 (1990).

[74] L. E. Kay, D. A. Torchia, and A. Bax, *Biochemistry* **28,** 8972 (1989).

[75] N. R. Nirmala and G. Wagner, *J. Am. Chem. Soc.* **110,** 7557 (1988).

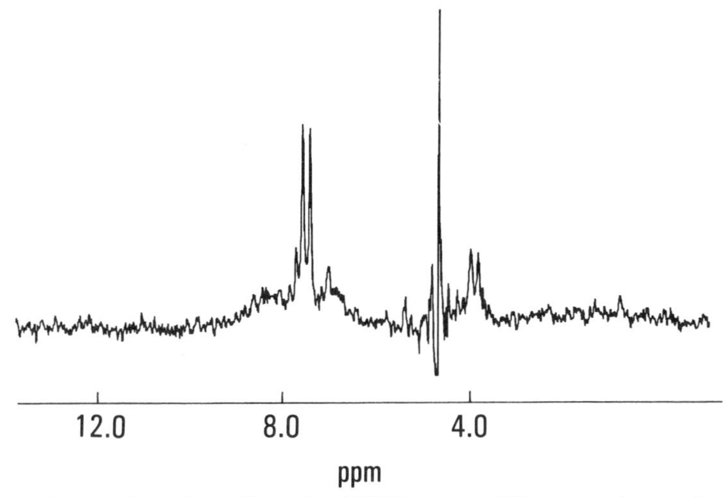

FIG. 7. Isotope-directed one-dimensional NOE spectrum (50-msec mixing time) recorded from a Fab complex of the peptide MHKDFLEKIGGL (see Fig. 2) labeled with ^{15}N at G85. Due to the lack of heternuclear decoupling during acquisition, the resonances at 7.58 and 7.43 ppm correspond to the G85 amide proton doublet. The remaining resonances at approximately 7.72, 7.03, 3.99, and 3.83 ppm correspond to NOEs to nearby protons of the complex. The spectrum was recorded at 600 MHz and 308 K, using a sequence similar to that shown in Fig. 6B with solvent presaturation and a homo-spoil pulse to suppress the residual water resonance intensity.

edited experiments.[7] This is illustrated in Fig. 7, which shows the full, 1D, isotope-directed NOE spectrum of the complex of an ^{15}N-labeled peptide and an antibody Fab fragment. A doublet from the amide proton coupled to the ^{15}N label at G85 is observed at 7.58 and 7.43 ppm (no heteronuclear decoupling was employed); the remaining resonances correspond to NOEs from this proton to neighboring protons within an approximate radius of 3 Å.[7] The protons that give rise to these NOEs correspond to nearby α, amide, and aromatic protons belonging to residues of either the peptide or the Fab′ combining site. Additional labeling of the peptide enables unambiguous identification of some of these resonances as arising from intrapeptide NOEs.[7] Thus, direct detection of a select group of NOEs from a specified amide proton to nearby protons is possible using the methods outlined here. At this time, the extent of conformational information obtainable for the bound peptide is limited to local, short-range interactions between protons.

From previous theoretical investigations of NOE behavior[76,77] of high-

[76] J. A. Glasel, *J. Mol. Biol.* **209,** 747 (1989).

[77] M. Madrid, J. E. Mace, and O. Jardetzky, *J. Magn. Reson.* **83,** 267 (1989).

molecular-weight systems and a specific deuteration study conducted on a Fab–peptide complex,[33] it appears that direct NOEs may be interpreted with reasonable accuracy for relatively short distances in the molecular weight and correlation time range of Fab–peptide complexes[7,17] (~56K and ~20 nsec, respectively). Specific deuteration of the peptide antigen or the Fab represents a method to improve the sensitivity and selectivity of the NOE experiments. An example is shown in Fig. 8, which compares the isotope-directed NOE spectra obtained from complexes containing either specifically deuterated or undeuterated peptide.[33] As this spectrum shows, enhancement of certain direct NOEs may be accomplished through the use of amino acids labeled with multiple isotopes such as $^{13}C/^{2}H$ or $^{15}N/^{2}H$. Further improvement in the measurement of NOEs by isotope-editing techniques should be possible in situations where both specific deuteration of the peptide antigen and selective biosynthetic deuteration of the Fab (using methods previously applied to Fab complexes[23]) are employed. The

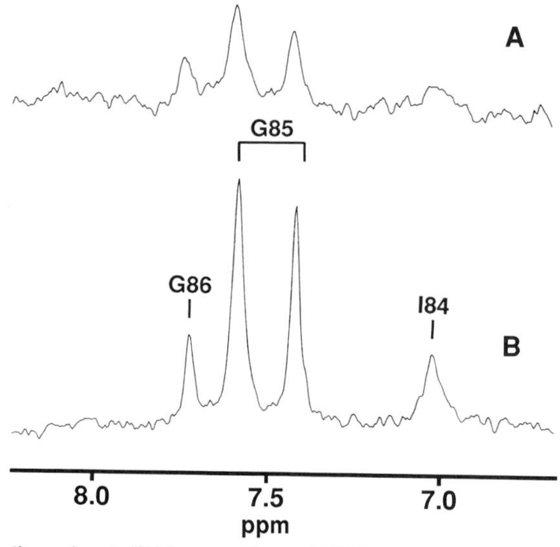

FIG. 8. One-dimensional, ^{15}N isotope-directed NOE spectra recorded with a mixing time of 50 msec from two Fab′–peptide complexes at 600 MHz and 308 K. (A) Peptide MHKDFLEKIGGL (see Fig. 2) labeled with ^{15}N at G85. (B) Complex with the same peptide ^{15}N-labeled at G85 and specifically deuterated at the adjacent residues, G86 and I84; the α protons of G85 were also replaced by deuterons. The concentrations and ratios of Fab′ : peptide in the two samples were similar. The resonances corresponding to the G85 amide doublet (no heteronuclear decoupling was employed during acquisition), and NOEs assigned to the amide protons of G86 and I84[7] are indicated. The pulse sequence used is shown in Fig. 6B. Reproduced with permission of the American Chemical Society from *J. Am. Chem. Soc.* **112,** 8183 (1990).

combined approach of isotope editing and specific deuteration of the antigen and antibody fragment should generally facilitate the measurement of specific and direct NOE intensities due to interactions between protons over both short- and long-distance ranges,[7] and hence provide important structural information on antigen–antibody complexes.

In conclusion, isotope-edited NMR techniques offer new possibilities for detailed studies of the structure and dynamics of peptide–Fab complexes and promise to provide important new insights into the molecular basis for antibody–antigen recognition. The specificity of these methods allows one to distinguish residues of the antigen that interact directly at the antibody combining site.[7] By use of appropriate isotope labeling, in conjunction with NOE experiments, detailed information can be obtained on both the conformation of the bound peptide and on the nature of the peptide–antibody interactions. The same NMR techniques that offer so much promise for studies of antibody–antigen interactions should also be more generally applicable to studies of peptide–receptor complexes.

Acknowledgments

This work was supported in part through grants from the National Institutes of Health (GM40089 and CA27489) and the National Science Foundation (DMB-8903777). The authors thank R. A. Lerner, R. A. Houghten, T. M. Fieser, J. Ostresh, and L. K. Tennant for their support and assistance throughout this project.

[12] Reductive Methylation and Carbon-13 Nuclear Magnetic Resonance in Structure–Function Studies of Fc Fragment and Its Subfragments

By JOYCE E. JENTOFT

Introduction

The Fc region of immunoglobulin G (IgG), made up of the disulfide linkage and the C_H2 and C_H3 domains of both heavy chains, mediates many of the biological functions of the humoral immune response, including complement fixation,[1,2] antibody-dependent cytotoxicity,[1,2] IgG bind-

[1] D. Yasmeen, J. R. Ellerson, K. J. Dorrington, and R. H. Painter, *J. Immunol.* **116,** 518 (1976).

[2] J. R. Ellerson, D. Yasmeen, R. H. Painter, and K. J. Dorrington, *FEBS Lett.* **24,** 318 (1972).

ing to macrophages,[3,4] and the release of histamine from mast cells.[5] The macromolecular interactions responsible for these effects are preserved in the Fc fragment obtained from papain digestion of IgG, and, as appropriate, in smaller fragments containing intact C_H2 or C_H3 domains. The specific subfragments discussed herein are: the C_H2 fragment; residues 225–338 of chain A and residues 225–248 of chain B (based on the Eu numbering system[6]); pFc′, residues 333–443; and tFc′, residues 339 (or 341)–443.

Despite the biological importance of the events mediated through the Fc region, few investigations have focused upon the structure–function properties of the Fc fragment and subfragments.

Rationale. The Fc fragment of human IgG (subclass 1) has 18 lysine residues, with 12 lysyl groups in the C_H2 domain and 6 in the C_H3 domain. The pK_a value of each of these residues reflects the microenvironment of the parent amino group, and any change in pK_a is indicative of a change in that microenvironment. Thus, if one could obtain and compare pK_a values for each of the lysine residues in, e.g., the Fc fragment and the pFc′ fragment containing the C_H3 domain, any lysine residue from the C_H3 domain of the Fc fragment that had a different pK_a value than its counterpart in the pFc′ fragment would be presumed to exist in a different conformation in the two macromolecules. A global conformational change would be indicated if all (or most) of the corresponding lysine residues had different pK_a values in the two fragments, while a localized conformational change would be indicated if only one or two corresponding lysine residues were different. Conclusions could be made regarding the existence (or lack) of a conformational change in a system even without complete assignment of resonances to residues by asking if the measured pK_a values changed or were altered, respectively, in going from the Fc fragment to a subfragment.

Advantages and Disadvantages. The primary advantage of nuclear magnetic resonance (NMR) as a probe for studying macromolecules is that the properties of individual atoms can be studied. However, applications of standard NMR methods to native immunoglobulins, or to their Fc fragments, has not been practical because the large number of, e.g., protons, in the 150-kDa IgG, or in the 50-kDa Fc fragment, make it difficult to both resolve and assign individual resonances. Direct detection of ionization events at amino groups is impossible because of rapid exchange of amino

[3] D. Yasmeen, J. R. Ellerson, K. J. Dorrington, and R. H. Painter, *J. Immunol.* **110,** 1706 (1973).

[4] G. O. Okafor, M. W. Turner, and F. C. Hay, *Nature (London)* **248,** 228 (1974).

[5] J. O. Minta and R. H. Painter, *Immunochemistry* **9,** 1041 (1972).

[6] G. M. Edelman, B. A. Cummingham, W. E. Gall, P. D. Gallieb, U. Rutishauser, and M. J. Wardal, *Proc. Natl. Acad. Sci. U.S.A.* **63,** 78 (1969).

protons with H_2O or D_2O. A further complication is that the NMR resonance linewidths generally broaden as the macromolecular size increases, due to a coupling between resonance linewidth and rotational motion of the atom giving rise to the resonance.

These difficulties are minimized by utilizing reductive ^{13}C-methylation and ^{13}C NMR. Reductive methylation with ^{13}C-enriched formaldehyde covalently methylates the free amino groups of proteins and provides a means, via ^{13}C NMR spectroscopy of the dimethylamino groups, to experimentally deduce the pK_a values of individual dimethylamino groups on N termini and on lysyl residues in proteins. Previous studies[7-10] have shown that free and dimethylated amino groups have essentially the same pK_a values. An exception to this correspondence between pK_a values occurs when the native amino group forms multiple hydrogen bond interactions, since the replacement of two protons by methyl groups removes two hydrogen bond donors.[9] In this case, the modification perturbs the amino group. In principle titrations of reductively methylated lysines followed by ^{13}C NMR spectroscopy can be used to determine the pK_a values of all the lysine residues in a protein.

The rate at which amino groups are reductively methylated seems to depend only on their respective pK_a values and not on their solvent accessibility or peculiarities of their microenvironment.[7,11] This is in contrast to classical modification reagents, where these factors frequently lead to selective reactivity of a single residue.

The difficulties associated with completely assigning the ^{13}C NMR resonances of dimethyllysyl groups to the parent amino acid residue constitute the major disadvantage of the reductive methylation technique. The application of reductive methylation to the Fc fragment and to its subfragments could have been considered problematic because of the large number of lysine residues and the unlikelihood of complete assignment of the dimethyllysyl resonances. We took the approach that the resonance behavior, at the worst, could be analyzed to determine if global or local conformational changes were occurring. At the best, we could combine the NMR data with strucutral information and information deduced from comparison of the Fc fragment with single-domain fragments in order to

[7] J. E. Jentoft, N. Jentoft, T. A. Gerken, and D. G. Dearborn, *J. Biol. Chem.* **254,** 4366 (1979).

[8] J. E. Jentoft, T. A. Gerken, N. Jentoft, and D. G. Dearborn, *J. Biol. Chem.* **256,** 231 (1981).

[9] T. A. Gerken, J. E. Jentoft, N. Jentoft, and D. G. Dearborn, *J. Biol. Chem.* **257,** 2894 (1982).

[10] T. A. Gerken, *Biochemistry* **23,** 4688 (1984).

[11] N. Jentoft and D. G. Dearborn, This series, Vol. 91, p. 570.

assign resonances and, more importantly, assess their conformational similarities and differences.

Procedures

Fragment Isolation. A monoclonal IgG_1 (Tu)[12,13] is used as the source of all fragments (although commercial Cohn fraction II has been used as a source of IgG in other, unpublished experiments). The Fc fragment is obtained by a modification of the method of Porter,[12–14] the pFc′ fragment is obtained after pepsin digestion,[15] the tFc′ fragment is obtained from the pFc′ fragment after trypsin treatment,[15] and the C_H2 fragment is isolated after low-pH trypsin treatment of IgG.[16] Samples used for reductive methylation are at least 90% homogeneous, as assessed by sodium dodecyl sulfate (SDS)-polyacrylamide gel electrophoresis and Ouchterlony double diffusion.

Reductive Methylation. The detailed methods for reductive methylation of proteins have been described previously.[10] The reaction is specific for N-terminal and lysyl amino groups. The only known side reactions occur occasionally with the use of cyanoborohydride as the reducing agent, where cyanomethyl adducts can form.[17] No resonances attributable to cyanomethyl adducts were detected in these studies.

NMR.[13] The proton-decoupled ^{13}C NMR data are obtained at 67.916 MHz on a Bruker WH180/270 Fourier transform NMR spectrometer equipped with a Nicolet 1180 computer.[12] The protein samples contain 20% D_2O, 0.1 *M* KCl, and 0.5% (v/v) methanol as an internal chemical shift reference [49.40 ppm relative to tetramethylsilane (TMS)]. The protein concentrations vary from 0.1 m*M* (for the C_H2 fragment) to 0.4 m*M* and require accumulation of 15,000 to 1000 scans/spectrum, respectively, requiring about 50 min of acquisition time/1000 scans. Chemical shift data are generally obtained from resolution-enhanced spectra, which are obtained by Gaussian multiplication of the free induction decay prior to Fourier transform. Resonance areas are obtained from nonenhanced spectra by integration.

pH Measurements.[13] Reported pH values represent the average of measurements taken before and after the acquisition of the NMR data; data are

[12] J. E. Jentoft, D. G. Dearborn, and G. M. Bernier, *Biochemistry* **21,** 289 (1982).

[13] J. E. Jentoft and R. Rayford, *Biochemistry* **28,** 3250 (1989).

[14] R. R. Porter, *Biochem. J.* **73,** 119 (1959).

[15] J. B. Natvig and M. W. Turner, *Clin. Exp. Immunol.* **8,** 685 (1971).

[16] J. R. Ellerson, D. Yasmeen, R. H. Painter, and K. J. Dorrington, *J. Immunol.* **116,** 510 (1976).

[17] N. Jentoft and D. G. Dearborn, *Anal. Biochem.* **106,** 186 (1980).

used for pH titration analyses only if the two pH values agreed within 0.03 pH units. The pH meter reading is taken as the pH value; no correction is made for the D_2O in the solution.

Analysis of Titration Data. The pH dependence of the chemical shift is followed between pH 7 and 11 for each resonance in the ^{13}C NMR spectrum. In the pH region where peaks are titrating, and thus shifting through each other, the pH increments are very close together, so that the pH dependence of the chemical shifts of the individual peaks can be followed. As a first stage in data analysis, all of the chemical shift versus pH data for resonances in a single fragment are plotted on a single graph. Both the shape of the individual preliminary titration curves and the changes in areas under all resonances during the pH titration are assessed for consistency, leading to some adjustment of the data points corresponding to the individual titration curves. In the second stage, individual sets of pH versus chemical shift data are fitted to a theoretical titration curve using an iterative program. The chemical shift limits for protonated and deprotonated forms derived from this analysis are used to construct a Hill plot of the data points, which is then subject to linear regression analysis and from which n_H and pK_a values are obtained.[13] Values of n_H are generally unity, within the range of experimental error (± 0.03).[13] The standard deviations on the pK_a values are on the order of ± 0.03 pH units.[13] The resulting parameters, protonated chemical shift in parts per million, change in chemical shift on deprotonation, and pK_a value, are summarized for all fragments in Table I.

Criteria for Application of Method

In order to utilize the dimethyllysyl resonances of a protein as conformational probes, the lysine residues of the protein must reside in multiple environments and give rise to multiple peaks in the ^{13}C NMR spectrum. A single resonance for all dimethyllysine residues, or a broad envelope, cannot provide the multiple probes of protein conformation implicit in the approach described here. This criterion is met for reductively ^{13}C-methylated Fc fragment, as shown by the seven resonances in its spectrum, shown in Fig. 1. Second, the protein must be stable at high pH values (up to pH 11) so that a pH titration curve can be defined for the dimethyllysyl groups (characteristic pK_a of 10.2). Stability is demonstrated experimentally by first titrating to high pH and then by titrating back to neutral pH; stability is presumed if the plot of chemical shift versus pH is independent of the order in which the data are obtained. This criterion was met for all IgG-derived fragments. Finally, data interpretation is greatly facilitated by access to a crystal structure for the protein(s) of interest. In the case of the

TABLE I
CHARACTERISTIC PARAMETERS OF DIMETHYLLYSYL RESONANCES OF Fc, C_H2, pFc′, AND tFc′ FRAGMENTS[a]

Fragment	Estimated number of lysines	Parameters		
		δ_{HA} (ppm)[b]	Δ (ppm)[b]	pK_a
		Resonance 1		
Fc[c]	5–6	43.10	1.29	10.13
pFc′	2	43.16	1.09	10.16
tFc′	1	43.10	1.01	10.18
C_H2	2	43.02	1.19	10.15
		Resonance 7		
pFc′	3	43.10	1.09	9.83
tFc′	1	43.26	1.05	10.02
C_H2	1	43.23	1.28	10.16
		Resonance 2		
Fc	1	42.60	1.06	9.93
C_H2	1	43.14	1.01	9.83
		Resonance 3		
Fc	3	43.11	1.16	9.74
pFc′	2	43.19	1.03	9.63
tFc′	2	43.16	0.97	9.68
C_H2	2	43.15	0.98	9.74
		Resonance 4		
Fc	3	43.13	1.10	10.50
C_H2	1	43.31	1.11	10.67
pFc′	1	43.06	1.09	10.68
tFc′	1	43.08	1.10	10.89
		Resonance 5		
Fc	2–3	43.12	1.43	10.66
		Resonance 6		
Fc	1	43.04	1.49	9.62
C_H2	1	43.12	1.17	9.32
		Resonance 6′		
pFc′	1	43.28	1.16	10.04
tFc′	1	43.28	1.16	10.18
C_H2	1	42.61	0.79	10.38
		Resonance 8		
Fc	1	42.44	1.54	10.25
C_H2	1	42.46	1.05	10.07

[a] Reprinted with permission from *Biochemistry* **28,** 3250. Copyright 1989 American Chemical Society.

[b] δ_{HA}, Chemical shift of the fully protonated dimethyllysyl resonance; Δ, increase in the chemical shift of the dimethyllysyl resonance on deprotonation.

[c] Resonances 1 and 7 are indistinguishable for the Fc (Tu) fragment in 0.1 *M* KCl. At other salt concentrations, resonance 7 can be seen as a separate resonance with a relative area of 1 (data not shown).

Fc fragment and subfragments, the crystal structure of the Fc fragment derived from human IgG_1 has been solved.[18]

Results

The reductively methylated Fc fragment, with 18 lysine residues in each heavy-chain fragment, has 7 distinct titrating dimethyllysyl resonances with relative areas and titration parameters given in Table I. Resonance 7 is identical to resonance 1 for the Fc fragment, but is included in Fig. 1 because it can be distinguished under other solvent conditions. Resonance 1 has the largest area, representing five to six lysine residues, while resonances 2, 6, and 8 have areas suggesting that they each arise from a single lysine residue. The chemical shift parameters for the Fc resonances are given in Table I. Note that all three chemical shift parameters vary from resonance to resonance. Among the resonances are those with both elevated and depressed pK_a values relative to the expected value for a solvent-exposed lysine residue (10.2). The properties of resonance 1 are consistent with those previously assigned to solvent-exposed lysine residues,[7-9] and we presume that the lysines giving rise to resonance 1 in the Fc fragment and its subfragments are solvent exposed and noninteracting.

The pFc′ fragment consists of the entire C_H3 domain, which has six lysines, and a nine-residue, probably unstructured, segment from the C_H2 domain, which contains three lysines. The reductively methylated pFc′ fragment gives rise to five distinct titrating resonances, the properties of which are summarized in Table I. As with the reductively methylated Fc fragment, resonance 1 is the most intense resonance, corresponding to two lysine residues.

The tFc′ fragment contains the intact C_H3 domain with its six lysine residues, but is missing six or eight of the C_H2 domain residues found in the pFc′ fragment and two or three of the lysines from that domain. The reductively methylated tFc′ fragment contains five titrating resonances with equivalent properties to those measured for the pFc′ fragment, but the areas of resonances 1 and 7 were reduced by an amount corresponding to approximately three resonances, suggesting that those resonances "missing" in the reductively methylated tFc′ fragment were from the lysine residues in the N-terminal segment of the pFc.

The C_H2 fragment contains a complete C_H2 domain, the hinge disulfide, and 24 residues from the second C_H2 domain (B chain). The reductively methylated C_H2 fragment has 8 well-resolved dimethyllysyl resonances, representing 11 lysine residues from the intact domain and 2 lysines from the partial domain. The area under resonance 1 corresponds to two lysine residues for the C_H2 fragment.

[18] J. Deisenhofer, *Biochemistry* **20,** 2361 (1981).

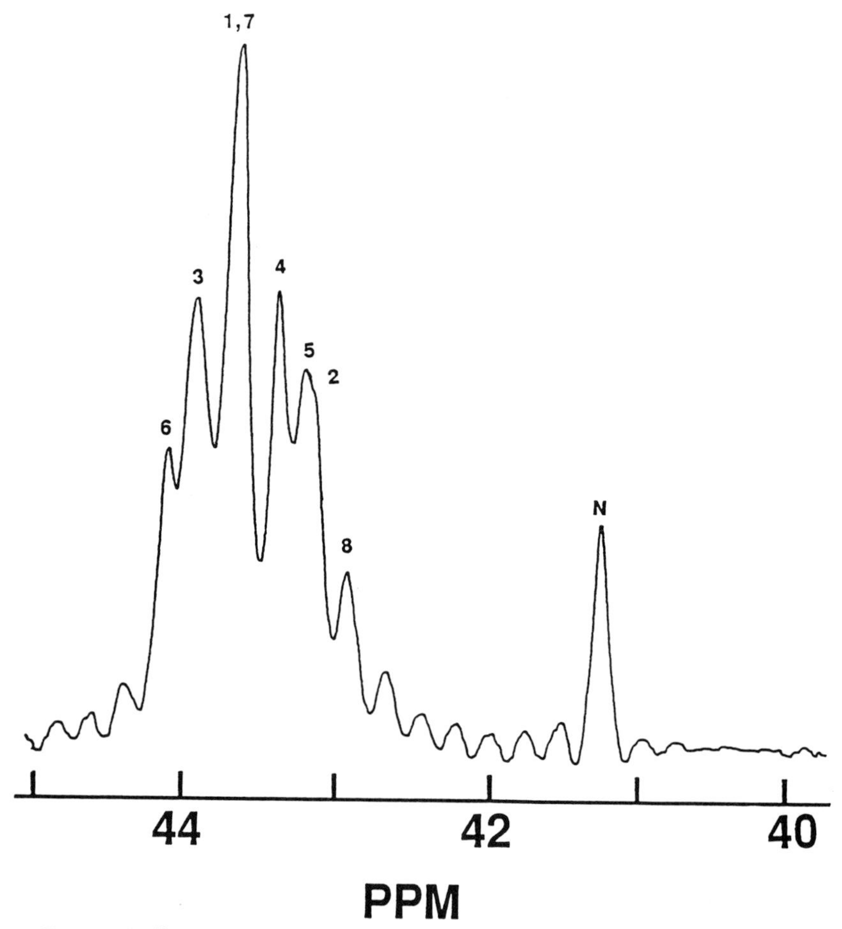

FIG. 1. The ^{13}C NMR spectrum of the dimethylamino resonances from the reductively ^{13}C-methylated Fc fragment [obtained from the monoclonal IgG (Tu)] at pH 10.0 in 0.1 *M* KCl. Conditions for the NMR experiment are given in Procedures. Dimethyllysyl resonances are numbered to correspond to the text. N, Resonance from the dimethylated N terminus. Reproduced, by permission, from J. E. Jentoft & R. Rayford, *Biochemistry,* **28,** 3250–3257 (1989).

Alignment of Resonances from Different Fragments

The alignment of the data in Table I implies correspondence between resonances from the Fc fragment and the various subfragments. Indeed, the resonances were numbered to reflect the correspondence after the alignment of resonances for the various fragments was established. The primary basis for this alignment was the pK_a value, as is evident by inspection of Table I. The secondary basis for the alignment was the protonated chemical shift value. This approach to alignment gave a consistent correspondence of pH-dependent chemical shift values for the various fragments.[13]

Following resonance alignment, the structural implications of the data in Table I can be evaluated. Resonances 1, 3, and 4 correspond in the Fc and all subfragments, indicating that these resonances in the Fc fragment arise from lysine residues from both domains, and that they arise from residues that are not perturbed in going from the Fc fragment to the smaller, single-domain subfragments. Resonances 2, 6, and 8 are unique to the Fc fragment and the C_H2 subfragment, indicating that the parent residues for these resonances are in the C_H2 domain, and that their respective microenvironments are the same in the Fc fragment and in the smaller C_H2 fragment. Interestingly, each of these three resonances arise from a single lysine residue, so that these residues have the potential of serving as sequence-specific probes. Resonance 5 is found only for the Fc fragment, indicating that the microenvironment of its parent residues was destroyed in its corresponding subfragment. Resonances 6′ and 7 are unique to the two subfragments, indicating that their parent residues are in changed environments in the subfragments relative to the Fc fragment.

This experimental information can now be combined with structural information derived from the X-ray crystal structure of the Fc fragment to deduce partial assignments for the resonances. Table II summarizes the solvent accessibility and the location and type of interaction seen for each lysine residue in crystalline Fc, and shows the possible resonances that could give rise to each residue. The solvent accessibility provides very basic information about the lysine residues likely to have perturbed properties (those with the least solvent exposure) and those likely to display properties similar to lysine residues in unstructured peptides (those with the greatest exposure). Similarly, those lysine residues forming hydrogen-bond and ion pair interactions in crystalline Fc are those most likely to have perturbed pK_a values; we presume that an amino group in an ion pair interaction is more likely to have a perturbed pK_a value than one in a hydrogen bond. Finally, lysine residues that form interactions across the C_H2-C_H3 domain interface are certain to have different properties in the Fc fragment and in single-domain subfragments.

Generally speaking, we would expect resonance 1 to correspond to noninteracting, solvent-exposed residues. Resonance 5 is a good candidate for lysine residues that form interactions across the C_H2-C_H3 domain that will be disrupted in the subfragments. It is known that pFc′ and tFc′ fragments are dimers,[19] so C_H3-C_H3 interdomain interactions should be

[19] M. W. Turner and H. Bennich, *Biochem. J.* **107,** 171 (1968).

TABLE II
ACCESSIBILITIES, INTERACTIONS, AND PARTIAL RESONANCE ASSIGNMENTS FOR LYSINE RESIDUES OF Fc FRAGMENT[a]

Lysine[b]	Solvent accessibility[c] (Å^2)	Site of interaction[b]	Type of interaction[b]	Possible resonances[d]
		C_H2 domain		
246	47	Intra-C_H2	H bond to NAG-456[e]	6
248	23	Inter-C_H2–C_H3	Ion pair with Glu-380	5 or 6
274	44	Noninteracting		1
288	48	Noninteracting		1
290	54	Noninteracting		1
317	41	Intra-C_H2	Ion pair with Asp-280	2, 3, 4, or 8
320	26	Intra-C_H2	Ion pair with Glu-333	2, 3, 4, or 8
322	35	Intra-C_H2	H bond to Tyr-278	2, 3, 6, or 8
326	54	Noninteracting		1
334	31	Intra-C_H2	H bond to Val-240	2, 3, or 6
338	5	Inter-C_H2–C_H3	Ion pair with Glu-430	5 or 6
340	44	Noninteracting		1
		C_H3 domain		
360	50	Noninteracting		1
370	17	Inter-C_H3	Ion pair with Glu-357[f]	4
392	33	Inter-C_H3	H bond to Leu′-398	6′ or 7
409	0.4	Intra-C_H3	H bond to Thr-441	3
		Inter-C_H3	Ion pair with Asp′-399	
414	38	Intra-C_H3	H bond to Met-358	7 or 6′
439	5	Intra-C_H3	H bond to Thr-350	3
		Inter-C_H3	Ion pair with Glu′-356	

[a] Adapted with permission from *Biochemistry* **28,** 3250. Copyright 1989 American Chemical Society.
[b] J. Deisenhofer, *Biochemistry* **20,** 2361 (1981).
[c] B. K. Lee and F. M. Richards, *J. Mol. Biol.* **55,** 379 (1971).
[d] J. E. Jentoft and R. Rayford, *Biochemistry* **28,** 3250 (1989).
[e] NAG, *N*-acetylglucosamine.
[f] Interactions between chains are indicated by a prime after the three-letter code for the nonlysine amino acid of the interacting pair.

preserved in both the Fc fragment and in these subfragments (resonances 3 and 4?).

tFc′ Fragment. Using Table II we can predict the properties of the lysine resonances in the tFc′ fragment, the smallest fragment for the C_H3 domain. Lysine-360 is noninteracting in crystalline Fc and so should correspond to resonance 1. Lysine-370, in an ion pair, is likely to be more strongly perturbed than are Lys-392 or Lys-414, which are in hydrogen bonds. Hence, Lys-370 most closely corresponds to the perturbed single lysine of resonance 4. The properties of dimethylated Lys-409 and Lys-439 are hard to predict since they form multiple interactions, a hydrogen bond within the domain and an ion pair with two hydrogen bonds across the C_H3-C_H3 interface. Both of these residues could be perturbed by dimethylation, since they appear to form three hydrogen bonds, and the presence of two methyl groups on the parent nitrogen only leaves room for one or two hydrogens as hydrogen-bond donors. Despite this uncertainty, we would expect the properties of these lysine residues to be similar to each other and to be unchanged in the dimethylated Fc, pFc′, and tFc′ fragments, by virtue of their location at the stable C_H3-C_H3 interface. This description corresponds most closely to resonance 3, since it has a relative area of two lysines and has a highly perturbed pK_a value. Finally, the remaining resonances, resonances 6′ and 7, can be assigned to the remaining lysines, Lys-392 or Lys-414 (one each). Note that resonances 6′ and 7 are not found in the Fc fragment, thus implying that Lys-392 and Lys-414 may be perturbed in the subfragment.

pFc′ Fragment. The spectra of the dimethylated pFc′ fragment contains areas corresponding to one additional lysine residue in its resonance 1 and two additional lysines in its resonance 7, relative to the spectra of the dimethylated tFc′ fragment. All other resonances correspond very closely for these two fragments, confirming that the structure of the C_H3 domain is conserved for these two fragments. These extra areas for the pFc′ fragment correspond to the three lysine residues (334, 338, and 340) in the nine additional N-terminal residues that distinguish the pFc′ fragment from the tFc′ fragment. The relatively unperturbed characteristics of the resonances are consistent with the expected loss of structure and, for Lys-334 and Lys-338, defined interactions for these lysines.

C_H3 Domain in Fc Fragment. Resonances 1, 3, and 4 in the Fc fragment correspond to those in the tFc′ and pFc′ fragments. Thus, Lys-360 is assigned to resonance 1, Lys-409 and Lys-439 are assigned to resonance 3, and Lys-370 is assigned to resonance 4 in the Fc fragment. The relative areas of these resonances indicate that the lysines from the C_H2 domain also contribute to each resonance in the Fc fragment. Resonances 6′ and 7 are not observed for the Fc fragment, implying that the microenvironments for Lys-392 and Lys-414 are different in the two domains. Fc resonances 4 and 5 have properties most similar to that of resonance 6′ for the subfragments, while Fc resonance 1 has properties most similar to that for resonance 7 of the subfragments, and therefore, these lysines, 392 and 414, are assigned to one of these three Fc resonances.

C_H2 Fragment. The structural data in Table II can be used to predict

that the C_H2 fragment would contain six noninteracting, solvent-exposed lysines (246B, 248B, 274, 288, 290, and 366) with properties corresponding to resonance 1 and/or 7, two lysines (317, 320) in intradomain ion pairs, and two lysines (322, 334) in intradomain hydrogen bonds, the properties of which should be similar in the Fc fragment and the C_H2 fragment, and three lysines (246, 248, and 338) that are expected to be perturbed in the C_H2 subdomain relative to the Fc fragment. Lysine-246 is hydrogen bonded to the carbohydrate chain that lies between the C_H2 domains in the intact Fc fragment, but the conformation of the carbohydrate side chain in the C_H2 subfragment could easily be altered. The ion pair interactions of Lys-248 and Lys-338 across the C_H2–C_H3 interdomain interface are certainly disrupted in the C_H2 subfragment.

Suprisingly only three or four lysines contribute to resonances 1 and 7 for the reductively methylated C_H2 fragment, indicating that more lysyl residues are structured (e.g., in perturbed microenvironments) than are expected. Of the four lysines expected to be in similar environments in the Fc and C_H2 fragments, resonances 2, 3, 4, 6 and 8 correspond reasonably well and account for six lysine residues in the C_H2 fragment. Of these, resonance 4, which arises from a single lysine, has the elevated pK_a value commonly associated with a lysine residue in an ion pair interaction, consistent with it arising from Lys-317 or Lys-320. Resonances 2 (from a single lysine) and 3 (from two lysines) have depressed pK_a values, which can occur for solvent-protected residues; any three of the interactive lysines predicted to be similar for the Fc and the C_H2 fragments could give rise to these resonances (Lys-322, -334, -320, or -317). Resonances 6 and 8 arise from single lysine residues in the C_H2 fragment whose pK_a values and chemical shift changes are altered relative to those in the Fc fragment, but whose chemical shift values during titration still most closely resemble those of the corresponding resonance in the Fc fragment (see Table I). These resonances could arise from lysine residues (Lys-246, -248, -334, -338) whose chemical shifts are perturbed by proximity to aromatic side chains and whose environments are altered somewhat in the C_H2 fragment relative to those in the Fc fragment. Resonance 6′ in the C_H2 fragment arises from a lysine unique to the fragment, and hence from a residue of altered environment relative to the Fc. Candidate lysines for resonance 6′ are Lys-246, -322, or -334. Note that the pool of potential candidate lysines for assignment to each resonance from the reductively methylated C_H2 fragment is confused by the unexpected excess of resonances from "structured" lysines.

C_H2 Domain in Fc Fragment. Resonance 5, unique to the Fc fragment, has the elevated pK_a value predicted for solvent-shielded residues and/or residues in ion pair interactions. Because all six of the lysine residues from the C_H3 domain in the Fc fragment could be associated with other reso-

nances, resonance 5 is presumed to arise in large part from Lys-248 and Lys-338, residues that form ion pairs across the C_H2-C_H3 interface and whose interactions therefore are certain to be destroyed on formation of the C_H2 fragment. The similarities between resonances 2, 3, 4, 6, and 8 from the Fc fragment and from the C_H2 fragment imply that a substantial portion of the tertiary structure of the C_H2 domain is unaffected by formation of the C_H2 fragment[13]; the differences suggest that the conservation of structure in the C_H2 fragment is not as great as that for the pFc′ and tFc′ fragments.

Summary and Extension of Approach

The titration data on the Fc fragment, the subfragments, and the information from the crystal structure of the Fc fragment were combined to make partial correlations between lysine residues and dimethyllysyl resonances in the three fragments (Table II). The comparisons between the Fc fragment and its subfragments allowed us to assess the detailed similarities (and hence conservation of structure) between the larger fragment and the single-domain fragments,[13] and to conclude that while the structure of the C_H3 domain is very well conserved in the pFc′ and tFc′ subfragments, the structural conservation in the C_H2 domain is much weaker. An important finding related to the C_H2 domain is the association of the single-residue resonances 2, 6, and 8 with the C_H2 domain; they provide single-lysine probes of events in the C_H2 domain that can be used to monitor their interactions in the Fc fragment. Further, the lysines that can give rise to these three resonances have been limited to only seven residues (246, 248, 317, 320, 322, 334, and 338).

Using the results of this study, we can further study the Fc fragment and distinguish between global and local conformational changes on the basis of the number of resonances perturbed. Potentially the ^{13}C NMR spectra of reductively methylated Fc could be used to define the extent and location of a local conformational change in the Fc fragment, perhaps as a result of binding an effector, or in respond to some change in solution conditions. It is important to note that another macromolecule will not interfere with the ^{13}C NMR spectrum of a reductivity methylated protein, except to broaden the resonance lines if the rotational correlation time of the resonance is lengthened by formation of a larger macromolecular complex.

Assignments can be further refined by other strategies. For example, resonances from residues in ion pair interactions may be identified by the ionic strength dependence of their pK_a values, since the strength of solvent-accessible ion pair interactions is attenuated in a predictable fashion by increasing ionic strength.[20] Another approach to identifying lysines in particular types of interactions is to determine the temperature dependence of the pK_a values. The perturbations of pK_a values for lysyl residues in hydrogen-bond interactions are expected to become stronger with decreasing temperature, while those in electrostatic interactions are expected to become stronger with increasing temperature.

If lysyl residues help to form an effector binding site on the Fc fragment, the addition of the effector would be expected to perturb the chemical shifts and/or pK_a values of the corresponding residues. The binding may also result in broadened resonance lines for the dimethyllysyl derivatives, depending on the exchange rate of the complex with free components. In either case, it should be possible to identify the resonances perturbed upon complex formation. An accompanying strategy for studying lysine residues at a binding site relies on the fact that these residues may be protected from methylation, if the modification reaction takes place under conditions where saturating amounts of effector are present. In such an experiment, the residues at the binding site would be missing from a ^{13}C NMR spectrum. To label lysine residues specifically at a binding site, one would, e.g., ^{12}C-methylate the complex, dissociate it, and remethylate the protected lysines with [^{13}C]formaldehyde. Note that this latter strategy for specific modification could be used to ^{14}C- or ^{3}H-methylate active site residues for subsequent chemical identification of the residues.

Acknowledgments

This work was supported by grants-in-aid from the American Heart Association, Northeast Ohio Affiliate, and by the Northeast Ohio Arthritis Center Grant AR 20618 from the National Institutes of Health.

[20] J. B. Matthew, M. A. Flanagan, B. Garcia-Moreno, K. L. March, S. J. Shire, and F. R. N. Gurd, *Crit. Rev. Biochem.* **18,** 91 (1985).

[13] Immunoelectron Microscopy and Image Processing for Epitope Mapping

By NICOLAS BOISSET and JEAN N. LAMY

Introduction

Immunoelectron microscopy (IEM) is a labeling method based on the analysis of immunocomplexes by electron microscopy (EM) that can be used to study a large variety of structures. However, specimen preparation and observation conditions (dehydration, exposure to the electron beam, etc.) may induce a broad range of deformations and partial destruction of biological molecules. Fortunately, these factors do not affect equally all molecules present in the preparation, so that a part of the lost information can be recovered by image processing. However, image processing cannot be used with all types of immunocomplexes. With undefined antigens, such as cells or organelles, it is often impossible to carry out statistical image processing because of the size and/or shape heterogeneity of the antigen structures. With defined antigens such as proteins, ribosomes, and certain viruses, the lost information can generally be recovered by statistical methods of image processing, provided that the population is randomly affected.

The following methodology, valuable exclusively for defined immunocomplexes composed of monoclonal antibodies (MAbs) and well-defined antigens, was developed for the localization of epitopes on hemocyanin, a copper-containing respiratory pigment of arthropods and mollusks.[1-3]

Hemocyanin from Androctonus: The Antigen Used in This Study

Androctonus australis hemocyanin (Hc), is a highly immunogenic, large-sized (25 × 25 × 16 nm) protein. In the electron microscope, it produces three different views: top, side, and 45°, that are easy to recognize and to align.[4,5] Its quaternary structure, determined by IEM with polyclonal antibodies, is well understood,[6,7] and the amino acid sequences and three-dimensional (3D) structures of evolutionarily related species are known.[8,9]

[1] N. Boisset, J. Frank, J. C. Taveau, P. Billiald, G. Motta, J. N. Lamy, P. Y. Sizaret, and J. N. Lamy, *Proteins* **3,** 161 (1988).

[2] J. Lamy, *in* "Biological Organization: Macromolecular Interactions at High Resolution" (R. M. Burnett and J. Vogel, eds.), p. 153. Academic Press, New York, 1987.

[3] J. Lamy, P. Billiald, J. C. Taveau, N. Boisset, G. Motta, and J. N. Lamy, *J. Struct. Biol.* **103,** 64 (1990).

[4] M. Van Heel and J. Frank, *Ultramicroscopy* **6,** 187 (1981).

[5] M. Van Heel, W. Keegstra, W. Schutter, and E. F. J. Van Bruggen, *Life Chem. Rep., Suppl.* **1,** 69 (1983).

[6] J. Lamy, M. M. C. Bijlholt, P. Y. Sizaret, J. N. Lamy, and E. F. J. Van Bruggen, *Biochemistry* **20,** 1849 (1981).

[7] J. Lamy, J. Lamy, P. Billiald, P. Y. Sizaret, G. Cavé, J. Frank, and G. Motta, *Biochemistry* **24,** 5532 (1985).

[8] B. Linzen, N. M. Soeter, A. F. Riggs, H. J. Schneider, W. Schartau, M. D. Moore, E. Yokota, P. Q. Behrens, H. Nakashima, T. Takagi, T. Nemoto, J. M. Vereijken, H. J. Bak, J. J. Beintema, A. Volbeda, W. P. J. Gaykema, and W. G. J. Hol, *Science* **229,** 519 (1985).

The native molecule is composed of 24 subunits belonging to 8 different 75-kDa polypeptide types organized in 4 groups of 6 subunits, improperly called hexamers.[10] The disposition of the subunits in the hexamer, initially supposed to be identical to that of the spiny lobster *Panulirus interruptus* hemocyanin,[11] and the organization of the hexamer in the 24-mer were confirmed by 3D reconstruction of randomly oriented isolated particles.[12]

Native hemocyanin presents two other interesting features. First, as shown by image processing,[4] the molecule in its top view (Fig. 1A and B) projects as a parallelogram and can lie on the support film on either of its flip and flop faces. The molecule stands on its flip face (Fig. 1A and E) when the long diagonal of the parallelogram passes by the lower left and upper right hexamers. The flop face is in contact with the support plane when the long diagonal passes by the top left and bottom right corners of the molecule (Fig. 1B and F). The second important structural feature is that the centers of the four hexamers are not coplanar, so that the molecule, when placed on a flat surface, is unstable and lies on three hexamers (the top left, the bottom right, and alternatively one of the other two). This phenomenon is known as the rocking effect.

The disposition of the four *Aa*6 subunits at the four corners of the molecule is such that two neighboring subunits always have opposed orientations with respect to the support plane. For example, the two copies located on the top right and bottom left corners (Fig. 1E and F) have their flat faces looking upward (toward the observer), while the flat faces of the other two subunits are looking toward the support plane. This later point explains why, when Fab fragments are bound to *Aa*6 subunits, their orientations depend on the type of subunit to which they are bound.

Method

Localizing an epitope by IEM is equivalent to the determination of the structure of an immunocomplex from EM images. Depending on the antigen and the objective in mind, different methods are used for the preparation of the materials (antibodies and immunocomplexes), the preparation and observation of the specimens, and the image processing to

[9] W. P. J. Gaykema, W. G. J. Hol, J. M. Vereijken, N. M. Soeter, H. J. Bak, and J. J. Beintema, *Nature (London)* **309,** 23 (1984).

[10] J. Lamy, J. Lamy, and J. Weill, *Arch. Biochem. Biophys.* **193,** 140 (1979).

[11] A. Volbeda and W. G. J. Hol, *J. Mol. Biol.* **209,** 249 (1989).

[12] N. Boisset, J. C. Taveau, J. Lamy, T. Wagenknecht, M. Radermacher, and J. Frank, *J. Mol. Biol.* **216,** 743 (1990).

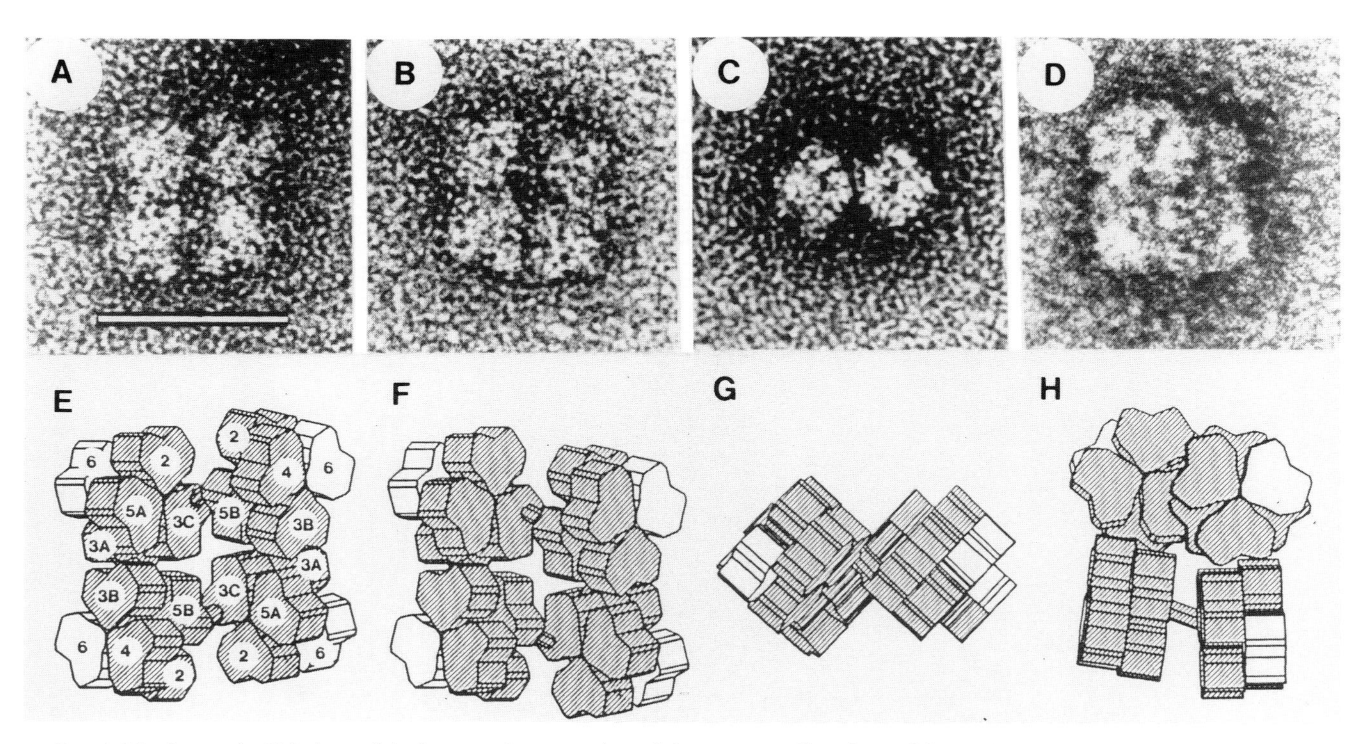

FIG. 1. The four main EM views of *Androctonus* hemocyanin and the corresponding views of the model. (A and E) Top view flip; (B and F) top view flop; (C and G) side view; (D and H) 45° view. The scale bar is 25 nm.

reinforce the significant details. In this section we describe the topological localization of several epitopes on a single subunit of *Androctonus* hemocyanin in order to illustrate the pertinent methodologies.

Preparation of Monoclonal Antibodies

Immunization. Different forms of the same antigen can be used for immunization. With a complex oligomeric protein such as hemocyanin, the whole oligomer, purified subunits, proteolytic peptides, or synthetic peptides can be given to the mouse, but the results will probably be different. Although there is no universal rule, a few guidelines may be of interest. For example, since most of the immunocomplexes are to be purified, the antigen–antibody binding must be strong. Because the anti-native protein antibodies generally have higher affinities than those directed toward peptides, it is often preferable to immunize the mouse with the whole protein or with purified subunits.

Another point to take into consideration is that the MAb must be highly specific for the epitope to be labeled. It may be more favorable to immunize with a subunit or a domain rather than to screen a large number of cross-reacting antigens. For example, because the subunits of *Androctonus* hemocyanin descend from a common ancestor, there are multiple cross-reactivities between them. Therefore, immunization with the whole oligomer produces far fewer subunit-specific clones than when the immunization is done with an isolated subunit.

The choice of the antibody class is much simpler. Although immunoglobulin M (IgM) produces immunocomplexes easily observed in EM, the images are generally difficult to interpret, because of the well-known polymorphism of IgM. In practice only IgG is used for the preparation of immunocomplexes.

Various immunization protocols have been published. The reader is referred to Volumes 92 and 121 of this series and to Harlow and Lane[13] for the preparation of hybridomas, and to Lamy *et al.*[7] for antihemocyanin antibodies.

Hybridoma Screening. The test used in the hybridoma screening must be the same type of test as that used for the Mab. However, a screening test based on IEM would be too time consuming, so that a specific test must be devised. With anti-hemocyanin antibodies we proceed in two steps. First, detection of the MAbs producing soluble immunocomplexes with the antigen is carried out by a classical enzyme-linked immunosorbent assay (ELISA) system using alkaline phosphatase-conjugated sheep anti-mouse

[13] E. Harlow and D. Lane, "Antibodies: A Laboratory Manual" Cold Spring Harbor Laboratory, Cold Spring Harbor, New York, 1988.

immunoglobulin. Second, the positive clones are subjected to crossed immunoelectrophoresis, which tests the strength of the antibody–antigen bond, as described by Weeke.[14] The procedure is as follows.

1. The antibody and the antigen are mixed and left in the deposit well of an agarose plate for 30 min at room temperature.
2. Electrophoresis is started. The conditions are such that the mobilities of the free antigen and the free antibody are high and low, respectively (0.15 *M* barbital buffer, pH 8.6). If the antigen mobility is low, it can be increased by covalent binding to a high-mobility protein such as serum albumin.
3. The plate is turned 90° and immunoelectroprecipitation is performed against a polyclonal antiserum specific for the antigen.
4. After completion of the electroimmunoprecipitation step, the plate is washed and stained with Coomassie Brillant Blue R250 and the precipitation pattern of the antigen is compared to a control plate without Mab.

When the antibody produces a stable immunocomplex, a precipitation peak with a low mobility is present between the deposit and the free antigen peak. Figure 2 shows an example of this test for the detection of MAb L102, an anti-*Aa*6 subunit antibody.

Storage and Purification of MAb. The MAb can be used as ascitic fluid or as purified IgG. For IgG purification the immunoglobulins are precipitated from the ascites fluids with 45% saturated ammonium sulfate, washed twice with 50% saturated ammonium sulfate, resuspended, and dialyzed against a 0.05 *M* Tris-Hcl buffer, pH 7.5, containing 0.15 *M* NaCl. Finally, they are purified by ion-exchange chromatography on a Mono Q HR 5/5 column [FPLC (fast protein liquid chromatography); Pharmacia, Uppsala, Sweden] using a 0.02 *M* triethanolamine buffer, pH 7.7, and a linear 0–0.35 *M* NaCl gradient. Purified IgG is then stored at −70°.

When Fab fragments are necessary, the purified MAb is subjected to a hydrolysis by papin.[15]

Characterization of Antibodies

Monoclonal antibodies are characterized by class and subclass, affinity, activity, and specificity.

1. The immunoglobulin class is determined by Ouchterlony double diffusion in agarose with rabbit anti-IgG subclass-specific antisera.

[14] B. Weeke, *Scand. J. Immunol.* **2**,(Suppl. 1), 47 (1973).

[15] R. R. Porter, *J. Biochem. Tokyo* **73,** 119 (1959).

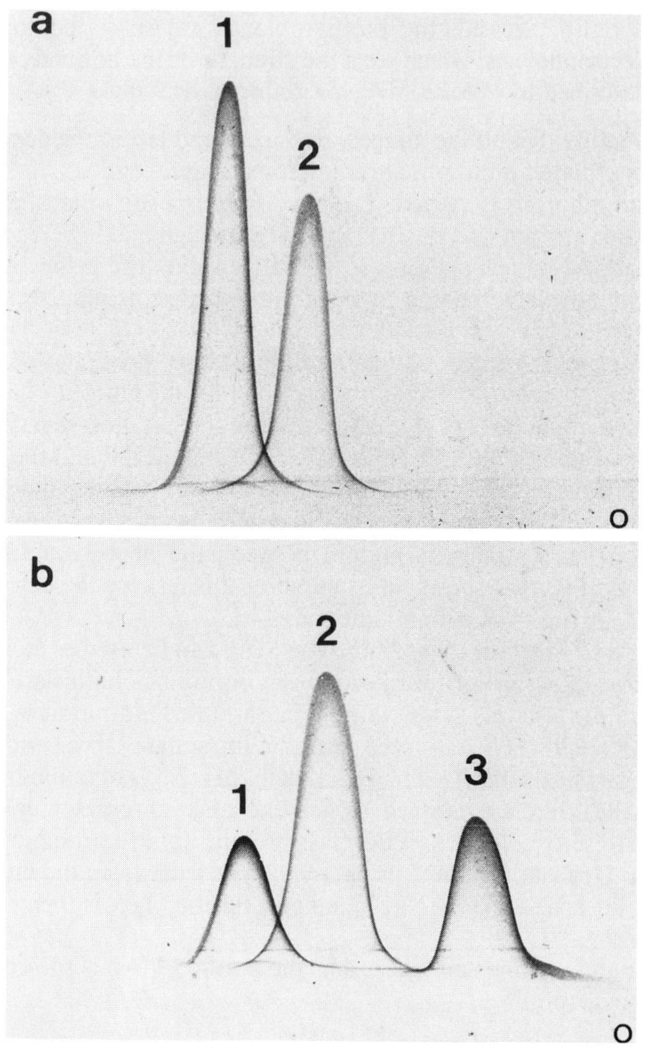

FIG. 2. Characterization of MAb L102 by crossed immunoelectrophoresis against a polyclonal anti-hemocyanin antiserum. (a) Control: the antigen (*Aa*6 subunit) and a reference substance to which the MAb does not bind (*Aa*4 subunit) are deposited in the same well. (b) Same as in (a), but MAb L102 has been added to the mixture. 1, *Aa*6 subunit; 2, *Aa*4 subunit; 3, immunocomplex. Electrophoretic migration from right to left; electroimmunoprecipitation from botton to top.

2. The affinity is determined by inhibition of binding MAb to ^{125}I-labeled subunit in a radioimmunoassay.[16]

3. The activity is expressed as the reciprocal of the highest dilution of ascites fluid producing a positive reaction in ELISA.

4. The result of an immunolabeling experiment is highly dependent on the antibody specificity. When cross-reactivities exist between the various parts of the whole antigen (subunits of an oligomer or domains of a subunit), the specificity of the various clones must be carefully determined.

Preparation of Immunocomplexes

Choice of Antigen. All the antigen visible in EM can be used for IEM. However, the choice depends on the final goal of the experiment. For a topological localization of epitopes the best approach is to use an antigen molecule as large as possible with as many asymmetry elements as possible in order to simplify the alignment step of the image processing. The fact that a large antigen bears several copies of an epitope (for example, located on several copies of a same subunit) is not inconvenient because the Fab fragments bound to these epitopes will be easily distinguished in the average images. Another point to take into consideration is the different accessibilities of the various subunits in an oligomer. For example, in *Androctonus* hemocyamin the internal subunits (*Aa*3A, 3B, 3C, and 5B) have only a small portion of their surface accessible to antibodies. Conversely, more external subunits are more easily accessed (*Aa*2, 4, 5A, and 6). *Aa*6 was choosen for the epitope localization experiment, because it is the most accessible subunit of the oligomer.

When IEM is used to investigate overlaps between epitopes it is critical to ensure that the antigen molecule bears a single copy of each epitope. In this case it is preferable to use the smallest antigen possible. However, these immunocomplexes are never submitted to image processing by multifactorial methods.

Choice of IgG or Fab. Choosing IgG or Fab is important. Because of the bivalence of the antibody molecule, the immunocomplex strings from IgG are usually long and contain several antigen molecules. Conversely, immunocomplexes made with Fab fragments never contain more than one antigen molecule. These immunocomplexes are easier to handle, but a part of the information, such as the torsion angles around the long axes of the Fab arms or the flexion angles at the hinge, is definitely lost.

Length of Immunocomplex String. The length of the immunocomplex string may also be important in view of the image processing. For example,

[16] J. A. Schroer, T. Bender, R. J. Feldmann, and K. J. Kim, *Eur. J. Immunol.* **13,** 693 (1983).

in long strings the antigen molecules may lie on the support film in an orientation different from those found in the uncomplexed antigen. In short strings the position of the antigen molecule is usually similar to a position observed in the free state.

Incubation. For isolated molecules the rule is to incubate the antigen with the antibody and then to purify the immunocomplexes if necessary. Common incubation conditions are 30 min at 37°, followed by an overnight incubation at 4°. The pH is often slightly alkaline; a pH 7.5 is common. Ionic strength is important to prevent the denaturation of the antibodies. A 0.05 *M* Tris-HCl buffer, pH 7.5, containing 0.15 *M* NaCl is a widely used system.

Purification. When the ascites fluid is used as such, the immunocomplexes must be purified in order to remove the ascites material, which could interfere with observation of the immunocomplexes. Conversely, if both the antigen and the antibody have been highly purified, if the affinity of the antibody for the antigen is high, and if antigen excess is used, all the antibody molecules are bound to the antigen so that purification of the immunocomplexes may be unnecessary. With oligomeric antigens an additional requirement is that the whole molecule does not dissociate on antibody binding. These requirements almost always necessitate purification.

All the purification methods convenient for proteins can be used with immunocomplexes. For hemocyamin, we use gel filtration on a microcolum (30 × 0.5 cm) of AcA 34 (Industrie Biologique Francaise, Villeneuve-la-Garenne, France) with an incubation buffer to remove the small-sized material. Because of their large size the immunocomplexes are excluded from the gel and are free from IgG, whose molecular weight is in the separation range of the gel. With smaller antigen, a good separation is required because of a higher risk of contamination and gel permeation on Superose 12 HR 10/30 (FPLC; Pharmacia) can be used.

Electron Microscopy

Although almost any method of specimen preparation can be used in IEM, negative staining is the most widely employed. The single-carbon layer technique is especially interesting in image processing because of preferential staining of the face of the molecule in contact with the support. In addition, the presence of stain menisci and the incomplete immersion of the molecules in the stain layer signal the presence of portions of the molecule that are located far above the support. Thus the stain-exclusion pattern of the Fab arms in the immunocomplexes may suggest their orientation with respect to the support film.

Because of differential staining of the faces of the molecule, the investigator must be very cautious with the handedness of the molecule. Obviously, when the grid is placed in the electron microscope with the carbon film facing the electron beam, the image obtained is enantiomorphic compared to that produced by the same molecule with the carbon film facing the photographic emulsion.

For the localization of the *Aa*6 subunit epitopes, 13 highly specific MAbs were selected, characterized, and incubated, 1 at a time, with an excess of native hemocyanin. Then, the short immunocomplex strings were purified by gel filtration, negatively stained by the method of Valentine[17] with 1–2% (w/v) uranyl acetate, and observed in EM under low-dose conditions at a magnification of ×50,000. As shown in Fig. 3, most of the strings are composed of two antigen molecules in their top view and of Fab arms bound to the corners of the molecule. Figure 3 also shows that the hemocyanin molecules are aligned, parallel or perpendicular, sometimes in the same immunocomplex string (e.g., MAb L102). This architecture is important because it reflects the degree of torsion and flexing of the Fab arms,[18] information lost in Fab-containing immunocomplexes.

Image Processing

Strategy. Image processing of MAb-containing immunocomplexes by a statistical method allows a considerably more precise localization of the antibody–antigen contacts than can be achieved by visual observation. The image-processing analysis begins with an evaluation of the quality of the micrographs by optical diffraction, followed by digitization of the negatives with a microdensitometer. Then, after a windowing step, the immunocomplexes are aligned and the whole image set is subjected to multivariate statistical analysis and classification, procedures that are used to find homogeneous subsets. Once these are obtained, the corresponding average and variance maps can be computed. Such a complex process requires sophisticated software. The results presented below were obtained using the SPIDER (system for processing image data in electron microscopy and related fields) program.[19] For general information, the reader is referred to a chapter in this series[20] and to other general papers.[21–23]

[17] R. C. Valentine, B. M. Shapiro, and E. R. Stadtman, *Biochemistry* **7,** 2143 (1968).

[18] R. H. Wade, J. C. Taveau, and J. N. Lamy, *J. Mol. Biol.* **206,** 349 (1989).

[19] J. Frank, B. Shimkin, and H. Dowse, *Ultramicroscopy* **6,** 343 (1981).

[20] J. Frank, M. Radermacher, T. Wagenknecht, and A. Verschoor, this series, Vol. 164, p. 3.

[21] J. Frank, A. Verschoor, and T. Wagenknecht, *in* "New Methodologies and Studies of Protein Configuration" (T. T. Wu, ed.), p. 36 Van Nostrand, New York, 1985.

[22] J. Frank, *Electron Microsc. Rev.* **2,** 53 (1989).

[23] M. Van Heel, *Ultramicroscopy* **13,** 165 (1984).

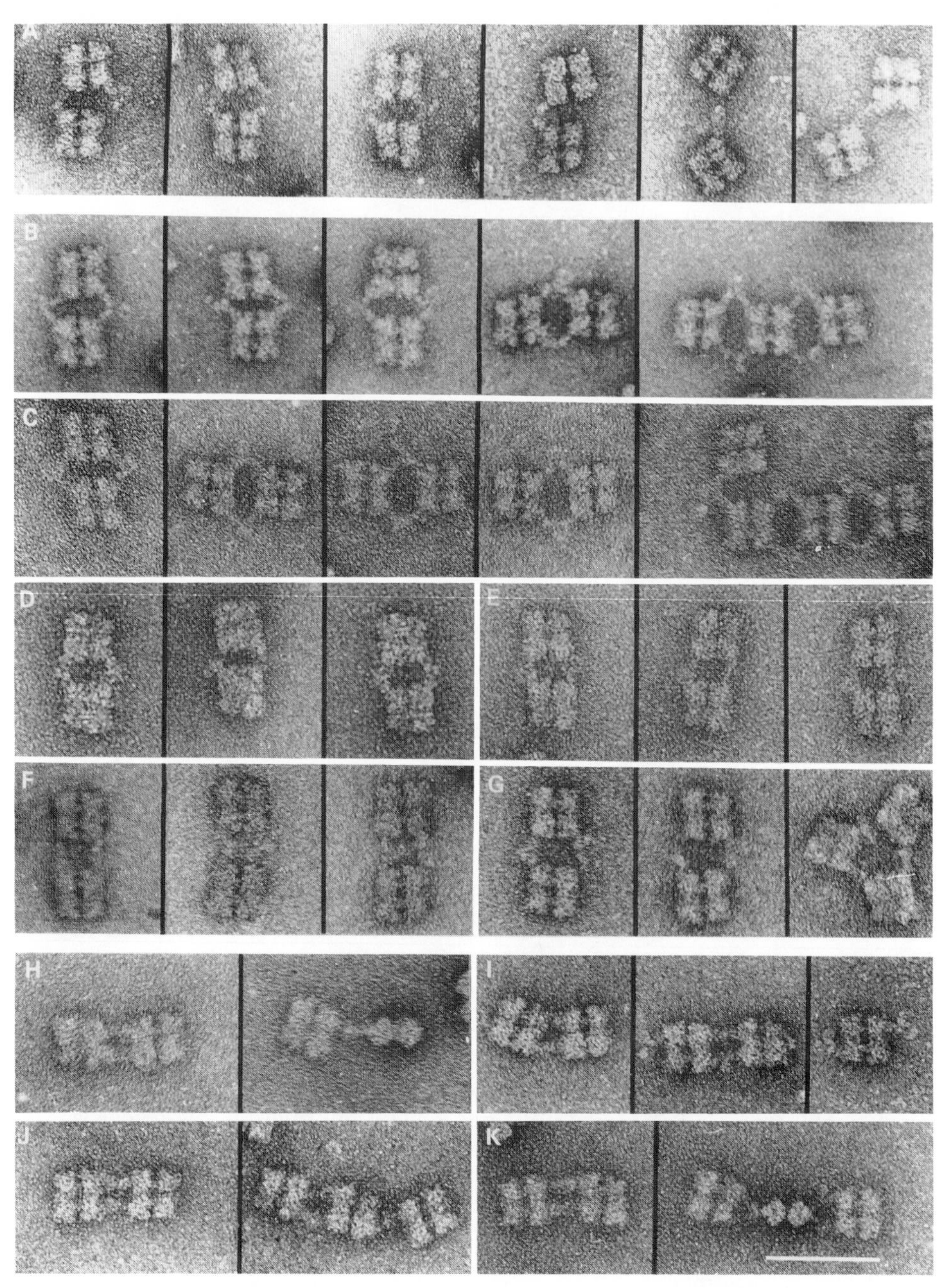

FIG. 3. Architecture of immunocomplex strings produced by 11 MAbs with whole *Androctonus* hemocyanin. (A–K) 5701, L8, L102, L100, L101, L103, L107, L104, L106, L114, L115. The scale bar is 50 nm. [Reproduced with permission from J. Lamy, P. Billiald, J. C. Taveau, N. Boisset, G. Motta, and J. N. Lamy, *J. Struct. Biol.* **103,** 64 (1990).]

The following section describes the aspects of the method that are specific for immunocomplexes, taking as an example localization of the epitope of MAb L102 on the *Aa*6 subunit.

Digitization, Image Selection, and Windowing. Before image processing the negatives must be evaluated by optical diffraction and digitized. Only negatives slightly underfocused (1000–2000 Å), allowing one to "see" details in the size range of 3 nm, devoid of drift and astigmatism, are selected for image processing. The digitization consists of a measurement of the optical density (OD) of small image elements called pixels (picture element) on a fine sampling grid, using a microdensitometer. The mesh size of the grid and hence the pixel size must be smaller than or at most equal to half the resolution. For example, at a resolution of 20 $Å^{-1}$, a pixel size equal to or smaller than 10 Å is required. In the case of the hemocyanin-containing immunocomplexes the image is composed of 128 × 128 pixels of 25 μm each. At a magnification factor of ×50,000 this field corresponds to a window of 640 Å on the scale of an object, which is large enough to contain a hemocyamin molecule with 4 Fab fragments, each bound to a corner. The various factors to be taken into account in the digitization have been reviewed in detail.[24]

Windowing is a process by which a portion of an EM field is extracted for the purpose of image processing. In the case of immunocomplexes, the windowing step is used to select small structural units, called elementary immunocomplexes (EICs), composed of an antigen molecule and one or several Fab fragments or Fab arms, depending on the nature of the antibody moiety of the immunocomplex (IgG or Fab fragment). Windowing requires the formulation of precise criteria for inclusion in the image population. The criteria we used are as follows.

1. An EM view of the antigen must be carefully chosen. Usually, the most asymmetrical view is the easiest to align. The position of the attachment point of the Fab to the antigen must also be taken into account. In the case of MAb L102, the top view, which shows the four copies of the *Aa*6 subunit, is obviously the most favorable.

2. With antigen bearing several copies of the epitope (e.g., *Androctonus* hemocyanin), the hemogeneity of the EIC population with respect to the number of Fab bound to the antigen must be verified. For example, in the case of anti-*Aa*6 subunit MAbs, there are 15 different possible EIC types, depending on the number (1, 2, 3, or 4) of Fab bound and on the location of the subunit to which they are bound in the oligomer. Therefore, it is easy

[24] J. Frank, *in* "Biological Electron Microscopy" (J. K. Koehler, ed.) p. 215. Springer-Verlag, Berlin, 1973.

to select a given EIC type at the windowing step; for example, two L102 Fab arms bound on the same side of the molecule.

3. The cleanliness of the image in the zone surrounding the EIC and the absence of deformation and disruption of the antigen molecule must also be checked. If deformed or contaminated molecules are present in the sample population, they will generally be recognized by multivariate statistical analysis, but the number of useful images will be proportionately reduced.

Alignment. The alignment procedure entails the construction of a reference image, followed by alignment of all images to this reference. The reference can be authentic image, an average image, such as the average resulting from a first alignment cycle, or a synthetic image. In addition, the reference can represent the antigen, the antibody, or the immunocomplex. Because of the various flexibilities of the antibody molecule (rotational flexibility around the long Fab axis, segmental flexibility in the hinge or in the elbow areas), it is generally better to choose the free antigen or an EIC as the reference image.

Strictly speaking, the alignment comprises calculation of the rotation angle and the translation vector, which must be successively applied to an image so that it is brought into the best coincidence with the reference. For this purpose we use the method based on autocorrelation and cross-correlation functions.[24-27]

In order to avoid the introduction of a bias due to the presence of Fabs, the antigen portion of the immunocomplex is centered for EICs containing MAb L102. Then, a centered circular mask is applied so that the Fab arms are completely hidden, but the entire antigen molecule is visible. The density in the hidden area is set as the average of the density in the visible area. Such a mask does not interfere with rotational alignment because it possesses circular symmetry, nor with translational alignment, because the density step at the mask boundary is kept small. Two or three alignment cycles are usually performed. The alignment reference used at the first cycle is generally an authentic EIC image, the averge map computed from the whole population of aligned molecules is used as a second reference for the second alignment cycle, the average of the second cycle for the third cycle, and so on. This iterative method reduces the bias introduced in the alignment by the first reference image.

Selection of Homogeneous Image Subsets by Correspondence Analysis. Averaging aligned images of isolated molecules reinforces significant de-

[25] W. O. Saxton and J. Frank, *Ultramicroscopy* **2,** 219 (1977).

[26] J. Frank, W. Goldfarb, D. Eisenberg, and T. S. Baker, *Ultramicroscopy* **3,** 283 (1978).

[27] M. Kessel, J. Frank, and W. Goldfarb, *J. Supramol. Struct.* **14,** 405 (1980).

tails only if the image series is homogenous with respect to the considered details. Therefore, the critical step of the procedure is the selection of homogeneous subsets from the whole image propulation. For this purpose, we use correspondence analysis (CORAN), a multifactorial statistical method applied for the first time to image processing of macromolecules.[4] For the principle and applications of the method the reader is referred to the books by Benzecri[28] and Lebart *et al.*,[29] and to review articles on this subject.[21,22] The following section is limited to the aspects specific to immunocomplexes.

1. At the end of the alignment step, when all the images appear in the same orientation as the reference, the size of the images can generally be cut down without loss of information. In the case of MAb L102-containing EICs, the size of the images can be reduced from 128×128 to 100×100 pixels. This operation substantially decreases the computation time required for the succeeding steps.
2. In general, the various immunocomplex images processed in one pass with CORAN cannot be extracted from a single micrograph. Since the mean optical density of the micrographs included in the analysis may vary over a broad range, it is necessary to rescale each image according to the micrograph zone from which it originates. Although rescaling of the images is not absolutely required by CORAN, it is desirable for the purpose of forming averages over selected subsets, so that all images are represented with equal weight. For a given image, the mean optical density of all pixels is first computed. Subsequently, the optical density of each pixel is divided by the mean optical density, resulting in a corrected optical density centered on a value of 1.0.
3. In order to increase the signal-to-noise ratio of the images, and thus to enhance signal-related clustering where present, low-pass filtration is applied using a Gaussian function with a radius dependent on the image resolution. Subsequently, the images are undersampled into arrays of 64×64 without loss of information.
4. In order to reduce the influence on the classification of the parts of the image external to the molecule, a mask is imposed on the images prior to CORAN. The definition of this mask is important. In the case of MAb L102-containing EICs a rectangular mask hiding the Fabs is used. The reason for this choice is that the separation is expected to be based only on the varying features of the antigen

[28] J. P. Benzecri, *Cah. Anal. Donnêes.* **1,** 1 (1976).

[29] L. Lebart, A. Morineau, and K. M. Warwick, "Multivariate Descriptive Statistical Analysis," p. 231. Wiley, New York, 1984.

molecule (flip and flop faces, rocking effect, etc.), two features that are usually associated with the first two factorial axes.

5. CORAN carried out according to Van Heel and Frank[4] and Frank *et al.*[30] produces three types of output: (1) a map corresponding to the projection of the data cloud onto planes spanned by the factors with the highest contributions to the total interimage variance, (2) a histogram of eigenvalues associated with the highest ranking factors, (3) a set of importance images or of reconstituted images used to identify the pattern of variation expressed by each factor.

Figure 4 shows the projection of the data cloud on the plane defined by factors 1 and 2. As expected, the cloud is divided into four well-defined image clusters.

Interpretation of Factorial Axes. Several methods can be used to identify the regions of the images associated with a given factor involved in the separation.

1. One can compare the average maps of image groups falling on the factorial axis. The image regions that progressively vary between extreme negative and extreme positive values of the factor are likely to be associated with the factor.

2. Importance images[31,32] are based on the fact that, in addition to the images, the pixels can be projected on the CORAN map, and the pixels with extreme coordinates are more involved in the separation than those projecting near the origin. For each factorial axis, a pair of importance images showing the pixels with negative and positive values is calculated. In those images, each pixel appears with a color code depending on its eigenvalue (those projecting near the origin have low color codes, whereas those with high positive or negative values have high codes).

3. Reconstituted images[33] are based on a partial reconstitution of the contingency table from the image and pixel coordinates on the most significant factorial axes (we use the first eight axes). The image with null coordinates on all the axes is the average map of the whole image set. A reconstituted image has null coordinates on all axes except on the axis to be considered, where the coordinates are extreme. These images, which provide essentially the same type of information as the importance images, are easier to interpret because of their resemblance to average maps.

[30] J. Frank, A. Verschoor, and M. Boublick, *J. Mol. Biol.* **161,** 107 (1982).

[31] M. Van Heel and J. Frank, *in* "Proceedings of the 7th European Congress of Electron Microscopy" (P. Brederoo and W. de Priester, eds.), Vol. 2, p. 692. 7th Eur. Congr. on E. M. Foundation, Leiden, The Netherlands, 1980.

[32] M. Van Heel, P.h.D. Thesis, University of Groningen, Groningen, The Netherlands, 1981.

[33] J. P. Bretaudière and J. Frank, *J. Microsc.* **144,** 1 (1986).

U

V

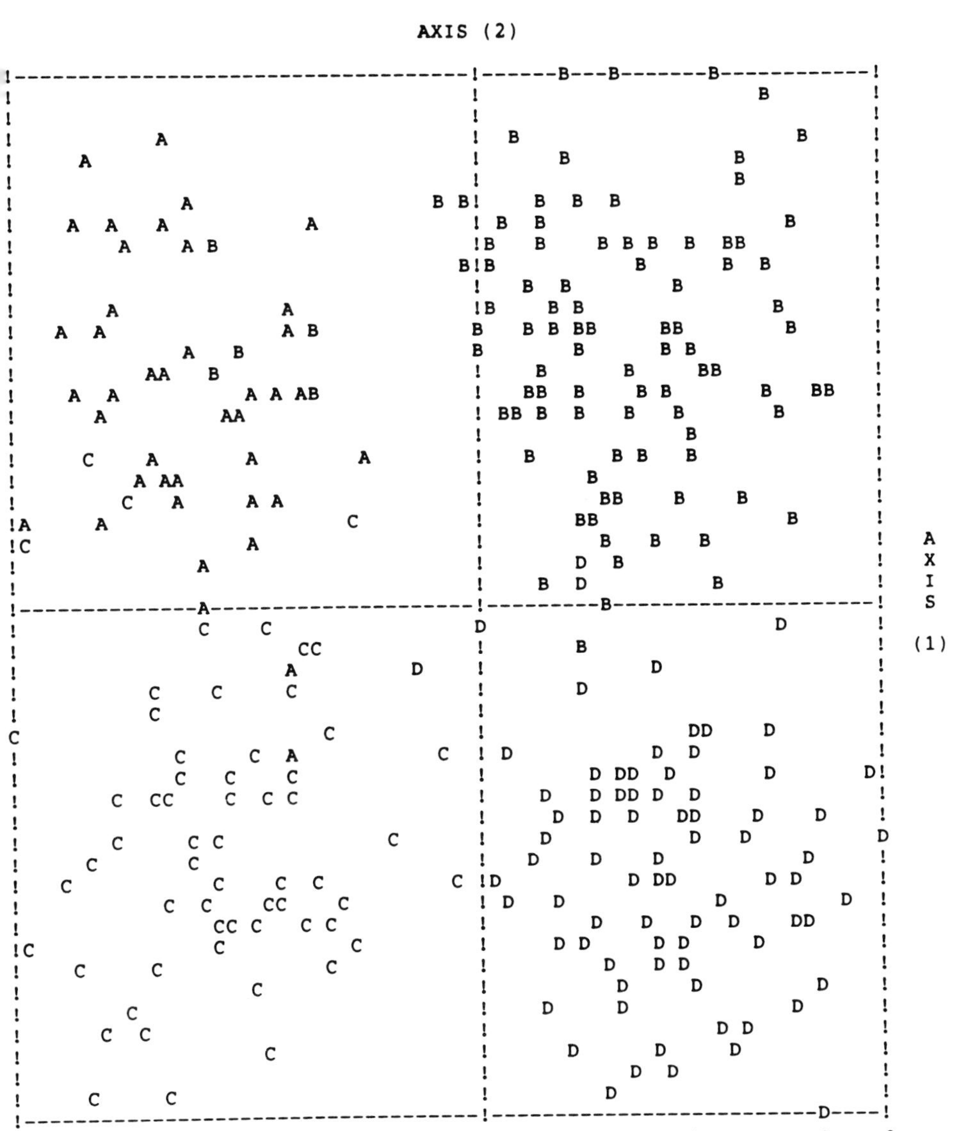

FIG. 4. CORAN map showing the separation of the elementary immunocomplexes of MAb L102 incubated with whole hemocyanin. (A–D) The four main clusters separated by hierarchical ascendant classification. [Reproduced with permission from J. Lamy, P. Billiald, J. C. Taveau, N. Boisset, G. Motta, and J. N. Lamy, *J. Struct. Biol.* **103,** 64 (1990).]

Clustering. When the data cloud is composed of well-defined subclouds, the projection on the map defined by the first two factorial axes produces several clusters of points that are easily distinguished from each other. This situation was encountered in the original work of Van Heel and Frank[4] with the four clusters produced by the top view of *Limulus polyphemus* hemocyanin and in the separation of L102-containing EICs.

It is not uncommon that the separation of the cloud is incomplete, so that the variation is somewhat continuous even along the first axes. In this case the centers of mass of the subclouds are distinct and situated on two opposite parts of the axis, but with no clear separation between them. To avoid a subjective definition of the image classes, an automatic method of classification must be used. The application of automatic classifications to single-particle image processing has been reviewed.[34]

The first application of automatic classification to images by Van Heel[23] was based on hierarchical ascendant classification (HAC). Since this first approach, various methods based on HAC, partitional classification,[35] or the clustering method around moving centers, have been proposed.[36]

For immunocomplexes we use HAC as a routine test. The principle of the method is an iterative aggregation of the points representing the images as a function of their coordinates on the main factorial axes of CORAN. The hierarchy of the aggregation is represented by a tree, where the most similar images are aggregated first and appear in the lower part of the tree, while the most different groups are aggregated only at the end of the classification. By cutting the tree at various heights one can individualize several subpopulations. Based on this general principle, several classification methods differing by the calculation of the distances and the aggregation criterion are possible. We use the reciprocal neighbors-chain search algorithm,[37] along with the aggregation cirterion of Ward.[38] The main inconvenience of this algorithm is that it always separates several classes even if the distribution is homogeneous. Therefore, several classification procedures should be used with the same data.

Average and Variance Maps. In the average and variance maps each pixel is equal to the arithmetic mean and the variance of that pixel in the images composing the subset. To prevent misinterpretation resulting from statistical fluctuations in the pixel values in the Fab areas, the reproducibil-

[34] J. Frank, *Q. Rev. Biophys.* **23,** 281 (1990).

[35] J. M. Carazo, F. F. Rivera, E. L. Zapata, M. Radermacher, and J. Frank, *J. Microsc.* **157,** 187 (1990).

[36] J. Frank, J. P. Brentaudière, J. M. Carazo, A. Verschoor, T. Wagenknecht, *J. Microsc.* **150,** 99 (1988).

[37] J. P. Benzecri, *Cah, Anal. Donnèes* **7,** 209 (1982).

[38] J. H. Ward, *J. Am. Stat. Assoc.* **58,** 236 (1963).

ity of the average and variance images is evaluated separately for each subset as follows. First, the subset is further divided, by random selection, into two groups of equal size. Average images are then computed for each group and compared. Based on the result of this comparison, the average of the complete subset is low-pass filtered for final presentation using the Fermi filter function in Fourier space.[21] When the number of images composing the subset is small (less than 25), a standard filtration radius of 0.125 nm^{-1} and a "temperature" parameter value of 0.0625 nm^{-1} is used for each group average, and the overall resemblance in characteristic regions of the molecule projection is visually confirmed. When the number of images is sufficient (≥ 25) the reproducible resolution of the subset averages is quantitatively determined by phase residual analysis,[39] and the corresponding Fourier radius is used in the Fermi filtration of the complete subset averages. The same filtration is applied to the variance map. A detailed description of the method is given in Frank *et al.*[20] Other methods have been proposed to compute the resolution of the averge maps, using (1) the Fourier ring correlation criterion,[5,40] and (2) the spectral signal-to-noise ratio.[41]

Figure 5 shows the average maps corresponding to the four clusters of the L102-containing EICs. A comparison of these images shows that the importance of the stain exclusion produced by a Fab depends on the corner of the molecule to which the Fab is bound, and on the rocking effect. Acutally, when a Fab is linked to the top right-hand (Figs. 5A and B) or the bottom left-hand corner (Fig. 5C and D), the hexamer to which it is bound excludes the stain more strongly than does the diagonally opposed hexamer. In other words the presence of a Fab arm forces the rocking. In addition, the stain exclusion due to the Fab arms bound to rocking (top right and bottom left) hexamers is slightly stronger than that of Fab arms bound to nonrocking hexamers.

Interpretation. There is no doubt that the most difficult step of image processing is the interpretation of the average maps in terms of structure. In the case of negatively stained EICs, the selection of homogeneous clusters by CORAN provides (1) a precise determination (in projection) of the contact point between the Fab arm and the antigen (2) a qualitative identification of the portions of the molecule located high above the support by the presence of the stain menisci, and (3) a qualitative determination of the Fab orientation in the stain layer.

Sometimes the simplest features can be visually interpreted, but in

[39] J. Frank, A. Verschoor, and M. Boublik, *Science* **214,** 1352 (1981).

[40] W. O. Saxton and W. Baumeister, *J. Microsc.* **127,** 127 (1982).

[41] M. Unser, B. L. Trus, and A. C. Steven, *Ultramicroscopy* **23,** 39 (1987).

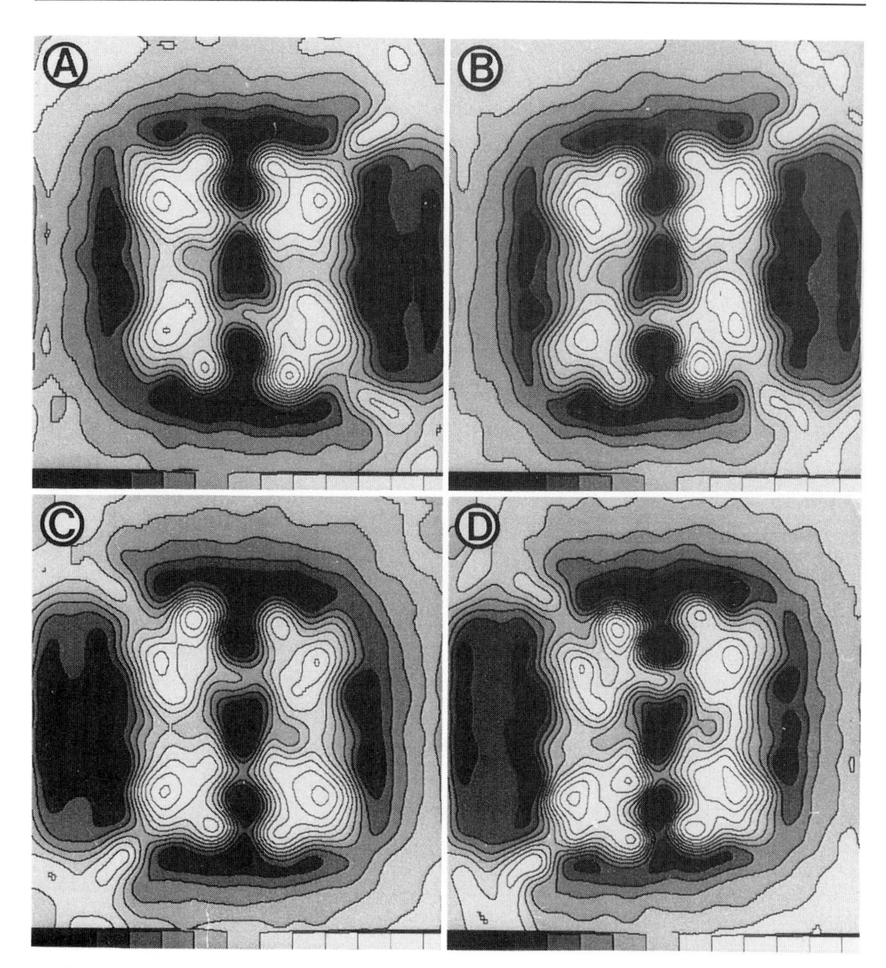

FIG. 5. Average images of the four main clusters detected by hierarchial ascendant classification in the CORAN separation of elementary immunocomplexes of L102 and *Androctonus* hemocyanin. The average maps of Fig. 5A through D correspond to the clusters of Fig. 4A through D, respectively. The scale bar is 10 nm. [Reproduced with permission from J. Lamy, P. Billiald, J. C. Taveau, N. Boisset, G. Motta, and J. N. Lamy, *J. Struct. Biol.* **103,** 64 (1990).]

most cases specific tools are required to understand three-dimensional (3D) structural details from two-dimensional projections. In the case of immunocomplex strings, particularly when the antigen contains several copies of the epitope, the architecture is so complex that 3D models are absolutely required. Other tools, such as the simulation of EM images obtained from 3D models and the comparison of simulated images to authentic images, may also help in the validation of a model. For this purpose we devised a software called SIGMA (software of imaging and graphics in molecular architecture), which contains programs of 3D construction (specifically immunocomplexes), visualization, stain simulation, etc., applied to the interpretation of molecular EM images.

Figure 6 shows an example of a 3D model of MAb L102-containing EIC, calculated by the SIGMA software from the results of CORAN, determined by minimizing the torsion around the Fab arm and the flexions in the hinge and elbow-bend areas of the IgG molecule. Comparison of the model with authentic images permits checking that the topological location deduced from the image processing of the EICs is compatible with the architecture of the whole immunocomplex string.

Immunological Approach

The topological localization of the epitopes of 13 MAbs specific for the *Aa*6 subunit produced two main patterns of Fab binding to hemocyamin, suggesting that most of the epitopes are located in domains 1 and 3 of the

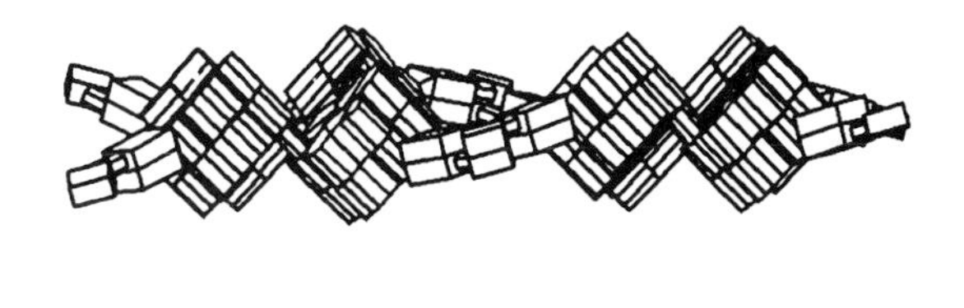

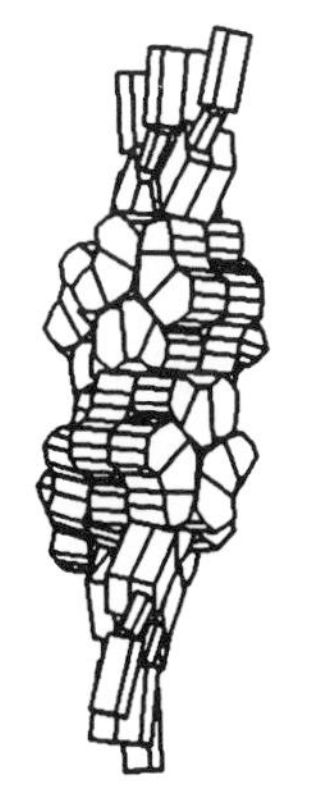

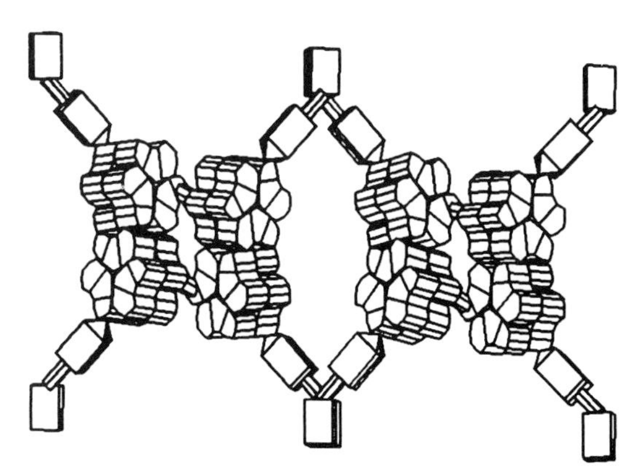

FIG. 6. Three perpendicular views of a model of immunocomplex strings containing MAb L102 and native *Androctonus* hemocyanin, computed from the data of Fig. 5 by the SIGMA program. Compare with the structure shown in Fig. 3C.

subunit. An immunological mapping was carried out in order to check this hypothesis.

The method is based on the fact that it is impossible to bind two different MAb molecules on the same antigen if their epitopes overlap.

1. The antigen (free hemocyanin subunit) is incubated with a mixture of the two MAbs, one of which (MAb_2) is labeled (for example, labeled with biotin).

2. The identification of ternary immunocomplexes (MAb_1 – antigen – MAb_2) is then based on their characteristic shapes (such as rings or strings in EM) sizes (i.e., larger than those of the corresponding MAb_1 – antigen or MAb_2 – antigen in polyacrylamide gel electrophoresis), and identification of the labeled MAbs in the immunocomplex [3] by a classical test of immunocompetition.

When present, a ternary immunocomplex definitely excludes the possibility of an overlap of the two epitopes. However, with this method overlap is obtained by default, since only nonoverlapping epitopes can be detected. For example, Fig. 7 shows a gallery of immunocomplexes obtained with MAbs L8 and 5701. The IgG molecules (anti-L8 and anti-5701) bound to

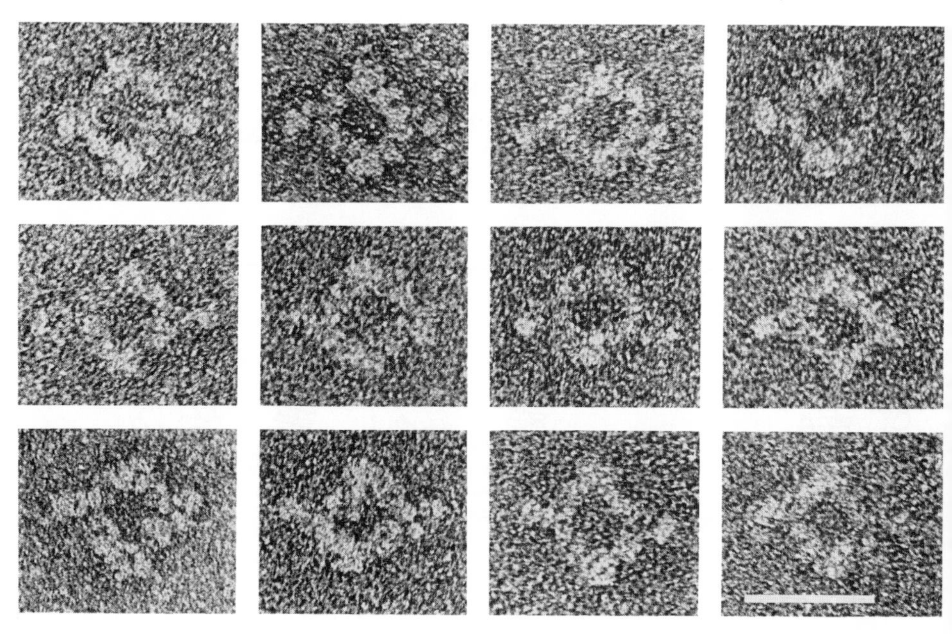

FIG. 7. Electron microscope views of ternary immunocomplexes composed of free *Aa*6 subunit and two different MAbs (5701 and L8), demonstrating an absence of overlap between the corresponding epitopes. The scale bar is 25 nm. [Reproduced with permission from N. Boisset, J. Frank, J. C. Taveau, P. Billiald, G. Motta, J. Lamy, P. Y. Sizaret, and J. Lamy, *Proteins* **3**, 161 (1988).]

the two copies of the *Aa*6 subunit are clearly visible and demonstrate the absence of overlap between the two epitopes.

The results of the immunological mapping are in perfect agreement with those of IEM and image processing. The 13 epitopes are distributed into 4 groups containing 1, 6, 4, and 2 epitopes, respectively. The fourth group (two epitopes) is accessible only in the free subunit, and the other three groups have different topological locations in the areas of domain 3 (first and second groups), and domain 1 (third group), respectively, of the *Aa*6 subunit.[3]

Thus the combination of immunological and topological mappings considerably improves the reliability of epitope localization.

Concluding Remarks

In the near future, 3D reconstruction of immunocomplex EM views will probably simplify direct epitope localization. In addition, the preliminary results obtained with specimens embedded in vitreous ice[18] suggest that the negative stain could be replaced by frozen hydrated specimens. However, simple negative staining combined with a well-standardized method of image processing produces average images with a resolution of 3–4 nm, a very reasonable result at present. Finally, topological location of epitopes refined by immunological mapping produces an epitope location that is surpassed only by the crystallographic determination of the structure of Fab-containing immunocomplexes by X-ray diffraction.

Acknowlegments

We thank all current and previous members of this laboratory for their skillfull contributions, Prof. J. Frank for many stimulating and helpful discussions on image processing during a long-term collaboration, and Prof. S. Vinogradov for reading and correcting this manuscript.

[14] Characterization of Antigenic Structures by Mapping on Resin-Bound Epitope Analogs

By WAN-JR SYU and LAWRENCE KAHAN

Epitope characterization is a sometimes tedious necessity for obtaining an understanding of the interactions between protein antigens and antibodies. Common approaches rely on differences in antibody binding to homologous antigens, mutagenized proteins, and synthetic epitope analogs.[1-3] These approaches offer differing advantages. The synthetic peptide approach offers both efficiency and versatility. Methods for simultaneous synthesis of a large number of peptides have been developed. One method follows the procedure of standard solid-phase peptide synthesis with the resin packed into mesh bags.[4] After synthesis, the free peptides are coated on a solid support and tested for antibody binding. A second approach involves peptide synthesis on solid supports followed by determination of antibody binding. This is accomplished by derivatizing solid supports with acrylic acid and use of a device formatted to accommodate multiple rods during simultaneous synthesis.[5] Although both approaches are useful in simultaneously generating many epitope analogs, they both require hardware that differs from that used in standard peptide synthesis. We have developed a method for epitope mapping, using standard peptide synthesis machinery to synthesize a series of peptides on solid supports, that may be used directly for the assay of antibody binding.

Principle

In standard solid-phase synthesis of a free peptide, the peptide is linked to the support resin by an ester bond.[6] This linkage is relatively labile and is destined to be cleaved by acidolysis after synthesis. Since we intend to use the resin that carries the synthetic peptides as the solid support for antibody binding, a more stable linkage is required. An amide linkage chemically similar to the peptide bond is stable to the deprotection condition and serves this purpose. Aminomethyl-cross-linked polystyrene resin provides the necessary linkage when reacted with the carboxyl group of the first (C-terminal) amino acid added to the resin. In the standard solid-phase synthesis, peptides are lengthened by one amino acid per step at the N terminus; sampling of the resin beads after each coupling cycle generates a series of epitope analogs of increasing length. Since these epitope analogs are resin bound, they can be easily manipulated and readily regenerated for the epitope mapping of multiple antibodies.

[1] R. Jemmerson and Y. Paterson, *BioTechniques* **4**, 18 (1985).

[2] Y. Paterson, *Biochemistry* **24**, 1048 (1985).

[3] L. A. Lasky, G. Nakamura, D. H. Smith, C. Fennie, C. Shimasaki, E. Patzer, P. Berman, T. Gregory, and D. J. Capon, *Cell (Cambridge, Mass.)* **50**, 975 (1987).

[4] R. A. Houghten, *Proc. Natl. Acad. Sci. U.S.A.* **82**, 5131 (1985).

[5] H. M. Geysen, R. H. Meloen, and S. J. Barteling, *Proc. Natl. Acad. Sci. U.S.A.* **81**, 3998 (1984).

[6] J. M. Stewart and J. D. Young, "Solid Phase Peptide Synthesis," 2nd Ed., p. 9. Pierce, Rockford, Illinois, 1984.

Method

Preparation of Resin-Bound Epitope Analogs

All materials and chemicals may be obtained from commercial vendors and used without further purification. Aminomethyl-cross-linked polystyrene resin is the precursor used for preparation of *N-tert*-butyloxycarbonyl (Boc)-aminoacyl phenylacetamidomethyl (Pam) resin.[6] To the aminomethyl resin (1 mmol equivalent), the first Boc-amino acid (3 mmol) is added, together with 3 mmol of the coupling reagent, *N,N'*-dicyclohexylcarbodiimide. For reaction, any vessel that can be tightly capped (to exclude atmospheric moisture) and is resistant to the reaction reagents can be used. Since these resin beads tend to stick to glass more than to polypropylene, a 50-ml polypropylene conical tube is an ideal reaction vessel. The reaction is carried out by agitating in a total volume of 10 ml dichloromethane at room temperature. Agitation can be done either by shaker or rocking table. After a 1- to 2-hr agitation, the resin beads are collected on a sintered glass filter. The completeness of the coupling reaction is monitored by the ninhydrin reaction, using a few sampled resin beads.[7] If a high degree of substitution is desired, the resin beads may be resuspended in fresh reagents and the coupling reaction repeated. The remaining unreacted amino groups on the resin should be blocked by rocking the resin in 15 ml dichloromethane containing 0.5 ml acetic anhydride and 0.7 ml triethylamine for 30 min, or until the ninhydrin test is negative. The resin is then thoroughly washed with dichloromethane and dried in a desiccator.

Comments. To reduce the possible steric hindrance by the resin, Boc-4-aminobutyric acid may be used as a spacer and coupled to the aminomethyl resin.

Approximately 1 g of aminomethyl resin containing 0.5 mmol of the first Boc-amino acid derivative is subjected to sequential addition of residues using an automatic peptide synthesizer. Sampling of resin beads can be automatically carried out by synthesizers such as the Applied Biosystems model 438 (Foster City, CA). If automatic sampling is not available, the resin should be sampled manually after each Boc-amino acid has been coupled.

A nearly quantitative yield in each reaction step can usually be achieved. Synthesis of 0.5 mmol of 20-mer peptide will generate about 1100 mg final products, which is far more than needed for assay of antibody binding. Sampling 1–2% of the resin after each cycle of Boc-amino

[7] V. K. Sarin, S. B. H. Kent, J. P. Tam, and R. B. Merrifield, *Anal. Biochem.* **117**, 147 (1981).

acid addition removes a total of 0.1–0.2 mmol of different resin-bound peptides; the resin left at the final stage still carries 0.3–0.4 mmol of the longest peptide. Thus, the scale of synthesis may be reduced if desired. Additionally, if peptides longer than those synthesized are needed after preliminary tests, the resin-bound Boc-peptide derivatives could be subjected to reactions for further extension of the necessary residues.

The resin beads collected after each residue addition are put into 1.5-ml polypropylene micro test tubes. To remove the side-chain protecting groups and the last Boc group on the peptides, the resin (2.0 mg) is treated with a large excess (0.33 ml) of 0.5 *M* boron tris(trifluoroacetate)[8] in trifluoroacetic acid (Pierce, Rockford, IL). The tubes are tightly capped and rocked and the reactions are allowed to proceed for 90 min at room temperature. Anhydrous methanol (1 ml) is then added to each tube; resin beads are spun down by centrifugation in a microcentrifuge for (11,600 rpm, 1 min, 22°). After the supernatant is withdrawn, the beads are washed twice more with methanol and air dried.

Boron tris(trifluoroacetate) does not remove the protecting groups as universally as liquid HF cleavage does. Stable protecting groups created for synthesis of long peptides may be resistant to deprotection with boron tris(trifluoroacetate). Examples are the 2-chlorobenzyloxycarbonyl group for the ε-amino group of Lys and the 2-bromobenzyloxycarbonyl group for the side chain of Tyr. These are hardly removed by boron tris(trifluoroacetate); they may be replaced by the more easily removed carbobenzoxyl and *o*-2,6-dichlorobenzyl groups, respectively.

Reaction with Antibodies

Binding of antibodies to the resin-bound epitope analogs is carried out in the micro test tubes without transfer of the resin. The peptide-containing resin beads are blocked with 10% (v/v) calf serum, 10% (w/v) bovine serum albumin, 1% (v/v) Tween 20 (Sigma, St. Louis, MO) in Tris/NaCl (0.02 m*M* Tris-HCl, 0.5 *M* NaCl, pH 7.5) for 1 hr at room temperature. Purified antibodies diluted in the preincubation mixture (above solution) are added to individual tubes (75 μl/tube) and incubated overnight at 4°. The peptide–resin beads are collected by centrifugation and the antibody-containing supernatant is aspirated. The beads are washed five times with Tris/NaCl containing 0.05% (v/v) Tween 20. Alkaline phosphatase-conjugated goat anti-mouse IgG diluted to 1 μg/ml in the blocking solution is added to individual tubes and incubated at room temperature for 2 hr. The excess conjugate is removed by washing five times with Tris/NaCl containing 0.05% (v/v) Tween 20. The immunocomplexes bound to the pep-

[8] J. Pless and W. Bauer, *Angew. Chem.* **85**, 142 (1973).

tide resin are then detected by reaction for 90 min with alkaline phosphatase substrate, 300 μl *p*-nitrophenyl phosphate (1.2 mg/ml) in 1 *M* 2-amino-2-methyl-1-propanol, 1.0 m*M* $MgCl_2$, 0.02 m*M* $ZnCl_2$, pH 10.3. The resin is pelleted and 200 μl of the reaction solution is transferred to the wells of 96-well plates; absorbance at 405 nm is measured with a 96-well plate reader, such as model EL310 from BioTek Instruments (Burlington, VT).

Regeneration of Peptide–Resin Beads

To regenerate the resin-bound epitope analogs by dissociating the bound immunocomplexes, the resin beads are soaked twice for 10 min with 8 *M* urea containing 0.1% (w/v) sodium dodecyl sulfate and 0.1% (v/v) 2-mercaptoethanol, followed by extensive washing with Tris/NaCl–Tween 20 buffer.

Comments. *p*-Nitrophenol, the yellowish product of the alkaline phosphatase reaction, is soluble in the above buffers and is removed during the regeneration. However, a very faint yellow color remains on highly reactive peptide resins; this presumably resin-trapped color does not affect reuse of the resin.

Example of Method

We have been studying monoclonal antibodies as probes of the structure of the *Escherichia coli* ribosome. Three monoclonal antibodies (MAbs 19, 20, and 21) bind to the C terminus approximately 34 residues of ribosomal protein S13. In an attempt to determine the differences between these three MAbs and to limit the epitopes defined by each, we synthesized a series of resin-bound epitope analogs of a 23-residue peptide (PVRGQRTKTNARTRKGPRKPIKK),[8A] based on other complementary data.[9] After removal of the protecting groups, these resin-bound peptides were tested for binding of MAbs 19, 20, 21, and a control antibody, MAb 22, which is directing to the N-terminal epitope of S13. Binding of MAbs 19, 20, and 21 increased when Arg was incorporated as the twenty-first residue of the S13 peptide, whereas that of the control MAb 22 remained unchanged (Fig. 1). The interactions of these MAbs with peptides of lengths greater than 18 residues were examined and results are

[8a] Single-letter abbreviations for amino acids: A, alanine; R, arginine; N, asparagine; D, aspartic acid; C, cysteine; Q, glutamine; E, glutamic acid; G, glycine; H, histidine; I, isoleucine; L, leucine; K, lysine; M, methionine; F, phenylalanine; P, proline; S, serine; T, threonine; W, tryptophan; Y, tyrosine; V, valine.

[9] W.-J. Syu and L. Kahan, *J. Immunol. Methods* **118**, 153 (1989).

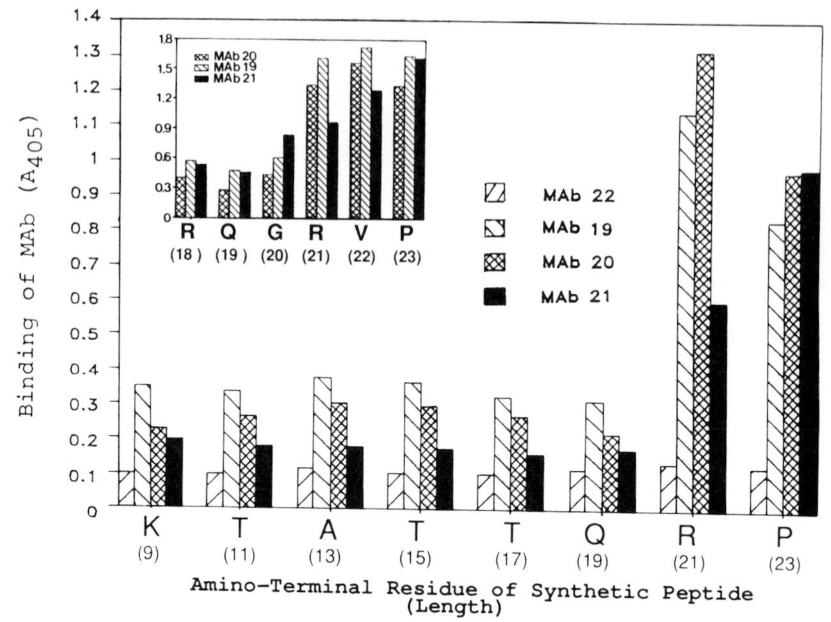

FIG. 1. Binding of MAbs to a set of resin-bound epitope segments of a 23-residue S13 peptide (PVRGQRTKTNARTRKGPRKPIKK).[9] Consecutive peptide segments differ in length by two residues, as indicated on the abscissa. Binding of MAb to resin-bound peptide was determined by enzyme immunoassay. The *p*-nitrophenol product generated in the immunoassay was monitored by absorbance at 405 nm. The inset illustrates the binding of the MAbs to a series of peptides differing by 1 residue in length, beginning with the 18 C-terminal residues of the above peptide.

shown in the inset in Fig. 1. Binding of MAbs 19 and 20 increased when Arg-21 was added to the peptide sequence, indicating the critical contribution of this Arg to this epitope. Addition of two more residues did not further increase the bindings of MAbs 19 and 20, suggesting that the C-terminal 21 residues of S13 contain all the amino acids involved in the interaction with MAbs 19 and 20. In contrast, MAb 21 not only bound to the peptide of 20 residues but bound more strongly when the sequence was lengthened. These results indicated that the sequence PVRG affects the structure of the epitope interacting with MAb 21. The epitope bound by MAb 21, therefore, is affected by at least two additional residues than that recognized by MAbs 19 and 20.

[15] Epitope Mapping of Allergens for Rapid Localization of Continuous Allergenic Determinants

By BRADLEY J. WALSH and MERLIN E. HOWDEN

Introduction

An antigen acts as an allergen when it stimulates receptors on B lymphocytes, causing them to differentiate and form mast cells, which in turn manufacture and secrete immunoglobulin E (IgE) specific to that allergen. Study of the binding between allergens and IgE is most important, as this interaction is the trigger for all type I (immediate) allergic reactions. Some of the symptoms of such reactions are asthma, urticaria, rhinitis, eczema, and conjunctivitis. Allergens occur in many different sources, such as dusts (e.g., grass pollens, animal danders, and dust mites), where the route of sensitization is by inhalation, and foods (e.g., milk, egg, fish, and wheat), where it is by ingestion. Allergen-specific IgE is usually measured by the radioallergosorbent test (RAST), which reacts patient serum samples with allergen bound to a solid-phase support. Any such IgE is then detected by reaction with ^{125}I-labeled anti-IgE. Results are expressed as a percentage ratio of radioactive label bound to label added.

The term allergenic epitope or allergenic determinant is used to denote a peptide sequence from a protein allergen that binds specific IgE. Such peptide allergenic determinants may be prepared by enzymatic or chemical degradation of the parent protein or by solid-phase peptide synthetic methods.[1] Allergenic determinants may be of two types: continuous or conformational. A continuous determinant refers to a linear amino acid sequence in the protein chain. A conformational allergenic determinant is formed from residues that are in close proximity due to the three-dimensional structure of the protein.[2] Thus a continuous allergenic determinant will endure after removal of tertiary structure, for example by reduction and carboxymethylation of cysteine residues, but a conformational allergenic determinant will not. Our experience is that continuous determinants appear to be important in binding to IgE. For example, reduction and carboxymethylation of a cow milk allergen (β-lactoglobulin, a dimer with a single disulfide bridge) resulted in IgE binding that was as high for the monomer as for the native protein.[3] Reduction and carboxymethylation of an allergen from wheat flour (the 0.28 α-amylase inhibitor,

[1] H. Atassi and M. Z. Atassi, *FEBS Lett.* **188,** 96 (1985).
[2] M. Z. Atassi and J. A. Smith, *Immunochemistry* **15,** 609 (1978).
[3] S. L. Adams, B. J. Walsh, and M. E. H. Howden, unpublished results (1987).

0.28 AI) also resulted in as high binding of specific IgE as that recorded for the native protein.[4] These results are consistent with the notion of linear determinants binding IgE from serum samples.

Only two allergens have been studied in the detail required to provide information on their IgE-binding sites. These are allergen M from cod and the ragweed pollen allergen Ra3. Allergen M is a glycoprotein of 113 amino acid residues and contains 1 glucose moiety.[5] The molecule is divided into three homologous domains called AB, CD, and EF. As denaturation does not affect the allergenicity of the native molecule, it is presumed to contain continuous IgE-binding sites. This was confirmed by cleaving the molecule into two fragments, both of which were antigenic[6] and allergenic.[7] Further study localized the allergenic determinants to four distinct sequences. The first allergenic determinant was residues 13–32 in the AB loop, which, it was concluded, functions as a divalent allergenic determinant.[8] The second and third were residues 33–44 and 65–74, which join the AB to CD and CD to EF loops, respectively. These allergenic determinants also bind IgE *in vitro* and have sequence homology in the regions 35–41 and 67–73.[9] The fourth allergenic determinant was residues 41–64 in the CD loop.[10]

Atassi and Atassi[1] identified continuous allergenic determinants on ragweed pollen allergen Ra3 by the use of solid-phase peptide synthesis. Ten overlapping pentadecapeptides spanning the whole sequence were synthesized. These were then tested separately for their ability to bind IgE in a pooled serum from patients allergic to ragweed pollen. The peptides 1–15, 21–35, 31–45, 51–65, and 71–85 bound specific IgE to varying degrees. It was suggested by the authors that the allergenic determinants fell within the regions encompassed by the synthetic peptides and that the allergenic determinants may be only part of the sequence of the synthetic peptide in each case.

The major drawbacks of the work on both allergen M and Ra3 were that tedious trial-and-error methods were used to localize the determinants

[4] B. J. Walsh, Ph. D. Thesis, "The Molecular Basis of Allergenicity and Cross-Allergenicity in Certain Wheat Proteins." Macquarie University, Sydney, 1988.

[5] S. Elsayed and H. Bennich, *Scand. J. Immunol.* **4,** 203 (1975).

[6] J. Apold, S. Elsayed, L. Aukrust, and H. Bennich, *Int. Arch. Allergy Appl. Immunol.* **58,** 337 (1979).

[7] S. Elsayed, J. Apold, K. Aas, and H. Bennich, *Int. Arch. Allergy Appl. Immunol.* **52,** 59 (1976); S. Elsayed and J. Apold, *Int. Arch. Allergy Appl. Immunol.* **54,** 171 (1977).

[8] S. Elsayed, U. Ragnarsson, and B. Nettland, *Scand. J. Immunol.* **17,** 291 (1983).

[9] S. Elsayed and J. Apold, *Allergy* **38,** 449 (1983).

[10] S. Elsayed, S. Sornes, J. Apold, K. Vik, and E. Floorvag, *Scand. J. Immunol.* **16,** 77 (1982).

and that the RAST and RAST inhibition studies were done only with pooled sera. This reduced the reliance on scarce individual sera. However, studies by Sutton *et al.*,[11] Tovey and Baldo,[12] and Ford *et al.*[13] indicated that there were considerable differences in the patterns of individual reactivities, in the various allergen sources examined by them. If this was so for allergen M and Ra3, then other allergenic determinants may exist that were not detected, since the serum pool may not contain IgE binding to them. This has major implications when considering an allergen or allergenic determinant, for use in standardization and immunotherapy.[12]

Epitope Mapping: Application to Localization of Allergenic Determinants

The need thus arose for a method that would allow the continuous allergenic determinants to be rapidly identified for allergic patients whose serum IgE bound to various purified allergens. A method was reported that allowed rapid localization of *antigenic* determinants (IgG binding) by using solid-phase peptide synthesis on an array of polypropylene pins.[14] The peptides synthesized are those corresponding to overlapping sequences covering the complete antigen being studied. The procedure is similar to that used for fluorenylmethoxycarbonyl (Fmoc) peptide synthesis with minor modifications. By using a direct enzyme-linked immunosorbent assay (ELISA) with enzyme-linked antisera the continuous antigenic determinants of the protein can be mapped.

Because of the small amount of IgE (range, 0–437 ng/ml) relative to IgG (range, 7–15 mg/ml) present in serum, an enzyme-linked anti-IgE was not suitable for direct probing of the pins, as the resulting signal would not be strong enough to distinguish it from background. We therefore developed a method using epitope mapping and employing an avidin–biotin system to increase the signal and thereby detect binding of allergen-specific IgE to the pins. Thus localization of continuous allergenic determinants only requires the sequence of the allergen to be known and samples of serum to be available from patients hypersensitive to that allergen. At least two sera can be probed per week to give a full map of IgE-binding peptides

[11] R. Sutton, J. H. Skerritt, B. A. Baldo, and C. W. Wrigley, *Clin. Allergy* **14,** 93 (1984).
[12] E. R. Tovey and B. A. Baldo, *Int. Arch. Allergy Appl. Immunol.* **75,** 322 (1984).
[13] S. A. Ford, E. R. Tovey, and B. A. Baldo, *Int. Arch. Allergy Appl. Immunol.* **78,** 15 (1985).
[14] H. M. Geysen, R. H. Meloen, and S. J. Barteling, *Proc. Natl. Acad. Sci. U.S.A.* **81,** 3998 (1984).

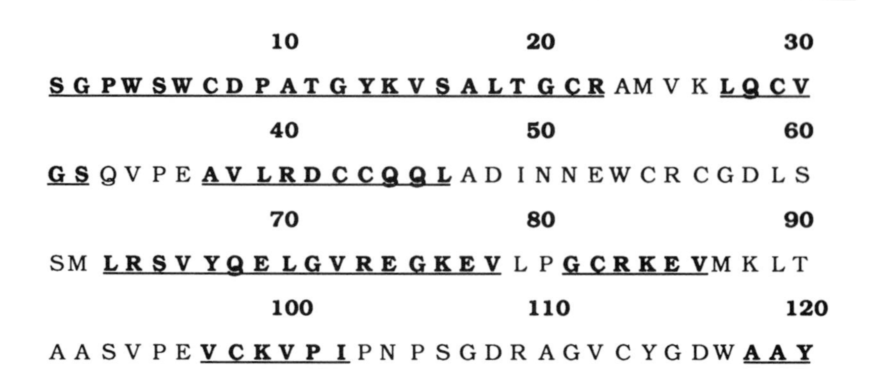

P D V

FIG. 1. Sequence of the wheat 0.28 AI.[16] Allergenic determinants selected by the epitope mapping method are shown in boldface and underlined.

for each individual. The peptide determinants selected by this method can then be synthesized on a larger scale and tested for their ability to inhibit IgE binding to the whole protein in the RAST.[15] The technique is cost effective, rapid, selective, and sensitive. A major advantage is that it is not necessary to purify large amounts of the allergen of interest, apart from that initially required for sequencing. If the DNA or protein sequence is known, no purification procedures are required at all, thus removing the need for an often long and arduous task.

The peptides produced on the pins in this work were from the sequence of an allergen of interest (the 0.28 AI).[16] To probe for the allergenic determinant(s), overlapping hexapeptides were produced. Synthesis began at the amino-terminal hexapeptide and then moved along, two amino acids at a time, to produce the next overlapping hexapeptide, and so on. This set of peptides, bound to pins, was probed with the sera of allergic patients, thus mapping all of their allergenic determinants (see Fig. 1).

Enzyme-Linked Immunosorbent Assay for Epitope Mapping

The principle of detecting IgE binding to the pins with biotin and avidin peroxidase is as follows[16]: IgE in serum is reacted with the peptides

[15] G. J. Gleich, J. B. Larsen, R. T. Jones, and H. Baer, *J. Allergy Clin. Immunol.* **53,** 158 (1974).

[16] B. J. Walsh and M. E. H. Howden, *J. Immunol. Methods* **121,** 275 (1989).

on the pins. The peptide–IgE complex is reacted with goat anti-human IgE. A third antibody, rabbit anti-goat IgG, is reacted with the peptide–IgE–anti-IgE complex. The rabbit anti-goat IgG is labeled with biotin. The final reagent that is bound to the complex is a peroxidase enzyme labeled with avidin. Avidin is a protein with four high-affinity binding sites for biotin. This high-affinity quarternary reagent binds tightly to each biotin molecule, resulting in a greatly increased sensitivity of the assay. The complex can then be detected spectrophotometrically following incubation with substrate, which gives a colored product on reaction with the enzyme.

Stock Solutions

Phosphate-buffered saline (PBS): 0.1 *M* phosphate and 0.15 *M* NaCl at pH 7.4

Wash buffer: 0.05% Tween 20 in PBS (w/v)

Dilution buffer: Used to dilute all reagents, consists of 2% bovine serum albumin and 0.05% Tween 20 (w/v in PBS)

Lyophilized goat anti-human IgE (Bio-Rad, Richmond, CA): Made up in dilution buffer at a concentration of 1 mg/ml

Biotinylated rabbit anti-goat IgG, avidin peroxidase, and *o*-phenylenediamine: From Sigma Chemical Co. (St. Louis, MO); the rabbit anti-goat IgG is supplied at a concentration of 0.6 mg/ml and the avidin peroxidase is made up by dissolving in dilution buffer at a concentration of 1 mg/ml

Note: All antibodies are aliquoted and stored at − 20°. Avidin peroxidase is stored at 4°.

The substrate reagent is made up just prior to use as 0.05% *o*-phenylenediamine (w/v), 0.01% H_2O_2 (v/v) in 0.1 *M* sodium citrate, 0.1 *M* Na_2HPO_4 (pH 5.0). The mixture is agitated and the reaction is allowed to proceed in the dark for 2 to 5 min. It is then stopped by removal of the pins and the addition of 0.05 ml 4 MH_2SO_4.

Microtiter plates are used (low-binding Immulon; Nunc, Denmark) to incubate the reagents with the pins. The plates and pins are blocked separately in dilution buffer for 1 hr at room temperature (0.2 ml/well). Following blocking, the plates are washed three times, 10 min each, with wash buffer. The dilution, time, and temperature of incubation for each antibody are given in Table I.

All antibody dilutions are made in dilution buffer. Following each incubation the plates are washed four times with wash buffer for 10 min/wash. For the incubations, 0.175 ml of each reagent is placed into the wells of a plate and the pins are lowered into it. The pins and plate are then

TABLE 1
ANTIBODY DILUTIONS FOR COMPONENTS USED IN IgE EPITOPE-MAPPING ELISA SYSTEM

Reagent	Dilution (v/v)	Incubation time	Temperature (°C)
Human IgE	1:7	22 hr	4
Goat anti-human IgE	1:2,000	20 hr	25
Biotinylated rabbit anti-goat IgG	1:10,000	2 hr	25
Avidin peroxidase	1:400	2 hr	25
o-Phenylenediamine and hydrogen peroxide	—	5 min	25

sealed in a polypropylene box for the duration of each incubation. A fresh plate is used with each solution. Between each addition of reagent the pins are washed four times with 300 ml of wash buffer for 10 min/wash. Substrate for the peroxidase enzyme is *o*-phenylenediamine (150 ul/well). The results are read at 492 nm in a Titertek Multiskan MCC plate reader (Flow Laboratories, McLean, VA).

Since any IgE binding nonspecifically to peptide on the pins is a possible source of error, the pins are also probed with a sample of serum

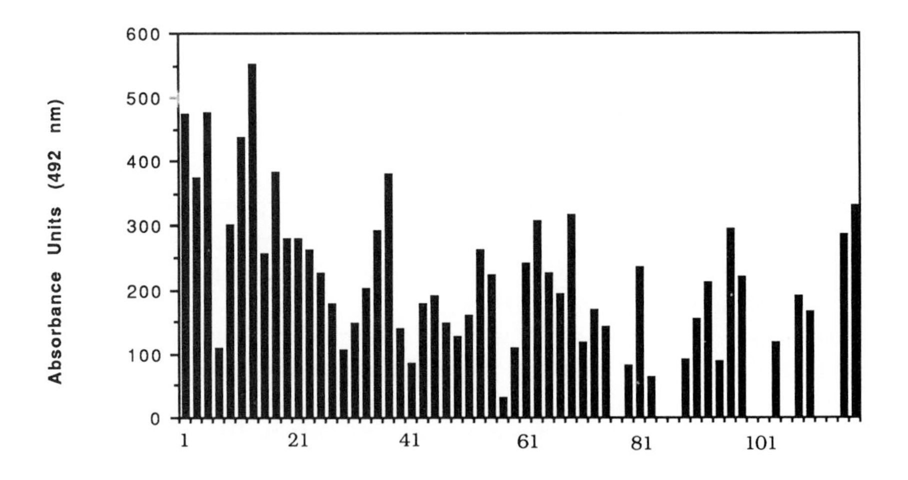

FIG. 2. Epitope maps for the 0.28 AI resulting from probing with IgE in the serum of patient Bro allergic to wheat. The results are expressed as absorbance of the product (× 1000) from the ELISA measured at 492 nm. Residue number refers to the number of the first residue in each hexapeptide with respect to the sequence of the allergen numbered from the N to C terminus.

containing 100 kilounits (kU)/liter of IgE nonspecific for wheat proteins. Results are corrected for nonspecific binding by subtracting the values obtained for this high-IgE serum test from the results obtained for sera from the patient. An example of the resultant epitope map for the 0.28 AI is given in Fig. 2. The map has been corrected for nonspecific binding. Table II gives the 10 highest peaks for two serum samples probed expressed as a percentage increase in absorbance over the background absorbance for each pin.

Procedure for Removal of Bound Antibodies from Pins

The blocks of pins are placed in polypropylene baths containing 300 ml 1% (w/v) sodium dodecyl sulfate (SDS), 0.1% (v/v) 2-mercaptoethanol, 0.1 *M* NaH_2PO_4 (pH 7.2) prewarmed to 60°. They are then sonicated for 30 min, removed, and fully immersed in water at 60°, three times. After the water washes the blocks are finally immersed in a bath of gently boiling methanol for 2 min before being air dried. After removal of bound antibodies by the disruption procedure, pins are probed with further sera in the same manner.

Preparation and Testing of Allergenic Determinant Peptides

Residues 9–25 from the 0.28 AI, which bound IgE in the epitope mapping experiments, are synthesized on a larger scale. These amino acid residues were selected because they are in the set of 10 highest peaks and they bind IgE in the sera of both of the individuals tested (Table II). Solid-phase peptide synthesis is performed essentially according to Atherton *et al.*,[17] either by using a simple, semiautomated instrument or a 9050 PepSynthesizer (MilliGen Division, Millipore, Burlington, MA). The sequences of the synthetic peptides are confirmed to be correct by gas-phase sequencing via Edman degradation and by amino acid analysis.

Coupling of Proteins to Disks

The 0.28 AI is coupled to nitrocellulose disks by the method of Walsh *et al.*[16] RAST buffer is made as follows: to PBS is added 3% bovine serum albumin (BSA) and 1% Tween 20 (w/v). Coupled and blocked nitrocellulose disks are stored dry at 4°.

[17] E. J. Atherton, C. J. Logan, and R. C. Sheppard, *J. Chem. Soc., Perkin Trans. 1* **38,** 538 (1981).

TABLE II
PERCENTAGE INCREASE OVER CONTROL FOR 10 HIGHEST PEAKS FROM 0.28 AI EPITOPE MAP[a]

Peptide sequence	Increase (%) over control for sera of patients[b]	
	Arm	Bro
SGPWSW	—	240
PWSWCD	—	341
SWCDPA	—	654
TGKYVS	—	395
YKVSAL	393	335
ALTGCR	312	247
LQCVGS	759	—
AVLRDC	—	1103
DCCQQL	675	—
LRSVYQ	280	186
ELGVRE	626	244
REGKEV	741	—
GCRKEV	400	—
VCKVPI	705	—
AAYPDV	880	—

[a] Above are given the 10 peaks with the highest absorbance at 492 nm for the results of epitope mapping using the sera from two subjects. They are listed in order of highest to lowest absorbance for each subject. The sequence for each hexapeptide is also given from N to C terminus. The percentage increase over background absorbance (i.e., the absorbance obtained with the control serum) for peptide is given for each serum.

[b] Arm and Bro are patient codes.

RAST

The RAST is employed to identify the presence of IgE binding to allergenic proteins. The ^{125}I-labeled anti-IgE is raised in rabbit and the IgG fraction is affinity purified. The antibody does not cross-react with other immunoglobulin classes and gives low levels of nonspecific binding. The antibody may be purchased from either Pharmacia Diagnostics (Uppsala, Sweden) or Kallestad Laboratories, Inc. (Austin, TX), and should be used within 1 month of receipt. The RAST method is performed according to the instructions of the manufacturer (Pharmacia Diagnostics, Uppsala, Sweden). Each disk is incubated with 0.05 ml of either fetal cord or test serum (diluted with PBS if necessary) in a 6-mm diameter tube for 3 hr at room temperature. The disks are washed three times in wash buffer (2.5 ml each, 10 min/wash). Then 0.05 ml ^{125}I-labeled anti-IgE is placed in each tube and allowed to react overnight. Unbound anti-IgE is removed with three more washes. The percentage of ^{125}I-labeled anti-IgE that remains bound to the disk is calculated as percentage radioactive uptake:

$$\text{Percentage uptake} = \left(\frac{\text{Total counts bound to disk}}{\text{Total counts in 0.05-ml aliquot anti-IgE}}\right) 100$$

Specific IgE binding is calculated by deducting the percentage radioactive uptake for cord serum (minus the value for its blank disk) from the value for the test serum (minus its blank).

In order to select sera for the epitope-mapping experiments, specimens collected from wheat allergic patients are tested in the RAST against the 0.28 AI coupled to nitrocellulose disks. Samples giving a high percentage radioactive uptake in the test are used for the epitope-mapping studies. The results for a group of patients allergic to wheat are given in Table III.

TABLE III
RAST RESULTS[a]

Serum code	0.28 AI
Arm	16
Bro	40
Pas	16
Spa	14
Wil	8
Cord[b]	0.2
High[c]	0.2

[a] RAST (radioallergosorbent test) results are expressed as percentage of total radioactive uptake. A result of 5 times background is considered to be positive.

[b] Cord refers to fetal cord serum, which is normally free of IgE.

[c] High refers to the serum containing 100 kU/liter IgE nonspecific for wheat proteins.

RAST Inhibition

The ability of the synthetic peptide to act as an allergenic determinant is tested by assessing its ability to inhibit IgE binding to the parent protein attached to nitrocellulose disks in RAST by the method of Gleich *et al.*[15] Serum (0.05 ml) is incubated for 3 hr at room temperature with 0.05 ml of inhibitor solution (1 mg/ml). Positive and negative control sera are also incubated with inhibitor to check for any nonspecific reactions. In our hands, none is found. The mixture is then added to nitrocellulose disks coupled to the purified 0.28 AI. The percentage of ^{125}I-labeled anti-IgE that remains bound to the disk is calculated as above for both inhibited *(i)* and uninhibited *(u)* samples. Percentage RAST inhibition is then calculated as

$$\text{Percentage RAST inhibition} = 100(u - i)/u$$

Table IV shows the results for the RAST inhibition studies using the 0.28 AI. In all cases the peptide is able to inhibit IgE binding to the protein from which it was derived. Since the peptide inhibits IgE binding to the parent allergen, it must contain allergenic determinant(s) present in the native conformation of the allergen. The inhibition of binding varies with each serum sample. This is to be expected, since RAST measures all IgEs directed against all of the determinants of the allergen and only one allergenic determinant was used in this RAST inhibition experiment. These results thus demonstrate that the linear peptide can act as an IgE-binding determinant. The control sera do not show any reactivity to the peptides or proteins, thus indicating that the results are not due to nonspecific reactions.

TABLE IV
RAST INHIBITIONS[a]

0.28 AI inhibited with	Degree of RAST inhibition (%) with serum code indicated				
	Bro	Pas	Spa	Wil	High[b]
SWCDPATGYKVSALTGCRAMV	44	13	27	17	0

[a] Shown as percentage decrease in radioactive uptake in the RAST of the 0.28 AI after incubation of sera with the synthetic peptide. The quantity of inhibitor was 50 μg/50 μl of PBS in each case.

[b] High refers to the serum containing 100 kU/liter IgE nonspecific for wheat proteins.

Discussion

This is the first time that the allergenic determinants of a cereal protein have been elucidated and a peptide containing a determinant synthesized. The method given here demonstrates that there are individual reactivities to the allergen. The serum from each patient contained allergen-specific IgE directed against a number of different allergenic determinants. This is consistent with the results for Ra3, for which five allergenic determinants were detected. Four of the determinants bound IgE in the sera of both patients (Table II) tested in the 0.28 AI epitope-mapping experiments.

Comments on Method

The method depends on the absorbance values from the negative control sample being low and those from the test samples being high. Therefore, the anti-IgE used must be affinity purified and also tested in epitope mapping without adding any serum containing IgE. This will indicate if there is any nonspecific binding by the anti-IgE antibody to the peptides on the pins. The problem resulting from such binding can, in principle, be eliminated by preabsorbing the antiserum with the allergen bound to a solid phase.

If serum samples are small, or the protein to be probed is large, a 1-in-10 dilution is possible for allergic sera that give RAST values of greater than 10%, approximately, against the allergen.

The value of this method is its speed and selectivity. When performed in conjunction with large-scale peptide synthesis and RAST inhibition to confirm the results, it is a powerful technique.

Acknowledgments

This work was supported by grants from the National Health and Medical Research Council of Australia. We thank the editors of the *Journal of Immunological Methods* (Elsevier) for permission to use material published in their journal.

[16] Design and Production of Bispecific Monoclonal Antibodies by Hybrid Hybridomas for Use in Immunoassay

By MIYOKO TAKAHASHI, STEVEN A. FULLER, and SCOTT WINSTON

Introduction

The production of novel antibody reagents has become possible through advances in hybridoma techniques, recombinant DNA manipulation, and gene transfection technology. One such novel reagent is the bispecific antibody, a hybrid immunoglobulin molecule that is formed of nonidentical heavy- and light-chain pairs expressing different antigen specificities. The ability of bispecific antibodies to cross-link two different antigens has led to applications in immunohistochemistry,[1] cell targeting,[2,3] complement-mediated cytotoxicity,[4] and immunoassays.[5,6] In conventional use, monoclonal antibodies (MAb) are covalently coupled to various reagents (e.g., enzymes, fluorochromes, radioisotopes, dyes, and toxins). Enzyme immunoassays (EIA) utilize MAb cross-linked to any of several marker enzymes. Such chemical modification of both antibody and enzyme has several limitations, including (1) an antibody–enzyme mixture that includes both active and inactive species in varying proportions, (2) a preparation that contains both antibody and enzyme coupled to contaminants as well as to each other, (3) loss of antibody function, and (4) antibody–enzyme conjugates that are often less stable than the unmodified components. Bispecific monoclonal antibodies (bsMAb), developed originally by Milstein and Cuello (1983),[1] can circumvent these problems by avoiding entirely the chemical modification.

Antibodies with bifunctional specificities have been generated by a number of methods: (1) cross-linking of heterologous MAb,[7,8] (2) reassociation of monovalent antibody fragments,[9–11] and (3) cell fusion of hybrid-

[1] C. Milstein and A. C. Cuello, *Nature (London)* **305,** 537 (1983).
[2] U. D. Staerz and M. J. Bevan, *Proc. Natl. Acad. Sci. U.S.A.* **83,** 1453 (1986).
[3] A. Lanzavecchia and D. Scheidegger, *Eur. J. Immunol.* **17,** 105 (1987).
[4] J. T. Wong and R. B. Colvin, *J. Immunol.* **139,** 1369 (1987).
[5] M. R. Suresh, A. C. Cuello, and C. Milstein, *Proc. Natl. Acad. Sci. U.S.A.* **83,** 7989 (1986).
[6] M. Takahashi and S. A. Fuller, *Clin. Chem.* **34,** 1693 (1988).
[7] U. D. Staerz, O. Kanagawa, and M. J. Bevan, *Nature (London)* **314,** 628 (1985).
[8] P. Perez, R. W. Hoffman, S. Shaw, J. A. Bluestone, and D. M. Segal, *Nature (London)* **316,** 354 (1985).
[9] A. Nisonoff and W. J. Mandy, *Nature (London)* **194,** 355 (1962).
[10] V. Raso and T. Griffin, *Cancer Res.* **41,** 2073 (1981).
[11] M. Brennan, P. F. Davison, and H. Paulus, *Science* **229,** 81 (1985).

omas or sensitized B lymphocytes.[1] The cell fusion approach has several advantages. Hybrid hybridomas form a high percentage of hybrid antibody molecules. The ease of the procedure allows any experienced cell culturist to perform the work. In addition, bsMAbs that are biologically produced by hybrid hybridomas often exhibit no observable changes in specificity or physical characteristics after prolonged storage at 4° (M. Takahashi, S. A. Fuller, and S. Winston, unpublished data, 1988).

Our research interests lie in the development of EIAs for a wide variety of analytes. In this chapter we describe methods we have utilized or developed to prepare a panel of fusion partners that simplifies production of bsMAbs, the preparation of hybrid hybridomas, and the production and use of bsMAbs. While the focus of our work is urease-based EIAs, these methods should be readily applicable to other enzyme/detector systems. A detailed report concerning the production of bsMAbs by hybrid hybridomas (theory, materials, and methods) was published in an earlier volume in this series.[12]

Hybrid Hybridoma Production

In our attempts to utilize bsMAbs in urease-based EIAs, we wished to set up a simple and flexible system that might be useful for any analyte or MAb of interest. Consequently, we have prepared a panel of anti-urease fusion partners that represent the major murine IgG subclasses and that possess two selectable markers. This panel has the following advantages:

1. bsMAbs of most homologous or heterologous heavy-chain subclasses can be readily prepared, which provides flexibility in stability, function, or purpose.
2. The hybridoma to be fused need not be modified, since the selectable markers of the anti-urease fusion partner allow both parental lines to be eliminated.

Figure 1 is a schematic representation of the processes involved in producing hybrid hybridomas. Each of these is described in the methods that follow.

Isolation of Hybridoma Lines for Use as Fusion Partners

A panel of anti-urease-producing hybridoma lines was established from a 42% (w/v) polyethylene glycol (PEG) (M_r 4000; Merck, Darmstadt, Germany) fusion of Sp2/0 cells and splenocytes isolated from BALB/c

[12] M. R. Suresh, A. C. Cuello, and C. Milstein, this series, Vol. 121, p. 210.

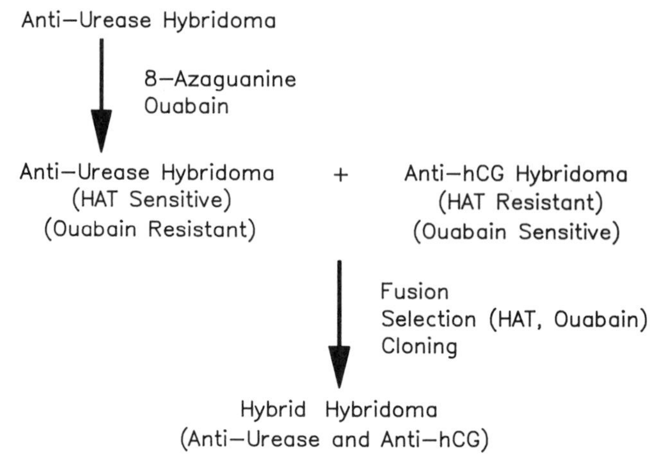

FIG. 1. Flow diagram for preparation of hybrid hybridomas. HAT, Hypoxanthine/aminopterin/thymidine medium; hCG, human chorionic gonadotropin.

mice immunized with jackbean urease.[6] In order to use these anti-urease-producing hybridoma lines as universal fusion partners with either immunocytes (lymphocytes) or other hybridomas, double-mutant lines resistant to both 8-azaguanine and ouabain were isolated.

Procedure to Isolate 8-Azaguanine-Resistant Hybridomas. In hybridoma technology, the most common selection method is the use of a parental cell line that is sensitive to hypoxanthine/aminopterin/thymidine (HAT) medium.[13,14] Aminopterin blocks *de novo* purine and pyrimidine synthesis by inhibition of the enzyme dihydrofolate reductase. However, if present, the enzymes hypoxanthine–guanine phosphoribosyltransferase (HGPRT) and thymidine kinase can utilize hypoxanthine and thymidine, respectively, via salvage pathways. Animal cell lines deficient for HGPRT (HAT sensitive) can be isolated by adaptive growth and selection in 8-azaguanine (8-AG). This antimetabolite is cytotoxic to normal cells and only $HGPRT^-$ cells can survive.

Selection of 8-AG-resistant variants involves adaptive growth in successively increasing concentrations of the drug (to 20 μg/ml).

1. Prepare a 2 mg/ml stock solution of 8-AG. Weigh 200 mg of the purine analog and add to 90 ml distilled water. Add 1 *N* NaOH dropwise

[13] S. A. Fuller, M. Takahashi, and J. G. R. Hurrell, *in* "Current Protocols in Molecular Biology" (F. Ausubel, B. Brent, R. Kingston, D. D. Moore, J. G. Seidman, J. A. Smith, and K. Struhl, eds.), Unit 11. Greene Publishing, New York, 1987.

[14] G. Galfrè and C. Milstein, this series, Vol. 73, p. 3.

until dissolved. Add distilled water to 100 ml. Filter sterilize the solution and then store frozen at −20° in convenient aliquots.

2. Prepare six wells (six-well cluster dish; Costar, Cambridge, MA) containing actively growing hybridomas (3 × 10^5 cells/ml, anti-urease cell lines in this case) in 10 ml of Dulbecco's modified Eagle's medium (DMEM) supplemented with 10% (v/v) fetal calf serum (FCS). Add the 8-AG stock solution to a final concentration of 1, 2, and 4 μg/ml, using two wells for each of the concentrations (we have found that different murine hybridoma lines show varied sensitivity to the drug). Grow cells in an incubator under 8% CO_2.

3. Make a 50% medium change every 2 days using growth medium containing appropriate amounts of 8-AG.

4. Select the culture growing at the highest drug concentration. Double the drug concentration and continue to culture for several generations.

5. Repeat step 4 until a viable cell population growing in the presence of 20 μg/ml 8-AG is obtained.

6. When cells are adapted to the 20 μg/ml drug concentration (approximately 3 weeks), clone by the limiting dilution method[13] using medium containing 20 μg/ml 8-AG.

7. When the clones are confluent, select 10 wells containing a single colony and expand them in duplicate in a 24-well plate (Costar). Culture one set in HAT selection medium[13,14] and discard the clones that survive.

8. Screen the remaining clones that are 8-AG resistant and HAT sensitive for antibody production (anti-urease).

9. Select clones with good growth and antibody production. Freeze aliquots for safe keeping. The remaining cells are subjected to another round of adaptive growth in the presence of ouabain.

Procedure for Isolation of Ouabain-Resistant Hybridomas. Ouabain is a plant glycoside that inhibits membrane-bound Na^+,K^+ − ATPase. Ouabain resistance is an additional property useful for the selection of desirable fusion products. An 8-AG-resistant hybridoma line that is also ouabain resistant can be fused with any hybridoma line. The fusion products are selectively cultured in medium containing HAT and ouabain. In this medium, the double-mutant parental cells do not survive because of their deficiency for HGPRT. The other parental hybridoma will die due to its lack of resistance to ouabain, and only the hybrid hybridomas will survive.

The procedure used to isolate ouabain-resistant hybridomas is similar to that described in the previous section for azaguanine.

1. Prepare 50 m*M* ouabain (Sigma, St. Louis, MO) stock solution in distilled water. Dissolve by heating to 70°. Mix with an equal

volume of 2× DMEM (room temperature), filter sterilize, and store frozen at −20° in convenient aliquots.

2. In a 75-cm^2 tissue culture flask, grow the 8-AG-resistant hybridoma line to a cell density of approximately 4×10^5 cells/ml in 30 ml medium with 15% FCS. Add the ouabain stock solution to a final concentration of 0.1 m*M*. Set up a small control flask and grow the cells in the same medium without the drug, in order to ensure that the cell growth retardation is due to the drug.
3. Make a 50% medium change by removing the spent medium and supplementing with fresh medium or by merely doubling the medium as required.
4. Allow cells to grow for several generations in medium containing 0.1 m*M* ouabain, then double the culture volume with fresh medium containing 0.2 m*M* ouabain. When the culture has reached a cell density of $4-6 \times 10^5$ cells/ml, make an appropriate split with medium containing 0.2 m*M* ouabain.
5. Repeat step 4, by doubling the culture volume with medium containing 0.4 m*M* ouabain. It is very important to increase the ouabain level gradually. In comparison to human cell lines, rodent cell lines have been reported to be far more resistant to ouabain. All of the nonadapted murine hybridoma lines are very sensitive to the drug at around 0.2–0.4 m*M*.
6. Continue the process and gradually increase the ouabain level up to 1 m*M*.
7. When the cells are well adapted to 1 m*M* ouabain, add 20 μg/ml 8-AG to the medium supplemented with 15% (v/v) FCS. Allow them to grow for several generations in the presence of the two drugs.
8. Clone by the limiting dilution method[13] in medium containing 1 m*M* ouabain and 20 μg/ml 8-AG, supplemented with 20% FCS.
9. Screen only those wells with a single colony for antibody activity.
10. Choose the clone(s) with the best growth characteristics and the best production of antibody.

Using the procedures described above, we have isolated numerous 8-AG/ouabain-resistant hybridoma lines that produce anti-urease of IgM or IgG (subtypes 1, 2a, and 2b) class. Each of these lines were continuously cultured for at least 8 weeks in DMEM containing 10% FCS, after which all lines still demonstrated resistance to 20 μg/ml 8-AG and 1 m*M* ouabain and were sensitive to HAT medium. Production of anti-urease in the variant lines was measured by enzyme-linked immunosorbent assay (ELISA) following continuous culture. Despite the stability of drug resistance characteristics, some hybridomas expressed reduced antibody production relative to wild type. Cell doubling times of the variant lines in the presence and absence of azaguanine and ouabain were as reported previously.[6]

Cell Fusions and Cloning

Two cell fusion procedures used to obtain hybrid hybridomas are described in the following sections. Most reagents and procedures are as for conventional hybridoma work.[13,14] Differences exist in the use of selection medium, the fusion cell ratio for hybridoma × hybridoma fusions, and screening for bispecific antibody activity. The latter is discussed separately.

Hybridoma × Splenocyte Cell Fusions. Fusions between the 8-AG-resistant hybridomas that produce anti-urease and splenocytes obtained from an immunized mouse, or splenocytes immunized *in vitro*, are basically as for Sp2/0 × splenocyte fusions. The ratio of splenocytes to hybridoma cells is 5 : 1. The polyethylene glycol (PEG)-mediated fusion products are resuspended in HAT selection medium and incubated for 1–3 hr in a CO_2 incubator. The cells are then plated out at $1-2 \times 10^5$ cells/well on 96-well Costar plates that were preseeded with mouse peritoneal exudate cells (PEC) as feeders in HAT selection medium.

Hybrid hybridomas are generally screened 2–3 weeks postfusion. Selected positive cultures are then subjected to cloning by the limiting dilution method.

We have successfully established hybrid hybridomas that produce bsMAb by fusions of anti-urease producing hybridomas with splenocytes sensitized *in vitro* with human luteinizing hormone as well as with splenocytes obtained from BALB/c mice immunized with various immunogens, including human chorionicgonadotropin (hCG).

The limitation of this approach, i.e., fusions of hybridoma × splenocytes, is that the isolation of hybrid hybridomas that produce bsMAbs of desired specificity and affinity depends on chance.

Hybridoma × Hybridoma Cell Fusions. We describe here the protocol used for the fusion experiments between anti-urease hybridomas and hybridomas secreting anti-hCG (schematically shown in Fig. 1).

1. From each of the two hybridoma lines to be fused, harvest 4×10^7 cells from a logarithmically growing culture.
2. Centrifuge both hybridoma lines. Pool the cell pellets, and wash three times in warm DMEM.
3. Subject the final cell pellet (which is a 1 : 1 cell mixture of the two parental hybridomas) to PEG cell fusion.[13]
4. Resuspend the fused cells in 40 ml DMEM supplemented with 20%

FCS, 2× HAT, and 2 m*M* ouabain. Incubate for 30 min in a CO_2 incubator prior to plating out at 2×10^5 cells/0.1 ml into each well on 96-well Costar plates. The plates should be preseeded with 0.1 ml/well mouse PEC in DMEM with 20% FCS, and without HAT/ouabain.

5. Prepare control cultures as in the experimental plates for both parental hybridomas using 10 wells each. PEC feeder cells are preseeded and each hybridoma line is plated at 2×10^5 cells/well. In the HAT/ouabain selection medium, the control anti-urease hybridomas should not survive because of their sensitivity to HAT, while anti-hCG hybridomas are killed due to ouabain sensitivity.
6. Leave plates undisturbed in the CO_2 incubator for 1 week, except for a brief check for microbial contamination a few days after the fusion. One week postfusion, examine the control cultures carefully. When both parental cells are dead in the HAT/ouabain medium, start feeding the experimental plates with DMEM supplemented with 20% FCS and HT (HAT medium from which aminopterin is omitted) and ouabain.
7. Feed cultures as required until the hybrid hybridomas become confluent and the culture medium becomes acidic (yellow in appearance).
8. Screen the culture supernatants for parental and bispecific antibody activities, as described in the following sections.
9. Select cultures indicating bispecific antibody activity, and clone by the limiting dilution method.
10. When the hybrid hybridomas are established and cloned, freeze a number of vials and confirm a good recovery from the liquid nitrogen freezer.

The procedure described above to prepare bispecific antibody-producing hybridomas by the fusion between two hybridoma lines has the advantage that the characteristics of antibody specificity and affinity as well as the heavy-chain isotypes of immunoglobulins produced by each of the parental hybridomas are already known. This allows for an easy selection of clones that meet the specific criteria desired.

Our initial bsMAb work[6] indicated that fusion of IgM, IgG_1, IgG_{2a}, and IgG_{2b} anti-urease hybridomas with an IgG_1 anti-hCG hybridoma resulted in efficient production of bispecific monoclonal antibodies only for the homologous $IgG_1 \times IgG_1$ fusions. Based on this observation, we have performed extensive fusion experiments to evaluate the outcome of homologous as well as heterologous heavy-chain combinations using our expanded panel of anti-urease variant hybridoma cell lines. Table I presents the

TABLE I
PRODUCTION OF BISPECIFIC ANTIBODY FROM FUSIONS BETWEEN HYBRIDOMAS PRODUCING HOMOLOGOUS AS WELL AS HETEROLOGOUS IMMUNOGLOBULIN SUBCLASSES OF IgG

Parental antibodies	Parental hybridoma lines[a]	Percentage of positive wells[b]			Number of wells screened
		Anti-urease	Anti-hCG	bsMAb	
Homologous fusions					
IgG_1-IgG_1	3U66^r × 2G131	96	57	57	130
$IgG_{2a}-IgG_{2a}$	7U114^r × 6G94	100	17	17	180
$IgG_{2b}-IgG_{2b}$	7U215^r × IG151	100	52	49	180
Heterlogous fusions					
IgG_1-IgG_{2a}	3U66^r × 6G94	100	53	6.9	160
IgG_1-IgG_{2b}	2G131 × 6U334^r	97	99	2.2	180
IgG_1-IgG_{2b}	3U66^r × 1G151	100	37	5.5	110

[a] Superscript r designates the anti-urease-producing variant hybridoma lines.
[b] Calculated as (number of positive wells/number of wells screened)100.

results obtained from cell fusions between anti-urease and anti-hCG hybridomas of homologous or heterologous heavy-chain subclass. In fusion experiments between homologous heavy chains of IgG_1, IgG_{2a}, and IgG_{2b} subclasses, bispecific antibody-positive wells were identified in 17–57% of wells screened. Compared to the homologous fusion results, the fusions between heterologous subclasses of IgG yielded only 2.2–6.9% positive wells for bispecific antibody activity. Moreover, use of IgM or IgG_3 hybridomas yielded no or only a few positive wells regardless of homologous or heterologous fusions. It is important to note that isolation of bsMAbs composed of heterologous heavy chains is possible despite the lower efficiency compared to homologous fusions.

Antibody Screening of Culture Supernatants of Hybrid Hybridomas

Enzyme immunoassays using 96-well microtiter plates are routinely used in our laboratory for the screening of culture supernatants for bsMAb activity as well as monospecific MAb activities of parental cells.

Materials

Polyvinyl or polystyrene 96-well microtiter plates
Coating buffer: 0.05 *M* carbonate buffer, pH 9.6
TEN buffer: 0.05 *M* Tris, 1 m*M* ethylenediaminetetraacetic acid (EDTA), 0.15 *M* NaCl, pH 7.3
Wash buffer (WB): Phosphate-buffered saline (PBS), pH 7.2, with 0.05% Tween 20

Diluting buffer (DB): PBS, pH 7.2, with 0.05% Tween 20 and 0.25% bovine serum albumin (BSA)
hCG (Sigma)
Urease and urease substrate solution (ADI Diagnostics, Inc., Rexdale, Canada)
o-Phenylenediamine (OPD) (Sigma)
Peroxidase-conjugated goat anti-mouse IgG (heavy and light chain) (Cappel Worthington, Malvern, PA)
Urease-conjugated rabbit anti-mouse $F(ab')_2$ of IgG (ADI Diagnostics)

Methods

Anti-hCG Screening

1. Coat 96-well polyvinyl microtiter plates (Dynatech, Chantilly, VA) with 100 μl/well of hCG (5 IU/ml) in 0.05 *M* carbonate buffer. Seal the plates and store at 4° for at least 24 hr before use.
2. Remove antigen solution and fill all wells with DB to block additional protein binding sites for 1 hr at room temperature.
3. Wash the plates three times in WB.
4. Add hybrid hybridoma culture supernatants (100 μl/well) into the plates and incubate for 60–90 min at 37°.
5. Wash plates three times in WB.
6. Add 100 μl/well of urease-conjugated rabbit anti-mouse $F(ab')_2$ fragment of IgG. Seal the plates and incubate for 30 min at 37°.
7. Wash the plates three times with WB, then three times with 0.85% NaCl.
8. Add 100 μl/well urease substrate solution, and after a 10- to 20-min incubation at room temperature, read the plates at 590 nm.

Anti-Urease Screening. The procedure is the same as for anti-hCG screening with minor modifications. The marker enzyme conjugated to the detector antibody must be one other than urease.

1. Coat polyvinyl plates with urease (10 μg/ml) in TEN buffer. Seal the plates and store at 4° until use.
2. Block plates for 1 hr at room temperature.
3. Wash three times in WB.
4. Add culture supernatants to the plate, and incubate for 60–90 min at 37°.
5. Wash plates three times.
6. Add peroxidase-conjugated goat anti-mouse IgG (heavy and light chains) and incubate for 30 min at 37°.
7. Wash plates four times with WB.

8. Add OPD in phosphate–citrate buffer (2.43 ml 0.1 *M* citric acid, 2.57 ml 0.2 *M* phosphate, 5 ml H_2O, and 0.015 ml 30% H_2O_2). Incubate for 30 min in the dark at room temperature.

9. Stop the reaction with 2 *M* H_2SO_4, and read the results at 492 nm.

Anti-Urease–Anti-hCG Screening (bsMAb Activity). The bispecific antibody activities can be screened either by direct assay involving the binding of hCG directly to microtiter wells, or by sandwich assay, in which hCG is attached to the solid phase by capturing it with an antibody that is bound to the microtiter well. In general, we find that the sandwich assay works better for bsMAb screening. The anti-urease–anti-hCG bispecific monoclonal antibody activities in culture supernatants of hybrid hybridomas were assayed by simultaneous sandwich immunoassay using 96-well plates coated with affinity purified capture antibody.

1. Coat polyvinyl plates with capture antibody (10 μg/ml, either polyclonal or monoclonal antibody specific for the antigen, hCG) in 0.05 *M* carbonate buffer, pH 9.6. When monoclonal capture antibodies are used, they should not compete with the bsMAbs for the binding site of analyte.
2. Prior to use, block the plate with DB for 1 hr at room temperature.
3. Wash plates three times in WB.
4. Add an appropriate amount of antigen (50 mIU hCG) to each well in a 20-μl volume. Mix the plate by gently tapping the side of plate. Then add culture supernatants (70 μl/well) and 400 ng urease in 10 μl PBS.
5. Incubate the plates for 30 min at 37°.
6. Wash the plates three times with WB, followed by three times with 0.85% NaCl.
7. Add 100 μl/well urease substrate solution and incubate for 30 min at room temperature prior to reading the reaction at 620 nm.
8. Repeat the assay on positive wells, and run control assays in parallel in which only culture supernatants and urease are added to the assay wells. If the sample contains functional bsMAb, the experimental wells should remain positive, while the control assays should be negative.

Production and Purification of Bispecific Monoclonal Antibodies

Production of bsMabs

In Vitro Production. Large-scale culture of hybridomas has been described in detail in this series.[14] For the production of murine bsMAbs, it is more practical and economical to use an *in vivo* system. However, if the derivation of bsMAbs utilizes species other than mouse, *in vitro* large-scale production of bsMAbs might be the method of choice. Culture *in vitro* has the advantage of obtaining purer antibody preparations. Contamination of the harvested end product is further reduced by the use of serum-free medium for short cultures of 24–48 hr using a high cell density of $1-2 \times 10^6$ cells/ml.

In Vivo Production. Procedures for the production of bsMAbs from ascitic tumors are similar to conventional hybridoma work.[13,14] It is important to remember that the animals used, e.g., mice or rats, for ascites production must be histocompatible with the parent cells. If, for example, two strains of mice are involved, an F_1 hybrid of the two strains should be used for *in vivo* propagation. Use of irradiated mice at 300–400 rad or nude mice may be considered for nonhistocompatible situations. However, such mice should be maintained in a germ-free environment.

Processing Antibodies Produced. Antibodies obtained from *in vitro* propagation are very dilute. They can be concentrated by precipitation at 40–50% saturation with ammonium sulfate, dissolving the precipitates in a smaller volume, and dialyzing against PBS. Antibodies in ascitic fluid are already concentrated, along with contaminating proteins such as albumin. Albumin levels can be greatly reduced by $(NH_4)_2SO_4$ precipitation at 50% saturation at 4°. Antibodies sensitive to $(NH_4)_2SO_4$ precipitation can be concentrated by dialysis against hygroscopic materials (e.g., PEG powder), by vacuum dialysis, or by filtration through selective Amicon (Danvers, MA) membranes. Partially purified or concentrated antibodies as described above are stored at 4° with preservatives such as NaN_3 at 0.1%, or frozen at −20° if a long period of storage is required prior to further purification by column chromatography.

Affinity chromatography using various immobilized staphylococcal protein A matrices is a common method used to prepare purified antibody. Procedures have been described elsewhere[13] or are available from manufacturers [e.g., Pharmacia (Piscataway, NJ) and Bio-Rad (Richmond, CA)]. Protein A chromatography of ascites fluid or culture supernatant will provide a preparation containing both bsMAbs and parental MAbs free of contaminants. Antibody should be concentrated and dialyzed against appropriate buffer prior to use in immunoassay or further purification.

Purification of Bispecific Monoclonal Antibodies

Functional bsMAbs may be purified by a number of techniques. Anion-exchange chromatography has been used successfully to prepare bsMAbs substantially free of contamination by parental antibodies when two different heavy-chain subclasses are present in the hybrid molecule.[5]

This technique has been discussed earlier in this series.[12] Others have used ion-exchange and hydroxylapatite chromatography with mixed results.[15,16]

Affinity chromatography may be used to purify bsMAbs when sufficient quantities of antigen are available to prepare affinity matrices.

1. Prepare antigen affinity columns.[13]
2. Ascitic fluid containing bsMAbs is diluted in 3 vol phosphate-buffered saline (PBS; 10 mM sodium phosphate, 150 mM sodium chloride, pH 7.2) and applied to the first antigen affinity column.
3. Wash the column with PBS and elute antibody with glycine–HCl buffer (50 mM glycine, 150 mM sodium chloride, pH 2.3).
4. Concentrate eluate by ultrafiltration and dialyze against PBS.
5. Apply dialyzed eluate to the second antigen affinity column and repeat steps 3 and 4.

Only bsMAbs can bind to both columns. Parental antibodies will bind only one column. When bsMAb is to be used to detect or quantitate an analyte by enzyme or other detector [e.g., detection of human chorionicgonadotropin (hCG) in an EIA by anti-hCG–anti-urease bsMAb], it may be necessary to eliminate only the anti-analyte parental antibody. Purification on only one antigen affinity column (that of the detector) will yield both bsMAb and one parental antibody (e.g., anti-urease), which should not interfere in detection of the analyte.

In the case of urease-based EIAs, the size of the urease molecule (M_r 540,000) can be utilized to separate bsMAb by size exclusion chromatography.

1. Antibody from bsMAb-containing ascitic fluid is purified by protein A-chromatography.[13]
2. Antibody (250 μg) is mixed with 800 μg urease in a total volume of 200 μl PBS for 10 min at room temperature.
3. This mixture is then loaded onto a 0.75 × 30 cm TSK G3000SW HPLC column (Varian, Lexington, MA). The sample is eluted at a flow rate of 1 ml/min using a mobile phase of 67 mM sodium phosphate, 300 mM sodium chloride, pH 7.0.
4. Fractions of 0.5 ml are collected and assayed for bsMAb activity as described earlier.

Figure 2 shows a set of elution profiles of antibody plus urease, urease

[15] M. R. Clark and H. Waldman, *J. Natl. Cancer Inst.* **79,** 1393 (1987).

[16] L. Karawajew, O. Behrsing, G. Kaiser, and B. Micheel, *J. Immunol. Methods* **111,** 95 (1988).

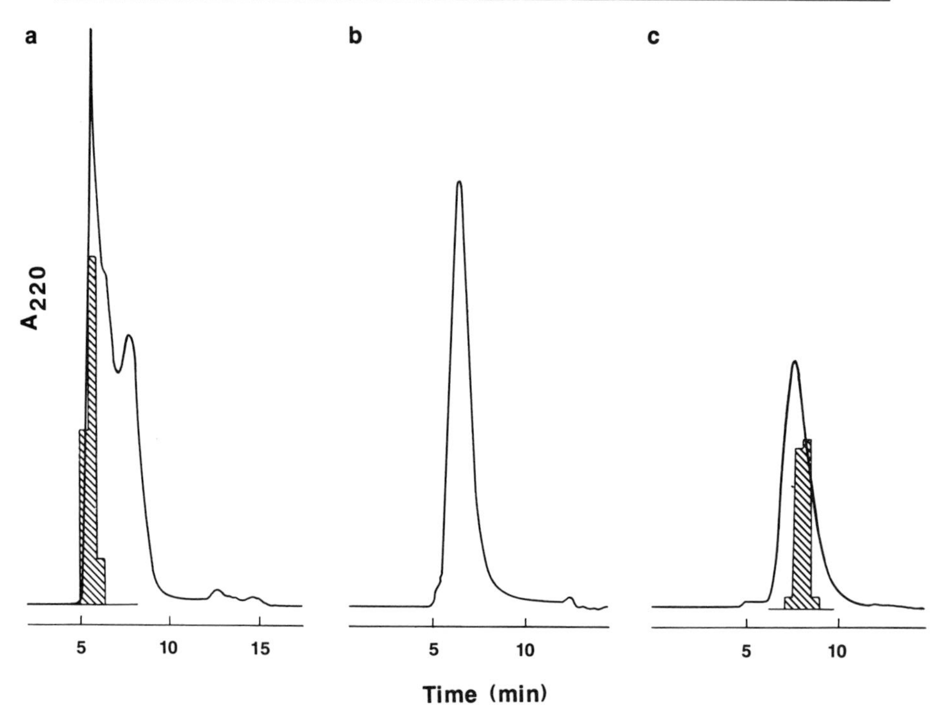

FIG. 2. HPLC TSK G3000SW size-exclusion chromatography of bispecific monoclonal antibodies. (a) Protein A-purified bsMAb (anti-hCG–anti-urease, 250 μg) + urease (800 μg); (b) urease (800 μg); and (c) protein A-purified bsMAb (anti-hCG–anti-urease, 250 μg). Hatched areas indicate proportional levels of bsMAb activity in column fractions (0.5 ml). Mobile phase is 0.067 *M* sodium phosphate, pH 7.0 with 0.3 *M* NaCl, flow rate of 1.0 ml/min.

alone, and antibody alone. Retention times are 5.5, 6.4, and 7.8 min, respectively. The 5.5-min peak contains both bsMAb and parental anti-urease complexed with urease, well separated from the other parental antibody. This fraction can be used directly in EIA for bsMAb detection of analyte.

Either of the latter two techniques can be used with bsMAbs formed of homologous or heterologous heavy-chain subclasses with similar results.

Bispecific Monoclonal Antibodies Used in Enzyme Immunoassays

A bsMAb with the specificities to link both marker enzyme and an analyte may possibly be a better alternative to chemically linked antibody–enzyme conjugates as immunoassay reagents, for reasons described earlier. A schematic comparison of simultaneous EIA formats using monospecific or bispecific antibodies is presented in Fig. 3. Each assay consists of the same number of steps. The use of a MAb–enzyme conjugate versus use of bsMAb and free enzyme marks the difference between the two assays. Successful applications of bsMAb on EIA have been reported. Suresh *et al.*[5] described a competitive dipstick immunoassay for substance P using an anti-substance P–anti-HRP bispecific MAb. It is a very simple and fast assay that may have an application in qualitative analysis of a substance where sensitivity is not critical. Karawajew *et al.*[16] applied purified bsMAb in two-site binding enzyme immunoassays as bridging agents between fluorescein isothiocyanate (FITC)-labeled MAb specific for the analyte and the marker enzyme horseradish peroxidase (HRP). The assay was performed sequentially, involving three incubation steps. The sensitivity of these assays to detect α-fetoprotein (AFP) and hCG were reported to be equivalent to those reported using other conventional assay methods. Another example of successful application of bsMAb in

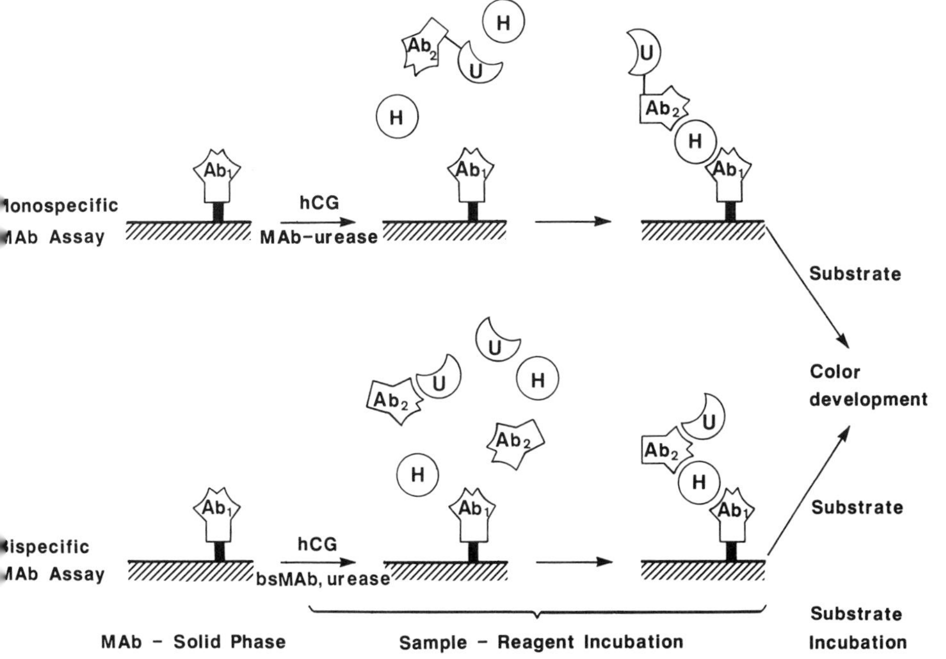

FIG. 3. Simultaneous EIAs for hCG. Ab_1, anti-hCG; Ab_2, anti-hCG monovalent MAb conjugated to urease for monospecific MAb assay (top structure), and anti-hCG-anti-urease bsMAb for bispecific MAb assay (bottom structure); H, human chorionicgonadotropin; U, urease.

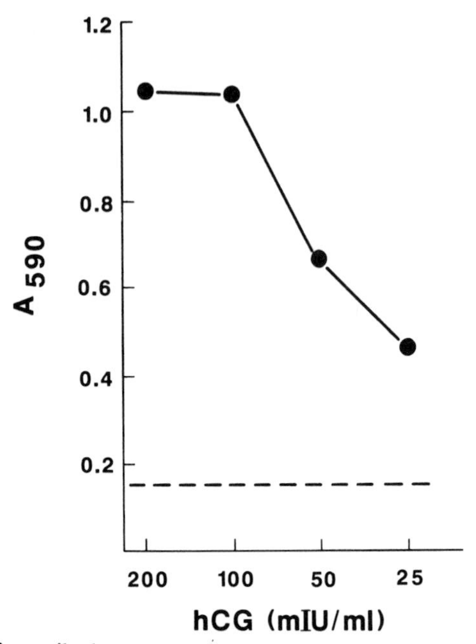

FIG. 4. Bispecific antibody–enzyme immunoassay for hCG. Anti-hCG–anti-urease-containing ascites fluid was incubated 15 min simultaneously with hCG and urease, followed by a 5-min incubation with urease substrate. Dashed line represents background absorbance with no hCG present.

EIA has been described by Tada *et al.*[17] for detection of human lymphotoxin (hLT). A one-step bsMAb sandwich EIA detected a 1 U/ml or 0.1–0.2 ng/ml hLT level within 2 hr, whereas a two-step sandwich EIA using biotinylated monospecific MAb required 3–4 hr to achieve a similar sensitivity.

Our interest is to utilize bsMAb in urease-based EIAs. One example of this form of EIA is the detection of human chorionicgonadotropin (hCG) using an anti-hCG–anti-urease bsMAb. The assay is performed in the following manner.

1. Prepare microtiter plates coated with anti-hCG monoclonal antibody (see section entitled "Antibody Screening of Culture Supernatants of Hybrid Hybridomas").

2. Incubate bsMAb antibody solution (ascites fluid, culture supernatant, or purified bsMAb preparations) with hCG solution and 50 μg of urease in a total volume of 100 μl DB for 15 min at room temperature.

[17] H. Tada, Y. Toyoda, and S. Iwasa, *Hybridoma* **8,** 73 (1989).

3. Wash the wells four times with a 0.85% sodium chloride solution.

4. Incubate with 100 μl urease substrate solution for 5 min at room temperature. Read absorbance at 590 nm.

Results of such an experiment are shown in Fig. 4. This assay detected 25 mIU/ml hCG using bsMAb-containing ascitic fluid or protein A-purified bsMAb. This level of detection is equivalent to the traditional MAb–urease assay, even though monospecific parental antibodies are still present. Using HPLC-purified bsMAb (which lacks the parental anti-hCG antibody), the level of detection is slightly enhanced to 15 mIU/ml.

[17] Catalytic Antibodies

By KEVAN M. SHOKAT and PETER G. SCHULTZ

Introduction

The specificity and affinity of antibody molecules have made them important tools in biology and medicine and as diagnostics. Our efforts have focused on exploiting the vast repertoire of unique antibody specificites for the immune system to generate novel catalysts. To date antibodies have been demonstrated to catalyze a wide variety of reactions, which have recently been reviewed (Table I).[1-4] Several general strategies for the production of antibody catalysts have emerged since the first reports of catalytic antibodies in 1986, and these are reviewed here.

Thus far, two major approaches have been applied to the design of antibody catalysts. The most widely used method exploits the electronic and steric complementarity of an antibody to its corresponding hapten. This approach allows for (1) generation of precisely positioned catalytic amino acid side chains in the combining site, (2) stabilization of high-energy transition states in reactions (3) reduction in the entropies of reactions by orienting reaction partners in reactive conformations, and (4) the incorporation of cofactor binding sites into antibody combining sites. The second general approach to the generation of catalytic antibodies involves direct introduction of catalytic groups into the antibody combining site by either selective chemical modification, site-directed mutagenesis, or genetic selections or screens. The development of a general set of "rules" for

[1] P. G. Schultz, *Angew. Chem., Int. Ed. Engl.* **28**, 1283 (1989).
[2] P. G. Schultz, *Acct. Chem. Res.* **22** 287 (1989).
[3] K. M. Shokat and P. G. Schultz, *Annu. Rev. Immunol.* **8**, 355 (1990).
[4] P. G. Schultz, R. A. Lerner, and S. J. Benkovic, *Chem. Eng. News* **68**, 26 (1990).

TABLE I
ANTIBODY-CATALYZED REACTIONS

Reaction	Ref.[a]
Ester, carbonate hydrolysis	*1–6*
Stereospecific lactonization	*7*
Amide bond formation	*7, 8*
Redox	*9*
Thymine dimer cleavage	*10*
Chorismate mutase	*11–13*
Stereospecific ester hydrolysis	*14, 15*
β-Elimination	*16*
Diels–Alder	*17, 18*
Porphyrin metallation	*19*
Porphyrin-mediated oxidations	*20*
Activated amide bond hydrolysis	*21*
Peptide bond hydrolysis	*22, 23*

[a] Key to references: (*1*) K. D. Janda, M. Weinhouse, D. M. Schloeder, R. A. Lerner, and S. J. Benkovic, *J. Am. Chem. Soc.* **112,** 1274 (1990); (*2*) A. Tramontano and D. Schloeder, this series, Vol. 178, p. 535; (*3*) K. M. Shokat, M. K. Ko, T. S. Scanlon, L. Kochersperger, S. Yonkovich, S. Thairivongs, and P. J. Schultz, *Angew. Chem. Int., Ed. Engl.* **29(11),** 1296 (1990); (*4*) C. N. Durfor, R. J. Bolin, R. J. Sugasawara, R. J. Massey, J. W. Jacobs, and P. G. Schultz, *J. Am. Chem. Soc.* **110,** 8713 (1988); (*5*) E. Baldwin and P. G. Schultz, *Science* **245,** 1104 (1989); (*6*) D. Y. Jackson, J. R. Prudent, E. P. Baldwin, and P. G. Schultz, *Proc. Natl. Acad. Sci. U.S.A.* in press (1990); (*7*) S. J. Benkovic, A. D. Napper, and K. A. Lerner, *Proc. Natl. Acad. Sci. U.S.A.* **85,** 5355 (1988); (*8*) K. D. Janda, R. A. Lerner, and A. Tramontano, *J. Am. Chem. Soc.* **110,** 4835 (1988); (*9*) K. M. Shokat, C. J. Leumann, R. Sugasawara, and P. G. Schultz, *Angew. Chem. Int. Ed. Engl.* **27,** 1172 (1988); (*10*) A. G. Cochran, K. Sugasawara, and P. G. Schultz, *J. Am. Chem. Soc.* **110,** 7888 (1988); (*11*) D. Y. Jackson *et al., J. Am. Chem. Soc.* **110,** 4841 (1988); (*12*) D. Hilvert, S. H. Carpenter, K. D. Nared, M.-T. M. Auditor, *Proc. Natl. Acad. Sci. U.S.A.* **85,** 4953 (1988); (*13*) D. Hilvert and K. D. Nared, *J. Am. Chem. Soc.* **110,** 5593 (1988); (*14*) K. D. Janda, S. J. Benkovic, and R. A. Lerner, *Science* **244,** 437 (1989); (*15*) S. J. Pollack, P. Hsuin, P. G. Schultz, *J. Am. Chem. Soc.* **111,** 5961 (1989); (*16*) K. M. Shokat, C. J. Leumann, R. Sugasawara, and P. G. Schultz, *Nature (London)* **338,** 269 (1989); (*17*) D. Hilvert, K. W. Hill, K. D. Nared, and M.-T. M. Auditor, *J. Am. Chem. Soc.* **111,** 9261 (1990); (*18*) A. Braisted and P. G. Schultz, *J. Am. Chem. Soc.* **112,** 7430 (1990); (*19*) A. G. Cochran and P. G. Schultz, *Science* **249,** 781; (*20*) A. Cochran and P. G. Schultz, *J. Am. Chem. Soc.* **112,** 9414 (1990); (*21*) K. D. Janda, D. Schloeder, S. J. Benkovic, and R. A. Lerner, *Science* **241,** 1188 (1988); (*22*) B. L. Iverson and R. A. Lerner, *Science* **243,** 1184 (1989); (*23*) P. Sudhir, S. Paul, D. J. Volle, C. M. Beach, D. R. Johnson, M. J. Powell, and R. J. Massey, *Science* **244,** 1158 (1989).

generating catalytic antibodies is essential in order to exploit fully the diversity of the humoral immune system. For example, once a general approach to the hydrolysis of peptides by an antibody has been developed, a large number of sequence-specific peptides should become available simply by altering hapten structure accordingly. Therein lies the enormous advantage of using antibodies as catalysts for chemical transformations.

The development of catalytic antibodies of defined specificity promises to be of considerable value to biology, chemistry, and medicine. Catalytic antibodies may find use as therapeutic agents to selectively hydrolyze protein or carbohydrate coats of viruses, cancer cells, or other physiological targets. It may also be possible to hydrolyze selectively or ligate complex biomolecules such as polynucleotides, carbohydrates, and proteins, thereby facilitating structure–function studies or allowing the synthesis of new biomolecules with novel properties. The ready availability of large quantities of monoclonal antibodies may allow for their use as synthetic tools for the production of pharmaceuticals or new materials. The ability to generate antibody combining sites with specific catalytic groups and/or microenvironments should also serve to test fundamental notions of enzymatic catalysis.

Strategies

The strategies used to generate catalytic antibodies are based largely on principles of enzymatic catalysis including transition state stabilization, general acid–base catalysis, nucleophilic and electrophilic catalysis, strain, and proximity effects. To date, antibody-catalyzed rate enhancements over background rates range from 10^2 to 10^6. The generation of antibodies with rate accelerations of the order of 10^8 or greater will likely involve the simultaneous application of two or more strategies for introducing catalytic activity into antibodies. Clearly, then, strategies which yield a high percentage of catalytic antibodies relative to the total number of hapten-specific antibodies isolated are the most desirable and generalizable. The development of such strategies will require a thorough characterization of the structure and mechanism of the catalytic antibodies being generated. Each strategy is discussed in the context of specific reactions which have been catalyzed to date.

Eliciting Catalytic Groups

The introduction of a general acid or base into an antibody combining site should be an effective method of catalyzing a variety of chemical reactions using antibodies. The high effective concentration of the catalytic group in the antibody combining site as well as favorable orbital alignment should lead to considerable lowering of the entropy ($\Delta S^{\ddagger}$) and enthalpy ($\Delta H^{\ddagger}$) of activation for reaction.[5] Model systems with effective molarities greater than 10^7 M have been realized, and theoretical arguments suggest

[5] W. P. Jencks, "Catalysis in Chemistry and Enzymology." McGraw-Hill, New York, 1969.

accelerations up to 10^8 *M* can be achieved by approximation of reactive groups in enzyme active sites.[6]

Experiments by Pressman and Siegel in 1953 suggested a strategy whereby the electrostatic complementarity between haptens and antibodies can be used to introduce a catalytic carboxylate in an antibody combining site.[7] Negatively charged aspartate or glutamate residues were found in the combining sites of antibodies raised toward *p*-azobenzenetrimethylammonium cation. **(1)**.[8] Conversely, positively charged arginine and lysine residues were identified in the combining sites of antibodies elicited against negatively charged *p*-azobenzoate **(2)**.[9] Besides electrostatic forces, hydrophobic and hydrogen bonding interactions can be used to induce certain amino acids in antibody combining sites. A tryptophan in MOPC 315 has been shown to π stack with the aryl ring of 2,4-dinitrophenyl-containing ligands.[10]

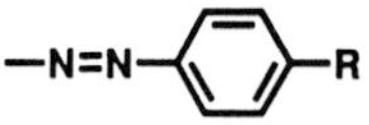

—N=N—C₆H₄—R

1: R= $-\overset{+}{N}(Me)_3$
2: R= $-CO_2^-$

β-Elimination. Isomerizations, eliminations, and many condensation reactions involve proton abstraction from carbon centers.[11-13] Such reactions are of great importance in many biological transformations. In general, enzymes which catalyze such reactions use carboxylate groups and imidazoles as the catalytic bases to deprotonate the substrate.[14] In order to generate an antibody capable of catalyzing a reaction of this type an Asp, Glu, or His residue needed to be positioned in the correct orientation to the bound substrate. Hapten **3** was used as an immunogen in order to generate an antibody that catalyzed the elimination of HF from *β*-fluoroketone **(4)**.[15] The position of the ammonium group corresponds to the position of

[6] M. J. Page and W. P. Jencks, *Proc. Natl. Acad. Sci. U.S.A.* **68**, 1678 (1971).
[7] D. Pressman and J. Siegel, *J. Am. Chem. Soc.* **75**, 686 (1953).
[8] A. L. Grossberg and D. Pressman, *J. Am. Chem. Soc.* **82**, 5478 (1960).
[9] M. H. Freeman, A. L. Grossberg, and D. Pressman, *Biochemistry* **7**, 1941 (1968).
[10] S. K. Dower and R. A. Dwek, *Biochemistry* **18**, 36687 (1979).
[11] K. J. Shray, E. L. O'Connell, and I. A. Rose, *J. Biol. Chem.* **248**, 2214 (1973).
[12] D. W. Banner, A. C. Bloomer, G. A. Petsko, D. C. Phillips, C. I. Pogson, I. A. Wilson, P. H. Corran, A. J. Furth, J. D. Milman, R. E. Offord, J. D. Priddle, and S. G. Waley, *Nature (London)* **255**, 609 (1975).
[13] H. M. Miziorko and M. D. Lane, *J. Biol. Chem.* **252**, 1414 (1977).
[14] S. M. Parsons and M. A. Rafferty, *Biochemistry* **11**, 1623 (1972).
[15] K. M. Shokat, C. J. Leumann, R. Sugasawara, and P. G. Schultz, *Nature (London)* **338**, 269 (1989).

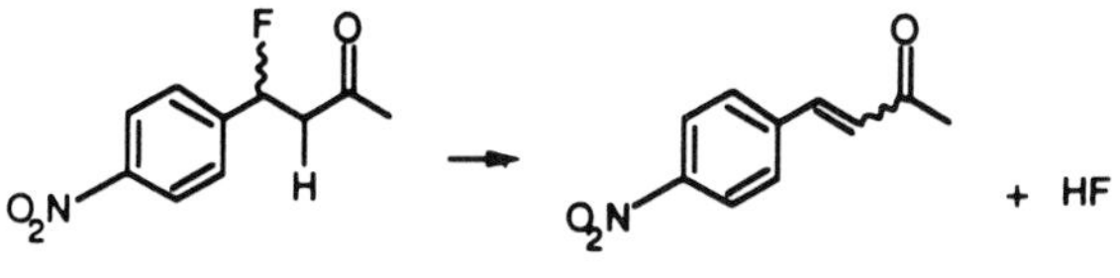

4 → 5 + HF

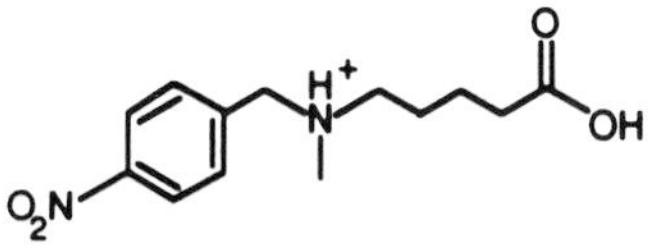

3

the abstractable proton in substrate **4** and should elicit a complementary catalytic carboxylate within bonding distance. The *p*-nitrophenyl ring was included to serve as a common recognition element between hapten and substrate. Moreover, replacement of hapten by substrate in the antibody combining site should lead to an increase in the pK_a of the catalytic carboxylate group (making it a better base) since a stabilizing salt bridge interaction is lost.

Of six antibodies tested which bound **3**, four accelerated the conversion of **4** to **5**, and were able to be inhibited by hapten. Hapten inhibition demonstrates that catalysis occurs in the binding site of the antibody. One of the four antibodies, 43D4-3D12, was further characterized. The kinetics of the 43D4-3D12-catalyzed reaction obeyed the Michaelis–Menten rate expression. The k_{cat} and K_m for substrate **4** were 0.2 sec^{-1} and 182 μM, respectively. In general the K_m values of antibody catalysts are in the same range as those of enzymes. The rate acceleration $((k_{cat}/K_m)/k_{uncat})$ by antibody compared to the background rate with acetate ion was 8.8×10^4, reflecting the contribution of proximity of substrate and a catalytic group in a protein binding site to rate enhancement. This value is similar to the rate acceleration of 10^4 attributable to the bases Glu-43 in staphylococcal nuclease[16] and Asp-102 in trypsin.[17] The antibody discriminated between the *p*-nitro substrate **4** and the *m*-nitro analog by a factor of 10 in overall rate (k_{cat}/K_m).

Chemical modification studies demonstrated that a carboxylate group was indeed responsible for catalysis. Selective chemical modification with the carboxylate-specific reagent diazoacetamide showed almost complete inactivation in the absence of inhibitor. When inhibitor was present activity was retained owing to protection of the carboxylate group responsible

[16] E. H. Serpersu, D. Shortle, and A. S. Mildvan, *Biochemistry* **26**, 1289 (1987).

[17] C. S. Craik, S. Roczniak, C. Largman, and W. J. Rutter, *Science* **237**, 909 (1987).

for catalysis. The pH dependence of catalysis indicated an essential residue with a pK_a of 6.3, active in the deprotonated form. Carboxylate groups on surfaces of proteins have pK_a values between 2.0 and 5.5.[18] However, in the hydrophobic cavities of enzymes they become stronger bases with p*K*a values in the range of 6.5 to 8.2.[18] Further mechanistic studies have shown there to be a kinetic isotope effect ($k_{cat,H}/k_{cat,D}$) for the antibody of 2.4.[19] This value can be compared to the kinetic isotope effect for the background reaction catalyzed by acetate ion, $k_H/k_D = 3.7$. The isotope effect argues against a mechanism involving rate-limiting S_N2 displacement of fluoride followed by elimination.

Charge–charge complementarity has also been applied by Janda *et al.* to design an esterolytic catalytic antibody.[20] These workers targeted the hydrolysis of the mildly activated substituted phenyl ester **6**. Hapten **7** catalyzed the desired reaction. Interestingly, the pH dependence of the catalyzed reaction indicates the catalysis is due to the presence of the basic form of a residue with pK_a = 6.23. This is identical to the pK_a dependence found in the β-elimination catalytic antibody.

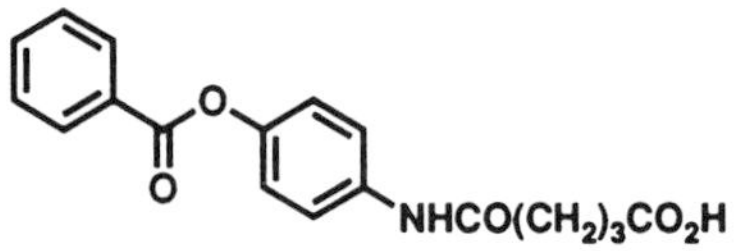

$NHCO(CH_2)_3CO_2H$

6

$NHCO(CH_2)_3CO_2H$

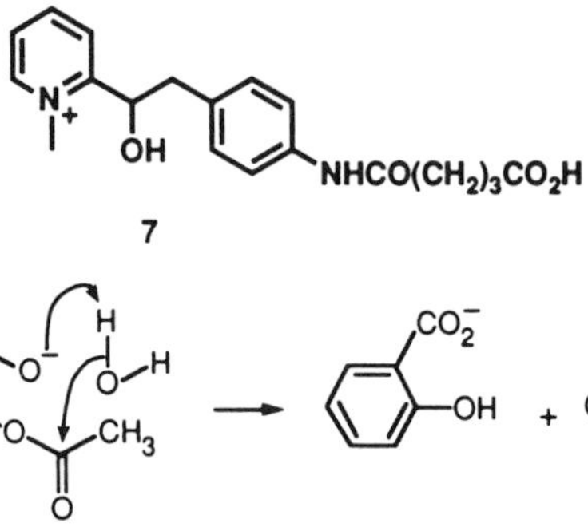

7

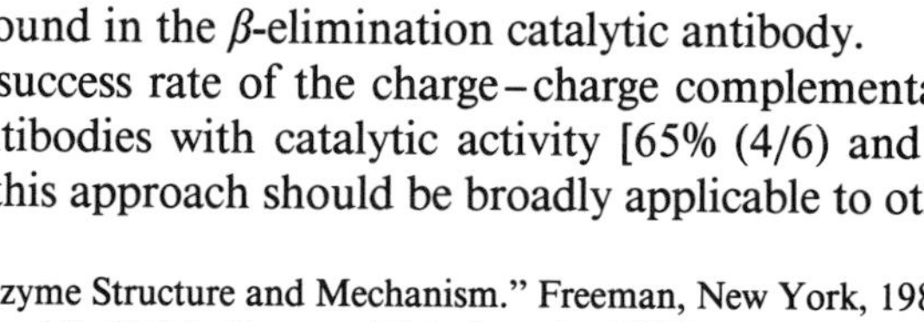

(1)

dependence found in the β-elimination catalytic antibody.

The high success rate of the charge–charge complementary design in generating antibodies with catalytic activity [65% (4/6) and 30% (7/23)] suggests that this approach should be broadly applicable to other reactions

[18] A. Fersht, "Enzyme Structure and Mechanism." Freeman, New York, 1985.

[19] K. M. Shokat and P. G. Schultz, unpublished results (1989).

[20] K. D. Janda, M. Weinhouse, D. M. Schloeder, R. A. Lerner, and S. J. Benkovic, *J. Am. Chem. Soc.* **112**, 1274 (1990).

in which negatively or positively charged amino acid side chains serve as catalysts. An important challenge will be the introduction of uncharged amino acids capable of nucleophilic catalysis into antibody combining sites. Perhaps haptens with reactive labeling groups such as α-halo ketones, epoxides, and maleimides can be exploited for this purpose. (A covalent bond between immunogen and antibody would be the basis for antibody selection rather than purely noncovalent interactions.)

Thymine Dimer Cleavage. Thymine dimers are the primary photolesion product of damage to DNA by UV light. The repair enzyme, DNA photolyase, catalyzes the light-dependent cycloreversion reaction to thymine, but its mechanism is not fully understood.[21] From model systems it is known that compounds such as indoles, quinones, and flavins can inefficiently photosensitize the reaction by reversibly accepting or donating an electron from or to the dimer to generate a radical cation or anion, respectively, both of which might be expected to undergo facile ring opening.[22-26] In order to generate a tryptophan residue in close proximity to the bound thymine dimer substrate antibodies were elicited toward the planar thymine dimer derivative **8**.[27] The extended π system of **8** should elicit a complementary π stacking aromatic amino acid (such as tryptophan) in the antibody combining site as was the case with MOPC 315. An appropriately positioned tryptophan should efficiently sensitize the cycloreversion reaction of **9** to thymine. In fact, five of six antibodies generated to hapten **8** showed light-dependent acceleration of the conversion of **8** to

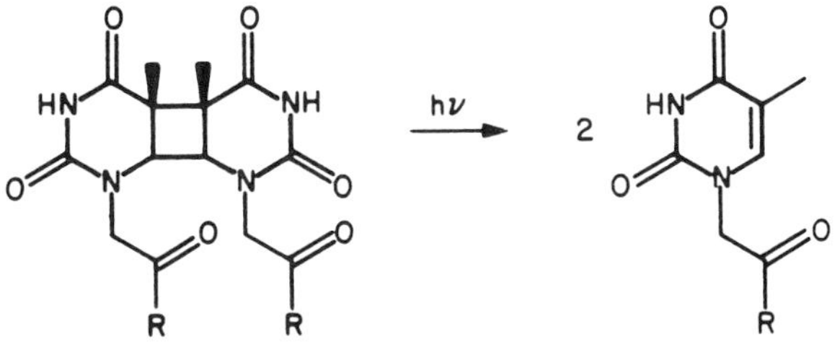

8: R = $NHCH_2COOH$
9: R = OH

thymine. One of these five, 15F1-B1, was further characterized. The substrate analog which is methylated at the two N-3 positions is not turned

[21] B. M. Sutherland, *Enzymes* **14**, 909 (1981).
[22] C. Helene and M. Charlier, *Photochem. Photobiol.* **25**, 429 (1977).
[23] J. R. Van Camp, T. Young, R. F. Hartman, and S. D. Rose, *Photochem. Photobiol.* **45**, 365 (1987).
[24] H. D. Roth and A. A. Lamola, *J. Am. Chem. Soc.* **94**, 1013 (1972).
[25] S. E. Rokita and C. T. Walsh, *J. Am. Chem. Soc.* **106**, 4589 (1984).

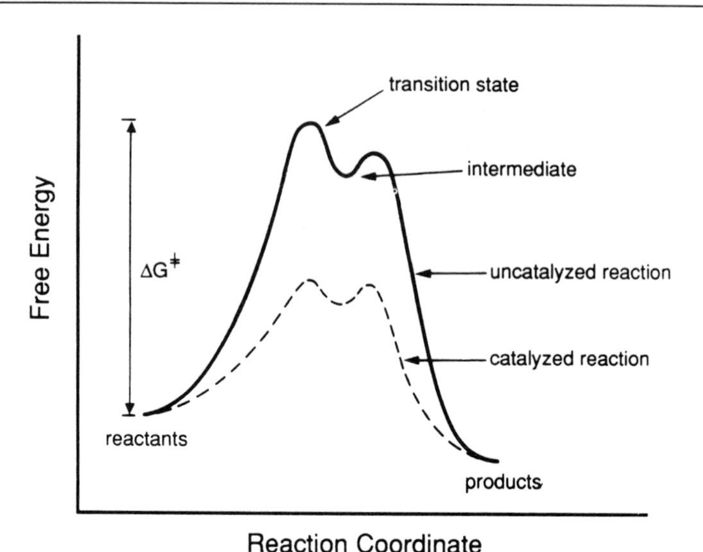

FIG. 1. Transition state stabilization.

over even at relatively high concentrations, consistent with the high specificity of antibodies. Antibodies not specific for **8** but which are known to contain binding site tryptophan residues do not catalyze the reaction. At nonsaturating light intensity the k_{cat} and K_m values were measured to be 1.2 min^{-1} and 6.5 μM, respectively. This k_{cat} value is quite close to the value for *Escherichia coli* DNA photolyase, 3.4 min^{-1}.[28]

The high success rate in using $\pi-\pi$ complementarity to generate catalytic antibodies suggests that this approach also should prove quite generalizable. The ability to generate combining sites with defined functional groups may lead to the generation of discrete microenvironments within antibody combining sites. These microenvionments could demonstrate unusual solvation, dielectric, or hydrophobic properties for studies of molecular recognition and other structure–function relationships.

Transition State Stabilization

The first successful approach used in the design of catalytic antibodies was that of transition state stabilization (Fig. 1), a proposal first made by Linus Pauling in 1948 to explain enzymatic catalysis.[29] Design of an antibody which takes advantage of transition state stabilization to acceler-

[26] M. S. Jorns, E. T. Baldwin, G. B. Sancar, and A. Sancar, *J. Biol. Chem.* **262**, 486 (1987).
[27] A. G. Cochran, R. Sugasawara, and P. G. Schultz, *J. Am. Chem. Soc.* **110**, 7888 (1988).
[28] M. S. Jorns, G. B. Sancar, and A. Sancar, *Biochemistry* **24**, 1856 (1985).
[29] L. Pauling, *Am. Sci.* **36**, 51 (1948).

ate a reaction requires an immunogen which closely resembles the transition state for the reaction. Since transition states are only transient species, stable analogs of the proposed transition state must be used as immunogens. These analogs are in many instances derived from known classes of enzyme inhibitors. Because antibodies are elicited toward transition state analogs, they should have lowered affinity for the reaction products and allow them to diffuse rapidly from the combining site.

Ester, Carbonate, and Amide Bond Hydrolysis.. The first antibodies characterized as having catalytic properties accelerated the hydrolysis rate of esters and carbonates **10–12.** These examples have been extensively

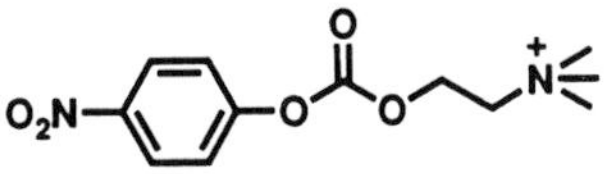

10

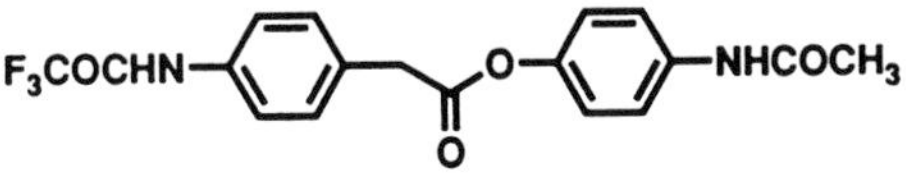

11

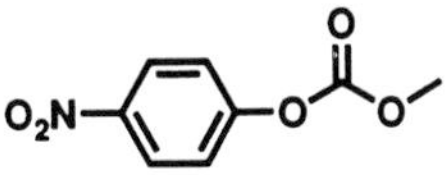

12

reviewed in a previous volume of this series.[30] In each of these cases negatively charged phosphonates or phosphates served as analogs of the tetrahedral negatively charged transition states. In addition to this well-known class of charged peptidase inhibitors, neutral inhibitors containing an hydroxyl group as an analog of the tetrahedral transition state are also potent protease inhibitors.[31] For example, pepstatin has a K_I of 50 nM for pepsin even though it lacks the negative charge which distinguishes the transition state from the products and substrate.[32]

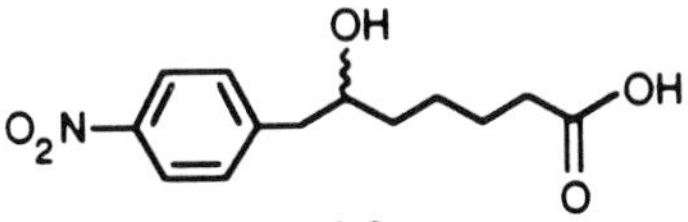

13

Antibodies elicited against alcohol **13** have been shown to catalyze the hydrolysis of **12** and **14**.[33] One out of four antibodies specific for **13** were catalytic with kinetic constants, $k_{cat,14}$ 0.72 min−1; $K_{m,14}$ 3.65 mM;K_I 140

[30] A. Tramoantano and D. Schloeder, this series, Vol. 178, p. 535.

[31] W. M. Kati, D. T. Pals, and S. Thaisrivongs, *Biochemistry* **26**, 7621 (1987).

[32] M. W. Holladay, F. G. Salituro, and D. H. Rich, *J. Med. Chem.* **30**, 374 (1987).

[33] K. M. Shokat, M. K. Ko, T. S. Scanlon, L. Kochersperger, S. Yonkovich, S. Thairivongs, and P. G. Schultz, *Angew. Chem., Int. Ed. Engl.* **29(11)**, 1296 (1990).

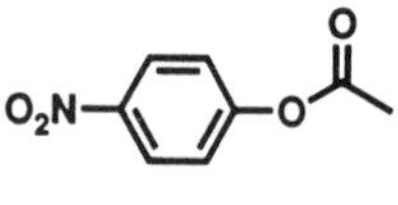

14

μM. The ratio K_m/K_I of 26 is several orders of magnitude lower than the rate acceleration achieved by the antibody, k_{cat}/k_{uncat} 2250. The disparity between the differential stabilization of the transition state relative to the substrate compared to the rate acceleration suggests that another factor such as acid–base catalysis is contributing to the catalysis observed. Chemical modification and affinity labeling experiments may suggest candidate residues in the combining site responsible for the additional catalytic activity.

Antibodies have been shown to catalyze the hydrolysis of water-insoluble substrates.[34] Antibodies specific for phenyl phosphonate **15** were shown to accelerate the hydrolysis of **16** in reverse miscelles. Reverse miscelles are

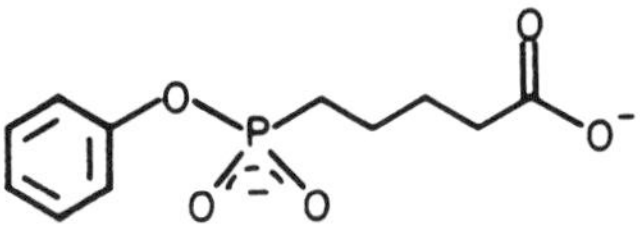

15

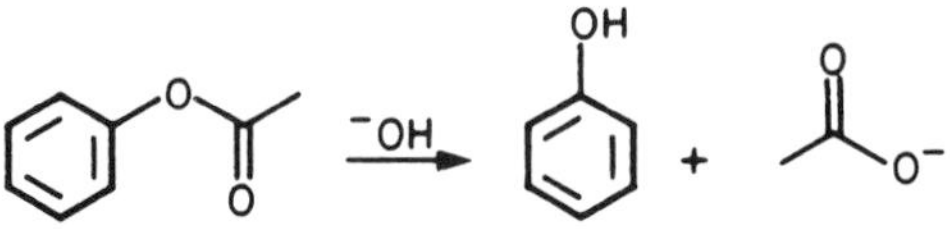

16

formed when water–sufactant mixtures are dissolved in water-immiscible solvents. The k_{cat} and K_m values of the antibody solubilized in isooctane in reverse miscelles were 3.89 min^{-1} and 569 μM, respectively. The optimal ratio of water to detergent (W_o) for antibody catalysis is significantly larger than values found for enzymes, consistent with the increased molecular weight of immunoglobulin G (IgG) molecules. The expansion of antibody catalysis to water-insoluble substrates in reverse miscelles or organic solvents should greatly expand the scope of antibody catalysis.

An antibody, NPN43C9, raised against aryl phosphonamidate **17** has been demonstrated to catalyze the hydrolysis of the activated amide bond in substrate **18**.[35] Hapten **17** is again an analog of the tetrahedral transition

[34] C. N. Durfor, R. J. Bolin, R. J. Sugasawara, R. J. Massey, J. W. Jacobs, and P. G. Schultz, *J. Am. Chem. Soc.* **110**, 8713 (1988).

[35] K. D. Janda, D. Schloeder, S. J. Benkovic, and R. A. Lerner, *Science* **241**, 1188 (1988).

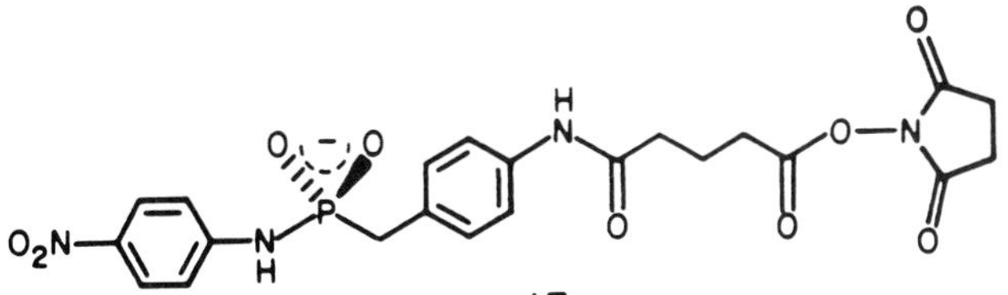

17

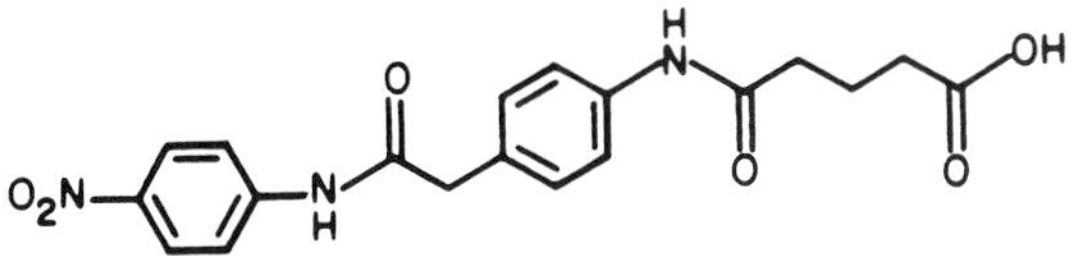

18

state in the hydrolysis of **18**. A rate acceleration by NPN43C9 over the background rate of 2.5×10^5 was reported. This rate acceleration is inconsistent with a mechanism which solely involves transition state stabilization since the difference in affinity for hapten **17** and substrate **18** is only $\Delta\Delta G^{\neq}$ $-$ 2.2 kcal/mol/. The difference in binding energy only accounts for a 100-fold rate acceleration. Other factors such as acid–base catalysis must be responsible for the observed rate acceleration. In this context it was found that 150 m*M* NaCl completely inhibited antibody catalysis. The structural feature of this hapten responsible for induction of such acid–base catalysis is not apparent since the phosphonamidate NH group is not charged in the immunogen. In fact the low success rate, 2% (1/44), suggests that immunological diversity may be responsible for the catalysis observed. Mechanistic analysis of this antibody catalyzed reaction should provide important insight into the generation of antibodies that catalyze the related but considerably more energetically demanding hydrolysis of peptide bonds.

The strategy of transition state stabilization should be applicable to the hydrolysis of other biopolymers such as carbohydrates and polynucleotides. Lysozyme[36] and RNase S[37] are classic examples of enzymes which function by transition state stabilization. The well-characterized transition state analog inhibitors of these enzymes, nojirimycin for lysozyme and uridine vandate complex for RNase S, if used as haptens may yield antibodies capable of catalyzing these important transformations.

Stereospecific Hydrolysis of Unactivated Esters. Antibodies elicited toward phosphonates **19**[38] and **20**[39] were shown to catalyze the hydrolysis

[36] T. I. Secemski and G. E. Liengard, *J. Biol. Chem.* **249**, 2932 (1974).

[37] F. M. Richards and H. W. Wycoff, "Bovine Pancreatic Ribonuclease." Academic Press, New York and London, 1970.

[38] K. D. Janda, S. J. Benkovic, and R. A. Lerner, *Science* **244**, 437 (1989).

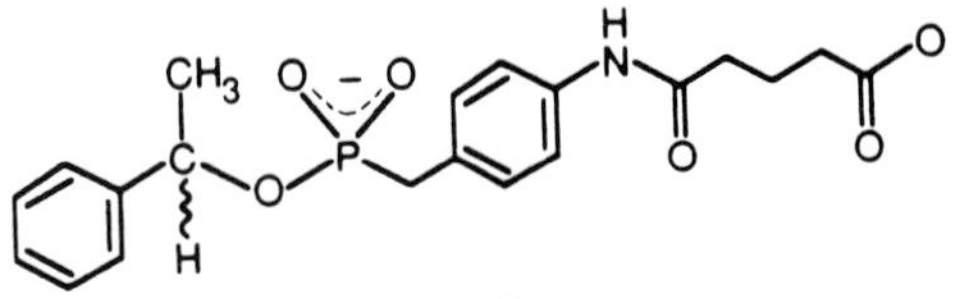

19

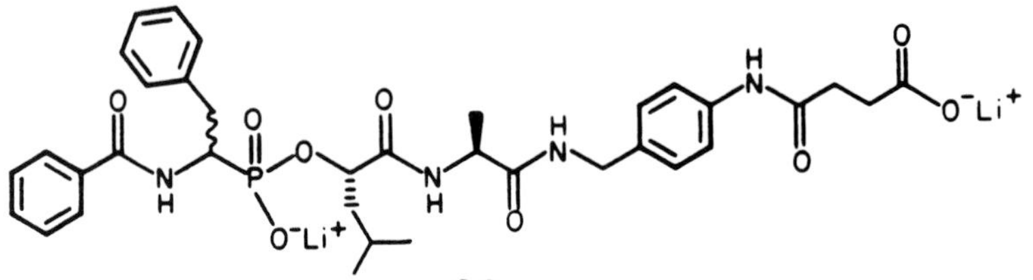

20

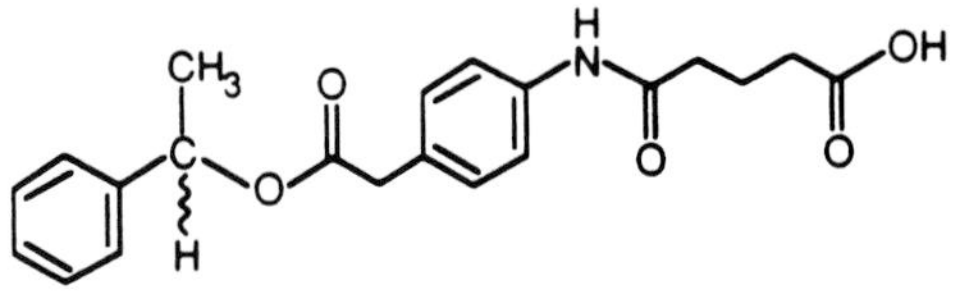

21

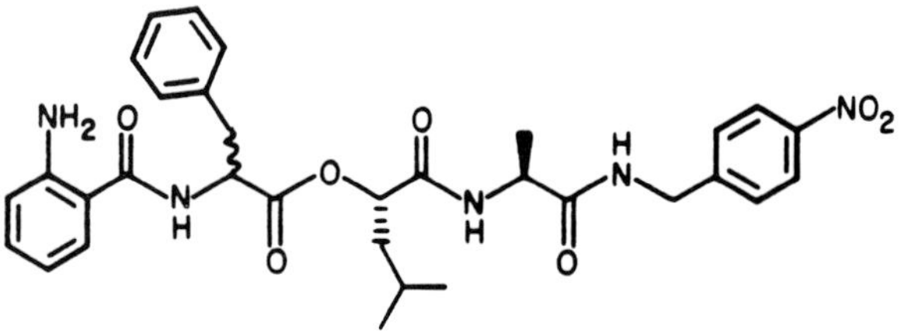

22

of **21** and **22**, respectively, in a stereospecific manner. In both cases antibodies were elicited against a 50/50 mixture of both stereoisomers. Of 18 antibodies specific for **19**, nine catalyzed the hydrolysis of (*R*)-**21** and two catalyzed the hydrolysis of (*S*)-**21**. Two antibodies, one of each specificity, were further characterized. The rate acceleration for the hydrolysis of (*R*)-**21** by antibody 2H6 was 80,000, whereas the (*S*)-**21** specific antibody 21H3 showed only a modest 1600-fold rate acceleration over background. The *R/S* selectivity is greater than 98%.

Of 25 antibodies isolated that were specific for the tripeptide phosphonate **20**, 18 were shown to accelerate the hydrolysis of the correspond-

[39] S. J. Pollack, P. Hsuin, and P. G. Schultz, *J. Am. Chem. Soc.* **111**, 5961 (1989).

ing ester substrates. Interestingly, an exclusive preference for antibodies of one specificity was observed. All 18 antibodies were selective for the D-phenlyalanine isomer of **22**. The selectivity for D-over L-phenylalanine at this position was greater than 99.5% for three of the five antibodies characterized. The modest rate accelerations of 50 to 300-fold may be a result of the size of the immunogen. The tetrahedral phosphonate contributes proportionally less to the overall binding energy of hapten to antibody than in smaller haptens. Note that the hapten also contains analogs of fluorogenic and quenching groups at the amino and carboxy termini of substrate **22**, respectively. These groups allow hydrolysis of the substrate to be monitored by observing the increase in fluorescence which occurs when the fluorescent 2-aminobenzoyl group is separated from the quenching 4-nitrobenzylamide group in the reaction.[40] This sensitive assay may allow direct screening of enzyme-linked immunosorbent assay (ELISA) plates for catalytic activity or the screening of λ libraries of antibody F_{ab} fragments,[41] facilitating the production of catalytic antibodies. These two examples suggest that excellent stereoselectivities can be achieved by antibodies with large or small substrates and chiral centers in the alcohol portion or the acyl group portion of the substrate. This strategy could be readily applied to the chiral resolution of racemic alcohols and esters in the production of pharmaceuticals.

Claisen Rearangement. The Claisen rearrangement of chorismate, **23**, to prephenate, **24**, formally involves the concerted cleavage of a carbon–oxygen bond and formation of a carbon–carbon bond. The enzyme which catalyzes this reaction, chrismate mutase, is at the branchpoint of aromatic amino acid biosynthesis in bacteria and plants. Bicyclic diacid **25** is the most potent transition state analog inhibitor of the enzyme known.[42] Two independent groups used **26**[43] and **27**[44,45] as haptens for generating antibody catalysts for the Claisen rearrangement of **23** to **24**. In one case one (11F1) out of eight antibodies specific for **26** was shown to catalyze the reaction with a k_{cat} of 2.7 min^{-1} and a K_m of 260 μM at 10°, pH7.0. This corresponds to a rate enhancement of 10^4 over the uncatalyzed rate and a decrease in $\Delta S^{\neq}$ from -12.9 e.u. for the uncatalyzed reaction to -1.2 e.u. for the 11F1 catalyzed reaction.[46] This factor compares with the 3×10^6-

[40] N. Nishino and J. C. Powers, *J. Biol. Chem.* **255**, 3482 (1980).

[41] W. D. Huse *et al., Science* **246**, 1275 (1989).

[42] P. A. Bartlett and C. R. Johnson, *J. Am. Chem. Soc.* **107**, 7792 (1965).

[43] D. Y. Jackson *et al., J. Am. Chem. Soc.* **110**, 4841 (1988).

[44] D. Hilvert, S. H. Carpenter, K. D. Nared, and M.-T. M. Auditor, *Proc. Natl. Acad. Sci. U.S.A.* **85**, 4953 (1988).

[45] D. Hilvert and K. D. Nared, *J. Am. Chem. Soc.* **110**, 5593 (1988).

[46] D. Y. Jackson and P. G. Schultz, unpublished results (1989).

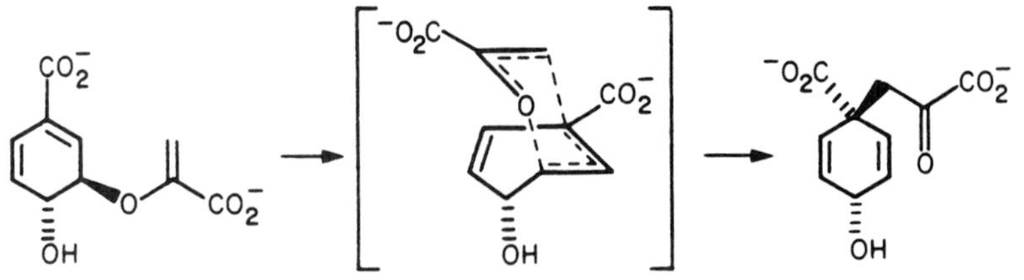

23

24

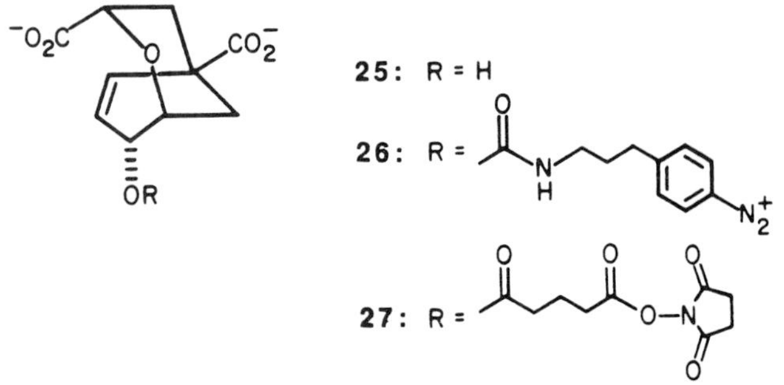

25: R = H

26: R =

27: R =

fold rate enhancement achieved by *E. coli* chorismate mutase assayed under the same conditions. Notice that this is the first example of a rate enhancement with respect to a unimolecular background reaction. In all the other examples the background reaction is bimolecular. With **27** as the immunogen, one (1F7) of 15 antibodies specific for **27** was catalytic, with a k_{cat} equal to 0.025 min^{-1} and a K_m of 22 μM at 13°, pH 7.5. Antibody 1F7 accelerates the reaction 190-fold over the uncatalyzed rate. Most of the rate acceleration is attributable to a decrease in $\Delta H^{\ddagger}$. Antibody 1F7 was also shown to be highly stereoselective, having a 90:1 preference for the (−) over the (+) isomer of chorismate. Complementation experiments underway in yeast and *E. coli* will provide formats to increase catalytic activity via genetic selection techniques.

Diels–Alder Reactions. Two groups have described the generation of antibodies capable of catalyzing Diels–Alder reactions. This reaction proceeds through a concerted transition state involving simultaneous formation and breakage of carbon–carbon bonds. It is one of the most important reactions in synthetic organic chemistry because it is a highly stereospecific route to cyclohexene derivatives. Hilvert and co-workers chose to catalyze the Diels–Alder reaction between tetrachlorothiophene 1,1-dioxide (**28**) and *N*-ethylmaleimide (**29**).[47] The hapten chosen was the bicyclic adduct

[47] D. Hilvert, K. W. Hill, K. D. Nared, and M.-T. M. Auditor, *J. Am. Chem. Soc.* **111,** 9261 (1990).

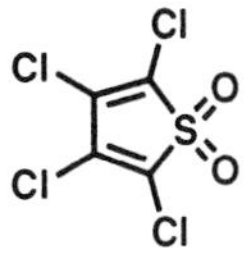

28

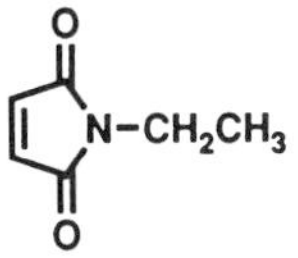

29

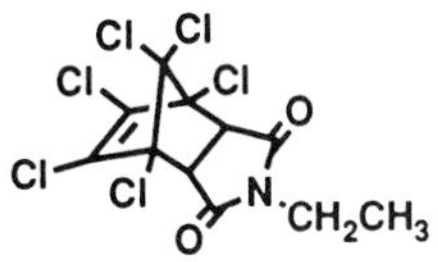

30

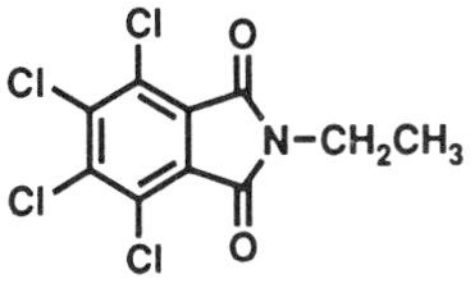

31

30. Tetrachlorothiophene 1,1-dioxide was used as the diene since extrusion of sulfur dioxide from the Diels–Alder adduct minimizes product inhibition. One of five antibodies specific for hapten **30** was shown to ctalyze the desired Diels–Alder reaction. Because of low solubility of the substrates the true k_{cat} value was not measurable.

An example of a more generalizable approach to catalyzing Diels–Alder reactions is the conversion of **32** and **33** to **34**.[48] This reaction does not depend on extrusion of sulfur dioxide. Hapten **35** contains an ethano

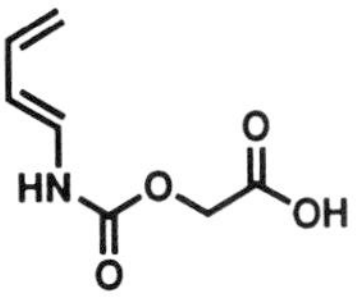

32

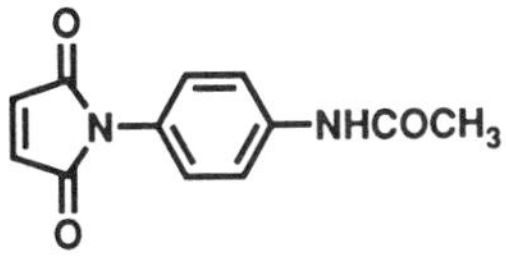

33

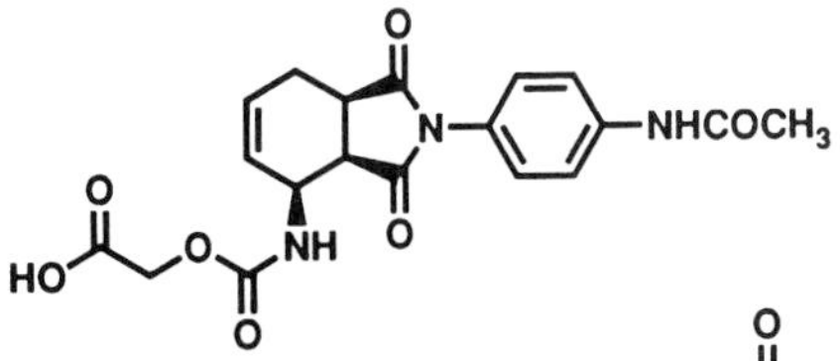

34

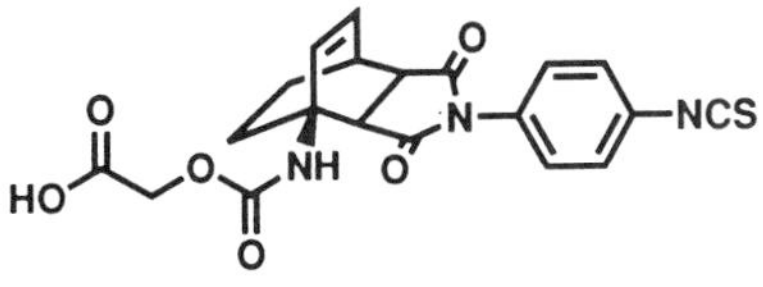

35

[48] A. Braisted and P. G. Schultz, *J. Am. Chem. Soc.* **112,** 7430 (1990).

bridge locking the cyclohexene ring into the proposed pericyclic transition state conformation. Furthermore, the high-energy boatlike conformation of the hapten should be distinct from the lowest energy conformer of the Diels–Alder product **34**, and hence product inhibition should be minimized. One of ten antibodies specific for **35** showed a large rate enhancement over the background rate: $k_{uncat,pH7.5}$ 1.9 M sec^{-1}, $k_{uncat,acetonitrile}$ 0.002 $M^{-1}sec^{-1}$. The catalytic rate constrants were measured to be k_{cat} 0.67 sec^{-1}, $K_{M,32}$ 1130 μM, and $K_{m,33}$ 740 μM. The K_D values for the hapten and product were determined to be 126 nM and 10 μM, respectively. The 100-fold difference in binding energy between the hapten and the substrates does lead to product inhibition after 5% reaction completion.

Combining strategies used to catalyze complex organic reactions such as Diels–Alder reactions with the ability to catalyze reactions in reverse miscelles provides intriguing possibilities for control of stereoselectivity.

Acyl-Group Transfer Reactions.. The first examples of antibody-catalyzed bimolecular reactions involved acyl-transfer reactions. Antibodies specific for bisubstrate analog **36** were shown to catalyze the bimolecular amide bond formation reactions of **37** to **38** involving phenylenediamine.[49]

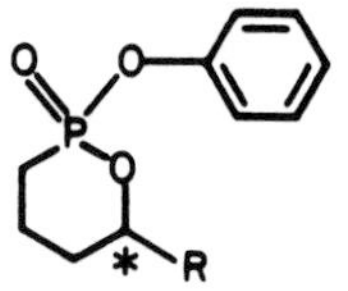

36

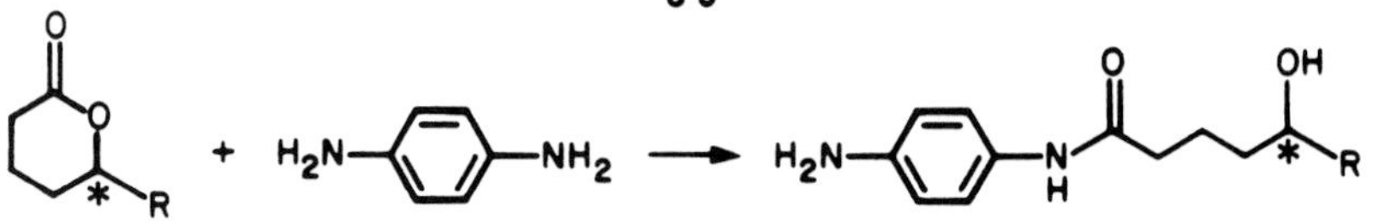

37 + H_2N–C$_6$H$_4$–NH_2 → 38

The antibody binds phenylenediamine and lactone **37** with K_m values of 1.2 and 4.9 mM, respectively. The inhibitor (**36**) was shown to be competitive for both reactions having a K_I of 75 nM for the bimolecular condensation reaction and 250 nM for the cyclization reaction. The effective molarity for the bimolecular reaction was measured to be 16 M, which is considerably below the theoretical limit of 10^8 M. This upper bound is based on the calculated value of approximately 45 e.u. for the translational and rotational entropy of activation of a bimolecular reaction.[6]

[49] S. J. Benkovic, A. D. Napper, and R. A. Lerner, *Proc. Natl. Acad. Sci. U.S.A.* **85**, 5355 (1988).

Another antibody generated against phosphonamidate **39** was shown to catalyze a bimolecular amide formation reaction [Eq.(2)][50].

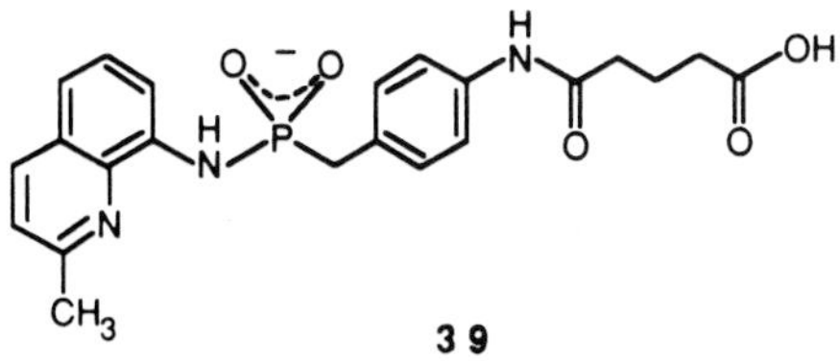

39

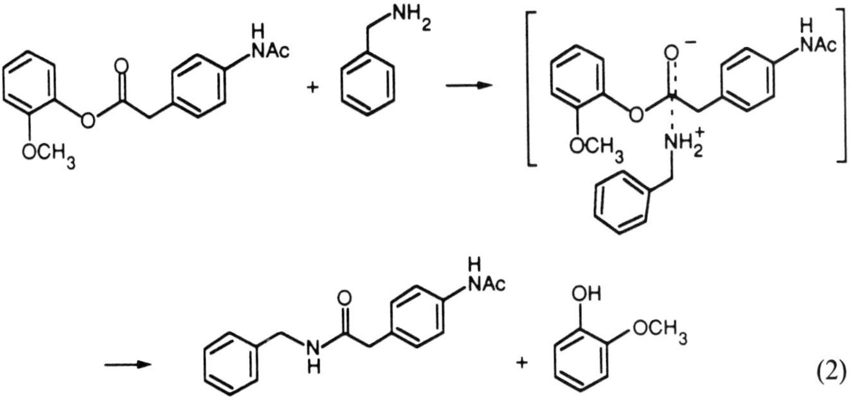

(2)

The antibody was shown to provide an effective molarity of 10.5 *M*. Here the hapten did not include the ester leaving group, an important element of the proposed transition state for the reaction. This design feature may explain why less than 2% (1/55) of antibodies specific for **39** were able to catalyze the reaction of interest.

The condensation of large peptide fragments synthesized on a peptide synthesizer is slow in solution and subject to many side reactions. Thus, the ligation of large peptide fragments ($>$ 50 amino acids) represents a significant obstacle to the *de novo* synthesis of large proteins. Antibodies could be designed using the strategy described above to bring the appropriate carboxyl and amino termini together (protecting groups would not be necessary) in the combining site and catalyze peptide bond formation. Another important class of bimolecular reactions that could be catalyzed by this strategy is aldol condensation. An example of this class of reaction is the formation of fructose 1,6-diphosphate from dihydroxyacetone phosphate and glyceraldehyde 3-phosphate catalyzed by aconitase.

[50] K. D. Janda, R. A. Lerner, and A. Tramontano, *J. Am. Chem. Soc.* **110**, 4835 (1988).

Cofactor Chemistry

Many enzymes utilize non-amino acid cofactors to catalyze reactions. Important members of this class of enzymes include cytochrome P_{450}(Fe-heme), α-keto-acid dehydrogenases (thiamin pyrophosphate), D-amino-acid oxidase (flavin), and alanine racemase (pyridoxal phosphate). In order to expand the scope of antibody catalysis to redox reactions and energetically demanding hydrolytic reactions, strategies that allow incorporation of cofactors into antibody combining sites will have to be used. To this end antibodies have been elicited against flavin cofactor **40**.[51] The three rings are coplanar in the oxidized flavin, **40**. The reduced dihydroflavin, **41**, has a substantially different electron distribution about the rings and is also conformationally distinct from the oxidized compound (**35**). The reduced form is butterfly shaped owing to a pucker about the two central nitrogen atoms in the central ring. Antibodies elicited toward oxidized **40** bind this form with a 4×10^4 higher affinity than **41**. This differential binding energy manifests itself in the reduction potential (E_m) of the bound flavin: the reduction potential of flavin in the absence of antibody is -206 mV, and in the antibody bound state the flavin E_m is -343 mV. Hence, bound flavin is a substantially stronger reducing agent than free flavin. Because the antibody–flavin complex is a stronger reductant than free flavin, the substrate safranine T (E_m -289 mV) could be rapidly reduced by the antibody flavin complex. This process does not occur with free flavin. The antibody–flavin complex is therefore able to mediate redox reactions not thermodynamically accessible to free flavin. Here antibody binding energy is used to destabilize the reduced form of a flavin, creating a more powerful donor of electrons to substrates. By incorporating substrate binding sites

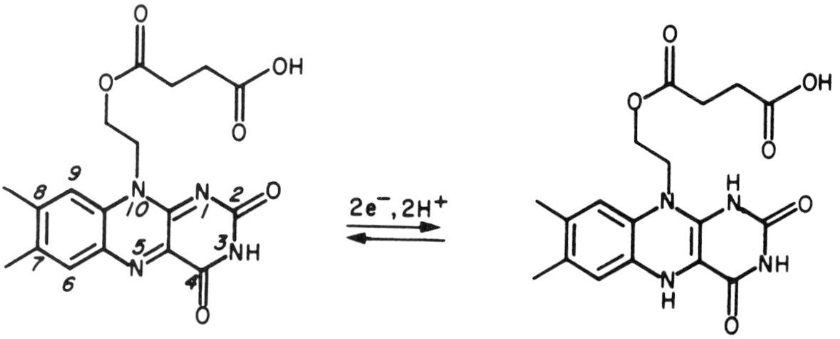

[51] K. M. Shokat, C. J. Leumann, R. Sugasawara, and P. G. Schultz, *Agnew. Chem. Int. Ed. Engl.* **27**, 1172 (1988).

adjacent to flavins, stereocontrolled chemical reductions could be carried out.

Antibodies have been generated against a peptide–cofactor complex (**42**) that are capable of catalyzing the hydrolysis of a Gly-Phe peptide

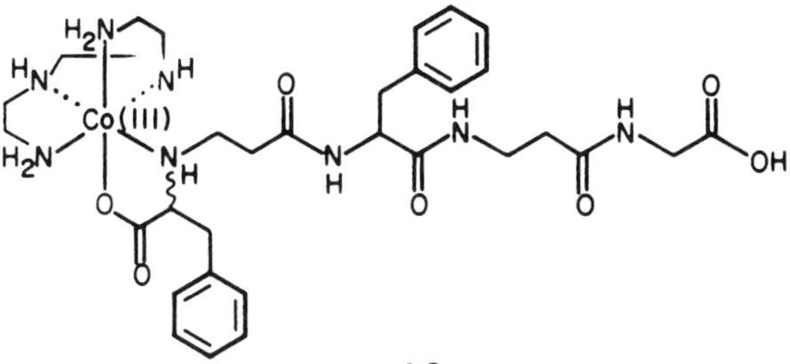

42

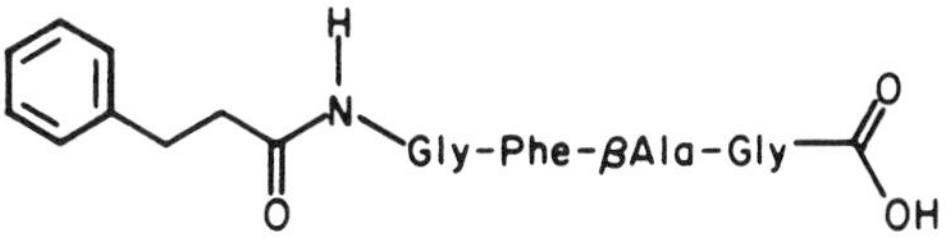

43

bond.[52] Although a structurally inert Co(III) trien [*N,N'*-bis(2-aminoethyl)-1,2-ethanediamine] complex was used as the cofactor in the immunogen, the 14 antibodies elicited could accommodate any of 13 different trien metal complexes. In the presence of a variety of metal trien complexes two of the antibodies were shown to catalyze the cleavage of the Gly-Phe peptide bond in **43**. A turnover number of 6×10^{-4} sec^{-1} was measured. Somewhat surprisingly, the Gly-Phe bond hydrolyzed is one removed from the bond one might have expected to have been cleaved based on model chemistry by Buckingham and others.[53] Nonetheless, the extension of antibody catalysis to biopolymers such as peptides is of great value. The general rules of using cofactors in such systems must be further developed.

A second example of antibody-catalyzed peptide hydrolysis has been reported. Two of six immune human IgG autoantibody preparations were shown to cleave ^{125}I labeled [Tyr^{10}] vasoactive intestinal peptide (VIP) between residues Gln-[16] and Met-[17].[54] The k_{cat} and K_m values reported were 15.6 min−1 and 37.9 n*M* respectively. Surprisingly the *background* rate for

[52] B. L. Iverson and R. A. Lerner, *Science* **243**, 1184 (1989).
[53] D. A. Buckingham, D. M. Foster, and A. M. Sargeson, *J. Am. Chem. Soc.* **92**, 6151 (1970).
[54] P. Sudhir, S. Paul, D. J. Volle, C. M. Beach, D. R. Johnson, M. J. Powell, and R. J. Massey, *Science* **244**, 1158 (1989).

the hydrolysis of the iodinated substrate used in the study is remarkably fast, considering a measurable amount of conversion could be observed in as little as 3 hrs. Radiolysis may contribute to the fast background rate and to the "antibody-catalyzed" reaction. The normal half-life ($t_{1/2}$) of peptide bonds is 7 years.[55] Clearly, it is essential that monoclonal antibodies with this activity be generated and carefully characterized.

One of the most interesting recent results in the field of catalytic antibodies has been catalysis of a porphyrin metallation reaction.[56] This system expands the scope of antibody-catalyzed reactions to ligand substitution reactions and is the first report of antibody binding of natural porphyrins. Two of three antibodies specific for the strained *N*-methymesoporphyrin (**44**) catalyze the metallation of mesoporphyrin IX (**45**). One of the antibodies catalyzes metallation with zinc(II) and copper(II). The kinetics of the zinc(II) metallation k_{cat} 80 hr^{-1}, compare favorably with the same reaction catalyzed by the enzyme ferrochelatase, k_{cat} 800 hr^{-1}. The antibody shows high substrate specificity in that protoporphyrin IX (**46**) and deuteroporphyrin IX (**47**) are not accepted as substrates in the metallation reactions.

Furthermore, this antibody also binds two biologically relevant metalloporphyrins, iron(III) mesoporphyrin and iron (III) protoporphyrin. The complex of antibody–iron(III) mesoporphyrin was recently shown to catalyze the reduction of hydrogen peroxide by several typical chromogenic peroxidase substrates: pyrogallol, hydroquinone, *o*-dianisidine, and 2,2′-azinobis(3-ethylbenzothiazoline-6-sulfonic acid).[57] Addition of a stoichiometric amount of hapten **44** completely inhibited antibody catalysis of the

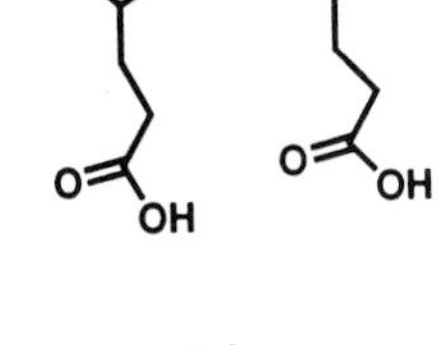

NH N= N N O OH O OH

44

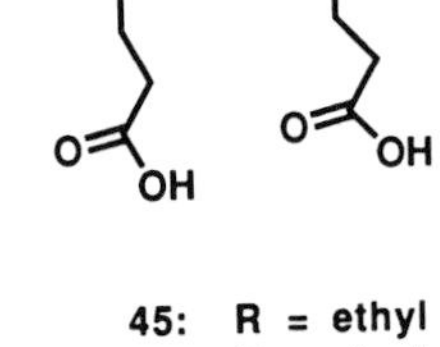

R R NH N= N HN O OH O OH

45: R = ethyl
46: R = vinyl
47: R = H

[55] D. Kahne and W. C. Still, *J. Am. Chem. Soc.* **110**, 7529 (1988).
[56] A. G. Cochran, and P. G. Schultz, *Science* **249**, 781 (1990).
[57] A. G. Cochran and P. G. Schultz, *J. Am. Chem. Soc.* **112,** 9414 (1990).

peroxidation reactions. Kinetic analysis of the oxidation of *o*-dianisidine catalyzed by the antibody afforded a $k_{cat}/k_m(H_2O_2)$ ratio of 274 $M^{-1}sec^{-1}$. For comparison, peroxidases are among the most efficient enzymes known, with typical $k_{cat}/K_m(H_2O_2)$ values of 10^7 $M^{-1}sec^{-1}$. The identity of axial ligands to the hemin iron in these complexes will be investigated in order to determine whether any are contributed by the antibody itself. By incorporating both a porphyrin and an oxidation substrate in an antibody combining site perhaps stereo and regioselective oxidations can be catalyzed.

Genetic and Chemical Modification of Existing Antibodies

A second major approach toward the design of catalytic antibodies involves introduction of catalytic activity into extant antibodies of the desired specificity either by selective chemical methods or by site-directed mutagenesis. Both approaches also allow for the incremental increase in rate enhancements achieved by antibodies generated via other strategies.

Site-Directed Mutagenesis. Site-directed mutagenesis has been successfully applied to the antibody MOPC 315.[58] MOPC 315 binds substituted 2,4-dinitrophenyl (DNP) ligands with association constants ranging from 10^3 to 10^7 M^{-1}.[59] The combining site has been characterized by spectroscopic methods (UV, fluorimetry, NMR), chemical modification, and amino acid sequencing of the variable region. Moreover, earlier affinity-labeling studies with reagents of varying structures defined a number of reactive amino acid side chains in the vicinity of the combining site.[60]

In the absence of an X-ray crystal structure of MOPC 315, chemical modification experiments were used to guide the mutagenesis. In spite of the fact that there are 14 potentially reactive side chains in the hypervariable region (2 histidines, 2 lysines, 3 arginines, and 7 tyrosines), DNP-containing affinity labels alkylate primarily two residues, tryrosine-34L and lysine-52H.[60] Tyrosine-34L was chosen as the initial target for mutagenesis since it appeared that histidine at residue 34 would be situated to catalyze the hydrolysis of esters. The role of tyrosine-34L in DNP binding has not been clearly established; affinity labeling of the tyrosine prevents binding of ligands,[60] perhaps by sterically blocking the entrance to the site. Nitration of the tyrosine ring has no effect.[61]

In order to generate mutants of MOPC 315, a recombinant Fv fragment was generated. MOPC 315 IgA can be proteolyzed with pepsin to

[58] E. Baldwin and P. G. Schultz, *Science* **245**, 1104 (1989).
[59] D. Haseldorn, S. Friedman, D. Givol, and I. Pecht, *Biochemistry* **13**, 2210 (1974).
[60] J. Haimovich, H. W. Eisen, E. Hurwitz, and E. Givol, *Biochemistry* **11**, 2389 (1972).
[61] R. J. Leatherbarrow, W. R. C. Jackson and R. A. Dwek, *Biochemistry* **21**, 5124 (1982).

yield functional Fab or Fv fragments.[62] The Fv fragment (26 kDa) contains all the sequences necessary for folding of the binding domain and recognition of the DNP hapten. Reduced and separated V_H and V_L can be efficiently formed by oxidation of the reduced forms prior to reconstitution.[62] This property has been exploited to produce a hybrid Fv, in which recombinant V_L produced in *E. coli* is reconstituted with V_H derived from MOPC 315 IgA. Functional Fv and Fab fragments of other antibodies have been previously expressed in *E. coli*.[63] A gene encoding the N-terminal 115 amino acids of MOPC 315 light chain was chemically synthesized and expressed in *E.coli* as a fusion with the λ *c*II gene.[64] The resulting hybrid protein was cleaved at one site with Factor Xa, the liberated V_L peptide was reconstituted with V_H, and the resulting Fv was purified to homogeneity by gel filtration and affinity chromatography.

Wild-type Fv [Fv(315)] and a phenylalanine mutant [Fv(Y34F_L)] bound DNP-*L*-lysine with similar affinities at pH 6.8 (K_D 250 n*M*), whereas the histidine mutant [Fv(Y34H_L)] bound DNP-lysine with 6-fold lower affinity (K_D 1400 n*M*). The histidine mutant Fv accelerated hydrolysis of **48** 6×10^5-fold over the background reaction with 4-methylimida-

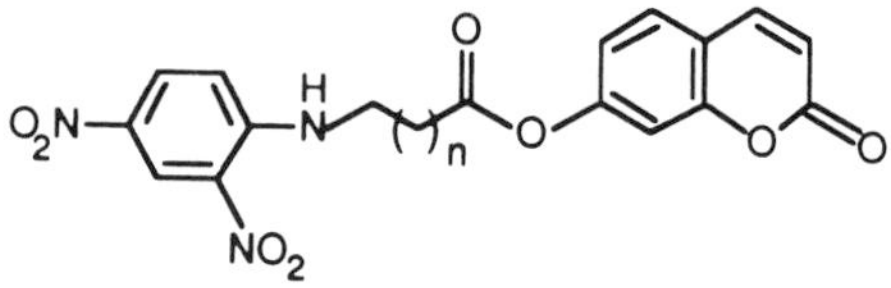

48: n = 1

zole at pH 6.8. The initial rate of ester cleavage is 50 times faster than that of wild-type Fv or the Fv(Y34F_L) mutant. In addition, the hydrolyses of aminopropanoic and aminohexanoic homologs were not significantly accelerated, consistent with the postulated location of the His-34 side chain in the active site. Site-directed mutagenesis should prove to be a powerful tool either for augmenting the rate enhancements of catatytic antibodies generated via other strategies or for the stepwise evolution of catalytic activity in antibodies to produce efficient selective catalysts.

Site-directed mutagenesis has also been used to probe the catalytic mechanism of phosphorylcholine (PC) binding antibody S107.[65] S107 is one member of a well-characterized class of PC binding antibodies. Two antibodies in this class, T15 and MOPC 167, have been shown to catalyze

[62] J. Hochman, D. Inbar, and D. Givol, *Biochemistry* **12**, 1130 (1973).

[63] A. Skerra and A. Pluckthun, *Science* **240**, 1038 (1988).

[64] K. Nagai and H. C. Thogersen, *Nature (London)* **309**, 810 (1984).

[65] D. Y. Jackson, J. R. Prudent, E. P. Baldwin, and P. G. Schultz, *Proc. Natl. Acad. Sci. U.S.A.* in press (1990).

the hydrolysis of PC *p*-nitrophenyl carbonate (**10**). Additionally, an X-ray crystal structure has been solved for the highly homologous PC binding antibody McPC 603.[66] Based on the three-dimensional structure and earlier kinetic and chemical studies, it is thought that T15 and MOPC 167 stabilize the tetrahedral negatively charged transition state for carbonate hydrolysis.[67] Two residues, Tyr-33 and Arg-52, are thought to play critical roles in the catalyzed reaction. Mutagenesis was performed in a pUC-fl plasmid containing the V_H binding domain of S107. After *in vitro* mutagenesis the V_H sequence was inserted back into a modified SV2 expression vector, (pSV2-S107). The vector allows coexpression of the S107 κ light chain and heavy chain variable region with a γ_{2b} constant region from a BALB/c liver cell. The wild-type and mutant genes were transfected into a myeloma cell line which produces no endogenous antibody and is gpt negative. Expression of up to 5 μg of antibody per milliliter of cell culture supernatant was obtained.

Mutations of Arg-52 to lysine, glutamine, and cysteine were made. Both the glutamine and cysteine mutants showed a large decrease in catalytic efficiency ($k_{cat,Glu}$ 0.04 min^{-1},$k_{cat,Cys}$ 0.02 min^{-1}), presumably because they lack a positive charge to stabilize the negatively charged transition state for hydrolysis of carbonate **10**. The lysine mutant displays catalytic activity similar to wild-type, consistent with molecular modeling which shows the ϵ-nitrogen of lysine to be in close proximity (3.4Å) to the phosphoryl oxygen of PC. These mutations emphasize the importance of electrostatic interactions in transition state stabilization. In light of this result it is interesting that uncharged transition state anaolgs successfully elicit catalytic antibodies for hydrolysis reactions. In contrast, mutations at Tyr-33 had little effect on catalytic activity with the exception of a Tyr-His mutant which showed a 6,700-fold rate enhancement of carbonate hydrolysis over the rate for 4-methylimidazole. No pre-steady-state burst of nitrophenolate was observed in the hydrolysis of **10** by the His mutant. This study successfully demonstrates that a general base catalysis can be combined with electrostatic catalysis in an additive fashion to improve the efficiency of an antibody by suing site-directed mutagenesis.

An important breakthrough in antibody technology is the recent development of a method to express the murine immunological repertoire in *E. coli*.[41] The method uses the polymerase chain reaction to make separate libraries of the gene segments encoding the antigen binding portion of the heavy-and light-chain proteins. The libraries are then recombined at ran-

[66] D. M. Segal, E. A. Padlan, G. H. Cohen, S. Rudikoff, D. Potter, and D. R. Davies, *Proc. Natl. Acad. Sci. U.S.A.* **71**, 4298 (1974).
[67] S. J. Pollack, J. W. Jacobs, and P. G. Schultz, *Science* **234**, 1570 (1986).

dom in a bacteriophage λ vector system. Infection of *E. coli* with the phage produces up to 10^{12} heavy–light chain combinations. Currently, several million clones can be screened in 2 days for antigen binding, compared with 6 months or longer needed to screen fewer than 1000 clones produced by conventional hybridoma technology. With this method, it may soon be possible to screen large numbers of antibody clones directly for catalytic activity. Such genetic screens can be combined with directed mutagenesis and chemical approaches to develop highly efficient catalytic antibodies.

Semisynthetic Antibodies. A comprehensive review of efforts in generating semisynthetic antibodies has recently appeared in this series.[68]

Antibody Generation

The protocols used to generate catalytic monoclonal antibodies are generally no different than those used for the production of monoclonal antibodies for other uses.[69,69a,70] The haptens described below have been conjugated to the carrier protein keyhole limpet hemocyanin (KLH) for immunization and to bovine serum albumin (BSA) for use in ELISA assays to identify hapten-specific antibodies.[71] The coupling strategies used have been designed to be compatible with hapten structure and *in vivo* stability. Couplings generally involve amide bond formation between carboxyl groups on haptens and ε-amino groups of surface lysine residues on carrier proteins.[72] Some haptens have been coupled via diazo linkages to surface tyrosine residues, via disulfide exchange reactions, and via reductive amination.[73] Typically, the length of the spacer arm between the hapten and carrier protein is greater than 6 Å, so as to preclude any steric interference from the carrier.[72] Epitope densities between 4 and 30 have been successfully used. Several strains of mice have been used including BALB/c, Swiss Webster, B10Q, and AJ1. Traditional polyethyleneglycol (PEG) fusions of splenocytes with the SP2/0 myeloma cell line have been used.[71,74] Antibodies were screened by ELISA for cross-reactivity with BSA–hapten conjugate, for inhibition of binding to the BSA–hapten conjugate by free hapten, and for lack of cross-reactivity with KLH. Those with maximum binding affinity for the hapten carrier conjugate were

[68] S. J. Pollack, G. R. Nakayama, and P. G. Schultz, this series, Vol. 178, p. 551.

[69] G. Kohler and C. Milstein, *Nature (London)* **256**, 495 (1975).

[69a] G. Galfrè, and C. Milstein, this series Vol. 73, p. 3. See also several chapters in Vols. 92 and 121. [Ed.]

[70] J. W. Goding, "Monoclonal Antibodies: Principles and Practice." Academic Press, New York, 1986.

[71] E. Engvall and P. Perlmann, *J. Immunol.* **109**, 129 (1972).

[72] T. Nishima, A. Tsuji, and D. K. Fukushima, *Steriods* **24**, 861 (1974).

[73] B. F. Erlanger, this series, Vol. 70, p. 85.

[74] R. Sugasawara, C. Prato, and J. Sippel, *Infect. Immun.* **42**, 863 (1983).

characterized. In our experience only IgG and IgA antibodies have been proved to function as catalysts, even though a large number of IgM antibodies have been tested. Either protein A affinity or ion-exchange chromatography methods have been used to purify antibodies to homogeneity.[75]

Characterization of the kinetic parameters, specificity, mechanism, and structural properties of catalytic antibodies is complicated and difficult to interpret unless a monoclonal antibody is generated. Moreover, reproducibility becomes an important concern when a polyclonal antibody preparation is generated. It should also be noted that over 15 years ago Raso and Stollar attempted to use pyridoxamine binding antibodies to catalyze Schiff base formation and transmination reactions.[76] Their lack of success may have been due to the use of polyclonal rather than monoclonal antibodies.

The purity of monoclonal antibodies is tremendously important, especially when an enzyme is known to catalyze the reaction of interest. For example, if the turnover number, k_{cat}, for a catalytic (IgG) antibody is 1 min^{-1} and a natural enzyme has a k_{cat} of $3 \times 10^4 min^{-1}$, contamination of the antibody with 6×10^{-5}% (on a mole/mole basis) of the natural enzyme would lead one to believe that the antibody is catalytic when in fact contaminating enzyme is responsible for the rate enhancement observed. Kinetic analysis, specificity, or inhibition data cannot always distinguish between catalysis by an antibody and catalysis by an enzyme impurity. Rigorous checks of specific activity must be performed after each purification step, along with a comparison between different methods of purification and comparison of whole antibody versus isolated Fab. For example, we have found it very difficult even after considerable purification (including affinity chromatography) to remove glycosidase, adenosine deaminase, and ribonuclease impurities from what appears by polyacrylamide gel electrophoresis (PAGE) to be homogeneous antibody.

Prospects

The rapid development of diverse strategies to elicit catalytic antibodies has led to the catalysis of a wide variety of reactions (see Table I). The concerted use of several of these strategies will be necessary in order to fully exploit the usefulness of catalytic antibodies.

Acknowledgements

This work was supported by the Office of Naval Research Grant No. N00014-87-K-0256.

[75] G. Kronvall, H. Grey and R. Williams, *J. Immunol.* **105**, 1116 (1972).
[76] V. Raso and B. Stollar, *Biochemistry* **14**, 584 (1975).

[18] Antibody Catalysis of Concerted, Carbon–Carbon Bond-Forming Reactions

By DONALD HILVERT and KENNETH W. HILL

Introduction

Linus Pauling recognized the similarity between enzymes and antibodies nearly 40 years ago.[1] While immunoglobulins do not normally catalyze reactions, they are well-characterized protein molecules that bind ligands with high affinity and exacting specificity.[2] Dissociation constants for typical antibody–antigen complexes fall in the range 10^{-4} to 10^{-12} M, and the same hydrophobic and electrostatic forces that typify the interactions of enzymes with substrates are utilized in antigen binding. Moreover, structural studies reveal that the basic sizes and shapes of the binding pockets of enzymes and antibodies are also similar. Consequently, given the spectacular diversity and specificity inherent within the immune system, it was not unreasonable to expect that antibodies might be found, or engineered, to possess intrinsic catalytic activity.

The immune system is certainly the most prolific source of specific receptor molecules known. This protein-generating system operates by natural selection[3] to produce antibody combining sites that precisely match the geometry and electrostatics of the immunizing antigen. High-affinity antibodies can be raised, on demand, against a wide range of substances, including virtually any small haptenic molecule synthesized by the organic chemist.[2] Because monoclonal technology[4] makes individual immunoglobulins available in large amounts and in homogeneous form, the key question with regard to the production of tailored antibody catalysts thus becomes to what extent the chemist can utilize this directable binding energy to do useful chemical work.

Successful strategies for enlisting the immune system for catalysis have been based on our understanding of how natural enzymes use binding

[1] L. Pauling, *Am. Sci.* **36**, 51 (1948).

[2] E. A. Kabat, *"Structural Concepts in Immunology and Immunochemistry."* Holt, Reinhart, and Winston, New York, 1976; A. Nisonoff, J. Hopper, and S. Spring, *"The Antibody Molecule."* Academic Press, New York, 1975; D. Pressman and A. Grossberg, *"The Structural Basis of Antibody Specificity."* Benjamin, New York, 1968.

[3] K. Rajewsky, I. Förster, and A. Cumano, *Science* **238**, 1088 (1987); D. L. French, R. Laskov, and M. D. Scharff, *Science* **244**, 1152 (1989).

[4] J. W. Goding, *"Monoclonal Antibodies: Principles and Practice."* Academic Press, New York, 1986; F. R. Seiler, P. Gronski, R. Kurrle, G. Lüben, H.-P. Harthus, W. Ax, K. Bosslet, and H.-G. Schwick, *Angew. Chem., Int. Ed. Engl.* **24**, 139 (1985).

energy to speed up a chemical reaction. According to the Haldane–Pauling principle of catalysis, enzymes lower the overall activation energy of a particular process by binding and stabilizing the ephemeral, high-energy transition state relative to the bound ground state. Jencks pointed out in 1969 that this principle could also be exploited to prepare catalytic antibodies,[5] proposing that stable molecules resembling the transition structure of specific chemical transformations could be employed as immunogens. If the transition state analog design is a good one, some fraction of the resulting antibodies will possess the desired catalytic activity.

In line with this suggestion, the first catalytic antibodies were raised against antigens that mimicked the stereoelectronic features of the transition state for acyl transfer reactions. Aryl phosphonate esters resemble the negatively charged tetrahedral intermediates and transition states that occur along the reaction coordinate for ester bond hydrolysis. Monoclonal antibodies that recognize these materials have been shown to hydrolyze aryl ester and carbonate substrates that are congruent with the immunizing antigen.[6,7] Impressive rate accelerations over background, with k_{cat}/k_{un} values as high as 10^6, have been observed in the best esterolytic systems. The availability of good transition state analogs, monoclonal antibody technology, and the ability to screen a sizable population of antibodies appear to be key components of these successful experiments.

In the last 3 years, this work has been extended and catalytic antibodies have been prepared that promote a range of mechanistically distinct chemical transformations, including amide hydrolysis,[8] acyl transfer to amines and alcohols,[9] a sigmatropic rearrangement,[10,11] photochemical processes,[12] a β elimination,[13] a Diels-Alder cycloaddition,[14] and redox reac-

[5] W. P. Jencks, *"Catalysis in Chemistry and Enzymology."* McGraw-Hill, New York, 1969.

[6] A. Tramontano, K. D. Janda, and R. A. Lerner, *Science* **234**, 1566 (1986); A. Tramontano, A. A. Ammann, and R. A. Lerner, *J. Am. Chem. Soc.* **110**, 2282 (1988).

[7] S. J. Pollack, J. W. Jacobs, and P. G. Schultz, *Science* **243**, 1570 (1986); J. W. Jacobs, P. G. Schultz, R. Sugasawara, and M. Powell, *J. Am. Chem. Soc.* **109**, 2174 (1987).

[8] K. D. Janda, D. Schloeder, S. J. Benkovic, and R. A. Lerner, *Science* **241**, 1188 (1988); B. L. Iverson and R. A. Lerner, *Science* **243**, 1184 (1989).

[9] S. J. Benkovic, A. D. Napper, and R. A. Lerner, *Proc. Natl. Acad. Sci. U.S.A.* **107**, 5355 (1988).

[10] D. Hilvert, S. H. Carpenter, K. D. Nared, and M.-T. M. Auditor, *Proc. Natl. Acad. Sci. U.S.A.* **85**, 4953 (1988); D. Hilvert and K. D. Nared, *J. Am. Chem. Soc.* **110**, 5593 (1988).

[11] D. Y. Jackson, J. W. Jacobs, R. Sugasawara, S. H. Reich, P. A. Bartlett, and P. G. Schultz, *J. Am. Chem. Soc.* **110**, 4841 (1988).

[12] A. Balan, B. P. Doctor, B. S. Green, M. Torten, and H. Ziffer, *J. Chem. Soc., Chem. Commun.*, 106 (1988); A. G. Cochran, R. Sugasawara, and P. G. Schultz, *J. Am. Chem. Soc.* **110**, 7888 (1988).

[13] K. M. Shokat, C. J. Leumann, R. Sugasawara, and P. G. Schultz, *Nature (London)* **338**, 269 (1989).

tions.[15] Like enzymatic processes, antibody-catalyzed reactions exhibit substantial rate accelerations, substrate specificity, and, more importantly, exacting regio- and stereoselectivity. The observation of saturation kinetics in these systems indicates that substrate binds to the antibody to form a Michaelis complex prior to undergoing chemical change. Moreover, the specificity and the selectivity of antibody catalysts correlate precisely with the structure of the antigen used to elicit the immune response and, therefore, can be set by the researcher by judicious design of the haptenic transition state analog. As a result of these favorable properties, antibody catalysts have great promise as tools for studying how natural enzymes work and as practical agents for accelerating reactions for which natural enzymes are unsuitable, unstable, or unavailable.

While the feasibility of generating antibody catalysts has been demonstrated, the generality and scope of this approach must still be shown. First-generation antibody catalysts are generally inferior to their naturally occurring enzymatic counterparts. Consequently, transition state analog design must be optimized and the rules that relate hapten design to catalytic mechanism need to be clearly defined. For many chemical processes of interest, appropriate transition state analogs have yet to be developed. In addition, approaches are needed for avoiding product inhibition and for introducing functional groups and cofactors site specifically into antibody combining sites. Finally, general techniques for screening large libraries of antibody molecules for the desired reactivity must be perfected so as to better utilize the full diversity available in the immune system. Efficient screening protocols will be especially important for energetically difficult reactions. In this chapter, we focus on some of these issues as they relate to the production of antibodies that catalyze concerted carbon–carbon bond-forming reactions.

Pericyclic Processes: General Considerations

Many chemical reactions require the participation of catalytic groups, including nucleophiles, general acids, general bases, metal ions, and other chemical cofactors, if they are to proceed with appreciable rates. Although individual catalytic residues have been induced in a number of antibody active sites (for example, by charge complementarity[13]), proper positioning of these groups can be achieved only by chance. In addition, the probabil-

[14] D. Hilvert, K. W. Hill, K. D. Nared, and M.-T. M. Auditor, *J. Am. Chem. Soc.* **111**, 9261 (1989).

[15] K. M. Shokat, C. J. Leumann, R. Sugasawara, and P. G. Schultz, *Angew. Chem., Int. Ed. Engl.* **27**, 1172 (1988); N. Janjić and A. Tramontano, *J. Am. Chem. Soc.* **111**, 9109 (1989).

ity of generating a constellation of *multiple* catalytic groups during immunization is likely to be low. For these reasons, reactions that do not require chemical catalysis may be more appropriate targets for antibody catalysis.

Pericyclic processes comprise a broad and important class of concerted reactions of particular theoretical and practical interest.[16] These transformations involve the making and breaking of carbon–carbon bonds and include electrocyclic reactions, sigmatropic rearrangements, and cycloadditions. They are unique in organic chemistry in that they proceed without formation of ionic or radical intermediates. For many years they were classified as "no mechanism" reactions. However, it is now clear that each of these reactions occurs via concerted, symmetry-allowed reorganization of cyclic overlapping π orbitals in the transition state. Pericyclic processes generally do not require chemical catalysis but can be highly sensitive to strain and proximity effects. The latter factors differentiate enzymes from ordinary chemical catalysts. Since they result from the direct utilization of binding energy, they are also the principal effects antibodies are likely to impart.

Although pericyclic reactions are rare in nature, these transformations are among the most valuable to the synthetic organic chemist for the construction of carbon–carbon bonds.[16] They are utilized extensively in syntheses of complex natural products, therapeutic agents, and synthetic materials of all kinds. Antibodies that catalyze pericyclic reactions with the high rates and specificities of enzymes would be extremely valuable, as they would simplify both the planning and execution of the total syntheses of many useful materials.

Development of general strategies for catalyzing concerted chemical reactions is also important, because the resulting antibodies would be unique tools for elucidating how noncovalent interactions between enzyme and substrate lead to enormous rate increases and to precise regio and stereoselectivities. By studying antibody-catalyzed sigmatropic rearrangements, cycloadditions, and electrocyclic reactions we may be able to sort out the relative importance of approximation and geometric distortion in enzymic catalysis. Such knowledge will be invaluable for the development of better transition state analog inhibitors for existing enzymes and also for the preparation of improved second-generation catalytic antibodies.

In the following sections we describe the successful production of antibody catalysts for two concerted pericyclic reactions, a unimolecular [3,3]-

[16] R. B. Woodward and R. Hoffmann, *Angew. Chem., Int. Ed. Engl.* **8**, 781 (1964); G. Desimoni, G. Tacconi, A. Barco, and G. P. Pollini, *"Natural Products Synthesis Through Pericyclic Reactions,"* ACS Monograph 180, American Chemical Society, Washington, D.C., 1983.

sigmatropic rearrangement and a bimolecular Diels–Alder cycloaddition. In each case, the discussion of hapten design will focus on our understanding of basic reaction mechanism, the stereoelectronic nature of the targeted transition state, and the problem of product inhibition. From what we have seen so far, exploitation of the immune system for catalysis is limited only by our mechanistic understanding of a given transformation and the extent to which a mimic of the relevant transition state can be devised. Consequently, the strategies that we have developed are likely to have broad generality for other pericyclic processes and perhaps for other classes of organic reactions as well.

Claisen Rearrangements

The unimolecular Claisen rearrangement of allyl enol ethers (Fig. 1) is an important example of a [3,3]-sigmatropic shift in which formation of a carbon–carbon bond is accompanied by breaking of a carbon–oxygen bond.[17] Together with the Cope rearrangement, it is one of the most valuable sigmatropic rearrangements in organic synthesis. Claisen rearrangements have been used to prepare diverse natural products, including C-nucleosides like the antibiotic showdomycin,[18] as well as numerous alkaloids and terpenes.[16,17] In addition, the only known enzyme-catalyzed sigmatropic process in primary metabolism, the conversion of chorismate into prephenate (Fig. 2), is formally a Claisen rearrangement.[19]

The sigmatropic rearrangement of chorismate is the committed step in the biosynthesis of the aromatic amino acids tyrosine and phenylalanine via the shikimate pathway in bacteria, fungi, and higher plants.[19] It is catalyzed by the enzyme chorismate mutase by a factor of 2 million-fold over the spontaneous thermal rearrangement. Because both the enzymatic and nonenzymatic processes have been extensively studied, this reaction is an ideal target for evaluating the ability of antibodies to promote pericyclic processes. The naturally occurring enzyme provides a suitable benchmark for judging the chemical efficiency of any abzymes that are produced.

In order to produce viable chorismate mutase antibodies, it is necessary to devise a suitably functionalized transition state analog. As indicated in Fig. 1, the chair geometry for the transition state of Claisen rearrangements is generally favored over the boat by at least 5 kcal/mol for unsubstituted

[17] R. P. Lutz, *Chem. Rev.* **84**, 205 (1984); F. E. Ziegler, *Chem. Rev.* **88**, 1423 (1988).

[18] D. B. Tulshian and B. Fraser-Reid, *J. Org. Chem.* **49**, 518 (1984).

[19] U. Weiss and J. M. Edwards, *"The Biosynthesis of Aromatic Amino Compounds,"* p. 134. Wiley, New York, 1980.

[20] W. V. E. Doering and W. R. Roth, *Tetrahedron* **18**, 67 (1962); A. Brown, M. J. S. Dewar, and W. Schoeller, *J. Am. Chem. Soc.* **92**, 5516 (1970).

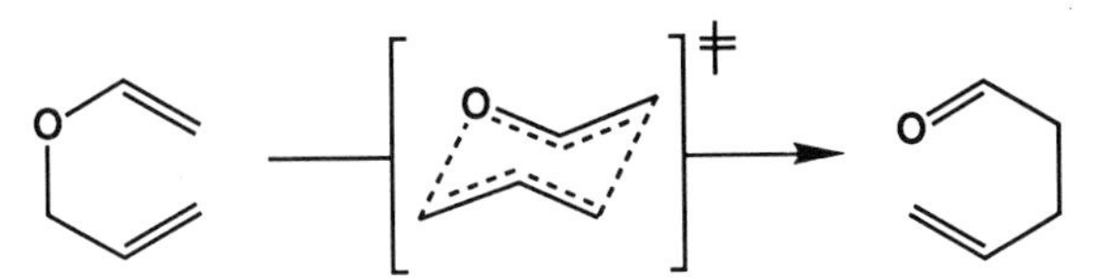

FIG. 1. Prototypic Claisen rearrangement of an allyl enol ether into 4-pentenal.

molecules.[20] *Ab initio* calculations performed at the MP-2/6-31G* level have helped to define the electronic characteristics of the rearrangement.[21] The partially formed carbon–carbon bond length is calculated to be 2.1–2.2 Å, while the breaking carbon–oxygen bond is 33% longer than its reactant value. Thus, appreciable carbon–oxygen bond breaking accompanies carbon–carbon bond formation.

Stereochemical studies[22] have shown that the enzymatic and nonenzymatic rearrangement of chorismate also proceeds by a chairlike transition state **1** (Fig. 2). This compact structure differs from that of substrate and product, which prefer to adopt an extended conformation. Secondary isotope effects have provided evidence that the transition state for the nonenzymatic reaction is also substantially asymmetric, with carbon–oxygen bond breaking preceding carbon–carbon bond formation.[23] In the case of the enzymatic reaction, however, the isotope effects are suppressed, possibly indicating that the rate-determining transition state occurs before the chemical rearrangement step.[23]

A number of research groups have designed successful inhibitors for chorismate mutase assuming that the enzymatic reaction proceeds through the bicyclic transition structure **1**. These include a number of adamantane derivatives and bicyclic diacids.[24,25] The oxabicyclic diacid **2a** (Fig. 2), synthesized by Bartlett and co-workers,[25] illustrates this strategy. The bicyclo[3.3.1]nonene skeleton is a straightforward mimic of the putative bicyclic transition structure, with normal carbon–carbon and carbon–oxygen bonds replacing the longer making and breaking bonds of the high-energy species, respectively. Although the carboxylate in the enol pyruvate side chain adopts a roughly planar configuration in the transition state **1**, it is

[21] R. L. Vance, N. G. Rondan, K. N. Houk, F. Jensen, W. T. Borden, A. Komornicki, and E. Wimmer, *J. Am. Chem. Soc.* **110**, 2314 (1988).

[22] S. G. Sogo, T. S. Widlanski, J. H. Hoare, C. E. Grimshaw, G. A. Berchtold, and J. R. Knowles, *J. Am. Chem. Soc.* **106**, 2701 (1984).

[23] L. Addadi, E. K. Jaffe, and J. R. Knowles, *Biochemistry* **22**, 4494 (1983).

[24] P. R. Andrews, E. N. Cain, E. Rizzardo, and G. D. Smith, *Biochemistry* **16**, 4848 (1977); H. S.-I. Chao and G. A. Berchtold, *Biochemistry* **21**, 2778 (1982).

[25] P. A. Bartlett and C. R. Johnson, *J. Am. Chem. Soc.* **107**, 7792 (1985); P. A. Bartlett, Y. Nakagawa, C. R. Johnson, S. H. Reich, and A. Luis, *J. Org. Chem.* **53**, 3195 (1988).

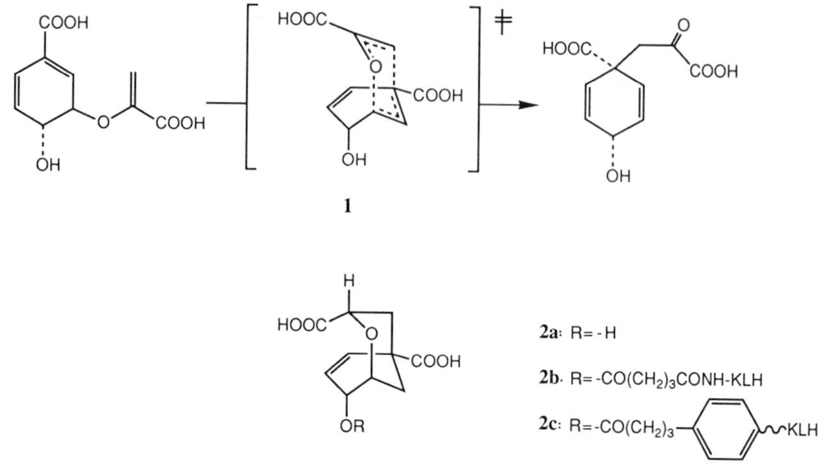

2a: R = -H

2b: R = -CO(CH$_2$)$_3$CONH-KLH

2c: R = -CO(CH$_2$)$_3$–C$_6$H$_4$∿KLH

FIG. 2. The Claisen rearrangement of chorismate into prephenate proceeds through the formation of the putative transition state **1**. The oxabicyclic molecule **2** was prepared as a stable mimic of this high-energy species and has been employed as a hapten to elicit chorismate mutase abzymes.

attached in endo configuration to an sp^3-hybridized carbon in the analog. Nevertheless, racemic **2a** is the most potent inhibitor ever developed for chorismate mutase. It binds to the enzyme with a dissociation constant (K_i) of 0.12 μM under conditions in which chorismate itself has a K_m of 34 μM.[25] Although **2a** captures only part of the binding affinity expected for a perfect transition state analog, it is an excellent first-generation template for constructing chorismate mutase antibodies. Antibodies raised against it should lower the activation energy for the reaction by differential binding of the transition state and the ground state. In addition, because **2a** has a much different conformation than the extended product molecule, product inhibition should not be a serious problem.

We synthesized **2a** according to published procedures and coupled it to the carrier protein keyhole limpet hemocyanin (KLH) through the pendant hydroxyl group using a glutaric acid linker.[10] The hydroxyl group is known to be nonessential for the rearrangement chemistry.[26] We hyperimmunized several mice with the KLH protein conjugate **2b** and prepared 45 high affinity monoclonal antibodies to the hapten with standard protocols.[4] The immunoglobulins were purified to homogeneity from ascites fluid by a three-step procedure involving ammonium sulfate precipitation,

[26] J. L. Pawlak, P. E. Padykula, J. D. Kronis, R. A. Aleksejczyk, and G. A. Berchtold, *J. Am. Chem. Soc.* **111**, 3374 (1989).

affinity chromatography on protein A, followed by FPLC (fast protein liquid chromatography) ion-exchange chromatography on a Mono Q HR 10/10 column from Pharmacia (Piscataway, NJ). Antibodies were assayed individually for chorismate mutase activity at 275 nm and by analytic high-performance liquid chromatography (HPLC). Two of them (1F7 and 27G5) catalyzed the reaction significantly over background.[10]

Control experiments with 1F7 and 27G5 established that prephenate was the authentic product of the catalyzed reaction, and that the rate of chorismate disappearance equalled the rate of prephenate appearance. We were also able to show that the reaction is specifically inhibited by **2a** ($K_i = 0.6\ \mu M$), indicating that the observed reaction takes place in the induced binding pocket. Saturation kinetics at high substrate concentrations provided further support for the formation of a Michaelis-type complex between substrate and antibody prior to the rearrangement event. The kinetic parameters determined for 1F7 at pH 8.0 and 14° were as follows: $k_{cat} = 0.024\ min^{-1}$ and $K_m = 49\ \mu M$. Because the k_{cat} parameter is the first-order rate constant for conversion of the bound starting material into the bound product, it could be compared directly to the first-order rate constant for the uncatalyzed thermal reaction to provide a direct measure of the chemical efficiency of the abzyme. Making this comparison (k_{cat}/k_{uncat}), we found that 1F7 accelerates the chorismate mutase reaction by a factor of roughly 250-fold. The kinetic profile of 27G5 was identical to that of 1F7 and recent cDNA sequencing data confirm that the two antibodies are derived from the same daughter clone (K. Bowdish, unpublished results, 1988).

Working independently, Schultz, Bartlett, and colleagues have identified another chorismate mutase antibody from a mouse hyperimmunized with the protein–hapten conjugate **2c**.[11] Their antibody exhibits a rate acceleration of 1×10^4-fold at 10°, pH 7, and is 50 to 100 times more efficient than 1F7. This result is significant for two reasons. First, the antibody generated against **2c** possesses roughly 1% of the activity of the natural enzyme, demonstrating that an imperfect transition state analog can be employed to create an abzyme that is chemically quite efficient. Second, the identification of multiple antibodies with a range of chemical efficiencies underscores the crucial importance of screening the immune response to a given antigen extensively in order to identify those molecules with the desired properties (specific activity and/or specificity). Although different linkers were used to couple **2** to KLH in the two experiments described, it is unlikely that linker geometry alone can account for the different properties of the induced abzymes. Rather, the observed differences presumably reflect the enormous structural diversity inherent within the immune system. By screening substantially larger numbers of immunoglobulins, even better chorismate mutase abzymes might be found. Work on the production of very large combinatorial libraries of Fab proteins in *Escherichia coli* is exciting in this regard.[27] Coupled with improved screening techniques, including *in vivo* activity assays, this new technology will have a major impact on our ability to identify highly efficient antibody catalysts.

To gain some insight into mechanism we have carried out a thermodynamic analysis of the antibody-catalyzed chorismate mutase reaction.[10] We found that the rate enhancement observed with 1F7 is due entirely to a 5.7-kcal/mol lowering of the enthalpy of activation. Surprisingly, the entropy of activation for the antibody reaction is 12 eu (entropy unit) less favorable than for the thermal rearrangement in solution. These results are consistent with the notion that induced strain might be an important component of catalysis in this case. In contrast, the antibody raised against **2c** is an "entropy trap," having a more favorable activation entropy (13 eu more than the uncatalyzed process).[28] Although activation parameters are notoriously difficult to interpret because of the complex contributions of solvation and desolvation, it is clear that the two chorismate mutase antibodies achieve their catalytic effects in mechanistically diverse ways, underscoring again the versatility of the immune system with respect to catalysis.

The exacting specificity and selectivity displayed by biocatalysts is of potentially greater practical importance than rate accelerations alone. Specificity is also a hallmark of antibody molecules.[2] In the chorismate mutase system, for example, Bartlett and Schultz have shown that the substrate carboxylate groups are essential for recognition by the antibody: the dimethyl ester of chorismate is not a substrate for the abzyme.[11] On the other hand, antibodies are chiral molecules and should exert considerable regio- and stereochemical control over the reactions they promote. We have shown that 1F7 exhibits high enantioselectivity, accepting only (−)-chorismate as a substrate (Fig. 3).[10] We exploited this property to carry out a kinetic resolution of racemic chorismate and obtained the unreacted (+)-isomer optically pure in high yield. Antibody (18 μM) was incubated with racemic chorismate (75 μM) in phosphate-buffered saline (PBS, pH 7.5) at 24°. The reaction was monitored spectroscopically at 275 nm until the rapidly reacting isomer was consumed. After 6.5 hr the reaction mixture was ultrafiltered at 4° (Centricon 30; Amicon, Danvers, MA) to remove protein and then lyophilized. Unreacted chorismate was isolated and purified by preparative HPLC on a Vydac (Hesperia, CA) C_{18} 218-TP-

[27] W. D. Huse, L. Sastry, S. A. Iverson, A. S. Kang, M. Alting-Mees, D. R. Burton, S. J. Benkovic, and R. A. Lerner, *Science* **246**, 1275 (1989).

[28] P. G. Schultz, *Acc. Chem. Res.* **22**, 287 (1989).

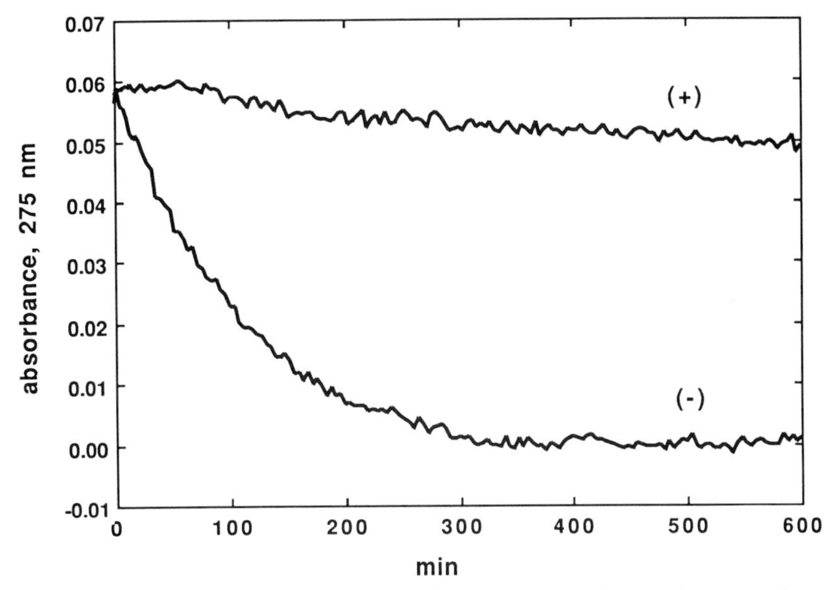

FIG. 3. Time course of the rearrangement of (−)- and (+)-chorismate with the chorismate mutase abzyme 1F7. Antibody (18 μM) was incubated with substrate (ca. 20 μM) at pH 7.5 and 14° and the ensuing reaction monitored spectroscopically at 275 nm. Adapted with permission from D. Hilvert and K. D. Nared, *J. Am. Chem. Soc.* **110**, 5593 (1988). Copyright 1988 American Chemical Society.

510 reversed-phase column (10 mm × 25 cm) with an isocratic elution with 93% aqueous trifluoroacetic acid (0.05%)/7% acetonitrile. The isolated material, which coeluted with authentic chorismate on analytic HPLC, was characterized by circular dichroism spectroscopy and shown to be optically pure (+)-chorismate. The high enantioselectivity of 1F7 and other antibodies is likely to be one of the most valuable properties of these inducible catalysts for practical applications in chemistry and biology.

Because the immunizing antigen **2** in the studies outlined above was racemic, it was conceivable that antibodies possessing (+)-chorismate mutase activity might have been produced. Although no such catalyst was found among the 45 monoclonals that were screened,[29] we anticipate that screening larger immunoglobulin libraries[27] will yield abzymes with the desired specificity. Indeed, in another system, two classes of antibody have been obtained with a racemic hapten that are able to hydrolyze either the R or S enantiomer of a homologous ester substrate.[30]

[29] D. Hilvert, *ACS Symp. Ser.* **389**, 14 (1989).
[30] K. D. Janda, S. J. Benkovic, and R. A. Lerner, *Science* **244**, 437 (1989).

We feel that the chorismate mutase system is ideal for studying protein structure–function relationships. In addition to the natural enzyme, we currently have two distinct antibody catalysts for this reaction and several immunoglobulins that bind the transition state analog tightly but that do not promote the sigmatropic rearrangement. Detailed structural and chemical characterization of these molecules and comparison of their properties is likely to result in a far better understanding of the mutase reaction and, perhaps, protein catalysis in general.

Future studies in this area will attempt to extend and generalize the findings obtained with the chorismate mutase abzymes. We can expect that the same design principles that produced **2** will be applied to the design of transition state analogs for sigmatropic shifts that have no physiological counterpart. Thus, in addition to other Claisen rearrangements, Cope rearrangements (including its oxyanionic version) and hydrogen atom shifts are likely candidates for investigation. Refinements in the design of these transition state mimics can also be expected. For example, although saturated six-membered rings successfully simulate the conformationally restricted transition state of a [3,3]-sigmatropic rearrangement, it may be possible to improve these first-generation analogs still further by incorporating large heteroatoms (e.g., S, P, or Si) into the ring systems to mimic the elongated making and breaking bonds in the high-energy species more effectively.

Diels–Alder Cycloadditions

To examine the ability of antibodies to exploit proximity effects for catalysis we have chosen a bimolecular reaction of considerable importance in organic synthesis. The Diels–Alder reaction is a thoroughly studied [2 + 4] cycloaddition reaction that involves concerted addition of an olefin (or another doubly or triply bonded molecule) across the 1,4-positions of a conjugated diene (Fig. 4).[31] As the reaction proceeds, two new σ bonds are formed at the expense of two π bonds in the starting materials.

The tremendous versatility of the Diels–Alder reaction has made it one of the most powerful tools in the armamentarium of the synthetic organic chemist for constructing carbon–carbon bonds, providing practical entry to important classes of compounds.[16,32] The Diels–Alder reaction has been utilized, for instance, as the key step in the preparation of steroids, alka-

[31] J. Sauer, *Angew. Chem., Int. Ed. Engl.* **6**, 16 (1967); A. Wassermann, *"Diels–Alder Reactions."* Elsevier, Amsterdam, 1965.

[32] J. Sauer, *Angew. Chem., Int. Ed. Engl.* **5**, 211 (1966); R. R. Schmidt, *Acc. Chem. Res.* **19**, 250 (1986).

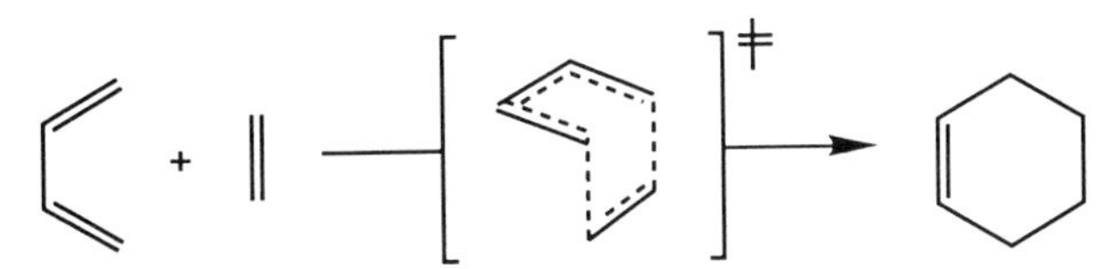

FIG. 4. Concerted [2 + 4] cycloaddition, or Diels–Alder reaction, of a conjugated diene to an unsaturated dienophile to give a cyclohexene derivative.

loids and terpenes, sugars, prostaglandins, and antineoplastic agents like adriamycin and daunomycin.[16]

While not usually subject to catalysis by nucleophiles, general acids, or general bases, the Diels–Alder reaction is enhanced when carried out in water,[33] under pressure,[34] and in the presence of Lewis acids[34] or cyclodextrins.[33] Interestingly, though, no enzyme has ever been isolated that catalyzes this process. Nevertheless, we felt that antibodies were ideally suited for this task. Bimolecular Diels–Alder reactions are entropically very unfavorable. Activation entropies are typically in the range of -30 to -40 entropy units due to the large loss of rotational and translational entropy that is incurred when the two substrates are brought together in the transition state.[31] In principle, it should be possible to avoid this enormous entropy loss by binding the reactants together in an antibody combining site in the proper orientation for reaction. Page and Jencks have calculated that by converting a bimolecular reaction into a unimolecular process on the surface of the protein, rate accelerations as high as 10^8 *M* can be achieved for 1 *M* standard states.[35]

The transition state for the Diels–Alder reaction has a highly ordered, boatlike cyclohexene conformation (Fig. 4).[31,36] It is this structure that must be mimicked in the design of haptens for the production of catalytic antibodies. Although the forming carbon–carbon bonds have been calculated to be 2.217 Å long,[36] stable transition state analogs containing normal carbon–carbon bonds might be expected to induce a combining site with the proper dimensions for preorganizing the substrates, as in the chorismate mutase system discussed above.

A more daunting problem with regard to the design of haptens for bimolecular reactions, however, is that of product inhibition. If the product of the reaction binds too tightly to an active site, turnover of the

[33] D. C. Rideout and R. Breslow, *J. Am. Chem. Soc.* **102**, 7816 (1980); D. D. Sternbach and D. M. Rossana, *J. Am. Chem. Soc.* **104**, 5853 (1984); P. A. Grieco, K. Yoshida, and Z.-M. He, *Tetrahedron Lett.* **25**, 5715 (1984).

[34] J. Sauer and R. Sustman, *Angew. Chem., Int. Ed. Engl.* **19**, 779 (1980).

[35] M. I. Page and W. P. Jencks, *Proc. Natl. Acad. Sci. U.S.A.* **68**, 1678 (1971).

[36] F. K. Brown and K. N. Houk, *Tetrahedron Lett.* **25**, 4609 (1984).

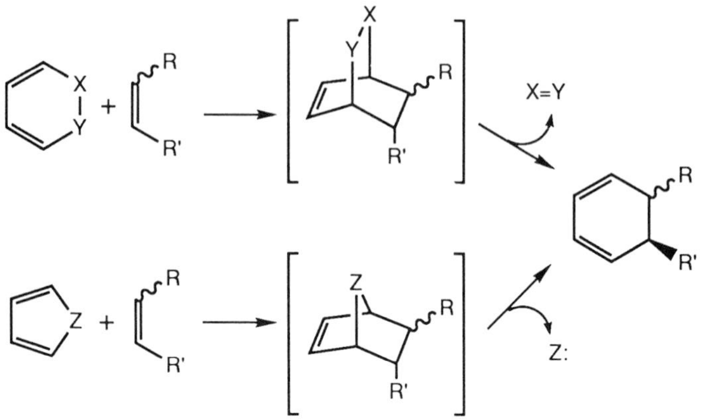

FIG. 5. Diels–Alder reaction of a suitably substituted cyclic diene and a dienophile to yield an unstable bicyclic intermediate which undergoes rapid chelotropic elimination of X=Y or Z:.

catalyst will be prevented. For this reason, the product of a cycloaddition reaction would be a poor choice for eliciting Diels–Alderase abzymes, even though it contains all of the elements present in the transition state. Antibodies raised against such a compound would likely be limited to a single turnover, unable to release the product once it had been assembled within the binding pocket (if they worked at all). In order to produce a multiturnover abzyme a transition state analog must therefore be devised that not only has the necessary boatlike geometry of the transition state but that does not resemble the final product of the reaction too closely.

One solution to this general problem is illustrated in Fig. 5. Product inhibition can be circumvented, if the initially synthesized adduct (which is roughly complementary in shape and charge to the induced active site) is structurally altered by a subsequent chemical reaction. In the case of cycloadditions, for example, a number of cyclic dienes undergo reaction with olefins to give bicyclic adducts that are intrinsically unstable and subsequently decompose by an alternate cycloreversion (e.g., by chelotropic elimination of nitrogen, carbon dioxide, sulfur dioxide, or nitriles).[16] In principle, other chemistries, in addition to cycloreversion, can also be employed to alter the structure of product molecules, including selective hydrolysis, oxidation, and reduction.

We have successfully used the strategy shown in Fig. 5 to catalyze the Diels–Alder reaction between *N*-alkylmaleimides and tetrachlorothiophene dioxide (Fig. 6).[37] The initially formed tricyclic compound **3** (Fig. 6)

[37] M. S. Raasch, *J. Org. Chem.* **45**, 856 (1980).

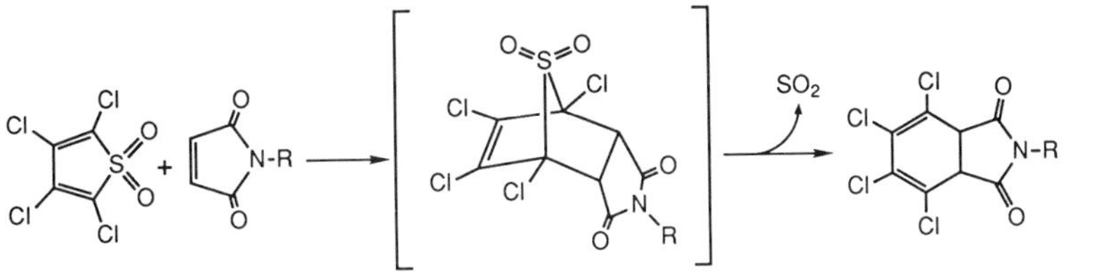

3

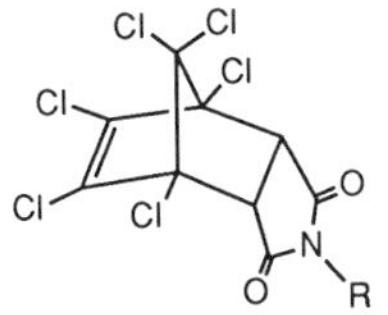

4a: R= $-(CH_2)_5COOH$

4b: R= $-(CH_2)_5CONH$-KLH

4c: R= $-CH_2CH_3$

FIG. 6. The hexachloronorbornene adduct **4a**, which mimicks the high-energy species **3**, has been used to raise catalytic antibodies for the Diels–Alder cycloaddition betweeen tetrachlorothiophene dioxide and *N*-ethylmaleimide and the subsequent chelotropic elimination of sulfur dioxide.

rapidly eliminates sulfur dioxide to give dihydro-(*N*-ethyl)tetrachlorophthalimide, which is subsequently oxidized to *N*-ethyltetrachlorophthalimide under the reaction conditions. The transition states for both the cycloaddition and the subsequent cycloreversion steps of this reaction resemble the high-energy intermediate **3**. We anticipated that antibodies raised against a stable analog of this material would possess the proper shape for bringing the two substrates together. In addition, because **3** has a much different shape than the planar starting materials and reaction products, the induced combining site would not be expected to be subject to severe product inhibition.

We chose the hexachloronorbornene derivative **4a** as a potential mimic of the transition states of the reaction depicted in Fig. 6. It is a close analog of **3**,containing a Cl-C-Cl moiety in place of the O=S=O bridgehead. Compound **4a** is readily prepared by refluxing 1,4,5,6,7,7-hexachloro-5-norbornene-2,3-dicarboxylic anhydride with an equimolar amount of 6-aminohexanoic acid in a dichloromethane–acetone mixture. An equivalent of acetyl chloride is added and the mixture is refluxed for 2 hr. Evaporation of solvent, followed by chromatography over silica gel, gives **4a** in 70% yield. This material is activated for coupling to the carrier proteins KLH and bovine serum albumin (BSA) by treating it with a slight excess of *N,N'*-disuccinimidyl carbonate for 45 min. After solvent evaporation and silica gel chromatography, an 89% yield of the *N*-hydroxysuccinimidyl ester of **4** is obtained. KLH and BSA conjugates are prepared by treating dilute solutions of each protein in borate buffer (pH 9.0) with an excess of the activated ester of **4**. After standing at room temperature for 1 hr, each protein–hapten conjugate is purified over a Sephadex G-25 column. Protein concentrations are determined by the method of Smith *et al.*[38] Epitope densities (KLH, 30; BSA, 15 haptens/protein) are determined by the method of Habeeb.[39]

We prepared eight monoclonal antibodies with the KLH conjugate **4b** using standard immunization and fusion protocols.[4] Hybridomas secreting high-affinity immunoglobulins are identified by enzyme-linked immunosorbent assay (ELISA). They are propagated individually in ascites and the resulting antibodies are purified to homogeneity by affinity and ion-exchange chromatography.

The purified antibodies are tested individually for their ability to catalyze the cycloaddition between *N*-ethylmaleimide and tetrachlorothiophene dioxide. Because the thiophene dioxide undergoes a rapid and undesired side reaction with antibody lysine residues, it is necessary to protect the ϵ-amino groups by reductive methylation with formaldehyde and sodium cyanoborohydride.[40] We find that the permethylated antibodies retain their ability to bind **4** but no longer react with substrate.[14] Chemical modification of antibodies in this way is likely to be a generally useful tactic, as it makes it possible to employ reactive substances as substrates.

A number of the antibodies tested substantially accelerate the reaction between *N*-ethylmaleimide and tetrachlorothiophene dioxide in spectroscopic and HPLC assays.[14] The best of these, 1E9, was investigated in some detail. By monitoring its fast reaction with Malachite Green, we are able to verify that SO_2 is produced in the course of the antibody-catalyzed reaction, albeit in less than stoichiometric amounts. Model studies have shown that SO_2 acts as an efficient oxidant of dihydrophthalimides, and we have isolated *N*-ethyltetrachlorophthalimide as the major product from the crude reaction mixture.

That the cycloaddition is antibody catalyzed, and occurs within the binding pocket induced by hapten **4**, is supported by several observations. We showed that the disappearance of thiophene dioxide is first order in antibody concentration and is also severely inhibited by **4c**. When 1E9 is mixed with an equimolar amount of **4c**, no catalysis above background is observed for the reaction, even in the presence of a nearly 10^3-fold molar

[38] P. K. Smith, R. I. Krohn, G. T. Hermanson, A. K. Mallia, F. H. Gartner, M.D. Provenzano, E. K. Fujimoto, N. M. Goeke, B. J. Olson, and D. C. Klenck, *Anal. Biochem.* **150**, 76 (1985).

[39] A.F.S.A. Habeeb, *Anal. Biochem.* **14**, 328 (1966).

[40] N. Jentoft and D. G. Dearborn, *J. Biol. Chem.* **254**, 4359 (1979).

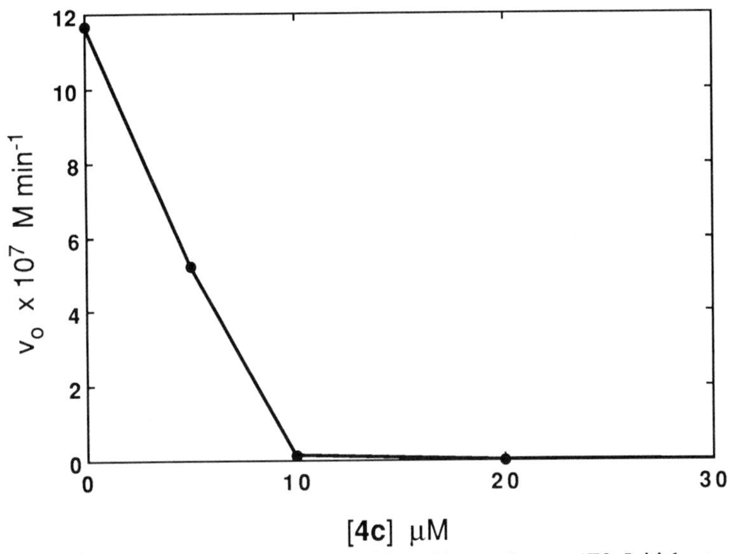

FIG. 7. Titration of the active site of the Diels–Alderase abzyme 1E9. Initial rates of the reaction between tetrachlorothiophene dioxide (0.18 mM) and N-ethylmaleimide (3.4 mM) were determined spectroscopically at pH 6 and 25° in the presence of antibody (5.2 μM, 10 μM binding site) and varying amounts of the hapten analog **4c**.

excess of N-ethylmaleimide. As shown in Fig. 7, this fact allows the available antibody active sites to be titrated with inhibitor.

The effects of minor structural variation in the substrates on reaction rate also accord with expectations based on the hapten structure. For example, maleimides with long N-alkyl groups are better substrates than those with shorter tails, presumably because they can better exploit binding interactions within that portion of the active site that was induced by the hexanoic acid linker of **4b**. Thus, the observed order of reactivity in the presence of the antibody is butyl $\approx$ propyl $>$ ethyl $>$ methyl $\gg$ H, even though each of these maleimide derivatives has comparable reactivity in the uncatalyzed process (K. Hill, unpublished results, 1990).

To gauge the chemical efficiency of 1E9 we carried out a series of kinetic experiments.[14] Holding the concentration of tetrachlorothiophene dioxide constant, initial rates of reaction are determined by HPLC as a function of N-alkylmaleimide concentration. Hyperbolic kinetics are observed in each case, indicating the formation of a Michaelis complex between the substrates and the antibody. Analysis of the rate data obtained with N-propylmaleimide (NPM) by standard methods allows us to calculate apparent values of k_{cat} and $(K_m)_{NPM}$ of 6.9 ± 0.5 min^{-1} and 36 ± 6 mM, respectively, at 0.5 mM tetrachlorothiophene dioxide. Comparison of

$(k_{cat})_{app}$ obtained under these conditions with the second-order rate constant for the uncatalyzed reaction ($k_{un} = 0.040\ M^{-1}\ \text{min}^{-1}$) gives an apparent effective molarity of 170 *M* for the Diels–Alderase antibody. Under identical conditions, an effective molarity of 105 *M* is obtained for *N*-ethylmaleimide, indicating that the extra methylene in the propyl derivative contributes roughly 0.28 kcal/mol in favorable binding interactions toward more efficient catalysis.

The effective molarity is the concentration of substrate that would be needed in the uncatalyzed reaction to achieve the same rate as in the antibody ternary complex.[41] Clearly, the binding site of 1E9 confers a significant kinetic advantage to this bimolecular Diels–Alder reaction: the observed effective molarities are several orders of magnitude larger than the maximum physically accessible concentration of substrate in aqueous buffer. However, the measured values underestimate the actual chemical efficiency of 1E9. The true effective molarities could not be measured because the antibody could not be saturated with tetrachlorothiophene dioxide due to its low solubility in the reaction medium. Under the conditions of the experiment, the value of $(k_{cat})_{app}$ is linearly dependent on thiophene dioxide concentration. Hence, for *N*-propylmaleimide the true effective molarity will be substantially (10 to 100 times) larger than the measured value of 170 *M*.

Catalysis of a bimolecular Diels–Alder cycloaddition by an antibody is notable because it demonstrates the feasibility of exploiting proximity effects in antibody catalysis of bimolecular reactions. In addition, the Diels–Alder reaction has no physiological counterpart. Our success at promoting this process therefore underscores both the power and versatility of the catalytic antibody approach to enzyme engineering. Significantly, we explicitly incorporated a mechanism for avoiding product inhibition into our hapten design. The fact that 1E9 undergoes multiple turnovers (>50) demonstrates that our strategy successfully solves this vexing problem. We were able to show by competition ELISA that the final reaction product, *N*-ethyltetrachlorothiophene dioxide, binds roughly 10^3 times less tightly to the combining site than does the hapten derivative **4b** (K. Hill, unpublished results, 1990).

As the Diels–Alder reaction is a prototype for a broad and important class of pericyclic processes, it should now be possible to apply our design strategy to many related reactions. The use of an antigen that resembles an unstable intermediate along the reaction coordinate is likely to be a general solution to the problem of product inhibition in antibody-catalyzed bimolecular reactions. In the future, we hope to exploit other chelotropic elimi-

[41] W. P. Jencks, *Adv. Enzymol.* **43**, 219 (1975).

nations, as well as acid–base and redox chemistry, to release a wide range of materials from antibody combining sites and thereby facilitate catalyst turnover. The ability to carry out bimolecular Diels–Alder reactions, dipolar cycloadditions, and the like with the high rates and specificities of enzymes will have considerable practical value in organic synthesis.

Conclusions

Catalytic antibody technology represents a revolutionary new approach to the study of enzyme mechanism and the development of practical protein catalysts for use in chemistry and biology. In the last few years, a variety of mechanistically distinct chemical reactions have been successfully catalyzed by rationally induced abzymes. These include hydrolysis of esters and amides, acyl transfer to amines and alcohols, photochemical processes, a β elimination, redox reactions, and the pericyclic processes discussed above. In the next few years, many additional transformations will be tackled, as this methodology is limited, in principle, only by our ability to create a suitable transition state analog and the compatibility of the target reaction with an aqueous milieu and the protein active site. Thus, we are likely to see abzymes that can site-specifically cleave peptides and proteins, modify complex oligosaccharides regio- and stereospecifically, carry out selective aldol condensations, and effect polyene cyclizations and a host of useful redox reactions. As we learn more about structure–function relationships in abzymes and optimize our strategies for preparing them, artificial protein catalysts that rival the efficacy of naturally occurring enzymes may well become commonplace.

Acknowledgments

This work was supported in part by a grant from the National Institutes of Health (GM38273 to D.H.), a Junior Faculty Research Award from the American Cancer Society (JFRA-182 to D.H.), and a postdoctoral fellowship from the American Heart Association (to K.W.H.).

[19] Selection of T Cell Epitopes and Vaccine Engineering

By FRANCESCO SINIGAGLIA, PAOLA ROMAGNOLI, MARIA GUTTINGER, BELA TAKACS, and J. RICHARD L. PINK

Introduction

It is currently impossible to predict with reasonable accuracy whether a particular peptide sequence will be recognized by mouse or human T cells in association with a particular major histocompatibility complex (MHC) molecule.[1] Empirical methods of assessing immunogenicity for T cells are therefore still an important tool in characterizing potential vaccine candidates. This is particularly true for synthetic peptide vaccines, where the inability of a peptide to bind to and be recognized in association with a variety of allelic forms of human MHC antigens may be an important limitation on its use.

In this chapter we discuss approaches to identifying and characterizing epitopes recognized by human $CD4^+$ T cells (work involving $CD8^+$ cells has been discussed elsewhere[2,3]). The first section describes *in vitro* stimulation of human peripheral blood mononuclear cells (PBMC) and subsequent cloning of T cells as an assay for immunogenicity of potential vaccine candidates. The second section gives methods for identifying the epitopes and restriction elements recognized by the T lymphocyte clones (TLC). We then describe assays for identifying peptides that bind to human MHC class II antigens. These assays can be used to select peptides that bind strongly to several different MHC allelic products and are therefore likely to be useful immunogens. Examples of results obtained with these methods, as applied to our work on malaria vaccine development, are given in each case.

In Vitro Stimulation of Human T Cells

The immunogenicity of vaccine candidates can be tested by using the candidate to stimulate human T cells *in vitro*. If the donors are already primed, a standard proliferation assay can be used.[4] If naive donors are

[1] J. L. Cornette, H. Margalit, C. DeLisi, and J. A. Berzofsky, this series, Vol. 178, p. 611.
[2] A. Townsend and H. Bodmer, *Annu. Rev. Immunol.* **7,** 601 (1989).
[3] D. F. Nixon, A. R. M. Townsend, J. G. Elvin, C. R. Rizza, J. Gallwey, and A. J. McMichael, *Nature (London)* **336,** 484 (1988).
[4] F. Sinigaglia, M. Guttinger, D. Gillessen, D. M. Doran, B. Takacs, H. Matile, and A. Trzeciak, and J. R. L. Pink, *Eur. J. Immunol.* **18,** 633 (1988).

used, a restimulation assay of the type described by Hensen and Elferink,[5] or direct cloning of T cells following bulk culture with antigen, may be necessary to demonstrate immunogenicity. The cloning method, although more time consuming, has several advantages over the stimulation–restimulation technique: experiments with clones are more reproducible than those with uncloned cells, and the clones can be used to define unambiguously both the human major histocompatibility complex (HLA) restriction elements involved and the precise peptide sequence constituting the epitope recognized by the T cells.

Media for Cell Stimulation and Growth. The culture medium is RPMI 1640 (GIBCO, Paisley, Scotland) supplemented with L-glutamine (2 mM), sodium pyruvate (1 mM), 5×10^{-5} M 2-mercaptoethanol, 1% (v/v) of a 100× mixture of nonessential amino acids (GIBCO), 50 U/ml penicillin, 50 U/ml streptomycin, and 10% (v/v) heat-inactivated fetal calf serum (RPMI–FCS) or 10% pooled human AB serum (RPMI–HS). To support the antigen-independent growth of T cell clones, RPMI-HS is supplemented with 100 U/ml recombinant human interleukin 2 (rIL-2; Hoffmann-La Roche, Inc., Nutley, NJ).

Restimulation Assay. Peripheral blood mononuclear cells (PBMC) are isolated from heparinized whole blood by centrifugation over Ficoll–Hypaque (Pharmacia Fine Chemicals, Uppsala, Sweden). The washed PBMC (2×10^6/ml) are cultured in 24-well flat-bottomed plates (2 ml/well) in RPMI-HS for 6 days with antigen at a suitable concentration. Several antigen concentrations (e.g., 0.03–30 μg/ml) should be tested in preliminary experiments to determine the optimal stimulatory dose. Cells are then washed twice and cultured for a further 6 days in IL-2-containing medium without antigen. Finally, cells from individual wells are washed, adjusted to a concentration of 4×10^5 cells/ml in RPMI–HS, and added in duplicate 100-μl aliquots to wells of 2 separate 96-well flat-bottomed plates (Costar, Cambridge, MA) with irradiated (5000 rad) autologous PBMC (1×10^6/ml), also in 100-μl aliquots. Antigen at a range of concentrations is added in 20 μl of RPMI–HS. After 3 days the cultures of one plate are pulsed with 1 μCi of [^{3}H]thymidine, and incorporation of labeled nucleotide is determined after another 16 hr. After assessment of antigen-induced proliferation, positive cultures from the duplicate plate are expanded by adding rIL-2 and, after another 3 to 7 days, cloned by limiting dilution.

Cloning of Antigen-Specific T Lymphocytes. For cloning, T blasts are seeded at 0.3 cell/well (20 μl) in Terasaki trays (Falcon Division, Becton Dickinson, Cockeysville, MD) in the presence of an optimal concentration

[5] E. J. Hensen and B. G. Elferink, *Hum. Immunol.* **10,** 95 (1984).

of antigen (typically 0.1 to 5 μg/ml for proteins), 10^4 autologous irradiated (5000 rads), freshly taken PBMC and rIL-2. Alternatively, PHA-P (Wellcome, Beckenham, England) at 2 μg/ml, rIL-2, and 10^4 allogeneic irradiated PBMC may be used. After 1 to 2 weeks, cell growth is detected microscopically and growing cells are expanded further in medium with antigen, autologous irradiated PBMC, and rIL-2. The clones are maintained in culture by periodic restimulation (2–6 weeks) in the presence of allogeneic irradiated PBMC, 2 μg/ml PHA-P, and rIL-2.

Immortalization of B Lymphocytes with Epstein–Barr Virus (EBV). PBMCs (10^7) are resuspended in 10 ml RPMI–FCS containing 30% supernatant of the EBV-producing marmoset cell line B95.8 and 600 ng/ml cyclosporin A (Sandoz, Basel, Switzerland), and distributed in the wells of 96-well microculture plates at 5×10^4/well. EBV-transformed B cells (EBV-B cells) are expanded in RPMI–FCS for at least 2 months before being tested for their ability to serve as antigen-presenting cells (APC). All the lines should be screened for mycoplasma contamination[6] before being used as APC.

Proliferation Assay. T lymphocyte clones (TLC) are tested for antigen specificity at least 2 weeks after restimulation with PHA to avoid problems of residual nonspecific proliferation. T lymphocyte clones (TLC; 2×10^4 cells) are cultured in 200 ul RPMI-HS in flat-bottomed microplates with 1×10^4 irradiated (7000 rads) autologous EBV-B cells in the presence or absence of serial dilutions of antigen. After 48 hr the cultures are pulsed with 1 μCi [^{3}H]thymidine, and incorporation of labeled nucleotide is determined after another 16 hr.

Cell Surface Analysis. Clones of T cells are analyzed by direct immunofluorescence on a flow cytometer (FACS-Star, Becton Dickinson, Sunnyvale, CA) by using $CD4^+$- and $CD8^+$-specific monoclonal antibodies (Simultest T "Helper/Suppressor" kit) purchased from Becton Dickinson.

Determination of T Cell Restriction Specificity. The isotype of class II molecules recognized by each T cell clone can be determined by antibody-blocking experiments. T cells are cultured with autologous EBV-B cells, limiting concentrations of antigen (e.g., 10- to 100-fold lower than optimal), and anti-DR (E.31),[7] anti-DQ (Tu22 or SVPL3),[8,9] or anti-DP

[6] T. R. Chen, *Exp. Cell. Res.* **104,** 255 (1977).

[7] M. M. Trucco, G. Garotta, J. W. Stocker, and R. Ceppellini, *Immunol. Rev.* **47,** 219 (1979).

[8] A. Ziegler, J. Heinig, C. Muller, H. Gotz, F. P. Thinnes, B. Uchanska-Ziegler, and P. Wernet, *Immunobiology* **171,** 77 (1986).

[9] H. Spitz, J. Borst, M. Giphart, J. Coligan, C. Terhorst, and J. E. De Vries, *Eur. J. Immunol.* **14,** 299 (1984).

(B7.21)[10] monoclonal antibodies (MAb) as 1 : 1000 ascites. To identify the restricting alleles we use as APC a panel of allogeneic HLA-homozygous EBV-B cells (available from the European Collection for Biomedical Research, Interlab Project, Istituto Nazionale per la Ricerche sul Cancro, Genoa, Italy). The cells are pulsed for 2 hr at 37° with the antigen or medium alone, washed four times, and irradiated before T cells are added and proliferation assayed as above.

Murine L cells transfected with MHC class II α and β chains can also be used as APC.[11] Transfectants grown to confluence are washed in Hanks' medium (GIBCO), resuspended at $5-10 \times 10^6$ cells/ml in mitomycin C (Sigma Chemicals, Poole, England) at a final concentration of 100 μg/ml, incubated at 37° for 1 hr, and then washed several times before the addition of antigen and T cells. The proliferative response is assayed as described above.

Immunogenicity of Synthetic Peptide and Recombinant Malaria Vaccine Candidates. The *Plasmodium falciparum* malaria circumsporozoite (CS) protein is the major sporozoite surface antigen and the best studied malaria vaccine candidate.[12] We searched for T cell epitopes in the nonrepetitive regions of the CS protein by testing the ability of different CS-derived synthetic peptides to stimulate T cell proliferation in donors without a history of malaria infection. Using a restimulation assay, we could show specific proliferation of cells from 18 out of 33 donors (Fig. 1) in the presence of a synthetic peptide, CS.T3, corresponding to CS protein residues 378-398 (but with cysteines 384 and 389 replaced by alanine).[4] These results suggested that CS.T3 can interact with a wide range of different MHC alleles, a finding confirmed by extensive analysis of cloned T cells from some of the donors.[13-15]

The p190 molecule is the precursor of the major merozoite surface proteins of the malaria parasite *P. falciparum.* Trials in monkeys have suggested that induced immunity to a polypeptide (190L) from a conserved region of p190 modifies the clinical course of the malaria disease,

[10] A. M. Watson, R. DeMars, I. S. Trowbridge, and F. H. Bach, *Nature (London)* **304,** 358 (1983).

[11] D. Wilkinson, R. R. P. De Vries, J. A. Madrigal, C. B. Lock, J. P. Morgenstern, J. Trowsdale, and D. M. Altman, *J. Exp. Med.* **167,** 1442 (1988).

[12] V. Nussenzweig and R. Nussenzweig, *Adv. Immunol.* **45,** 283 (1989).

[13] M. Guttinger, P. Caspers, B. Takacs, A. Trzeciak, D. Gillessen, J. R. L. Pink, and F. Sinigaglia, *EMBO J.* **7,** 2555 (1988).

[14] F. Sinigaglia, M. Guttinger, J. Kilgus, D. M. Doran, H. Matile, H. Etlinger, A. Trzeciak, D. Gillessen, and J. R. L. Pink, *Nature (London)* **336,** 778 (1988).

[15] J. Kilgus, T. Jardetzky, J. C. Gorga, A. Trzeciak, D. Gillessen, and F. Sinigaglia, *J. Immunol.* **146,** 307 (1991).

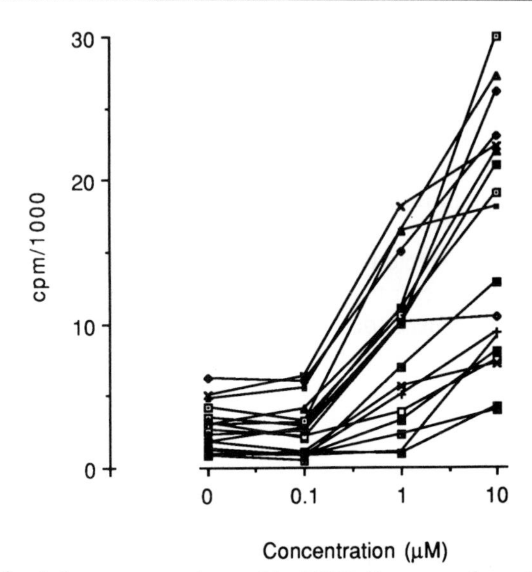

FIG. 1. Restimulation responses to peptide CS.T3. Responses (cpm [^{3}H]thymidine uptake) of CS.T3-sensitized PBMCs from 18 donors restimulated with 3 concentrations (0.1, 1, and 10 μM) of CS.T3 peptide in the presence of autologous irradiated PBMCs. No response was obtained from 15 other donors tested.

both in Aotus[16] and Saimiri monkeys.[16a] We asked whether the 190L recombinant malaria polypeptide is immunogenic for T cells of donors with no history of exposure to malaria. For this purpose we obtained T cell clones from a panel of such donors with different MHC haplotypes by *in vitro* priming as described above.[17] Of eight tested clones, all except one also proliferated in response to parasitized red blood cells (Table I), indicating that the clones are parasite specific and that processing of the whole parasite and of 190L generates similar or identical peptides. These results support the idea that *in vitro* priming can be used to identify and characterize pathogen-specific T cell epitopes.

The isotype and allele of the class II-restricting molecule for each clone was determined as described above. The 190L-specific clones were DR, DQ, or DP restricted, although the DR isotype seems to be used preferentially. At least seven DR molecules [DRw52b, DRw53, DR1, DRw15(2),

[16] S. Herrera, M. A. Herrera, B. L. Perlaza, Y. Burki, P. Caspers, H. Döbeli, D. Rotmann, and U. Certa, *Proc. Natl. Acad. Sci. U.S.A.* **87,** 4017 (1990).

[16a] H. Etlinger, P. Caspers, H. Matile, H.-J. Schönfeld, D. Stüber, and B. Takacs, *Infect. Immun.* (in press).

[17] F. Sinigaglia, B. Takacs, H. Jacot, H. Matile, J. R. L. Pink, A. Crisanti, and H. Bujard, *J. Immunol.* **140,** 3568 (1988).

TABLE I
ANTIGEN SPECIFICITY OF 190L-INDUCED T CELL CLONES[a]

	T cell clone (cpm × 10^{-3})							
Antigen	AH.L16	DP.L31	JK.L20	UB.L8	HK.L41	LG.L7	WS.L6	HH.L11
190L	101	13	90	26	22	34	33	180
I-RBC	9	51	4	<1	12	8	6	11
N-RBC	<1	<1	<1	<1	<1	<1	<1	<1
Medium	<1	<1	<1	<1	<1	<1	<1	<1

[a] T cells were stimulated in the presence of autologous irradiated PBMC with medium alone, with recombinant 190L (1 μg/ml), with *P. falciparum*-infected red blood cells (I-RBC), or noninfected RBC (N-RBC). Results are given as cpm × 10^{-3} [^{3}H]thymidine uptake of triplicate cultures. Positive responses are underlined.

DRw17(3), DR4, and DRw13(w6)] can present 190L to T cells (Table II).[17a]

Epitope Identification

If the location of a T cell epitope in a large protein is unknown, it is most efficient to localize the epitope first to a small (< about 100 residues) portion of the protein by testing the reactivity of the appropriate TLC with fragments of the molecule. These can be produced by proteolytic digestion or by recombinant DNA techniques. Once the approximate location of the epitope is known, it can be better defined either by using controlled exonuclease digestion to progressively shorten the coding DNA as described below, or by an adaptation of the Pepscan technique, in which overlapping synthetic peptides are synthesized on plastic pins and then cleaved from the solid support, so that their ability to stimulate TLC can be tested in solution.[18]

T Cell Epitope Mapping by Recombinant Technique. Many methods can be used to systematically generate shortened DNA coding sequences; we briefly describe one. The DNA fragment containing the coding sequence is ligated to suitable linkers and cloned into the polylinker site of the expression vector pUH130,RBSII, a derivative of plasmid pDS5, as

[17a] M. Guttinger, P. Romagnoli, L. Vandel, R. Meloen, B. Takacs, J. R. L. Pink, and F. Sinigaglia, submitted (1991).

[18] R. Van der Zee, W. Van Eden, R. H. Meloen, A. Noordzij, and J. D. A. Van Embden, *Eur. J. Immunol.* **19,** 43 (1989).

TABLE II
190L CONTAINING SEVERAL T CELL EPITOPES RECOGNIZED IN ASSOCIATION WITH DIFFERENT HLA CLASS II MOLECULES

		Antigen specificity	
Clone	Restriction	Peptide	Sequence[a]
AC69	DQw1	211–224	YKLNFYFDLLRAKL
HH.L11	DRw53	239–249	LKIRANELDVL
WS.L15	DR4	241–255	IRANELDVLKKLVFG
AC129	DRw13(w6)	260–273	LDNIKDNVGKMEDY
HH.L2	DRw15(2)	311–327	YQAQYDLFIYNKQLEEA
HH.L27	DQw6	330–341	LISVLEKRIDTL
JK.L20	DRw52b	338–350	IDTLKKNENIKEL
LG.L7	DRw52b	338–350	IDTLKKNENIKEL
AH.L16	DR1	344–355	ENIKELLDKIN
HK.L41	DRw17(3)	ni[b]	
UB.L8	DPw2	ni	

[a] Single-letter abbreviations are used for amino acid residues: A, alanine; R, arginine; N, asparagine; D, aspartic acid; C, cysteine; Q, glutamine; E, glutamic acid; G, glycine; H, histidine; I, isoleucine; L, leucine; K, lysine; M, methionine; F, phenylalanine; P, proline; S, serine; T, threonine; W, tryptophan; Y, tyrosine; V, valine.

[b] ni, nonidentified.

described.[19,20] On induction with IPTG (isopropyl-β-D-thiogalactopyranoside), fusion proteins containing the expressed sequence at the amino terminus of chloramphenicol acetyltransferase are obtained. After cleavage of the plasmid DNA at a unique 5′ linker site, the coding fragment is shortened from the 5′ end by treatment with exonuclease III and nuclease S1, and blunt ends are generated with Klenow enzyme.[20] Suitable linkers (see example below) are ligated to the blunted ends and, after digestion with appropriate restriction enzymes, fragments of the desired length are isolated from polyacrylamide gels and recloned into pUH130,RBSII. The progressively shortened fusion proteins, which differ from each other in size by a few amino acids at the amino-terminus, are expressed in *E. coli.* The induced *E. coli* cultures are lysed in 6 *M* urea (10^{10} cells/ml) and added to the T cell proliferation assay as crude lysates (10 μl/ml).

[19] A. Crisanti, H. M. Muller, C. Hilbich, F. Sinigaglia, H. Matile, M. McKay, J. Scaife, K. Beyreuther, and H. Bujard, *Science* **240,** 1324 (1988).

[20] H. Bujard, R. Gentz, M. Lanzer, D. Stueber, M. Mueller, I. Ibrahimi, M.-T. Haeuptle, and B. Dobberstein, this series, Vol. **155,** p. 416.

Pepscan for T Cell Epitope Mapping. Peptides are synthesized on polyethylene pins as described by Van der Zee *et al.*[18] Polyacrylic acid-coated polyethylene pins are derivatized to provide a coating of amino groups. Amino acids are added to the pins by deprotection and washing steps, coupling of the next protected amino acid, further washing, and then repetition of the cycle until the synthesis is complete. The peptides are synthesized with a C-terminal extension of three amino acids (Asp-Pro-Gly) to allow acid cleavage of the Asp-Pro bond. Cleavage is done using 70% formic acid in a 0.2-ml vol for 20 hr at 37°. The solvent is then removed by evaporation/lyophilization. Phosphate-buffered saline (PBS; 100 μl) is used to dissolve the peptides; PBS containing urea can also be used, provided that the final urea concentration in the proliferation assay does not exceed 60 m*M*.

Approximately 10 μg of peptide, about 10% of the total present on the pins, is released by cleavage with formic acid. This amount is sufficient for 5–10 T cell proliferation assays.

190L Containing Multiple Epitopes Recognized in Association with Different HLA Class II Molecules. To determine whether 190L-specific T cell clones recognize different regions of the antigen in association with different HLA class II molecules, two different methods are used. In the first method, a set of truncated recombinant 190L proteins progressively shortened at the amino terminus are produced. The coding sequence of p190 between the unique *Pvu*II and *Hae*III sites [base pairs (bp) 316–964] is fused to *Bam*HI (5′ end) and *Hin*dIII (3′ end) linkers and cloned into the polylinker site of pUH130, RBSII. After cleavage of the plasmid DNA at the *Bam*HI site, the sequence was shortened from the 5′ end by exonuclease III (digestion times 40, 60, 80, 100, and 120 seconds) and then treated with S1 and Klenow enzyme. *Bam*HI linkers are ligated to the blunted ends, and *Bam*HI and *Hin*dIII are used to isolate DNA fragments coding for a series of truncated fusion proteins. Each is then tested for its ability to stimulate 190L-specific TLC. Two T cell epitopes (211–224 and 260–273) were thus identified.[19]

In the second method we use the Pepscan system as modified by Van der Zee *et al.*[18] Overlapping nonapeptides covering the whole protein are synthesized and tested for their capacity to stimulate T cells. In the first screening, three different epitopes were identified: 241–249 recognized by a DRw53-restricted clone, 340–348 recognized by two DRw52b-restricted clones, and 245–253 recognized by a clone restricted to DR4. None of the other clones tested responded to any peptide. Consequently a second series of dodecapeptides was synthesized. In the second screening three additional epitopes were identified: 314–325 recognized in association with DRw15(2), 330–341 recognized in association with DQw6, and 344–355 recognized in association with DR1.[17a] All epitopes were confirmed by conventional synthesis of identical or longer peptides (Table II). We were not able to identify specific determinants for six clones. In these cases, the antigenic peptide might be insoluble at a concentration sufficient to stimulate the cells; alternatively, the length and/or conformation of the peptides used in the Pepscan procedure might not correspond to those of the stimulatory peptides generated during 190L processing.

Peptide–MHC Binding Assays

Although the affinity of peptide binding to MHC molecules is low (ca. micromolar), it is possible to measure binding to purified MHC class II molecules by equilibrium dialysis[21] or by filtration or nondenaturing gel electrophoresis assays, such as those described below. Alternatively, binding of peptides to HLA-DR molecules can be determined in a cellular competition assay, which measures the presentation of antigen by MHC molecules through specific proliferation.[22] The tools required for this assay are a panel of cell lines that can present defined peptides to a panel of T cell clones with different known restriction specificities.

Affinity Purification of DR Molecules. DR molecules are isolated in solution[23] from detergent-solubilized DR-homozygous human EBV-B cell lines NOL (DR1) and BSMII (DR4). Cells are harvested, washed twice with Hanks' balanced salt solution by centrifugation at 750 *g* for 15 min, and lysed at a density of 2×10^7/ml on ice for 30 min in 1% (v/v) Nonidet NP-40 (NP-40), 25 m*M* iodoacetamide, 1 m*M* phenylmethylsulfonyl fluoride (PMSF), 10 μl of a saturated solution of ϵ-aminocaproic acid/ml, and 10 μg/ml each of soybean trypsin inhibitor, antipain, pepstatin, leupeptin, and chymostatin in 0.05 *M* Na/PO_4 buffer, 0.15 *M* NaCl, pH 7.5 (0.05 *M* PBS). The lysates are cleared of nuclei and debris by centrifugation at 27,000 *g* for 30 min. If not used immediately, detergent extracts can be stored at −70°.

For the immunoaffinity purification of HLA-DR antigens, detergent extracts obtained from about 10^{10} cells are applied using a recycling system to two successive columns, the first containing Sepharose 4B-coupled nonspecific mouse MAb, and the second specific mouse monoclonal anti-human HLA-DR (E.31)[7] MAb. The specific MAb column is prewashed with

[21] B. P. Babbit, P. M. Allen, G. Matsueda, E. Haber, and E. Unanue, *Nature (London)* **317,** 359 (1985).

[22] J. Kilgus, P. Romagnoli, M. Guttinger, D. Stuber, L. Adorini, and F. Sinigaglia, *Proc. Natl. Acad. Sci. U.S.A.* **86,** 1629 (1989).

[23] J. C. Gorga, V. Horejsi, D. R. Johnson, R. Raghupathy, and J. Strominger, *J. Biol. Chem.* **262,** 16087 (1987).

elution buffer [1% octyl-β-D-glucopyranoside, 50 mM diethylamine hydrochloride (pH 11.5), 0.15 M NaCl, 1 mM EDTA, 1 mM EGTA, 10 mM iodoacetamide] and both columns are equilibrated with lysis buffer. After application of the sample, the specific immunoadsorption column is washed successively with three column volumes of lysis buffer, 10–20 column volumes of washing buffer 1 [0.5% NP-40, 0.02% (w/v) sodium dodecyl sulfate (SDS), 0.05 M PBS (pH 7.5), 0.02% (w/v) NaN_3], and 3–5 column volumes of washing buffer 2 [1% octyl-β-D-glucopyranoside, 0.05 M PBS (pH 7.5), 0.02% NaN_3]. Specifically bound HLA-DR antigens are then eluted with elution buffer (see above). Four-milliliter fractions are

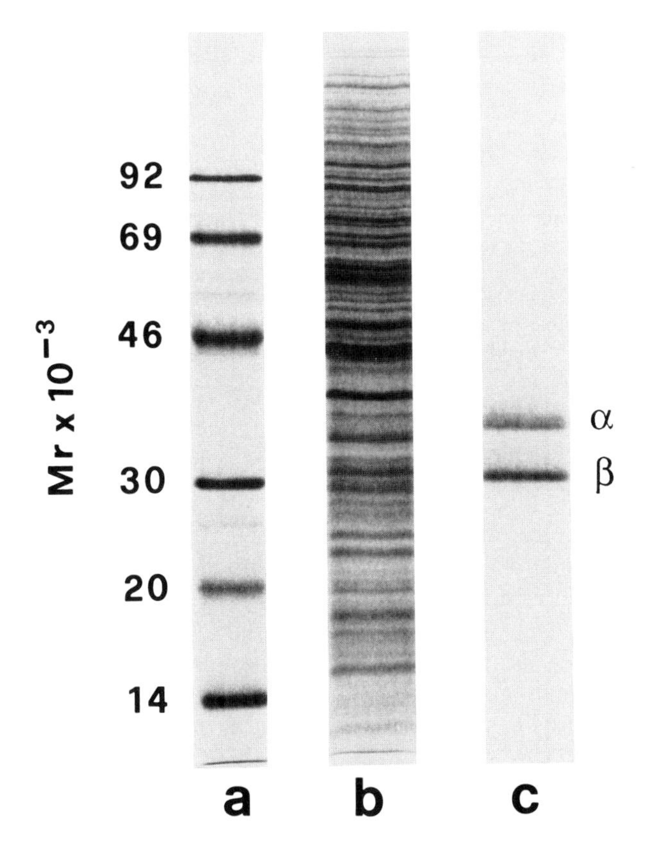

FIG. 2. SDS-PAGE analysis of immunoaffinity-purified HLA-DR1 molecules. For the immunoaffinity purification of DR antigens (lane c), detergent extracts (lane b) obtained from about 10^{10} DR1-homozygous EBV-B cells were applied to a column containing Sepharose 4B-coupled specific mouse monoclonal anti-HLA-DR Ab. Lane a contains Coomassie Blue-stained molecular weight standards. The two noncovalently associated α (M_r 34,000) and β (M_r 29,000) chains are indicated.

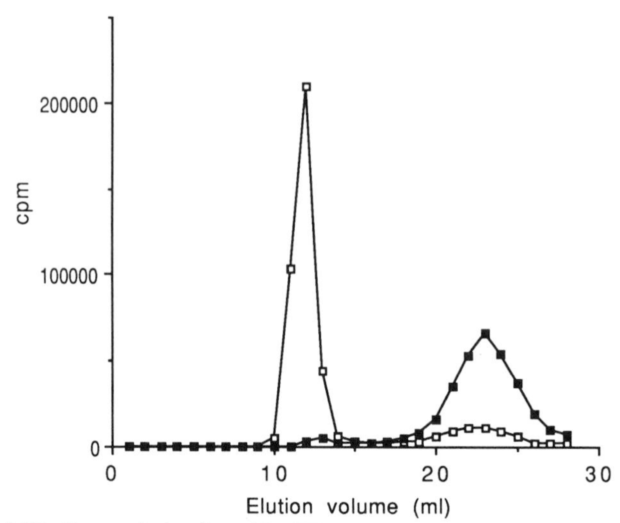

FIG. 3. Gel-filtration analysis of peptide–DR1 complexes. The DR1-binding influenza virus hemagglutinin peptide Ha(307–319) was modified by the addition of a tyrosine residue to the amino terminus to allow ^{125}I labeling. The incubation mixture contained 100–300 pmol of purified DR1 and 1–3 pmol of ^{125}I-radiolabeled peptide alone (□), or in the presence of a 100-fold excess of unlabeled Ha(307–319) (■). After incubation at room temperature for 48 hr, the DR–peptide complexes were separated from free peptide by gel filtration on a Sephadex G-50 column and 2-ml fractions were collected and assayed for radioactivity in a gamma spectrometer.

collected into tubes containing 250 μl of 1 *M* Tris-HCl (pH 6.8), 0.15 *M* NaCl solution in order to neutralize the pH. Fractions containing DR antigens are pooled, and either used immediately for binding studies or concentrated under N_2 pressure in microcollodion bags (Sartorius, Göttingen, Germany) and stored at −70°.

Peptide Synthesis. Peptides are synthesized manually using the simultaneous multiple solid-phase peptide synthesis of Houghton,[24] which is adapted to the base-labile *N*-fluorenylmethoxycarbonylamino acids (Fmoc strategy), *tert*-butyl-based side-chain protecting groups, and a *p*-benzyloxybenzyl alcohol polystyrene resin, as described by Atherton and Sheppard.[25] As a coupling reagent the *O*-benzotriazolyl-*N,N,N,N*-tetramethyluronium hexafluorophosphate is used.[26,27] Each peptide is purified using reversed-

[24] R. A. Houghton, *Proc. Natl. Acad. Sci. U.S.A.* **82,** 5131 (1985).

[25] A. Atherton and R. C. Sheppard, *in* "The Peptides: Analysis, Synthesis, Biology" (S. Udenfriend and J. Meierhofer, eds.), Vol. 9, p. 1. Academic Press, New York, 1987.

[26] V. Dourtoglou, B. Gross, V. Lambropoulou, and C. Zioudrou, *Synthesis* **15,** 572 (1984).

[27] R. Knorr, A. Trzeciak, W. Bannwarth, and D. Gillessen, *Tetrahedron Lett.* **30,** 1927 (1989).

TABLE III
BINDING OF CIRCUMSPOROZOITE PEPTIDES TO DR1 AND DR4

Peptide	Sequence	50% inhibition (μM)[a]	
		DR1	DR4
P. falciparum			
CS(23–43)	YQCYGSSSNTRVLNELNYDNA	—[b]	—
CS(103–122)	EKLRKPKHKKLKQPGDGNPD	—	—
CS(120–135)	NPDPNANPNVDPNANP	—	—
CS(124–135)	NANPNVDPNANP	—	—
CS(148–159)	NANPNANPNANP	—	—
CS(297–309)	GHNMPNDPNRNVD	—	—
CS(310–318)	ENANANNAV	—	—
CS(325–341)	EPSDKHIEQYLKKIKNS	—	—
CS(378–398)	DIEKKIAKMEKASSVFNVVNS	1.2 ± 0.3	21.7 ± 2.2
Influenza virus			
Ha(307–319)	PKYVKQNTLKLAT	0.5 ± 0.2	3.2 ± 1.1

[a] Average of at least two independent determinations.
[b] Dashes indicate >200 μM.

phase, high-performance liquid chromatography[28] and is routinely >95% pure. The composition (molecular weight) of each peptide is verified by fast-atom bombardment mass spectroscopy.

Gel-Filtration Assay. Binding studies are done using an inhibition assay by a gel-filtration procedure, essentially as described by Buus *et al.*[29] with some modifications. The incubation mixture (100 μl) contains 100–300 pmol of purified DR1, inhibitory peptide (1–100 pmol), and 1–3 pmol of ^{125}I-radiolabeled peptide [influenza hemagglutinin (Ha) 307–319] in 1% NP-40/0.05 *M* PBS, pH7.5 containing 1 m*M* PMSF, 1.6 m*M* EDTA, 6 m*M* *N*-ethylmaleimide, and 10 μg/ml each of soybean trypsin inhibitor, antipain, pepstatin, leupeptin, and chymostatin. After incubation at room temperature for 48 hr, the DR1–peptide complexes are separated from free peptide by gel filtration on a Sephadex G-50 (Pharmacia, Uppsala, Sweden) column equilibrated with 0.5% NP-40 in 0.05 *M* PBS, pH 7.5, containing 0.02% NaN_3. Two-milliliter fractions are collected and assayed for radioactivity in a gamma spectrometer. The fraction of peptide bound to DR is calculated as the radioactivity eluting in the void volume of the column divided by the total radioactivity recovered.

Native PAGE Binding Assay. Nondenaturing polyacrylamide gel elec-

[28] R. C. Merrifield, *Science* **150,** 178 (1965).
[29] S. Buus, A. Sette, S. M. Colon, D. Jenis, and H. M. Grey, *Cell (Cambridge, Mass.)* **47,** 1071 (1986).

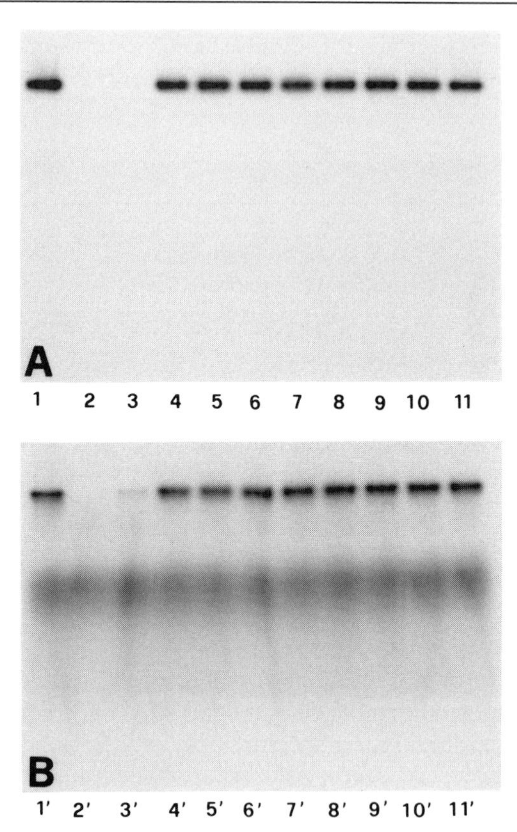

FIG. 4. Autoradiographic analysis of native PAGE binding assay. ^{125}I-Labeled influenza virus hemagglutinin peptide [Ha(307–319)], 0.1 μg, was incubated with 10 μg of immunoaffinity-purified HLA–DR1 (A) or HLA-DR4 (B), at room temperature for 48 hr. The incubation mixture was diluted with an equal volume of PAGE sample buffer and one-eighth of the sample was analyzed by native PAGE (lanes 1 and 1′), as described in the section, Native PAGE Binding Assay. In the other lanes the same incubation mixtures were analyzed in the presence of an excess (10 μg) of nonradioactive test peptides. Samples analyzed in lanes 2 and 2′ contained peptides Ha(307–319), lanes 3 and 3′ CS(378–398), lanes 4 and 4′ CS(23–43), lanes 5 and 5′ CS(103–122), lanes 6 and 6′ CS(120–135), lanes 7 and 7′ CS(124–135), lanes 8 and 8′ CS(148–159), lanes 9 and 9′ CS(297–309), lanes 10 and 10′ CS(310–318), and lanes 11 and 11′ CS(325–341).

trophoresis (PAGE) is performed in slab gels containing 12% acrylamide and the buffer system described previously,[30] with some modifications. SDS is replaced in all the buffers by 3-[(3-cholamidopropyl)dimethylammonio]-1-propane sulfonate (CHAPS; Calbiochem, San Diego, CA). The

[30] B. Takacs, *in* "Immunological Methods" (I. Lefkovits and B. Pernis, eds.), p. 81. Academic Press, New York, 1979.

separating gel, stacking gel, and the tank buffers contain 0.00215% (w/v) CHAPS, whereas the sample buffer contains 0.036% CHAPS and no 2-mercaptoethanol. All stock solutions are vacuum filtered through a 0.22-μm membrane filter (Millipore, Bedford, MA) and stored at 4°.

The binding assay is based on that of Jardetzky *et al.*[31] Samples are usually prepared by mixing, in a final volume of 200 μl, 100μl of twice-concentrated incubation buffer [0.05 *M* PBS (pH 7.5), 10 m*M* EDTA, 10 m*M* *N*-ethylmaleimide, 2% NP-40, 2 m*M* PMSF, and 20 μg/ml each of soybean trypsin inhibitor, antipain, pepstatin, leupeptin, and chymostatin], 10 μg of immunoaffinity-purified DR antigen, 0.1 μg of ^{125}I-labeled immunogenic peptide, and either none of 10 μg of nonradioactive test peptide. Samples are incubated at room temperature for 48 hr and diluted twofold with PAGE-sample buffer [0.1 *M* Tris-HCl (pH 6.8), 17% (v/v) glycerol, 8% (w/v) Ficoll, 0.036% CHAPS, and 0.001% (w/v) bromphenol blue]. Up to 100 μl of this incubation mixture is applied in each sample well of 1.5-mm slab gels. Electrophoresis is carried out at room temperature at constant current (30 mA/slab), until the marker dye reaches the bottom of the gel, usually after 2.5–3 hr. Proteins are fixed by incubating the gels in an ethanolic formaldehyde solution [25% (v/v) ethanol, 5% (w/v) formaldehyde] for 45 min at room temperature with agitation. Gels are rinsed in distilled water and dried on filter paper at reduced pressure, prior to exposure to Fuji RX medical X-ray films (Fuji Photo Film Co., Tokyo, Japan) with Cronex (Du Pont, Wilmington, DE) intensifying screens at −70°.

Peptide Binding to Multiple Allelic Forms of DR Molecules. We have analyzed, for DR1 and DR4 binding, a set of peptides encompassing more than 50% of the *P. falciparum* CS protein sequence. Affinity-purified preparations of DR proteins (Fig. 2) are used in these experiments. The influenza virus hemagglutinin peptide Ha(307–319)[32] is radioiodinated by the chloramine-T method,[33] incubated for 48 hr with purified HLA–DR, and the binding assayed by Sephadex G-50 gel filtration as described above. Binding of labeled peptide is saturable and can be inhibited by an excess of cold Ha peptide (Fig. 3). We then examined the interaction of DR1 and DR4 with CS peptides by competition with the labeled Ha(307–319) peptide (Table III). Peptide CS.T3, corresponding to CS sequence 378–398 but with Cys-384 and -389 replaced by Ala,[15] competes very

[31] T. S. Jardetzky, J. C. Gorga, R. Busch, J. Rothbard, J. L. Strominger, and D. C. Wiley, *EMBO J.* **6,** 1797 (1990).

[32] J. B. Rothbard, R. Lechler, K. Howland, V. Bal, D. D. Eckels, R. Sekaly, E. Long, W. R. Taylor, and J. Lamb, *Cell (Cambridge, Mass.)* **52,** 515 (1988).

[33] F. C. Greenwood, W. M. Hunter, and J. S. Glover, *Biochem. J.* **89,** 114 (1963).

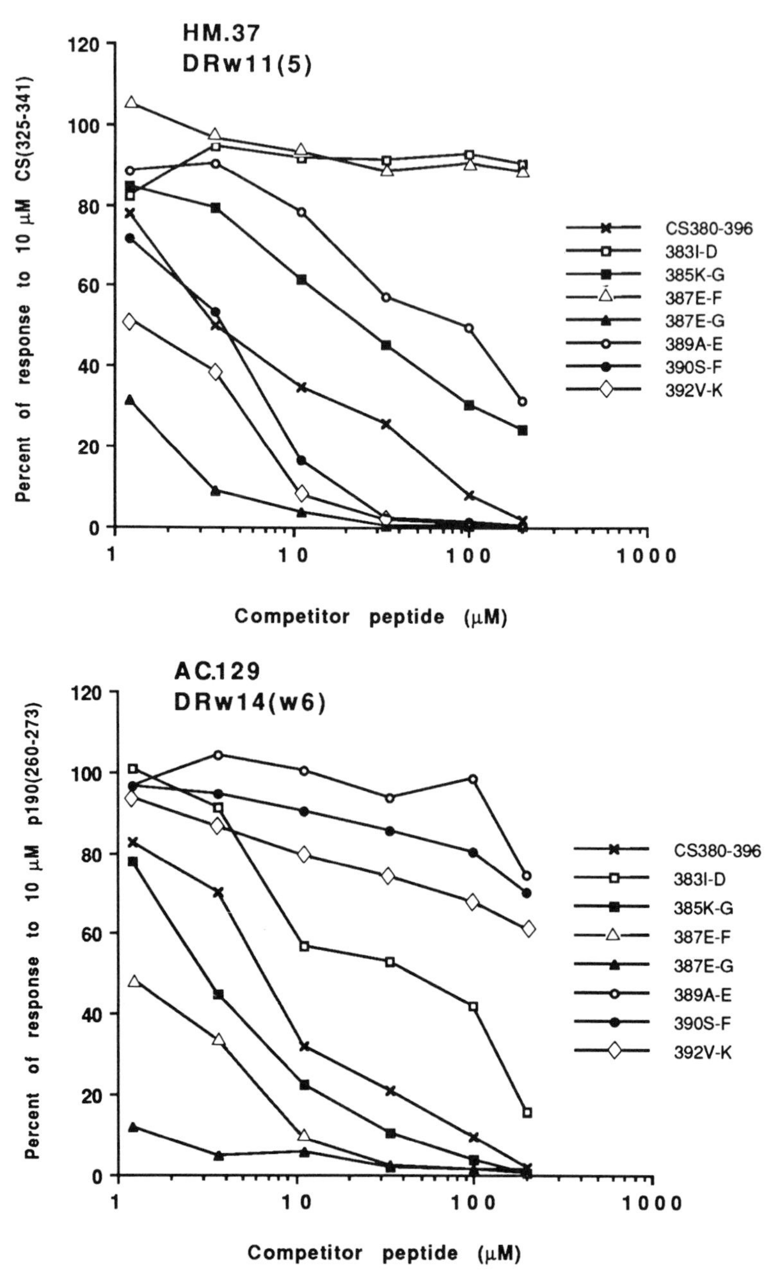

FIG. 5. Differential binding of CS.T3 peptide analogs to DRw11(5) and DRw14(w6). Inhibition of antigen presentation was determined by incubation of glutaraldehyde-fixed EBV-B cells (5×10^4 cells/well) with 10 μM peptide CS(325–341) (TLC-HM.37) or peptide p190(260–273) (TLC-AC.129) and 1–200 μM competitor peptides for 18 hr. APCs were then washed and HM.37 or AC.129 T cells (4×10^4 cells/well) were added. After 48 hr, the cultures were pulsed with [^{3}H]thymidine. Data are presented as percentages of the [^{3}H]thymidine incorporation obtained in response to 10 μM stimulator peptide alone.

efficiently with the binding of the labeled peptide for both DR molecules. The other peptides tested do not significantly inhibit the binding of Ha peptide to DR1 and DR4.

An example of such an assay is given in Fig. 4, where the same series of CS peptides, previously tested in the gel-filtration assay, are analyzed in the native gel-binding assay. There is good agreement between the two methods, and in addition the native gel assay has the advantage that many samples can be analyzed in parallel (Fig. 4).

Peptide Competition for Antigen Presentation. The cellular competition assay is based on the proliferative response of a given T cell clone to a "test peptide," and asks whether other peptides can specifically inhibit this reaction by competing with the test peptide for a single binding site on the MHC restriction element of the antigen-presenting cells (APCs) in culture. In the first step of the competition assay, fixed APCs are pulsed with suboptimal concentrations of stimulator peptide in the presence of varying concentrations of peptides to be examined for binding. In the second step the amount of stimulator peptide that is present on the surface of the APC is monitored in a proliferation assay, using the appropriate T cell test clones as responder cells.

Procedure. Competition for antigen presentation is performed with fixed EBV-B cell lines as APC. After four washing steps in serum-free medium, cells are fixed by resuspending them in 0.025% (v/v) glutaraldehyde in Hanks' balanced salt solution for 90 sec. The reaction is stopped by addition of 0.2 *M* glycine and the cells are then washed and resuspended in complete RPMI 1640 medium containing PMSF (final concentration, 1 m*M*), aprotinin (100 units/ml), leupeptin (10 μg/ml), aminocaproic acid (10 μl/ml of a saturated solution), *N*-α-(*p*-tosyl)-L-lysine-chloromethyl ketone (0.1 m*M*), and 10% human serum. Fixed APCs (5×10^4 cells/well) are incubated with various concentrations of the competitor peptides (1–200 μM) and suboptimal concentrations (1–10 μM) of the stimulator peptides that are recognized by the test clones. After incubation for 20 hr at 37°, the cells are washed three times and cultured with 4×10^4 T cells from the test clones. The cultures are pulsed with 1 μCi of [^{3}H]thymidine after 48 hr and the incorporation of the labeled nucleotide is determined after another 16 hr.

CS Peptides Showing Enhanced Binding to DR Antigens. Using the competition assay,[22] we tested a large series of singly substituted analogs as well as truncated forms of the CS.T3 peptide, and analyzed their binding capacity to two DR alleles.[15] Of the 51 analogs tested, the peptide substituted at position 387(E $\rightarrow$ G) showed enhanced binding to both the DR molecules tested (Fig. 5). Because the CS.T3 peptide is able to induce helper T cells which could be boosted by immunization with the whole malaria parasite,[34] these results could be relevant for the construction of a vaccine inducing cellular immunity to sporozoites. Such a result is interesting, as it would suggest that the naturally occurring sequences might be improved as vaccine candidates by selection of cross-reactive, more immunogenic variant peptides.

[34] F. Sinigaglia, M. Guttinger, H. Matile, and J. R. L. Pink, *Bull. WHO* **68,** 94 (1990).

[20] Design, Construction, and Characterization of Poliovirus Antigen Chimeras

By DAVID J. EVANS and JEFFREY W. ALMOND

Introduction

Poliovirus, the causative agent of paralytic poliomyelitis, is one of the best characterized of all human viruses. Over the last decade, the combined approaches of molecular genetics, protein crystallography, and molecular immunology, have led to a detailed understanding of the molecular determinants of many of the biological properties of this virus. We have exploited the detailed knowledge of the structure of the poliovirus in experiments aimed at redesigning the antigenic characteristics of the virus particle. This work started as part of our research into the design of new improved polio vaccines but has also opened up the possibility of constructing vaccines against other infectious diseases, based on the poliovirus particle. The studies have resulted in the development of poliovirus as an *epitope expression vector,* into which short, well-defined, antigenic peptides can be inserted. We present here an overview of the techniques involved in the construction and use of poliovirus cDNA cassette vectors, and in the recovery and exploitation of poliovirus antigen chimeras.

Poliovirus

The live-attenuated oral poliovirus vaccines (OPV) currently in use were developed by Albert Sabin during the 1950s by multiple passage in culture of wild-type strains of each of the three serotypes.[1] The resulting attenuated strains are now established as extremely effective vaccines; they induce good levels of immunity in recipients, they are easy to administer

[1] A. B. Sabin and L. R. Boulger, *J. Biol. Stand.* **1,** 115 (1973).

(usually on a piece of sugar or as a drop on the tongue), and inexpensive to produce. However, although OPV is one of the safest vaccines in current medical use, a small risk of vaccine-associated paralytic poliomyelitis does exist for vaccinees and/or their close contacts. In the United States, this vaccine-associated disease has accounted for all cases of poliomyelitis since 1981, and has been variously estimated at around 1 case of paralysis per 1.2 million recipients under the age of 3 years.[2] Characterization of the viruses isolated from vaccine-associated cases has demonstrated convincingly that the disease is due to reversion to neurovirulence of mainly the type 2 and type 3 vaccine strains (approximately 20 and 80% of cases, respectively[2]). In contrast, the Sabin type 1 strain of the vaccine appears to be safer by at least an order of magnitude. Molecular genetic studies have shown that the Sabin type 2 and 3 vaccine strains contain fewer mutations than the Sabin 1 strain as compared to their wild-type progenitors, and this probably accounts for their poorer genetic stability.[3,4]

The major advances in the molecular characterization of polioviruses that have enabled their exploitation was outlined in this chapter include the following:

1. determination of the nucleotide sequences of numerous enteroviruses, including attenuated and neurovirulent representatives of each serotype of poliovirus (reviewed by Palmenberg[5]). This has indicated the extent of natural sequence variation among the genus and has identified the regions that are highly divergent
2. demonstration that infectious virus can be recovered from full-length cDNA copies of the genome.[6] This opened up the possibility of manipulating the genome using recombinant DNA techniques
3. determination of the three-dimensional structure of the poliovirus particle at near atomic resolution[7,8]
4. identification of the regions of the caspid responsible for induction of neutralizing antibodies for each serotype[9]

[2] F. Assaad and W. C. Cockburn, *Bull. WHO* **60,** 231 (1982).

[3] J. W. Almond, *Annu. Rev. Microbiol.* **41,** 153 (1987).

[4] V. R. Racaniello, *Adv. Virus Res.* **34,** 217 (1987).

[5] A. C. Palmenberg, *in* "Molecular Aspects of Picornavirus Infection and Detection" (B. L. Semler and E. Ehrenfield, eds.), p. 211. American Society for Microbiology, Washington, D.C., 1990.

[6] V. R. Racaniello and D. Baltimore, *Science* **214,** 916 (1981).

[7] J. M. Hogle, M. Chow, and D. J. Filman, *Science* **229,** 1358 (1985).

[8] D. J. Filman, R. Syed, M. Chow, A. J. Macadam, P. D. Minor, and J. M. Hogle, *EMBO J.* **8,** 1567 (1989).

[9] P. D. Minor, M. Ferguson, D. M. A. Evans, J. W. Almond, and J. P. Icenogle, *J. Gen. Virol.* **67,** 1283 (1986).

The poliovirus capsid is composed of 60 copies each of 4 virus proteins, VP1–4, which are assembled to form the icosahedral virus particle. Of these, VP4 is entirely internal, and does not contribute to the antigenicity of the mature virion. Virus proteins VP1–3 have a similar core structure characterized by an eight-stranded antiparallel β-pleated sheet flanked by two short α helices.[7] A series of techniques have been used to map the neutralizing antigenic sites of the virus (reviewed by Minor *et al.*[9,10]), which are localized to four regions of the surface of the particle, and can be visualized as forming surface-projecting loops on the three-dimensional structures of serotypes 1 and 3.[7,8] Three of these antigenic sites (2, 3, and 4) are discontinuous or conformational epitopes, being composed of more than one region of the capsid proteins. In contrast, antigenic site 1 seems to be predominantly a linear epitope,[9] being largely composed of the region of VP1 separating the βB and βC strands.[7] This antigenic site is therefore very amenable to genetic manipulation.

Intertypic Poliovirus Antigen Chimeras

The construction of intertypic poliovirus antigen chimeras was first attempted as an approach to improving the current type 2 and type 3 vaccines. The rationale was that such chimeras would be based on the very stable Sabin 1 vaccine strain, which would have its own antigenic determinants substituted by those of the type 2 and/or type 3 strains. Since the antigenic determinants of Sabin 1 probably do not contain attenuation mutations, the resulting chimeric virus should retain the stable attenuation phenotype of this strain. The region encoding antigenic site 1 in a full-length cDNA clone of Sabin 1 was therefore replaced with the corresponding sequence from a type 3 strain by site-directed oligonucleotide mutagenesis. Following transfection of Hep2c cells a viable chimeric virus, designated S1/3.10, was recovered.[11] As expected, S1/3.10 was shown to be still neutralized by polyclonal type 1 antisera, and by monoclonal antibodies directed against antigenic sites 2 and 3, but not by antibodies against site 1 of serotype 1. The virus was also neutralized by type 3 polyclonal antisera and by most of the monoclonal antibodies directed against antigenic site 1 of serotype 3, thereby demonstrating that the virus was a genuine chimera exhibiting dual antigenicity. Further studies demonstrated that S1/3.10 also possessed composite immunogenicity, since it elicited neutralizing antibodies against both type 1 and type 3 strains after

[10] P. D. Minor, *in* "Current Topics in Microbiology and Immunology," Vol. 161 (V. R. Racaniello, ed.), p. 121. Springer-Verlag, New York.

[11] K. L. Burke, G. Dunn, M. Ferguson, P. D. Minor, and J. W. Almond, *Nature (London)* **322,** 81 (1988).

administration to mice, rabbits, and a single monkey.[11] These and similar studies from other groups[12-14] demonstrated convincingly that antigenic site 1 of the poliovirus particle could be modified by replacement with related, poliovirus-derived sequences, and suggested that this region may be relatively flexible in terms of the sequences it can accommodate. These results prompted us to exploit Sabin 1 for the expression of antigenic sites from more distantly related sources.

Poliovirus Cassette Vectors

pCAS1

To facilitate the construction of further poliovirus antigen chimeras—both to improve the immunogenicity of the site 1 intertypic chimeras, and to extend the use of poliovirus as a vector on which to present foreign antigenic determinants—a "cassette system" for antigenic site 1 was designed. The purpose was to enable the rapid replacement of the cDNA region encoding antigenic site 1 with complementary oligonucleotides encoding sequences of choice. A vector was therefore designed based on a full-length infectious cDNA clone of poliovirus (to avoid any subcloning steps) that contained unique restriction sites flanking the region of the cDNA encoding antigenic site 1.

The full construction details of this Sabin 1-based antigenic site 1 cassette vector, designated pCAS1, have been published elsewhere.[15] Briefly, *Sal*I, and *Dra*I restriction sites were introduced into a full-length infectious cDNA clone of Sabin 1, at nucleotides 2753 and 2783, respectively (Fig. 1). The engineered Sabin 1 cDNA was subcloned into a pBR322-derived vector (pFB1[2]; Pharmacia, Piscataway, NJ) that had been modified by removal of the three *Dra*I sites from the ampicillin resistance gene. A T7 promoter was introduced at the extreme 5′ end of the poliovirus cDNA, so that virus could be recovered by transfection of RNA transcripts rather than cDNA, which is comparatively inefficient.[16] The

[12] A. Martin, C. Wychowski, T. Couderc, R. Crainic, J. M. Hogle, and M. Girard, *EMBO J.* **7,** 2839 (1988).

[13] M. G. Murray, R. J. Kuhn, M. Arita, N. Kawamura, A. Nomoto, and E. Wimmer, *Proc. Natl. Acad. Sci. U.S.A.* **85,** 3202 (1988).

[14] P. D. Minor, M. Ferguson, K. Katrak, D. Wood, A. John, J. Howlett, G. Dunn, K. L. Burke, and J. W. Almond, *J. Gen. Virol.* **71,** 2543 (1990).

[15] K. L. Burke, D. J. Evans, O. Jenkins, J. M. Meredith, E. D. A. S'Souza, and J. W. Almond, *J. Gen. Virol.* **70,** 2475 (1989).

[16] S. van der Werf, J. Bradley, E. Wimmer, F. W. Studier, and J. J. Dunn, *Proc. Natl. Acad. Sci. U.S.A.* **83,** 2330 (1986).

a

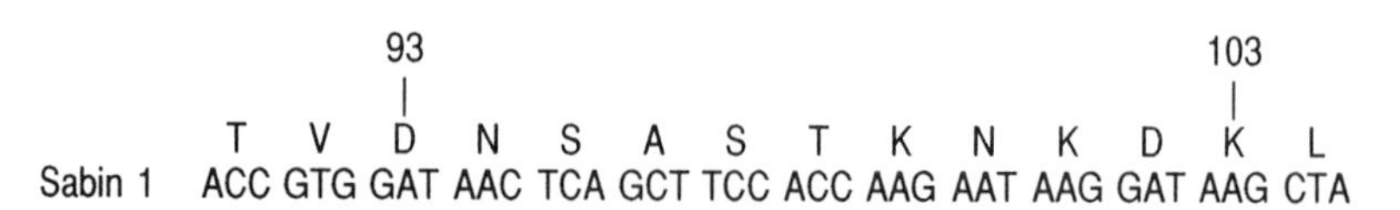

b

T V D N S A S T K N K F K L
pCAS1 ACC GTC GAC AAC TCA GCT TCC ACC AAG AAT AAG TTT AAA CTA
SalI DraI

FIG. 1. Poliovirus cassette vectors: the construction of pCAS1. (a) Nucleotides 2749–2791 of Sabin type 1 poliovirus (P1/LSc 2ab), encoding residues 91 to 104 of capsid protein VP1, are shown. The cassette was constructed by introducing unique *Sal*I and *Dra*I sites at nucleotides 2753 and 2783 using site-directed oligonucleotide mutagenesis. The resulting vector (b), designated pCAS1, is illustrated with the single amino acid substitution of Asp-102 for Phe highlighted.

introduction of the *Dra*I site at position 2783 resulted in the substitution of the aspartic acid at VP1 position 102 by phenylalanine (Fig. 1). However, since *Dra*I cuts in the center of its recognition sequence, digestion and religation with suitable oligonucleotides can restore the aspartate if required, although the substituted phenylalanine at position 102 was shown not to affect virus viability adversely.

The pCAS1 cassette vector has been extensively used in the construction of intertypic poliovirus chimeras, and chimeras expressing epitopes from other pathogens, including viruses such as human immunodeficiency virus type 1 (HIV-1),[17] human papillomavirus type 16 (HPV-16),[18] and hepatitis A,[19] and bacterial pathogens such as *Chlamydia trachomatis*.[20] Other poliovirus cassette vectors have also been described. Girard and colleagues introduced *Hpa*I/*Hin*dIII or *Eco*RV/*Hin*dIII restriction site pairs to facilitate the construction of recombinants expressing antigenic site 1 of the mouse-adapted Lansing strain.[12] Similarly, Wimmer and colleagues introduced a *Hin*dIII site and adopted a naturally occurring *Sph*I

[17] D. J. Evans, J. McKeating, J. M. Meredith, K. L. Burke, K. Katrak, A. John, M. Ferguson, P. D. Minor, and J. W. Almond, *Nature (London)* **339,** 385 (1989).

[18] O. Jenkins, J. Cason, K. L. Burke, D. Lunney, A. Gillen, D. Patel, D. J. McCance, and J. W. Almond, *J. Virol.* **64,** 1201 (1990).

[19] K. L. Burke, M. Ferguson, P. D. Minor, S. Lemmon, and J. W. Almond, unpublished data (1989).

[20] K. L. Burke, H. Caldwell, and J. W. Almond, unpublished data (1989).

site to construct a cassette system into which poliovirus type 2 and type 3 sequences were introduced.[13,21] In both cases these cassette systems were based on subcloned restriction fragments of the neurovirulent type 1 Mahoney strain of poliovirus, thereby necessitating an additional rebuilding step for the construction of a full-length infectious cDNA.

Other Poliovirus Cassette Vectors

Antigenic site 2b of poliovirus, although forming part of a well-defined conformational epitope (with site 2a), is also a contiguous stretch of amino acids and was considered to be a potentially good site for insertion of foreign substances. Cassette systems have therefore been constructed with restriction sites flanking this region of a poliovirus cDNA.[22,23] Experience has shown, however, that the flexibility of this region of the poliovirus capsid is significantly less than that of site 1, and this limits the usefulness of this region of the capsid for the expression of foreign epitopes.[23] A mutagenesis cassette has also been used in the study of the genome-linked protein VPg of poliovirus.[24]

pCAS7: An Improved Version of pCAS1

The pCAS1 cassette vector has been used for the construction of a large number of viable antigen chimeras.[15,17–19,25] However, extensive use of pCAS1 highlighted a series of problems inherent to the design of the vector. *Escherichia coli* strains harboring the vector grew particularly slowly and exhibited a marked instability, both of which we ascribed to the use of the modified pFB1[2] plasmid as the basis for the construct. In addition, because pCAS1 could itself give rise to viable virus, it was important to be certain that a given cDNA was a recombinant before proceeding to the transfection step. We have therefore constructed an improved vector (designated pCAS7) that overcomes these drawbacks. The construction details of pCAS7 are illustrated in Fig. 2, with the nucleotide sequences of relevant regions shown in Fig. 3. The 3.7-kilobase (kb) plasmid (pJM1) used for the construction of pCAS7 is a pAT153-derived vector created by the replacement of the entire tetracyclin-resistance gene of pAT153 with a gene

[21] M. G. Murray, J. Bradley, X.-F. Yang, E. Wimmer, E. Moss, and V. R. Racaniello, *Science* **241,** 213 (1988).

[22] A. D. Murdin and E. Wimmer, *J. Virol.* **63,** 5251 (1989).

[23] K. L. Burke and J. W. Almond, unpublished results (1989).

[24] R. J. Kuhn, H. Tada, M. F. Ypma-Wong, J. J. Dunn, B. L. Semler, and E. Wimmer, *Proc. Natl. Acad. Sci. U.S.A.* **85,** 519 (1988).

[25] C. S. P. Rose, S. P. Whelan, and D. J. Evans, unpublished results (1990).

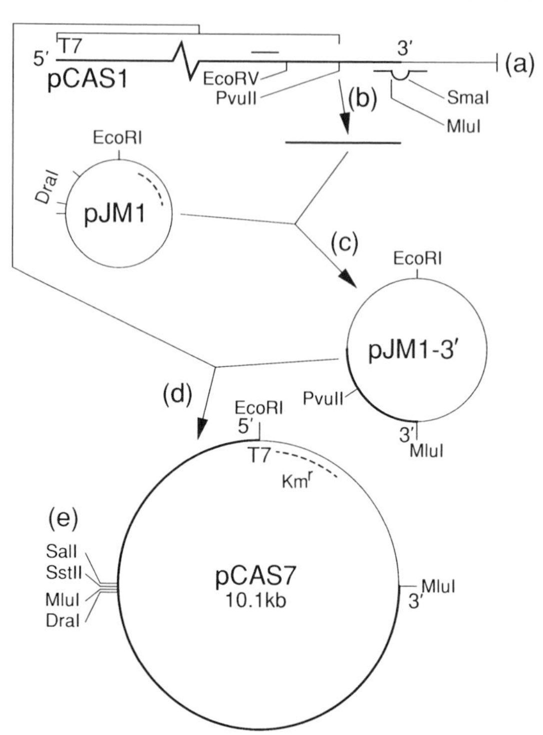

FIG. 2. Construction of pCAS7. (a) Linear representation of pCAS1 opened at the *Eco*RI site, which immediately precedes the T7 promoter. (b) Synthetic oligonucleotide primers complementary to nucleotides 6340–6360 and the junction between the poly(A) and G/C tails of pCAS1 (respectively, 5′-CCCTTATGTAGCAATGGG-3′ and 5′-GGGGGGGGG<u>CCCGGGACGCGT</u>TTTTTTTTT-3′) were used to amplify a 1.1-kb fragment using the polymerase chain reaction, and to introduce unique *Sma*I and *Mlu*I (underlined) sites at the extreme 3′ end of the poly(A) tract. (c) The purified reaction product was cleaved with *Eco*RV (which cuts at nucleotide 6373) and *Sma*I and cloned into *Dra*I-cut pJM1 to generate the plasmid pJM1-3′, in which the introduced poliovirus sequence was in the same orientation as the ampicillin-resistance gene. (d) The 7.1-kb *Eco*RI–*Pvu*II fragment from pJM1 was cloned into *Eco*RI–*Pvu*II-digested pJM1-3′ to generate pCAS6, which was modified (e) by replacement of poliovirus antigenic site 1 with complementary oligonucleotides containing *Mlu*I and *Sst*II restriction sites. The resulting "dead" vector was designated pCAS7. Plasmid vector regions are represented by thin lines, and Sabin 1 by thick lines. T7 indicates the location of the bacteriophage T7 promoter; 5′ and 3′ the ends of the Sabin 1 cDNA. The kanamycin-resistance gene of pJM1 is indicated by a dotted line. See Fig. 3 for further details on the construction of pCAS7.

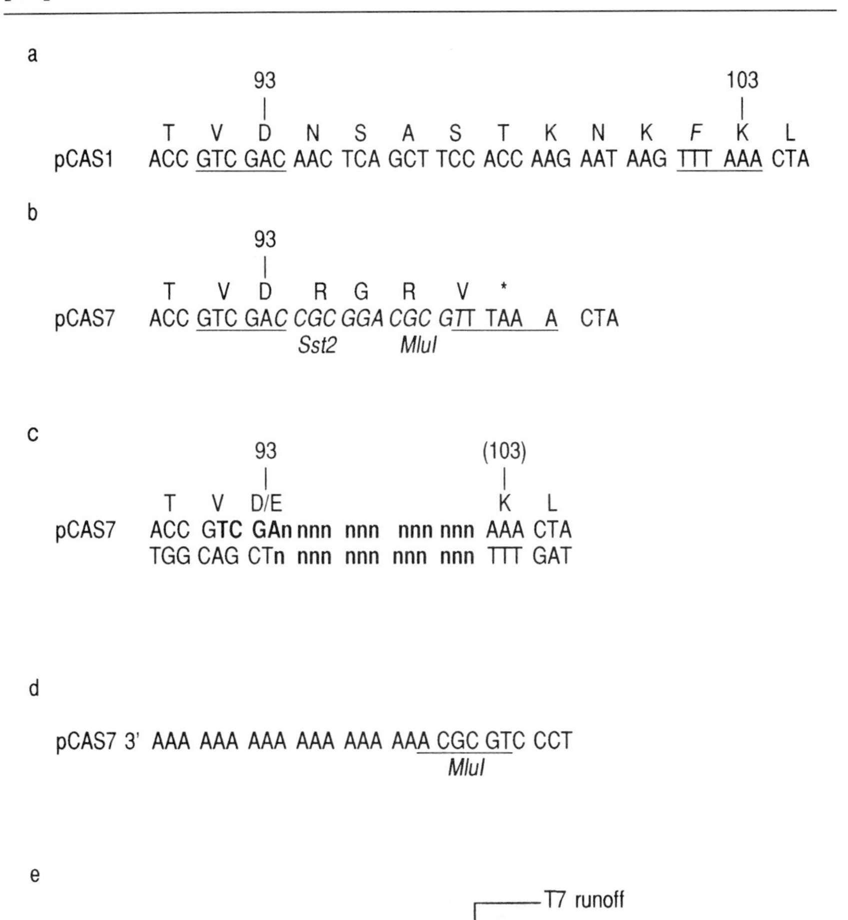

FIG. 3. Construction and use of pCAS7. (a) The original pCAS1 vector served as the basis for the construction of pCAS7. The cassette region of pCAS1 was replaced with oligonucleotides encoding *Sst*II and *Mlu*I sites, thereby introducing a frameshift to the poliovirus polyprotein (b) that results in premature termination within the *Dra*I site forming the 3′ extent of the cassette. (c) Oligonucleotides encoding sequences of choice (indicated by nnn), and with compatible 5′ and 3′ termini are ligated with *Sal*I–*Dra*I-digested pCAS7, thereby restoring the correct reading frame. (d) The extreme 3′ end of pCAS7 has been modified by the introduction of a *Mlu*I site following the 26-nucleotide poly(A) tail. This serves as a convenient site with which to linearize recombinant cDNA constructs prior to use in a T7 reaction. (e) A promoter (underlined) for the DNA-dependent RNA polymerase of bacteriophage T7 has been introduced at the extreme 5′ end of the Sabin 1 cDNA of pCAS1 and pCAS7. The transcription start point is indicated, as is the first nucleotide of the Sabin 1 cDNA.

cartridge encoding kanamycin resistance.[26] Recombinants derived from pCAS7 grow well and exhibit minimal instability problems. The deletion of the 3′ GC tails, and provision of an *Mlu*I site immediately after the 26-nucleotide poly(A) tail (Fig. 3), allows linearization of recombinant cDNAs prior to T7 transcription so that they do not have additional nonvirus nucleotides at the 3′ end. This has been reported to improve the infectivity of T7 transcripts.[27] Practically, the most significant differences between pCAS1 and pCAS7 are in the cassette region. The original sequences encoding antigenic site 1 of Sabin 1 have been replaced with complementary oligonucleotides containing two restriction sites (*Sst*II and *Mlu*I), which introduce a frameshift to the poliovirus polyprotein (Fig. 3). This results in premature termination of the polyprotein within the recognition sequences of *Dra*I that forms the 3′ site of the cassette region [TT(TAA)A]. This makes pCAS7 incapable of yielding infectious poliovirus unless the cassette region has been replaced, and ensures that any viable virus recovered must be a recombinant. It also allows the cassette to be used for the generation of multiple antigenic site sequence variants in a single step, selecting for the viable chimeras by transfection and plaquing.

The explosion of novel amplification and cloning techniques based on the polymerase chain reaction (PCR)[28] enables the rapid construction of cassette vectors without recourse to standard methods of site-directed mutagenesis. We currently use recombinant PCR[29] technology to rapidly generate cassette vectors similar in concept to pCAS7. In addition, we routinely use the University of Wisconsin Genetics Computing Group program MAPSORT, run with the command line qualifiers /SILENT or /MISMATCH to identify suitable restriction sites within a region of interest.[30] To realize the full potential of this approach, all blunt restriction enzyme "half-sites" (e.g., GAT and ATC representing the 5′ and 3′ "half-sites" of *Eco*RV) should be listed in the associated ENZYME.DAT file.

Construction and Recovery of Poliovirus Antigen Chimeras

The poliovirus epitope expression vectors pCAS1 and pCAS7 have been used in our laboratory to generate more than 70 viable poliovirus antigen chimeras. The 9-amino acid sequence of antigenic site 1 (Asn-94–

[26] J. M. Meredith and D. J. Evans, unpublished results (1989).

[27] P. Sarnow, *J. Virol.* **63,** 467 (1989).

[28] R. K. Saiki, D. H. Gelfand, S. Stoffel, S. J. Scharf, R. Higuchi, G. T. Horn, K. B. Mullis, and H. A. Erlich, *Science* **239,** 487 (1988).

[29] R. Higuchi, *in* "PCR Protocols" (M. A. Innis, D. H. Gelfand, J. J. Sninsky, and T. J. White, eds.), p. 177. Academic Press, San Diego, California, 1990.

[30] J. Devereaux, P. Haeberli, and O. Smithies, *Nucleic Acids Res.* **12,** 387 (1984).

Phe-102) has been replaced by sequences ranging from 3 to 26 amino acids in length.[15,25] These results have demonstrated the considerable flexibility of this part of the virus particle. However, only about 80% of all recombinant cDNA constructs encode viable virus. In the remainder the insertion presumably disrupts an essential stage of the virus replication cycle e.g., translation, polyprotein processing, particle assembly, receptor binding, or particle uncoating. Different, nonviable constructs may be blocked at different stages of the replication cycle. We are currently comparing our extensive panel of viable and nonviable recombinants in an attempt to identify features that characterize both categories. Preliminary analysis suggests that the charge distribution at the N terminus of the insertion may be critical—the presence of excessive positive or negative charges seems to result in nonviability for reasons that are not clear. Analysis of some closely related viable and nonviable constructs using computer-based molecular modeling techniques also suggests that the overall width of the introduced loop, determined by measuring the predicted separation of the α-carbons of Asp-92 and Lys-103, is also important. Those in which the width is predicted to be significantly wider than the native loop tend to be nonviable, suggesting that the flanking βB and βC strands may be distorted and so result in the disruption of folding or assembly. Alternatively, the increased width of the loop may lead to unfavorable interactions with other regions of the capsid.[31]

Epitope Selection and Oligonucleotide Design

Our experience of the poliovirus system based on the expression of antigenic determinants from HIV-1,[17] HIV-2 and simian immunodeficiency virus (SIV), HPV16,[18] *Chlamydia trachomatic*,[20] and hepatitis A,[19] suggest that the system is best suited to the expression of linear epitopes that have previously been identified by other means. However, the system can also be used to express a range of sequences in an attempt to identify those that constitute recognized epitopes in the original protein. There are many approaches published for identifying candidate antigenic domains[32,33] and computer programs helpful for these purposes are available.[30] For pCAS7, complementary oligonucleotides are designed so that, when annealed, the 5′ end is compatible with a *Sal*I site, and the 3′ end is blunt and can therefore be ligated with *Dra*I-digested vector (Fig. 3). Although the usage of certain codons in the poliovirus polyprotein is

[31] M. J. C. Crabbe, D. J. Evans, and J. W. Almond, *FEBS Lett.* **271,** 194 (1990).
[32] T. P. Hopp and K. R. Woods, *Proc. Natl. Acad. Sci. U.S.A.* **78,** 3824 (1981).
[33] J. A. Berzofsky, *Science* **229,** 932 (1985).

biased, we have yet to demonstrate an effect on the growth of the virus by inclusion of such codons in a construct. The precise sequence chosen for the oligonucleotides is obviously determined by the epitope to be generated; however, it is often helpful to exploit the degeneracy of the code to include restriction sites that may aid the subsequent screening of cDNA recombinants.

Construction Details

The construction and characterization of recombinant poliovirus cDNA's uses the standard techniques of molecular genetics.[34] The pCAS7 vector is digested with *Sal*I and *Dra*I, phenol extracted, and ethanol precipitated. There is no need to separate and purify the larger of the two restriction fragments by agarose gel electrophoresis, or to prevent religation of the vector by phosphatase treatment. Approximately 300 ng of each of the two complementary oligonucleotides is annealed by boiling for 3 min and slowly cooling to room temperature. We routinely ligate one-sixth of this annealed mix with 50 ng of prepared vector, and transform *E. coli* MC1061. Prior to transformation the ligation mix is digested with either *Sst*II or *Dra*I (assuming the latter site is destroyed on ligation of the oligonucleotides) to prevent the subsequent recovery of religated pCAS7 vector. Recombinants are selected on medium containing kanamycin at 50 ng/μl. Plasmid DNA extracted by alkaline lysis miniprep can be rapidly screened with *Mlu*I; heterologous recombinants are linearized, whereas reformed pCAS7 forms a doublet of 4.7 and 5.4 kb. The sequence through the modified region of cDNA constructs that appear correct by restriction analysis can be rapidly verified using a protocol suitable for double-stranded plasmids.[35]

Recovery of Poliovirus Chimeras

As mentioned above, the recovery of viable virus from poliovirus cDNA by transfection of permissive cells is more efficient using RNA,[16] rather than DNA. We have therefore engineered a promoter for the DNA-dependent RNA polymerase of bacteriophage T7 immediately preceding the 5′ end of the Sabin 1 cDNA in pCAS7 (Fig. 3). The recombinant cDNA construct is linearized with *Mlu*I, phenol extracted, ethanol precipitated, and resuspended in diethyl pyrocarbonate-treated water, prior to use

[34] J. Sambrook, E. F. Fritsch, and T. Maniatis, "Molecular Cloning: A Laboratory Manual." Cold Spring Harbor Laboratory, Cold Spring Harbor, New York, 1989.

[35] G. Murphy and T. Kavanagh, *Nucleic Acids Res.* **16,** 5198 (1988).

as the template in a T7 reaction. Approximately 100 ng of *Mlu*I-linearized DNA is ample for use in the T7 reaction. We have observed that template DNA produced by alkaline lysis protocols generally produces better T7 transcripts than plasmids purified by methods involving the use of cesium chloride. Before transfection of permissive cells, the products of the T7 reaction are visualized following gel electrophoresis. RNA transcripts produced from *Mlu*I-linearized pCAS7-derived plasmids have an apparent size of 2.1 kb on 1% agarose gels. Extensive smearing of the products suggests contamination with RNAses, although limited degradation is often visible and should not prevent recovery of viable chimeric viruses. Subconfluent Hep2c or HeLa cells are transfected with the T7 reaction using DEAE-dextran according to published protocols.[16] Transfected cells should be incubated at 34° until a cytopathic effect (cpe) is observed. This usually takes between 3 and 4 days, although some slowly growing chimeras may require as long as 10 days. It is advisable to perform parallel positive and negative control transfections, for which we routinely use T7 runoff transcripts produced from a full-length Sabin 1 cDNA and pCAS7, respectively. The tissue culture supernatant from transfections that fail to yield viable virus after 7 days should be passaged on fresh monolayers and monitored for a further 7 days. The recovered master stock of a viable chimera should be stored as aliquots at −70° in cell-free tissue culture supernatant, and used to raise submaster stocks for further characterization.

Safety Aspects of Poliovirus Antigen Chimera Characterization

According to the "canyon hypothesis" proposed by Rossman *et al.*, the cleft surrounding the pentameric apex of the rhinovirus (and closely related poliovirus) capsid is the region of the virus that interacts with, and determines the specificity for, the cellular receptor.[36] This is supported by site-directed mutagenesis studies of the canyon of rhinovirus.[37] However, there is also some evidence that the sequences present in antigenic site 1 of poliovirus may be involved in virus tropism. La Monica *et al.*[38] demonstrated that mouse neurovirulence of the type 2 Lansing strain of poliovirus is dependent on sequences within site 1. This was supported by the observation that a chimera based on the mouse avirulent strain P1/Ma-

[36] M. G. Rossman, E. Arnold, J. W. Erickson, E. A. Frankenburger, J. P. Griffith, H. J. Hecht, J. E. Johnson, G. Kamer, M. Luo, A. G. Mosser, R. R. Rueckert, B. Sherry, and G. Vriend, *Nature (London)* **317,** 145 (1985).

[37] R. Colonno, J. Condra, S. Mizutani, P. Callahan, M. Davies, and M. Murcko, *Proc. Natl. Acad. Sci. U.S.A.* **85,** 5449 (1988).

[38] N. La Monica, W. J. Kupsky, and V. R. Racaniello, *Virology* **161,** 429 (1987).

honey became fully virulent for mice when its site 1 sequence was derived from the mouse virulent P2/Lansing strain.[12,21] This suggested that sequences inserted at antigenic site 1 may affect the tropism of the virus, and therefore that certain chimeras may acquire novel cell tropisms and thus constitute a biohazard. Routinely, therefore, we check that each new chimera is neutralized by pooled human serum containing poliovirus antibodies before preparing or purifying significant amounts of the virus. We additionally determine whether infection of susceptible cell lines with a chimera is blocked by a monoclonal antibody directed against the poliovirus receptor.[39] Chimeras with novel tropisms might be expected to overcome such a block. We regard both these safety checks as important preliminaries to handling the virus under the containment conditions appropriate for poliovirus.

Following the safety precautions we routinely determine the infectious titer of the chimera using either a standard plaque assay or a quantal infectivity assay ($TCID_{50}$), both of which are described elsewhere.[40] Before further characterization of the antigenicity or immunogenicity of a particular chimera, the RNA sequence of a region spanning the modified antigenic site(s) should be determined to verify that no spurious deletions or substitutions have occurred during manipulations. RNA extracted and purified directly from virions can be directly sequenced using ^{35}S and the dideoxy chain-termination method described by Rico-Hesse *et al.*[41] However, this involves the use of relatively large amounts of purified virus,and we prefer to amplify by PCR and sequence a 300-nucleotide (nt) region spanning the modified site.

RNA Extraction, Reverse Transcription, and PCR Amplification

The *positive-* and *negative-*sense primers used in the following protocol are complementary to nucleotides 2627–2649 and 2924–2905 of Sabin 1, respectively, and are stored at 100 ng/μl.

1. Mix 25 μl or vanadyl ribonucleoside complexes (Sigma, St. Louis, MO) with 250 μl of cell-free tissue culture supernatant containing the viable antigen chimera.
2. Add 14 μl of 10% sodium dodecyl sulfate (SDS) and incubate for 10 min on ice.

[39] P. D. Minor, P. A. Pipkin, D. Hockley, G. C. Schild, and J. W. Almond, *Virus Res.* **1,** 203 (1984).

[40] P. D. Minor, *in* "Virology: A Practical Approach" (B. W. J. Mahy, ed.), p. 25. IRL Press, Oxford, 1984.

[41] R. Rico-Hesse, M. A. Pallanasch, B. K. Nottay, and O. M. Kew, *Virology* **160,** 311 (1987).

3. Add 20 μl of 20 mg/ml proteinase K (Sigma) and incubate at 37° for 20 min.

4. Taking care to avoid the interface, phenol/chloroform extract. Repeat. Precipitate nucleic acids by addition of 4 μl 2 *M* sodium acetate with 750 μl ice-cold ethanol and centrifugation at 12,000 *g*. Resuspend the black pellet in 40 μl 10 m*M* Tris-HCl (pH 8)/1 m*M* EDTA.

5. Prepare the following in a 500-μl microcentrifuge tube: 2 μl of RNA (from step 5), 1 μl *negative*-sense primer, 2 μl Moloney Murine Leukemia Virus (MMLV) RT buffer (5× concentrate), 2.5 μl 2 m*M* dNTPs, 1.5 μl H_2O. Add 1 μl MMLV reverse transcriptase (BRL, Bethesda Research Laboratory, Gaithersburg, MD), and incubate at 37° for 5 min.

6. Using the entire reaction from the preceding step, add to the same microfuge tube: 10 μl 2 m*M* dNTPs, 10 μl 10× concentrated Taq buffer (as specified by manufacturer), 500 ng of both primers, H_2O to a total volume of 100 μl, and 2 U of Cetus Amplitaq-cloned Taq polymerase. Cover reaction with 60 μl of mineral oil and amplify by completing 30 cycles of 95° for 1 min, 55° for 2 min, and 72° for 2 min.

The 300-nt product of the PCR reaction is separated by electrophoresis on a 1.5% agarose gel, isolated using Prep-A-Gene (Bio-Rad, Richmond, CA), and sequenced directly using the dideoxy chain-termination method[42] as modified by Winship.[43] Alternatively, the 300-base pair (bp) fragment amplified using the primers described above contains unique *Sph*I and *Xba*I restriction sites, thereby allowing the direct cloning into a suitable vector.

Antigenic and Immunogenic Characterization of Chimeras

The detailed protocols used to characterize the antigenic and immunogenic characteristics of poliovirus antigen chimeras are beyond the scope of this chapter. In practice, the antigenic characterization involves two well-documented approaches: (1) single radial immunodiffusion, or "antigen-blocking" assays, and (2) neutralization of virus infectivity.[40] Both assays can be used to demonstrate an interaction between the chimeric virus and antibodies raised against the foreign protein from which the expressed epitope is derived. The immunogenicity of a chimera is determined by analysis of sera following administration to a suitable small animal. Immunizations are generally performed by parenteral administration of approximately 10^7 $TCID_{50}$ of virus in Freund's complete or incomplete adjuvant.

[42] F. Sanger, S. Nicklen, and A. Coulson, *Proc. Natl. Acad. Sci. U.S.A.* **74,** 5463 (1977).
[43] P. Winship, *Nucleic Acids Res.* **17,** 1266 (1989).

We have also immunized primates via the oral route and further studies to characterize the immune responses obtained are in progress. One of the attractions of poliovirus as a vector is that oral immunization is known to induce a good secretory as well as circulatory immune response.

Conclusion

Our success in producing live chimeras with novel antigenicity based on the stably attenuated vaccine P1/Sabin suggests that the primary application of such viruses will be in the field of vaccination. However, other work has demonstrated that poliovirus antigen chimeras may have wider applications, such as in the study of picornavirus structure and function, and in the antigenic characterization of other pathogens or of any foreign protein. The system provides an attractive alternative to the use of free peptides for eliciting antibodies against a defined region of a given protein. The expression of a normally immunosilent epitope in a well-defined immunoexposed location on the poliovirus particle may elicit antibodies that have valuable diagnostic or therapeutic applications. We are also exploiting the novel antigenicity of chimeric virus particles in HIV serological studies.

Acknowledgments

We acknowledge the significant contributions made to these studies by our colleagues at Reading University (J. M. Meredith, C. S. P. Rose, S. P. Whelan, and Drs. K. L. Burke and M. J. C. Crabbe), the National Institute for Biological Standards and Control (Drs. M. Ferguson, P. D. Minor, and C. Vella), and Birkbeck College (H. Stirk and Dr. J. Thornton). This work was funded by the Medical Research Council and the MRC AIDS Directed Programme.

Section II

Nucleic Acids and Polysaccharides

[21] Computer Simulation of DNA Supercoiling

By WILMA K. OLSON and PEISEN ZHANG

Introduction

Superimposed on the right-handed coiling of the familiar DNA double helix is a higher order of coiling of the helix axis itself, called supercoiling.[1] This coiling is analogous to the tertiary folding of helical segments in protein structures, but is spread over a much larger molecular scale. In naturally occurring DNA there is roughly 1 superhelical turn for every 20 local turns (e.g., 200 bp) of the approximately 10-fold B-DNA helix. In long proteins, in contrast, a corresponding 360° turn of the polypeptide backbone can be accommodated by 10 or fewer chain residues. Moreover, like the proteins, which fold in both a right-handed and a left-handed manner, the supercoiling of native DNA is also right and left handed. The two complementary strands of the supercoiled DNA duplex are respectively over and underwound compared to the number of times they would intertwine around one another in a relaxed (nonsupercoiled) structure and are consequently subjected to a persistent internal strain tending to wind and unwind local regions of the double helix. Not surprisingly, supercoiling is an important facet of biological processes that entail local helical winding and unwinding, such as replication, transcription, and recombination.[2-5]

While first invoked to account for the peculiar physicochemical properties of closed circular DNA,[6] supercoiling can be manifested by any spatially constrained sequence, including, for example, a linear chain segment tightly bound to a rigid protein scaffold. In such systems the ends of the DNA are required to adopt specific three-dimensional arrangements and the secondary and tertiary structures are forced to vary in an interdependent fashion. The basic topological parameter characterizing the constrained structure is the so-called linking number, Lk, the total number of revolutions of one single strand about the other when the DNA lies in a plane. In the absence of strand breaks Lk has a fixed integer value that may

[1] W. R. Bauer, *Annu. Rev. Biophys. Bioeng.* **7,** 287 (1978).
[2] N. R. Cozzarelli, *Science* **207,** 953 (1980).
[3] L. F. Liu and J. C. Wang, *Proc. Natl. Acad. Sci. U.S.A.* **84,** 7024 (1987).
[4] D. M. J. Lilley, *Nature (London)* **305,** 276 (1983).
[5] L. M. Fisher, *Nature (London)* **307,** 686 (1984).
[6] J. Vinograd, J. Lebowitz, R. Radloff, R. Watson, and P. Laipis, *Proc. Natl. Acad. Sci. U.S.A.* **53,** 1104 (1965).

be decomposed into a contribution, Wr, which describes the locus of the DNA helix and Tw, which describes the twisting of the strands about the helical axis.[7,8]

$$\text{Lk} = \text{Wr} + \text{Tw} \quad (1)$$

Wr is called the writhing number of the macromolecule and Tw the total twist. Wr and Tw are differential geometric parameters that vary continuously with the shape of the duplex, so that when the strands are covalently closed the linking number is fixed. Their values can be obtained by various mathematical procedures (see below).

According to Eq. (1), the supercoiling of DNA, in which the linking number is altered by ΔLk from its value in the relaxed planar form, involves either writhing (ΔWr) or twisting (ΔTw) or both. Changes in Tw are generally thought to involve torsional deformations of the DNA duplex and/or local transitions to alternate conformational forms. Processes such as duplex denaturation,[9,10] cruciform formation,[10-14] and right- to left-handed transitions[15,16] are well known to be facilitated by DNA supercoiling. Changes in the tertiary structural parameter Wr entail some sort of chain bending. Supercoils are found to adopt compact interwound forms in hydrodynamic and electron microscopic studies,[17-19] while there are additional suggestions that toroidal-like conformations may be favored at low ionic strengths.[20,21] A toroidal trajectory is that of a closed circular DNA superhelix with the axis of the double helix winding around the surface of a torus (e.g., a donut) and the bases of the DNA turning around the deformed secondary helical axis.

[7] J. H. White, *Am. J. Math.* **91,** 693 (1969).

[8] W. B. Fuller, *Proc. Natl. Acad. Sci. U.S.A.* **68,** 815 (1971).

[9] J. Vinograd, J. Lebowitz, and R. Watson, *J. Mol. Biol.* **33,** 173 (1968).

[10] T.-S. Hsieh and J. C. Wang, *Biochemistry* **14,** 527 (1975).

[11] M. Gellert, K. Mizuuchi, M. H. O'Dea, H. Ohmori, and J. Tomizawa, *Cold Spring Harbor Symp. Quant. Biol.* **43,** 35 (1979).

[12] K. Mizuuchi, M. Mizuuchi, and M. Gellert, *J. Mol. Biol.* **156,** 229 (1982).

[13] D. M. J. Lilley, *Proc. Natl. Acad. Sci. U.S.A.* **77,** 6468 (1980).

[14] N. Panayotatos and R. D. Wells, *Nature (London)* **289,** 466 (1981).

[15] L. J. Peck, A. Nordheim, A. Rich, and J. C. Wang, *Proc. Natl. Acad. Sci. U.S.A.* **79,** 4560 (1982).

[16] L. J. Peck and J. C. Wang, *Proc. Natl. Acad. Sci. U.S.A.* **80,** 6206 (1983).

[17] W. Bauer and J. Vinograd, *J. Mol. Biol.* **33,** 141 (1968).

[18] J. C. Wang, *J. Mol. Biol.* **43,** 25 (1969).

[19] R. T. Espejo, E. S. Canelo, and R. L. Sinsheimer, *Proc. Natl. Acad. Sci. U.S.A.* **63,** 1164 (1969).

[20] N. B. Gray, *Biopolymers* **5,** 1009 (1967).

[21] G. W. Brady, D. B. Fein, J. Lambertson, V. Grassian, D. Foos, an C. J. Benham, *Proc. Natl. Acad. Sci. U.S.A.* **80,** 741 (1983).

Polymer Models

Elastic Energy Models

Theoretical understanding of DNA supercoiling at the all-atom level is not computationally feasible. The long chain must be described by simpler phenomenological models that account for certain macroscopic properties of the double helix, but which ignore details of its chemical structure and environment. The DNA duplex has been idealized in a number of studies, for example, as a symmetric, linearly elastic rod that, like a rubber hose, exhibits no preferential directions of bending and twisting.[22-33] The forces opposing deformation of the rod are partitioned locally into independent harmonic contributions from the changes of curvature κ (e.g., local bending) and the rate of local twist ω.

$$E = E_{\mathrm{bend}} + E_{\mathrm{twist}} = \frac{A}{2}\oint \kappa(s)^2 ds + \frac{C}{2}\oint (\omega - \omega_0)^2 ds \qquad (2)$$

The parameter ω_0 is the intrinsic rate of twist; A, the coefficient of flexural (e.g., bending) rigidity; C, the twisting stiffness constant; and s, the arc length. Values of A, typically $2.0-3.0 \times 10^{-19}$ erg, are deduced from the observed persistence length of DNA,[34-36] while values of C, of the order of $1.0-2.0 \times 10^{-19}$ erg, are related to various optical measurements,[37-49] the

[22] C. J. Benham, *Proc. Natl. Acad. Sci. U.S.A.* **74,** 2397 (1977).
[23] C. J. Benham, *J. Mol. Biol.* **123,** 361 (1978).
[24] C. J. Benham, *Biopolymers* **18,** 609 (1979).
[25] C. J. Benham, *Biopolymers* **22,** 2477 (1983).
[26] M. Le Bret, *Biopolymers* **17,** 1939 (1978).
[27] M. Le Bret, *Biopolymers* **18,** 1709 (1979).
[28] M. Le Bret, *Biopolymers* **19,** 619 (1980).
[29] M. Le Bret, *Biopolymers* **23,** 1835 (1984).
[30] M. D. Frank-Kamenetskii, A. V. Lukashin, and A. V. Vologodskii, *Nature (London)* **258,** 398 (1975).
[31] A. V. Vologodskii, V. V. Anshelevich, A. V. Lukashin, and M. D. Frank-Kamenetskii, *Nature (London)* **280,** 294 (1979).
[32] M. D. Frank-Kamenetskii and A. V. Vologodskii, *Sov. Phys. Usp. (Engl. Transl.)* **24,** 679 (1981).
[33] M. D. Frank-Kamenetskii, A. V. Lukashin, V. V. Anshelevich, and A. V. Vologodskii, *J. Biomol. Struct. Dyn.* **2,** 1005 (1985).
[34] P. J. Hagerman, *Annu. Rev. Biophys. Biophys. Chem.* **17,** 265 (1988).
[35] H. Eisenberg, *Acc. Chem. Res.* **20,** 276 (1987).
[36] H. Eisenberg, *Eur. J. Biochem.* **187,** 7 (1990).
[37] M. D. Barkley and B. H. Zimm, *J. Chem. Phys.* **70,** 2991 (1979).
[38] J. C. Thomas, S. A. Allison, C. J. Appellof, and J. M. Schurr, *Biophys. Chem.* **12,** 177 (1980).

rates of cyclization of small DNA chains,[50–52] and the equilibrium distribution of helical twist in circular DNA.[53–56]

According to elasticity theory, the local twist density ω is independent of s for a normally straight rod of circular cross section. It then follows that for given linking number Lk the total energy in Eq. (2) can be reduced to[8,57]

$$E = \frac{A}{2} \oint \kappa(s)^2 ds + \frac{2\pi^2 C}{L} (\Delta\mathrm{Lk} - \mathrm{Wr})^2 \tag{3}$$

where L is the total contour length of the DNA and $\Delta\mathrm{Lk} = \mathrm{Lk} - \mathrm{Lk}_0$, the linking number difference, with Lk_0, the so-called intrinsic linking number, equal to $(L/2\pi)\omega_0$. The curvature integral and the writhing number appearing in the above expression are evaluated from the equations or points describing the spatial trajectory of the helix axis as outlined below. Because the local twisting energy is related to the overall shape of the curve in this approximation, there is no need to locate the positions of individual base pairs.

This simple elastic model ignores the well-known preferential modes of bending of the DNA double helix. Various X-ray crystallographic[58–69] and

---

[39] D. P. Millar, R. J. Robbins, and A. H. Zewail, *Proc. Natl. Acad. Sci. U.S.A.* **77,** 5593 (1980).

[40] D. P. Millar, R. J. Robbins, and A. H. Zewail, *J. Chem. Phys.* **74,** 4200 (1981).

[41] D. P. Millar, R. J. Robbins, and A. H. Zewail, *J. Chem. Phys.* **76,** 2080 (1982).

[42] I. Hurley, P. Osei-Gyimah, S. Archer, C. P. Scholes, and L. S. Lerman, *Biochemistry* **21,** 4999 (1982).

[43] M. Hogan, J. Le Grange, and B. Austin, *Nature (London)* **304,** 752 (1983).

[44] J. C. Thomas and J. M. Schurr, *Biochemistry* **22,** 6194 (1983).

[45] J. H. Shibata, J. Wilcoxon, J. M. Schurr, and V. Knauf, *Biochemistry* **23,** 1188 (1984).

[46] B. S. Fujimoto, J. H. Shibata, R. L. Schurr, and J. M. Schurr, *Biopolymers* **24,** 1009 (1985).

[47] B. Berkoff, M. Hogan, J. Le Grange, and B. Austin, *Biopolymers* **25,** 307 (1986).

[48] P. Wu, L. Song, J. B. Clendenning, B. S. Fujimoto, A. S. Benight, and J. M. Schurr, *Biochemistry* **27,** 8128 (1988).

[49] B. S. Fujimoto and J. M. Schurr, *Nature (London)* **344,** 175 (1990).

[50] D. Shore, J. Langowski, and R. L. Baldwin, *Proc. Natl. Acad. Sci. U.S.A.* **78,** 4833 (1981).

[51] D. Shore and R. L. Baldwin, *J. Mol. Biol.* **170,** 957 (1983).

[52] D. Shore and R. L. Baldwin, *J. Mol. Biol.* **170,** 983 (1983).

[53] R. E. Depew and J. C. Wang, *Proc. Natl. Acad. Sci. U.S.A.* **72,** 4275 (1975).

[54] D. E. Pulleyblank, M. Shure, D. Tang, J. Vinograd, and H.-P. Vosberg, *Proc. Natl. Acad. Sci. U.S.A.* **72,** 4280 (1975).

[55] J. C. Wang, L. J. Peck, and K. Becherer, *Cold Spring Harbor Symp. Quant. Biol.* **47,** 85 (1982).

[56] D. S. Horowitz and J. C. Wang, *J. Mol. Biol.* **173,** 75 (1984).

[57] F. Tanaka and H. Takahashi, *J. Chem. Phys.* **83,** 6017 (1985).

[58] R. E. Dickerson and H. R. Drew, *J. Mol. Biol.* **149,** 761 (1981).

theoretical[70-80] models of oligonucleotide helices find the chain bends more easily into its major and minor grooves than in other directions. Individual base pairs in these structures exhibit more pronounced rolling about their long axes, and hence into the grooves of the duplex, than tilting about their short (dyad) axes. Moreover, according to these models, the rolling is base sequence dependent, with certain base pairs more likely to bend in an anisotropic fashion into one groove and others able to flex with comparable likelihood into either groove. These properties can influence overall chain bending, particularly when the sequence of base pairs is regularly repeated in phase with the 10-fold helical repeat.[76,77,80] The discussion in this chapter, however, is limited to chains without these special features.

---

[59] A. V. Fratini, M. L. Kopka, H. R. Drew, and R. E. Dickerson, *J. Biol. Chem.* **257,** 14686 (1982).

[60] B. N. Conner, C. Yoon, J. L. Dickerson, and R. E. Dickerson, *J. Mol. Biol.* **174,** 663 (1984).

[61] A. H.-J. Wang, S. Fujii, J. H. van Boom, and A. Rich, *Proc. Natl. Acad. Sci. U.S.A.* **79,** 3968 (1982).

[62] A. H.-J. Wang, S. Fujii, J. H. van Boom, G. A. van der Marel, S. A. A. van Boeckel, and A. Rich, *Nature (London)* **299,** 601 (1982).

[63] Z. Shakked, D. Rabinovich, O. Kennard, W. B. T. Cruse, S. A. Salisbury, and M. A. Viswamitra, *J. Mol. Biol.* **166,** 183 (1983).

[64] M. McCall, T. Brown, and O. Kennard, *J. Mol. Biol.* **183,** 385 (1985).

[65] G. Kneale, T. Brown, O. Kennard, and D. Rabinovich, *J. Mol. Biol.* **186,** 805 (1985).

[66] M. McCall, T. Brown, W. N. Hunter, and O. Kennard, *Nature (London)* **322,** 661 (1986).

[67] Z. Shakked and D. Rabinovich, *Prog. Biophys. Mol. Biol.* **41,** 159 (1986).

[68] T. E. Haran and Z. Shakked, *J. Mol. Struct. (THEOCHEM.)* **179,** 367 (1988).

[69] O. Kennard and W. N. Hunter, *Q. Rev. Biophys.* **22,** 327 (1990).

[70] V. B. Zhurkin, Y. P. Lysov, and V. I. Ivanov, *Nucleic Acids Res.* **6,** 1081 (1979).

[71] V. B. Zhurkin, Y. P. Lysov, V. L. Florentiev, and V. I. Ivanov, *Nucleic Acids Res.* **10,** 1811 (1982).

[72] N. B. Ulyanov and V. B. Zhurkin, *J. Biomol. Struct. Dyn.* **2,** 361 (1984).

[73] W. K. Olson, A. R. Srinivasan, M. A. Cueto, R. Torres, R. C. Maroun, J. Cicariello, and J. L. Nauss, *in* "Biomolecular Stereodynamics IV" (R. H. Sarma and M. H. Sarma, eds.), p. 75. Adenine Press, Guilderland, New York, 1985.

[74] C.-S. Tung and S. C. Harvey, *Nucleic Acids Res.* **12,** 3343 (1984).

[75] C.-S. Tung and S. C. Harvey, *J. Biol. Chem.* **261,** 3700 (1986).

[76] A. R. Srinivasan, R. Torres, W. Clark, and W. K. Olson, *J. Biomol. Struct. Dyn.* **5,** 459 (1987).

[77] W. K. Olson, A. R. Srinivasan, R. C. Maroun, R. Torres, and W. Clark, *in* "Unusual DNA Structures" (R. D. Wells and S. C. Harvey, eds.), p. 207. Springer-Verlag, New York, 1987.

[78] A. Sarai, J. Mazur, R. Nussinov, and R. L. Jernigan, *Biochemistry* **27,** 8498 (1988).

[79] A. Sarai, J. Mazur, R. Nussinov, and R. L. Jernigan, *Biochemistry* **28,** 7842 (1989).

[80] W. K. Olson and A. R. Srinivasan, *Comput. Appl. Biosci.* **4,** 133 (1988).

Excluded Volume Effect

The DNA is also subject to the so-called excluded volume effect.[81] Since each repeating segment has finite volume, other segments cannot occupy the same space. The excluded volume contribution represents the effect of interactions between segments that are far apart along the chain contour. In real DNA, the nature of the excluded volume interaction is quite complicated, involving both long-range steric contacts and electrostatic repulsions of the negatively charged phosphate groups. This effect can be crudely mimicked by use of a hard sphere potential that restricts the contacts of chain contour points to distances greater than some arbitrary limit. Specifically, if the distance r_{ij} between chain contour points i and j is greater than the cutoff distance D, the excluded volume contribution V_{ij} is zero, and if less than or equal to the limit, the energy is infinite.

$$V_{ij} = \begin{Bmatrix} \infty & r_{ij} \leq D \\ 0 & r_{ij} > D \end{Bmatrix} \tag{4}$$

Alternatively, the self-interaction energy can be expressed by a van der Waals term of the form

$$V_{ij} = \begin{Bmatrix} \epsilon\left[\left(\dfrac{d}{r_{ij}}\right)^{12} - \left(\dfrac{d}{r_{ij}}\right)^{6}\right] & r_{ij} \leq d \\ 0 & r_{ij} > d \end{Bmatrix} \tag{5}$$

where ϵ is an empirical parameter equal to the depth of the pairwise energy well and d the effective interaction radius of the chain. The latter distance does not necessarily correspond to the hard sphere cutoff distance D used in Eq. (4).

Both Eqs. (4) and (5) are costly steps in a computer simulation. The number of nonbonded terms can be reduced, however, if chain configuration is perturbed locally rather than globally (see below).

Constrained DNA Models

General Computational Approaches

The description of closed chain molecules (or, alternatively, open chains with ends confined to a fixed end-to-end separation and orientation) is a long-standing problem in polymer physical chemistry that has

[81] P. J. Flory, *J. Chem. Phys.* **17,** 303 (1949).

been attacked from two basic points of view. In one approach the configurations of open linear chains that meet specified spatial criteria are collected through exhaustive simulation studies.[82-85] The method, however, is not generally practical due to limitations on computer availability. The probability of identifying specific configurations from the random sampling of an unconstrained chain is so low that it is difficult to draw conclusions regarding the conformations and flexibility of the constrained system of interest. The method, however, is very useful in the simulation of the kinetics of cyclization or loop formation of open chains. The second approach to the problem is to start with a constrained chain configuration, such as the closed, 100-residue, figure-eight trajectory in Fig. 1, and to allow the system to deform subject to some potential function.[30-33] The major difficulty in such computer simulations is the preservation of the desired constraints on the positions and orientations of chain ends during the simulation. Individual Cartesian coordinates must be moved in small concerted steps or internal torsion angles in highly correlated pairwise increments to maintain the fixed positions of chain ends.

Until recently these two approaches to the closed polymer problem have been limited to systems of relatively small size by the number of variables needed to describe the chain trajectory. As a result, it has not been possible to simulate the transformation of relaxed closed circular DNA to the elaborate interwound forms observed experimentally. The treatment of larger DNAs has necessitated simplifications of the chain model beyond the simple elastic approximation.[33]

Work from this laboratory shows that the size-limitation problem in the simulation of supercoiled DNA can be overcome, at least in part, by taking advantage of curve-fitting methods commonly used in computer-aided design and engineering.[86-89] Both piecewise B-spline curves and Fourier series representations have been successfully used to describe the complex trajectories of constrained DNA polymers. These two approaches automatically satisfy the ring-closure constraints of cyclic DNA and/or the end-to-end limitations on constrained open DNA. As outlined below, the formu-

[82] U. W. Suter, M. Mutter, and P. J. Flory, *J. Am. Chem. Soc.* **98,** 5733 (1976).

[83] S. D. Levene and D. M. Crothers, *J. Mol. Biol.* **189,** 61 (1986).

[84] N. L. Marky and W. K. Olson, *Biopolymers* **26,** 415 (1987).

[85] P. J. Hagerman and V. A. Ramadevi, *J. Mol. Biol.* **212,** 351 (1990).

[86] M.-H. Hao and W. K. Olson, *Biopolymers* **28,** 873 (1989).

[87] M.-H. Hao and W. K. Olson, *Macromolecules* **22,** 3292 (1989).

[88] W. K. Olson, A. R. Srinivasan, and M.-H. Hao, *in* "Proceedings Supercomputing '89: Supercomputer Applications" (L. P. Kartashev and S. I. Kartashev, eds.), p. 361. International Supercomputing Institute, Inc., St. Petersburg, Florida, 1989.

[89] P. Zhang, W. K. Olson, and I. Tobias, *Comput. Polymer Sci.* **1,** 3 (1991).

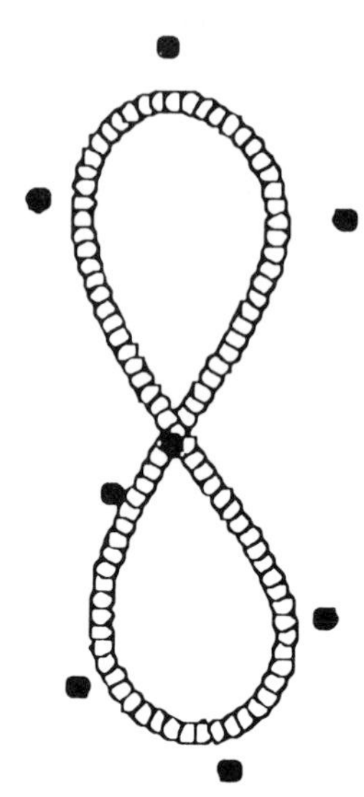

FIG. 1. Planar projection of a closed figure-eight trajectory of 100 residues generated with the 8 controlling points denoted by solid circles and the order four cyclic B-spline formulation of Eq. (6).

lations used to describe complex curves involve a relatively limited number of variables. In addition, chains of any desired length can be studied. The closed figure-eight trajectory in Fig. 1, for example, is described by eight B-spline controlling points (e.g., $3 \times 8 = 24$ spatial variables) noted by solid spheres. The number of points on the overall curve, however, is independent of the number of controlling points. There are 100 evenly spaced points in this example, corresponding to a DNA of 100×3.4 Å = 340 Å total contour length. Chains of different lengths can be generated by simply rescaling the controlling points.

The B-spline formulation of a closed chain trajectory is local in the sense that the change of a single variable alters only part of the curve. The Fourier representation of a cyclic DNA, on the other hand, is global in that an individual variable affects the entire chain trajectory. The principal advantage of the B-spline representation is the direct control of the chain trajectory provided by the choice of controlling points. As evident from Fig. 1, the controlling points guide but do not fall along the B-spline curve. The B-spline method, however, is limited by the number of controlling points required to describe elaborately folded trajectories. The more complicated the trajectory, the more controlling points are required. The latter problem can be partially overcome with the Fourier series representation.

Many known curves are also expressed exactly by a finite Fourier series. The exact trajectory of a known curve, in contrast, cannot be described by a finite number of B-spline controlling points. Other advantages and disadvantages of the two representations of structure will become apparent in the discussion that follows.

B-Spline Representations

The B-splines are a series of polynomial expressions that smoothly connect a sequence of arbitrary controlling points **p** in space.[90] The curves are regionally defined functions that are evaluated separately over different intervals of the chain trajectory. The polynomials can be chosen with as many continuous derivatives as needed for the mathematical analyses to be performed. The order four (e.g., cubic) cyclic B-spline with continuous first and second derivatives is given by

$$\mathbf{r}_i(u) = F_1(u)\mathbf{p}_{i-1} + F_2(u)\mathbf{p}_i + F_3(u)\mathbf{p}_{i+1} + F_4(u)\mathbf{p}_{i+2} \tag{6}$$

where $\mathbf{r}_i$ is the ith segment of the closed curve located in the vicinity of points i and $i + 1$. The $F_j(u)$, where $j = 1$ to 4, are basis functions that weight the relative contributions of $\mathbf{p}_{i-1}$, $\mathbf{p}_i$, $\mathbf{p}_{i+1}$, and $\mathbf{p}_{i+2}$ at different points along $\mathbf{r}_i$. The number of points along the curve is determined by the increments of the parameter u, which varies between zero and unity. The $F_j(u)$ values are defined such that successive pieces of the curve and their first- and second-order derivatives are smoothly connected.

$$\begin{aligned} F_1(u) &= (-u^3 + 3u^2 - 3u + 1)/6 \\ F_2(u) &= (3u^3 - 6u^2 + 4)/6 \\ F_3(u) &= (-3u^3 + 3u^2 - 3u + 1)/6 \\ F_4(u) &= u^3/6 \end{aligned} \tag{7}$$

Specifically, at $u = 0$, $\mathbf{r}_{i-1}(1) = \mathbf{r}_i(0)$, $d\mathbf{r}_{i-1}(1)/du = d\mathbf{r}_i(0)/du$, and $d^2\mathbf{r}_{i-1}(1)/du^2 = d^2\mathbf{r}_i(0)/du^2$ and at $u = 1$ $\mathbf{r}_i(1) = \mathbf{r}_{i+1}(0)$, $d\mathbf{r}_i(1)/du = d\mathbf{r}_{i+1}(0)/du$, and $d^2\mathbf{r}_i(1)/du^2 = d^2\mathbf{r}_{i+1}(0)/du^2$. The complete chain pathway is described by M controlling points and therefore M smoothly connected piecewise curves. As can be shown from Eq. (6), variation of the components of $\mathbf{p}_i$ will alter four of the piecewise fragments, $\mathbf{r}_{i-2}$, $\mathbf{r}_{i-1}$, $\mathbf{r}_i$, and $\mathbf{r}_{i+1}$, and their derivatives.

The order four cyclic B-spline described in Eqs. (6) and (7) is sufficient for the calculation of topological and energetic parameters of an ideal

[90] W. J. Gordon and R. E. Riesenfeld, *in* "Computer Aided Geometric Design" (R. E. Barnhill and R. E. Riesenfeld, eds.), p. 95. Academic Press, New York, 1974.

elastic closed circular chain molecule. Higher order B-spline functions with additional derivatives of the curve are required in the treatment of more realistic and elaborate models of the potential energy of DNA.

Fourier Series Representations

A finite Fourier series representation of a closed curve is given by the expression

$$\mathbf{r}(u) = \sum_{n=1}^{N} \mathbf{a}_n \sin(2\pi nu) + \sum_{n=1}^{N} \mathbf{b}_n \cos(2\pi nu) \tag{8}$$

where the Fourier coefficients $\mathbf{a}_n$ and $\mathbf{b}_n$ are vectors in three-dimensional space and the parameter u varies between zero and unity. The Fourier coefficients are analogous to the controlling points of the B-spline representation, while the scalar sine and cosine terms in Eq. (8) are comparable to the B-spline weighting functions. Because of the nature of the Fourier basis functions [e.g., $\sin(2\pi nu)$ and $\cos(2\pi nu)$], the variation of the components of $\mathbf{a}_n$ or $\mathbf{b}_n$ alters the complete chain trajectory. In contrast, the configurational variations caused by changes in the B-spline controlling points are local. On the other hand, the sine and cosine functions in Eq. (8) can be differentiated infinitely, so that unlike the cubic B-spline, the formalism does not need to be changed for more elaborate models of the chain energy. The Fourier representation also ensures that the center of gravity of the chain trajectory, $\langle \mathbf{r} \rangle = \int_0^1 \mathbf{r}(u)du$, is fixed at $\mathbf{0}$.

Many standard curves can be expressed as a finite Fourier series. The toroidal pathway of a closed circular superhelix of radius r_1 and k turns deformed around a larger circle of radius r_2,

$$\begin{aligned}\mathbf{r} = {} & [r_2 + r_1 \sin(2\pi ku)]\cos(2\pi u)\mathbf{e}_1 + [r_2 + r_1 \sin(2\pi ku)]\sin(2\pi u)\mathbf{e}_2 \\ & + r_1 \cos(2\pi ku)\mathbf{e}_3\end{aligned} \tag{9}$$

where $\mathbf{e}_1, \mathbf{e}_2, \mathbf{e}_3$ are orthogonal unit vectors in three-dimensional Cartesian space, can be reexpressed, for example, by a Fourier series of $k + 1$ terms with the following coefficients:

$$\begin{aligned} \mathbf{a}_1 &= (0, r_2, 0) & \mathbf{b}_1 &= (r_2, 0, 0), \\ \mathbf{a}_{k-1} &= \left(\frac{r_1}{2}, 0, 0\right) & \mathbf{b}_{k-1} &= \left(0, \frac{r_1}{2}, 0\right) \\ \mathbf{a}_k &= (0, 0, 0) & \mathbf{b}_k &= (0, 0, r_1), \\ \mathbf{a}_{k+1} &= \left(\frac{r_1}{2}, 0, 0\right) & \mathbf{b}_{k+1} &= \left(0, -\frac{r_1}{2}, 0\right) \end{aligned}$$

$$\mathbf{a}_n = \mathbf{b}_n = \mathbf{0},\ n \neq 1, k-1, k, k+1 \tag{10}$$

The Fourier series of a circle is a special case of the above summation with both r_1 and k equal to zero.

Computational Methodology

Stochastic Energy Minimization

The global minimum of the elastic energy potential in the above-described closed curves has been identified with two different algorithms. One combines Metropolis–Monte Carlo sampling with a simulated annealing procedure,[87] while the other combines a fixed temperature Metropolis–Monte Carlo calculation with a modified simplex acceleration.[89] B-spline controlling points or Fourier coefficients are moved at random with new configurations of lower energy automatically accepted and those of higher energy accepted on the basis of the Boltzmann factor of the increase in energy.[91] A move of higher energy is accepted only if a randomly generated number between zero and unity is less than or equal to $\exp(-\Delta E/k_B T)$, where ΔE is the increase in energy of the proposed configurational move relative to the current state, k_B the Boltzmann constant, and T the absolute temperature. If the increase in energy is small, the probability of acceptance is high and vice versa. The actual configurational moves are small increases or decreases of one component of a randomly selected B-spline controlling point or Fourier coefficient. The step size is chosen so that the rate of configurational acceptances is in the range 0.40–0.60.

In the case of simulated annealing,[92] as the system approaches an equilibrium distribution at a given temperature, the temperature is lowered by a small amount. The rate of configurational acceptances is consequently decreased at the new temperature. The system is again moved at random according to the same procedure until a new equilibrium distribution is reached. The cooling process is repeated several times until the temperature is below the "freezing point," where the configuration is no longer changed. The procedure is basically a reversible process with a similar energy trajectory obtained when the chain is "heated" (see below).

The modified simplex acceleration involves a record-keeping Monte Carlo simulation at a fixed (high) temperature combined with a downhill acceleration modification. Configurational moves that are found to lower the chain energy are retained for up to 30–50 steps. As illustrated below,

[91] N. N. Metropolis, A. Rosenbluth, M. Rosenbluth, A. Teller, and E. Teller, *J. Chem. Phys.* **21,** 1087 (1953).

[92] S. Kirkpatrick, C. D. Gelatt, and M. P. Vecchi, *Science* **220,** 671 (1983).

the attainment of the configurational minimum is enhanced in this case by a factor of approximately 20 over the time required by simulated annealing. The global minimum is identified from the recorded lowest energy configuration in the sample. This differs from the standard annealing procedure, where the minimum is represented by the last configuration obtained from a series of simulations at decreasing temperatures.

As illustrated below, the two minimization and modeling approaches produce similar results. Minimization of the composite elastic/long-range potential under the constraints of ring closure and chain length is found to produce structures that are consistent with both the configurations of supercoiled DNA observed in electron micrographs and the macroscopic properties of elastic rods. The chain configurations at small linking number difference are very much like those found in the classic elastica problem, in which an elastic thin rod is allowed to bend in a plane subject to isolated forces or torques acting at its ends.[93] Unlike the planar elastica trajectory, the energy-minimized configurations do not self-intersect. The most stable structures of the closed chain are interwound configurations that are critically dependent on the specified linking number difference. Toroidal configurations are found to be unstable and, on energy minimization, are transformed to interwound forms.

Curvature Integral and Writhing Number

The curvature integral and the writhing number used to evaluate the elastic energy [Eq. (3)] of an arbitrary closed curve $\mathbf{r}(u)$ are obtained with the standard formulas:

$$\oint \kappa(s)^2 ds = \int_0^1 \left[\frac{|\mathbf{r}'(u) \times \mathbf{r}''(u)|^2}{|\mathbf{r}'(u)|^5} \right] du \tag{11}$$

and

$$\mathrm{Wr}(\mathbf{r}) = \frac{1}{4\pi} \oint \oint \left[\mathbf{r}'(u_1) \times \mathbf{r}'(u_2) \cdot \frac{\mathbf{r}(u_1) - \mathbf{r}(u_2)}{|\mathbf{r}(u_1) - \mathbf{r}(u_2)|^3} \right] du_1 du_2 \tag{12}$$

where $\mathbf{r}(u)$ is the equation of the closed curve, $\mathbf{r}'(u)$ the first derivative $d\mathbf{r}/du$, and $\mathbf{r}''(u)$ the second derivative $d^2\mathbf{r}/du^2$.

Numerical integration of the writhing number with Eq. (12), the so-called Gauss double integral, is very time consuming. The calculation is simplified by taking advantage of a theorem first proposed by Fuller to evaluate the difference in writhing numbers between two curves.[94] If the

[93] A. E. H. Love, "Treatise on the Mathematical Theory of Elasticity," 4th Ed., Chap. 18 and 19. Dover, New York, 1944.

[94] F. B. Fuller, *Proc. Natl. Acad. Sci. U.S.A.* **75**, 3557 (1978).

writhing number of an initial nonself-intersecting space curve $\mathbf{r}_1(s)$ is known, the writhing number of a second such curve $\mathbf{r}_2(s)$ can be obtained from the single integral

$$\mathrm{Wr}(\mathbf{r}_2) - \mathrm{Wr}(\mathbf{r}_1) = \frac{1}{2\pi} \int_0^1 \frac{\mathbf{r}_1'(u) \times \mathbf{r}_2'(u)}{|\mathbf{r}_1'(u)||\mathbf{r}_2'(u)| + \mathbf{r}_1'(u) \cdot \mathbf{r}'_2(u)} \cdot$$

$$\left[\frac{\mathbf{r}_1''(u)}{|\mathbf{r}_1'(u)|} - \frac{\mathbf{r}_1'(u)[\mathbf{r}_1''(u) \cdot \mathbf{r}_1'(u)]}{|\mathbf{r}_1'(u)|^3} + \frac{\mathbf{r}_2''(u)}{|\mathbf{r}_2'(u)|} - \frac{\mathbf{r}_2'(u)[\mathbf{r}_2''(u) \cdot \mathbf{r}_2'(u)]}{|\mathbf{r}_2'(u)|^3}\right] du \quad (13)$$

given that $\mathbf{r}_1(u)$ can be deformed into $\mathbf{r}_2(u)$ through the nonself-intersecting curves $\mathbf{r}_\lambda(u)$, $1 \leq \lambda \leq 2$, in such a way that $\mathbf{r}_1'(s)$ and $\mathbf{r}_\lambda'(s)$ are never oppositely directed. The latter prerequisite is usually satisfied in systems with an excluded volume potential like that in Eq. (4) or (5) preventing the self-intersection of the chain.

Contour Length

The variation of B-spline controlling points and Fourier coefficients with the above Monte Carlo procedure does not necessarily preserve the contour length of a closed circular DNA. If the total contour length of the DNA is kept strictly constant, few moves can be made, even at high temperatures. This constraint has therefore been relaxed in the following two ways. The first approach involves the introduction of a pseudopotential of the form

$$V_\mathrm{L} = K_\mathrm{L}(L - L_0)^2 \quad (14)$$

where K_L is a force constant and L the actual contour length, that allows the chain to fluctuate to a small degree about its given or starting value L_0 during the simulation. The second approach involves scaling each new chain configuration $\mathbf{r}$ of length L by the factor L_0/L in order to keep the length exactly fixed at L_0.

Results and Discussion

Simulated Annealing and Melting

The simulated annealing procedure is basically a reversible process. If the cooling has been carried out so that an equilibrium is attained at each temperature, a comparable trajectory is obtained when the process is reversed (i.e., the system is heated up at a similar rate). The results of such an annealing and melting cycle of a B-spline chain representation are illustrated in Fig. 2, where the energy and the writhing number are plotted as

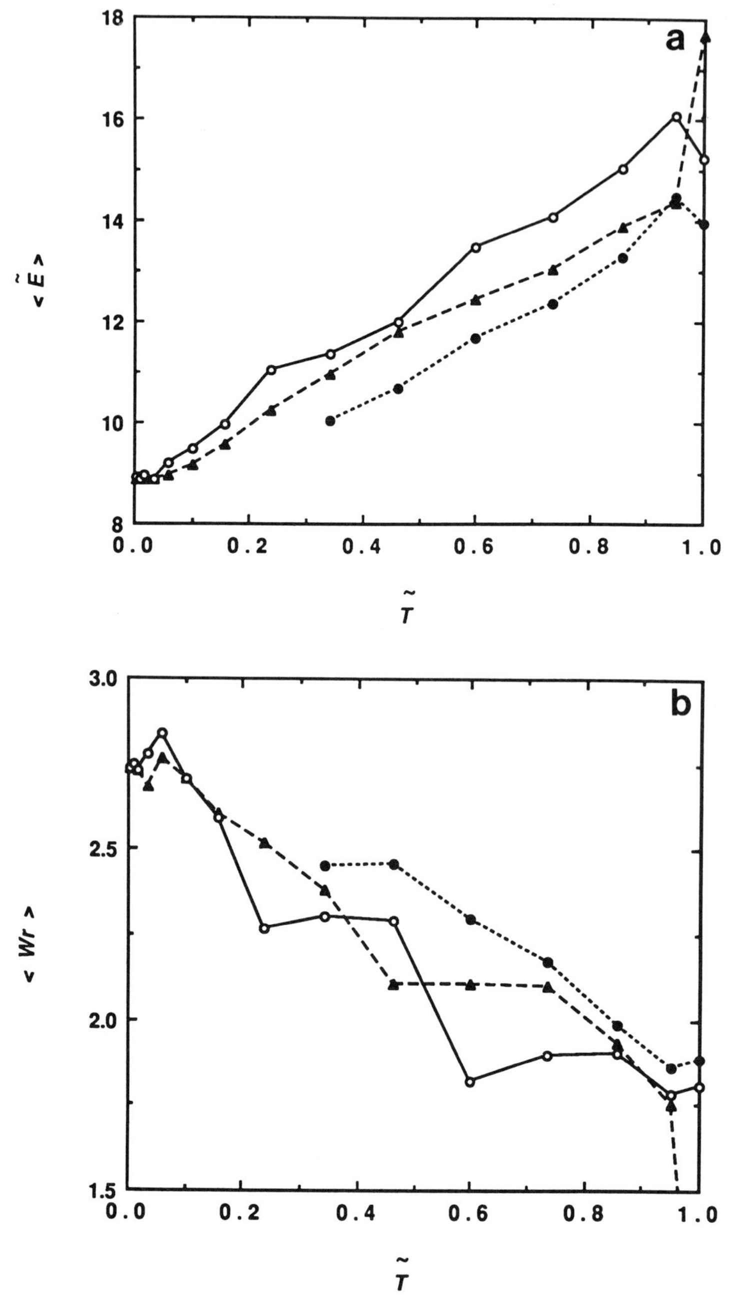

FIG. 2. The simulated annealing and melting trajectories of a closed circular DNA with imposed linking difference $\Delta Lk = 4$ as illustrated by the variation of (a) average reduced energy $\langle \tilde{E} \rangle$ and (b) average writhing number $\langle Wr \rangle$ with respect to reduced temperature $\tilde{T}$.

functions of the temperature for a set of heating and cooling trajectories. The starting configuration is a near-circular configuration of radius r described by 12 evenly spaced controlling points $\mathbf{p}_i$ located at [$r \cos(2\pi i/12)$, $r \sin(2\pi i/12)$, 0), where $i = 1$ to 12 and r is chosen so that the total contour length equals 3400 Å (e.g., 1000 bp). The energy is reported in reduced form as $\tilde{E}$, a unitless quantity given by the quotient of the total elastic and excluded volume energy $E + V_{ij}$ [Eqs. (3) and (4), respectively] and the factor $2\pi^2 C/L$, and the temperature as $\tilde{T} = T/T_0$, where T_0 is the starting temperature. The value of T_0 is obtained from the starting energy E_0 through the expression $k_B T_0 = \sigma E_0$, where k_B is the Boltzmann constant and σ (here equal to 0.045) is a variable that governs the configurational acceptance rate in the Monte Carlo simulation. The linking number difference ΔLk is fixed at 4, the ratio of bending to twisting force constants A/C at 1.35, and the hard sphere contact distance D at $L_0/90$ (e.g., 38 Å). The contour length is preserved using the harmonic expression in Eq. (14) with the stretching force constant K_L equal to $20/L_0$. The cooling schedule adopted is an accelerative cooling procedure in which the temperature is lowered by a factor of 0.95^p at the pth temperature-lowering step and the configuration relaxed by up to 30,000 Monte Carlo moves at each temperature stage of the calculation. (For mere minimization purposes, however, fewer Monte Carlo runs are sufficient.) The heating process (open circles connected by solid lines) is started from the final frozen configuration of the cooling sequence with the heating factor given by the reciprocal of the cooling factor at the same temperature. A second partial annealing is then carried out starting from the final state of the heating process. The energies and writhing numbers along all three pathways are average values accumulated at the specified temperatures. The global minimum is identified from the energy value at the lowest temperature.

As observed from Fig. 2, the average reduced energy $\langle \tilde{E} \rangle$ along the melting simulation is very similar to the corresponding values along the two cooling (e.g., annealing) trajectories. The variation of the average writhing number ⟨Wr⟩ with $\tilde{T}$ in the figure is also comparable along the heating and cooling curves. During the annealing process, an irregular coil is frozen into a regular interwound helix with roughly three chain crossings, and the process is reversed during the heating process. The observed hysteresis in $\langle \tilde{E} \rangle$ and ⟨Wr⟩ in Fig. 2 is attributed to the random features of the algorithm and the finite number of configurations sampled. Indeed, the

---

The calculations are started from a circle approximation defined by 12 B-spline controlling points located at [$r \cos(2\pi i/12)$, $r \sin(2\pi i/12)$, 0], where the radius r is assigned so the total contour length L_0 equals 3400 Å (e.g., 1000 bp). The ratio of the elastic force constants A/C is set at 1.35 and the hard sphere contact distance D at $L_0/90$. (▲), Anneal 1; (○), heat; (●), anneal 2.

average chain configuration as measured by the root-mean-square radius of gyration is quite different at several of the higher temperatures along the annealing and heating pathways (data not shown). The writing number is not a unique descriptor of chain configuration. Very different three-dimensional forms are described by the same value of Wr, thereby accounting for the observed discrepancy in the radii of gyration. Longer sampling times are required to achieve the same average radii of gyration at corresponding high temperatures along the annealing and cooling pathway. The average chain dimensions, however, are similar in the vicinity of the low-temperature global energy minimum in the reported samples.

Representative equilibrium configurations obtained at the end of the above Monte Carlo runs at selected temperatures are illustrated in Fig. 3. Note that these are individual configurations and are subject to fluctuations during the simulation. The energy and writhing number fluctuations ($\langle \delta \tilde{E}^2 \rangle^{1/2} = 1.0-4.0$ and $\langle \delta \mathrm{Wr}^2 \rangle^{1/2} = 0.10-0.30$) are largest at the start of the annealing trajectories and at the conclusion of the melting trajectory, consistent with increased flexibility of the superhelix at high temperature. At the lowest temperatures the chain is frozen in the apparent global

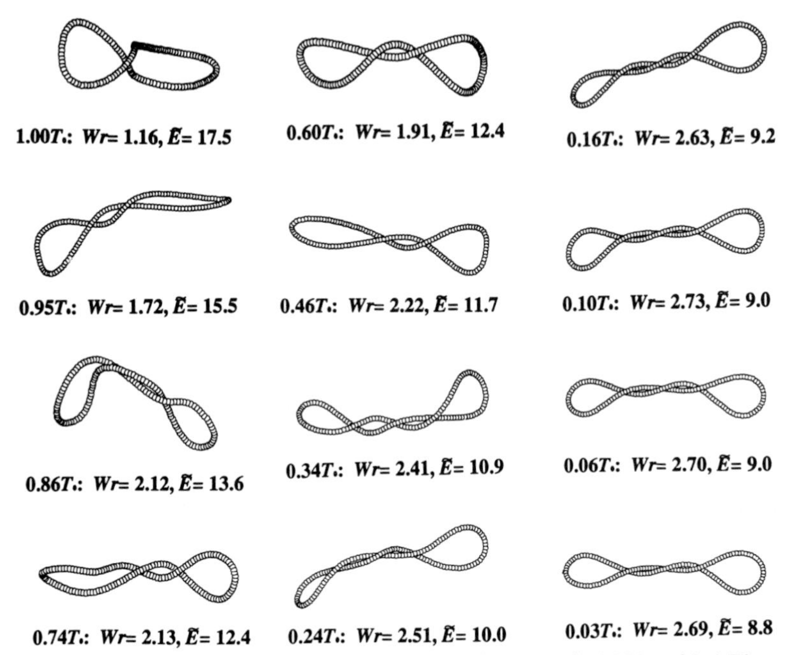

FIG. 3. Representative equilibrium configurations of a supercoiled DNA with Δ Lk = 4 at selected temperatures during the simulated annealing trajectory described in Fig. 2.

minimum energy configuration ($\tilde{E} = 8.8$ and $\mathrm{Wr} = 2.69$). At the early stages of the simulation the open circle collapses into an asymmetric figure-eight form and by $T/T_0 = 0.95$ the chain is loosely intertwined. The remainder of the simulation involves further winding and unwinding of the central superhelical core of the structure as well as bending and straightening of the interwound polymer. The energy minimum is achieved during the annealing cycle after 11 temperature lowerings at $T/T_0 = 0.03$. The final interwound configuration is nearly straight with the end loops propeller twisted by $\sim 30°$ with respect to one another and the central core containing three chain crossings.

Attainment of Global Minimum

The simulated annealing of the folding of a Fourier series representation of a DNA circle of 1000 bp into an interwound configuration with $\Delta\mathrm{Lk} = 4$ is compared with the corresponding B-spline treatment in Fig. 4. The total reduced energy $\tilde{E}$ and its bending $\tilde{E}_{\mathrm{bend}}$ and twisting $\tilde{E}_{\mathrm{twist}}$ components are plotted as functions of the number of simulation steps for the two schemes. The Fourier chain pathway is defined by the first four coefficients in each term of Eq. (8), corresponding to $2 \times 4 \times 3 = 24$ independent variables, with the hard sphere contact radius D set at $L_0/170$ (e.g., 20 Å) and the radius of the starting circle chosen so that the contour length L_0 equals 3400 Å. The intricate folding of the interwound configuration involves fewer variables in the Fourier series representation than in the B-spline formulation, where there are 12 controlling points or $12 \times 3 = 36$ independent parameters. The two representations, however, converge to comparable energies and writhing numbers. The Fourier series calculation takes a total of 200,000 steps and 10 temperature lowerings to level off to a final minimum energy configuration with $\tilde{E} = 9.8$ and $\mathrm{Wr} = 2.71$. The B-spline calculation with the same number of steps and temperature lowerings attains a somewhat more favorable configuration with $\tilde{E} = 7.9$ and $\mathrm{Wr} = 3.32$. The rough similarity of the two final states, together with the dramatically different starting configurations and the very different representations of the chain trajectory, suggest that the global energy minimum has been identified. A similar minimum ($\tilde{E} = 8.8$ and $\mathrm{Wr} = 2.70$) is also attained when starting from a toroidal starting state of the same contour length and linking number difference and using the Fourier chain representation.[89] As evident from Fig. 4, the decrease in total energy in both the Fourier and the B-spline simulations primarily reflects decreases in the twisting component of the elastic energy ($\Delta\tilde{E}_{\mathrm{twist}} \approx -15$) brought about by the increase of the writhing number. The increase in bending energy caused by the collapse and folding of the final interwound structure is of

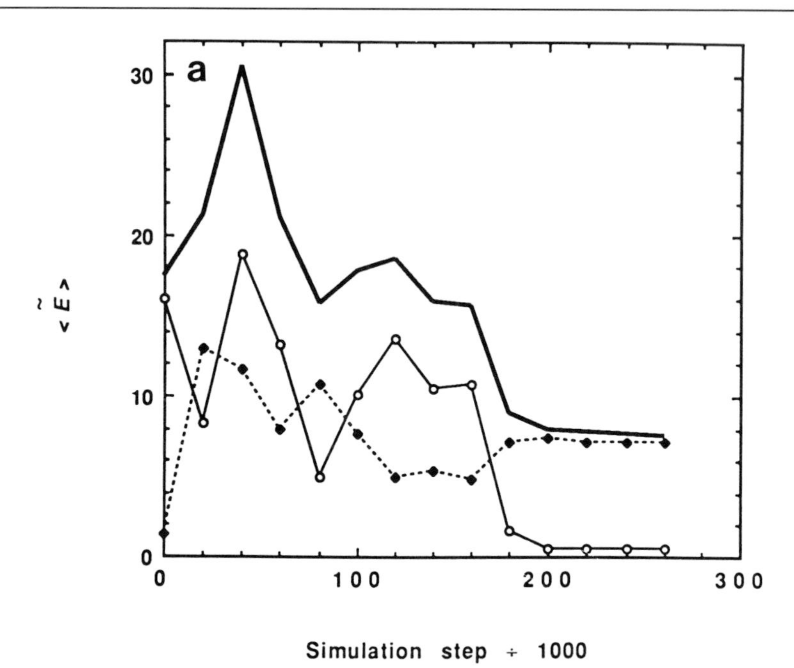

FIG. 4. The variation of the reduced energy $\tilde{E}$ and its bending $\tilde{E}_{\text{bend}}$ and twisting $\tilde{E}_{\text{twist}}$ components during the simulation of an elastic DNA with $\Delta\text{Lk} = 4$ during the simulated annealing of (a) B-spline, (b) Fourier series chain representations, and (c) the fixed high-temperature simulation of a Fourier series chain trajectory. The B-spline simulation is carried out with the parameter set at $L_0 = 3400$ Å, $A/C = 1.35$, and $D = L_0/50$. The Fourier pathways, defined by the first four coefficients in Eq. (8), are initiated from a circular starting state of contour length 3400 Å with the same energy parameters.(—), Total; (●), bending; (○), twisting.

much lesser magnitude ($\Delta E_{\text{bend}} \approx 5-6$). While there are no disallowed contacts in the final minimized structures, the excluded volume contribution plays an important role in determining the relative proportions of the extended internal core and the large external loops of the interwound structure (see below).

The record of energy lowering found in the accelerated Monte Carlo simulation at fixed (high) temperature of the circle to interwound transition of a Fourier series chain representation with $\Delta\text{Lk} = 4$ is also presented in Fig. 4. The calculation is initiated from the same circular starting state used in the simulated annealing Fourier series calculation with up to 50 energy-lowering configurational moves of the same coordinate permitted. The Fourier chain pathway is again defined by the first four coefficients in each term of Eq. (8). As evident from Fig. 4, the global energy minimum is

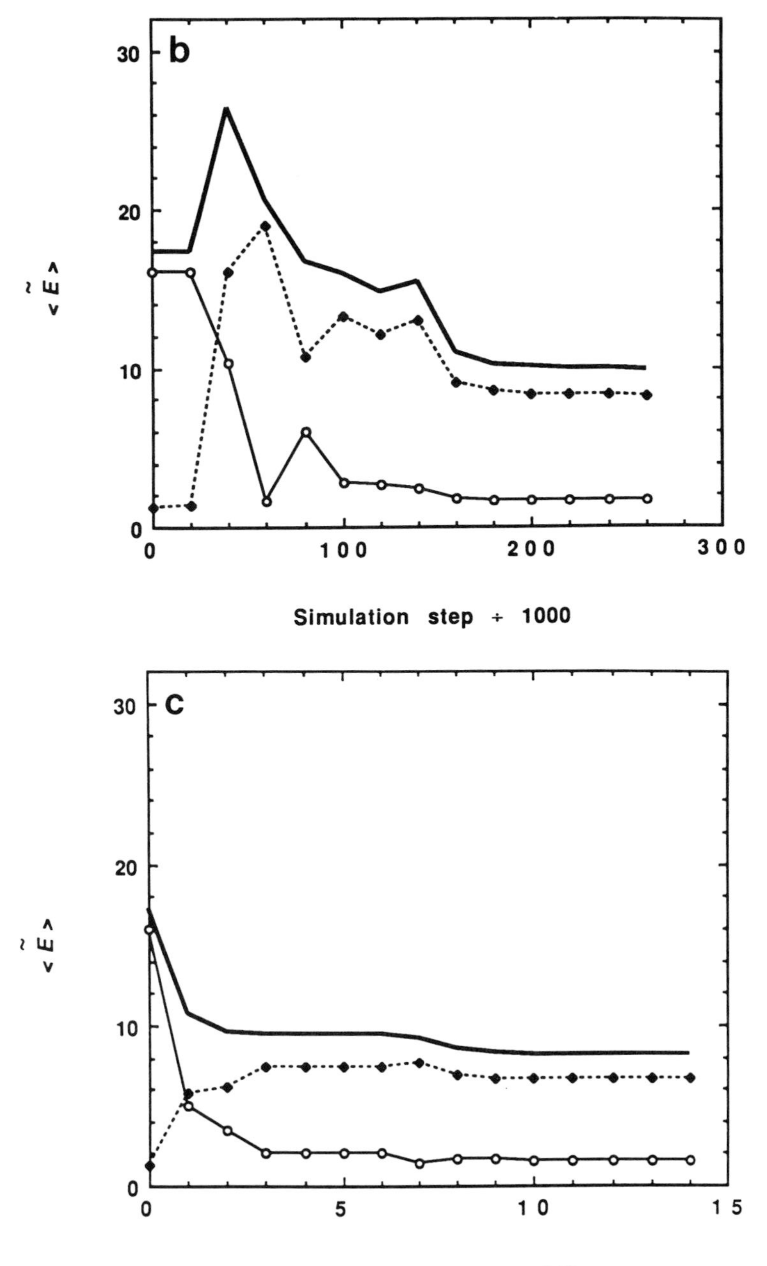

b
30
20
10
0
0
100
200
300
<$\tilde{E}$>
Simulation step ÷ 1000
c
30
20
10
0
0
5
10
15
<$\tilde{E}$>
Simulation steps ÷ 1000

FIG. 4b and c.

achieved much more rapidly than with the simulated annealing procedure (~10,000 compared to ~200,000 steps). The minimum energy arising from the record-keeping procedure ($\Delta\tilde{E} = 8.2$) is comparable to that resulting from the B-spline-simulated annealing calculations ($\Delta\tilde{E} = 7.9$). The writhing number at the global minima of the accelerated high-temperature simulations (Wr = 2.74), however, is closer to that obtained in the Fourier-simulated annealing (Wr = 2.70).

Simulated versus Exact Solutions of the Elastica Problem

The closed figure-eight configuration is found in the classic elastica problem, in which a thin elastic rod is allowed to bend in a plane subject to isolated forces or torques acting at its ends.[93] The problem can be solved analytically with the closed figure-eight configuration appearing at a specific bending angle. This problem, however, is not exactly the same as the DNA supercoiling problem. Ring closure is an intrinsic constraint in the DNA problem and supercoiling is due to a linking number difference in the molecule. Nevertheless, it has been shown by Le Bret[29] that the same figure-eight form is a local minimum energy configuration that becomes more stable than the planar circle when the linking number difference is greater than a critical value determined by the ratio A/C of the bending to twisting force constants.

As a simple test of the calculations, minimum-energy figure-eight configurations obtained through computer simulations are compared in Fig. 5 with the exact solution to the elastica problem. One trajectory is obtained from the simulated annealing of a B-spline chain representation described by 14 controlling points with a van der Waals excluded volume term,[87] while the other is obtained from the high-temperature simulation of a Fourier series representation with three Fourier coefficients and a hard sphere energy term.[89] The starting B-spline configuration is an arbitrarily chosen figure-eight shape generated with the controlling points listed in the figure legend. The chain is annealed by an accelerative procedure with a temperature lowering factor of 0.95^p. The system is relaxed with 4000 moves at each cooling temperature, ΔLk fixed at 1.8, the ratio A/C at 1.5, d at $L_0/225$ (e.g., 15 Å), $\tilde{\epsilon} = 3$, and K_L at 0.3. The initial Fourier state is a closed circle of 1000 residues (e.g., $L_0 = 3400$ Å) with the same choices of ΔLk and A/C but $D = L_0/170$ (e.g., 20 Å). The chain is kept at fixed length in this case with a multiplicative scale factor.

Because of the self-contact energy contribution, both energy-minimized structures are three-dimensional, in contrast to the planar elastica. The two trajectories are essentially identical to the elastica in the main plane projection but are different from the exact solution when viewed in a perpendic-

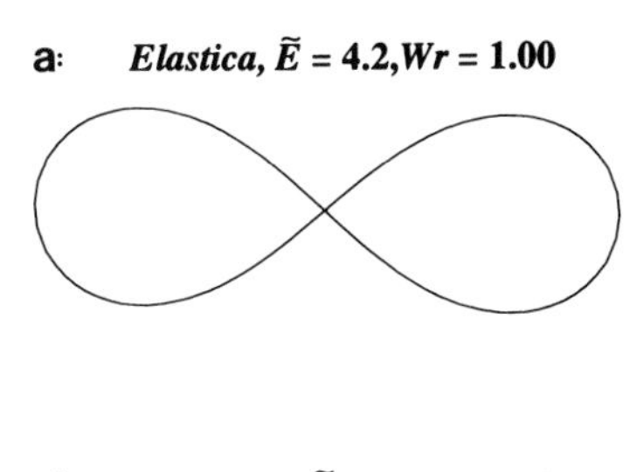

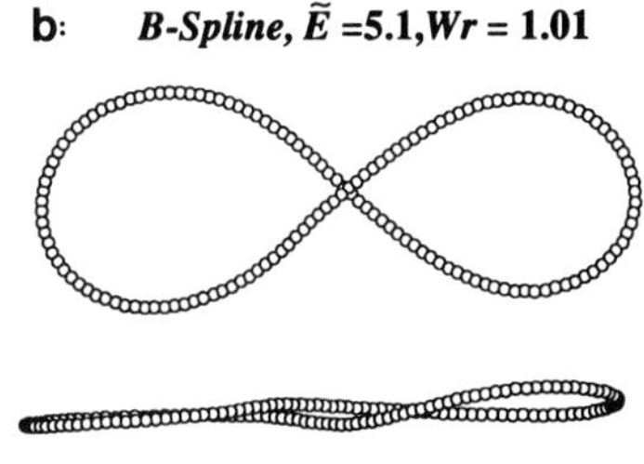

FIG. 5. A comparison of the exact elastic equilibrium configuration[29,93] of a thin rod in the figure-eight form (a) with the minimum energy configurations obtained with a 14-point B-spline approximation with van der Waals' excluded volume contributions (b) and a three coefficient Fourier series model with hard sphere energy terms (c). The controlling points of the starting B-spline model are (0.6, 0.6, −1.0), (−1.4, −0.6, −1.0), (−3.0, −3.0, −0.60), (−2.8, −4.8, −0.2), (0.0, −6.0, 0.0), (2.0, −5.2, 0.2), (3.0, −3.0, 0.6), (1.4, −1.4, 1.0), (−1.4, 0.6, 1.0), (−3.6, 3.0, 0.6), (−2.8, 4.8, 0.2), (1.0, 6.0, 0.0), (2.0, 4.8, −0.2), (3.0, 3.0, −0.6). The ratio of the elastic force constants A/C is set at 1.5 in both simulations. The nonbonded contact limit d is set at $L_0/225$ and the van der Waals energy well $\tilde{\epsilon}$ at 3 in (b), and the excluded volume cutoff distance D at $L_0/170$ in (c).

ular projection. As expected from their nonplanarity, the reduced energies of the simulated curves (5.1 and 4.8 for the B-spline and Fourier, respectively) are slightly higher than that of the elastica (4.2) found at $A/C = 1.5$. The respective writhing numbers of the optimized closed curves are 1.01 and 1.08, compared to a writhing number of unity for the elastica. The configurations, however, are quite similar to the elastica solution despite the approximate nature of the calculated structures.

Dependence of Supercoiled Configuration on Linking Number Difference

The simulated annealing method has also been used to estimate the critical value of ΔLk at which the equilibrium figure-eight configuration becomes more stable than the planar circle. The latter is a local minimum-energy configuration over a certain range of ΔLk.[29] The results of the minimization of a series of closed DNA curves of variable linking number difference are reported as a graph of $\tilde{E}$ versus ΔLk in Fig. 6. The total reduced energy and its bending and twisting components are plotted as functions of ΔLk over the range 0–7. Unless noted otherwise, the starting trajectory is the 14-point B-spline curve detailed in Fig. 5 and the energy parameters are the values previously specified. The simulations up to ΔLk = 2.0 are performed with a 12-step accelerative cooling procedure

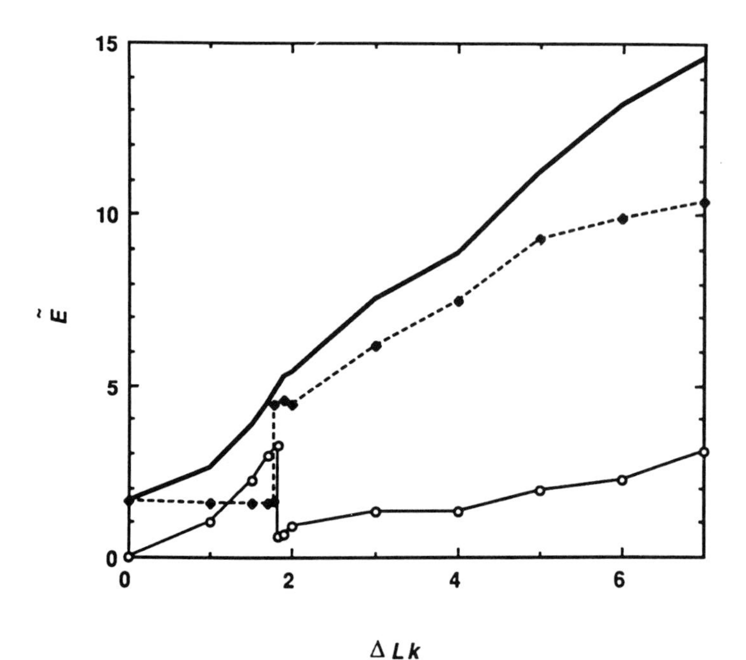

FIG. 6. The reduced energy $\tilde{E}$ and its bending $\tilde{E}_{bend}$ and twisting components $\tilde{E}_{twist}$ at the global minimum energy configuration plotted as functions of the imposed linking number difference ΔLk of an elastic DNA. Energy parameters are given in the caption to Fig. 5. (—), Total; (●), bending; (○), twisting.

with a temperature lowering factor of 0.95^p, where p refers to the temperature step, and up to 12,000 Monte Carlo moves per step. According to Fig. 6, the energy-minimized configurations in the range $\Delta Lk = 0-2$ are circular with constant bending energy up to a point where the chain folds abruptly into a figure-eight form. The structures with $\Delta Lk \leq 1.7$ are planar circles with writhing numbers of approximately zero, while those with $\Delta Lk \geq 1.9$ have writhing numbers greater than unity. This increase in writhing number accounts for the sudden drop in the twisting energy at $\Delta Lk \sim 1.8$. Repeated simulations with $\Delta Lk = 1.8$ are found to yield both circular and figure-eight structures. The reduced energy of the annealed circle (4.9) is somewhat lower than that (5.1) of the figure-eight with $\Delta Lk = 1.8$, suggesting that the critical linking number difference at which the circle becomes less stable is between 1.8 and 1.9. This value is in good agreement with the analytical results of Le Bret,[29] who finds with the same ratio of A/C (1.5), that the circular configuration is less stable than the figure-eight configuration when the critical linking number difference is approximately 1.9. The correspondence between the numerical simulation and the analytical solution is further improved if the self-contact energy contribution is reduced in value (data not shown).

As evident from Fig. 6, the numerical procedure has also been extended to chains of ΔLk greater than those that can be studied by exact analytical methods. The simulated annealing procedure was repeated on the 14-point starting structure detailed above for values of the linking number difference up to $\Delta Lk = 5$. These trajectories were relaxed with an accelerative cooling scheme involving 2000–4000 moves per temperature step and a temperature-lowering factor at step p of 0.95^p. The most complicated interwound trajectories with $\Delta Lk = 6$ and 7 were obtained with Fourier series chain representations because of the limitations of the B-spline approach. These structures were identified using fixed high temperature record keeping Monte Carlo simulations. Elastic and excluded volume energy parameters were held at the value noted above.

The final minimized configurations at selected linking number differences are illustrated in Fig. 7. As evident from the figure, the simulated chains are found to relax to nearly perfect circles when $\Delta Lk = 0-1$, but to form various interwound helices for $\Delta Lk \geq 2$. At $\Delta Lk = 2$, the minimized configuration is a figure-eight, at $\Delta Lk = 3$ an interwound form with two chain crossings, and at $\Delta Lk = 4$, 5, 6, and 7 interwound helices with three, four, five, and six chain crossings, respectively, in the projections shown. It is very clear that as ΔLk increases, the energetically favored configurations are superhelices with additional helical turns. It is also noteworthy that the minimized structures at the larger values of ΔLk are slightly curved. Such

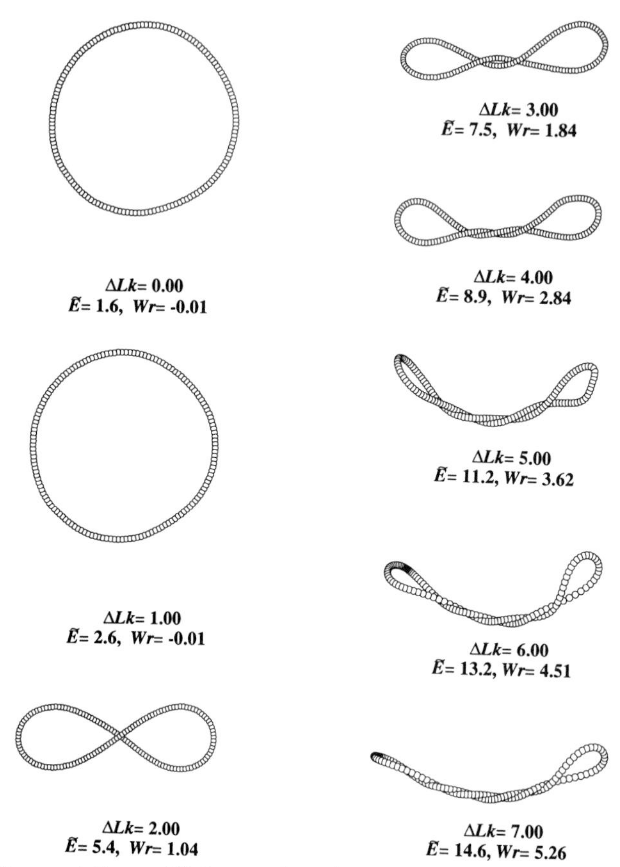

FIG. 7. Representative equilibrium configurations of supercoiled DNA as a function of the imposed linking number difference ΔLk. Structures correspond to data plotted in Fig. 6.

bending may be a preliminary step to the branching of supercoiled DNA observed experimentally at higher linking number differences.[95]

Two other points can be made from the energy minimized curves in Fig. 7. First, the final configurations at $\Delta\text{Lk} > 1.8$ are dramatically different from the circular starting structures. It is therefore unlikely that local minima have trapped the algorithm and prevented identification of the global minimum at each ΔLk. Second, the critical influence of ΔLk on the supercoiled configurations is manifested through the various optimized supercoiled structures. When ΔLk is less than or equal to unity, the most stable configurations are planar circles with near-zero writhe. This conclu-

[95] T. C. Boles, J. H. White, and N. R. Cozzarelli, *J. Mol. Biol.* **213,** 931 (1990).

sion has been reached by others.[29,57] When $\Delta Lk \geq 2$, however, the most stable configurations are exclusively interwound supercoils. This conclusion has not been directly reached in other simulation studies. The experimentally observed supercoiling of closed circular DNA with linking number deviations is correctly predicted by the present approaches,[6] as is the observed dependence of writhing number of ΔLk. Beyond the critical transition between the circle and figure-eight, the computed writhing number is linearly correlated to the imposed linking number difference with a correlation coefficient of 0.99. The computed slope of the Wr vs. ΔLk simulation data in this range (0.85), however, is somewhat greater than that (0.72) deduced from electron microscopic measurements of plectonemically interwound DNA combined with an assumed model of the local structure of an interwound supercoiled segment.[89,95]

Excluded Volume Effects

Besides the linking number difference, the optimized configuration of a supercoiled DNA is affected by a number of other factors. One of these is the self-interaction of the chain as characterized by the excluded volume contact distance D in Eq. (4). This parameter is determined to a large extent by physical interactions of the sugar-phosphate backbone of the DNA double helix.

Optimized B-spline supercoiled configurations derived from the same starting structure with different repulsive contact distances are compared in Fig. 8. In all cases ΔLk is fixed at 4 and A/C at 1.35. The starting structure is the 12-point circular B-spline structure used in Fig. 2. The ratio D/L_0 is set equal to 1/179, 1/89, 1/45, and 1/22 (e.g., $D = 19$, 38, 76, and 152 Å in the 1000-bp DNA model) with all other energy parameters held at the values cited previously. It is observed that as D increases and repulsive interactions force the two strands of the interwound structures greater distances apart, the pitch angles of the interwound helices become smaller and their radii larger. These contact limits crudely mimic the effects of reduced ionic concentration on DNA structure, approximating the enhanced interactions of the negatively charged phosphate groups at large distances. The writhing numbers of the optimized configurations also becomes smaller with increase in D, while the reduced energies become greater. The computed values of Wr with increasing values of D are, respectively, 3.32, 2.69, 2.38, and 1.73, while the corresponding values of $\tilde{E}$ are 7.9, 8.8, 10.1, and 12.0. It is noted that the configuration with the most severe repulsive contacts unwinds by more than one superhelical turn relative to the configuration with the smallest self-contact limit. The bending in the terminal loops is increased as D varies from $L/179$ to $L/45$, but is

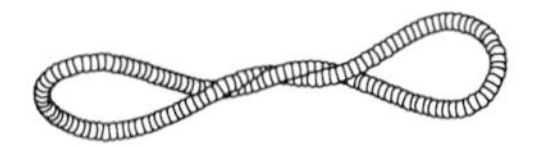

$D = L/179$, $\tilde{E} = 7.9$, $Wr = 3.32$

$D = L/89$, $\tilde{E} = 8.8$, $Wr = 2.69$

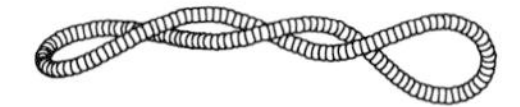

$D = L/45$, $\tilde{E} = 10.1$, $Wr = 2.38$

$D = L/22$, $\tilde{E} = 12.0$, $Wr = 1.73$

FIG. 8. Structural examples illustrating the dependence of the optimized supercoiled configuration of an elastic DNA with $\Delta Lk = 4$ on the ratio of the hard sphere radius D to the initial contour length L_0. Remaining energy parameters are held at the values listed in the caption to Fig. 2.

relieved by unwinding of the DNA at $D = L/22$. It is not yet clear whether this is a sudden jump in configuration or a gradual unwinding of the helix in the range $L/45 \leq D \leq L/22$. Both twisting and bending energy components are increased in the altered structures, illustrating the balance between excluded volume and elastic energy contributions.

Bending-to-Twisting Ratio

The ratio of flexural rigidity to twisting stiffness A/C is another factor that affects the optimized supercoiled configuration. Both the critical point of the circle to figure-eight transition and the magnitude of the elastic energy are determined by this ratio. The greater the value of A/C, the more difficult it is to deform the circle into a figure-eight and the higher the linking number difference at the critical transition point. Optimized configurations resulting from the simulated annealing of the same 14-point B-spline starting structure used in Fig. 5 under constraints of varying ΔLk and A/C are depicted in Fig. 9. Structures of constant ΔLk are shown in the separate columns, and those of constant A/C in the different rows. The reduced energies and writhing numbers are listed below each configura-

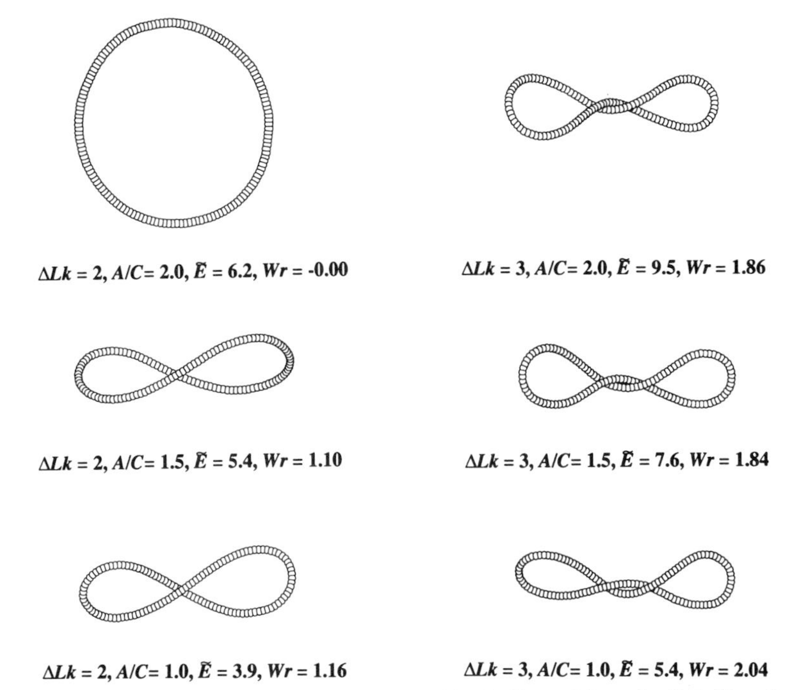

FIG. 9. Structural examples illustrating the combined effects of the ratio A/C of bending to twisting coefficients and ΔLk. The starting configuration and energy parameters are held at the values listed in the caption to Fig. 5.

tion. When $\Delta\mathrm{Lk} = 2$ (column 1), the optimized configurations fall in the critical range between circular and figure-eight forms. The choice of A/C is crucial to the computed configurations. With $A/C = 2$, the optimized structure is a deformed circle, but with $A/C = 1.5$ or 1.0, the optimized structure is a figure-eight. When $\Delta\mathrm{Lk} = 3$ (column 2), however, the supercoiling energy is relieved during the minimization process with all three choices of A/C by increasing the writhing number to values of approximately two. The resulting structures are characterized by two chain crossings in the projections shown. In this case the optimized configurations are not greatly affected by the different choices of A/C. At larger ΔLk, the effect of A/C is even smaller, and the minimized supercoiled configurations are closely related interwound forms.

Conformational Constraints within Circular DNA

It is not necessary to vary all of the B-spline controlling points during a computer simulation. Indeed, it is possible to take advantage of the fixed positions of some controlling points to simulate the effects of local rigidity within a circular DNA. A set of points describing a superhelix of the correct proportions can be used, for example, to model nucleosomes on small DNA rings such as those reconstituted by Prunell and associates[96] or the local bending of (A·T)-rich stretches within the intrinsically circular kinetoplast DNA of trypanosomes.[97,98]

Preliminary runs on a circular DNA of 500 bp containing a right-handed superhelix of ~1.5 turns demonstrate how a nucleosome or a curved DNA fragment might be incorporated into the hairpin loops of supercoiled DNA (Fig. 10). (The nucleosome, however, is a left-handed superhelix with linking number effects of the opposite sense from those predicted here.) Perturbed closed circular chains are represented by 7 fixed B-spline controlling points defining the internal superhelical constraint and 21 variable points describing the remainder of the chain. The starting configuration is the open structure illustrated at the top of Fig. 10 and the minimized forms correspond to structures with $\Delta\text{Lk} = \pm 4$. The chains are annealed by an accelerative procedure with a temperature lowering factor of 0.95^p and up to 30,000 moves at each cooling temperature. The ratio A/C is fixed at 1.5, $\tilde{\epsilon}$ at 3, d at $L_0/90$, and K_L at $20/L_0$.

The three minimized structures are depicted in stereo in Fig. 10, with the superhelical constraint fixed in a common orientation. As evident from Fig. 10, the perturbation of structure increases the energy and alters the usual symmetric influence of ΔLk on chain configuration. The increase in energy found in the perturbed chains relative to that in the unconstrained closed chain is a consequence of the tight bending of the ~1.5-turn superhelical constraint. The apparent asymmetry in both the energy ($\tilde{E} = 21.4$ at $\Delta\text{Lk} = 4$ and 27.3 at $\Delta\text{Lk} = -4$) and the writhing number ($\text{Wr} = 3.03$ at $\Delta\text{Lk} = 4$ and $\text{Wr} = -1.94$ at $\Delta\text{Lk} = -4$) in chains with linking number differences of the same magnitude arises because the lowest energy configuration containing the superhelical constraint is found at a linking number difference close to unity. The energy ($\tilde{E} = 8.3$) and writhing numbers ($\text{Wr} = \pm 2.7$), in contrast, are equivalent at $\Delta\text{Lk} = \pm 4$ in the corresponding unperturbed elastic circular DNA.

Summary

Major goals of this research are to comprehend and visualize the detailed three-dimensional arrangements of supercoiled DNA. Attention has been focused in the initial stages on mathematical procedures to generate the spatial coordinates of the B-DNA double helix constrained to specific

[96] Y. Zivanovic, I. Goulet, B. Revet, M. Le Bret, and A. Prunell, *J. Mol. Biol.* **200,** 267 (1988).
[97] P. T. England, S. L. Hajduk, and J. C. Marini, *Annu. Rev. Biochem.* **51,** 695 (1982).
[98] M. Linial and J. Shlomai, *Nucleic Acids Res.* **16,** 6477 (1988).

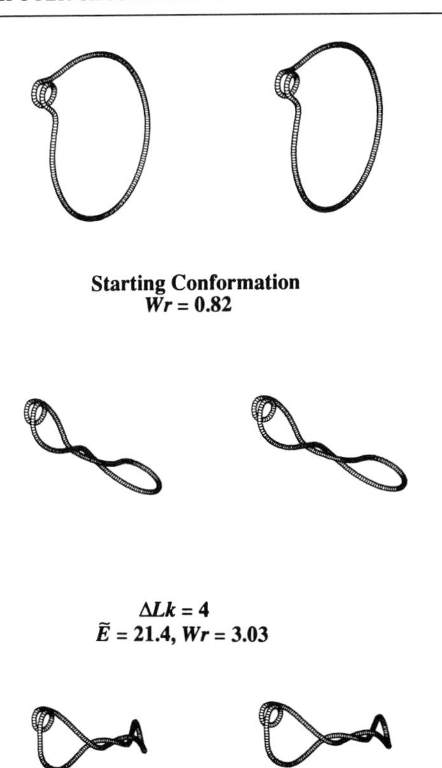

FIG. 10. Starting and optimized supercoiled configurations of closed circular elastic DNA (~500 bp) constrained by the presence of a right-handed superhelix of ~1.5 helical turns (~140 bp) with $\Delta Lk = \pm 4$, $A/C = 1.35$, $d = L_0/90$, $\tilde{\epsilon} = 3.0$, $K_L = 20/L_0$.

spatial pathways and on simple energy models of chain conformation. The new treatment of superhelical DNA in terms of parametric curves is an important first step in being able to generate and examine tertiary structure systematically. The location of every residue is implicitly determined by the equation of the closed curve, with the number of computational variables sharply reduced compared to the number required for explicit specification of all chain units. Furthermore, the constraints of ring closure in cyclic chains and/or the end-to-end limitations on constrained open chains are automatically satisfied by the formulations (cubic B-splines and finite Fourier series) chosen in this work.

The predicted conformations of elastic DNA do not appear to be tied to either the form of chain representation or the computer simulation method. Significantly, two very different minimization and modeling approaches come to the same structural conclusions. The most stable configurations of the closed circular elastic DNA model are found to be interwound superhelices that are critically dependent on the specified linking number difference. The total elastic energy is proportional to the imposed linking number difference, and beyond the critical linking number difference separating the circular and figure-eight forms, the writhing number of the DNA superhelices is directly proportional to ΔLk. The measured proportionality constant between Wr and ΔLk, however, is somewhat greater than that deduced from experimental observations of plectonemically interwound DNA chains and an assumed structural model.[95] Furthermore, at large ΔLk, the interwound structures appear to curve.

The treatment of the DNA double helix as an ideal elastic rod is clearly incorrect. The chain cannot bend with the same ease in all directions. The degree of bending observed in atomic level models is also tied to the angular twist so that the presumed partitioning of bending and twisting components is in error. Furthermore, the local chain bending and twisting are base sequence dependent, with certain residues able to flex more symmetrically than others. The polyelectrolyte character of the DNA is additionally expected to govern the overall folding of the chain and to influence the local secondary structure. The next step in this work is to compare the properties of such "real" DNA with conventional elastic models. These features are currently being incorporated into the simulations by use of analytical expressions that mimic the observed bending, twisting, and long-range electrostatic behavior of the double helix.

Acknowledgments

The authors are grateful to the U.S. Public Health Service for support of this work under research Grant GM34809, the Rutgers Center for Computational Chemistry and the Pittsburgh Supercomputer Center for computational facilities, Andrew J. Olson for preparation of molecular graphics, and Irwin Tobias for helpful discussions.

[22] Molecular Modeling to Study DNA Intercalation by Anti-tumor Drugs

By STEPHEN NEIDLE and TERENCE C. JENKINS

Introduction

The concept of intercalation was first introduced by Lerman[1] to explain the reversible, noncovalent binding of compounds with planar aromatic (most often heteroaromatic) extended chromophores to double-helical DNA. The essence of intercalation is the insertion of such a chromophore between adjacent base pairs of the DNA, thereby extending and stabilizing the double helix; base-pair separation at an intercalation site thus increases from 3.4 to 6.8 Å.[2,3] Processes of DNA transcription, translation, and replication can all be affected by intercalation; these effects have been rationalized by the decreased ability of the double helix to unwind consequent to drug binding. A number of cancers, especially leukemias, which are characterized by rapid cell proliferation, can have their growth preferentially inhibited by a variety of intercalating compounds as a consequence of their inhibition of DNA function. The effects produced by, in particular, the anthracyclines, are such that they have a clinically exploitable window of therapeutic index, thereby discriminating between normal and some tumor cells. The intercalating compounds doxorubicin (Adriamycin), mitoxantrone, amsacrine, and actinomycin (Fig. 1) are all in current clinical use, with the anthracene-9,10-dione doxorubicin having the widest spectrum of activity of any current anticancer drug. Thus the major interest in DNA-intercalating compounds has focused on their anticancer activity.[4-6]

In general, intercalating drugs have a polarizable, electron-deficient chromophore that is typically composed of three fused six-membered rings. A formal positive charge is common, located either on the central chromophore or on attached substituent groups. The presence of a substituent moiety appears to be a necessary precondition for antitumor action, although not for intercalative binding. Thus, unsubstituted acridines are often good intercalators and yet are devoid of antitumor activity. Considerable variation in the nature of the substituent is possible, ranging from

[1] L. Lerman, *J. Mol. Biol.* **3,** 18 (1961).

[2] H. M. Berman and P. R. Young, *Annu. Rev. Biophys.* **10,** 87 (1981).

[3] W. D. Wilson and R. L. Jones, *Adv. Pharmacol. Chemother.* **18,** 177 (1981).

[4] M. J. Waring, *Annu. Rev. Biochem.* **50,** 159 (1981).

[5] L. P. G. Wakelin *Med. Chem. Rev.* **6,** 275 (1986).

[6] L. H. Hurley, *J. Med. Chem.* **32,** 2027 (1989).

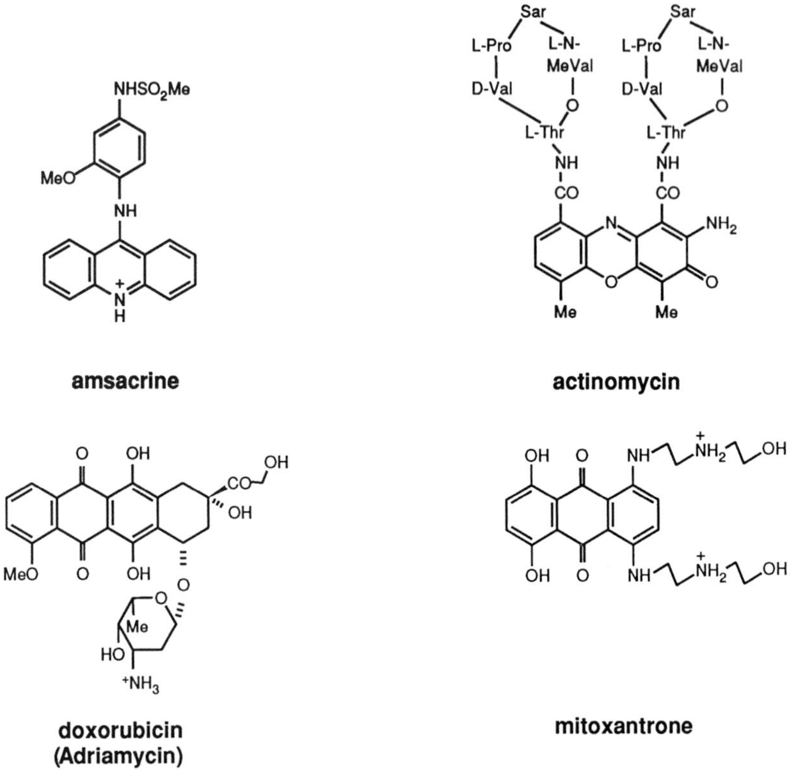

amsacrine

actinomycin

doxorubicin (Adriamycin)

mitoxantrone

FIG. 1. Chemical formulas of several intercalating anticancer drugs.

uncharged cyclic polypeptides (in actinomycin) to protonated amino sugars (in the anthracycline antibiotics).

There is little evidence that these intercalating drugs have a DNA sequence selectivity that extends beyond the 2- or 3-bp level. Thus, the sensitivity of some tumor cells to these compounds cannot readily be explained in terms of selective inhibition of tumor-specific DNA sequences (oncogenes). Rather, evidence has accumulated that the antitumor effects of such agents are related to a drug-induced stabilization of a cleavable DNA complex involving the enzyme DNA topoisomerase II, which results in lethal double-strand breaks.[7] Thus, intercalation is currently considered to be an essential (although by itself insufficient) component of the anticancer action of these drugs,[8] and is probably the primary recognition event prior to the formation of a ternary drug–DNA–enzyme complex.

[7] E. M. Nelson, K. M. Tewey, and L. F. Lui, *Proc. Natl. Acad. Sci. U.S.A.* **81,** 1361 (1984).

[8] W. A. Denny, *Anti-Cancer Drug Des.* **4,** 241 (1989).

Although many simple acridines are at least as effective DNA intercalators as amsacrine (as measured by DNA binding in solution), they are invariably devoid of anticancer activity, since they are unable to induce DNA double-strand breaks *in vivo.* It is nonetheless clear that some structure–activity relationships do exist between *in vitro* cytotoxicity and strength of intercalative DNA binding; these correlations are not general, however, but are confined to series of structurally related analogs in which biological activity is manifest.[8]

The primary role of molecular modeling in the study of drug–DNA intercalation has historically been to rationalize, explain, and extend the observed solution binding behavior in both structural and dynamic terms. These agents can therefore be considered as effective probes of DNA behavior. Rather less attention has focused on the application of modeling to anticancer drug discovery and as yet few novel compounds of potential clinical utility have been developed from such analyses. This situation is largely attributable to the fact that it is not at present possible to relate studies of intercalation at the molecular level to those design features required for improved efficiency in DNA topoisomerase II inhibition, and hence to improvements in tumor versus normal cell selectivity.

Goals of DNA–Drug Modeling

The discovery of clinically useful activity by several intercalating agents, albeit with a lack of true selectivity toward tumor cells, has led to a quest for superior agents, usually on the presumption that increased DNA affinity is a worthwhile goal. This, it can be argued, may be a factor in improvements in therapeutic index. Molecular modeling can identify and define the possible key details of the molecular interaction and, using the graphics/computational approach, decide on optimal structural modifications likely to enhance either general binding affinity or recognition of a particular nucleotide-binding site. A number of studies have revealed the converse, with the discovery that particular patterns of chemical modification on a given drug (e.g., doxorubicin, Fig. 1) can result in both low biological activity and low-ranking interaction energy.

Structural Models for Intercalation

The structures of a large number of polymorphs of DNA at the polymer level have been established by X-ray fiber diffraction. The consensus is that the B-DNA form represents the physiologically important structure.[9] Ex-

[9] J. M. Gottesfield, J. Bianco, and L. L. Tennant, *Nature (London)* **329,** 460 (1987).

tensive single-crystal studies on short (≤12-bp length) oligonucleotides during the past decade have shown the importance of sequence-dependent structural effects, although general rules for them remain to be established.[10-13] Sequence dependency is undoubtedly of prime importance in DNA recognition by regulatory and functional proteins, although its role in the mechanism of action of intercalating drugs is imperfectly understood at present. The relatively weak and short-run sequence preferences shown by doxorubicin,[14] for example, are probably of secondary importance compared to any preferred sites of double-strand breakage produced by DNA topoisomerase II (although the two factors may be related).

Crystallographic data on intercalation complexes fall into three categories:

1. The first category includes deoxydinucleoside monophosphate intercalation complexes, exemplified by that between $d(CG)_2$ and proflavine,[15] with the two strands forming an antiparallel, 2-bp length of double helix. Some 20 such complexes have been described,[16] each having closely similar nucleoside conformations regardless of the nature of the intercalated species (e.g., proflavine, 9-aminoacridine, or ethidium).
2. The second category includes deoxyhexanucleoside pentaphosphate complexes of the anthracycline daunomycin and several of its derivatives,[17-19] where the nucleic acid sequences are of the type $d(CGXYCG)_2$ and XY is either TA or AT. The structures are antiparallel duplexes with two drug molecules bound per duplex at the terminal CpG sites, with the two binding sites having broadly equivalent conformations.
3. The third category, bis intercalators echinomycin and triostin, have been cocrystallized with octanucleotide and hexanucleotide sequences.[20,21]

[10] H. R. Drew and R. E. Dickerson, *J. Mol. Biol.* **151,** 535 (1981).
[11] H. R. Drew and A. A. Travers, *Nucleic Acids Res.* **13,** 4445 (1985).
[12] U. Heinemann, C. Alings, and H. Laube, *Nucleosides Nucleotides* **9,** 349 (1990).
[13] R. E. Dickerson, *J. Biomol. Struct. Dyn.* **6,** 627 (1989).
[14] J. B. Chaires, J. E. Herrera, and M. J. Waring, *Biochemistry* **29,** 6145 (1990).
[15] H.-S. Shieh, H. M. Berman, M. Dabrow, and S. Neidle, *Nucleic Acids Res.* **8,** 85 (1980).
[16] S. Neidle, *in* "Landolt-Bornstein, Nucleic Acids" (W. Saenger, ed.), Vol. 7/1b, p. 247. Springer-Verlag, Berlin and Heidelberg, 1989.
[17] A. H.-J. Wang, G. Ughetto, G. J. Quigley, and A. Rich, *Biochemistry* **26,** 1152 (1987).
[18] M. H. Moore, W. N. Hunter, B. Langlois d'Estaintot, and O. Kennard, *J. Mol. Biol.* **206,** 693 (1989).
[19] C. A. Frederick, L. D. Williams, G. Ughetto, G. A. van der Marel, J. H. van Boom, A. Rich, and A. H.-J. Wang, *Biochemistry* **29,** 2538 (1990).
[20] A. H.-J. Wang, G. Ughetto, G. J. Quigley, T. Hakoshima, G. A. van der Marel, J. H. van Boom, and A. Rich, *Science* **225,** 1115 (1984).
[21] G. J. Quigley, G. Ughetto, G. A. van der Marel, J. H. van Boom, A. H.-J. Wang, and A. Rich, *Science* **232,** 1255 (1986).

These drugs have received little attention by molecular modelers[22]; the conformational features of bisintercalation will not be discussed in this chapter, although the interested reader will be able to extrapolate much of the detail for monointercalation to these systems. Experimental studies on bis(acridines), which were predicted to have high DNA affinity, have shown uniformly disappointing antitumor activity.

The experimental structural data available for DNA intercalation are in principle insufficient to provide a full description of an intercalation site that is embedded within a DNA sequence, rather than located at either end of it. Both the dinucleotide and hexanucleotide complexes are subject to end effects with 5′- and 3′-end nucleosides having conformations that are unconstrained, since there are no succeeding residues. Basic physicochemical parameters of drug intercalation, such as the extent of helix unwinding, cannot therefore be obtained from such systems.

A number of theoretical models have been derived for intercalation within a DNA sequence at sites distant from either end of the sequence. These provide a plausible basis for further modeling studies. Few of these models directly relate to the single-crystal results and the majority force helicity throughout the structure by restricting conformational change solely to the intercalation site by means, for example, of *trans* conformations around the P–0-5′ and C-4′–C-5′ bonds. Their relative stabilities have not been evaluated, however, due to the diversity of force fields used in their derivation and hence poor comparability between cited energies. The model derived in our laboratory (the ICR model),[23] and those from Pullman and associates,[24–26] have been extensively used as the basis for structure–activity studies. The satisfactory results obtained with them in terms of agreement with experimental data strongly suggest that, in spite of significant differences in many of their features, there is not a unique geometry for intercalation, but rather a number of plausible drug–DNA complexes associated with local minima of closely similar energy. No study to date has systematically compared these and other models.

ICR Intercalation Model

This model has been derived with the crystal structure conformation of the $d(CG)_2$ duplex from the $d(CG)_2$–proflavine intercalation complex as a

[22] U. C. Singh, N. Pattabiraman, R. Langridge, and P. A. Kollman, *Proc. Natl. Acad. Sci. U.S.A.* **83,** 6402 (1986).

[23] S. Neidle, L. H. Pearl, P. Herzyk, and H. M. Berman, *Nucleic Acids Res.* **16,** 8999 (1988).

[24] K. X. Chen, N. Gresh, and B. Pullman, *FEBS Lett.* **224,** 361 (1987).

[25] K. X. Chen, N. Gresh, and B. Pullman, *Anti-Cancer Drug Design* **2,** 79 (1987).

[26] N. Gresh, B. Pullman, F. Arcamone, M. Menozzi, and R. Tonani, *Mol. Pharmacol.* **35,** 251 (1989).

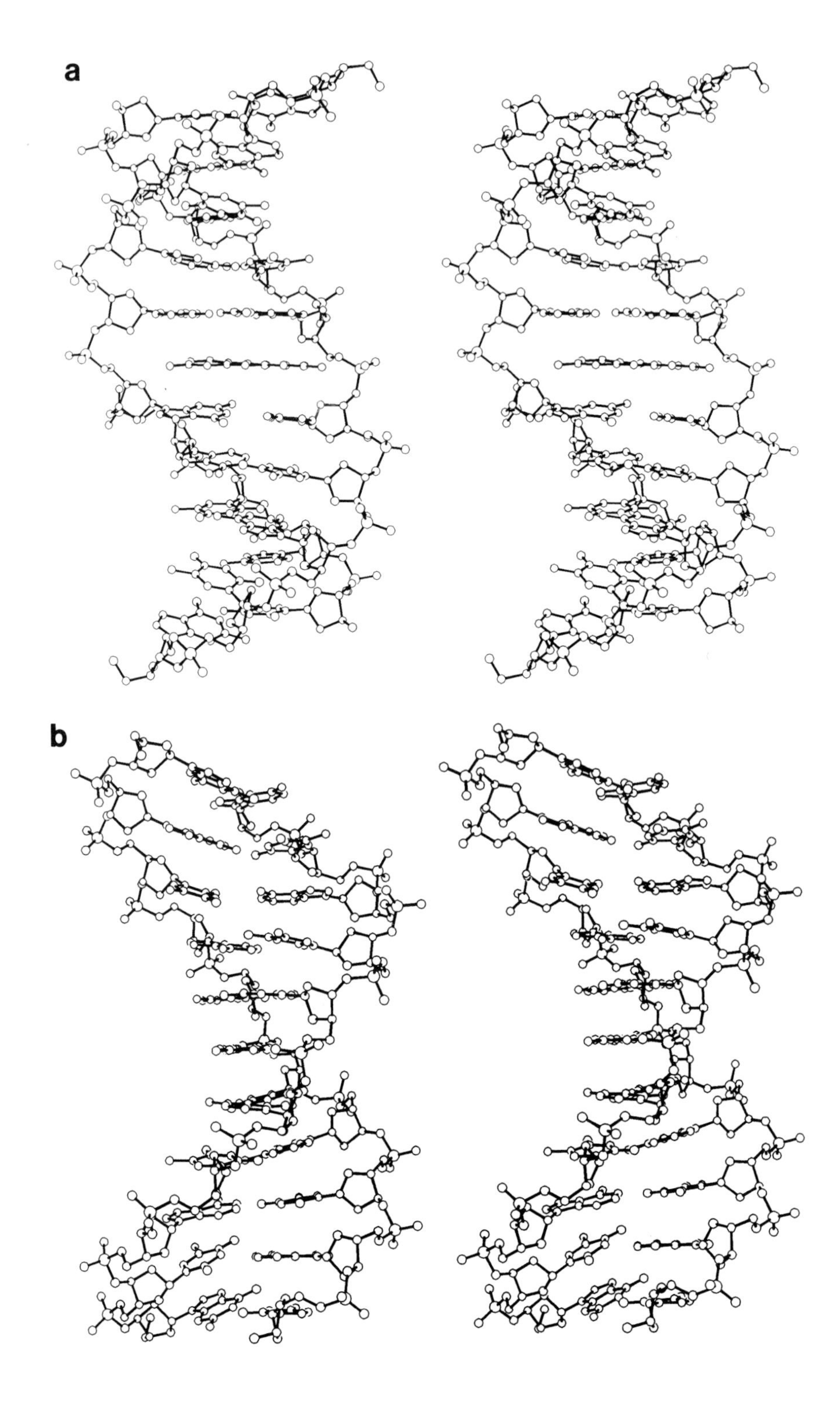
a
b

starting point (which itself has been used directly in a number of earlier simulations of drug–DNA intercalation), with two tetranucleotide duplexes in classic B-DNA form attached at the ends to achieve a full decamer sequence of $d(GATACGATAC)_2$. The 5′- and 3′-terminal GC base pairs were used in order to minimize the possibility of strand separation or "fraying" during subsequent energy refinement. An archetypal intercalator molecule, the acridine proflavine, was inserted into the central CpG site using computer graphics molecular modeling; energy refinement in the absence of an intercalated chromophore would result in the collapse of the structure surrounding the 6.8 Å-separated intercalation site. This initial model was subsequently energy refined using molecular mechanics with the "Kollman" all-atom force field (see Methods for Modeling Intercalation).[23]

The refined model (Figs. 2 and 3) has a smoothly deformed structure with both major and minor grooves widened compared to classical B-DNA. The marked asymmetry of the two nucleosides in each CpG intercalation site revealed in their crystal structures has been only partially retained in the final model. Instead, the refinement has forced conformational changes in most of the residues of the decamer, so that the structure has many features of A-form polynucleotide duplexes (Table I). Details of base-pair deformations in terms of helical parameters (i.e., propeller twist, roll, tilt, and helical twist) have been described.[23] The implication from this model that medium and even long-range structural perturbations are induced following intercalative binding is in accord with solution data accumulated from extensive DNA footprinting studies.[27] The decamer model has been modified to a hexamer length in several studies, largely for reasons of computational cost; the majority of DNA-intercalating compounds have a binding site that spans fewer than 4 bp.

Methods for Modeling Intercalation

The computational of low-energy oligo- and polynucleotide conformations and both inter- and intramolecular energies is in general a more complex problems than that for polypeptides, in part because of the much greater number of degrees of freedom per nucleotide repeating unit. In addition to the five backbone torsion angles the deoxyribose sugar has pseudorotational flexibility and can adopt a number of distinct puckers;

[27] P. E. Nielsen, *J. Mol. Recognition* **3,** 1 (1990).

FIG. 2. Stereo views of the ICR intercalation model of sequence $d(GATACGATAC)_2$ with proflavine. Hydrogen atoms have been omitted for clarity. (a) Looking into the major groove around the proflavine; (b) rotated by 90° around the helix axis.

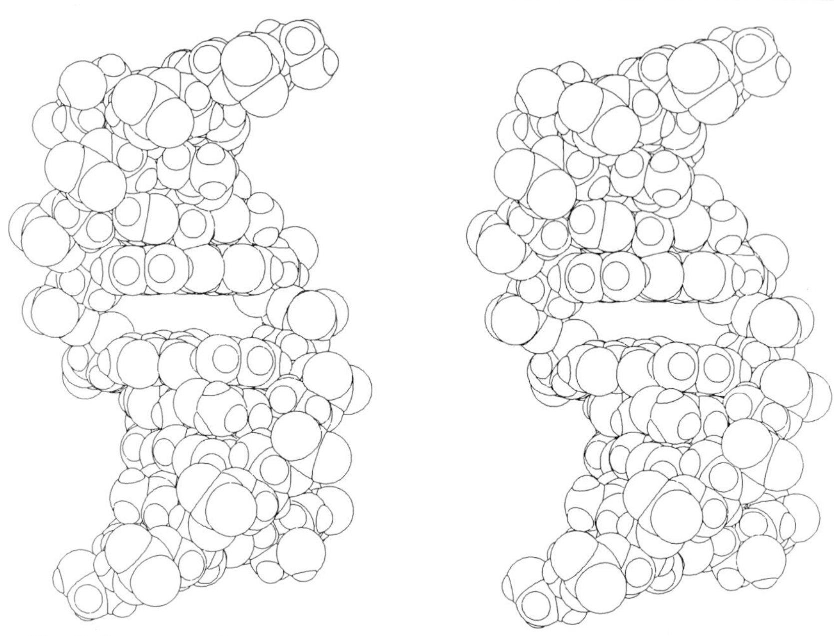

FIG. 3. Stereo space-filling view of the intercalated sequence $d(GATACGATAC)_2$ in the absence of ligand.

the major types C-2′(*endo*) and C-3′(*endo*) are separated by a shallow energy well.

The force field developed by Kollman, Weiner, and colleagues for DNA (and RNA) units is the most extensive, and consequently the most frequently employed,[28,29] using comprehensive parameterization based on the force field:

$$V_{\text{total}} = \sum_{\text{bonds}} K_r(r - r_{\text{eq}})^2 + \sum_{\text{angles}} K_\Theta(\Theta - \Theta_{\text{eq}})^2 + \sum_{\text{dihedrals}} \frac{Vn}{2}[1 + \cos(n\phi - \gamma)] + \sum_{i<j} \{[(A_{ij}/R_{ij}^{12}) - (B_{ij}/R_{ij}^{6})] + (q_i q_j/\epsilon R_{ij}\} + \sum_{\text{H bonds}} [(C_{ij}/R_{ij}^{12}) - (D_{ij}/R_{ij}^{10})] \quad (1)$$

[28] S. J. Weiner, P. A. Kollman, D. A. Case, U. C. Singh, C. Ghio, G. Alagona, S. Profeta, and P. Weiner, *J. Amer. Chem. Soc.* **106,** 765 (1984).

[29] S. J. Weiner, P. A. Kollman, D. T. Nguyen, and D. A. Case, *J. Comput. Chem.* **7,** 230 (1986).

TABLE I
BACKBONE CONFORMATIONAL ANGLES AND SUGAR PUCKER PSEUDOROTATION PARAMETERS (Q,P)[a]

Number	Base	Q	P	χ	α	β	γ	δ	ϵ	ζ
1	G5T	0.410	78.90	189.65	0.00	185.23	59.35	84.36	190.57	285.13
2	ADE	0.420	139.45	236.82	289.34	179.48	59.54	129.49	180.00	266.23
3	THY	0.399	93.89	218.27	290.03	178.37	57.65	92.54	177.26	273.05
4	ADE	0.378	128.91	236.31	299.54	171.26	60.83	120.37	186.98	272.91
5	CYT	0.402	64.18	230.83	287.54	167.09	55.84	80.89	189.71	281.02
6	GUA	0.418	82.88	243.88	293.04	183.30	70.25	82.16	44.92	52.16
7	ADE	0.433	130.61	241.11	224.33	187.02	57.49	121.47	175.26	268.70
8	THY	0.444	133.81	238.81	292.91	178.23	57.52	126.75	189.67	228.94
9	ADE	0.376	162.91	246.43	291.31	165.51	65.53	144.83	179.87	271.55
10	C3T	0.434	117.46	224.88	290.68	171.47	60.50	113.37	182.77	265.29
11	G5T	0.441	129.70	225.29	128.42	13.07	60.24	121.35	187.67	272.10
12	THY	0.416	70.24	214.41	278.23	174.59	55.84	80.72	177.54	275.75
13	ADE	0.369	160.51	247.22	298.01	172.05	66.52	142.03	182.54	254.89
14	THY	0.409	128.96	234.57	288.59	176.25	59.22	121.75	181.43	259.62
15	CYT	0.438	26.53	213.19	276.78	167.21	61.66	75.49	198.17	288.89
16	GUA	0.454	149.20	274.51	296.61	201.02	68.26	145.73	271.33	165.57
17	THY	0.424	68.07	211.26	284.08	118.18	54.80	77.58	174.98	271.51
18	ADE	0.364	161.48	246.52	297.06	177.52	63.23	142.19	178.26	268.24
19	THY	0.407	119.30	227.15	293.32	172.02	61.78	112.78	182.04	274.00
20	C3T	0.419	118.81	226.53	294.65	167.63	60.06	116.08	296.32	68.67

[a] For the two strands of the ICR intercalation model, each of sequence 5′-d(GATACGATAC).

where V_{total} represents the potential energy of the system. The CHARMm force field, developed by Karplus and colleagues,[30] and subsequently employed within the X-PLOR macromolecular refinement package,[31,32] uses the same formalism and a similar parameterization, although it has not been as extensively tested on either nucleic acids or drug–nucleic acid complexes. The total energy is obtained by minimization, usually with a conjugate-gradient approach. Convergence is normally achievable for drug–DNA complexes with a root-mean-square (rms) derivative of less than 0.1 kcal $Å^{-1}$ and a negligible energy change in consecutive cycles of refinement.

The 6–12 van der Waals nonbonded term is of conventional Lennard–Jones type, and the electrostatic term utilizes partial atomic charges derived, for example, from CNDO/2, MNDO, or AM-1 molecular

[30] B. R. Brooks, R. E. Bruccoleri, B. D. Olafson, D. J. States, S. Swaminathan, and M. Karplus, *J. Comput. Chem.* **4,** 187 (1983).

[31] A. T. Brünger, "X-PLOR Manual, Version 2.1," Yale University, New Haven, Connecticut, 1990.

[32] A. T. Brünger, *J. Mol. Biol.* **203,** 803 (1988).

orbital calculations (and a Mulliken population analysis) or, better, from quantum mechanically derived electrostatic potentials fitted to a point charge model. This force field, as incorporated into the AMBER[33] package, has been used by a number of workers and has been remarkably successful in both reproducing and predicting low-energy conformations for a number of nucleic acid duplex sequences, especially when an all-atom rather than a united-atom representation is used. Other less extensive (and computationally cheaper) force fields can be used satisfactorily for describing the geometric features of base–base stacking interactions. Thus, a systematic examination of possible orientations for proflavine in a CpG intercalation site[34] found low-energy positions in excellent agreement with the crystallographic structures, using a simple nonbonded potential. If only such qualitative geometric data are required, then the problem of parameterizing the full force field does not need to be approached. However, if accurate relative energies (or even ultimately thermodynamic energies) are desired, then full parameters for the nonstandard molecular components in the system (i.e., those for the ligand molecule) must be derived. A full discussion of this important topic is beyond the scope of this chapter. In practice, choosing parameters by analogy is often effective, with geometric values obtained from standard tabulations.[35] Force-field parameters and atom-type designations for several intercalating drugs are given below.

The use of united-atom parameterizations of the AMBER force field has been examined in the case of both ethidium–dinucleotide and ethidium–hexanucleotide intercalation models.[36] Although similar final conformations were obtained with the two formalisms, the united-atom relative energies for various models examined of each type were inconsistent and not at all in accord with experimental data. Thus, although the united-atom approach is computationally more attractive, and may even be genuinely useful for nucleic acid and proteins in the absence of ligands, its use for drug–DNA complexes is not to be recommended.

Nucleic acids are highly charged anionic molecules, with the phosphate groups on the exterior of a double helix carrying most of the formal negative charge. The extensive studies of the Pullmans has focused attention on their electrostatic potential and field, and has, for example, shown that, contrary to expectation, the minor groove is the region of greatest negative electrostatic potential, in a sequence-specific manner. Electrostatic potential considerations are clearly of importance in the binding of

[33] P. K. Weiner and P. A. Kollman, *J. Comput. Chem.* **2,** 287 (1981).

[34] S. A. Islam and S. Neidle, *Acta Crystallogr., Sect. B* **40,** 424 (1984).

[35] F. H. Allen, O. Kennard, D. G. Watson, L. Brammer, A. G. Orpen, and R. Taylor, *J. Chem. Soc., Perkin Trans. 2,* S1 (1987).

[36] T. Lybrand and P. Kollman, *Biopolymers* **24,** 1863 (1985).

intercalating drugs (most of which carry cationic charge). Pullman has developed an intermolecular energy force field that explicitly incorporates several factors that are electrostatic in origin[37]:

$$E_{\text{INTER}} = E_{\text{MTP}} + E_{\text{POL}} + E_{\text{REP}} + E_{\text{DISP}} + E_{\text{CT}} \quad (2)$$

where E_{MTP} and E_{POL} represent electrostatic and polarization contributions to the total energy E_{INTER}, calculated by means of a multipolar expansion of the *ab initio* wavefunctions of the individual structural fragments. The E_{REP} and E_{DISP} terms correspond to the standard Lennard–Jones potentials, and the E_{CT} contribution explicitly takes account of charge-transfer factors. These terms are implicitly incorporated in the 10–12 terms of the Kollman *et al.* force-field in Eq. (1), which have been extensively parameterized for nucleic acid constituents as such, but not in a variety of interacting situations, such as when polarization and charge-transfer factors may be significant. The inability of the simple Coulombic law to treat induced polarization effects is probably its most serious defect when considering drug–nucleic acid complexes, with their combination of high formal charges and very polar groupings.

These force fields can be used to calculate (1) E_{D}, the internal energy of a drug, (2) E_{H}, the internal energy of the oligonucleotide sequence, usually as B-form DNA, (3) E_{HC}, the internal energy of the nucleic acid duplex in an intercalated complex, (4) E_{HD}, the drug–helix intermolecular interaction energy, and (5) E_{COM}, the summation of intermolecular and internal energies in the drug–DNA complex, together with individual component contributions to the totals.

Helix destabilization energy is then defined as

$$E_{\text{D}} = E_{\text{HC}} - E_{\text{HD}} \quad (3)$$

and represents the energy lost as a result of affecting the distortions in DNA necessary to form an intercalative binding site. The net binding energy ΔE is then the energy of the drug–DNA complex minus the summed energies of the drug and the B-DNA helix:

$$\Delta E = E_{\text{COM}} - (E_{\text{D}} + E_{\text{H}}) \quad (4)$$

The problem of solvent treatment is common to all these molecular mechanics methods. In general, explicit structured solvent or hydration contributions have only very rarely been considered for drug–DNA complexes. Thus, molecular mechanics calculations on these systems in principle afford potential energies and cannot normally provide data on energy values that have direct relevance to thermodynamic measurements of free

[37] B. Pullman, *Adv. Drug Res.* **18,** 1 (1989).

energy and entropy. However, relative binding energies calculated for a closely related series of drugs that do not displace different numbers of water molecules should correlate with experimental relative enthalpies and can indicate changes in free energy. Exclusion of explicit solvent causes a further serious problem that is particular to nucleic acids: the large formal negative phosphate charges are in reality shielded by solvent, and thus neglect of this factor will greatly overemphasize any electrostatic contributions in the force field. The problem has been tackled in various ways: (1) the phosphate charges have been modified so as to neutralize the group as whole, while still retaining the phosphate dipole (e.g., see Refs. 38 and 39); (2) explicit cationic counterions, usually sodium, are placed between the phosphorus and its two monosubstituted oxygen atoms; and (3) the most straightforward (and probably most widely used) procedure is to employ a distance-dependent dielectric constant in the Coulombic electrostatic term, of the form $\epsilon = cR_{ij}$, thereby damping this energy component.

We have shown[40] that in the situation with anionic phosphate charges are no explicit solvent, a constant $c = 4$ in the distance-dependent term gives structural results after full molecular mechanics energy minimization that compare significantly better with a simulation on a fully solvated model than is the case with $c = 1$. DNA helical parameters such as base-pair tilt, roll, twist, and propeller twist were found to alter, depending on the value of c, and were most consistent with $c = 4$ rather than with lower values; the structures themselves also had superior convergence properties and more acceptable structural features.

The major problem of the conformational pluralism of nucleic acids cannot adequately be explored by molecular mechanics, with its inherent limitation of being restricted to the minimization of local potential energy wells on the total multidimensional energy surface. Molecular dynamics techniques have much promise in being able to explore many more features of molecular flexibility as well as being able to calculate thermodynamic free energy changes. These methods still have the problems involved in the treatment of solvent and electrostatic factors. It has been pointed out that these problems are more acute in the case of nucleic acids than for proteins, due to the inherent polyelectrolyte nature of nucleic acids and their larger surface-to-volume ratio. To date, only a few molecular dynamics simulations of drug–DNA complexes have been reported,[41–43] and

[38] B. Tidor, K. Irikura, B. R. Brooks, and M. Karplus, *J. Biomol. Struct. Dyn.* **1,** 231 (1983).

[39] R. F. Tilton, P. K. Weiner, and P. A. Kollman, *Biopolymers* **22,** 969 (1983).

[40] M. Orozco, C. A. Laughton, P. Herzyk, and S. Neidle, *J. Biomol. Struct. Dyn.* **8,** 359 (1990).

[41] M. Prabhakaran and S. C. Harvey, *J. Phys. Chem.* **89,** 5767 (1985).

very recently a free energy perturbation analysis of a daunomycin complex.[44] Molecular dynamics simulations are typically of up to 50–100 psec; thus, they are not at present able to model the dynamics of drug–DNA association, which is typically in the 10^{-4}- to 10^{-1}-sec range.

The principal alternative method for the study of multidimensional degrees of conformational freedom involves the use of interactive computer graphics, for more than mere display and analysis purposes. It has found special use in the estimation of plausible low-energy positions for a drug in an intercalation binding site, usually prior to molecular mechanics regularization. Although such an approach is usually unable to systematically study all conformational and positional possibilities, especially when the ligand itself has flexibility, it can be remarkably speedy in locating sterically (and therefore energetically) reasonable structures, when in the hands of a skilled molecular modeler. The graphics method has been the method of choice for the location of starting structures prior to full energy minimization and refinement. The use of color-coded surface representations has often been found to aid drug–DNA maneuvers, especially when incorporating an electrostatic potential display to show, for example, the negative electrostatic potential associated with an AT-rich minor groove surface.[37] Several more systematic procedures have been developed to examine speedily all possible locations for a rigid drug molecule. The procedure developed in our laboratory[45] produces nonbonded potential maps by alteration of ligand orientation and two-dimensional translation in the intercalation plane. This approach highlights the importance of chromophore size and shape in relation to its flexibility within the binding site, and furthermore that the nature of the substituents plays a major role in defining major versus minor groove preferences. It appears that the majority of biologically active intercalating drugs have their substituent groups in the minor groove; this may well be a stereoelectronic requirement for effective recognition by DNA topoisomerase II prior to formation of the ternary "cleavable complex."

Modeling Strategy

Plausible initial structures for the complex formed between a candidate intercalator molecule and a chosen DNA sequence must first be generated, normally using a computer graphics "docking" approach. These are judged

---

[42] M. Prabhakaran and S. C. Harvey, *Biopolymers* **27,** 1239 (1988).

[43] S. N. Rao and P. A. Kollman, *Proc. Natl. Acad. Sci. U.S.A.* **84,** 5735 (1987).

[44] P. Cieplak, S. N. Rao, P. D. J. Grootenhuis, and P. A. Kollman, *Biopolymers* **29,** 717 (1990).

[45] L. H. Pearl and S. Neidle, FEBS Lett. **209,** 269 (1986).

on the basis of minimal nonbonded steric clashes, possibly with some attractive interactions and maximal chromophore/base-pair overlap. Coordinates for the ligand may be obtained directly from crystallographic data, where available, or by adaptation of those derived for a closely related molecule. The Cambridge Crystallographic Database is the principal source of information for this.[45a] Where this is not feasible, there is a wide choice of graphics packages (e.g., GEMINI, QUANTA, ALCHEMY, or INSIGHT/DISCOVER) that facilitate molecular construction using standard libraries of bond lengths and bond angles. Although crystallographic data is often found to provide reliable initial coordinates, other structures may require optimization to take care of stereoelectronic effects and untoward intramolecular clashes. Molecular mechanics or possibly *ab initio* or semiempirical quantum mechanical optimization methods may be applied (using, for example, GAUSSIAN or AMPAC).

Molecules with likely modes of conformational or torsional flexibility may exert a profound influence on the ease with which a chromophore may be inserted within a model DNA intercalation site. Preliminary investigations of such modes by, for example, searching conformational space for the isolated ligand and determining low-energy geometries for the flexible regions, will assist the subsequent docking procedure and reduce the net computation time by eliminating unfavorable structures. This approach has been adopted to probe, for example, the conformational flexibilities of daunomycin[46] and *N*-substituted 9-aminoacridine-4-carboxamides[47] with regard to their interactions with DNA.

The docking procedure involves interactive graphics manipulation of both the drug and DNA molecules such that the ligand becomes located within the intercalation site in plausible orientations. This process may require alterations in drug conformation to eliminate or to minimize unfavorable intermolecular or van der Waals close contacts ("bumping"), as judged from displays of interatomic distances and angles, molecular surfaces, or simple computed energies. The influence of favorable electrostatic or hydrogen-bonded interactions between substituents or side-chain moieties may also be investigated. Intercalated chromophores are normally aligned or stacked within the DNA helix to effect optimal π overlap with the two base pairs that constitute the binding site. Automated alignment techniques, usually employing a combination of translation and rotation processes together with rigid-body energy refinement, may also be used to generate plausible structures for the complex.

[45a] F. H. Allen, O. Kennard, and R. Taylor, *Acc. Chem. Res.* **16,** 146 (1983).

[46] S. A. Islam and S. Neidle, *Acta Crystallogr., Sect. B* **39,** 114 (1983).

[47] B. D. Hudson, R. Kuroda, W. A. Denny, and S. Neidle, *J. Biomol. Struct. Dyn.* **5,** 145 (1987).

Alternative orientations of the chromophores within the DNA binding site are possible, especially if the molecule is asymmetric, since substituents may then be accommodated within either the major or minor groove. In addition, alternative partial modes of binding such as "spear" intercalation (in which a doubly substituted molecule has one substituent in each groove) may be possible. This molecular modeling approach requires systematic assessment of each feasible alignment of the ligand within the drug–DNA complex (see section entitled "Amido-Functionalized 9,10-Anthraquinones").

Subsequent energy refinement of such structures requires further information pertaining to the introduced ligand, since the available force fields (e.g., AMBER, CHARMm, COSMIC, MM2) exploit details of (1) explicit bond connectivity, (2) the various atom types that define the molecular complexion, and (3) their associated atom-centered charges. In addition, parameters relating to those atom types not available within the force field [see Eq. (1)] must be provided; such parameters are often derived by interpolation from data available for other configurations or deduced from crystallographic or spectroscopic information.[35] Designations of atom type for representative intercalator molecules examined at the Institute of Cancer Research (Sutton, England) are given in Fig. 4. The charges must reflect the likely protonation status (i.e., neutral versus cationic) of the drug, if appropriate, so that comparison may be made with available solution or biological data involving a relevant species.

Once parameterized, the complexes may be refined by energy minimization using the appropriate force field as outlined in the previous section. Defined constraints and restraints may be applied at this stage, if required,

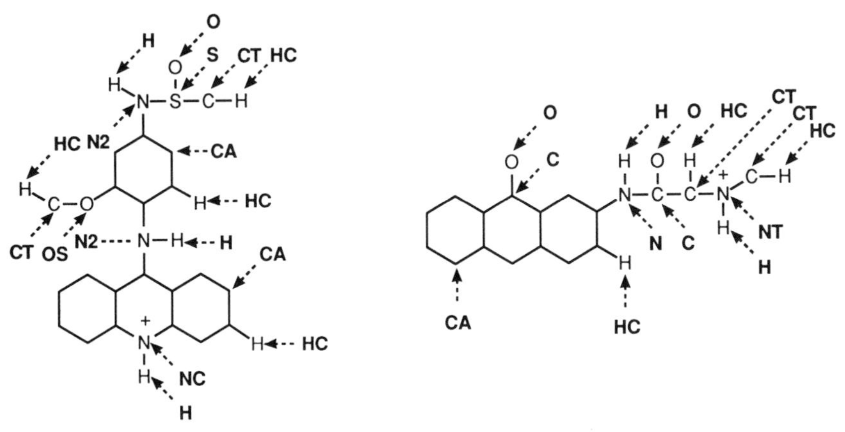

FIG. 4. Atom-type designations for amsacrine and a 2,6-bis(aminoalkanamido)anthracene-9,10-dione, based on the AMBER force-field formalism.[28,29,33]

to maintain planarity of the chromophore, favorable interatomic contacts or hydrogen bonding, or system integrity. Typically, the phosphorus atoms of the DNA backbone are fixed in position during early cycles of refinement and such constraints may be relaxed, as necessary, once the system converges to a low-energy or stable geometry. The convergence achieved will reflect the validity of the docked complex, but will be sensitive to the problem associated with multiple local minima situated on a global potential energy surface. Caution must be exercised in "overrefining" the structure to, for example, an unrealistic energy gradient.

Following satisfactory refinement, as judged by graphics visualization of the final structure, the energetics of the model may be determined as already described. Thus, the perturbation energies (enthalpies) induced on both drug and DNA may be calculated, if the separate or unbound molecules have been subjected to energy refinement. Further, the binding energy and its component terms (i.e., bonded, nonbonded, hydrogen bonded or electrostatic) can be analyzed for each of the models. The computed energies cannot, however, be formally compared with experimentally derived free energies, for example, those determined by differential calorimetry or calculated from solution binding data using the following expression:

$$\Delta G = -RT \ln K$$

where K is the equilibrium binding constant for the process. The entropic contribution to the binding is not easily derived from molecular modeling, hence the computed binding energies derived from such models normally relate only to the enthalpic component. Thermodynamic energies would require modeling of explicit hydrated structures for both the unbound free molecules and the drug–DNA complex, using either a full Monte Carlo or molecular dynamics refinement procedure.

Analysis of the final structures provides information likely to be useful for the design of improved DNA-intercalating compounds and may indicate those features of the molecule that enhance or interfere with the modeled binding process (such as particular hydrogen bonding between bases and ligand side chains). Further, details of the structure may indicate whether substituent groups are stereoselectively or preferentially accommodated within either the major or minor groove(s) of the DNA, and whether site or sequence selectivity effects are feasible. Structural analysis of the DNA in the final drug–DNA complex in terms of conformational helical parameters will highlight those perturbations associated with the intercalation process, and whether such effects are confined locally to the binding site or propagated through the DNA helix. Similarly, the influence of any sequence-dependent effects on binding can be probed.

Examples of Modeled Complexes

This section discusses several drug–DNA intercalative systems studied in the authors' laboratory, with one area described in detail.

Porphyrin–DNA Interactions

The intercalative properties and possible binding site preference of tetra-(4-*N*-methylpyridyl)porphyrin have been studied using both $d(CG)_2$ and $d(TA)_2$ sequences and an automated rigid-body molecular mechanics approach.[48] Minimum-energy configurations for the symmetric porphyrin in the two intercalation sites indicate that the principal contributions to the stabilization of the complex are electrostatic. Further, intercalation into the CpG rather than the TpA site is likely to be preferred since the methyl group of the thymine hinders a low-energy geometry for the complex. Indeed, the modeling suggests that intercalation of the bulky ligand can only be partial at TpA sites, whereas full intercalation is possible at CpG-binding sites. These predictions are consistent with physicochemical measurements of porphyrin–DNA binding.

Acridine–DNA and Related Interactions

This laboratory has accumulated a wealth of data relating to the DNA-binding properties of acridine derivatives, which behave as "classical" intercalators and often possess useful antitumor properties. The model complex formed between proflavine and the decamer sequence $d(GATACGATAC)_2$ has been described above and provides a useful starting geometry for other studies of DNA intercalation. The clinical activity displayed by many acridines has prompted structure–activity analyses of several acridine derivatives in the search for compounds with improved therapeutic utility. Thus, analogs of the antitumor drug amsacrine (Fig. 1) have been extensively studied using molecular modeling techniques. Studies of 1-methylamsacrine, using crystallographic data to model its intercalation with the $d(CG)_2$ dinucleotide sequence, suggested[49] that the energy of interaction is greater than for the parent compound, in accord with biological data. It was not possible, however, to establish whether intercalation was favored from either groove direction. However, the influence of

[48] K. G. Ford, L. H. Pearl, and S. Neidle, *Nucleic Acids Res.* **15,** 6553 (1987).

[49] S. Neidle, G. D. Webster, B. C. Baguley, and W. A. Denny, *Biochem. Pharmacol.* **35,** 3915 (1986).

a methyl group upon the planarity of the chromophore and hence the likely effect upon *N*-functionalized 9-aminoacridine 4-carboxamides has revealed implications for their potential DNA-binding behaviors.[47] Thus, the influence of conformational flexibility and any potential intramolecular hydrogen bonding involving the acridinium chromophore on binding to a model CpG site has been described.

The DNA-binding properties of amsacrine and an analog substituted at the 4-position with a 4-methanesulfonanilide group have revealed that the derivative is at least as effective an intercalator as the parent compound,[50] although its anticancer properties are markedly inferior. Modeling of intercalation used a hexanucleotide duplex of sequence $d(TACGTA)_2$ that had the same backbone conformation as the central region of the ICR intercalation model described above. For amsacrine itself, minor groove intercalation was indicated to be energetically preferred.

Amido-Functionalized 9,10-Anthraquinones

The evaluation of amino-functionalized anthraquinones for antitumor potency has led to the 1,4-bis-substituted compound mitoxantrone (Fig. 1). This agent is a DNA intercalator *in vitro,* an effective inhibitor of nucleic acid synthesis, and produces lethal double-strand beaks via DNA topoisomerase II complex formation. The clinical utility of this compound has prompted the quest for analogous compounds with improved therapeutic indices, and to examine the influence of (1) ring substitution of the chromophore, (2) net molecular charge, and (3) alternative functionalization on potency.

Early studies in this laboratory of 9,10-anthraquinone (or, more correctly, anthracene-9,10-diones) agents, particularly those which are amido-functionalized, have shown that computer modeling can rationalize much of the solution DNA-binding behavior, and that both can illuminate the available structure–activity data. Other studies[51] have shown direct relationships between DNA-binding parameters and some biological effects for a series of monosubstituted amidoanthraquinones. This work has been elaborated by an extensive study of 1-mono- and 1,4-disubstituted alkylamido-functionalized analogs.[52] The inclusion of side chains containing basic nitrogen atoms has a marked effect on the DNA-binding properties and initial calculations (with a restricted force field) revealed that

[50] Z. H. L. Abraham, M. Agbandje, S. Neidle, and R. M. Acheson, *J. Biomol. Struct. Dyn.* **6,** 471 (1988).

[51] G. Palii, M. Palumbo, C. Antonello, G. A. Meloni, and S. M. Magno, *Mol. Pharmacol.* **29,** 211 (1986).

[52] D. A. Collier and S. Neidle, *J. Med. Chem.* **31,** 847 (1988).

intercalative interaction with $d(CpG)_2$ is dominated by specific electrostatic energy terms involving the formal protonated center. It was concluded that the positive charge associated with the side-chain moiety can more than compensate for the lack of charge on the anthraquinone chromophore itself.

A detailed study of the effects of varying positions of anthraquinone disubstitution with protonatable side chains[53] also suggested that isomers with groups attached at the 1,4-, 1,5-, or 1,8-positions result in severe constraints on the geometries of the low-energy complexes. Thus, for example, minor groove entry to the intercalation site is prevented by steric hindrance for both the isomeric 1,4- and 1,8-derivatives, whereas the binding of the 1,5-compound is enhanced by favorable nonbonded energy terms.

Bis-substitution at both the 2- and 6-positions was originally envisaged by us to result in compounds with well-defined DNA interaction properties, since geometric introduction of two bulky groups at the distal positions would result in superior kinetic stabilities for the drug–DNA complexes. In particular, 2,6-disubstitution of the chromophore requires that each of the two substituent groups must be simultaneously accommodated within the major and minor grooves adjacent to the intercalation site. It is not feasible to construct a model in which both side chains of the drug are localized in only one of the DNA grooves.

A series of seven ω-aminoalkanamido 2,6-disubstituted compounds has been synthesized (Fig. 5) in which (1) the length of the side chain was altered by the inclusion of either one or two methylene groups, and (2) where the protonation status of the pendant basic groups, and hence the overall molecule, was modulated by appropriate choice of terminal amine moiety as determined solely by pK_a values (e.g., morpholino < piperidino ≅ diethylamino < piperazino).[54] Thus, a family of related analogs was generated where the side-chain functionalization is expected to influence and determine the DNA-binding behavior of the intercalating chromophore. In this chapter we restrict our discussion to the most biologically potent compounds, where the pendant amine functions have pK_a values such that the agents are effectively protonated at physiological pH.

Molecular modeling of these novel agents involved the use of a consensus dinucleoside d(CpG) intercalation site geometry (the ICR model, see earlier), where the CpG site was contained in the center of an alternating hexanucleotide duplex of sequence $d(TACGTA)_2$. The coordinates of the

[53] S. A. Islam, S. Neidle, B. M. Gandecha, M. Partridge, L. H. Patterson, and J. R. Brown, *J. Med. Chem.* **28,** 857 (1985).

[54] M. Agbandje, T. C. Jenkins, and S. Neidle, submitted for publication.

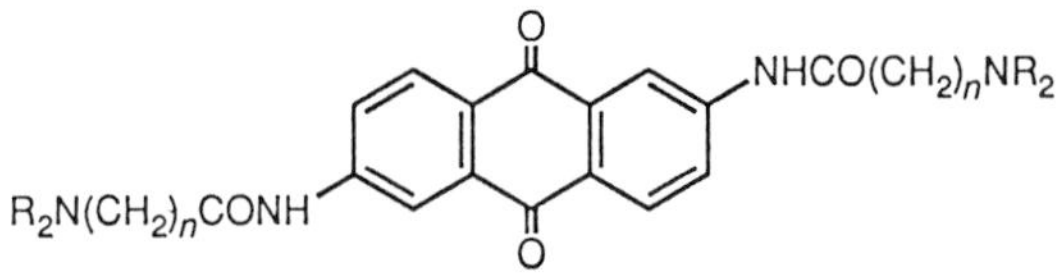

O
$NHCO(CH_2)_nNR_2$
$R_2N(CH_2)_nCONH$
O

Compound	*n*	$-NR_2$
1	1	$-NEt_2$
2	1	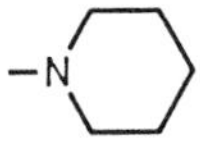 –N
3	1	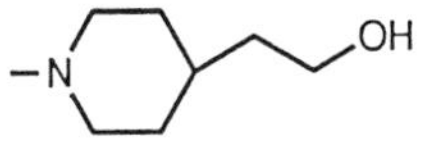 –N OH
4	2	$-NEt_2$
5	2	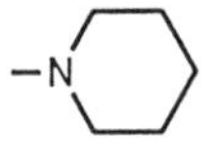 –N
6	2	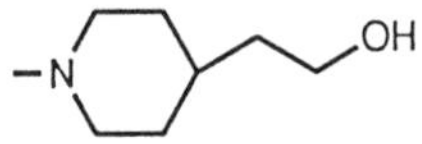 –N OH
7	2	$-N(CH_2CH_2OH)_2$

FIG. 5. Chemical formulas of 2,6-bis(aminoalkanamido)anthracene-9,10-diones (see text).

anthraquinone chromophore and the functionalized amido groups forming the 2,6-substituents were taken from the X-ray crystal structure of the piperidino compound itself (**2** in Fig. 5). The alternative side chains were generated by computer graphics modeling using standard bond lengths and angles; all molecules were generated with *N*-protonated terminal amine moieties for comparative purposes.

Interactive graphics modeling using the program GEMINI was used to dock each of the molecules into the intercalation site of the hexamer sequence with acceptable nonbonded contacts and minimal steric clash. The oligonucleotide residues were held fixed throughout the docking maneuvers, and the initial objective was to intercalate the planar chromophore at an average distance of 3.4 Å from the adjacent base pairs. Attempts to dock the molecules with both side chains positioned in either the major or minor grooves of the DNA revealed that plausible models could not be achieved due to unacceptable repulsive close intermolecular contacts involving the DNA backbone. In each case, the docking maneuvers showed that effective DNA intercalation required simultaneous positioning in both the major and minor grooves.

The initial position of the intercalated chromophore was retained for all of the $n = 2$ compounds (Fig. 5) to ensure that the energy differences subsequently calculated for the drug–DNA complexes were due to the side-chain substitutions rather than approaches to a radically different local energy minimum. The same procedure was employed for the $n = 1$ analogs, but involving a different initial chromophore position that reflects the inherent inflexibility of the shorter side chains. In general, better overlap or π stacking between the base pairs at the intercalation site and the chromophore was achieved with the $n = 2$ derivatives, with the chromophore positioned in a plane almost parallel to the long axis of the base pairs. With the $n = 1$ derivatives, parallel positioning of the chromophore resulted in unfavorable repulsive contacts; these derivatives were thus intercalated diagonally between the long axes of the base pairs. For both series of compounds the side-chain amide groups were positioned with the N—H pointed toward the sugar-phosphate backbone and the C=O directed into the groove; this amide conformation was retained for each model.

The successful docking of the drug chromophores was followed by a search for possible hydrogen-bonded interactions involving the protonated amine (i.e., N^+—H) residues and the phosphate oxygens on the DNA backbone. Thus, by systematically changing all side-chain torsion angles it was possible to evolve model structures with both maximal hydrogen-bonded interactions and optimized van der Waals hydrophobic contacts. Repulsive intra- and intermolecular contacts could thus be minimized during this search of local conformational space.

The coordinates for the intercalated drug–DNA complexes were subjected to molecular mechanics energy minimization using the AMBER all-atom force field. Force-field parameters for the new molecules were adapted from values derived in previous studies reported from this laboratory; the atom types used for a typical molecule are shown in Fig. 4, and additional parameters are listed in Table II. Atom-centered charges were calculated for the anthracene-9,10-diones in their intercalated conformations using the MNDO procedure within the AMPAC program. A distance-dependent dielectric constant of the form $\epsilon = 4R_{ij}$ was used in the

TABLE II
ADDITIONAL FORCE-FIELD PARAMETERS FOR ENERGY REFINEMENT OF 2,6-BIS(AMINOALKANAMIDO)ANTHRACENE-9,10-DIONES AND COMPLEXES[a]

Atom types	Force constant	Equilibrium value	
Bonds			
C N	490[b]	1.335 Å	
CA N	481	1.340 Å	
CT NT	367	1.471 Å	
Angles			
C CT NT	63[c]	110.1°	
C N H	35	120.0°	
CA CA N	75	120.0°	
CT CT NT	80	111.2°	
CT NT H	38	118.4°	
CT NT CT	50	110.0°	
	Fourier components	Energy barrier	Fold
Torsions			
XX CA N XX	1	−6.80[d]	2
XX CA NT XX	1	−6.80	2
XX CT NT XX	1	1.40	3
	Force constant	Fold	
Improper torsions (wags)			
XX C N XX	10.5[d]	2	
XX CA N XX	10.5	2	
XX CT NT XX	14.0	3	

[a] Atom types, parameters, and values are as defined by Kollman *et al.* (Refs. 28 and 29).
[b] In kilocalories $mole^{-1}$ $angstrom^{-1}$.
[c] In kilocalories $mole^{-1}$ $radian^{-2}$.
[d] In kilocalories $mole^{-1}$.

refinement, and energy minimization was judged to have reached convergence when the root-mean-square (rms) value of the first derivative was less than 0.15 kcal mol^{-1} $Å^{-1}$.

In addition, the isolated $d(TACGTA)_2$ hexanucleotide duplex (in the lower-energy, canonical, B-DNA double-helical form) and each of the drugs were separately subjected to energy refinement in order to determine the net binding energy involved in forming the DNA–drug complex [see Eq. (4)]. The computed energies for each drug and their simulated drug–DNA intercalation complexes are collected in Table III.

TABLE III
ENERGIES (ENTHALPIES) CALCULATED FOR INTERACTIONS OF 2,6-BIS(AMINOALKANAMIDO)-SUBSTITUTED ANTHRACENE-9,10-DIONES WITH CpG INTERCALATION SITE IN $d(TACGTA)_2$[a]

Compound	Drug–DNA complex energy, E_{COM}	Isolated drug energy, E_D	Interaction energy, ΔE
2	−310.4	−3.2	−35.3
1	−299.8	1.8	−29.8
5	−325.0	−3.4	−49.7
4	−302.4	4.7	−35.2
6	−331.6	−6.2	−53.6
7	−301.5	9.5	−39.1
3	−316.8	−4.4	−40.6

[a] All energies are given in kilocalories $mole^{-1}$. The computed enthalpy for the duplex (E_H) energy refined in the absence of ligand is −271.9 kcal mol^{-1}. Energies calculated using Eq. (4).

Visualization of the different models following energy refinement reveals only slight alterations in the stacking interactions between the chromophores and the base pairs. The observed differences in binding energies calculated for compounds with similar side-chain lengths indicate that classical intercalative binding is not the sole determinant of DNA-binding affinity. Major roles are clearly played by intermolecular contacts of the side chains with the DNA strands (i.e., grooves). The first observation to be made from Table III is that compounds with longer side chains (i.e., $n = 2$ derivatives) form more favorable complexes than their $n = 1$ counterparts with the same functionalization, since the computed interaction enthalpies are more negative. The presence of an additional methylene group serves primarily to relieve undesirable clashes between the backbone and the bulky side-chain groups. The importance of side-chain intermolecular contacts is clearly demonstrated for compounds **2** and **5** (Fig. 5, i.e., where $n = 1$ or 2 and $-N^{+}R_2H$ is piperidinium) if models are constructed without considerations of close van der Waals contacts. Subsequent refinement in each case affords structures that are less exothermic by 7–9 kcal mol^{-1}.

The refined structures of the complexes are in accord (e.g., see Figs. 6 and 7 for compounds **2** and **5**) with data obtained in solution binding studies that indicate that the ligands occlude three or four base pairs at the binding site. Further, the ranking order of binding energies calculated for the anthracene-9,10-diones are in qualitative agreement with both the determined equilibrium binding constants and the observed cytotoxic behaviors *in vitro*. That is, compounds with $n = 1$ are less DNA affinic and

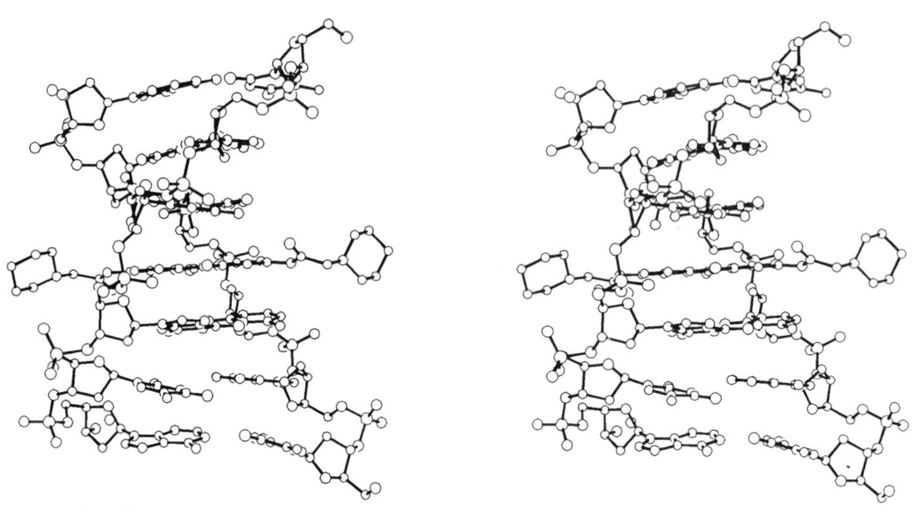

FIG. 6. Stereo view of a simulated intercalation complex between compound **2** ($n = 1$; see text) and the sequence d(TACGTA)$_2$. Hydrogen atoms have been removed for clarity.

generally less cytotoxic than their $n = 2$ equivalents. Further, the presence of hydroxyl groups in the side chains is observed to markedly improve the binding energy of the complex, by ~4 kcal mol^{-1}, largely as a result of secondary hydrogen bonding involving the phosphate backbone. It is evident from Table III that bulky terminal $-NR_2$ amine groups also lead to favorable DNA binding (e.g., $n = 1$: **1** < **2** < **3**; $n = 2$: **4** < **5** < **6**), largely due to the greater number of close van der Waals contacts involving the hydrophobic walls of the DNA strands.

For compound **6**, which has the greatest computed binding enthalpy of this series, there is considerable π overlap of the chromophore with G3 of strand 2 (the Crick or 3′–5′ strand), with none from its complementary C3 on strand 1 (the Watson or 5′–3′ strand). Partial stacking of the chromophore with the six-membered ring of G4 on strand 1 and its complementary C4 is also observed. The chromophore is thus positioned parallel to the long axis of the base pairs. The side chain in the major groove (1) closely follows the helical twist of strand 2 toward the 5′ end, (2) extends to T6 spanning 3 bp, and (3) presents a possible hydrogen bond from the terminal hydroxyl group in the 4-(2-hydroxyethyl) substituent to the A5pT6 phosphate oxygen, with an O—H $\cdots$ O—P separation of 2.66 Å. The side chain in the minor groove (1) follows the helical twist of strand 1 toward the 5′ end, (b) extends to A6, spanning 3 bp, (3) presents a large number of favorable van der Waals contacts involving the sugar hydrogens, at distances of 2.2–2.4 Å, (4) affords the possibility of a weak

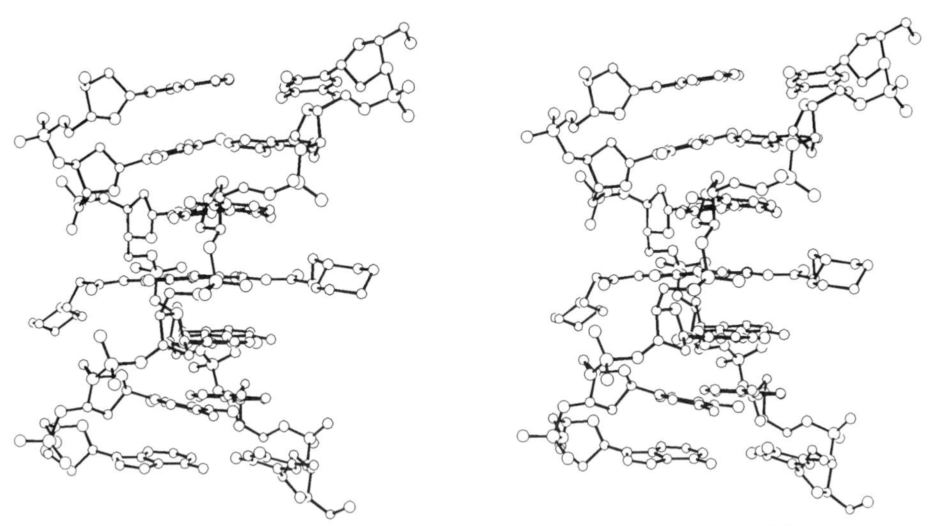

FIG. 7. Stereo view of a simulated intercalation complex between compound **5** ($n = 2$; see text) and the sequence d(TACGTA)$_2$. Hydrogen atoms have been removed for clarity.

hydrogen bond betweem the amide hydrogen and the oxygen of G4 sugar, with an N—H···O-1′ separation of 2.47 Å, and (5) gives little or no interaction with strand 2.

Conclusions

The undoubted success of a number of intercalating drugs in the treatment of at least some human cancers is due at least as much to their particular pharmacokinetic properties as to their DNA-binding ones. The former factors are important determinants in conferring cell and tissue selectivity, since the agents themselves are not specific to genomic DNA sequences of tumor cells. This chapter has not discussed in detail the aspects of ternary complex formation with DNA topoisomerase II, which is probably the major determinant of selectivity at the cellular level.[7] As yet, lack of direct structural data on the enzyme precludes detailed study of the molecular factors involved. However, indirect molecular modeling approaches could play an important role in this in the future, thereby rationally optimizing the role played by this enzyme. Comparative examination of models for drug–DNA complexes with different degrees of ability to form the ternary complex should enable structural and electronic features that may be important for ternary complex recognition to be highlighted and ultimately exploited.

It has been shown[55] that intercalating agents can be very usefully exploited to enhance the value of antisense and antigene oligonucleotides. These have the capability of selectivity at the genome level and thus hold out considerable promise for the future as truly specific anticancer agents active against defined loci such as oncogene sequences. It has been shown that attachment of an acridine moiety to the 3′ end of an oligonucleotide confers increased resistance to nuclease attack, improved cellular uptake, and, crucially, much enhanced DNA-binding properties for the oligonucleotide. Rational analysis and exploitation of these findings will undoubtedly be a fruitful area for future molecular modeling studies of drug–DNA intercalation.

[55] C. Hélène and J.-J. Toulmé, *Biochim. Biophys. Acta* **1049,** 99 (1990).

[23] Molecular Modeling in Mutagenesis and Carcinogenesis

By EDWARD L. LOECHLER

For chemicals and radiation to cause cancer, a large body of literature suggests that (1) they should effectively react with DNA to generate DNA adducts and (2) these adducts (or their breakdown products, such as apurinic/apyrimidinic sites) must be efficient premutagenic lesions. Thus, the study of the principles of adduct formation and the principles of adduct-induced mutagenesis is fundamental to the study of the carcinogenic process initiated by mutagens/carcinogens. Much has been learned about the adducts formed in DNA by mutagens/carcinogens[1] and the mutational specificity of these mutagens/carcinogens,[2–8] but little is known about the relationship between these two; i.e., which adducts induce which

[1] B. Singer and D. Grunberger, "Molecular Biology of Carcinogenesis and Mutagenesis." Plenum, New York, 1983.

[2] J. H. Miller, *in* "The Operon" (J. H. Miller and W. S. Reznikoff, eds.), p. 31. Cold Spring Harbor Laboratory, Cold Spring Harbor, New York, 1980.

[3] P. L. Foster, E. Eisenstadt, and J. H. Miller, *Proc. Natl. Acad. Sci. U.S.A.* **80,** 2695 (1983).

[4] M. Bichara and R. P. P. Fuchs, *J. Mol. Biol.* **183,** 341 (1985).

[5] E. Eisenstadt, A. J. Warren, J. Porter, D. Atkins, and J. H. Miller, *Proc. Natl. Acad. Sci. U.S.A.* **79,** 1945 (1982).

[6] T. A. Kunkle, *Proc. Natl. Acad. Sci. U.S.A.* **81,** 1494 (1984).

[7] C. Coulondre, J. H. Miller, P. J. Farbaugh, and W. Gilbert, *Nature (London)* **274,** 775 (1978).

[8] T. A. Kunkle, *J. Biol. Chem.* **260,** 12866 (1985).

mutations and why.[9,10] In most cases even less is known about the conformations that adducts adopt in relation to biological end points, such as mutation.[9-12] For example, more is known about O^6-methylguanine (O^6-MeGua) than any other premutagenic lesion,[13-16] and yet the structural basis for its induction of Gua to adenine (Ade) mutations is controversial.[9] An adequate O^6-MeGua:T base pair can be drawn, but various experiments suggest that the actual DNA structure may be sufficiently disrupted by the presense of the methyl moiety in O^6-MeGua that this simple base pair may not be the structural basis for mutation.[17]

The study of the relationship between adducts and mutation and other biological end points is being conducted in many laboratories.[9-11,18,19] Although there are drawbacks in studying structures using molecular modeling techniques,[20] this technique can prove very useful in this effort.[21,22] In this chapter the term *molecular modeling* includes both computer graphics techniques and molecular mechanical calculations. Indeed, in certain cases no other technique can realistically be used; e.g., in the evaluation of transient intermediates. In other cases molecular modeling is probably the best technique currently available; e.g., in probing for subtle changes in structure that may not be possible to evaluate by physical techniques, such as nuclear magnetic resonance (NMR). Molecular modeling can also be used to aid in the refinement of structures derived by other techniques, such as NMR. Last, molecular modeling represents a relatively rapid means of obtaining structurally reasonable information that may suggest experiments to be conducted.

[9] A. K. Basu and J. M. Essigmann, *Chem. Res. Toxicol.* **1,** 1 (1988).

[10] A. K. Basu and J. M. Essigman, *Mutat. Res.* **233,** 189 (1990).

[11] E. L. Loechler, M. Benasutti, A. K. Basu, C. L. Green, and J. M. Essigmann, *in* "Progress in Clinical and Biological Research, Volume 340A, Mutations and the Environment, Part A: Basic Mechanisms" (M. L. Mendelsohn and R. J. Albertini, eds.), p. 51. Wiley-Liss, New York, (1990).

[12] E. L. Loechler, *Biopolymers* **28,** 909 (1989).

[13] E. L. Loechler, C. L. Green, K. W. and J. M. Essigmann, *Proc. Natl. Acad. Sci. U.S.A.* **81,** 6271 (1984).

[14] R. W. Chambers, E. Sledziewska-Gojska, S. Hirani-Hojatti, and H. Borowy-Borowski, *Proc. Natl. Acad. Sci. U.S.A.* **82,** 7173 (1985).

[15] O. S. Bhanot and A. Ray, *Proc. Natl. Acad. Sci. U.S.A.* **83,** 7348 (1986).

[16] M. Hill-Perkins, M. D. Jones, and P. Karran, *Mutat. Res.* **162,** 153 (1986).

[17] D. J. Patel, L. Shapiro, S. A. Kozlowski, B. L. Gaffney, and R. A. Jones, *Biochemistry* **25,** 1036 (1986).

[18] M. Benasutti, Z. D. Ezzedine, and E. L. Loechler, *Chem. Res. Toxicol.* **1,** 160 (1988).

[19] J. O. Ojwang, D. Grueneberg, and E. L. Loechler, *Cancer Res.* **49,** 6529 (1989).

[20] E. L. Loechler, M. M. Teeter, and M. D. Whitlow, *J. Biomol. Struct. Dyn.* **6,** 1237 (1988).

[21] P. A. Kollman, *Acc. Chem. Res.* **18,** 105 (1985).

[22] S. J. Weiner, P. A. Kollman, D. T. Nguyen, and D. A. Chase, *J. Comput. Chem.* **7,** 230 (1986).

Molecular Mechanical Calculations

Several different methods of performing molecular modeling studies have been developed. Frequently, this involves the generation of an initial, unrefined structure, which uses idealized coordinates for B-DNA (or A-DNA, Z-DNA, etc.) into which the chemical adduct of interest is introduced. This structure is then refined by calculations that effectively move the positions of its atoms, such that favorable interactions (e.g., Coulombic attractions) are maximized, while unfavorable interactions (e.g., van der Waals contacts) are minimized.[21-24] Favorable and unfavorable interactions are evaluated assuming a classical mechanical description of the interactions of atoms within the structure.

Several programs have been developed to conduct this refinement. AMBER[21-23] and CHARMM,[24] which are fundamentally similar, were developed under the leadership of Drs. Peter Kollman (UC San Francisco) and Martin Karplus (Harvard Univ.), respectively. These programs consider atoms in Cartesian space. A second approach, exemplified by the program, DUPLEX, developed by Drs. Brian Hingerty (Oak Ridge Natl. Lab.) and Suse Broyde (New York Univ.), fixes bond lengths and bond angles, and searches for low-energy structures in torsional space by allowing dihedral (tortion) angles to vary.[25-29] Searches in Cartesian space have the advantage that bond lengths and angles can be perturbed in the location of a minimum-energy structure, while the disadvantage is that the energy-minimized structure that emerges is frequently similar to the initial structure. The latter can be overcome to some extent either by systematically changing the input structure in order to locate the lowest energy structure or by employing techniques that allow structures to escape local minima, such as molecular dynamics or Boltzmann jump techniques. Searches in torsional space are more likely to find lower energy structures that are more distantly related to the initial structure, but suffer from not permitting perturbations in bond lengths and bond angles.

A detailed description of how one does molecular modeling is not the focus of this chapter; rather, what is addressed here is how molecular

[23] P. Kollman, *Annu. Rev. Phys. Chem.* **38,** 303 (1987).

[24] C. L. Brooks III, M. Karplus, and B. M. Pettitt, *Adv. Chem. Phys.* **71,** 1 (1988).

[25] B. E. Hingerty and S. Broyde, *Biochemistry* **21,** 3243 (1982).

[26] B. E. Hingerty and S. Broyde, *J. Biomol. Struct. Dyn.* **4,** 365 (1986).

[27] R. Shapiro, G. R. Underwood, H. Zawadzka, S. Broyde, and B. E. Hingerty, *Biochemistry* **25,** 2198 (1986).

[28] D. Norman, P. Abuaf, B. E. Hingerty, D. Live, D. Grunberger, S. Broyde, and D. Patel, *Biochemistry* **28,** 7462 (1989).

[29] R. Shapiro, B. E. Hingerty, and S. Broyde, *J. Biomol. Struct. Dyn.* **7,** 493 (1989).

modeling can be best harnessed to answer questions of relevance to the carcinogenesis process.

Carcinogenesis Paradigm

The question that many of us are trying to answer is, what makes a chemical a carcinogen? To evaluate this, Fig. 1 shows the carcinogenesis paradigm using aflatoxin B_1 (AFB_1) as an example. Arrows pointing horizontally lead to mutation, while arrows pointing vertically lead to detoxification. At each junction, factors that favor partitioning horizontally vs vertically will improve the mutagenic potency of a mutagen/carcinogen. Mutagens/carcinogens are frequently metabolized; in some cases to less potent derivatives (e.g., AFP_1, which is demethyl-AFB_1), but importantly

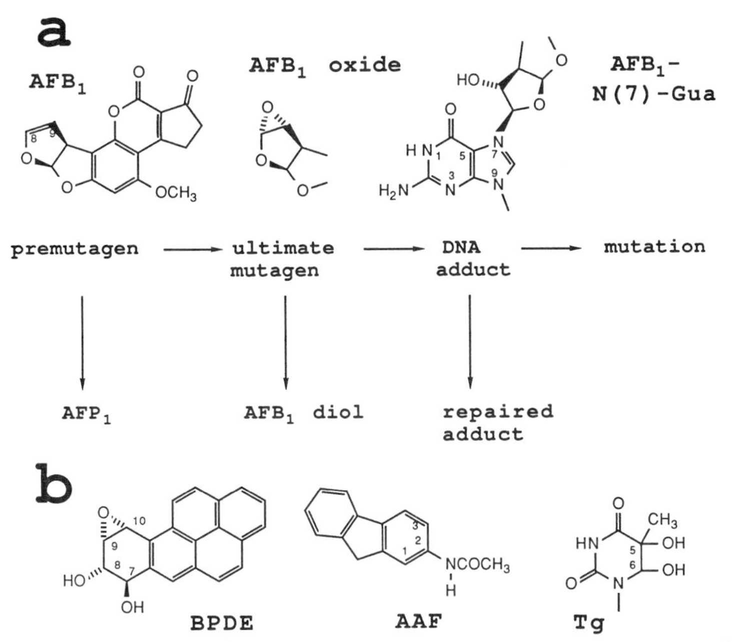

FIG. 1. Mutagenesis paradigm (a), using AFB_1 as an example, and mutagen/carcinogen structures (b). (a) AFB_1 can partition toward either mutagenesis (horizontally) or detoxification (vertically). Cellular enzymes can activate AFB_1 either to its epoxide (AFB_1 oxide) or to AFP_1 (demethyl-AFB_1); the latter is considerably less mutagenic and carcinogenic. AFB_1 oxide reacts readily with nucleophiles, such as water, to give the less toxic AFB_1 diol, and DNA to give the major DNA adduct [AFB_1-N(7)-Gua]. This species could be accurately repaired, but if not, could be encountered by a polymerase and a mutation induced. (b) Structures of the (+)-*anti*-7,8-diol-9,10-epoxide of benzo[*a*]pyrene (BPDE), which is one of four enantiomers; 2-acetylaminofluorene (AAF), which can be activated at the exocyclic amino group for reaction with DNA by several means; and thymine glycol (Tg).

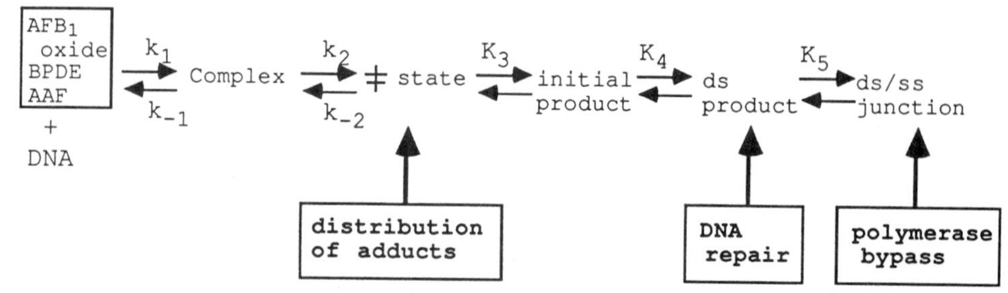

FIG. 2. The interrelationship between molecular modeling and the mutagenesis paradigm. Mutagens/carcinogens, such as AFB_1 oxide, BPDE, or an activated derivative of AAF, form noncovalent complexes with DNA (K_1), from which reaction can occur via a transition state (k_2). Following reaction (K_3), the adduct adopts an initial conformation, which may not be lowest in energy, whereupon rearrangement may occur (K_4) to reach the lowest energy structure in double-stranded (ds) DNA. Finally, when the adduct is encountered by a DNA polymerase during replication, it is located at a double strand/single strand (ds/ss) junction.

for mutagenesis also to more potent derivatives (i.e., AFB_1 oxide). These activated derivatives, or ultimate mutagens, react with nucleophiles, such as water (detoxification) or DNA to form DNA adducts [e.g., AFB_1-N(7)-Gua], which are generally thought to be misreplicated during DNA synthesis, such that mutations result. Of course, accurate DNA repair of a DNA adduct leads to no mutation. Finally, a mutation can potentially activate a protooncogene to an oncogene.[30]

This chapter focuses on relating this paradigm to the structures of mutagens/carcinogens and their interaction with DNA (Fig. 2), because it is within this framework that molecular modeling can best be utilized to contribute to our understanding of the process of carcinogenesis. Importantly, it is imperative to be performing molecular modeling studies on structures that are relevant to the experimental data to be rationalized.

Prelude to Adduction: Noncovalent Carcinogen–DNA Interactions

Frequently activated mutagens/carcinogens are thought to form complexes with DNA prior to adduction ($K_1 = k_1/k_{-1}$; Fig. 2) because they are hydrophobic, which is the case for the aflatoxins, such as AFB_1 oxide; for the polycyclic aromatic hydrocarbons, such as benzo[*a*]pyrene, which reacts with DNA following activation to its corresponding diol epoxide (BPDE; Fig. 1b); and the aromatic amines, such as 2-acetylaminofluorene (AAF; Fig. 1b), which react with DNA following activation in several ways.

[30] A. Balmain and K. Brown, *Adv. Cancer Res.* **51,** 147 (1988).

Proper orientation within this complex may help remove the entropic barrier to reaction by improving the probability of reaching the transition state[31] in the reaction with DNA, especially in comparison to the corresponding reaction with other cellular nucleophiles, such as water and glutathione, where complexation does not occur.

Noncovalent complex formation may not guarantee efficient reaction with DNA; e.g., favorable noncovalent complexes may form from which no covalent reaction can occur because the mutagen/carcinogen is not properly oriented for reaction with a nucleophilic atom in DNA. This point may be of relevance in the reaction of AFB_1 oxide with DNA. Studies have suggested that the noncovalent interaction of AFB_1 with DNA involves intercalation,[32,33] and that AFB_1-N(7)-Gua has an intercalated AFB_1 moiety.[34,35] This appears reasonable based on molecular modeling studies, where reasonable intercalated species have been considered.[20,36] It is of interest to note that AFB_1 appears to have a slightly higher affinity (< twofold) for the formation of noncovalent (i.e., presumably intercalated) complexes with poly(dA-dT) than with poly(dG-dC), in spite of the fact that numerous studies have failed to demonstrate AFB_1 oxide reaction at the corresponding N(7) position of Ade.[35,37–39] This does not imply that the intercalated species is irrelevant to covalent adduction at N(7)-Gua, but rather that formation of the intercalated complex is not sufficient, although it may still be necessary, for covalent reaction. For example, if $k_{-1} > k_2$ and AFB_1 oxide concentration is below the association constant ($K_1 \sim 10^{-3}$ M),[34] which is likely, and if $k_2^{\text{poly(dA-dT)}}/k_2^{\text{poly(dG-dC)}}$ is small, then much less adduction would occur at N(7)-Ade. The nucleophilicity of N(7)-Ade is $\sim$10 to 50 times lower than N(7)-Gua, based on the reactions of simple alkylating agents.[1] Taking $k_2^{\text{poly(dA-dT)}}/k_2^{\text{poly(dG-dC)}} \simeq 30$, and $K_1^{\text{poly(dA-dT)}}/K_1^{\text{poly(dG-dC)}} \simeq 2$, then $\sim$15 times fewer N(7)-Ade than N(7)-Gua adducts would be expected. N(7)-Ade adducts must be formed many orders of magnitude less efficiently.[35,37,38] This may suggest that the complex (pre-

[31] W. P. Jencks, *Adv. Enzym. Mol. Biol.* **43,** 219 (1975).

[32] S. Gopalakrishnan, S. Byrd, M. P. Stone, and T. M. Harris, *Biochemistry* **28,** 726 (1989).

[33] M. P. Stone, S. Gopalakrishnan, T. M. Harris, and D. E. Graves, *J. Biomol. Struct. Dyn.* **5,** 1025 (1988).

[34] K. D. Raney, S. Gopalakrishnan, S. Byrd, M. P. Stone, and T. M. Harris, *Chem. Res. Toxicol.* **3,** 254 (1990).

[35] M. P. Stone and T. M. Harris, personal communication (1990).

[36] M. Bonnett and E. R. Taylor, *J. Biomol. Struct. Dyn.* **7,** 127 (1989).

[37] J. M. Essigmann, R. G. Croy, A. M. Nadzan, W. F. Busby, Jr., V. N. Reinhold, G. N. Buchi, and G. N. Wogan, *Proc. Natl. Acad. Sci. U.S.A.* **74,** 1870 (1977).

[38] J. M. Essigmann, C. L. Green, R. G. Croy, K. W. Fowler, G. W. Buchi, and G. N. Wogan, *Cold Spring Harbor Symp. Quant. Biol.* **47,** 327 (1983).

[39] F. L. Yu, W. Bender, and I. Geronimo, *Carcinogenesis* **11,** 475 (1990).

sumably intercalated) between aflatoxin B_1 and poly(dA-dT), while strong, does not permit reaction at the N(7)-Ade.

Molecular modeling studies may provide a rationale for this observation. Figure 3 shows a portion of an intercalated adduct between AFB_1 and both N(7)-Gua and N(7)-Ade. A steric problem for reaction is apparent in both cases, but is more severe in the latter case. If the adduct bond between AFB_1 oxide and Gua or Ade is formed to give a reasonable intercalated structure assuming reasonable bond angles, then H^9 on the AFB_1 moiety is

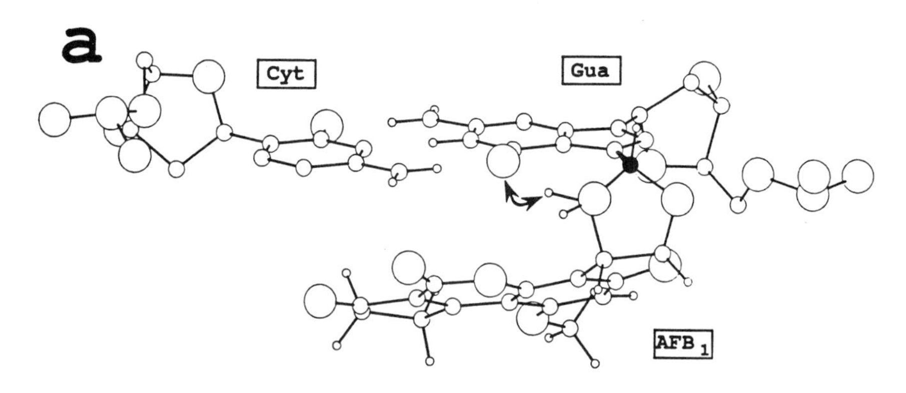

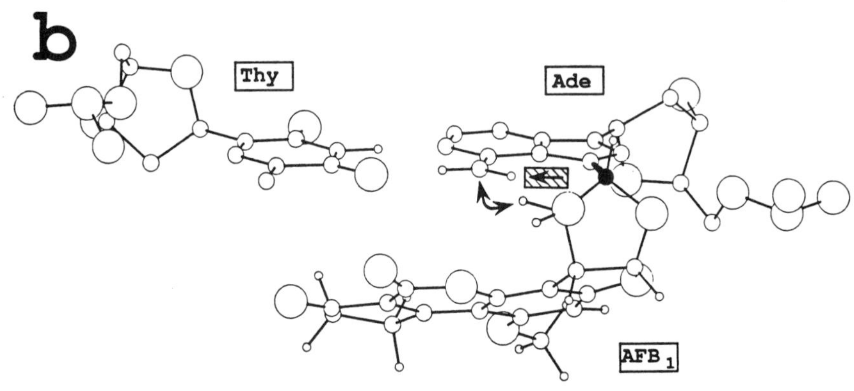

FIG. 3. Hypothetical structures of AFB_1-N(7)-Gua (a) and AFB_1-N(7)-Ade (b), when the AFB_1 moiety is intercalated. The AFB_1 moiety was minimized as described previously,[33] and then docked at the N(7) position of either Gua or Ade. The bonds (bold) between N(7)-Gua-C(8)AFB_1 and N(7)Ade-C(8)AFB_1 were set at a length of 1.48 Å;[33] the C(8)-AFB_1 atoms are shown as solid spheres. The positioning of the AFB_1 moiety was dictated by the assignment of appropriate sp^3 or sp^2 geometries to bond angles and keeping the plane of the base pair and the aromatic portion of the AFB_1 moiety as planar as possible given the angle constraints. The double-headed arrow indicates the close van der Waals contact between H^9-AFB_1 and O^6-Gua (a, 1.43 Å), and between H^9-AFB_1 and N^6Ade (b, 1.40 Å). The cross-hatched arrow indicates the hydrogen at N^6Ade, which is 0.69 Å from H^9-AFB_1.

~1.41 Å (double-headed arrows) from the O^6 position of Gua and the N^6 position of Ade (Fig. 3); this distance is less than what it must be based on van der Waals radii (~2.6 Å) in the actual adduct structure. In fact, this issue is more acute in the case of structures involving Ade, because the hydrogen attached to N^6-Ade is a mere ~0.7 Å (cross-hatched arrow) from H^9 of the AFB_1 moiety. Thus, this potential steric problem, which must be resolved during adduction, is much more severe in the case of reaction of AFB_1 oxide with the N(7) position of Ade than with the N(7) position of Gua. It is worth noting that this rationale for the lack of reaction between AFB_1 oxide and N(7)-Ade was originally noted by merely inspecting a reasonable intercalated complex between AFB_1 and Gua, and did not require any computational chemistry.

Adduction and Transition State Structures

When an experimental result involves the distribution of adducts in DNA, then the structure of the transition states vs starting material must be considered. Thus, if one is trying to understand why an adduct is formed more frequently at N(7)-Gua with AFB_1 oxide in a 5′-G<u>G</u>G-3′ than in a 5′-T<u>G</u>A-3′ sequence,[40] then the respective values for K_1 and K_2 (Fig. 2) must be considered. Relative structures of transition states must also be considered for relative reactivity at, e.g., O^6-Gua vs N(7)-Gua, although chemical effects (e.g., bond energy considerations) must also be considered in this case. As discussed above, the distribution of adducts may not be a simple function of the formation of noncovalent adducts, and it is for this reason that the modeling of noncovalent complexes may only be relevant when it is coupled to the modeling of the corresponding transition state. Of course, the description of a transition state is not a simple matter, which has been discussed with respect to the reaction of AFB_1 oxide with N(7)-Gua.[20] The author's preference has been to use the initial adduct between the mutagen/carcinogen and DNA as a model for the transition state structure, both because it is likely to be close to the actual structure of the transition state and because it certainly must be formed whatever the transition state is. Clearly, more sophisticated approaches to the structures of transition states must be pursued in the future.

Adduct Structures: To Rearrange or Not?

Some bulky mutagens/carcinogens, such as AAF, have the potential to induce dramatic distortional changes in DNA structure following adduc-

[40] M. Benasutti, S. Ejadi, M. D. Whitlow, and E. L. Loechler, *Biochemistry* 27, 472 (1988).

TABLE I
CLASSIFICATION OF NUCLEOPHILIC SITES IN DOUBLE-STRANDED DNA

Class	π orbitals	Nonbonding orbitals	Attached protons	β protons	Nucleophilic sites[a]
I	1	0	1 or 2	*b*	N²G, N⁶A, N⁴C, C(8)G, C(8)A
II	1	1 or 2	0	0	N(7)G, N(7)A, N(3)G, N(3)A, O²C
III	1	2	0	1	O⁶G, O²T, O⁴T

[a] Numbering is standard.[1]
[b] Because protons are attached to these nucleophilic sites, the presence or absence of β protons is irrelevant.

tion (see below).[1,41,42] Thus, an apparent contradiction emerges: if adduct formation is preferred at a particular atom [e.g., at C(8)-Gua with AAF], then why does adduction result in such a dramatic change in DNA structure, which must require a significant input of energy? This (and other apparent) contradiction(s) can be rationalized by considering that nucleophilic atoms in DNA can be classified into three categories (Table I) according to whether they have nonbonding electrons in addition to π electrons, and whether protons are located at the nucleophilic atom, β to that atom, or neither.

Following the formation of the adduct bond between the mutagen/carcinogen and DNA, the adduct adopts an initial conformation (i.e., following K_3 in Fig. 2). This structure may or may not be the lowest energy structure, and if it is not, then adduct rearrangement may occur (i.e., K_4 in Fig. 2) to give the structure observed in double-stranded (ds) DNA. This is thought to occur with BPDE, which forms an adduct at N²-Gua. Evidence suggests that the initial complex between BPDE and DNA is intercalated,[43-46] while the product has the BPDE moiety bound externally, probably in the minor groove.[47-52] What could possibly be the rationale for this seemingly unlikely scenario? Molecular modeling offers some insights.

[41] R. Fuchs and M. Daune, *Biochemistry* **11,** 2659 (1972).
[42] I. B. Weinstein and D. Grunberger, *in* "World Symposium on Model Studies in Chemical Carcinogenesis (P. O. P. Ts'o and J. DiPaolo, eds.), p. 217. Dekker, New York, 1974.
[43] N. E. Geacintov, H. Yoshida, V. Ibanez, and R. G. Harvey, *Biochem. Biophys. Res. Commun.* **100,** 1569 (1981).
[44] M. C. MacLeod and J. K. Selkirk, *Carcinogenesis* **3,** 287 (1982).
[45] T. H. Meehan and K. Straub, *Nature (London)* **277,** 410 (1977).
[46] N. E. Geacintov *Carcinogenesis* **7,** 759 (1986).
[47] H. Yoshida, C. E. Swenberg, and N. E. Geacintov, *Biochemistry* **26,** 1351 (1987).
[48] N. E. Geacintov, H. Yoshida, V. Ibanez, S. A. Jacobs, and R. G. Harvey, *Biochem. Biophys. Res. Commun.* **122,** 33 (1984).

The only free electron pair at N^2-Gua is perpendicular to the plane of the G:C base pair and conjugated with the π electron system of the purine. Molecular modeling studies show that this electron pair is not normally accessible in B-DNA because it is sandwiched between the base pairs that are found on either side (Fig. 4a). Thus, for reaction to occur at N^2-Gua, a deformation of B-DNA structure must occur. (The alternative is ionization of N^2-Gua followed by reaction of the anion. This is not favored because of the high pK_a of the N^2 position and the low β_{nuc} that is expected for the adduction reaction.) Intercalation is the most obvious perturbation that would permit the reactive C(10) position of BPDE to gain access to this electron pair (K_{in}; Fig. 4b). For BPDE there is evidence that noncovalent intercalation precedes covalent reaction (K_{ad}; Fig. 4b), and this certainly is a logical step toward adduct formation.[43-52] The following observation suggests that face-to-face stacking of aromatic systems may be important for improving reaction at N^2-Gua. Most aliphatic alkylating agents do not react appreciably at the N^2 position of guanosine,[1] but several benzylating agents do.[53,54] The benzyl functionality is capable of face stacking with the guanine moiety, which may be essential for efficient reaction at N^2-Gua. The authors of Refs. 53 and 54 have offered an alternative chemical rationale for their observations.

Following adduction, a new set of forces are operative in the case of N^2-Gua adducts. Initially, the N^2 position of the adduct is expected to have sp^3 geometry (III; Fig. 4b) and be positively charged; however, the proton at N^2-Gua will ionize because of its low pK_a (K_{io}; Fig. 4b). This ionization will leave an electron pair protruding into the minor groove at an angle that is not optimal for conjugation with the π electrons of the guanine moiety (Fig. 4b and c). Thus, there will be energetic pressure to reconjugate this electron pair with the π electrons of the purine. In addition, while the pyrene moiety of BPDE can lie approximately parallel to the plane of the base pair in the noncovalent complex, in the adduct it cannot unless bond angles are severely distorted, which was first noted by Miller.[55,56] (This is depicted in III and IV in Fig. 4b.) Thus, a wedge-shaped adduct structure is

[49] N. E. Geacintov, H. Hibshoosh, V. Ibanez, M. J. Benjamin, and R. G. Harvey, *Biophys. Chem.* **20,** 121 (1984).

[50] N. E. Geacintov, I. B. Ivanovich, and R. G. Harvey, *Biochemistry* **17,** 5256 (1978).

[51] T. Prusik, N. E. Geacintov, C. Tobiasz, I. B. Ivanovich, and I. B. Weinstein, *Photochem. Photobiol.* **29,** 223 (1979).

[52] F.-M. Chen, *J. Biomol. Struct. Dyn.* **4,** 401 (1986).

[53] R. C. Moschel, A. Hudgeins, and A. J. Dipple, *J. Org. Chem.* **44,** 3324 (1979).

[54] R. C. Moschel, A. Hudgeins, and A. J. Dipple, *J. Org. Chem.* **45,** 533 (1980).

[55] M. E. Hogan, N. Dattagupta, and J. P. Whitlock, Jr., *J. Biol. Chem.* **256,** 4504 (1981).

[56] E. R. Taylor, K. J. Miller, and A. J. Bleyer, *J. Biomol. Struct. Dyn.* **1,** 883 (1983).

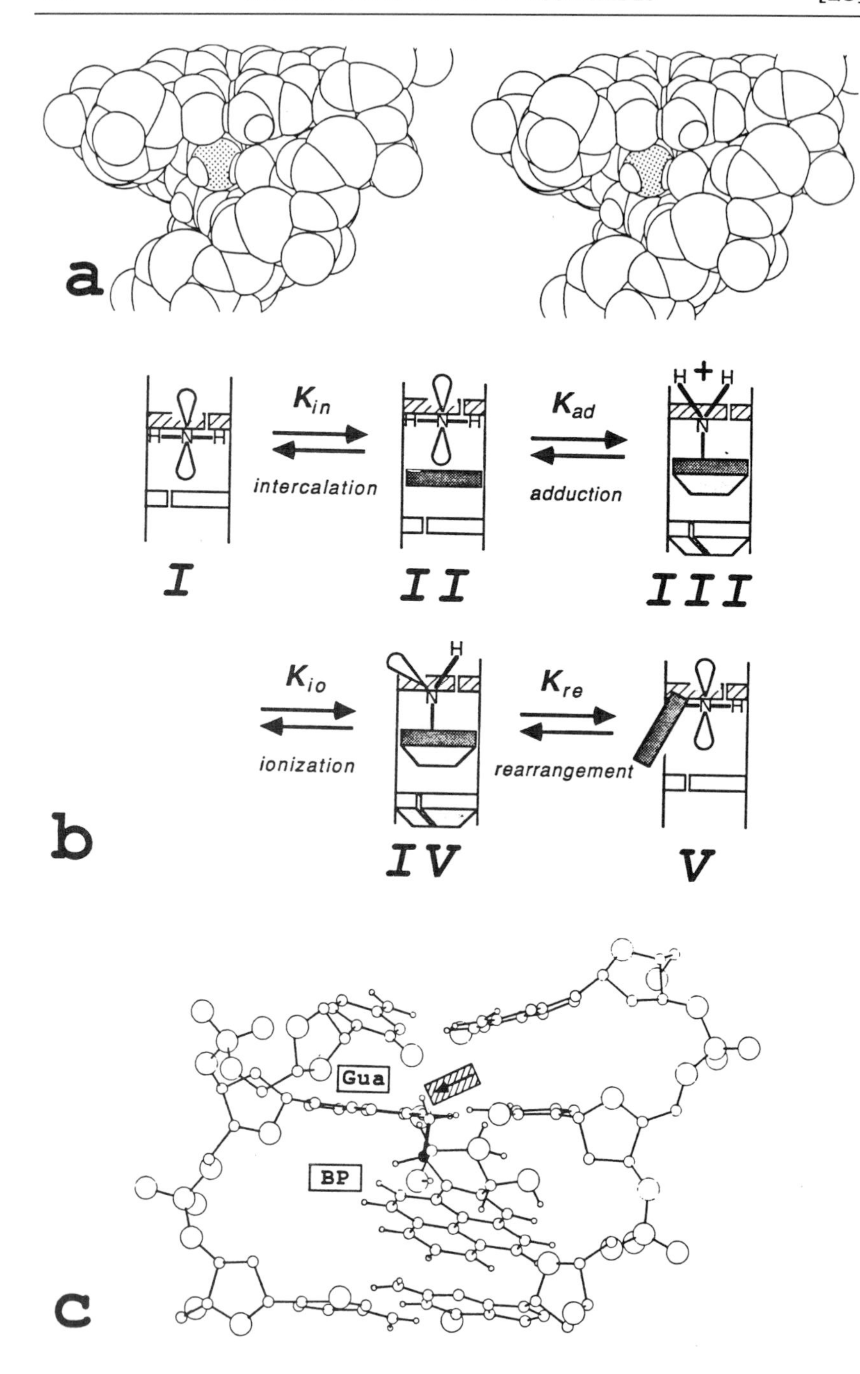

a
K_{in}
intercalation
K_{ad}
adduction
I
II
III
K_{io}
ionization
K_{re}
rearrangement
IV
V
b
Gua
BP
c

likely (Fig. 4c), and the face-stacking energy must be correspondingly diminished. If the wedge-shaped, intercalated adduct is not sufficiently stable to compensate for the energy that could be gained by reconjugating the N^2 electron pair into the π electron system of the Gua moiety of the adduct, then rotation about the C(2)–N^2 bond of the Gua moiety would be expected (K_{re}; Fig. 4b). Such a rotation would reposition the pyrene moiety into either the major or the minor groove. Although the final conformation of BP-N^2-Gua in dsDNA is controversial, there is mounting evidence to indicate that it is externally bound in one of the grooves.[47–52]

Based on this discussion, the proposition that BPDE intercalates prior to adduction at the N^2 position of guanine, and then subsequently rearranges, is chemically reasonable. All class I nucleophilic atoms in DNA (Table I) should react with electrophilic mutagens/carcinogens similarly, because the only electron pair associated with class I atoms is conjugated with a π system and is buried by surrounding base pairs. Thus, all of these sites are expected to undergo this same series of events, namely, intercalation, adduction, ionization, and, finally, rearrangement. In part, this may explain why AAF adducts form at C(8)-Gua even though they induce such dramatic DNA rearrangements.[1,41,42]

Class II sites have both π and nonbonding orbitals with which (in principle) a mutagen/carcinogen could react, and have no proton either attached or β to (i.e., in vinylogous conjugation with) the nucleophilic atom. The N(7) position of guanine is a class II site and is important in

FIG. 4. Potential steps in the reaction of BPDE with the N^2 position of guanine: intercalation, adduction, ionization, and rearrangement. (a) A view (stereo pair) of a portion of double-stranded B-DNA with an N^2-Gua atom shaded. The only free electron pair at N^2-Gua can be thought of as having p orbitals lying toward the top and bottom the shaded sphere. This orbital is sterically inaccessible. (b) A scheme of the putative cycle of reaction between BPDE and guanine. The N^2 position of Gua with its free electron pair is indicated (H-N-H) in a G:C base pair (cross-hatched). The BPDE moiety (shaded) intercalates (K_{in}), which permits the C(10) position of BPDE to gain access to the N^2 position of BPDE and covalent reaction to occur (K_{ad}). The initial product of this reaction (III) is expected to be protonated at N^2-Gua and have the BPDE moiety of the adduct no longer perfectly planar to the Gua moiety of the adduct due to angle constraints. [This structure can be approximated by molecular modeling techniques[12] and is shown in (c). The G:C base pair is shown approximately horizontally with the adduct bond between N^2-Gua and C(10)-BP (solid sphere) shown more boldly. Bond angle constraints prevent the BP moiety of the adduct from lying perfectly planar to the G:C base pair. The proton that will ionize is indicted by the cross-hatched arrow; following ionization the electron pair associated with this proton is not properly oriented for reconjugation with the π electrons in the Gua moiety of the adduct.] One of the protons attached at N^2Gua in the adduct is expected to ionize (K_{io}). If face-stacking energy between the BPDE and Gua moieties has decreased sufficiently (c), and the energy gained from reconjugation of the electron pair at N^2-Gua is sufficient, then rotation (K_{re}) about the adduct bond is expected such that the BPDE moiety emerges into the major or minor groove.

adduct formation with the bulky carcinogens, BPDE and AFB_1,[1] and if reaction occurred from an intercalated species utilizing π electrons, then the same principles as described for class I nucleophilic sites would be operative, except that class II sites have neither a proton attached to the site of adduction nor β to the site of adduction. Thus, the adduct cannot ionize and must remain positively charged. In spite of this, these sites would have considerable energetic pressure to reorganize following adduction. In contrast, following reaction with the nonbonding orbitals of class II sites [e.g., N(7)Gua], significant rearrangements following adduction would not be expected.

Class III nucleophilic sites are similar to class II sites except that they have a conjugated β proton which, following ionization, allows the adduct to shift from the amino (i.e., amide) to the imino (i.e., imidate ester) tautomer. The adducts derived from class III sites are, thus, uncharged. An example of a class III site is O^6-Gua.

Adducts in Double-Stranded DNA

Following rearrangements (K_3; Fig. 2) the adduct structure in dsDNA results, which may be relevant to an adduct undergoing DNA repair. Frequently, physical studies are done on adducts in dsDNA as well.

Adduct-Induced Mutations: At Replication Fork

In most cases it is likely that mutation occurs during replication, which means that a replication fork is involved, and the adduct is located at a junction between double-stranded and single-stranded (ss) DNA.[12] DNA polymerase will also be present and may have an important influence upon structure.

A study on the mutations induced by thymine glycol (Tg; Fig. 1) and a corresponding molecular modeling study are instructive in trying to relate molecular modeling to a mutagenic outcome.[57] Doctor Ashis Basu and Dr. John Essigmann (M.I.T.) have succeeded in determining the mutations induced by Tg. Thymine glycol is produced in DNA when reactive oxygen species encounter DNA; one pertinent example is the hydroxyl radical, which is a by-product of normal oxidative metabolism and is produced by ionizing radiation. The fact that thymine glycol is the major lesion formed in DNA exposed to ionizing radiation, and that ~40% of mutations induced by ionizing radiation are at A:T base pairs, led to the study of the

[57] A. K. Basu, E. L. Loechler, S. A. Leadon, and J. M. Essigmann, *Proc. Natl. Acad. Sci. U.S.A.* **86**, 7677 (1989).

genetic consequences of Tg adducts in DNA. A vector was constructed, by site-directed methods, that contained a single *cis*-5,6-dihydroxy-5,6-dihydrothymine (*cis*-Tg) moiety at a unique site. This vector was transfected into *Escherichia coli,* progeny vectors were isolated and shown to contain Thy at the original position of Tg at a frequency of 99.7%; however, cytosine (Cyt) was found at this genome location 0.3% of the time. Thus, the mechanism of mutation must provide a reason why Tg principally appears as thymidine (Thy) and only occasionally as Cyt.

Our working hypothesis can be best understood by anticipating the result. The most straightforward means for Tg to induce Thy to Cyt mutations would be for it to become paired with Gua (i.e., dGTP) during replication. The structure of the analogous T:G mismatch is known, where Thy and Gua interact via a wobble base pair.[58] Each base is displaced ~1 Å toward the major and minor grooves, respectively, compared to a normal T:A base pair. Thus, displacement of a template Tg toward the major groove might be expected to enhance the likelihood that DNA polymerase would incorporate dGTP opposite a template Tg compared to the situation with a template Thy. Molecular modeling in conjunction with molecular mechanical calculations were performed to test this model. Because the result to be explained involved replication errors, we chose to model the Tg lesion at a single-stranded/double-stranded (ss/ds) junction in DNA.

The base, Tg (or Thy), was placed in an initial, unrefined DNA structure at a ss/ds junction in such a position that the Tg (or Thy) would be the next base to be copied by DNA polymerase (Scheme 1).

```
3′-CTAGGGGCGAXCGC-5′    X = Tg or Thy
5′-GATCCCCGCT     -3′
```

SCHEME 1

This structure was then refined by molecular mechanical calculations.[57] A structure refined by this procedure has a minimized energy, and in this case might be a reasonable approximation of the structure that DNA polymerase encounters when it is about to copy a Tg (or Thy) residue at a ss/ds junction (i.e., at X in Scheme 1). The minimized structure in the case of Tg is shown in Fig. 5a. The approximation procedure unfortunately (but necessarily) ignores any effect DNA polymerase may have on local architecture.

The starting coordinates for the structures containing either Tg or Thy were *identical;* thus, any difference in the final structures must be reflective of differences in the ways that Tg and Thy are accommodated in the DNA

[58] T. Brown, O. Kennard, G. Kneale, and D. Rabinovich, *Nature (London)* **315,** 604 (1985).

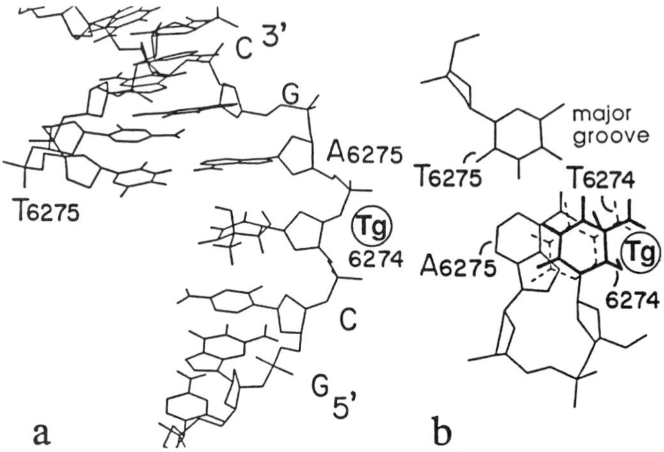

FIG. 5. Structure of thymine glycol (Tg) in DNA based on molecular mechanical calculation.[57] (a) Tg (5*R*,6*S* isomer) was located at position 6274 in the genome of the vector, M13-*Nhe*I, the sequence of which was used for the molecular modeling studies (Scheme 1). Four base pairs that are double stranded appear at the top of the panel, and four bases that are single stranded appear at the bottom. Tg is located in the template strand adjacent to the double-stranded domain; it is the next base to be copied by DNA polymerase. The view is from the major groove. (b) A portion of the structure from (a) when viewed along the helix axis from below. Tg (bold lines) is displaced toward the major groove compared to Thy (broken lines; from a structure computed separately); the base pair on the immediate 3′ side of Tg (i.e., $A_{6275}:T_{6275}$) is included for reference.

molecule. Following minimization the structure containing Thy was superimposed on the structure containing Tg in order to establish the relative positions of the bases Thy and Tg, in relation to their surrounding DNA sequences. As seen in Fig. 5b, Tg (bold) has been displaced toward the major groove compared to Thy (broken), an orientation that would be expected to facilitate mispairing with Gua (i.e., dGTP) during replication, based on the previously noted findings for the T:G wobble base pair.

This potential mechanism of inducing mutations is referred to as adduct-induced base wobble,[12] and its cause is readily understandable. Thy is aromatic, while the C(5) and C(6) positions of Tg are saturated, with hydroxyl and methyl groups projecting above and below the plane of the base. These groups form unfavorable van der Waals contacts with the base on the immediate 3′ side of Tg, which was Ade in this case. These unfavorable contacts were resolved during the minimization procedure by moving the Tg moiety toward the major groove.

It is important to point out that merely studying the structure of Tg in double-stranded DNA would not have revealed that Tg is more likely to

TABLE II
POTENTIAL MUTAGENIC MECHANISMS

1. Misinformational mechanisms
   a. Chemical perturbations
      i. Adduct-induced base tautomerization
      ii. Adduct-induced base ionization
   b. Structural perturbations
      i. Adduct-induced base rotation
      ii. Adduct-induced base wobble
2. Noninformational mechanisms
3. Other mechanisms

adopt a wobble conformation than Thy itself. When Tg is modeled opposite Ade, the structure appears approximately the same as when Thy is modeled opposite Ade; i.e., Watson–Crick-type pairing is observed. Similarly, when Tg is modeled opposite Gua, the structure appears approximately the same as when Thy is modeled opposite Gua; i.e., both show wobble. Thus, studies in double-stranded DNA provided no hint that Tg might have a greater proclivity than Thy itself for the formation of wobble base pairs with Gua.

Mechanisms of Mutagenesis

The previous section describes one mechanism by which an adduct might induce mutations, adduct-induced base wobble. Other mechanisms have been proposed, and this section discusses them systematically.[11]

Mechanisms of mutagenesis involving DNA lesions can be classified into three categories (Table II). (1) A *misinformational mechanism* is operative when a DNA polymerase attempts to "read" the base moiety in a DNA lesion, but misinterprets it. O^6-Methylguanine[9,13–16] and O^4-methylthymine,[59] which are formed from carcinogenic methylating agents, appear to belong in this class, and are most frequently misread as Ade and Cyt, respectively, during DNA replication. (2) A *noninformational mechanism* is operative when a DNA polymerase encounters an adduct that is uninterpretable and chooses to incorporate a particular deoxynucleoside triphosphate (dNTP) for reasons other than its attempt to "read" the lesion.[60] Reasons for choosing a particular dNTP could range from it being dictated by an inherent property of the DNA polymerase to the formation

[59] B. D. Preston, B. Singer, and L. A. Loeb, *Proc. Natl. Acad. Sci. U.S.A.* **83,** 8501 (1986).
[60] B. Strauss, S. Rabkin, D. Sagher, and P. Moore, *Biochemie* **64,** 829 (1982).

of an overall structure of DNA that can best be accommodated by the active site of a DNA polymerase. Apurinic/apyrimidinic sites (AP sites), which are formed spontaneously or are chemically induced by various means, appear to belong in this class,[61] and adducts derived from bulky mutagens/carcinogens may also belong in this class.[60,62] (3) *Other mechanisms* is a general category, and includes (a) the possibility that many lesions are processed to a common intermediate, which is then the mutagenic species, such as AP sites (see below),[31] and (b) mechanisms where DNA polymerase is not involved in the fixation of the mutation (e.g., mutagenic mechanisms involving topoisomerases have been suggested in several cases[63,64]).

Mutations involving misinformational lesions can in principle be divided into several categories (Table II). First, the carcinogen moiety of an adduct may induce a *chemical perturbation* in the base moiety of the adduct, which may improve the probability of misreplication. One category is adduct-induced base tautomerization.[65] The alkyl group in O^6-AlkGua adducts chemically lock the Gua moiety of the adduct in the enol (i.e., imidate ester) tautomeric form, which can base pair with Thy; this is the commonly accepted explanation for why O^6-AlkGua adducts induce G → A transition mutations. Adduct-induced base ionization would occur if a particular lesion significantly perturbed the pK_a of the base moiety of the adduct. For example, N(7)-Gua adducts are positively charged, which lowers the pK_a of the N(1) proton. This in turn increases the probability that N(1) is deprotonated, which might increase the probability that base pairing with the zwitterion will occur and a mutation result. There is no evidence for this mechanism.

The carcinogen moiety of an adduct could also affect the position of the base moiety of the adduct, resulting in a *structural perturbation,* which may improve the probability of misreplication. Adduct-induced base wobble is one example of this mechanism,[12,57] and was discussed in the previous section. Adduct-induced base rotation, in particular *anti* to *syn* base rotation,[65] is a second mechanism in this category. Recently, the major 2-aminofluorene adduct [AF-C(8)-Gua] was shown to form a base pair with adenine following an *anti* to *syn* base rotation,[28] and this may be related to

[61] L. A. Loeb and B. D. Preston, *Annu. Rev. Genet.* **20,** 201 (1986).

[62] J. McCann, N. E. Spingard, J. Kobori, and B. N. Ames, *Proc. Natl. Acad. Sci. U.S.A.* **72,** 979 (1975).

[63] D. Burnouf, P. Koehl, and R. P. P. Fuchs, *Proc. Natl. Acad. Sci. U.S.A.* **86,** 4147 (1989).

[64] L. S. Ripley, J. S. Dubins, J. G. deBoer, D. M. DeMarini, A. M. Bogerd, and K. N. Kreuzer, *J. Mol. Biol.* **200,** 665 (1988).

[65] M. D. Topal and J. R. Fresco, *Nature (London)* **263,** 283 (1976).

the mechanism by which AF induces GC → TA transversion mutations.[4,66,67] This is discussed at greater length below.

The exact definition of a noninformational lesions is elusive, but the term implies (1) that such a lesion is uninterpretable when encountered by DNA polymerase, and (2) that rules other than simple base-pairing schemes govern which base is incorporated opposite the lesion. It has been suggested that noninformational lesions share several characteristics: (1) Noninformational lesions block the progress of DNA polymerase in most primer extension studies *in vitro.*[60] (2) Noninformational lesions are not mutagenic in bacteria in the absence of the induction of the SOS response.[61,62,67,68] (3) Adenine (i.e., dATP) appears to be preferentially incorporated opposite noninformational lesions.[60,61,67,69] One unifying, but indirect, hypothesis has been offered to account for these observations; namely, that all mutations involving noninformational lesions share an AP site as a requisite intermediate.[69] No satisfying direct mechanism(s) has been proposed to account for the apparent characteristics of noninformational lesions.

In light of this discussion, does the fact that AF-C(8)-Gua adducts adopt a *syn* conformation when base paired with Ade in dsDNA[28,29] and that AF-C(8)-Gua adducts induce GC → TA transversion mutations[4,66,67] imply that a DNA polymerase reads an AF-C(8)-Gua adduct as Thy during replication; i.e., that AF-C(8)-Gua adducts are misinformational lesions? In fact, AF-C(8)-Gua adducts appear to satisfy all the criteria for being noninformational lesions (i.e., they block DNA replication *in vitro,*[60] their ability to induce mutation requires SOS induction,[4,67] and dATP appears to be preferentially incorporated).[4,66,67] Only further experimentation can resolve whether AF-C(8)-Gua, as well as other adducts of bulky carcinogens, are noninformational lesions, or are merely misinformational lesions that block DNA replication except following SOS induction, and that happen to adopt conformations that lead to dATP incorporation when misreplication occurs. If noninformational lesions truly exist, then gaining an understanding of their mechanism of mutagenesis, including by molecular modeling techniques, may be difficult.

Conclusion

This chapter has outlined some of the issues that relate carcinogen/DNA structure to the carcinogenesis process, and discusses how molecular

[66] A. M. Carothers, R. W. Streigerwalt, G. Urlaub, L. A. Chasin, and D. Grunberger, *J. Mol. Biol.* **208,** 417 (1989).

[67] T. H. Reid, M.-S. Lee, and C. M. King, *Biochemistry* **29,** 6153 (1990).

[68] G. C. Walker, *Microbiol. Rev.* **48,** 60 (1984).

[69] L. A. Loeb, *Cell (Cambridge, Mass.)* **40,** 483 (1984).

modeling may be used to provide some insights into these lesions. The main conclusion is that molecular modeling is of greatest utility when it is coupled with experimental results, and it is essential that the carcinogen/DNA structure to be modeled by relevant to this experimental data.

Acknowledgments

I am indebted to my many collaborators, including Marc Whitlow, Martha Teeter, John Essigmann, Ashis Basu, Thomas Harris, Michael Stone, Bea Singer, and Anthony Dipple. This work arose during research supported by the National Institutes of Health (CA-50432 and ES-03375).

[24] *In Situ* Hybridization and Immunodetection Techniques for Simultaneous Localization of Messenger RNAs and Protein Epitopes in Tissue Sections and Cultured Cells

By JUHA PELTONEN, SIRKKU JAAKKOLA, and JOUNI UITTO

Introduction

In situ hybridization (ISH) technique is used to localize specific cellular RNA or DNA sequences in tissues or in cell cultures by molecular hybridizations with complementary nucleotide probes. Since the half-life of functional mRNA molecules within cells is relatively short, varying from minutes to hours, the *in situ* detection of specific mRNAs allows both the temporal and spatial examination of cellular gene expression.

Well-established and standardized immunohistochemical and immunocytochemical methods allow the localization of antigenic protein or carbohydrate epitopes *in vivo* and *in vitro.* The steady state level of a given antigen within tissue or cell culture samples reflects the balance between its local synthesis and degradation, or its deposition from other sources, such as plasma. Thus, the presence of an antigen does not necessarily imply its synthesis by the resident cells.

In this chapter we describe the simultaneous use of ISH (with radiolabeled cDNA probes) to detect cellular mRNAs, and peroxidase–antiperoxidase (PAP) method to identify and localize protein epitopes in tissue samples and in cell cultures.

Procedures

The following is the general outline for simultaneous demonstration of mRNAs by *in situ* hybridization and protein epitopes by immunodetection in tissue sections and cultured cells. It should be noted that this protocol has been adapted to examine normal and pathological human skin specimens, as well as human skin fibroblasts and nerve-derived connective tissue cells (Schwann cells, perineurial cells, and fibroblasts) in culture. Application of this procedure to other tissue and cell types may require further optimization of specific steps during sample preparation.

The general outline is given in the following steps:

Step 1. Preparation and hybridization of tissue sections or cultured cells in combined use of ISH and PAP techniques (Table I and Refs. 1–3)

Step 2. Posthybridization treatments, including washes and S1 nuclease digestion (Table II).

Step 3. Peroxidase–anti-peroxidase immunostaining with specific antibodies (Table III and Ref. 4).

Step 4. Autoradiography for detection of radiolabeled cDNA–mRNA hybrids (Table IV).

Step 5. Counterstaining of specimens to visualize cell nuclei (if desired) (Table V).

Comments and Strategies

Sample Preparation

The sample preparation plays a critical role toward preserving the easily degradable cellular mRNAs. For this purpose, the tissue fixation must be immediate and complete, in particular since the half-lives of different mRNA species show considerable variation. Also, quick fixation of tissues rich in RNA-degrading enzymes (for example, liver and pancreas) is essential for good results.

When tissue specimens are fixed for *in situ* hybridization, best results are achieved by rapid immersion of the specimen immediately following

[1] J. Peltonen, S. Jaakkola, M. Lebwohl, S. Renvall, L. Risteli, I. Virtanen, and J. Uitto, *Lab. Invest.* **59,** 760 (1988).

[2] J. Peltonen, S. Jaakkola, K. Gay, D. R. Olsen, M.-L. Chu, and J. Uitto, *Anal. Biochem.* **178,** 184 (1989).

[3] J. Sambrook, E. F. Fritsch, and T. Maniatis, "Molecular Cloning," Vol. 2, Sect. P10.6. Cold Spring Harbor Laboratory, Cold Spring Harbor, New York, 1989.

[4] L. A. Sternberger, *in* "Immunocytochemistry," 3rd Ed., p. 90. Wiley, New York, 1986.

TABLE I
PREPARATION OF TISSUE SECTIONS OR CULTURED CELLS FOR COMBINED ISH–PAP METHOD[a]

Tissue sections	Cultured cells
1. Fixing of sample	
Fix immediately after excision by freezing in liquid nitrogen or in isopentane cooled on liquid nitrogen (samples can be stored for several months at −70° prior to sectioning)	Dip culture slide into PBS
Cut 5-μm sections with cryomicrotome	Incubate in cold (−20°) 100% ethanol in order to permeabilize and fix the cells; 15 min
Postfix in freshly prepared 4% paraformaldehyde in PBS at room temperature; 20 min	Rehydrate in 95 and 70% ethanol at room temperature; 1 min each
Rinse briefly in PBS	Postfix in freshly prepared 4% paraformaldehyde in PBS; 20 min
2. Pretreatments to increase penetration and reduce unspecific binding of probe	
Incubate in:	
0.2 M HCl containing 0.15 M NACl; 20 min	Not needed
Twice in 1 × SSC; 2 min/each to neutralize the sample prior to next step	
1 μg/ml of proteinase K (preheat buffer to 37°; 10 m*M* Tris, pH 7.4 and 2 m*M* $CaCl_2$); time of incubation (0–15 min) is to be tested in order to maintain the antigenic epitopes and yet to allow penetration of the probe into tissue sample	Not needed
PBS containing 2 mg/ml glycine; 3 and 10 min	Same as for tissue sections
Acetic anhydride in 0.1 *M* triethanolamine, pH 8.0; add the acetic anhydride just prior to incubation, mix well; 10 min	Same as for tissue sections
1 × SSC; 2 × 5 min	Same as for tissue sections
Dip twice into distilled H_2O	Same as for tissue sections
Dehydrate successively in 70, 95, and 100% ethanol; 1 min each	Same as for tissue sections
Prepare incubation chamber around the sample area using strips of microscope coverslips cut with a diamond pen	Same as for tissue sections
3. Prehybridization	Optional for cultured cells
Heat the slides on a hot plate at 90°; 5 min	
Cool rapidly on foil placed on ice/water bath	
Heat prehybridization solution[b] at 100°; 10 min; cool rapidly on ice	
Incubate with prehybridization solution; overnight	
4. Hybridization	
Empty the hybridization chamber by suction	Heat the slides on a hot plate at 90°; 5 min

TABLE I *(continued)*

Tissue sections	Cultured cells
Heat hybridization solution[b] at 100°; 10 min; cool rapidly on ice	Cool rapidly on foil placed on ice/water bath
Incubate with hybridization solution at 42°; overnight	Heat hybridization solution at 100°; 10 min; cool rapidly on ice
Remove coverslips and frames with forceps	Add hybridization solution, cover with a glass coverslip, and seal with rubber cement
	Incubate with hybridization solution at 42°; overnight
	Remove coverslips and frames with forceps

[a] ISH, *In situ* hybridization; PNP, peroxidase–anti-peroxidase; PBS, phosphate-buffered saline; SSC, standard saline citrate (1 × SSC = 0.15 NaCl/0.015 *M* sodium citrate, pH 6.8).

[b] Prehybridization and hybridization solution:

	Per 2 ml	Per 5 ml
Formamide	1.0 ml	2.5 ml
5.0 *M* NaCl	240 μl	600 μl
2.0 *M* Tris-HCl (pH 7.4)	10 μl	25 μl
0.5 *M* EDTA	2 μl	5 μl
50 mg/ml BSA (bovine serum albumin)	40 μl	100 μl
10% PVP (polyvinylpyrrolidone)	4 μl	10 μl
10% Ficoll	4 μl	10 μl
1.0 *M* DTT (dithiothreitol)	20 μl	50 μl
5.0 mg/ml ssDNA	80 μl	200 μl
Dextran sulfate	0.2 g	0.5 g
Probe (omit in prehybridization)	0.2 μg	0.5 μg
Diethyl pyrocarbonate (DEPC)-treated H_2O (1 : 1000, followed by autoclaving, three cycles)	600 μl*	1500 μl

* In hybridizations this includes the volume of the probe solution.

excision in liquid nitrogen or in isopentane that has been cooled on liquid nitrogen. Cultured cells grown on appropriate slides (see below) can be fixed by immersing the samples in cold (−20°) ethanol. Postfixation of both tissue specimens and cultured cells with a relatively high concentration (4%) of freshly prepared paraformaldehyde further serves toward optimal preservation of mRNAs and also protein epitopes (see Table I).

Pretreatment and hybridization solutions are preferentially made RNase free by autoclaving or boiling (see Table I). Also, use sterile containers and instruments before and during the hybridization steps.

In our experience, cryofixed tissue samples, such as skin specimens, can

TABLE II
POSTHYBRIDIZATION TREATMENT PROCEDURES FOR TISSUE SPECIMENS OR CULTURED CELLS

Treatment	Time	Temperature
1. Posthybridization washes		
0.5 × SSC,[a] 1.0 mM EDTA, 10 mM DTT	2 × 5 min	Room temperature
0.5 × SSC, 1.0 mM EDTA	2 × 5 min	Room temperature
50% formamide, 0.15 M NaCl, 5 mM Tris (pH 7.4), 0.5 mM EDTA	10 min	Room temperature
0.5 × SSC	2 × 5 min	55°
0.2 × SSC	2 × 5 min	55°
0.2 × SSC	5 min	Room temperature
2. S1 nuclease incubation[b]: Not needed for cell cultures (modified from Ref. 3)		
3. Pick up the slides from the washing solution and dry the excess SSC around the samples. Place the slides flat on wet paper towels or sponges and cover the sections with S1 nuclease buffer[b] for 30 min at room temperature, in a closed container. Wash the samples with 0.5 × SSC three times for 5 min each at room temperature		

[a] Standard saline citrate (1 × SSC = 0.15 M NaCl/0.015 M sodium citrate, pH 6.8).
[b] S1 buffer: 0.28 M NaCl; 5 mM sodium acetate, pH 4.6; 4.5 mM $ZnSO_4$; 1000 units/ml S1 nuclease; add just prior to use.

TABLE III
PEROXIDASE–ANTI-PEROXIDASE IMMUNOSTAINING[a]

Incubate tissue specimens or cultured cells in:

1. Tris-buffered saline (TBS: 0.15 M NaCl/0.05 M Tris-HCl, pH 7.6); three times, 5 min each
2. 1% Bovine serum albumin (BSA)/TBS; 30 min to block unspecific binding
3. Primary antibody diluted in 1% BSA/TBS; overnight at 4° or 1 hr at room temperature.
4. TBS; three times, 10 min each
5. Linking antibody[b] [e.g., swine anti-rabbit immunoglobulin G (IgG); 1 : 10 dilution]
6. TBS; three times, 10 min each
7. Rabbit PAP[b]; 1 : 80 dilution; 30 min
8. TBS; three times, 10 min each.
9. 0.05% 3,3′-Diaminobenzidine tetrahydrochloride (DAB), in freshly prepared 0.03% H_2O_2 in TBS; 5 min
10. Dehydrate with 70, 95, and 100% ethanol containing 0.3 M ammonium acetate

[a] Modified from Ref. 4.
[b] All antibodies diluted in 1% BSA/TBS.

TABLE IV
AUTORADIOGRAPHY

1. Warm Kodak NTB-3 autoradiography emulsion in 40° water bath until it becomes liquid

In darkroom with red light illumination:

2. Dilute emulsion with equal volume of 0.6 *M* ammonium acetate; to avoid air bubbles do not shake

Do not use red light during steps 3–7!

3. Dip slides into diluted emulsion; let excess emulsion run off, dry back side of slide
4. Place samples into desiccant-containing box large enough to allow drying of emulsion at room temperature for 5 hr
5. Return samples to desiccant-containing slide holder box and store at 4° until developed
6. Develop samples in Kodak D-19 developer at 15° for 90 sec
7. Rinse in tap water for 30 sec
8. Fix in Kodak Universal fixer for 5 min (light can be turned on)
9. Rinse in tap water for 5 min

TABLE V
COUNTERSTAINING[a]

1. Stain with Harris's hematoxylin for 30–60 sec
2. Rinse in tap water for 5 min; dip briefly into distilled water
3. Dehydrate in 70, 95, and 100% ethanol for 2 min each
4. Clear in xylene for two 5-min periods
5. Mount with Permount

[a] Optional; it is sometimes useful to observe and document findings without counterstaining.

be stored at −70° for several months, but storing samples after the sectioning can result in considerable reduction in the hybridization signal.

Primary cell cultures are conveniently produced by implantation of tissue pieces on a Petri dish under an acetylated[5] object glass. Passaged cells can be grown directly on acetylated object glasses.

When cell cultures are analyzed by *in situ* hybridization, the use of ethanol as fixative permeabilizes the cells by dissolving the lipid bilayer of the plasma membrane, and thus allows the penetration of nucleic acid probes into the intracellular compartment. Consequently, annealing of the cDNA with corresponding cellular mRNA sequences results in a good hybridization signal. In contrast, the penetration of the cDNA probe may

[5] M. Brahic, A. T. Haase, and E. Cash, *Proc. Natl. Acad. Sci. U.S.A.* **81,** 5445 (1984).

be a major problem in tissue sections. This point has been well demonstrated by studies in which keloids or neurofibromas, lesions with extensive collagenous extracellular matrix, have been used for *in situ* hybridizations.[1,6] In such a case, treatment of the tissue sections with proteolytic enzymes is necessary to obtain a detectable hybridization signal. The length of incubation with proteinase K must be optimized for each different tissue in order to allow hybridization, and yet to preserve the protein epitopes to be recognized by immunostaining (see Table I).

Hybridization Probe

The chemical stability and high specificity make radiolabeled cDNAs a preferable tool to detect cellular mRNAs, in comparison with relatively unstable cRNA probes. The cDNA probes can be radiolabeled to high specific activity, whereas, for example, biotinylated cDNAs may not be sensitive enough to detect rare or low-abundance mRNA species. Use of insert probe (double purified by electroelution), as opposed to entire plasmid, clearly results in an improved signal-to-background ratio. The use of probes labeled to a high specific activity ($\sim 10^9$cpm/μg) by nick translation[3] or by random priming either with [^{35}S]dATP, or double-labeled with [^{32}P]dCTP and [^{32}P]dGTP, results in a good hybridization signal which can be detected after a relatively short (2–7 days) exposure.

Development of Combined Procedure

To develop the combined methodology utilizing *in situ* hybridization and immunodetection techniques for simultaneous localization of mRNA and protein epitopes, these two techniques were first individually optimized.[2,6] The order of procedures selected for the final methodology, i.e., the *in situ* hybridization technique preceding the PAP immunodetection, was based on several considerations that were clearly advantageous. First, the hybridization efficiency was found to be much improved if the *in situ* hybridization was performed first. This observation is consistent with the preservation of the RNA content of cells. The enhanced hybridization efficiency may reflect the fact that once the cDNA has been fixed into its hybridization site on mRNA within the cells, the detection of the corresponding signal by autoradiography cannot be obliterated by residual RNase activity. Second, *in situ* hybridization prior to PAP avoids the nonspecific binding of cDNA probes to 3,3′-diaminobenzidine tetrahydrochloride (DAB) precipitates, a problem that can obscure the results.

[6] S. Sollberg, J. Peltonen, and J. Uitto, *Lab. Invest.* **64,** 125 (1991).

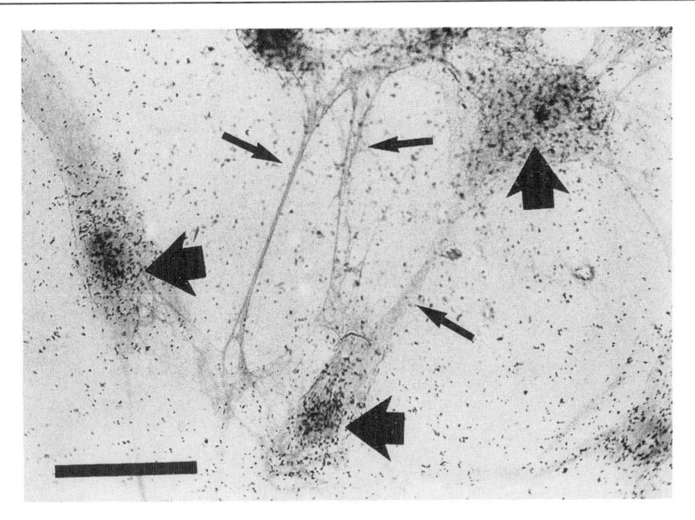

FIG. 1. Simultaneous detection of fibronectin mRNAs and the corresponding protein epitopes on the same cells in human skin fibroblast cultures. Autoradiographic grains represent ^{35}S-labeled cDNA–mRNA hybrids (large arrows), and the protein epitopes are visualized by brownish 3,3′-diaminobenzidine tetrahydrochloride (DAB) precipitates (small arrows). Cell preparations were counterstained by hematoxylin to visualize the nuclei. Bar: 50 μm.

Potential Utilization of Combined Technique

The methodology described in this chapter was initially developed in our laboratory in order to examine fibronectin gene expression in cultured cells both at mRNA and protein levels.[2] This application ensures that the presence of mRNA in cells, as detected by *in situ* hybridization, in fact translates to the presence of the corresponding protein epitopes in the same cells (Fig. 1). Obviously, this application is feasible with any mRNA–protein system for which specific cDNA probes and specific antibodies are available. An extension of this strategy is to examine the heterogeneity/homogeneity of gene expression in cell cultures. In particular, this technology is applicable to situations where mixed cell cultures consisting of several different cell types are examined. Utilization of immunodetection, together with *in situ* hybridization using cDNA for cellular mRNAs, allows identification of the cell type responsible for specific gene expression. For example, Schwann cells can be identified as S-100-positive cells by immunostaining in mixed neurofibroma cell cultures. Such Schwann cells have been shown to express, e.g., type VI collagen genes, while no expression of fibronectin could be detected.[7] Similarly, in tissue sections, the combina-

[7] S. Jaakkola, J. Peltonen, V. Riccardi, M.-L. Chu, and J. Uitto, *J. Clin. Invest.* **84,** 253 (1989).

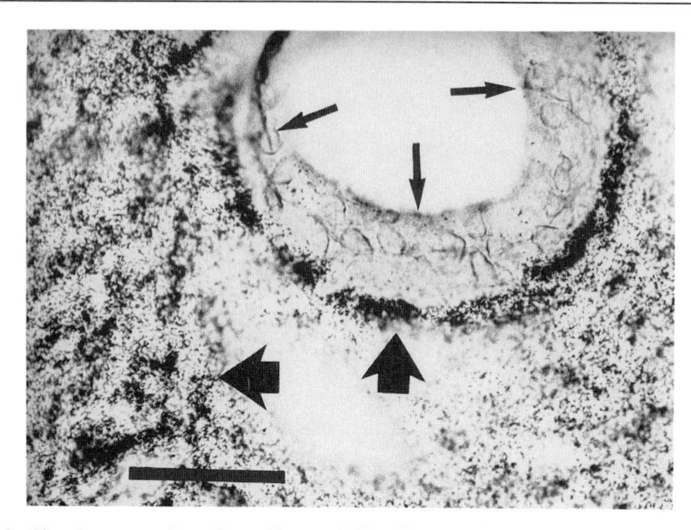

FIG. 2. Simultaneous detection of pro-α1(I) collagen mRNAs and factor VIII-related antigen epitopes in neurofibroma tissue. Endothelial cells can be identified by a positive staining reaction for factor VIII-related antigen (small arrows), while autoradiographic grains represent ^{32}P-labeled cDNA–mRNA hybrids. Note the strong hybridization signal for type I procollagen in the close proximity of, but apparently outside, the endothelium (large arrows). The sample was examined without counterstain. Bar, 50 μm.

tion of *in situ* hybridization and immunodetection allows identification of, e.g., factor VIII-related antigen-positive endothelial cells,[8] and localization of those cells that are responsible for expression of pro-α1(I) collagen genes (Fig. 2). Such observations have allowed elucidation of the relationship between the vascular component and the activation of type I collagen gene expression in cutaneous neurofibromas (Fig. 2). Clearly, this combined technique can be applied to study a variety of clinical situations, as well as normal tissue development, homeostasis, and repair.

Acknowledgments

The authors thank Charlene D. Aranda, Debra Pawlicki, and Eileen O'Shaughnessy for technical assistance. The original studies by the authors were supported by U.S. Public Health Service, National Institutes of Health Grants GM28833, AR35297, and AR41439. Additional support was provided by the Finnish Academy of Sciences and the Finnish Cultural Fund.

[8] K. Mukai, J. Rosai, and W. H. C. Burgdorf, *Am. J. Surg. Pathol.* **4,** 273 (1980).

[25] Transfer RNA with Double Identity for *in Vitro* Kinetic Modeling of Transfer RNA Identity *in Vivo*

By PAUL SCHIMMEL and JONATHAN J. BURBAUM

Introduction

The interpretation of trinucleotide codons as amino acids is due entirely to the specific recognition of anticodon-bearing transfer RNAs by aminoacyl-tRNA synthetases. Because there are approximately 60 different transfer RNAs and 20 aminoacyl-tRNA synthetases in the cytoplasm of a prokaryotic or eukaryotic cell,[1,2] the aminoacylation of a particular tRNA by its cognate enzyme is only 1 of approximately 1200 potential amino acid–tRNA combinations.[3] These combinations are accessible in principle because tRNAs have a similar L-shaped three-dimensional structure and have conserved nucleotides that stabilize the spatial arrangement of bases and phosphate groups. These conserved features facilitate the cross-interactions between synthetases and noncognate tRNAs that have been extensively demonstrated *in vitro.*[4] Notwithstanding the potential for these interactions and the ability to study them *in vitro,* the accuracy of the genetic code is maintained because misacylations *in vivo* are rare.

In addition to specific editing reactions that prevent or correct misacylations,[4-6] the high accuracy of the code *in vivo* is due in significant part to binding interactions that sequester tRNAs with their cognate aminoacyl-tRNA synthetases and thereby diminish the pool of each that is available to interact with a noncognate partner. This possibility was demonstrated *in vitro* with *Escherichia coli* isoleucine–tRNA synthetase (ligase), which aminoacylates $tRNA^{Ile}$ and misacylates $tRNA^{Phe}$. The misacylation reaction is suppressed by addition of $tRNA^{Ile}$, which competitively displaces $tRNA^{Phe}$ from Ile–tRNA synthetase.[3] Subsequently, mischarging of the *E. coli supF* $tRNA^{Tyr}$ amber suppressor *in vivo* by glutamine tRNA synthetase was shown to be prevented by elevation of the concentration of the $tRNA^{Gln_2}$ isoacceptor.[7] In this case, change of the $tRNA^{Tyr}$ GUA anti-

[1] M. Sprinzl, T. Hartmann, J. Weber, J. Blank, and R. Zeidler, *Nucleic Acids Res.* **17,** r1 (1989).
[2] P. Schimmel, *Annu. Rev. Biochem.* **56,** 125 (1987).
[3] M. Yarus, *Nature (London) New Biol.* **239,** 106 (1972).
[4] P. R. Schimmel and D. Soll, *Annu. Rev. Biochem.* **48,** 601 (1979).
[5] A. R. Fersch, "Enzyme Structure and Mechanism," 2nd Ed., p. 347. Freeman, New York, 1985.
[6] W. Freist, *Biochemistry* **28,** 6787 (1989).
[7] R. Swanson, P. Hoben, M. Sumner-Smith, H. Uemura, L. Watson, and D. Soll, *Science* **242,** 1548 (1988).

codon to CUA (to give the *supF* amber suppressor) creates a recognition site for glutamine–tRNA synthetase that facilitates mischarging. Other examples of misacylation of amber suppressors *in vivo* with glutamine or lysine have also been described,[8] although prevention of misacylation by elevated levels of a competing enzyme or tRNA has not been investigated in these instances.

The recent effort to define determinants for the recognition of tRNAs[8,9] has provided motivation to establish a quantitative framework to evaluate cross-competition and its effect on tRNA identity. Many studies have relied on *in vivo* screens or selections for suppression of amber codons that require a specific amino acid for manifestation of a conditional growth phenotype. Alternatively, the gene product of a suppressed amber-encoding mRNA is sequenced directly to identify the inserted amino acid and thereby establish the specificity of aminoacylation *in vivo.* With these experimental systems, nucleotides that confer a specific aminoacylation are identified. Those nucleotides are then transferred into other tRNA frameworks to determine whether the charging specificity follows the transferred nucleotides.[8-10] However, the interpretation of results that are obtained with this approach is hazardous. In particular, cross-competition *in vivo* can completely obscure the recognition of a major determinant for the identity of a specific tRNA that has been transferred into another tRNA, and thus suggest (falsely) that there is no such determinant.

A straightforward calculation with kinetic parameters obtained *in vitro* can predict circumstances that would obscure or reveal the amino acid-specific charging of a tRNA *in vivo.* To test the reliability of these calculations and their predictive value, a mutant tRNA that is charged completely and efficiently by either of two aminoacyl-tRNA synthetases acids was designed.[11] These two enzymes also aminoacylate their respective cognate tRNAs. Kinetic parameters for the four aminoacylation reactions were obtained and then used to predict the specificity of aminoacylation of the mutant tRNA *in vivo* under different circumstances (i.e., with different relative amounts of the two enzymes). While these calculations did not take account of all 1200 potential cross-interactions, consideration of the major set of interactions between the competing synthetases and their tRNAs is sufficient to recapitulate the main features of the specificity of charging that is observed *in vivo.* An important conclusion from these

[8] J. Normanly and J. Abelson, *Annu. Rev. Biochem.* **58,** 1029 (1989).

[9] P. Schimmel, *Biochemistry* **28,** 2747 (1989).

[10] J. Normanly, R. C. Odgen, S. J. Horvath, and J. Abelson, *Nature (London)* **321,** 213 (1986).

[11] Y. M. Hou and P. Schimmel, *Biochemistry* **28,** 4942 (1989).

studies is that a complete switch in amino acid specificity *in vivo* can be obtained by adjusting the relative enzyme levels by only 20-fold or less.[11]

Basic Considerations for Calculations of tRNA Identity

The overall aminoacylation reaction catalyzed by a specific enzyme E_i attaches amino acid i (AA_i) to its cognate tRNA ($tRNA^i$) according to the reaction

$$AA_i + ATP + tRNA^i \rightarrow AA_i\text{-}tRNA^i + AMP + PP$$

where AMP and PP are adenosine 5′-monophosphate and pyrophosphate, respectively. The overall reaction is the sum of two steps: activation of the amino acid by condensation with ATP to form a tightly bound aminoacyl adenylate, followed by transfer of the aminoacyl group from AMP to tRNA. *In vivo* the reaction may be considered as irreversible, because the charged tRNA is sequestered by translation factors and subsequently delivered to the ribosomes.

Kinetic parameters for aminoacylation *in vitro* consist of K_m parameters for the three substrates—amino acid, ATP, and tRNA—and k_{cat} (turnover number). These parameters are obtained by analysis of the dependence of the reaction velocity on the concentration of each of the three substrates, with catalytic amounts of enzyme. The problem is to calculate from these parameters the effective velocity of aminoacylation *in vivo*. For this calculation it is necessary to estimate the intracellular concentrations of the relevant synthetases, and of each of their substrates, under the specific conditions *in vivo* that are used for evaluating specificity of aminoacylation.

Methods for measuring intracellular concentrations of synthetases, tRNAs, and amino acids in *E. coli* have been described. Neidhardt *et al.* reported that, under a given set of growth conditions with defined media, concentrations of 10 synthetases that were measured by two-dimensional gel electrophoresis differ among themselves by two- to threefold and that the amount of each of these synthetases is increased in response to a faster growth rate.[12] For example, a 2.5- to 4-fold increase in the concentration of a specific enzyme accompanies a change in growth from 0.38 to 1.98 division/hr. In general, depending on the enzyme and growth conditions, the typical intracellular concentration of a synthetase is 0.2 to 2 μM.

There are roughly 1 to 20 specific tRNA molecules per cognate synthetase, with some variation among the different tRNAs.[13] Thus, both synthe-

[12] F. C. Neidhardt, P. L. Bloch, S. Pedersen, and S. Ree, *J. Bacteriol.* **129,** 378 (1977).
[13] H. Jakubowski and E. Goldman, *J. Bacteriol.* **158,** 769 (1984).

tases and tRNAs are in the micromolar concentration range. This means that *in vivo* the aminoacyl-tRNA synthetases are present at substrate levels of concentrations. Consequently, the Michaelis–Menten equations (which pertain to catalytic enzyme concentrations) for calculating reaction velocities should be applied with caution. Alternatively, parameters obtained from a standard kinetic assay (i.e., one that has been done with catalytic amounts of enzyme) can be used to estimate numerically the amount of tRNA that is bound to enzyme–aminoacyl adenylate complex. This concentration in turn is multiplied by k_{cat} to give an estimate of reaction velocity. In the kinetic modeling that is described below, calculations of tRNA identity are illustrated with Michaelis–Menten equations and more accurate numerical methods, and with different relative enzyme levels and numbers of competing reactions taken into account.

Kinetic Modeling of Behavior of tRNA with Double Identity

The major determinant for the identity of an alanine tRNA is a single G3:U70 base pair in the amino acid acceptor helix.[14] When this base pair is introduced into $tRNA^{Cys}$,[14] $tRNA^{Phe}$,[14,15] and $tRNA^{Tyr}$,[11] the resulting tRNAs can be charged with alanine. Moreover, alanine–tRNA synthetase aminoacylates a 7-bp hairpin microhelix that represents the acceptor helix, provided that the helix encodes G3:U70.[16] This illustrates the concentration of the primary site for recognition in a small element of the tRNA structure.

Introduction of G3:U70 into a $tRNA^{Tyr}$ amber suppressor [to give G3:U70 $tRNA^{Tyr}_{CUA}$ (Fig. 1)] results in a substrate that charges efficiently *in vitro* with either alanine or tyrosine.[11] Because G3:U70 $tRNA^{Tyr}_{CUA}$ can be charged by either enzyme, it can be used to investigate competitive factors that influence tRNA identity *in vivo.* Kinetic parameters for aminoacylation of $tRNA^{Ala}$ and G3:U70 $tRNA^{Tyr}_{CUA}$ with alanine–tRNA synthetase, and of $tRNA^{Tyr}_{CUA}$ and G3:U70 $tRNA^{Tyr}_{CUA}$ with tyrosine–tRNA synthetase, are given in Table I for the near-physiological conditions of pH 7.5, 37°. These parameters describe the dependence of the initial velocity V of the aminoacylation according to the Michaelis–Menten equation:

$$V = k_{cat}[\text{enzyme}]_0[\text{tRNA}]_0/(K_m + [\text{tRNA}]_0) \quad (1)$$

where $[\text{enzyme}]_0$ and $[\text{tRNA}]_0$ are concentrations of *total* enzyme and tRNA, respectively.

[14] Y. M. Hou and P. Schimmel, *Nature (London)* **333,** 140 (1988).
[15] W. H. McClain and K. Foss, *Science* **240,** 793 (1988).
[16] C. Francklyn and P. Schimmel, *Nature (London)* **337,** 478 (1989).

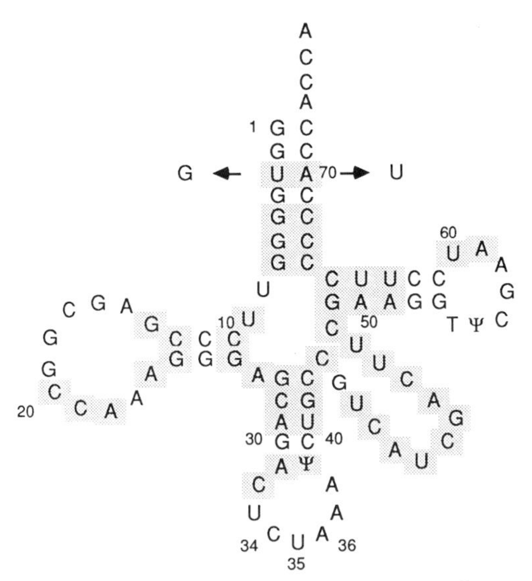

FIG. 1. Nucleotide sequence and cloverleaf structure of $tRNA^{Tyr}_{CUA}$. This transfer RNA is aminoacylated with tyrosine. Introduction of G3:U70 (shown by arrows) enables this tRNA also to be aminoacylated by alanine–tRNA synthetase with no diminution in the activity for tyrosine–tRNA synthetase. The shaded nucleotides are ones that are different between $tRNA^{Tyr}_{CUA}$ and $tRNA^{Ala}_{CUA}$. (Adapted from Ref. 11.)

It should be recognized that the terms that multiply k_{cat} in Eq. (1) give the concentration of enzyme–tRNA complex, under conditions where the enzyme is present in catalytic amounts compared to the tRNA substrate. Equation (1) also expresses only the dependence of the velocity of aminoacylation on enzyme and tRNA concentration. The concentrations of amino acid and ATP are fixed, so that the k_{cat} parameter pertains to specific concentrations of these ligands. The k_{cat} parameters given in Table I are for saturating concentrations of ATP and subsaturating concentrations (20 μM) of the respective amino acids.

As stated above, the concentrations of enzymes and tRNAs are usually comparable *in vivo,* so that the approximation given by Eq. (1) is inaccurate. More generally, the aminoacylation velocity is given by

$$V = k_{cat}[\text{enzyme} \cdot \text{tRNA}] = (k_{cat}/K_m)[\text{enzyme}][\text{tRNA}] \qquad (2)$$

where [enzyme], [tRNA], and [enzyme · tRNA] refer to the concentrations of *free* enzyme and tRNA, and of enzyme–tRNA complex, respectively. The concentrations of free enzyme and tRNA must be computed by numerical methods and then introduced into Eq. (2).

TABLE I
KINETIC PARAMETERS FOR AMINOACYLATION OF *E. coli* tRNAs WITH ALANINE– AND TYROSINE–tRNA SYNTHETASES[a]

Enzyme	tRNA	k_{cat} (sec^{-1})	K_m (μM)
Alanine	$tRNA^{Ala}$	1.4	2.8
Alanine	G3:U70 $tRNA^{Tyr}_{CUA}$	0.6	14.0
Tyrosine	G3:U70 $tRNA^{Tyr}_{CUA}$	1.24	8.3
Tyrosine[b]	$tRNA^{Tyr}_{CUA}$ [b]	1.4[b]	8.2[b]
Tyrosine	$tRNA^{Tyr}$	20.5	5.0

[a] At pH 7.5, 37°. Data are adapted from Ref. 11.
[b] Data for aminoacylation of $tRNA^{Tyr}_{CUA}$ with tyrosine–tRNA synthetase are shown for comparative purposes only, and are not required for the calculations that are described in the text.

In the experimental system used to investigate the parameters that determine alternative aminoacylations of G3:U70 $tRNA^{Tyr}_{CUA}$, the alanine enzyme not only interacts with G3:U70 $tRNA^{Tyr}_{CUA}$, but also with its own substrate ($tRNA^{Ala}$); similarly, the tyrosine enzyme also binds and aminoacylates $tRNA^{Tyr}$. The cognate enzyme–tRNA interactions diminish the amount of each enzyme that is available to aminoacylate G3:U70 $tRNA^{Tyr}_{CUA}$. Thus, the amount of free G3:U70 $tRNA^{Tyr}_{CUA}$ is influenced by a minimum of four synthetase–tRNA associations:

$$E_{Tyr} + tRNA^{Tyr} \rightleftharpoons E_{Tyr} \cdot tRNA^{Tyr} \tag{3a}$$

$$E_{Tyr} + \text{G3:U70 } tRNA^{Tyr}_{CUA} \rightleftharpoons E_{Tyr} \cdot \text{G3:U70 } tRNA^{Tyr}_{CUA} \tag{3b}$$

$$E_{Ala} + tRNA^{Ala} \rightleftharpoons E_{Ala} \cdot tRNA^{Ala} \tag{3c}$$

$$E_{Ala} + \text{G3:U70 } tRNA^{Tyr}_{CUA} \rightleftharpoons E_{Ala} \cdot \text{G3:U70 } tRNA^{Tyr}_{CUA} \tag{3d}$$

where E_{Tyr} and E_{Ala} refer to tyrosine- and alanine-tRNA synthetase, respectively, and each equilibrium is assumed to be characterized by a dissociation constant that is given by the appropriate K_m in Table I.

Once the velocities V_{Tyr} and V_{Ala} of aminoacylation of G3:U70 $tRNA^{Tyr}_{CUA}$ are computed, the fraction of aminoacylation with tyrosine is given by Eq. (4).

$$\text{Fraction tyrosine} = V_{Tyr}/(V_{Tyr} + V_{Ala}) \tag{4}$$

In the calculations that follow, three successive approximations are given for computing from the parameters of Table I the fraction of aminoacylation *in vivo* of G3:U70 $tRNA^{Tyr}_{CUA}$ with tyrosine and alanine, at different relative levels of the two enzymes that compete for G3:U70 $tRNA^{Tyr}_{CUA}$. These calculations also illustrate the effect of including progressively more competing synthetase–tRNA equilibria in accounting for the specificity of aminoacylation.

Numerical Methods

To use Eq. (2), the concentrations of free enzyme and tRNA species, or of enzyme–tRNA complex, must be computed for a given set of total enzyme and tRNA concentrations. For a simple equilibrium of a single enzyme E and a tRNA (T), the concentrations of individual species can be obtained from the expression for the dissociation constant K for the enzyme–tRNA (E·T) complex and the two mass conservation relationships:

$$[E]_0 = [E] + [E \cdot T] \quad (5a)$$

$$[T]_0 = [T] + [E \cdot T] \quad (5b)$$

With these relationships, the concentrations of the various species can be expressed as

$$[E \cdot T] = [E]_0[T]/(K + [T]) \quad (6a)$$

$$[E] = K[E]_0/(K + [T]) \quad (6b)$$

$$[T] = K[T]_0/(K + [E]) \quad (6c)$$

For a single enzyme–tRNA equilibrium, there are three equations (two mass conservation relationships and one equilibrium constant) in three unknowns (concentrations of enzyme, tRNA, and enzyme–tRNA complex). From these relationships, an exact solution for each of the concentrations can be obtained by solving a quadratic equation. For example, for [E·T] the exact solution is

$$[E \cdot T] = \{([E]_0 + [T]_0 + K)/2\} \{1 - [1 - 4([E]_0[T]_0)/([E]_0 + [T]_0 + K)^2]^{1/2}\} \quad (7)$$

However, exact solutions analogous to Eq. (7) cannot be obtained for more complex cases, without the introduction of simplifying approximations that are only valid under certain conditions (e.g., substrate concentrations in great excess of enzyme concentrations). For the most complex situation considered here, where two enzymes (E_1 and E_2) compete for the same tRNA species ($T_{1,2}$) and also bind to their cognate tRNAs (T_1 and T_2), there are nine equations in nine unknowns (five mass conservation relationships and four equilibrium constants [Eqs. (3a)–(3d)].

The concentrations of free $E_1 \cdot T_{1,2}$ complex and of free E_1 are given by

$$[E_1 \cdot T_{1,2}] = [E_1]_0[T_{1,2}]/\{[T_{1,2}] + K_{1,2}(1 + [T_1]/K_1)\} \quad (8a)$$

$$[E_1] = K_{1,2}[E_1]_0/\{[T_{1,2}] + K_{1,2}[1 + K_{1,2}(1 + [T_1]/K_1)]\} \quad (8b)$$

where K_1 is the dissociation constant for the cognate $E_1 \cdot T_1$ complex. An equation analogous to that for $[E_1 \cdot T_{1,2}]$ was used to compute $[E_2 \cdot T_{1,2}]$. The concentration of $[T_1]$ and $[T_2]$ can be calculated from an equation analogous to Eq. (5b), with K_1 and K_2, respectively, substituted for K.

Numerical calculations of free concentrations were performed on an Apple MacIntosh IIci computer with Microsoft Excel (version 2.2) software. For these calculations, a value is assumed for one of the concentration variables (e.g., [T]) and the others ([E] and $[E \cdot T]$) are then calculated from the above equations. This process is iterated until successive values of each of the various concentrations differ by no more than one part in 10^6. The convergence of numerical solutions was checked by back calculations of the total concentrations of each enzyme and tRNA species, and of the various equilibrium constants, from the numerical values of the concentrations of each species.

Calculations with No Competition

Estimation of Concentrations. The greatest uncertainty in applying the above equations is the estimate of the concentration *in vivo* of individual tRNA and enzyme species. For these estimations we have taken the volume of one cell of *E. coli* to be 10^{-15} liters[17] and use our own or other measurements of the number of molecules per cell to estimate concentrations. According to earlier estimates, the intracellular concentrations of alanine- and tyrosine-tRNA synthetases are 0.4[18] and 2 μM,[19] respectively. The concentrations of uncharged tRNA molecules can be derived from data of Jakubowski and Goldman, who measured charged tRNA molecules/cells for each of the 20 aminoacyl-tRNAs.[13] The aminoacylated tRNAs are believed to represent about 70% of the total tRNA.[13] From their data we calculate that total concentrations of uncharged alanine and tyrosine tRNAs are 2 and 0.5 μM, respectively. The total concentration of G3:U70 $tRNA^{Tyr}_{CUA}$ is estimated as 10 μM, which in turn corresponds to 3 μM of the uncharged species.[11]

The total intracellular concentrations of alanine (2.5 mM) and tyrosine

[17] R. Schleif, "Genetics and Molecular Biology," p. 4. Addison-Wesley, Reading, Massachusetts, 1986.

[18] S. D. Putney and P. Schimmel, *Nature (London)* **291,** 632 (1981).

[19] R. Calendar and P. Berg, *Biochemistry* **5,** 1681 (1966).

(1 mM)[20] are at least 10 times above the K_m values for the respective enzymes (240 μM for alanine[21] and 30 μM for tyrosine[22]). This raises the operational k_{cat} above the values reported in Table I (that are for subsaturating concentrations of amino acids) to 6.5 and 2.9 sec^{-1} for alanine- and tyrosine–tRNA synthetase, respectively. (We assumed that k_{cat} is proportional to the fractional saturation with amino acid; at least for alanine-tRNA synthetase, there is evidence to support this assumption.[23])

Michaelis–Menten Approximation. In the simplest approximation, Eq. (1) is used directly to calculate V_{Tyr} and V_{Ala} for aminoacylation of G3:U70 $tRNA^{Tyr}_{CUA}$. The above values for each of the respective total enzyme concentrations and for k_{cat} are substituted. However, a calculation of this sort treats each aminoacylation reaction as an independent event and ignores the effective reduction in concentration of free G3:U70 $tRNA^{Tyr}_{CUA}$ available to the alanine enzyme by virtue of the binding of G3:U70 $tRNA^{Tyr}_{CUA}$ to the tyrosine enzyme, and vice versa. In addition, the reduction in the free concentration of each enzyme caused by the sequestering of each one with its cognate tRNA [Eqs. (3b) and (3d)] is not taken into account.

Notwithstanding these limitations, the calculated fraction of aminoacylation with tyrosine is 77%. The observed value is >95% tyrosine and essentially no alanine. Thus, the tyrosine enzyme completely out-competes the alanine enzyme *in vivo.* The simplest calculation indicates that tyrosine is preferred, although not to the degree that is observed. The preference for tyrosine is due to the higher concentration of tyrosine–tRNA synthetase and its lower K_m for G3:U70 $tRNA^{Tyr}_{CUA}$.

On the other hand, if the concentration of the alanine enzyme is raised just 17-fold, then the simple application of Eq. (1) predicts that the aminoacylation of G3:U70 $tRNA^{Tyr}_{CUA}$ will switch to 84% alanine and 16% tyrosine. The observed values *in vivo* are >95% alanine and no detectable tyrosine. Thus, a complete switch in identity is achieved with only a 17-fold change in relative enzyme concentrations. The simple application of the Michaelis–Menten equation accounts for much, although not all, of the observed specificity of aminoacylation *in vivo* and its switch between two amino acids.

Numerical Calculations with Eq. (2). If the more accurate Eq. (2) is used to compute the respective aminoacylation velocities, then the fraction of aminoacylation with tyrosine and alanine is predicted to be 75 and 25%, respectively. With a 17-fold overproduction of alanine–tRNA synthetase,

[20] J. R. Piperno and D. L. Oxender, *J. Biol. Chem.* **243,** 5914 (1968).
[21] K. A. Kill and P. Schimmel, *Biochemistry* **28,** 2577 (1989).
[22] R. Calendar and P. Berg, *Biochemistry* **5,** 1690 (1966).
[23] M. Jasin, L. Regan, and P. Schimmel, *J. Biol. Chem.* **260,** 2226 (1985).

the proportions change to 19% (tyrosine) and 81% (alanine) (Table II). Thus, even though the Michaelis–Menten approximation [Eq. (1)] should not be used when substrate and enzyme concentrations are comparable, with the particular combination of parameters that have been used here, the approximation gives a reasonably accurate estimate of aminoacylation specificity.[11]

As stated above, these calculations do not treat the system as the coupled competitive equilibria that are described by Eqs. (3a)–(3d). Instead, the velocity of aminoacylation of G3:U70 $tRNA^{Tyr}_{CUA}$, with each of the amino acids, is computed as if the reactions were done in separate vessels. With this model the ratio of one amino acid to the other reaches a theoretical limit. One limit is obtained at low G3:U70 $tRNA^{Tyr}$ concentrations, where virtually all of either enzyme is in the free form. Another occurs at high concentrations of G3:U70 $tRNA^{Tyr}$, where each enzyme is saturated. Here, the concentration of enzyme–tRNA complex is equal to total enzyme concentration [Eq. (2)]. The magnitude of the effect of the concentration of G3:U70 $tRNA^{Tyr}_{CUA}$ on aminoacylation specificity can be substantial *(vide infra)*.

Numerical Calculations that Account for Direct Competition between Enzymes for G3:U70 $tRNA^{Tyr}_{CUA}$

The previous calculations treated the reactions of each enzyme with G3:U70 $tRNA^{Tyr}$ as independent and unlinked. In fact, the binding equilibria [Eqs. (3b) and 3d)] are mutually exclusive. Thus, the amount of G3:U70 $tRNA^{Tyr}$ available to react with alanine-tRNA synthetase is reduced by that fraction of the substrate that is bound to the tyrosine enzyme, and vice versa. In the simple two-step competition, whichever enzyme–G3:U70 $tRNA^{Tyr}$ complex is most favored in the absence of competition must become still more favored when competition is accounted for. This is because the preferred enzyme–G3:U70 $tRNA^{Tyr}$ complex has the greatest impact on lowering the free substrate concentration that is available to the alternative enzyme. Thus, numerical calculations that consider Eqs. (3b) and (3d) raise the proportion of tyrosine from 75 to 77% and, with overproduction of alanine-tRNA synthetase, raise the proportion of alanine from 81 to 83% (Table II).

Numerical Calculations that Account for All Competing Equilibria

The effect of including Eqs. (3a) and (3c) is to reduce the amount of each enzyme that is available to interact with G3:U70 $tRNA^{Tyr}$. The lowering of concentration is greatest for alanine–tRNA synthetase, because the concentration of $tRNA^{Ala}$ is four-fold higher than that of

TABLE II
CALCULATIONS OF IDENTITY OF G3:U70 tRNA^{Tyr} BASED ON *in Vitro* PARAMETERS[a]

Parameter	Eq. (2) (with no competition)		Eq. (3b) and (3d) (partial competition)		Eq. (3a)–(3d) (complete competition)		Eq. (3a)–(3d) (higher effective concentrations)[b]		Observed[c] (*in vivo*)	
	Tyr	Ala	Tyr	Ala	Tyr	Ala	Tyr	Ala	Tyr	Ala
Constitutive enzyme levels	0.75	0.25	0.77	0.23	0.83	0.17	0.90	0.10	≥0.95	≤0.05
Overproduction of Ala-tRNA synthetase[d]	0.19	0.81	0.17	0.83	0.19	0.81	0.17 (0.12)[e]	0.83 (0.88)[e]	≤0.05	≥0.95

[a] Comparison with results *in vivo*.
[b] The effective concentrations of enzymes and tRNAs have uniformly been raised 10-fold.
[c] Data from Ref. 11.
[d] A 17-fold overproduction of Ala-tRNA synthetase.
[e] Assuming a 25-fold overproduction of Ala-tRNA synthetase.

$tRNA^{Tyr}$ and the K_m for $tRNA^{Ala}$ is nearly two-fold smaller (Table I). Thus, the percentage of alanine acceptance drops when the additional equilibria are included. However, the extent of the reduction in alanine acceptance depends on the total concentration of the enzyme. When the enzyme is overproduced, the effect of including Eq. (3c) on the concentration of free alanine-tRNA synthetase is less significant.

The calculated proportion of acceptance of tyrosine thus rises to 83%.When alanine-tRNA synthetase is overproduced, there is 19% tyrosine and 81% alanine (Table II). This is the most accurate estimate of the identity of G3:U70 $tRNA^{Tyr}$ that can be obtained with the model. In general, the calculated effect of competition is to sharpen the specificity of charging and bring the predicted values somewhat closer to the observation of apparent exclusive charging with one amino acid or the other.

Effect of Adjusting Level of Test tRNA

For these experiments and calculations, the G3:U70 $tRNA^{Tyr}_{CUA}$ amber suppressor is used to test for aminoacylation specificity *in vivo.* The equilibria in Eqs. (3a)–(3d) are so tightly linked that, while both E_{Tyr} and E_{Ala} efficiently aminoacylate G3:U70 $tRNA^{Tyr}_{CUA}$, the proportion of aminoacylation with each amino acid depends on the absolute concentration of this tRNA, even with the concentrations of all other species (E_{Tyr}, E_{Ala}, $tRNA^{Tyr}$, $tRNA^{Ala}$) held fixed. This is illustrated in Fig. 2A and B, where the percentage of aminoacylation with tyrosine and with alanine is plotted as a function of (G3:U70 $tRNA^{Tyr}_{CUA}$), for constitutive and elevated levels of alanine-tRNA synthetase. As the concentration of G3:U70 $tRNA^{Tyr}_{CUA}$ is raised, the proportion of aminoacylation with tyrosine falls from one plateau to another. The most sensitive part of each curve occurs at the K_m (14 μM) for the interaction of G3:U70 $tRNA^{Tyr}_{CUA}$ with alanine-tRNA synthetase. Thus, at a concentration of G3:U70 $tRNA^{Tyr}_{CUA}$, where binding to alanine-tRNA synthetase becomes most competitive, the proportion of tyrosine acceptance drops. At constitutive levels of alanine-tRNA synthetase, the proportion of tyrosine acceptance is most affected and drops from 85 to 69%. The final plateau represents the partitioning of aminoacylation when both enzymes are saturated with the test tRNA.

This illustrates the sensitivity of the specificity of aminoacylation of a given tRNA species to its absolute concentration, even with all other concentration variables and kinetic parameters held fixed. The important variable is the concentration of a tRNA relative to the K_m values for the enzymes that aminoacylate the particular tRNA species. In a particular example, the K_m of G3:U70 $tRNA^{Tyr}_{CUA}$ for tyrosine-tRNA synthetase is lower than that for alanine-tRNA synthetase and, in addition, the K_m of

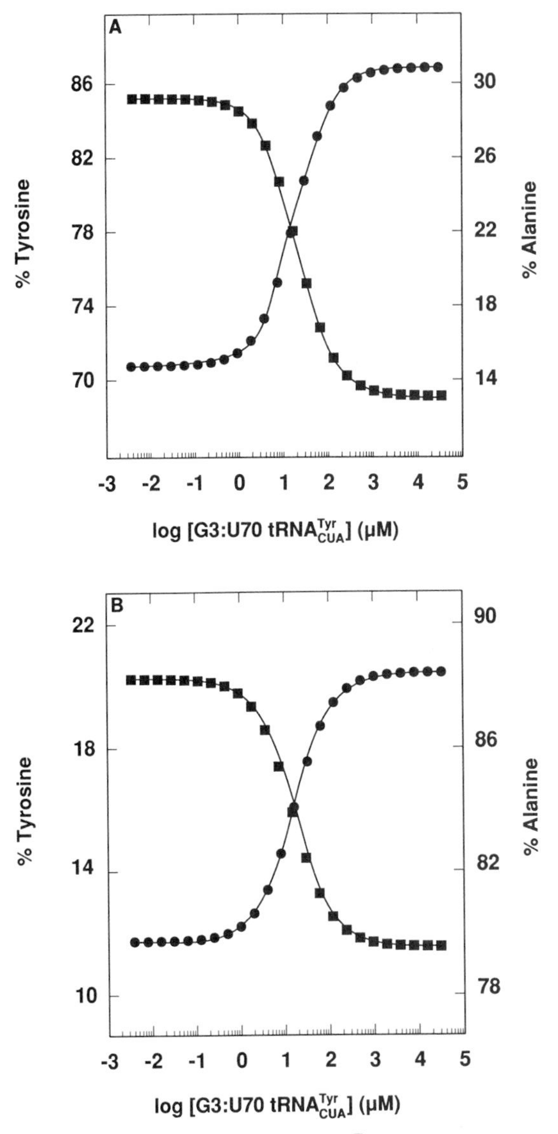

FIG. 2. Effect of concentration of G3:U70 tRNA$^{Tyr}_{CUA}$ on aminoacylation specificity for constitutive (A) and elevated (17-fold) (B) levels of alanine-tRNA synthetase. Numerical calculations, based on the equilibria of Eqs. (3a)–(3d) with Eq. (2) together with Eq. (8a) were done at a series of concentrations and the individual points were plotted and joined by a smooth curve. Kinetic parameters given in Table I and in the text were used for the calculations. (■), Tyrosine; (●), alanine.

the $tRNA^{Ala}$ that competes for alanine-tRNA synthetase is lower than the K_m of $tRNA^{Tyr}$ for tyrosine-tRNA synthetase (Table I). Thus, at lower concentrations of G3:U70 $tRNA^{Tyr}_{CUA}$, the equilibria are heavily weighted in favor of aminoacylation with tyrosine. However, as the concentration of G3:U70 $tRNA^{Tyr}_{CUA}$ is raised, the relative ability of alanine-tRNA synthetase to compete for this tRNA becomes more favorable as the differences between K_m values are overwhelmed by the higher concentration of the test tRNA.

Further Considerations

As shown above, with normal enzyme levels and a fixed value of 3 μM for G3:U70 $tRNA^{Tyr}_{CUA}$, there is a progressive improvement of the calculated specificity from 75 to 83% tyrosine as more and more competition is taken into account. The effect of this competition is to reduce the amount of alanine enzyme available to interact with G3:U70 $tRNA^{Tyr}$. This suggests that, were the system to come closer to conditions for maximal removal of alanine-tRNA synthetase by interaction with $tRNA^{Ala}$, then an even greater proportion of tyrosine might be achieved. One way to achieve greater degree of saturation of the enzymes is to raise uniformly all of the *effective* enzyme and tRNA concentrations. Because there is more $tRNA^{Ala}$ than $tRNA^{Tyr}$, complete saturation of each enzyme with the cognate tRNA will remove more alanine- than tyrosine-tRNA synthetase.

The rationale for uniformly raising concentrations is that *in vivo* the free volume available to the reacting molecules in one cell is less than the total volume of 10^{-15} liters. The available volume is reduced because large regions of intracellular space are occupied by other components (ribosomes, DNA, etc.) that exclude macromolecules like synthetases and tRNAs. Thus, if concentrations are calculated by dividing the total number of molecules per cell by the cell volume, then the total protein concentration in *E. coli* estimated from data tabulated by Neidhardt[24] is 4 mM, while the lipid concentration is about 35 mM. This results in an environment in which the synthetase–tRNA reactants have much higher local concentrations.

If concentrations are raised by 10-fold for each species, then the fraction of tyrosine rises to 0.90 (Table I). A similar effect could be achieved by lowering each K_m value by the same proportion so as to keep their relative values, or by a combination of lower K_m values *in vivo* and higher local

[24] F. C. Neidhardt, *in* "*Escherichia coli* and *Salmonella typhimurium:* Cellular and Molecular Biology, Volume 1" (F. C. Neidhardt, J. L. Ingraham, K. B. Low, B. Magasanik, M. Shaechter, and H. E. Umbarger, eds.), p. 3. American Society for Microbiology, Washington, D.C., 1987.

concentrations. Thus, without changing the *relative* values for any of the kinetic or concentration parameters, the model can bring the proportion of tyrosine acceptance for G3:U70 $tRNA^{Tyr}$ closer to the observed value.

With overproduction of alanine-tRNA synthetase, the calculated fraction of alanine (0.83) is still a little below the observed value. However, this fraction is sufficiently sensitive to the extent to which the concentration of alanine-tRNA synthetase is elevated that the gap between observed and calculated values could be due in part to a modest experimental error in the estimation of the degree of overproduction. For example, if overproduction is assumed to be 25-fold instead of 17-fold, then the recalculated fraction of alanine rises from 0.83 to 0.88 (Table II).

Comments

The G3:U70 $tRNA^{Tyr}_{CUA}$ system that was designed to investigate parameters that affect charging specificity *in vivo* is artificial because natural tRNAs do not have strong determinants for recognition by two different enzymes. Nonetheless, in the absence of overproduction of Ala-tRNA synthetase, G3:U70 $tRNA^{Tyr}_{CUA}$ is predicted and observed to be charged predominantly with tyrosine. Thus, a tRNA can, through mutation, acquire a major determinant for another enzyme and still retain its original specificity. At least part of the retained specificity depends on the degree to which a mutation disturbs the cognate enzyme-tRNA interaction. In the case studied here, the G3:U70 substitution into $tRNA^{Tyr}_{CUA}$ had no significant effect on the kinetic parameters for charging by tyrosine-tRNA synthetase (Table I). Another factor that works toward retention of the original specificity is the sequestering of the "second" enzyme with its cognate tRNA.

For the calculations we have rigidly fixed *in vitro* kinetic parameters and values for intracellular concentrations of enzymes and tRNAs and applied them to an *in vivo* system of amber suppression. The agreement between calculated and observed quantities is sufficiently close to suggest that the model is a realistic approximation of most of the features of the synthetase-tRNA interactions *in vivo*. The model would be more realistic if additional equilibria were taken into account, such as binding of the enzymes to other tRNAs and, conversely, the interaction of G3:U70 $tRNA^{Tyr}$ with other enzymes. Thus, it is known that the $tRNA^{Tyr}$ amber suppressor can interact with glutamine-tRNA synthetase[7] and there is a little misacylation of G3:U70 $tRNA^{Tyr}_{CUA}$ with glutamine *in vivo*.[11] The result of these additional interactions will be to lower the effective concentrations of free enzymes and substrates, although not necessarily by the same proportion for each. While perturbations of enzyme and substrate concentrations will affect the proportion of charging of G3:U70 $tRNA^{Tyr}$ with alanine and tyrosine, the extent of these changes depends entirely on the parameters for the cross-interactions. These additional interactions cannot be factored into the calculations without more experimental data. However, were such parameters available, the numerical methods that have been used here can accommodate additional reactions by a straightforward generalization of the procedures that are outlined here.

Acknowledgment

This work was supported by Grant GM 23562 from the National Institutes of Health (NIH). J. J. B. holds NIH postdoctoral fellowship GM 12122.

[26] Design of Simplified Ribonuclease P RNA by Phylogenetic Comparison

By NORMAN R. PACE and DAVID S. WAUGH

Introduction

Complex biological macromolecules commonly are multifunctional and consist of discrete structural domains with different functions. The fabrication of abbreviated versions of such complex molecules allows study or use of a particular function in isolation from the rest of the natural molecule. The design and testing of simplified enzymes is also an approach to identifying elements of molecular structure required for catalysis. The goal is to construct the smallest active version of an enzyme and thereby to determine the minimum structure required for catalysis. We have used this approach to outline the catalytic core of the RNA enzyme ribonuclease P (RNase P).[1]

RNase P, an endonuclease, forms the 5′ ends of tRNAs. In eubacteria RNase P consists of essential protein (ca. 14 kDa) and RNA (ca. 400 nucleotides) components.[2,3] However, the RNase P RNA alone is an efficient and accurate catalyst at high salt concentrations *in vitro.* The high ionic strength evidently screens anionic repulsion between the enzyme and substrate RNAs. In order to identify potentially important structure in

[1] D. S. Waugh, C. J. Green, and N. R. Pace, *Science* **244,** 1569 (1989).
[2] S. Altman, *Adv. Enzymol. Relat. Areas Mol. Biol.* **62,** 1 (1989).
[3] N. R. Pace and D. Smith, *J. Biol. Chem.* **265,** 3587 (1990).

RNase P RNA, we undertook to delete sequences that are not required for activity, seeking the minimum functional structure. This approach has been useful in localizing the portion of tRNA that is recognized by RNase P[4] and in the study of other catalytic RNAs.[5,6]

Design of *Min* 1 RNA

The RNase P RNAs are too complex for an efficient identification of nonessential sequences by random deletion mutagenesis. We therefore used a phylogenetic comparison of secondary structure models to identify potentially dispensable sequences. Secondary structure models of nine RNase P RNAs from two eubacterial phyla, the gram-positive bacteria (e.g., *Bacillus megaterium* in Fig. 1) and the "purple bacteria," (e.g., *Escherichia coli* in Fig. 1), have been derived using a comparative approach to identify base-paired sequences.[7] Differences between the models of the RNAs from the two phyla are due substantially to the phylum-specific occurrence of discrete helical elements at various positions in a core of homologous primary and secondary structure.[7,8] Since the phylum-specific helical elements are not present in the RNase P RNAs of all organisms, they are not expected to be crucial for the enzymatic activity of the RNA. We relied on the structure models to design a simplified RNase P RNA that consists only of phylogenetically conserved features.

The *E. coli* RNase P RNA (M1 RNA) was used as the foundation of the design because its structure model has fewer interruptions in the conserved core than do those of the RNase P RNAs from gram-positive bacteria. There are four proposed helices in the *E. coli* RNase P RNA model that have no counterparts in the gram-positive versions and so are not expected to be essential for enzymatic activity. These helices involve nucleotides 28–52, 148–167, 183–227, and 260–290 of the folded *E. coli* sequence in Fig. 1. They were excluded from the design of the simplified RNA using two strategies: simple omission from the design, and replacement of *E. coli* sequences with homologous segments of a gram-positive P RNA that lacks the *E. coli*-specific helical elements.

Nucleotides 28–52 in the *E. coli* RNase P RNA structure model form the apical part of a longer helix that involves nucleotides 20–61. Since only

[4] W. H. McClain, C. Guerrier-Takada, and S. Altman, *Science* **23,** 527 (1987).

[5] J. W. Szostak, *Nature (London)* **322,** 83 (1986).

[6] O. C. Uhlenbeck, *Nature (London)* **328,** 596 (1987).

[7] B. D. Jones, G. J. Olsen, J. Liu, and N. R. Pace, *Cell (Cambridge, Mass.)* **52,** 19 (1988).

[8] We use the term homologous in its strictest sense: homologous sequences have common ancestry and function. Homologous sequences are not necessarily identical; identical sequences are not necessarily homologous.

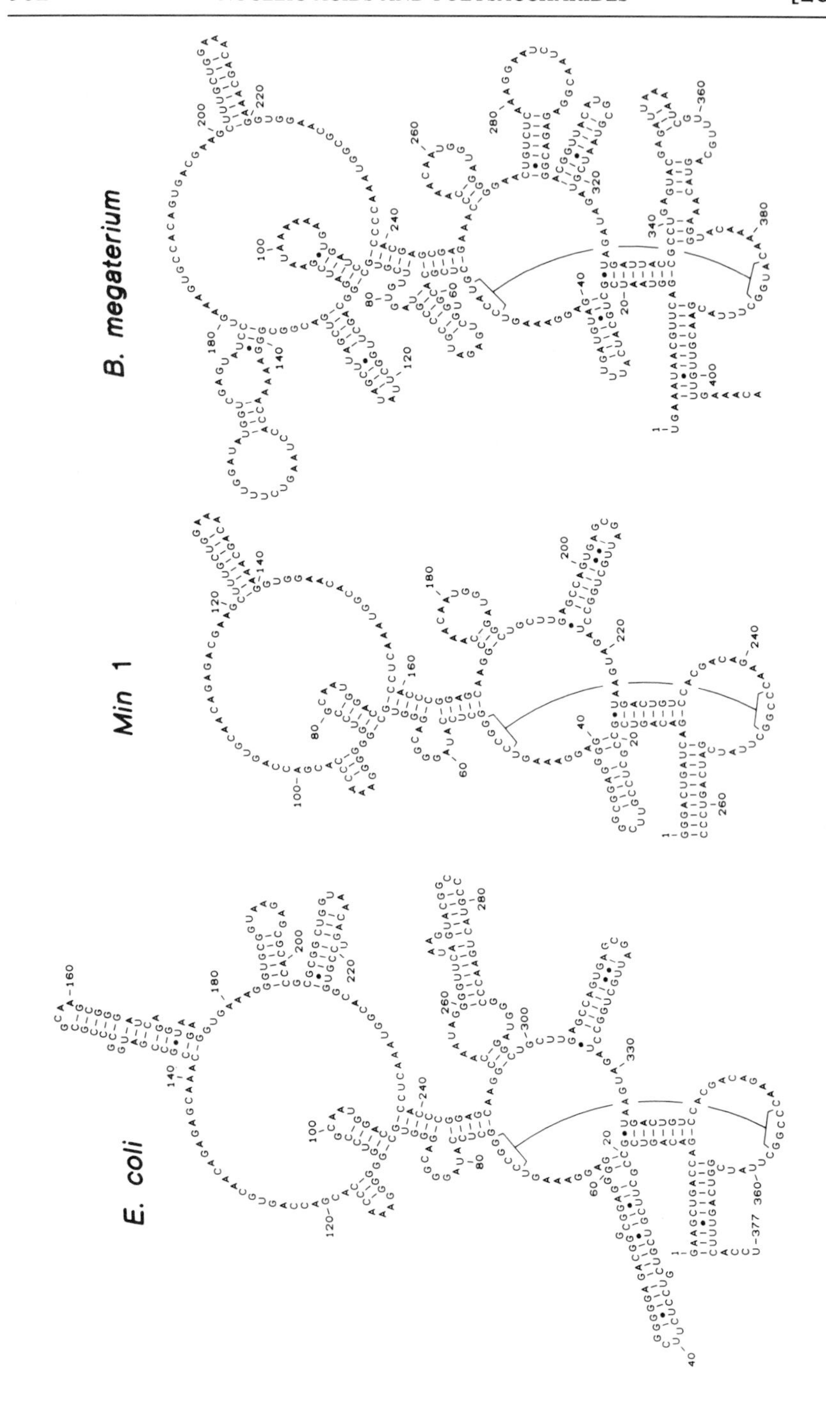
E. coli
Min 1
B. megaterium

the core-proximal portion of this helix (nucleotides 20–27 and 53–61) has a homolog in the gram-positive RNA models (e.g., *B. megaterium* in Fig. 1), nucleotides 28–52 were omitted in the design of the simplified RNA. The residual stem was capped with the former loop (UUCG, nucleotides 39–42). This sequence commonly occurs at the apex of RNA hairpins and confers particular stability.[9,10] We were, however, concerned that the shortened helix might be less stable than the original one, so U25 in the *E. coli* sequence was replaced with C in the simplified RNA. This converts a G/U pair to an ostensibly more stable G/C pair in the proposed, residual helix.

We were reluctant simply to omit from the design the three remaining phylum-specific helices present in the *E. coli* RNase P RNA because of concern that, in the absence of detailed knowledge of their structures, we might not properly restore the remnant secondary and tertiary structure. Consequently, the regions of the *E. coli* RNA sequence containing those helices were replaced with blocks of gram-positive sequence that occupy homologous positions in the proposed structure models, yet contain fewer nucleotides. We presumed that the superstructure of the synthetic RNA would accommodate the smaller homolog without steric constraint. Nucleotides 137–227 and 257–291 in the *E. coli* RNase P RNA were, respectively, replaced in the designed sequence with nucleotides 196–226 and 258–261 from the *B. megaterium* RNA. Thus, the simplified RNA design is a composite of RNase P RNA sequences from two different organisms. Portions of the *B. megaterium* RNase P RNA sequence were used instead of another gram-positive RNase P RNA sequence because this would result in the presence of a convenient restriction site in the synthetic gene.

A few additional changes from the native *E. coli* sequence were made near the ends of the designed RNA in order to create useful restriction sites and to accommodate a bacteriophage T7 RNA polymerase promoter. Nucleotides 2–4 in the *E. coli* sequence were changed from AAG to GGA in order to position a bacteriophage T7 promoter[11] as closely as possible to

[9] C. Tuerk, P. Gauss, C. Thermes, D. R. Groebe, M. Gayle, N. Guild, G. Stormo, Y. d'Aubenton-Carafa, O. C. Uhlenbeck, I. Tinoco, Jr., E. Brody, and L. Gold, *Proc. Natl. Acad. Sci. U.S.A.* **85,** 1364 (1988).

[10] D. R. Groebe and O. C. Uhlenbeck, *Biochemistry* **28,** 742 (1989).

[11] J. F. Milligan and O. C. Uhlenbeck, this series, Vol. 180, p. 51.

FIG. 1. Proposed secondary structures of the *Bacillus megaterium* and *Escherichia coli* RNase P RNAs based on phylogenetic comparisons,[7] and the proposed secondary structure of the *Min* 1 RNase P RNA. The 5′-terminal nucleotide in each sequence corresponds to position 1. Base-paired residues are indicated by hyphens (canonical Watson–Crick pairs) or by solid dots (noncanonical pairs, e.g., G · U)[1].

the synthetic gene. Nucleotides 371–372 in the *E. coli* sequence were therefore also changed from UU to CC in the designed sequence in order to maintain Watson–Crick complementarity between the termini. The base pair (bp) involving nucleotides 9 and 365 in the *E. coli* sequence was replaced by a different canonical base pair in order to create useful restriction sites (*Bcl*I, *Cla*I) near the termini of the synthetic RNase P RNA gene. These latter alterations maintained the potential for base pairing between the termini of the synthetic RNA. The proposed secondary structure of the simplified RNase P RNA, called *Min* 1 RNA, is shown in Fig. 1.

Methods

Assembly and Molecular Cloning of Min 1 RNase P RNA Gene. Both strands of a gene encoding the *Min* 1 RNase P RNA and a bacteriophage T7 promoter were assembled by annealing 12 synthetic oligodeoxyribonucleotides ranging from 41 to 62 residues in length. The oligonucleotides, provided by Dr. Chris Green of SRI International (Palo Alto, CA), were phosphorylated prior to annealing. The nicks were then sealed with DNA ligase, and the synthetic gene was cloned in a pUC[12]-derived plasmid vector to create pDW133.[13]

Synthesis of RNA Enzymes. Recombinant plasmids carrying either the *E. coli* (pDW27)[1] or the *Min* 1 (pDW133) RNase P RNA genes adjacent to phage T7 promoters were linearized with the appropriate restriction endonucleases and transcribed *in vitro*.[11] The (100 μl) transcriptions were carried out at 37° for 2 hr in 20 mM $NaPO_4$ (pH 7.7), 10 mM dithiothreitol, 8 mM $MgCl_2$, 4 mM spermidine trihydrochloride. The reactions also included each of the four ribonucleoside triphosphates at 1 mM, 10 μg of linear plasmid DNA (ca. 3 kbp), and a saturating amount of T7 RNA polymerase. The reaction products were resolved by electrophoresis through 5% (w/v) polyacrylamide gels containing 8 M urea. The transcription products were located in the gels by UV shadow[14] and excised. The RNA was recovered from the gel slices by soaking them overnight in 10 mM Tris-HCl (pH 8.0), 1 mM EDTA (TE) containing 0.1% (w/v) sodium dodecyl sulfate (SDS). The eluates were filtered, and then the RNA was precipitated by the addition of 0.1 vol of 3 M sodium acetate and 3 vol of ethanol. After centrifugation, the pellets were resuspended in 100 μl TE and the RNA was precipitated again with sodium acetate and ethanol. The RNA was pelleted by centrifugation and resuspended in TE. The concen-

[12] C. Yanisch-Perron, J. Vieira, and J. Messing, *Gene* **33,** 103 (1985).
[13] D. S. Waugh, Ph.D. Thesis, Indiana University, Bloomington, Indiana, 1989.
[14] S. M. Hassur and H. W. Whitlock, *Anal. Biochem.* **59,** 162 (1974).

trations of RNAs were estimated by ethidium bromide spot tests,[15] using *Bacillus subtilis* rRNA as the concentration standard.

Synthesis of Pre-tRNA Substrate. The tRNA precursor was synthesized by *in vitro* transcription, in the presence of [α-^{32}P]GTP, of a *B. subtilis* $tRNA^{Asp}$ gene cloned adjacent to a phage T7 promoter.[13] The synthetic pre-tRNA substrate has a mature 3′ end (CCA) and a 35-nucleotide, 5′ precursor segment that is removed by RNase P. The procedure for the synthesis and purification of ^{32}P-labeled pre-$tRNA^{Asp}$ was the same as for the runoff transcripts of RNase P RNA (above), except that transcription reactions contained only 100 μM unlabeled GTP (other nucleotides 1 mM) and 8% (w/v) polyacrylamide gels containing 8 M urea were used to resolve the transcription products. The concentration of ^{32}P-labeled pre-$tRNA^{Asp}$ was calculated using the specific activity of the labeled ribonucleotide reported by the manufacturer.

Assay of Activity of RNase P RNA. All (20 μl) reactions contained 50 mM Tris–acetate (pH 8.0), 50 mM magnesium acetate, 0.05% Nonidet P-40 (NP-40; Sigma, St. Louis, MO), [^{32}P]pre-$tRNA^{Asp}$ (10^{-7} M for the *E. coli* reactions and 10^{-5} M for reactions with the *Min* 1 RNase P RNA), and either the *E. coli* or the *Min* 1 RNase P RNA at 10^{-9}–10^{-8} M. Various concentrations of ammonium acetate were used, as discussed below. The reactions were incubated at the indicated temperatures for 20 min, then stopped by the addition of 3 vol of ethanol. The reaction components were recovered by centrifugation, resuspended in 10 μl or 8 M urea, 20 mM EDTA, 0.1% SDS, and resolved by electrophoresis in an 8% polyacrylamide gel containing 8 M urea. After fixing and drying the gel, the product bands were located by autoradiography (see Fig. 2), excised, added to scintillation fluid, and counted to determine the fraction of substrate cleaved.

Properties of *Min* 1 RNA

The *Min* 1 RNA was produced by runoff transcription and assayed for activity under a variety of conditions, all as described. The *Min* 1 RNA is catalytically active with a cleavage specificity for pre-tRNA that is indistinguishable from that of the naturally occurring RNase P RNAs (Fig. 2). However, its kinetic properties (Table I) and optimum reaction conditions differ from those of the native (*E. coli*) RNA (Fig. 3). The Michaelis constant (K_m) of the *Min* 1 RNA is 100-fold larger than that of the native RNA, indicating that the *Min* 1 RNA has less affinity for the substrate than

[15] T. F. Maniatis, E. F. Fritsch, and J. Sambrook, "Molecular Cloning: A Laboratory Manual," p. 468. Cold Spring Harbor Laboratory, Cold Spring Harbor, New York, 1982.

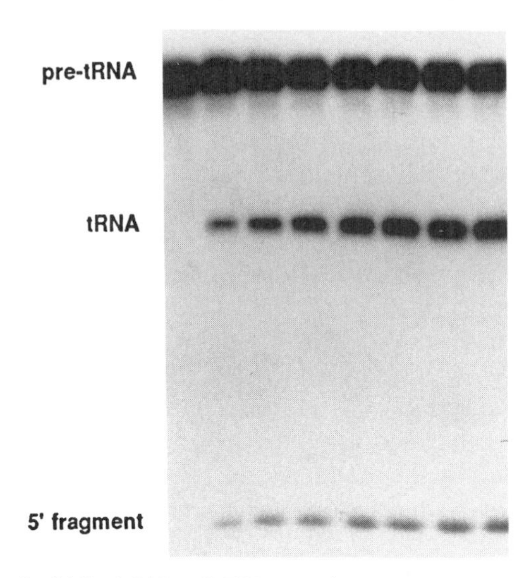

FIG. 2. Assay gel of *Min* 1 RNase P RNA reaction products. Assay of *Min* 1 RNase P RNA activity was carried out as described in the text using as substrate uniformly ^{32}P-labeled pre-tRNA^{Asp}. Samples were withdrawn from an assay at (left to right) 0, 4, 8, 12, 16, 20, 24, and 28 min; resolved by polyacrylamide gel electrophoresis; and visualized by autoradiography as described. An autoradiogram is shown. The positions of the substrate (pre-tRNA) and products in the gel are indicated. The homogeneity in sizes of the products is noteworthy, indicating the precision of the cleavage, at the identical site cleaved by the native RNA or holoenzyme (data not shown).

does the native RNA. On the other hand, at their respective optimal ionic strengths, the maximum velocity (k_{cat}) of the *Min* 1 RNA is 20-fold greater than that of the native RNA and about the same as that of the holoenzyme under physiological conditions.[16] The enhanced k_{cat} of the *Min* 1 RNA probably reflects a more rapid release of product than occurs with the native RNA, for which product release is rate limiting.[17] Since RNase P RNA binds to the mature domain of the pre-tRNA, the reduced affinity (higher K_m) of the *Min* 1 RNA for the substrate would result in more rapid dissociation of enzyme and product following cleavage. Overall, the cata-

[16] The k_{cat} of the *E. coli* and *B. subtilis* holoenzymes depends on reaction conditions and the character of the substrate. Examples of values for k_{cat} of the *E. coli* holoenzyme with pre-tRNA substrates are 0.35 min^{-1} [N. Lumelsky and S. Altman, *J. Mol. Biol.* **202,** 443 (1988)], 2 min^{-1},[19] 6.6 min^{-1} [C. Guerrier-Takada, A. van Belkum, C. W. A. Pleij, and S. Altman, *Cell (Cambridge, Mass.)* **53,** 267 (1988)], and 18.3 min^{-1}.[4] With the substrate and reaction conditions employed in this work k_{cat} for the holoenzymes of both *E. coli* (data not shown) and *B. subtilis*[17] is about 10 min^{-1}.

[17] C. Reich, G. J. Olsen, B. Pace, and N. R. Pace, *Science* **239,** 178 (1988).

TABLE I
KINETIC PARAMETERS FOR PROCESSING OF PRE-tRNA^{Asp} *in Vitro* BY *E. coli* AND *Min* 1 RNASE P RNAs[a]

RNase P RNA	K_m (M)	k_{cat} (min^{-1})	k_{cat}/K_m (M^{-1} min^{-1})
Min 1	5×10^{-6}	10	2×10^{6}
E. coli	4×10^{-8}	0.4	1×10^{7}

[a] Activity of both RNase P RNAs was assayed at 37° in 50 mM magnesium acetate, 50 mM Tris–acetate (pH 8.0), and 0.05% Nonidet P-40 (Sigma). Reactions containing the *E. coli* RNase P RNA (10^{-8} M) also included 0.5 M ammonium acetate, and reactions containing the *Min* 1 RNase P RNA (10^{-9} M) also included 3 M ammonium acetate. For each RNA enzyme, assays were conducted at various substrate concentrations above and below the K_m value. Aliquots were withdrawn from the reactions at regular intervals and processed as described (text and Fig. 2).

lytic efficiencies, or specificity constants (k_{cat}/K_m), of the *Min* 1 and *E. coli* RNase P RNAs are similar.

The design strategy was successful in that the synthetic RNA is highly active. However, the *Min* 1 RNase P RNA activity is significantly more sensitive to temperature and requires higher concentrations of monovalent (but not divalent) salt than the native RNA activity (Fig. 3). The temperature sensitivity of the *Min* 1 RNA activity suggests that the intramolecular forces responsible for its folding may be weaker than in the native RNA, possibly due to the absence of the phylum-specific structures or to a defect in the design of the RNA. The resulting instability could also be responsible for the increased salt dependence of the *Min* 1 activity. If the intramolecular packing forces of the *Min* 1 RNA are weaker than those of the native RNA, then electrostatic repulsion between phosphates could distort the structure into an active form. The high salt concentration required for activity of the *Min* 1 RNA could provide counterions to titrate such destabilizing repulsion and allow the RNA to assume its active conformation.

The addition of the RNase P protein from either *E. coli* or *B. subtilis* does not stimulate the activity of the *Min* 1 RNase P RNA at physiological

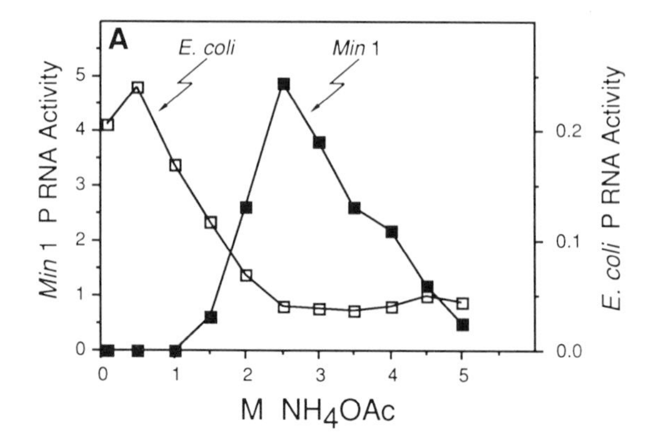

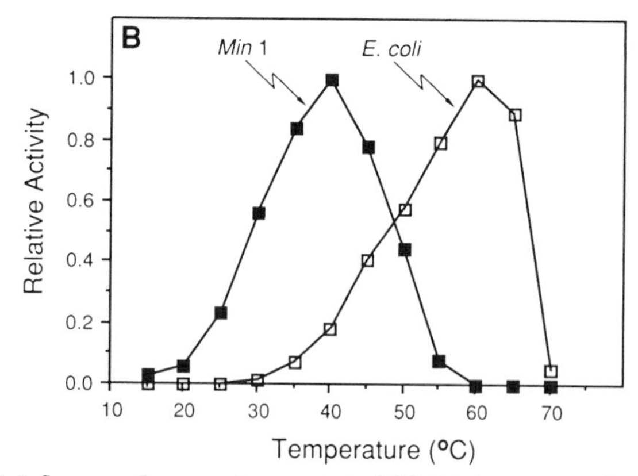

FIG. 3. (A) Influence of ammonium acetate (NH_4OAc) concentration on the *in vitro* tRNA processing activity of the *E. coli* and *Min* 1 RNase P RNAs. All reactions (20 μl) contained 50 m*M* Tris–acetate (pH 8.0), 50 m*M* magnesium acetate, 0.05% Nonidet P-40 (Sigma), [^{32}P]pre-tRNA^{Asp} (10^{-7} and 10^{-5} *M* for the *E. coli* and *Min* 1 RNase P RNA reactions, respectively), and either the *E. coli* or the *Min* 1 RNase P RNA at 10^{-8} *M*. The reactions were incubated at 37° for 20 min and stopped by the addition of 3 vol of ethanol. The products were recovered as ethanol precipitates, resuspended in 10 μl of 8 *M* urea, 20 m*M* EDTA, and 0.1% SDS, and resolved by electrophoresis in 8% polyacrylamide gels containing 8 *M* urea. After fixing and drying the gels, the bands were located by autoradiography, excised, and counted. Activity is expressed as moles of substrate cleaved per mole of enzyme per minute. (B) Influence of temperature on the *in vitro* tRNA processing activity of the *E. coli* and *Min* 1 RNase P RNAs. All reactions (20 μl) contained 50 m*M* HEPES (pH 8.0), 50 m*M* magnesium acetate, and 0.05% Nonidet P-40 (Sigma). Reactions containing *E. coli* RNase P RNA (10^{-8} *M*) also included 0.5 *M* ammonium acetate and 10^{-7} *M* [^{32}P]pre-tRNA^{Asp}. Reactions containing the *Min* 1 RNase P RNA (10^{-8} *M*) also included 3 *M* ammonium acetate and 10^{-5} *M* [^{32}P]pre-tRNA^{Asp}. Incubations were carried out at 37°, and the reactions were processed as described above. For each RNA enzyme, "Relative Activity" denotes the fraction of the maximum activity observed over the indicated temperature range.[1]

ionic strength.[18] This observation might suggest that one or more of the omitted structural elements are required for interaction with the proteins. However, the RNAs from gram-positive organisms also lack these structures, yet are able to form active holoenzymes with the *E. coli* RNase P protein.[19] Consequently, we believe that the omitted structures are not directly involved in the binding of the proteins. The inability of the proteins to stimulate catalysis by the *Min* 1 RNA at physiological ionic strength may be a consequence of the requirement by the synthetic RNA for particularly high salt concentrations in order to achieve appreciable activity. The conditions required to activate the *Min* 1 RNA may be incompatible with the function or assembly of the holoenzyme. The native holoenzyme also is inactive at high salt concentrations.[20]

The near-native catalytic efficiency of the *Min* 1 RNA (under the appropriate conditions) proves that the phylum-specific structures that were excluded from the design are not necessary for the specific cleavage of pre-tRNA. Moreover, the *E. coli*-specific structures evidently are not absolutely required *in vivo:* the RNase P RNA gene from *Bacillus subtilis,* which is similar to that of *B. megaterium,* can replace a deleted *E. coli* gene and maintain viability.[21] The salt dependence and temperature sensitivity of the *Min* 1 RNA may indicate that one or more of the variable elements contribute to the global stability of the RNA. If so, then structures that occur at different positions in the two types of RNAs (Fig. 1) must have similar effects on global stability of the ribozyme. Alternatively, some of the *B. megaterium*-specific nucleotides used to construct the *Min* 1 RNA may not have fulfilled the role of their *E. coli* homologs in supporting the tertiary structure. In any case, it is clear that the catalytic function of the naturally occurring RNase P RNAs resides in their phylogenetically conserved structures. The fact that the structure models could be used to predict the dispensability of extensive sequences, scattered throughout the interior of the natural RNAs, supports the accuracy of the models.

The *Min* 1 RNA may prove more useful than the natural RNase P RNAs for some studies. Because it is smaller than the natural molecules, the *Min* 1 RNA may be more amenable to spectroscopic studies. The reduced affinity of the *Min* 1 RNA for the substrate, yet holoenzyme-like maximum velocity, makes it more useful than the natural RNAs for steady

[18] We are grateful to Drs. Madeline Baer and Sidney Altman (Yale University, New Haven, CT) for providing *E. coli* RNase P protein, and to Dr. Bernadette Pace (Indiana University, Bloomington, IN) for providing *B. subtilis* RNase P protein.

[19] C. Guerrier-Takada, K. Gardiner, T. Marsh, N. R. Pace, and S. Altman, *Cell (Cambridge, Mass.)* **35,** 849 (1983).

[20] K. J. Gardiner, T. L. Marsh, and N. R. Pace, *J. Biol. Chem.* **260,** 5415 (1985).

[21] D. S. Waugh and N. R. Pace, *J. Bacteriol.* **172,** 6316 (1990).

state kinetic analyses that are not complicated by the rate-limiting step of product release.[17] Finally, the *Min* 1 RNA sequence is a new starting point for the elimination of additional sequence elements. We anticipate that further reductions in the size of the active molecule will be possible, since many random deletion mutants of the natural RNAs lack even conserved features and yet retain some (usually very low) activity.[13,22,23] The inspection of RNase P RNAs from more diverse eubacteria should provide further perspective on the minimal active structure, and allow the rational design of RNase P RNAs that are even simpler than the *Min* 1 RNA.

Acknowledgments

The authors' work is supported by NIH Grant GM34527 and Office of Naval Research Contract N14-87-K-0813. We thank Dr. Gary J. Olsen (University of Illinois) for assistance with artwork and many fruitful discussions.

[22] C. Guerrier-Takada and S. Altman, *Cell (Cambridge, Mass.)* **45,** 177 (1986).
[23] N. P. Lawrence and S. Altman, *J. Mol. Biol.* **191,** 163 (1986).

[27] Molecular Modeling and Electron Diffraction of Polysaccharides

By SERGE PÉREZ

Introduction

Polysaccharides constitute one of the most abundant and diverse families of biopolymers. With several hundred known examples, polysaccharides offer a great diversity of chemical structure, ranging from simple linear homopolymers to branched heteropolymers with repeating octasaccharide units.[1] Polysaccharides are distinguished by a fundamental involvement in the basic processes of life, and their ubiquitous presence is reflected in the spectrum of structural and functional roles they play. More recently, carbohydrate components of macromolecules such as glycopeptides and glycolipids have been shown to be essential as "signals," and their implication in biological recognition has been established. Such a spectrum of roles for carbohydrate polymers arises from their ability to generate a

[1] M. Yalpani, "Polysaccharides: Synthesis, Modifications and Structure/Property Relationships." Elsevier, Amsterdam, 1988.

large number of highly specific three-dimensional structures from only a relatively small number of monomeric residues.[2] Optimized storage, as in the starch granule, or constitution of cell walls, appears to be achieved through native semicrystalline arrangements of macromolecules.

In order to understand the molecular basis of such native arrangements, three-dimensional structures must be determined. The most important method for structure determination of crystalline polymers has been X-ray fiber diffraction. It has been observed that linear polysaccharides prefer to exist as long helices rather than as more complex folded structures. After solubilization, one usually can produce samples in which such helical macromolecules are aligned with their long axes either parallel or antiparallel. Further lateral organization may occur, but rarely to the degree of a three-dimensionally ordered single crystal. Fibrous structures typically provide diffraction data of low resolution. Many helical polysaccharides yield no more than 50 independent X-ray reflections that can be used to determine the molecular geometry of the crystallographic asymmetric unit. Other shortcomings in the use of this X-ray methodology are the difficulty in assigning the unit-cell parameters and ambiguities regarding the choice of the space group. Consequently advances in understanding the biomolecular structure of polysaccharides have been more recent than parallel developments with nucleic acids and proteins, although progress has been made in the development of experimental and computational methods. However, despite efforts of the fiber diffraction community, it has not always been possible to mimic classical crystallography studies and to reach credible solutions of structures using noncontroversial methods. It is the aim of the present chapter to illustrate how electron diffraction data coupled with realistic molecular modeling can yield, in a quantitative fashion, unambiguous structural information.

Before beginning a review and analysis of recent structural studies, a brief overview of the experimental methods and aspects of the theoretical basis of electron crystallography will be presented. Electron crystallography is the use of a coherent electron beam to obtain molecular and crystallographic information by scattering from a crystalline specimen. The electron microscope is still not used extensively for this purpose, and the methods and theory differ from those of conventional X-ray crystallography.[3] Only a brief summary of the main points is presented. The final section of this chapter also considers the power of molecular modeling in predicting three-dimensional arrangements of polymeric materials.

[2] D. A. Rees, "Polysaccharide Shapes." Chapman & Hall, London, 1977.

[3] B. O. P. Beeston, R. W. Horne, and R. Markham, "Electron Diffraction and Optical Diffraction Techniques." North-Holland/American Elsevier, Amsterdam, 1972.

Sample Preparation and Electron Diffraction

The crystallization behavior of polysaccharides, as seen through the light of modern concepts of polymer crystallization, has been the subject of an excellent review.[4] As with many other stereoregular polymers, simple linear polysaccharides, once dissolved and recrystallized, can yield single crystals. Crystal quality depends on proper nucleation followed by crystal growth. Usually, materials having a low molecular degree of polymerization (DP) and narrow molecular weight distribution give the best results. Crystallization is achieved from dilute solution by temperature modification or by the addition of solvent to induce precipitation (nonsolvent). Temperatures as high as possible and compatible with the system under investigation are preferred. Typical temperatures range from 100 to 200°, and the experiments are performed in pressure vessels. For electron microscopy investigations, a suspension of these crystals is evaporated on carbon-coated grids.

Polysaccharides often crystallize with the incorporation of water or solvent molecules. In most cases, a well-defined morphology is obtained, the most popular being platelike. These thin lamellae surfaces have lateral dimensions of several micrometers for only a few tenths of an angstrom in thickness. Usually, the polymer chain axis lies normal to the lamellae surface. Crystalline domains of such dimensions are well suited to examination by transmission electron microscopy in both imaging and diffraction modes.

Electron scattering is determined by the Coulomb potential of the individual atoms that make up the specimen. In the case of a single scattering called the kinematic approximation, the scattered wave function is proportional to the Fourier transform of the Coulomb potential. In the case of X-ray diffraction the scattered wave is proportional to the electron charge density. For both methods, the real space function is made up of cusp-shaped peaks located at the centers of every atom in the structure. The atomic scattering factors for electrons show a pronounced variation due to changes in the electronic shell structure as one advances through the periodic table.

For electron diffraction experiments, the accelerating voltage is at least 100 kV, which corresponds to a relativistic wavelength of 0.037 Å. The atomic cross-section for 100-kV electrons is in the range of 10^5 to 10^6 times greater than the cross-section for 1-Å wavelength X-rays. The very large scattering cross-section therefore requires that the specimens used for electron crystallography have a physical thickness in the range of 100 Å rather

[4] H. Chanzy and R. Vuong, *in* "Polysaccharides—Topics in Structure and Morphology" (E. D. T. Atkins, Ed.), p. 41. MacMillen, London, 1985.

than in the range of 10^7 Å. With the wavelength of the electron being small, the radius of the Ewald diffraction sphere is quite large. Hence, it is possible to record many diffraction maxima simultaneously. The diffraction pattern is recorded in two dimensions. In most cases, the polymer chain axes (c) are perpendicular (or nearly so) to the platelet face. Therefore, the *hk0* reflections can be readily collected from crystals aligned perpendicular to the electron beam; not only the **a***, **b***, and γ^* unit-cell parameters are available, but also the two-dimensional symmetry of the base plane. Depending on the perfection of the crystalline domains, information to a resolution of 1 Å can be obtained. Because typical unit-cell dimensions of crystalline linear polysaccharides are 10–30 Å, one can expect to collect between 50 to 100 $F(hk0)$ amplitudes. Some examples of typical polysaccharide single crystals, together with their electron diffraction diagrams, are presented in Fig. 1.

A problem frequently encountered when studying crystalline biopolymers with the electron microscope relates to the vacuum dehydration of the specimen when inserted inside the instrument column. This is particularly critical when water or solvent are part of the crystal structure. In such an instance, total or partial decrystallization takes place in a matter of minutes, accompanied by drastic distortion of the sample. Several methods have been developed whereby the sample is either viewed inside a hydration chamber[5,6] or quenched in a cryogenic bath prior to insertion into the electron microscope.[7-9] In these instances, the observations are performed at temperatures close to that of boiling liquid nitrogen, where the water of crystallization is stable in high vacuum. With such a technique, frozen wet electron diffractograms are readily recorded.

With the availability of accurately controllable electron microscope goniometer stages, the combination of data from tilt series offers the possibility of gaining insights into the third dimension of the unit cell. Determination of the remaining crystallographic parameters can be made after tilting the crystal around **a*** and **b*** (rotations about any of the reciprocal axis are also possible). Three-dimensional symmetry can be assigned by comparing the diffractograms obtained when the crystals are rotated by an angle μ about either **a*** or **b*** (Fig. 2).[10]

Because the primary recording medium used in electron crystallography is photographic film, it is necessary to use some type of densitom-

[5] U. R. Matricardi, R. C. Moretz, and D. F. Parsons, *Science* **177,** 278 (1972).
[6] S. W. Hui and D. F. Parsons, *Science* **184,** 77 (1974).
[7] H. Chanzy, E. Roche, and R. Vuong, *Kolloid Z. Z. Polym.* **198,** 1034 (1971).
[8] K. A. Taylor and R. M. Glaeser, *Science* **186,** 1036 (1974).
[9] K. J. Taylor, H. Chanzy, and R. H. Marchessault, *J. Mol. Biol.* **92,** 165 (1975).
[10] C. Guizard, D.Sc. Thesis, University of Grenoble, France (1981).

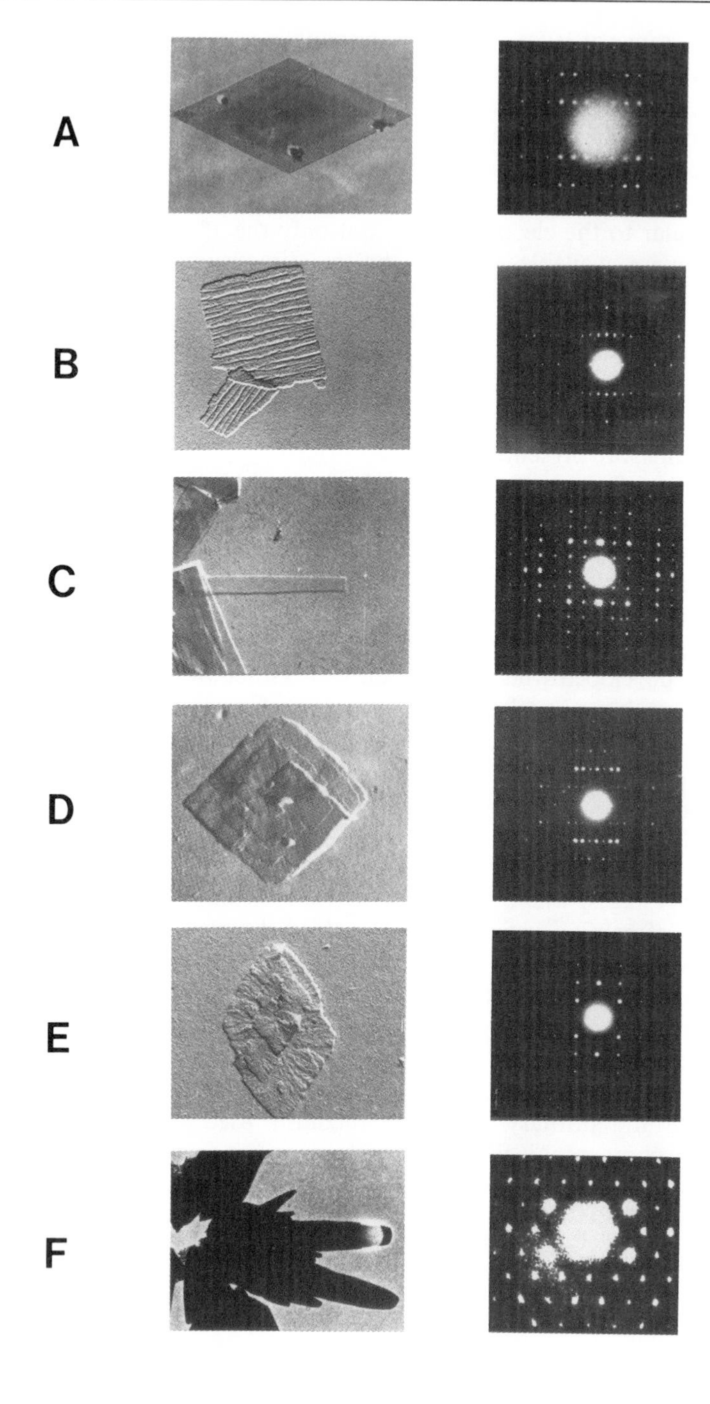
A
B
C
D
E
F

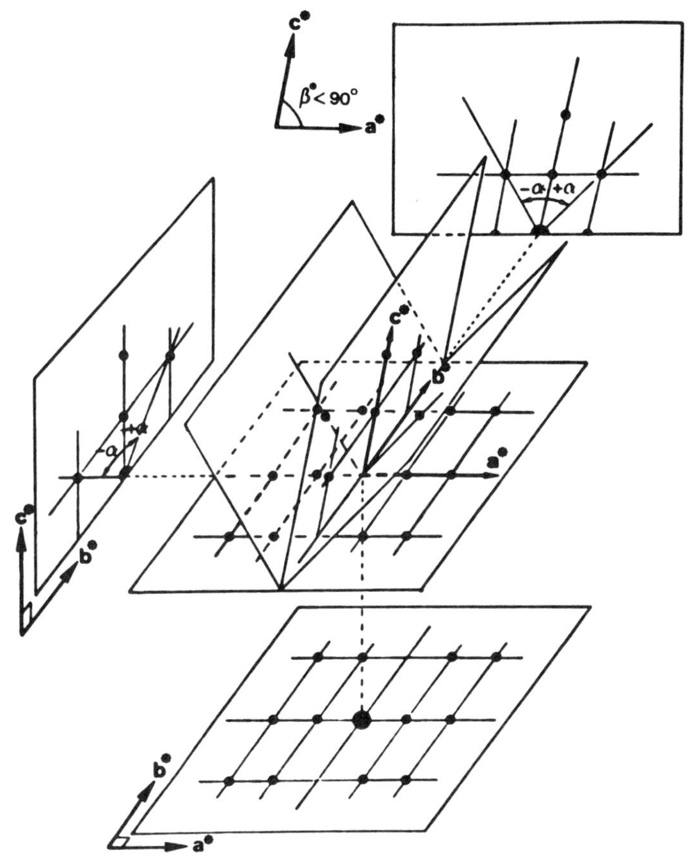

FIG. 2. Schematic representation of several electron diffraction reciprocal planes that can be recorded by proper rotation about the reciprocal axis. From Guizard.[10]

FIG. 1. Examples of typical polysaccharide single crystals, together with their diffraction diagrams. (A) Cellulose triacetate II. From [E. Roche, H. Chanzy, M. Boudeulle, R. H. Marchessault, and P. R. Sundararajan, *Macromolecules* **11,** 86 (1978)]. Reprinted with permission from the American Chemical Society. (B) Anhydrous nigeran. From [S. Pérez, M. Roux, J. F. Revol, and R. H. Marchessault, *J. Mol. Biol.* **129,** 113 (1979)]. (C) Anhydrous high-temperature polymorph of dextran. From [H. Chanzy, G. Excoffier, and C. Guizard, *Carbohydr. Polym.* **1,** 67 (1981)]. Reprinted with permission from App. Science Pub. (D) Hydrated low-temperature polymorph of dextran. From [C. Guizard, H. Chanzy, and A. Sarko, *J. Mol. Biol.* **183,** 397 (1985)]. (E) Mannan I. From [H. Chanzy, S. Pérez, D. P. Miller, G. Paradosi, and W. T. Winter, *Macromolecules* **20,** 2407 (1987)]. (F) A-type polymorph of starch. From [A. Imberty, H. Chanzy, S. Pérez A. Bulion, and J. Tran, *J. Mol. Biol.,* **201,** 365 (1988)].

etry system to measure diffracted intensities. Despite the small number of layers that can be expected to contribute to scattering from the tilted specimen, there are no detectable subsidiary minima, as might be expected to occur from small crystallites. Organic microcrystals may fulfill the requirements for near-kinematic intensities. They are usually extremely thin and composed of light elements. Within such a scheme, structure amplitudes that are the square root of the observed intensities can be used.

Some precautions and corrections may be applied in interpreting the data. The very thinness of the crystals presents a problem that compounds the difficulty. As evidenced by bend contours in diffraction contrast micrographs, the crystals are usually bent. The microcrystals are poor electrical conductors, so the beam damage is rapid. The correction for beam damage by measurements of intensities are various exposure times and their extrapolation to zero time is certainly critical (Fig. 3). Furthermore, exposure must be made quickly, resulting in considerable noise, both in the background and in the diffraction maxima. Preliminary trials have been

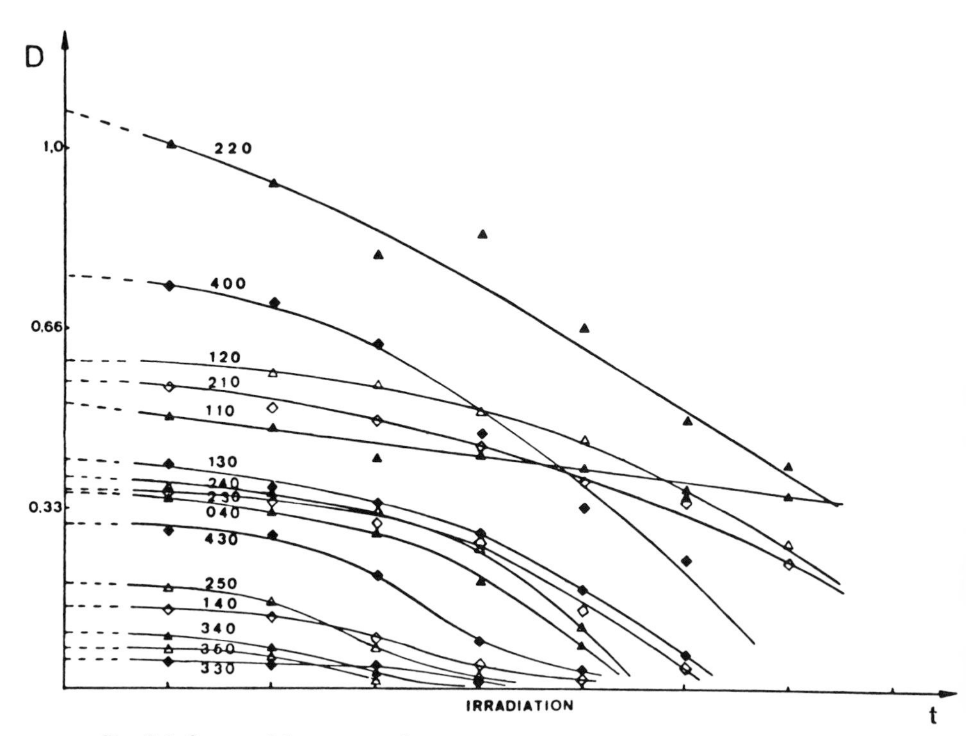

FIG. 3. Influence of the exposure times on the relative intensities of ($hk0$) reflections in the high-temperature polymorph of dextran. From Guizard.[10]

made to incorporate the intensities from higher layer reflections into the data set. Reviews[11,12] discuss the design of optimum diffraction experiments to yield intensity suitable for *ab initio* structure analysis. Methods and computer programs for evaluating electron diffraction data from films have become available.[13] The procedure starts with scanning the electron diffractogram, an area of 20×20 mm, by 10-μm steps through a 5-μm, circular aperture. A data array is thus produced and processed. A Gaussian function is fitted to the background, which is then deducted from the intensity points, and the intensity of each individual diffraction spot is integrated (Fig. 4). Scaling of the intensities on each *hkl* section (recorded using the tilt option) to those of the *hk0* section is done by using the intensities of some reflections that are common to both electron diffractograms. Once the intensities have been evaluated, the structure factors are derived by the relation

$$F(hkl)^2 = I(hkl) \tag{1}$$

described by Dorset.[14] For thin crystals ($d < 100$ Å), made up of light atoms (H, C, N, O), the diffraction conforms to the kinematic approximation.[15] An overall isotropic temperature factor is also included, so that

$$F(hkl) = \exp[-B(\sin \theta/\lambda)^2] \sum f_{\text{eli}} \exp[2\Pi i(r_i S)] \tag{2}$$

The development of very high-resolution electron microscopes of the transmission type has made it possible to obtain images with a point-to-point resolution of about 2 Å. In theory, a combination of the phase information from the image and the diffracted intensities should provide structural information at better than 2 Å. When the crystalline structures are composed of molecular arrays, the resolution requirements are less stringent. However, there are limitations due to several basic difficulties that are inherent to the radiation damage caused by the electron beam and specimen preparation.[16] Nevertheless, it is now possible to maintain high resolution at low magnification with very weak electron exposure. This has the advantage of preserving the molecular features. The images formed by the relatively small number of electrons (in the order of $1e^-/\text{Å}^2$) are statistically poorly defined. Since the molecular structure is periodic and repeats over relatively large distances, averaging techniques or summation procedures can be used to overcome the low signal-to-noise ratio limita-

[11] D. L. Dorset, *J. Electron Microsc. Tech.* **2,** 89 (1985).
[12] D. L. Dorset, *J. Electron Microsc. Tech.* **7,** 35 (1987).
[13] D. Miller, H. Chanzy, and G. Paradosi, *J. Phys.* **45,** C5 (1986).
[14] D. L. Dorset, *Acta Crystallogr. Sect. A* **36,** 592 (1980).
[15] J. M. Crowley and A. F. Moodie, *J. Phys. Soc. Jpn.* **17** (Suppl. BII), 86 (1962).
[16] J. R. Frier, *J. Electron Microsc. Tech.* **11,** 310 (1989).

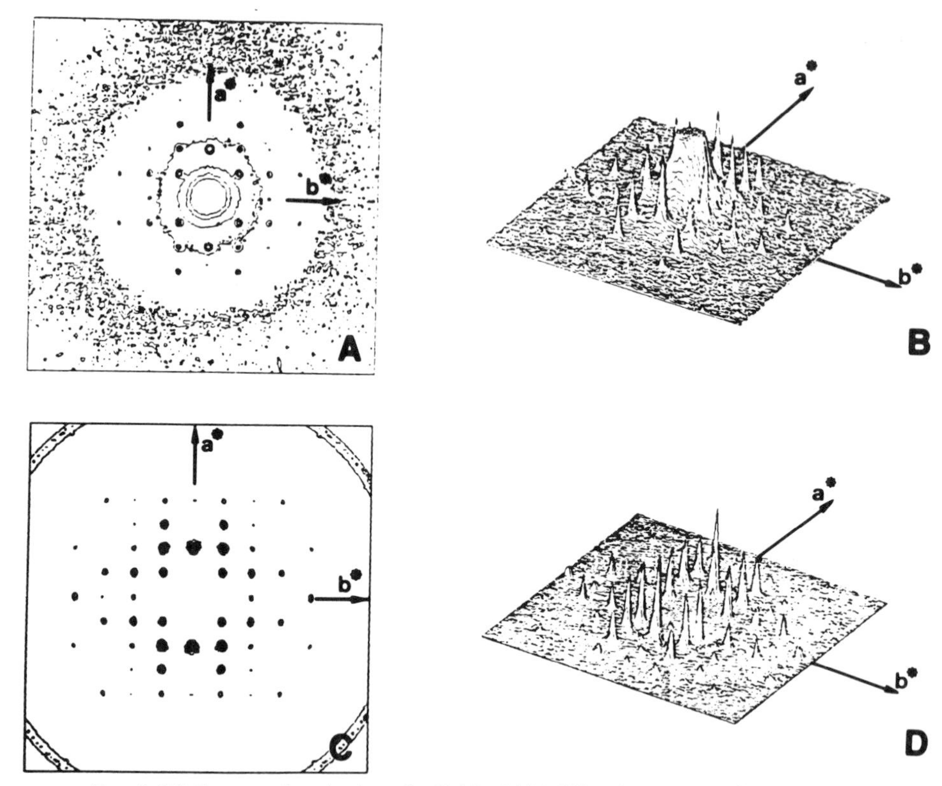

FIG. 4. (A) Contoured projection of a digitized *hk0* diffraction pattern of mannan I before background and noise removal. From [H. Chanzy, S. Pérez, D. P. Miller, G. Paradosi, and W. T. Winter, *Macromolecules* **20**, 2407 (1987)]. Reprinted with permission from the American Chemical Society. (B) As in (A), but in three-dimensional representation. (C) As in (A), but after background removal, and noise reduction. (D) As in (C), but in three-dimensional representation.

tion. These techniques of lattice imaging were first applied to the structural elucidation of the purple membrane protein, bacteriorhodopsin.[17] Their uses have been extended to the investigation of the morphology of crystalline synthetic and natural polymers, where they yield a wealth of structural information such as the local orientation of microcrystals or crystallites for fibrous polymers, and the detection of crystalline blocks with their respective orientation and their crystalline form in polymer films. Reports of

[17] P. N. T. Unwin and R. Henderson, *J. Mol. Biol.* **94**, 425 (1975).

crystal lattice fringes in polymers extremely sensitive to damage by electron irradiation have appeared.[18]

Molecular Modeling

Building Blocks

Carbohydrate building units usually have cyclic structures. One of the ring atoms is almost always an oxygen and the remaining atoms are carbon atoms. In polysaccharides most of the sugar residues are hexoses, which exist in the pyranose form and belong primarily to the D series. In theory the possibility of rotations about single bonds of the pyranose ring can generate numerous conformers. However, it has been established from X-ray structural and nuclear magnetic resonance (NMR) studies that the pyranose ring in monosaccharides, oligosaccharides, and polysaccharides exists generally in the chair conformation. Theoretical studies also indicate that the chair is slightly flexible and that small alterations of the ring torsional angles (up to 10°) and the ring bond angles (up to 3°) are possible. Molecular mechanics (MM) methods[19] using appropriate parameters to account for stereoelectronic effects such as the anomeric and exoanomeric effects (MM1[20], MM2CARB[21]) can be used to give accurate structures and energies for carbohydrates. Optimized geometries and conformations for a set of 35 monosaccharide units, which constitute more than 95% of the building blocks occurring in complex carbohydrates, has been reported and organized in a database.[22]

Junctions

Conformational analysis of disaccharides yields useful information for modeling polysaccharides. When two monosaccharide units are joined, they can rotate about the interglycosidic junctions (Fig. 5). The torsional angles of rotation about the glycosidic linkage are designated by ϕ and ψ; in principle they can take any value between -180 and 180°. Structural information derived from crystal structure determination of oligosaccharides have justified that, as a first approximation, the internal parameters

[18] J. F. Revol, *in* "Intermediate Voltage Microscopy, and Its Application to Materials Sciences" (K. Rajan, ed.), p. 126. Electron Optics Publishing Group Philips Electronic Instruments, Inc., Malwah, The Netherlands, 1987.

[19] N. L. Allinger, *J. Am. Chem. Soc.* **99,** 8127 (1977).

[20] G. A. Jeffrey and R. Taylor, *J. Comput. Chem.* **1,** 99 (1980).

[21] I. Tvaroska and S. Pérez, *Carbohydr. Res.* **149,** 389 (1986).

[22] S. Pérez and M. M. Delage, *Carbohydr. Res.* in press (1991).

A

B

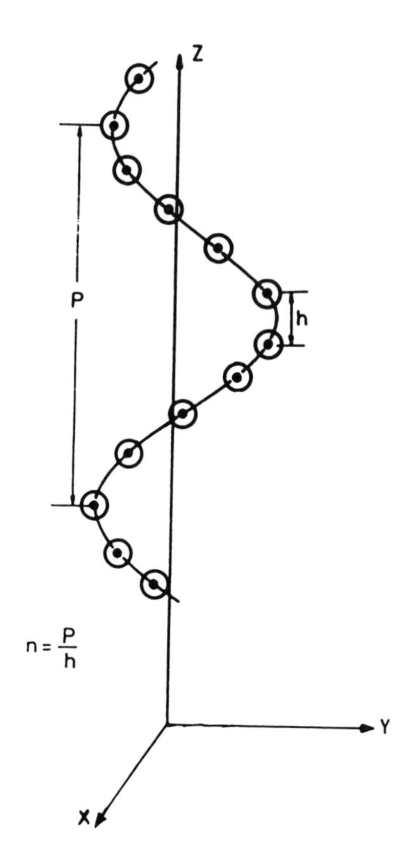

could be divided into rigid monomeric groups and flexible glycosidic linkage.[23] Also, because of the "spatial" separation of these "rigid" entities, which are interposed between the flexible linkages, there is an almost total independence between successive sets of glycosidic torsional angles.[24] For these reasons, the assessment of preferred conformations is usually performed on disaccharides by assessing the space that is available for rotation about the glycosidic junction.[25] The range of values that can be taken by the glycosidic torsion angles is limited due to steric restrictions. Set of limiting distances for different pairs of atoms to be used in the construction of steric maps have been proposed. These maps show the allowed and disallowed conformations for a particular disaccharide unit (Fig. 6A). More precise information about the potential energy of a given conformation can be gained by using "empirical methods." The potential energy E_{tot} is partitioned into a number of discrete contributors:

$$E_{tot} = E_{nb} + E_{tor} + E_{hb} + E_{exo} \dots$$

where E_{nb} is the nonbonded interaction energy,[26] E_{tor} is the energy due to the torsional strain about the glycosidic bonds, E_{hb} is the energy due to the formation of hydrogen bonds, and E_{exo} is the energy due to the exoanomeric effect. Detailed descriptions of these functions have been reported,[21] and are included in the PFOS (potential functions for oligosaccharide structures) computer program.[21,27]

Conformational analyses of disaccharides give different results depending on whether the residues are allowed to adjust internally at each increment in ϕ and ψ. Compared to conventional approaches where sugar residues are treated as rigid, optimization of all the internal parameters i.e., bond lengths, valence angles, and all the torsional angles is an important step toward more realistic information. Calculations of potential energy surfaces where all the internal coordinates of the molecules are "relaxed" and minimized through an extensive molecular mechanics scheme have

[23] D. A. Brant, *Q. Rev. Biophys.* **9,** 527 (1976).

[24] D. Gagnaire, S. Pérez, and V. Tran, *Carbohydr. Polym.* **2,** 171 (1982).

[25] V. S. R. Rao, P. R. Sundararajan, C. Ramakrishnan, and G. N. Ramachandran, *in* "Conformation in Biopolymers" (G. N. Ramachandran, ed.), Vol. 2, p. 721. Academic Press, London, 1967.

[26] R. A. Scott and H. A. Scheraga, *J. Chem. Phys.* **45,** 2091 (1966).

[27] S. Pérez, D.Sc. Thesis, University of Grenoble, France (1978).

FIG. 5. (A) Schematic representation of a maltose unit [Glc-α(1-4)-Glc] along with labeling of atoms and the torsional angles of interest. (B) Schematic representation of a helical chain, along with the helical parameters n and h.

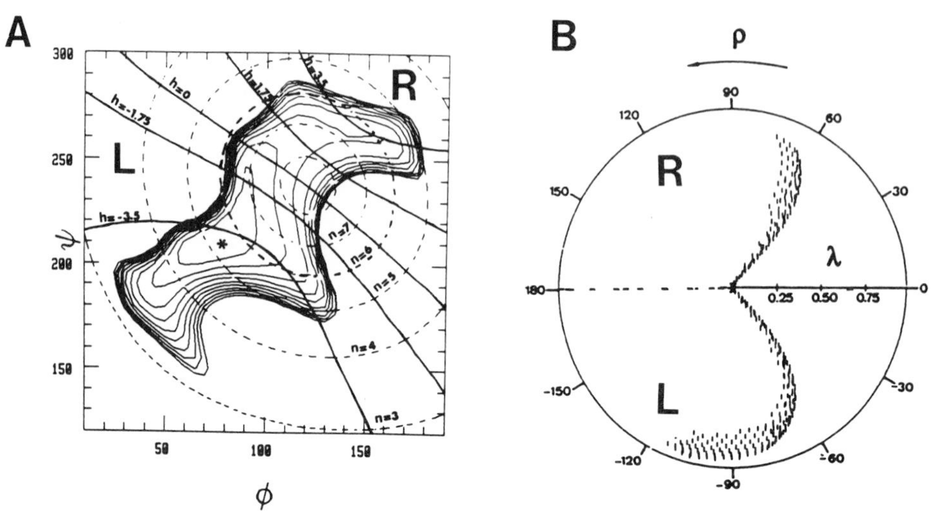

FIG. 6. (A) Selected iso-n and iso-h contours for a maltose unit, superimposed on the low-energy region of the classical isoenergy map of the α-(1 → 4)-linkage. The two-dimensional isoenergy map was calculated with the PFOS program [S. Pérez, D.Sc. thesis, University of Grenoble, France (1978)] using 5° increments for ϕ and ψ. With respect to the relative energy minimum (*), the isoenergy contours are drawn by interpolation of 1 kcal/mol. The iso-$h = 0$ contour divides the map into two regions corresponding to right-handed (R) and left-handed (L) chirality. From [A. Imberty, H. Chanzy, S. Pérez, A. Buléon, and V. Tran, *J. Mol. Biol.* **201**, 365 (1988)]. (B) Polar (ρ, λ) representation of the distribution of the local (n, h) helical parameters consistent with the low-energy domain of (A). The length of virtual bond is 4.410 Å. From [S. Pérez and C. Vergelati, *Biopolymers* **24**, 1809 (1985)]. Reprinted with permission from Wiley & Sons.

been reported for several disaccharides.[28-33] It was found that inclusion of the relaxed principle into conformational description of disaccharides does not generally alter the overall shape of the allowed low-energy regions, or the position of the local minima. However, flexibility within the ring plays a crucial role. Its principal effect is to lower energy barriers to conforma-

[28] S. N. Ha, L. Madsen, and J. W. Brady, *Biopolymers* **27**, 1927 (1988).

[29] A. D. French, *Biopolymers* **27**, 1519 (1988).

[30] V. Tran, A. Buléon, A. Imberty, and S. Pérez, *Biopolymers* **28**, 679 (1989).

[31] A. Imberty, V. Tran, and S. Pérez, *J. Comput. Chem.* **11**, 205 (1989).

[32] V. Tran and J. W. Brady, *Biopolymers* **29**, 961 (1990).

[33] A. D. French, V. Tran, and S. Pérez, *in* "Computer Modeling of Carbohydrate Molecules (A. D. French and J. W. Brady, eds.), p. 191. American Chemical Society, Washington D.C., 1990.

tional transitions about the glycosidic bonds, permitting pathways among the low-energy minima.

Helical Parameters

When subjected to the constraints imposed by the helical symmetry of macromolecular chains, equivalent monomeric units should occupy equivalent positions about the molecular axis.[34] This is achieved when the (ϕ, ψ) angles are the same at every linkage; the secondary structure is described in terms of a set of helical parameters n and h, where n is the number of residues per turn of the helix, and h is the translation of the corresponding residue along the helix axis (Fig. 5B). The chirality of the helix is defined by the sign of h. Arbitrarily, a positive value of h will designate a right-handed helix, and a negative value will correspond to a left-handed chirality. Whenever the values $h = 0$ or $n = 2$ are intercrossed, the screw sense of the helix changes to the opposite sign. In practice, the helical parameters are readily observable on the fiber diffraction pattern, where the spacing of the layer lines provides the pitch of the helical structure. The helix type is deduced from the position of the meridianal reflections. In conformational analysis studies, the helical parameters can be plotted as iso-n and iso-h contours as a function of torsional angles (ϕ, ψ) (Fig. 6A).[35,36] Values of the torsional angles consistent with the observed parameters are found at the intersection of the corresponding iso-h and iso-n contours. Discrimination between possible solutions is based on the magnitude of the potential energy.

Polysaccharides often crystallize in various shapes depending on slight alterations in the geometry of the constituent monosaccharides.[37] Therefore, a thorough conformational analysis for a polysaccharide requires either the use of flexible residues or parallel studies with several residues that span the range of residue variation. When flexibility of the residues and linkages is incorporated, the values of ϕ and ψ cannot specify n and h. Thus, an alternate representation of conformational space must be used.[38]

Whereas the number of residues per turn of the helix n has a universal meaning, the advance per residue h is strongly dependent on the nature of the repeating unit. The advance per residue, h, can be reduced to a dimensionless descriptor by considering the length of the virtual bond (L) be-

[34] G. Natta and P. Corradini, *Nuevo Cimento Suppl.* **15,** 9 (1960).
[35] T. Miyazawa, *J. Polym. Sci.* **55,** 215 (1961).
[36] D. Gagnaire, S. Pérez, and V. Tran, *Carbohydr. Res.* **78,** 89 (1980).
[37] A. D. French and V. G. Murphy, *Carbohydr. Res.* **27,** 391 (1973).
[38] A. D. French and V. G. Murphy, *Polymer* **18,** 489 (1977).

tween two adjacent linking atoms, which represents the maximum of the possible extension. The quotient $\lambda = h/L$ yields the percentage of total extension. An alternative use of the n parameter can be made by considering $\rho = 360°/n$. The purpose of doing this is to restrict the allowed region to finite areas for polymers that can assume large values of n. Also, it avoids any discontinuity on going from a right- to a left-handed helix. This unified representation of helical parameters[39] based on a polar mapping has been applied to some polysaccharides; this allowed for a straightforward comparison of all the secondary structures displayed in the solid state. It was found that despite the fact that polysaccharides display a wide range of helical structures, a tendency toward maximum extension is exhibited, along with a preference towards left-handed chirality.

Quite interestingly, this approach can be used to gain some insight into the disordered state of polysaccharides. There is increasing evidence that most of their physical properties in solution may be due to specific conformational behavior. Several attempts[24,40–43] have been made to describe the solution conformations of polysaccharides at the molecular level. To do so, spatial chain propagations were constructed based on the occurrence of conformational states (characterized by ϕ and ψ) within the allowed energy space calculated for the repeating unit. An alternative approach can be formulated based on the use of the helical parameters associated with each linkage conformation. In this case the helical parameters must be considered on a local basis. The helical parameters (ρ, λ) are determined for each ϕ, ψ grid point belonging to the low-energy region, and their distribution plotted using the polar representation (Fig. 6B).

Some interesting conclusions emerge from these results. Starting from only the primary structure of a stereoregular polysaccharide, it is quite straightforward to identify all the low-energy conformers that are likely to occur in vacuum. Taking into account solvation, even in a qualitative way, the equilibrium between these conformational states may be assessed, shedding some light into the inherent conformational characteristics of individual macromolecules free of interactions with other like or unlike species. Modeling of solution conformations of polysaccharides provides some insight not only into the overall shape of the random coil, but also

[39] S. Pérez and G. Vergelati, *Biopolymers* **24,** 1809 (1985).
[40] E. R. Morris, D. A. Rees, D. Thom, and E. J. Welsh, *J. Supramol. Struct.* **6,** 259 (1977).
[41] R. C. Jordan, D. A. Brant, and A. Cesaro, *Biopolymers* **17,** 2617 (1978).
[42] D. A. Brant and B. A. Burton, *in* "Solution Properties of Polysaccharides" (D. A. Brant, ed.), p. 81. American Chemical Society, Washington, D.C., 1981.
[43] B. A. Burton and D. A. Brant, *Biopolymers* **22,** 1769 (1983).

into the occurrence of local helical regions that might be appropriate for further ordering.[44]

Crystal Packing

Providing that the data set is of sufficient quality and/or the unit-cell dimensions and space group symmetry are well assigned, the final stage involves a complete structural determination of the unit-cell content. A linked-atom description similar to that reported by Smith and Arnott[45] or Zugenmaier and Sarko[46] may be used. Optimization procedures are used to fit observed and calculated structure amplitudes with simultaneous optimization of the nonbonded interactions and preservation of helix pitch and symmetry as well as ring closure. Interatomic energy functions, mimicking intermolecular nonbonded interactions,[47] are used extensively for this purpose.

In the linked-atom least-squares (LALS) procedure,[45] packing of the chains in the unit cell is described by the variable (μ) for orientation, and ($\overline{\omega}$) for translation along the helix axis. The procedure is aimed at minimizing the discrepancy function (Ω), where Ω is given by

$$\Omega = \sum W_m(|_{o}F_m|^2 - K^2|_{c}F_m|^2) + \sum \epsilon_{ij} + \sum \lambda_q G_q \qquad (3)$$

The first term seeks to optimize agreement between the observed amplitudes $_{o}F_m$ and those calculated from the models $_{c}F_m$, K is a scaling factor that places the observed amplitudes on the same scale as the calculated values, W_m is weight, and the summation is carried out over the m independent reflections. The second term provides a simple quadratic approximation to the nonbonded repulsion and is applied to all pairwise interactions between atoms i and j, when the distance between them is less than some "standard" values. The final term includes a set of coordinate constraint equations G_q, together with the initially undefined Langrangian multipliers λ_q, and is used to preserve helix pitch and symmetry as well as, in the case of flexible ring, the ring closure.

In another method,[46] both the monomer residue and the chain conformations are varied in such a way that all conformational and chain-packing features of the structures are optimized relative to the superimposed

[44] J. Jimenez-Barbero, C. Bouffar-Roupe, C. Rochas, and S. Pérez, *Int. J. Biol. Macromol.* **11,** 265 (1989).

[45] P. J. C. Smith and S. Arnott, *Acta Crystallogr. Sect. A* **34,** 3 (1978).

[46] P. Zugenmaier and A. Sarko, *Biopolymers* **15,** 2121 (1976).

[47] D. E. Williams, *Acta Crystallogr. Sect. A* **25,** 464 (1969).

constraints and available data. In stereochemical refinements, the function contains terms for bond length strain, bond and conformational angle strain, and nonbonded contacts. It has the following form:

$$Y = \sum_{i=1}^{l} \left(\frac{r_i - r_{o_i}}{\mathrm{SD}_i^r} \right)^2 + \sum_{i=1}^{m} \left(\frac{\theta_i - \theta_{o_i}\mathrm{SD}_i^\phi}{\mathrm{SD}_i^\theta} \right)^2 + \sum_{i=1}^{n} \left(\frac{\phi_i - \phi_{o_i}}{\mathrm{SD}_i^\phi} \right)^2 + \frac{1}{W^2} \sum_{i=1,j=1}^{N} w_{ij}(d_{ij} - d_{o_{ij}}) \quad (4)$$

where r_i, θ_i, and ϕ_i are, respectively, the bond lengths, bond and conformational angles for the l, m, and n corresponding variables in the residue, with r_{o_i}, θ_{o_i}, ϕ_{o_i}, and SD_i^r, SD_i^θ, SD_i^ϕ the corresponding best (or average) values and their standard deviations. The fourth term approximates nonbonded repulsion, with d_{ij} the nonbonded distance between the atoms i and j, $d_{o_{ij}}$ the corresponding equilibrium distance, w_{ij} the weight of that contact in the function Y. The summation is over all N nonbonded distances taken pairwise. When the refinement is against observed X-ray intensity data, the function is simply the equation for any of the crystallographic R factors, the simplest choice of which is

$$R = \Sigma||F_o| - |F_c||)/\Sigma|F_o| \quad (5)$$

where F_o and F_c are the observed and calculated structure amplitudes, respectively. It is desirable to perform x-ray refinement while maintaining some stereochemical constraints. To do so, a linear combination of both functions (4) and (5) should be minimized, in the form

$$\zeta = f_x R + (1 - f_x) Y \quad (6)$$

where f_x is the fractional weight of the R factor in the linear combination of R and Y.

These computational approaches have been used to produce a phasing model in the case of a disaccharide, which was further refined through conventional methods.[48] Similarly, the method was applied to a tetrasaccharide structure for which crystals good enough for conventional crystallographic work could not be grown.[49]

The various steps to arrive at the crystal structures of polysaccharides[50] are summarized in Fig. 7.

[48] S. Pérez, C. Vergelati, and V. Tran, *Acta Crystallogr. Sect. B* **41,** 262 (1985).

[49] B. Henrissat, S. Pérez, I. Tvaroska, and W. Winter, *in* "Solid State Characterization of Cellulose, Paper, and Wood" (R. H. Attala, ed.), Advanced Series, 340, p. 38. American Chemical Society, Washington, D.C., 1987.

[50] F. Brisse, *J. Electron Microsc. Tech.* **11,** 272 (1989).

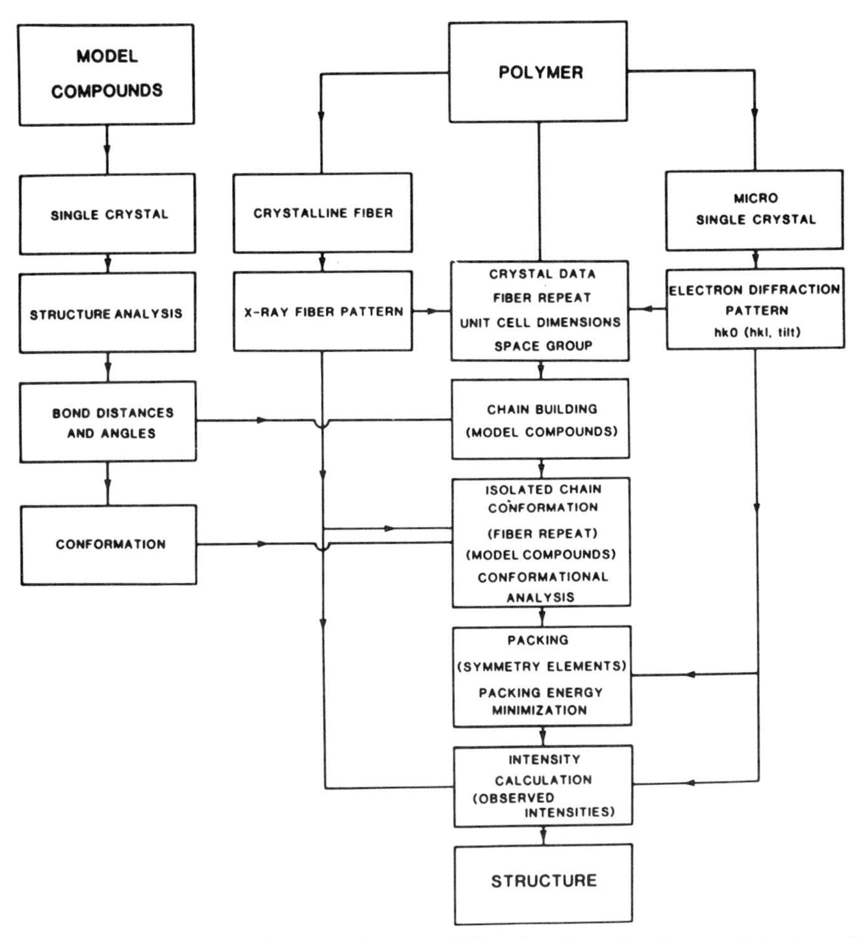

FIG. 7. Flow chart showing the various steps followed to determine the crystal structure of a polymer. From Brisse.[50] Reprinted with permission from A. R. Liss.

Applications

The combined use of molecular modeling and electron diffraction data has been invaluable in quantitative elucidations of the crystal and molecular structures of seven linear polysaccharides: cellulose triacetate II,[51] anhydrous nigeran,[52] anhydrous dextran,[53] hydrated dextran,[54] mannan I,[55]

[51] E. Roche, H. Chanzy, M. Boudeulle, R. H. Marchessault, and P. R. Sundararajan, *Macromolecules* **11,** 86 (1978).

[52] S. Pérez, M. Roux, J. F. Revol, and R. H. Marchessault, *J. Mol. Biol.* **129,** 113 (1979).

TABLE I
EXPERIMENTAL CONDITIONS FOR PREPARATION OF SINGLE CRYSTALS OF POLYSACCHARIDES[a]

Polysaccharide	DP	Method	Solvent	Precipitating agent	Polymer concentration (%)	Growth temperature	Crystallization rate	Ref.
Cellulose triacetate II poly[(1-4)-β-D-Glcp-AC3]	60	Addition of nonsolvent	Dibenzyl ether	*n*-Tetradecane	0.005	245°	4–6 hr	*b*
Nigeran, poly[(1-4)[α-D-Glcp-(1-3)-α-D-Glcp]	3000	Evaporation	Water		0.01	75°	12–24 hr	*c*
Dextran (high-temperature form), poly[(1-6)-α-D-Glcp]	60	Addition of nonsolvent	Water (35%)	PEG 300 (65%)	0.05	160°	3–4 hr	*d*
Dextran (low-temperature form), poly[(1-6)-α-DGlcp]	60	Addition of nonsolvent	Water (55%)	PEG 300 (45%)	0.02	95°	3–4 hr	*e*
Mannan I, poly[(1-4)-β-D-Manp]	15	Addition of nonsolvent	Water (50%)	DMSO (50%)	0.01	120°	10 min–10 hr	*f*
A-type starch, poly[(1-4)-α-D-Glcp]	15	Addition of nonsolvent	Water	Ethanol	0.1	60°	3–6 hr	*g*
B-type starch, poly[(1-4)-α-D-Glcp]	35	"Aging"	Water		0.05	4°	6 months–2 years	*h*

[a] DP, Degree of polymerization; PEG, polyethylene glycol; DMSO, dimethyl sulfoxide.
[b] E. Roche, H. Chanzy, M. Boudeulle, R. H. Marchessault, and P. R. Sundararajan, *Macromolecules* **11,** 86 (1978).
[c] S. Pérez, M. Roux, J. F. Revol, and R. H. Marchessault, *J. Mol. Biol.* **129,** 113 (1979).
[d] C. Guizard, H. Chanzy, and A. Sarko, *Macromolecules* **17,** 100 (1984).
[e] C. Guizard, H. Chanzy, and A. Sarko, *J. Mol. Biol.* **183,** 397 (1985).
[f] H. Chanzy, S. Pérez, D. P. Miller, G. Paradosi, and W. T. Winter, *Macromolecules* **20,** 2407 (1987).
[g] A. Imberty, H. Chanzy, S. Pérez, A. Buléon, and V. Tran, *J. Mol. Biol.* **201,** 365 (1988).
[h] A. Buléon, F. Duprat, F. Booy, and H. Chanzy, *Carbohydr. Polym.* **4,** 161 (1984).

TABLE II
CRYSTAL DATA OF POLYSACCHARIDE STRUCTURES SOLVED FROM ELECTRON DIFFRACTION DATA

Name	Unit cell	Space group	Density	Helical parameters	Motif	Packing	Ref.[a]
Cellulose triacetate II	$a = 24.68$ $b = 11.52$ $c = 10.54$	$P2_12_12_1$	1.29	$n \cong 2$ $h = 5.72$	Single	Antiparallel	*a*
Nigeran anhydrous	$a = 17.76$ $b = 6.00$ $c = 14.62$	$P2_12_12_1$	1.39	$n = 2$ $h = 7.31$	Single	Antiparallel	*b*
Dextran, high temperature anhydrous	$a = 9.22$ $b = 9.22$ $c = 7.78$ $\beta = 91.3°$	$P2_1$ *b* unique axis	1.69	$n = 2$ $h = 3.91$	Single	Antiparallel	*c*
Dextran, low temperature hydrated	$a = 25.71$ $b = 10.21$ $c = 7.76$ $\beta = 91.3°$	$P2_1$ *b* unique axis	1.67	$n = 2$ $h = 3.91$	Single	Antiparallel	*d*
Mannan I, anhydrous	$a = 8.92$ $b = 7.21$ $c = 10.27$	$P2_12_12_1$	1.57	$n = 2$ $h = 5.135$	Single	Antiparallel	*e*
A-type starch, hydrated	$a = 21.24$ $b = 11.72$ $c = 10.69$ $\gamma = 123.5°$	B_2	1.48	$n = 6$ $h = -3.55$	Parallel stranded double helix	Parallel	*f*
B-type starch, hydrated	$a = 18.5$ $b = 18.5$ $c = 10.4$ $\gamma = 120.0°$	$P6_1$	1.41	$n = 6$ $h = -3.47$	Parallel stranded double helix	Parallel	*g*

[a] E. Roche, H. Chanzy, M. Boudeulle, R. H. Marchessault, and P. R. Sundararajan, *Macromolecules* **11,** 86 (1978).
[b] S. Pérez, M. Roux, J. F. Revol, and R. H. Marchessault, *J. Mol. Biol.* **129,** 113 (1979).
[c] C. Guizard, H. Chanzy, and A. Sarko, *Macromolecules* **17,** 100 (1984).
[d] C. Guizard, H. Chanzy, and A. Sarko, *J. Mol. Biol.* **183,** 397 (1985).
[e] H. Chanzy, S. Pérez, D. P. Miller, G. Paradosi, and W. T. Winter, *Macromolecules* **20,** 2407 (1987).
[f] A. Imberty, H. Chanzy, S. Pérez, A. Buléon, and V. Tran, *J. Mol. Biol.* **201,** 365 (1988).
[g] A. Imberty and S. Pérez, *Biopolymers* **27,** 1205 (1988).

A-type amylose,[56] and B-type amylose.[57] THis set of structures was resolved over a 10-year period, during which most of the methodological aspects described above were established and used. The application of direct electron microscopic image data to visualize crystalline arrays of β-chitin[58] is described.

Table I summarizes the experimental conditions used for crystallizing these polysaccharides, whereas their crystal data are listed in Table II. Other micron-sized single crystals have been obtained from polysaccharides such as cellulose II,[59] curdlan[4], β-(1 → 4)-D-xylan,[60] inulin,[61] glucomannan,[62] and V-type amylose,[63,64] but have not been used for three-dimensional structural determination.

Cellulose Triacetate II

Cellulose occurs in several polymorphic forms, the most important being cellulose I (native state) and cellulose II (mercerized or recrystallized state). This polymorphism is found also in cellulose derivatives.[65] A familiar case is cellulose triacetate (CTA). CTA I and CTA II are named after the polymorphic state of their underivatized analogs. CTA I and CTA II were found to have specific crystalline morphologies closely related to those of cellulose I and cellulose II. This was an indication that the polarity of the chains was preserved in going from cellulose I to CTA I or cellulose II to CTA II. The crystal structure of CTA II[51] was derived from an X-ray and electron diffraction analysis using intensities from fiber and single-crystal data. A model was built from the known structure of the central residue of

[53] C. Guizard, H. Chanzy, and A. Sarko, *Macromolecules* **17,** 100 (1984).
[54] C. Guizard, H. Chanzy, and A. Sarko, *J. Mol. Biol.* **183,** 397 (1985).
[55] H. Chanzy, S. Pérez, D. P. Miller, G. Paradosi, and W. T. Winter, *Macromolecules* **20,** 2407 (1987).
[56] A. Imberty, H. Chanzy, S. Pérez, A. Buléon, and V. Tran, *J. Mol. Biol.* **201,** 365 (1988).
[57] A. Imberty and S. Pérez, *Biopolymers* **27,** 1205 (1988).
[58] J. F. Revol, K. H. Gardner, and H. Chanzy, *Biopolymers* **27,** 345 (1988).
[59] A. Buléon and H. Chanzy, *J. Polym. Sci. Polym. Phys. Ed.* **16,** 833 (1978).
[60] H. Chanzy, J. Comtat, M. Dubé, and R. H. Marchessault, *Biopolymers* **18,** 2459 (1979).
[61] R. H. Marchessault, T. Bleha, Y. Deslandes, and J. F. Revol, *Can. J. Chem.* **58,** 2415 (1980).
[62] H. Chanzy, A. Grosrenaud, J. P. Joseleau, M. Dubé, and R. H. Marchessault, *Biopolymers* **21,** 301 (1982).
[63] J. Brisson, H. Chanzy, and R. Vuong, *Food Hydrocolloids* **1,** 523 (1987).
[64] A. Buléon, M. M. Delage, J. Brisson, and H. Chanzy, *Int. J. Biol. Macromol.* **12,** 25 (1990).
[65] N. M. Bikales and L. Segal, "Cellulose and Cellulose Derivatives" (L. Segal, ed.), Part 4. Wiley-Interscience, New York, 1971.

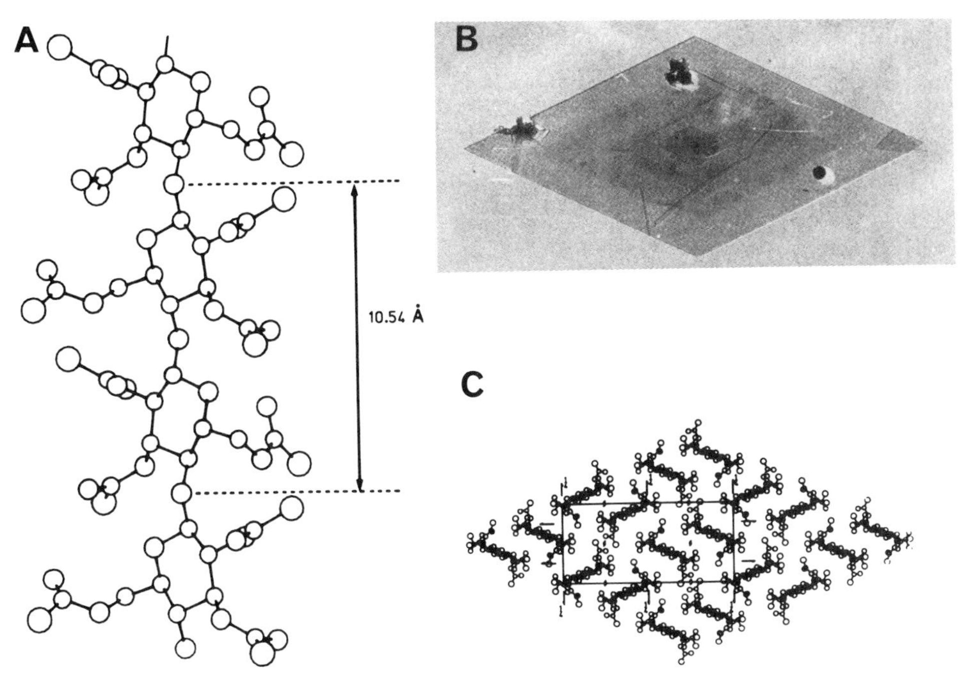

FIG. 8. Crystalline features of cellulose triacetate II. From Roche *et al.*[51] Reprinted with permission from the American Chemical Society. (A) Molecular drawing of the cellulose triacetate chain with its almost perfect twofold symmetry. (B) Structure of cellulose triacetate II viewed as a projection of the unit cell onto the (*a*, *b*) plane. In this structure, chains are alternatively "up" and "down." (C) Single crystal of cellulose triacetate II properly oriented with respect to the projection shown in (B).

cellotriose acetate[66] and further refined through packing energy refinement (Fig. 8A). The crystallographic reliability index was $R_e = 0.26$ and $R_x = 0.30$ for electron and X-ray diffraction data, respectively. The asymmetric unit contains two peracetylated β-(1 → 4)-linked glucopyranose units. The conformation of the chain is extended and close to a twofold screw symmetry. The resulting structure consists of antiparallel pairs of parallel CTA II chains, and there are no hydrogen bonds in the crystal. The packing analysis (Fig. 8C) can help rationalize characteristics of the crystal, such as the (110) growth plane, which is indeed parallel to the direction of the highest atomic density, or the occurrence of certain twinning planes (Fig. 8B).

[66] S. Pérez and F. Brisse, *Acta Crystallogr. Sect. B* **33,** 2578 (1977).

Nigeran

Nigeran is a polysaccharide found in the cell walls of lower fungi. It was first isolated from *Penicillium expansum* and *Aspergillus niger*[67] and has been shown to be synthesized by only a few species of *Aspergillus* and *Penicillium.* Other studies have shown that nigeran is located throughout the cell walls.[68] The presence of hydrated crystalline nigeran *in vivo* has been experimentally demonstrated.[69] Chemically, nigeran is a regular linear poly(disaccharide) of type $(A-B-)_p$, where A and B are glucose residues, and p can be as large as 3000. The linkages A–B and B–A are α-(1 → 3) and α-(1 → 4), respectively. Nigeran is unusual in its propensity to accommodate or lose water molecules in a reversible transition that occurs in the crystalline state. With proper control of relative humidity and temperature, nigeran can be induced to assume different packing modes characterized by changes in the base-plane parameters, as shown using electron diffraction on solution-grown crystals that mimic *in vivo* observations.[9] The crystal structure of anhydrous nigeran[52] was determined by a combined electron diffraction, X-ray diffraction, and molecular modeling analysis. The crystallographic reliability index was $R_e = 0.25$ and $R_x = 0.30$, when the model was tested against electron and X-ray diffraction data, respectively. The poly(disaccharide) chain is a twofold helix stabilized by an intrachain hydrogen bond between contiguous α(1 → 4)-linked glucose residues (Fig. 9A). Two such chains pack with antiparallel polarity and the twofold screw axis coincides with the macromolecular axis. A dense network of hydrogen bonds holds the chain together in the crystal (Fig. 9B). Calculation of dynamic structure factors from bent crystals has been carried out by using a supercell. An artificial superlattice was constructed containing several unit cells arranged to model the deformed curved foil.[70] Whereas the crystallographic reliability index was improved, the molecular and structural features remained unaltered.

Nigeran provides an interesting illustration of the mechanism of interaction of hydrolytic enzymes with polysaccharide crystals. When treated with endomycodextranase, they tend to break into "jigsaw" fragments, in which the relation to the original morphology is still visible.[71] Nigeran crystals are composed of folded chains, as evidenced from their growth

[67] A. W. Dox and R. E. Neidig, *J. Biol. Chem.* **18,** 167 (1914).

[68] I. R. Johnston, *Biochem. J.* **96,** 651 (1965).

[69] T. F. Bobbit, J. H. Nordin, M. Roux, J. F. Revol, and R. H. Marchessault, *J. Bacteriol.* **132,** 691 (1977).

[70] B. K. Moss and D. L. Dorset, *Acta Crystallogr. Sect. A* **39,** 609 (1983).

[71] R. H. Marchessault, J. F. Revol, F. Bobbit, and J. H. Nordin, *Biopolymers* **19,** 1069 (1980).

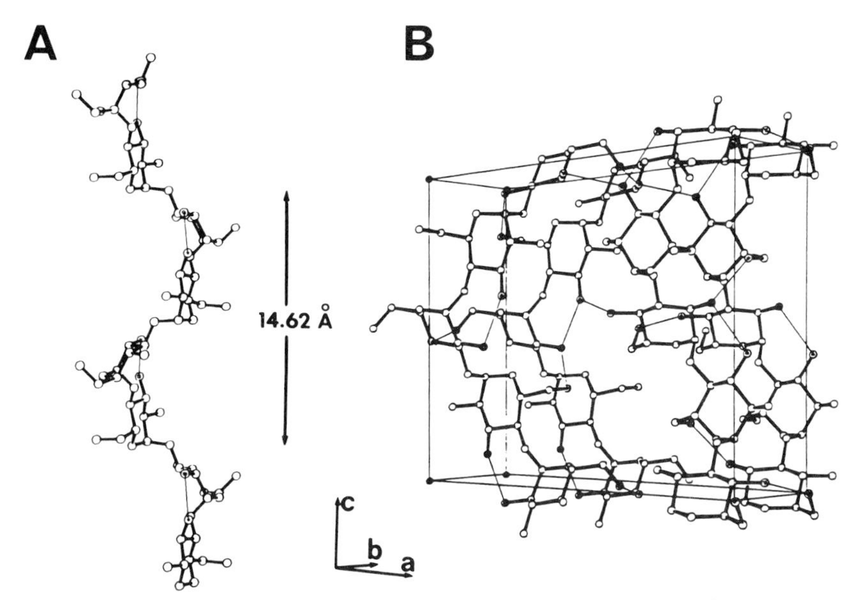

FIG. 9. Crystalline features of anhydrous nigeran. From Pérez *et al.*[52] (A) Molecular representation of the nigeran chain, with its twofold symmetry. Hydrogen bonds occurring between 1,4-linked α-D-Glc residues are shown as thin lines. (B) Molecular representation of the packing content of the unit cell of anhydrous nigeran with an antiparallel arrangement of the chains.

from high-molecular-weight material. They occur as a mosaic of tightly folded blocks, linked together by loosely folded, connecting nigeran chains. When performed at 20°, the enzymatic digestion will operate on the loose folds, leaving the components of the mosaic unaltered. This indicates that the erosion starts at the crystalline platelet surface, where long "amorphous" stems are accessible to the enzyme active sites.

Dextran

Dextran is a generic term used to describe D-glucans that contain a substantial number of 1,6-linked α-D-glucopyranosyl residues.[72] The great majority of dextrans are produced by bacteria growing on sucrose as a substrate, although some have been synthesized from other substrates. In addition, chemical synthesis of essentially unbranched dextran has been reported. The abundant literature related to dextran in chemical and medi-

[72] R. L. Sidebotham, *in* "Advances in Carbohydrate Chemistry and Biochemistry" (R. S. Tipson and D. Horton, eds.), Vol. 30, p. 371. Academic Press, San Francisco, California, 1974.

cal research is explained by the remarkable compatibility of this class of polysaccharides with biological systems. Dextran, like many other polysaccharides, exhibits polymorphism. Among the parameters that are critical for control of polymorphism are the temperature of crystallization and the crystallization medium. From *in vitro* growth of single crystals, two polymorphs have been identified, i.e., a high-termperature anhydrous form[73] and a low-temperature hydrated form.[74] The powder diffraction pattern of the high-temperature polymorph is very similar to that of a natural dextran. Similarly, the low-termpeature polymorph was shown to give powder diffraction patterns that were identical to those obtained from both bulk-crystallized synthetic dextrans and natural, bacterial dextrans crystallized from dilute solution.[74]

The crystal and molecular structure of the anhydrous polymorph[53] of synthetic dextran was determined through a combined electron and X-ray diffraction analysis and stereochemical model refinement. Electron diffraction data corrected for beam damage were obtained from lamellar single crystals grown from dilute solution and included *hk0* data to a resolution of 1.1 Å, as well as selected higher layer reflections. The X-ray data in the form of powder patterns were recorded from a collection of the same crystals. The crystallographic reliability index was $R_e = 0.189$ and $R_x = 0.101$ when the model was tested against electron and X-ray diffraction data, respectively. The asymmetric unit contains two α-(1 $\rightarrow$ 6)-linked glucopyranose residues. The conformation of the chain is relatively extended and ribbon-like, with successive residues in a near-twofold screw relationship (Fig. 10A). The chain is stabilized by an intrachain hydrogen bond between glucose residues i and $i + 2$. This feature explains why the observed conformation does not correspond to the lowest energy minimum calculated for the disaccharide isomaltose,[75] although it does reside within the theoretically allowed conformational space. Also, it does not correspond to the crystal conformation observed about the α-(1 $\rightarrow$ 6) linkage in the trisaccharide panose.[76] Two antiparallel-packed chains pass through the unit cell. The chains of like polarity pack into sheets with extensive intrasheet hydrogen bonding. The sheets, packed antiparallel, are extensively bonded together by intersheet hydrogen bonding (Fig. 10B). The fact that the structure is nearly maximally hydrogen bonded explains the quite dense packing of crystalline dextran and its high density.

The crystal and molecular structure of the low-temperature hydrated dextran polymorph[54] was established using the same strategy as in the case

[73] H. Chanzy, G. Excoffier, and C. Guizard, *Carbohydr. Polym.* **1,** 67 (1981).
[74] H. Chanzy, H. Guizard, and A. Sarko, *Int. J. Biol. Macromol.* **2,** 149 (1980).
[75] I. Tvaroska, A. Imberty, and S. Pérez, *Biopolymers* **30,** 369 (1990).
[76] A. Imberty and S. Pérez, *Carbohydr. Res.* **181,** 41 (1988).

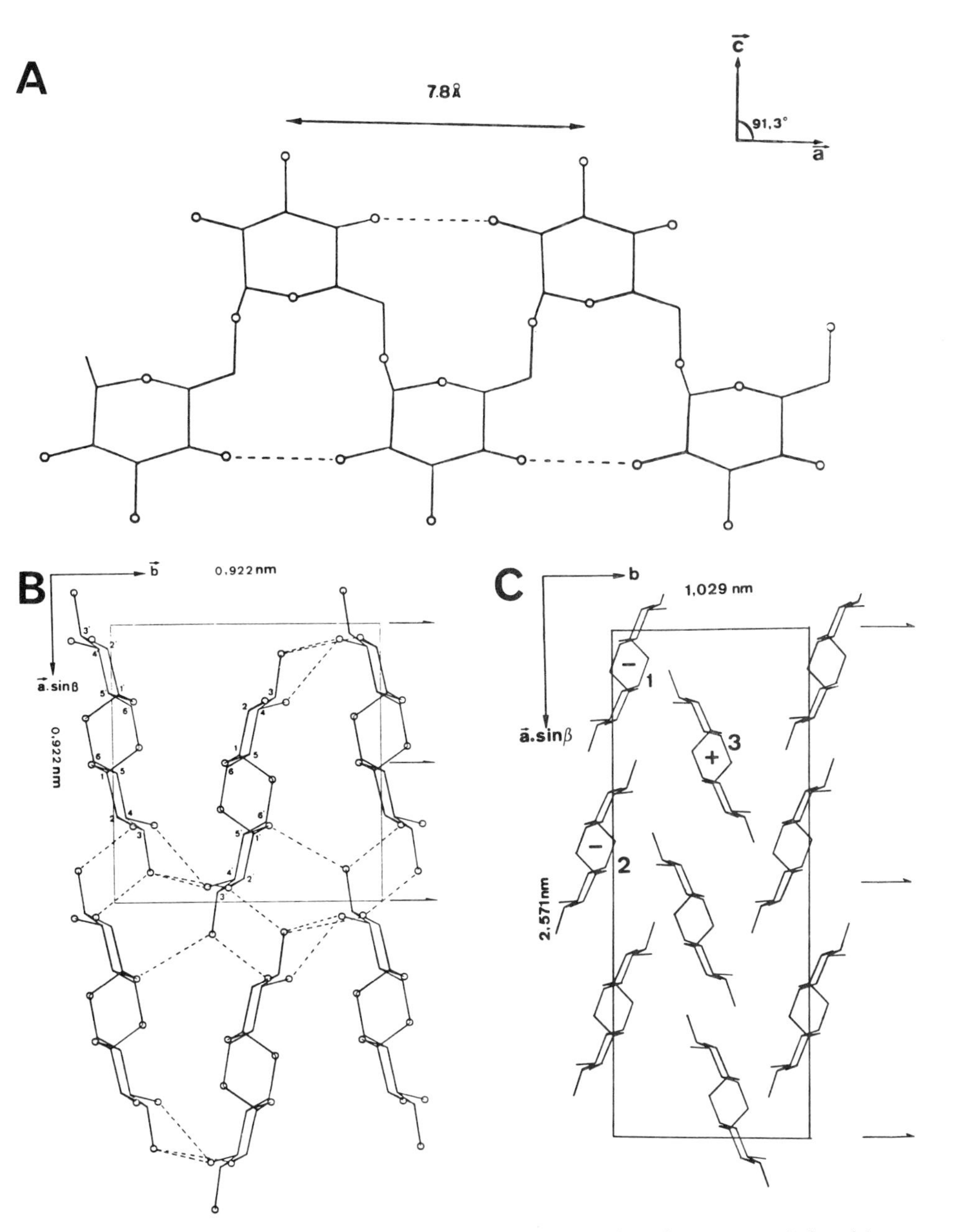

FIG. 10. Crystalline features of dextran. (A) Molecular drawing of the dextran chain, with its twofold symmetry. Intrachain hydrogen bonding (represented as broken lines) occurs between residue i and $i+2$. (B) Anhydrous high-temperature polymorph of dextran. Projection of the unit cell down the c axis. Hydrogen bonds are indicated as dashed lines. Corner chain has "up" polarity; center chain is "down." From Guizard *et al.*[53] Reprinted with permission from the American Chemical Society. (C) Low-temperature, hydrated polymorph of dextran. Projection of the unit cell down the c axis, with three crystallographically independent chains. Chain polarity is indicated by the + and − signs. From Guizard *et al.*[54]

of the anhydrous polymorph. Electron diffraction intensities were collected from frozen single crystals held at the temperature of liquid nitrogen, to a resolution of 1.6 Å. The crystallographic reliability index was $R_e = 0.258$ and $R_x = 0.127$ for electron and X-ray diffraction data, respectively. The unit cell contains six chains and eight water molecules, with three chains of the same polarity and four water molecules constituting the asymmetric unit. As in the anhydrous polymorph, the chain is extended and has close to a twofold screw symmetry. However, chain packing is different in the two structures in that rotational positions of the chains about the helix axes are considerably different. The chain packing in the hydrated structure is still quite tight (Fig. 10C) as is evident from the intermolecular contacts, numerous hydrogen bonds, and the high crystalline density. Water molecules occupy positions between sheets of chains and play an essential role in hydrogen bonding. The fact that these water molecules are tightly held in the crystal lattice explains the experimental observation that crystals of dextran hydrate do not deteriorate rapidly in the electron beam of the microscope.

Mannan

Mannan, a poly[(1 → 4)-β-D-mannose] can be obtained as a pure homopolysaccharide from the endosperm of certain plants, notably ivory nut,[77] *Phytelephas macrocarpa,* as well as from the walls of algae belonging to the families Codiaceae and Dasycladaceae.[78-80] Both the plant and algal materials exhibit *in situ* crystallinity, although the latter are optically anisotropic in the light microscope. The algal mannans have been reported to exist in two different polymorphs, the first similar to that observed with ivory nut mannan, being referred to as mannan I, and the second, in analogy with cellulose polymorphism, as mannan II.[81] The second structure is thought to be hydrated. The crystal and molecular structure of mannan I has been established[55] by a constrained linked-atom least-squares refinement using intensities derived from electron diffraction and stereochemical constraints. Intensities were measured from diffraction patterns produced by tilted specimens; several zones were collected, digitized, and reduced to integral intensities. The best model obtained using the base plane diffraction data coupled with a stereochemical refinement yields $R_e = 0.245$. It corresponds to a system of highly extended twofold

[77] H. Meier, *Biochim. Biophys. Acta* **28,** 229 (1958).
[78] E. Frei and R. D. Preston, *Proc. R. Soc. London, Ser. B* **169,** 124 (1968).
[79] Y. Iriki and T. Miwa, *Nature (London)* **185,** 178 (1960).
[80] J. Love and E. Percival, *J. Chem. Soc.,* 3345 (1964).
[81] H. Chanzy, A. Grosrenaud, R. Vuong, and W. Mackie, *Planta* **161,** 320 (1984).

helices stabilized by intramolecular hydrogen bonds (Fig. 11A) in agreement with the stereochemical features of its model compounds: mannotriose[82] and packing sheets of alternatively "up" and "down" chains along the apparent growth plane. The association of adjacent chains is stabilized by intermolecular hydrogen bonds (Fig. 11B).

Crystalline Part of Starches: A and B Types

Starch is the form of carbohydrate storage in all green plants. Starch granules contain two polysaccharides: amylose and amylopectin. Amylose is a linear polymer made of α-D-glucopyranose units linked through 1,4-bonds; amylopectin is a large branched molecule with side chains grafted to the linear α-(1 → 4)-polymer by a single α-(1 → 6) junction. The short chains are thought to be arranged in crystallites that would be responsible for the crystallinity of the native granule.[83] Depending on its botanical origin, native starch exhibits two types of X-ray diffraction patterns that are associated with two polymorphic forms: the A-type in cereal starches and the B-type in tuber starches.[84] There is much experimental evidence that although differing in their crystalline arrangements, the molecular conformations of the amylose chains are nearly identical in A- and B-type polymorphs. For both polymorphs, synthetic single crystals were grown in water or a mixture of water and ethanol from low-molecular-weight materials having respective DP values of 15 and 35.[85]

The structure of the crystalline part of A-type starch was established[56] from joint use of electron diffraction on single crystals, X-ray powder diffraction resolved into individual peaks,[86] and previously reported X-ray fiber diffraction[87] after appropriate reindexing. A thorough molecular modeling was performed[88] using information derived from structural studies of model compounds,[89] which helped to establish quite unusual features. The three-dimensional crystallographic reliability index was $R_x = 0.27$ and $R_x = 0.24$ for powder and fiber diffraction intensities, respectively. The intensities from electron diffraction were not used quantitatively. The content of the asymmetric unit consists of a maltotriose moiety and one

[82] W. Mackie, B. Scheldrick, D. Akrigg, and S. Pérez, *Int. J. Biol. Macromol.* **8,** 43 (1986).

[83] D. French, *in* "Starch: Chemistry and Technology" (R. L. Whistler, J. N. BeMiller, and E. D. Paschall, eds.), 2nd Ed., p. 183. Academic Press, New York, 1984.

[84] F. Duprat, D. Gallant, A. Guilbot, C. Mercier, and J. P. Robin, *in* "Les Polymères Végétaux" (B. Monties, ed.), p. 176. Gauthier-Villars, Paris, 1980.

[85] A. Buléon, F. Duprat, F. Booy, and H. Chanzy, *Carbohydr. Polym.* **4,** 161 (1984).

[86] V. Tran and A. Buléon, *J. Appl. Crystallogr.* **21,** 1887 (1987).

[87] H. C. H. Wu and A. Sarko, *Carbohydr. Res.* **61,** 27 (1978).

[88] S. Pérez and C. Vergelati, *Polym. Bull.* **17,** 141 (1987).

[89] W. Pangborn, D. Langs, and S. Pérez, *Int. J. Biol. Macromol.* **7,** 363 (1985).

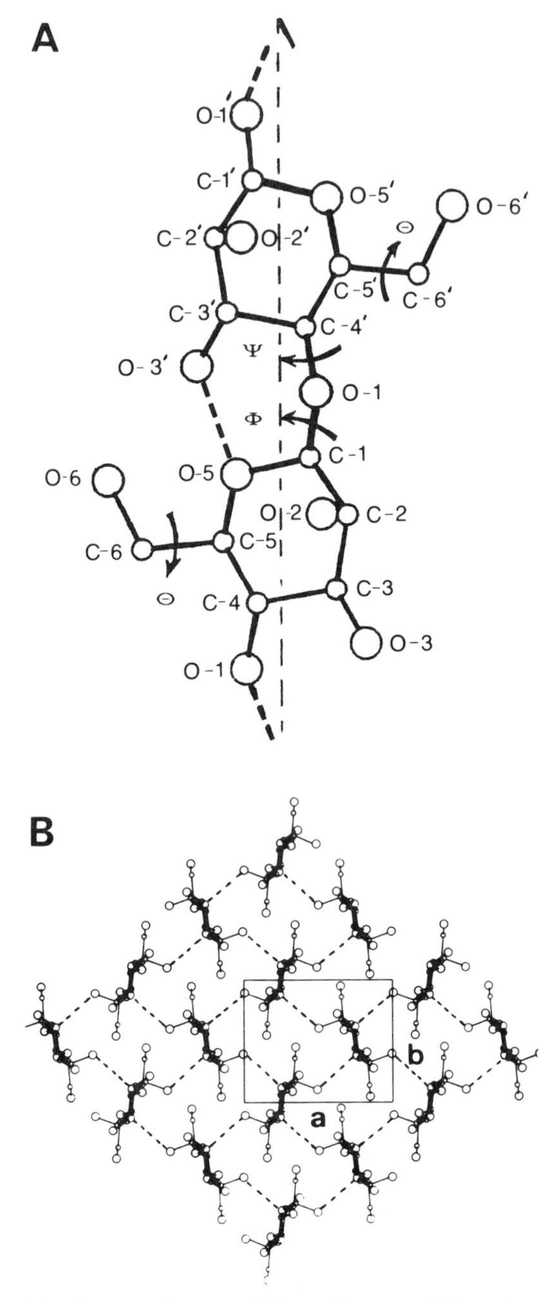

FIG. 11. Crystalline features of mannan I. From Chanzy *et al.*[55] Reprinted with permission from the American Chemical Society. (A) Molecular drawing of the mannan chain with its twofold symmetry. (B) Structure of mannan I viewed as a projection of the unit cell onto the (*a*, *b*) plane and outlining the shape of the crystals with their (110) growth plane. In this structure, chains are alternatively "up" and "down."

water molecule. This forms the basis of the structural motif, which is a parallel-stranded double helix where each strand is left handed (Fig. 12A). There are no intramolecular hydrogen bonds. Within the double helix, interstrand stabilization is achieved without any steric conflict through the occurrence of one type of hydrogen bond. In the unit cell double helices are packed in a parallel fashion; four water molecules are located between these double helices (Fig. 12B).

The structure of the crystalline part of B-type starch was solved[57] using information obtained from the base-plane electron diffractogram, i.e., unit cell parameters and two-dimensional symmetry, and the diffraction data that were taken from the previously reported X-ray fiber diffractogram.[90] The final R factor is 0.145 for the three-dimensional data. The unit cell contains 12 glucose residues located in 2 left-handed, parallel-stranded double helices packed in a parallel register; 36 water molecules are located between these helices (Fig. 12C). The asymmetric unit contains a maltose moiety and six water molecules. As in the A-type structure, there is no intramolecular hydrogen bond, and within the double helix, interstrand stabilization is achieved without any steric conflict and through the occurrence of one type of hydrogen bond. With hydration approximately 27% (w/w), the structure corresponds to a well-ordered crystalline sample, since all the water molecules could be located with no apparent sign of a disorder. Half of the water molecules are tightly bound to the double helices; the remainder forms a complex network centered around the sixfold screw axis of the unit cell.

Advantages and Possible Pitfalls

In all the cases studied, electron diffraction data provided highly reliable lattice constants, tests for symmetry elements, and base-plane structure amplitudes. An enlightening example was found in the course of studying the structure of the anhydrous form of dextran. It was observed that the base-plane diffraction pattern (**a***, **b***) displayed an *mm* symmetry with $\gamma = 90°$. However, it was found that the diagram obtained from a crystal rotated clockwise showed different intensities compared with the diagram obtained from the same crystal in a counterclockwise tilting of the same magnitude. A calibration of the tilted and untilted diagrams yielded a monoclinic unit characterized by $\gamma = 90°$, space group $P2_1$. Therefore, the (a,b) plane of the cell is orthogonal to the chain axis. Dextran was the first polysaccharide studied so far to exhibit such a feature. Because the β angle at 91.3° differs very little from 90°, it is doubtful that its value would have

[90] H. C. H. Wu and A. Sarko, *Carbohydr. Res.* **61,** 7 (1978).

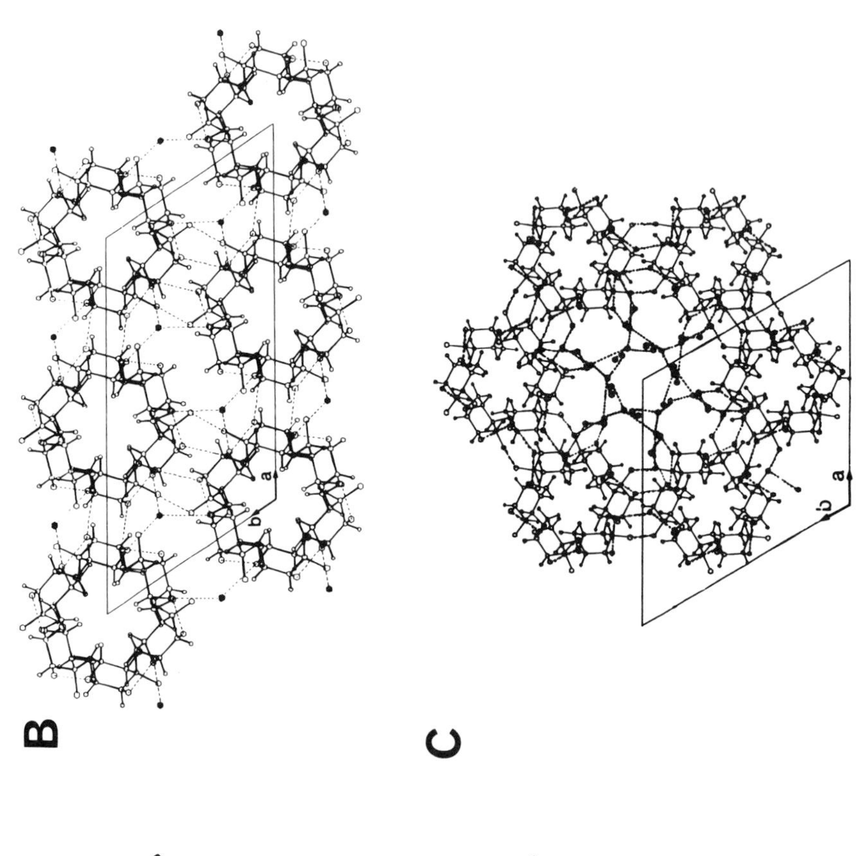
B
C
a
b
a
b

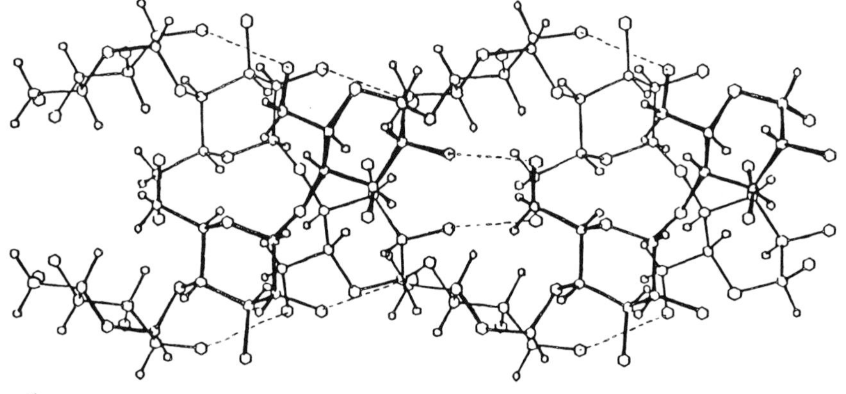
A

been determined correctly from a fiber X-ray diagram, had the latter been available.

Another illustration was provided by the structural determination of A-type amylose for which cell parameters and a space group symmetry were suggested by X-ray fiber and X-ray powder diffraction experiments. An orthorhombic-like unit cell, having $\alpha = \beta = \gamma = 90°$ was highly probable. Molecular modeling coupled with experimental values of helical parameters indicated the presence of a double-helical structure. Electron diffraction data provided some hints about the symmetry elements. Throughout the experimentally accessible reciprocal space, the electron diffractogram exhibited systematic absences of reflections with indices $h + k + l = 2n + 1$, (Fig. 13), indicating a body-centered lattice. It was rationalized that the possible orthorhombic space groups had too many symmetry requirements to accommodate two such chiral double helixes in the unit cell. The only remaining solution was the face-centered monoclinic space group B_2 (with the fiber axis c as the unique axis) after adequate transformation of the cell axis a, b, and c. Later, electron-diffraction data provided proof of monoclinic symmetry. By sequential tilting about an axis perpendicular to the crystal axis, some hkl and $-hkl$ (in the orthorhombic reference) reflection intensities could be compared and were found to be unequal, contrary to expectations for an orthorhombic space group.

Combination of data from tilt series offers the possibility of a three-dimensional structure refinement. The analyses reported to date show considerable variations in the degree of consistency between the different zonal data sets. Guizard[10] reported data for two zones in dextran with the upper layer data used qualitatively. In the structure refinement of mannan I, the base plane data were refined in conjunction with a stereochemical model in a least-squares procedure. The inclusion of several additional zones of data, obtained from tilt series, did not improve the refined model. Again, radiation damage was not considered to be a significant problem.

On the X-ray fiber diffractogram of anhydrous nigeran, a total of four layer lines were observed, along with systematic absences of odd meridi-

FIG. 12. Crystalline features of A- and B-type starch. (A) Molecular drawing of the double helix generated by the association of left-handed amylose strands, each one being a sixfold helix repeating in $2c = 21.38$ Å. Hydrogen bonds are represented as broken lines. From Imberty *et al.*[56] (B) Projection of the A-type structure onto the (a, b) plane. Hydrogen bonds are indicated as broken lines; (●), water molecules. From Imberty *et al.*[56] (C) Projection of the B-type structure onto the (a, b) plane. The unit cell and some neighboring double helices are represented in order to show the localization of the water molecules (●) in a channel. Hydrogen bonds are indicated as dashed lines. From Imberty and Pérez.[57] Reprinted with permission from Wiley & Sons.

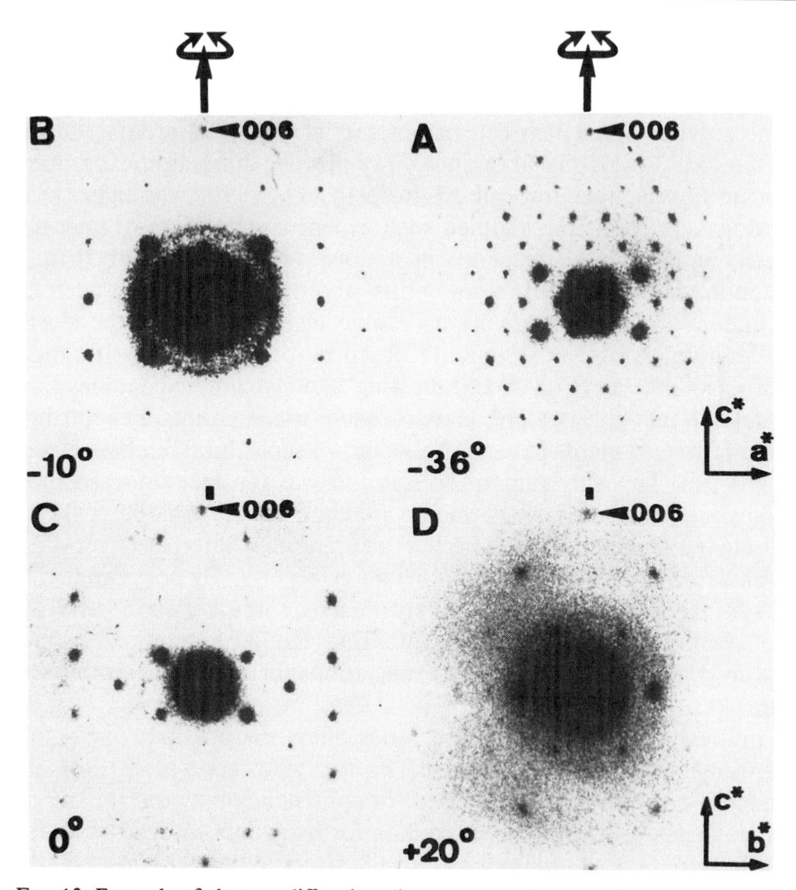

FIG. 13. Example of electron diffraction diagrams recorded under frozen wet conditions from crystals of A-type starch rotated about their long axis. The microscope goniometer setting was −36° in (A), −10° in (C), and +20° in (D). Indices in (B) are consistent with conditions for existing reflections of the type $h + k + l = 2n$, indicative of a body-centered lattice. From Imberty *et al.*[56]

anal reflections, which indicates a 2_1 crystallographic axis of symmetry along the c axis. Examination of the overall electron diffraction pattern indicates an *mm* symmetry of the base plane (*hk0*). Therefore, an orthorhombic space group was assigned. However, careful investigation of the systematic absences along **a*** and **b*** in the electron diffraction pattern yielded puzzling observations. The general features of the first layer-lines of reflections for small values of h are arced. Close examination revealed that two rows of reflections (separated by a distance of about 0.9 Å) were present. The strongest reflections were found to belong to the upper row, and no reflection having corresponding *d* spacing was observed on the zero-line of the fiber diagram. Furthermore, the strongest reflections in the fiber diagram were observed on the first-layer line. Eventually, it was established that some of the strongest reflections (*hkl*) were observed along with the (*hk0*) reflections on the electron diffractogram.

In vitro growth of micron-sized single crystals of amylose displaying a B-type structure has been described and the corresponding base plane electron diffractogram has been reported.[85] Electron diffraction from this sample in the frozen hydrated state resulted in sharp diffraction patterns from which the hexagonal symmetry and the dimensions of the base plane parameters could be ascertained. Close examination of the diffractogram indicated a 6*mm* symmetry (Fig. 14) that cannot be compatible with any of the noncentrosymmetrical space groups. It may, however, reflect the occurrence of two orientations of polar microcrystalline domains, both having the same structural features but growing in opposite *c* direction (Fig. 14). Similar features are commonly observed for minerals with hexagonal symmetry; they result from "twinning by merohedry."[91] This may occur with biological polymers grown from solution, where no template exists to orient the growth habit in a particular direction. Actually, the occurrences of domains in polymeric single crystals have been observed and characterized for poly-4-methyl-1-pentene.[92] Therefore, it was considered that the apparent 6*mm* symmetry observed on the base-plane diffractogram was likely to result from the existence of several narrowly delimited crystalline domains.

Lattice Imaging

This very promising technique has been applied to the molecular imaging of β-chitin[58] at 3.5-Å resolution, and compared with the known X-ray structure.[93] Chitin is a naturally occurring fiber-forming polymer and is an essential component of the skeletal materials of many "lower" animals, most notably the arthropods. Chemically it is a repeating polymer of β-(1 $\rightarrow$ 4)-linked anhydro-2-acetamido-D-glucose. Polymorphs have been recognized, of which the α and β forms are the best characterized.[94] Both have apparently the same 2_1 helical conformation. Highly crystalline β-chitin crystallizes in the monoclinic $P2_1$ space group with cell parameters

[91] H. Klapper, T. Hahn, and S. J. Chung, *Acta Crystallogr. Sect. B* **43,** 147 (1987).

[92] P. Pradère, J. F. Revol, and R. St. John Manley, *in* "Proceedings of the 45th Annual Meeting of the Electron Microscopy Society of America" (G. W. Bailey ed.), p. 488. San Francisco Press, San Francisco, 1987.

[93] K. H. Gardner and J. Blackwell, *Biopolymers* **14,** 1581 (1975).

[94] J. Blackwell, this series, Vol. 161, p. 435.

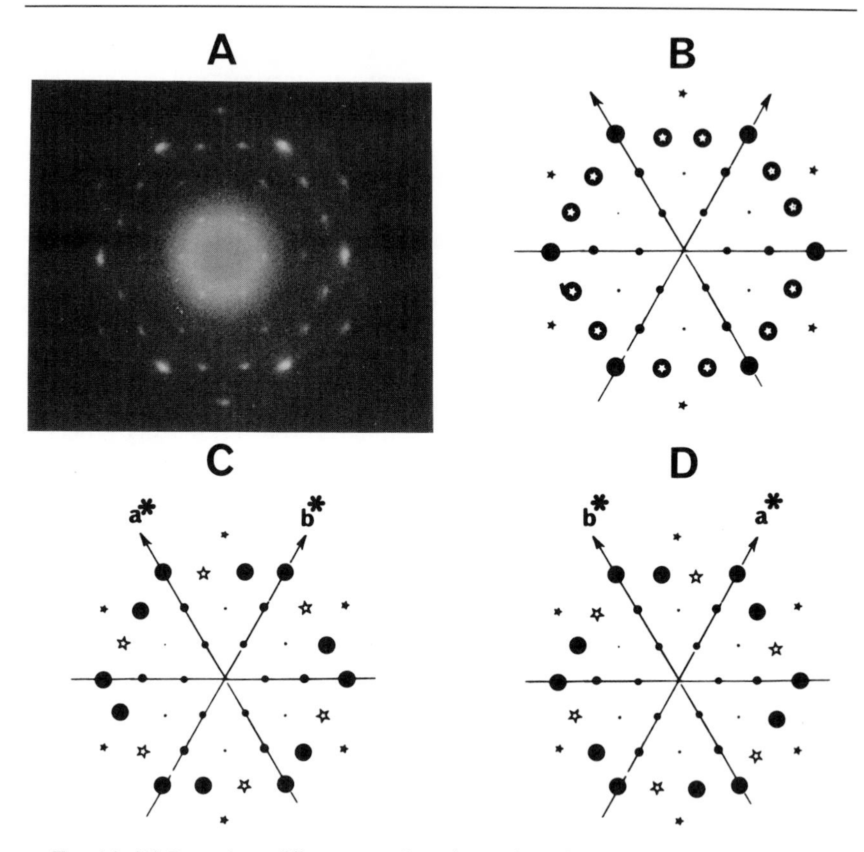

FIG. 14. (A) Base plane diffractogram for micron-sized single crystals of B-type amylose grown *in vitro,* along with its schematic representation (B) where the $6mm$ symmetry is apparent. Such a symmetry may result from the superimposition of the two diffractograms of (C) and (D). They correspond to the same structural arrangement, with parallel double helices in the $P6_1$ space group, but result from crystalline domains with opposite c direction (c^* would be "up" in (C) and would be "down" in (D)]. (A) From Buléon *et al.*[85] Reprinted with permission from Elsevier Science Publishers. (B), (C), and (D) From Imberty and Pérez.[57] Reprinted with permission from Wiley & Sons.

$a = 4.85$ Å, $b = 9.26$ Å, $c = 10.38$ Å (fiber axis), and $\gamma = 97.5°$; it is composed of an array of parallel chains.[93] An antiparallel arrangement of chains characterizes the α form.[95] Samples from the spines of *Thalassiosira fluviatilis* were selected for study, since they display a high crystallinity together with the absence of contaminating proteins. Selected-area diffraction on individual microfibrils gave rise to reflections that belong to the

[95] R. Minje and J. Blackwell, *J. Mol. Biol.* **120,** 167 (1978).

a*c* reciprocal lattice plane of β-chitin. This was an indication that the microfibrils are single crystals of high perfection, and that the macromolecular axis, c, is aligned along the microfibril axis. Conversely, the crystallographic b axis is normal to the plane of the microfibril. High-resolution, low-dose electron micrographs were obtained on similar preparations. They were digitized and subdivided into arrays of 200 × 200 pixels. These arrays were Fourier transformed and the results were examined to determine the regions of the original array that produced transforms with the highest symmetry (Fig. 15A). The square of the Fourier transform is identical to that produced by optical diffraction (inset in Fig. 15A). The observation of ($h0l$) reflections in the transform to a resolution of 3.5 Å indicates the presence of lattice lines to the same resolution. From the optically filtered image of Fig. 15A, the unit cell parameters and unit cell orientation were determined. At this stage, some information from the three-dimensional structure was added, i.e., symmetrization of the cells by

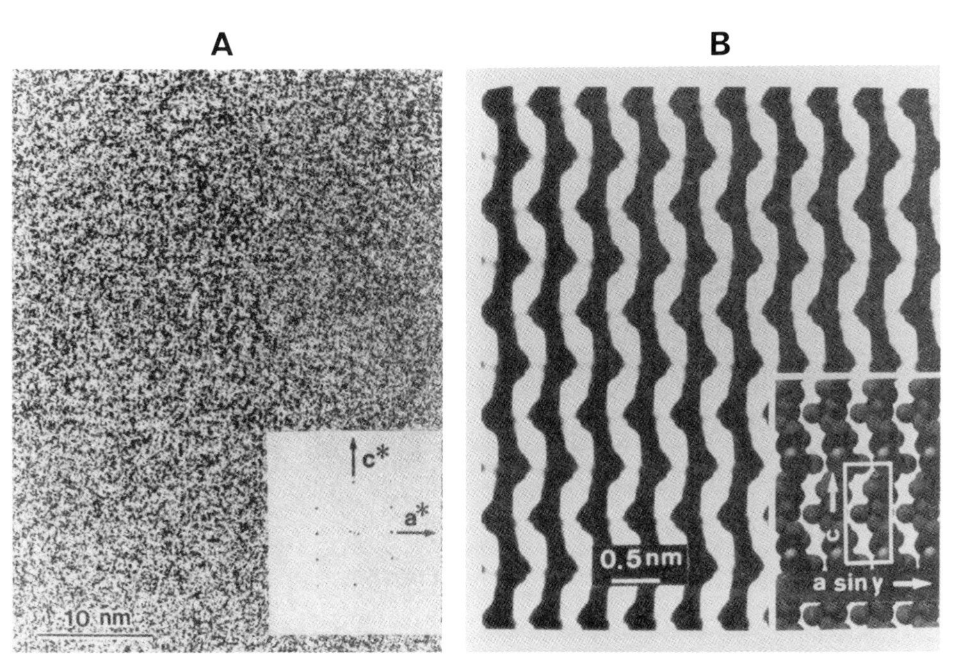

FIG. 15. Molecular imaging of β-chitin. From Revol *et al.*[58] Reprinted with permission from Wiley & Sons. (A) High-resolution, low-dose electron micrograph of a β-chitin microfibril. *Inset:* Corresponding computer-calculated Fourier transform. (B) Computer-reconstructed image showing the projection of the β-chitin crystal along the b axis. The inset represents such a projection according to the know crystal structure.[93]

applying a glide plane parallel to the c axis. This is the two-dimensional equivalent to the 2_1 screw axis lying along the chain.[93] This cell has been replicated to form the molecular image shown in Fig. 15B, from which the parallel arrangement of the chains could be assessed. Lattice imaging of α-chitin has been reported[96] without any attempt to solve the structure.

Structure Prediction

As amply demonstrated in the previous section, the modeling technique can be combined with information derived from electron diffraction, i.e., accurate unit cell parameters, and unambiguous determination of the space group symmetry, to help solve three-dimensional structures. When such constraints are used, one does not obtain an explanation of why the crystal structure is the stable form. There is therefore a need to understand and develop general rules for the stability of interhelical arrangements. Methods for investigating interhelix structure and energy through nonbonded forces have been suggested by a number of workers.[97-102] Basically, all these procedures minimize the interhelix energy. A method for predicting the packing relationships of two polymer chains have been developed[103-105] and its application to the case of polysaccharides is presented below.

Principle

In the following analysis, it is assumed that the two polymeric chains, referred to as chain A and chain B, are regular helices, i.e., that they have screw symmetry, with a repeat distance t_A and t_B, respectively. It is clear that, in an ordered periodic arrangement, the periods of the two interacting chains must be commensurate. This requirement of commensurability can be expressed as: $pt_A = st_B = t$, where p and s are integers. Chains can only be parallel or antiparallel to one another. A set of four interhelical parameters is required in order to define the geometric orientation of chain A

[96] J. F. Revol, *Int. J. Biol. Macromol.* **11,** 23 (1989).
[97] A. J. Hopfinger and A. G. Walton, *J. Macromol. Sci. Phys. B* **3,** 195 (1969).
[98] A. J. Hopfinger and A. G. Walton, *J. Macromol. Sci. Phys. B* **4,** 185 (1970).
[99] A. J. Hopfinger, *Biopolymers* **10,** 1299 (1971).
[100] K. Tai, M. Kobayashi, and H. Tadokoro, *J. Polym. Sci. Polym. Phys. Eds.* **14,** 783 (1976).
[101] K. C. Chou, G. Némethy, and H. A. Scherage, *J. Phys. Chem.* **87,** 2869 (1983).
[102] K. C. Chou, G. Némethy, and H. A. Scheraga, *J. Am. Chem. Soc.* **104,** 3161 (1984).
[103] R. P. Scaringe and S. Pérez, *J. Phys. Chem.* **91,** 2394 (1987).
[104] R. P. Scaringe and S. Pérez, unpublished results 1990.
[105] S. Pérez, *in* "Electron Crystallography of Organic Molecules" NATO ASI Ser. (J. Frier and D. L. Dorset, eds.), p. 33. Kluger, Amsterdam, 1991.

relative to chain B. They are μ_A, a rotation of the chain A about its axis, from 0 to 360°, and μ_B, a rotation of the chain B about its axis, from 0 to 360°.

Δx and Δz, which are taken as positive, represent positional shifts normal and parallel to the identity axes, respectively. Δz is bounded between 0 and t. The spatial description of these parameters is shown in Fig. 16.

The minimum energy arrangements of the two polymeric chains with respect to a displacement will tend to bring the molecules as close as possible without interpenetration. In reality, a small amount of repulsive energy resulting from interpenetration of some number of atom pairs can be compensated by additional attractions from the remaining atom pairs. However, nonbonded "contact distances" deviate by only 10% (0.20, 0.30 Å) in molecular solids. The principle of the contacting procedure has been described in details elsewhere.[103] In this procedure, the surface of each chain is described by circumscribing a hard sphere of the appropriate van der Waals radius R, around each of the constituent atoms. Then for a given

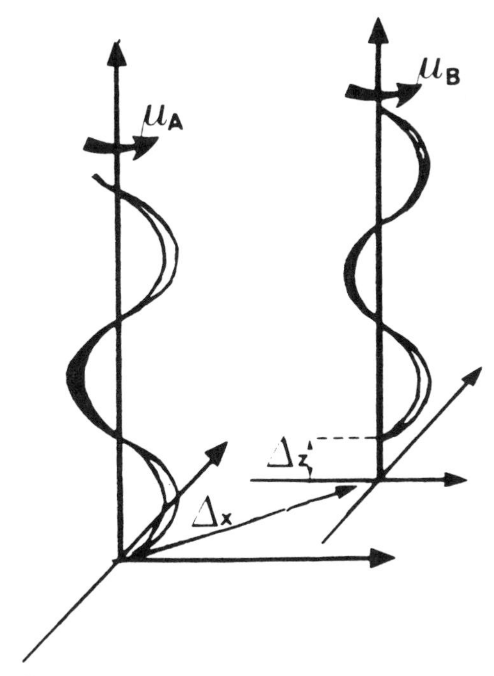

FIG. 16. Interhelical parameters required to define the geometric orientation of chain A relative to chain B.

orientation of the chains (as dictated by rotation angles μ_A, μ_B and an increment along the chain axis, Δz) one must find the relative translation, Δx, which will bring the interpenetrating surfaces into a position where they are in contact, without any interpenetration. In general the final position of the two initially penetrating polymers is characterized by the following conditions:

1. For at least one atom pair (i, j), the ith atom of chain A is separated from the jth atom of chain B by the sum of the associated van der Waals radii ($R_{ij} = R_i + R_j$). The atom pair i, j that satisfies this condition is referred to as the determining contact.
2. For all atom pairs in the two separate chains, there is no pair at a distance closer than the appropriate van der Waals radii sum ($D_{ij} > R_i + R_j$).
3. Condition (2) is violated by any atom pair i, j involved in hydrogen bonding for which the interatomic distances span 0.5 Å with no *a priori* optimum value. This limiting case is treated as follows. All the potential couples of atoms eligible to participate in an interchain hydrogen bond are identified and omitted from the contacting procedure. This implicitly means that hydrogen bonding will not violate principle (1) for the van der Waals bonded atoms. It is clear that if the shape of the molecule is not too complicated, there should be only one solution to the interaction problem.

If a contacting procedure is used, chain–chain construction requires only geometric information and, in principle, one can subsequently calculate the energy of the resulting interactions (E_{AB}) to any degree of approximation. The interaction energy of the two chains is considered to be the sum over all pairwise atom–atom interactions. Such an interaction is calculated according to 6–12 potential functions.[26] To these may be added Coulombic interactions between the atoms and the on charges. As for the energy of stabilization arising from hydrogen bonding, an extra term must be included. In the present simulations on polysaccharide chains, none of the hydroxyl hydrogen atoms is considered. Therefore, a simple energy criterion is used, based on the distance between the oxygen and/or the nitrogen atoms that can interact through hydrogen bonding.[88]

In practice a procedure is adopted in which Δx and E_{AB} are mapped as a function of the structural variables μ_A, μ_B, and Δz. The analysis is performed by rotating μ_A and μ_B over the whole angular range by increments of a few degrees and the relative translation (Δz) between the two chains is studied over the length of the whole fiber repeat, typically by 0.5-Å increments. For each setting of the chain as a function of μ_A, μ_B and Δz, the magnitude of the perpendicular offset (Δx) is derived according to the procedure outline above. Then the value of the energy E_{AB} corre-

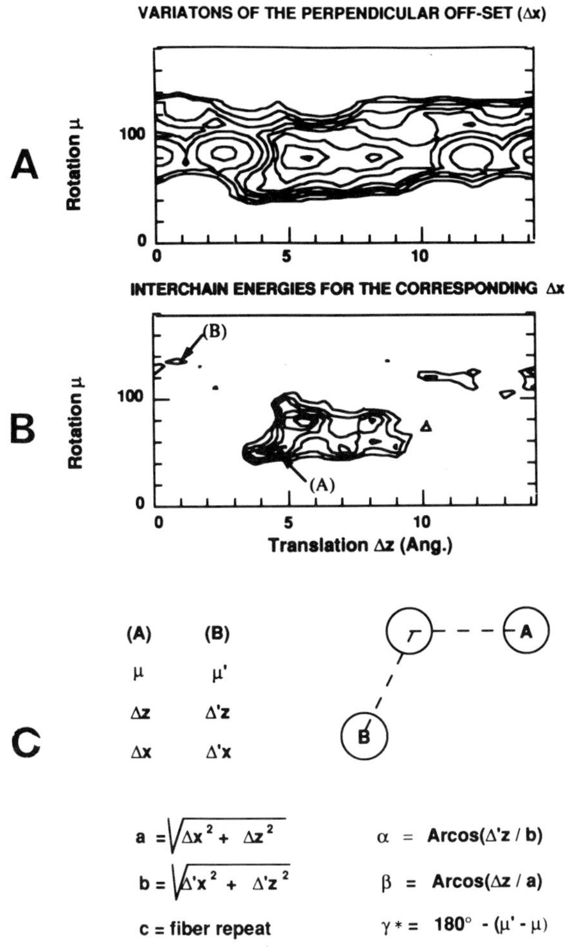

FIG. 17. *A priori* modeling of polymer–polymer interactions. The contour maps are drawn as a function of the translation: Δz, along the polymer fiber axis, and four coupled rotation angles $\mu_A = \mu_B$. From Pérez.[105] Reprinted with permission from "Electron Crystallography of Organic Molecules." Copyright 1991, Kluger. (A) Variations of the perpendicular off-set (Δx). Contours correspond to 4.1, 4.3, 4.5, 4.7, 4.9, 5.1, and 3.3 Å. (B) Interchain energies calculated for the corresponding Δx. Contours correspond to −11, −10, −9, −8, −7, −6 kcal/mol. (C) Derivation of unit cell parameters when two orientations of chain B with respect to chain A are known.

sponding to each set of chain orientations is computed. The mapping procedure is used to search for energy minima. This is adequate, because it provides a complete overview of the symmetry (or lack of symmetry) of the chain–chain interactions. The set of interhelical parameters relates directly to the symmetry operations and the type of lattices that characterize the solid organization (for detailed information, see Pérez[105]). Similarly, information about the unit cell parameters can be obtained, providing that two orientations of chain B with respect to chain A are known, i.e., μ_A, μ_B, Δx, Δz, and μ'_A, μ'_B, $\Delta' x$ and $\Delta' z$ (Fig. 17).

The structures derived by this process have been compared to experimentally observed structures in the case of synthetic polymers.[105] It was found that the predicted arrangements were surprisingly good approximations. The following section shows the application to reproducing known structures and understanding polymorphic transitions for polysaccharides.

Illustration

Chitin. A schematic representation of the disaccharide repeating unit of chitin is shown in Fig. 18. In the first step, the conformation of the chain was assessed through conformational analysis using the PFOS program.[21,27] The following parameters were determined:

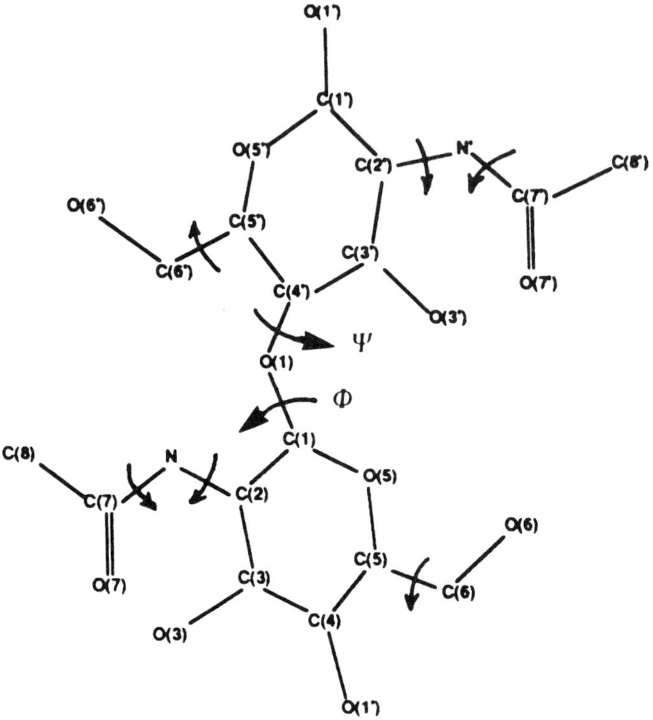

FIG. 18. Schematic representation of the disaccharide repeating unit of chitin.

1. rotation of O-6 about the exocyclic C-5–C-6 bond
2. rotation of the entire amide side chain about the C-2–N bond
3. rotation of the acetate group about the N–C-7 bond.

All the possible arrangements for polysaccharide chains were examined. It became clear that the significant energy minima were occurring for μ_A and μ_B and Δz such that $\mu_A = \mu_B = \mu$ and $\Delta z = 0$. This was an indication that with their given conformations, packing of the chitin chains is best achieved through translation symmetry operations. This allows for a straightforward one-dimensional study. Figure 19A is a representation of the variation of Δx, as derived from the contacting procedure, as a function of rotation of μ from 0 to 180°, and for $\Delta z = 0$. Figure 19B represents the variations of the interchain energy. Essentially, only one energy minimum is found at $\mu = 120°$ and $\Delta z = 0$ Å with $\Delta x = 4.88$ Å. Close examination also reveals that interchain hydrogen bonds are formed between O-6 and O-7 (2.967 Å), and N and O-7 (2.811 Å). In Fig. 20 are shown two representations of the chain pairing resulting from the *a priori* calculation.

The β-polymorph of chitin crystallizes in the monoclinic unit cell, space group $P2_1$; it contains a single chain disaccharide unit and hence the structure is an array of parallel chains.[93] One of the relative orientations of two neighboring chitin chains in the crystal structure is characterized by $\mu = 119.3°$, $\Delta z = 0$ Å and $\Delta x = 4.85$ Å. Also, the network of observed interchain hydrogen bonds corresponds very closely to the one which results from the method from predicting the packing relationship of two

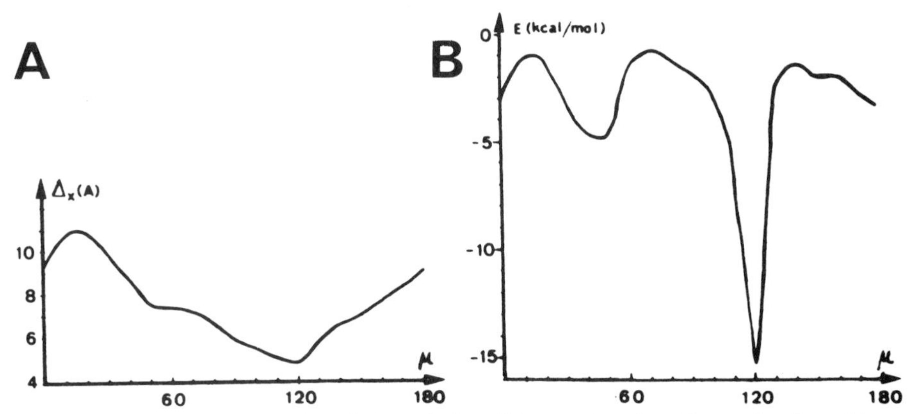

FIG. 19. (A) Representation of the variation of Δx as a function of μ ranging from 0 to 180°, and with a constant magnitude of Δz (=0), for chitin chains. (B) Representation of the interchain energy as a function of μ ranging from 0 to 180°, and with a constant magnitude of Δz (=0), for chitin chains. From Pérez.[105] Reprinted with permission from "Electron Crystallography of Organic Molecules." Copyright 1991, Kluger.

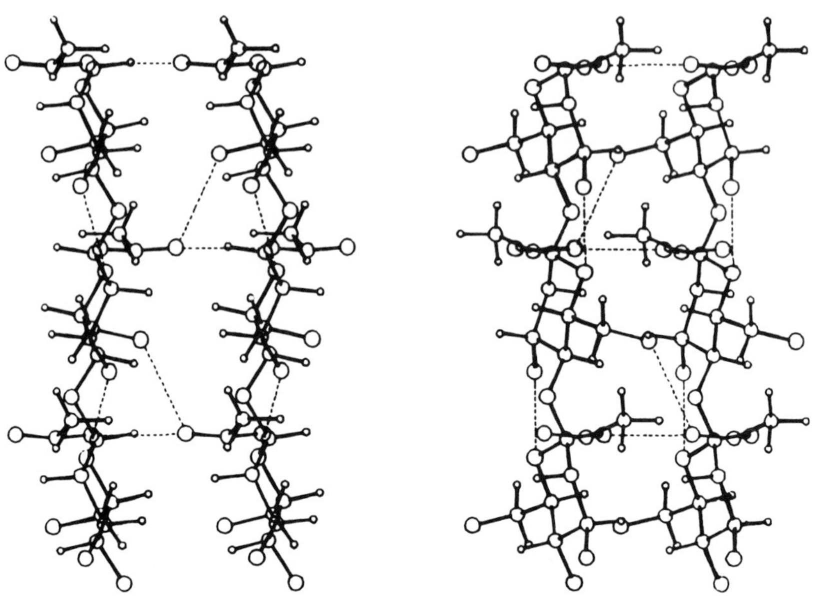

FIG. 20. Molecular drawings of the best parallel pairing of chitin chains. From Pérez.[105] Reprinted with permission from "Electron Crystallography of Organic Molecules." Copyright 1991, Kluger.

polymer chains. Therefore, the agreement between observed and interchain orientations is excellent.

Crystalline Polymorphism of Starch. Starting from the structural feature derived for the two crystalline polymorphs of starch,[56,57] an ideal parallel-stranded double helix was constructed. Each strand has a left-handed sixfold symmetry. All possible arrangements occurring between parallel and antiparallel double helices were examined. The details of these investigations have been reported elsewhere,[106] so only the essential features are summarized here. The calculated interchain parameters for the lowest energy minima for parallel and antiparallel arrangements are summarized in Table III. The best pairing in energy terms is found for a parallel arrangement of the double helices (Para 1). Another stable, parallel chain pairing is found that has an interaction energy 6.8 kcal/mol above the previous one (Para 2). The corresponding arrangements are shown in Fig. 21. As for the antiparallel case, two stable chain pairings are also

[106] S. Pérez, A. Imberty, and R. P. Scaringe, *in* "Computer Modeling of Carbohydrate Molecules" (A. D. French and J. W. Brady, eds.), Series 430, p. 281. American Chemical Society, Washington, D.C., 1990.

TABLE III
CALCULATED INTERCHAIN PARAMETERS FOR BEST ENERGY MINIMA FOR PARALLEL AND ANTIPARALLEL OF DOUBLE HELICES OF AMYLOSE

Parameter	Para 1	Para 2	Anti 1	Anti 2
μ_A (°)	11.5	26.0	78.5	47.0
μ_B (°)	11.5	167.5	41.5	13.0
Δz (Å)	5.25	3.22	7.86	7.39
Δx (Å)	10.77	11.20	10.77	11.16
E (kcal/mol)	−26.6	−19.8	−23.7	−20.9

found. The more stable arrangements for both parallel and antiparallel cases are characterized by a distance between the center of mass of the two double helices of 10.77 Å. The crystal structure of both polymorphs is based on a parallel arrangement of double helices that are slightly different since small variations away from the perfect sixfold symmetry are found. Nevertheless, they correspond closely to the idealized double helix studied in this work. The essential result is that in these two observed structures, the closest interactions between two neighboring double helices correspond closely to the duplex described as Para 1. In the crystal structures, neighboring double helices have the same rotational orientation and the same translation of half a fiber repeat as in the Para 1 model. Only the Δx vector is slightly larger in the calculated interactions (10.77 Å) than in the observed ones: 10.62 and 10.68 Å in the A type and B type, respectively. This may be due to the fact that in the crystal structures the helices depart slightly from perfect sixfold symmetry. Also, no interpenetration of the van der Waals surfaces is allowed in the modeling procedure, whereas some of them may occur in the crystal structures. It is interesting to note that the network of inter-double helix hydrogen bonds found in the calculated Para 1 model reproduces those found in the crystalline structures.

The differences between the two polymorphs occur from other effects. In its crystalline arrangement, the B-type polymorph has hexagonal symmetry; each double helix has only three neighbors corresponding to the interaction described above as Para 1. The channel created by six double helices packed in a hexagonal arrangement is occupied by a column of water molecules. In the less hydrated A-type structure, each double helix is surrounded by six neighboring ones. The chain pairing described by the Para 1 model corresponds to four out of six of these interactions, the two others being looser. This type of loose arrangement which is generated by translational symmetry ($\Delta x = 11.72$ Å, $\Delta z = 0$ Å) is also predicted by the

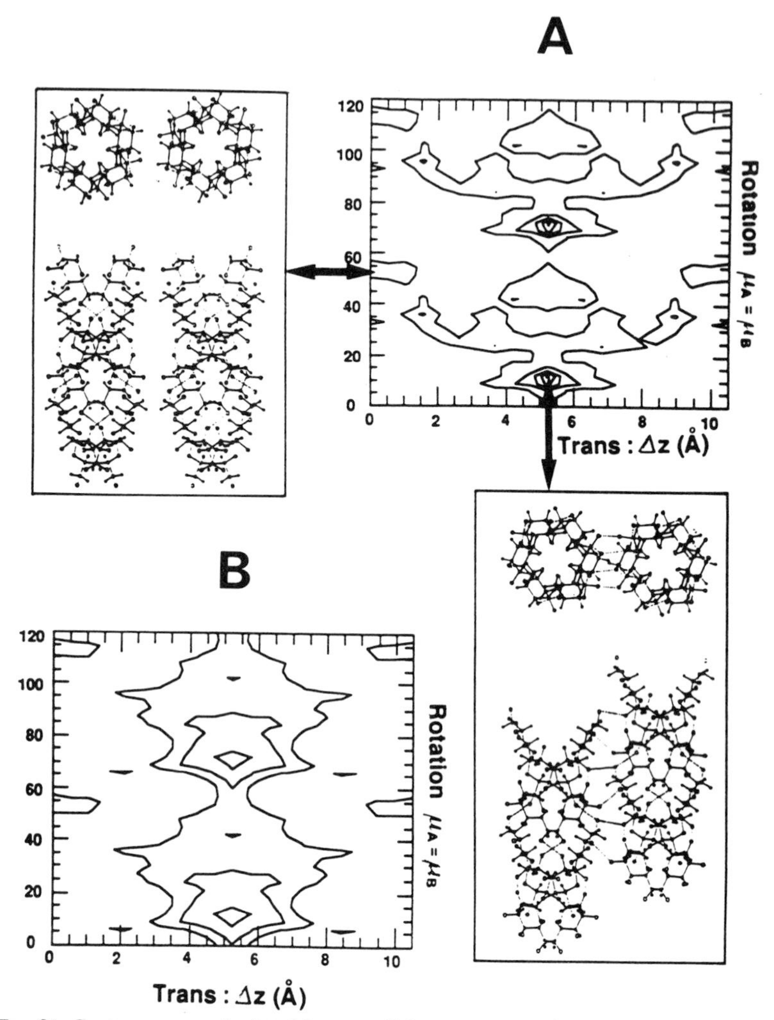

FIG. 21. Contour maps calculated for a parallel arrangement of double helices of amylose, as a function of the translation Δz along the fiber axis and of the coupled rotation angles $\mu_A = \mu_B$. From Pérez[105] (reprinted with permission from "Electron Crystallography of Organic Molecules." Copyright 1991, Kluger). (A) Interchain energies calculated for the corresponding Δx [see (B)]. Contours correspond to $-25, -20, -15, -10$, and -5 kcal/mol. (B) Variations of the perpendicular off-set (Δx). Contours correspond to 11.0, 11.5, and 12.0 Å. The arrangement occurring at $\Delta z = 5.25$ Å and $\mu_A = 11.5°$ corresponds to the lowest interchain energy and generates the Para 1 model. The one occurring at $\Delta z = 0$ Å and $\mu_A = 55°$ corresponds to a loose interchain interaction as found in the A-type crystal structure.[56]

modeling procedure and corresponds to a secondary energy minimum for double-helix pairings; it occurs for $\mu_A = \mu_B = 55°$, $\Delta z = 0$ Å, $\Delta x = 11.76$ Å. The agreement between experimental and calculated chain arrangements is obtained despite the fact that no water molecules were considered in the modeling. This would tend to indicate that these water molecules do not play an important role in establishing the three-dimensional packing of double helices.

Conclusion

Under appropriate conditions micron-sized crystals of linear polysaccharides can be grown. Electron diffraction data collected on such crystals provide highly reliable lattice constants, tests for symmetry elements, and base plane structure amplitude. The joint use of molecular modeling and electron diffraction has been invaluable in the quantitative elucidation of crystals and molecular structures. The modeling technique can be combined with information derived from electron diffraction along with fiber diffraction results for quantitatively and unambiguously solving three-dimensional polysaccharide structures. Essential molecular features, including chain conformation and polarity, packing structure, intermolecular interactions, and placement and role of water molecules, have all been characterized. The relationships between structure and crystal morphology and twinning planes have been established, along with information regarding chain folding at the crystal surface. This detailed understanding provides an essential basis for further investigations, such as the nature of the interactions of carbohydrases on solid substrates. The procedure also can be used for structure elucidation of large oligosaccharides, which are not likely to yield crystals good enough for conventional crystallographic work. Electron diffraction data may also be used to index powder diffraction patterns accurately and help resolve partially overlapping lines into individual intensities. Future efforts should be devoted to a better use of electron diffraction intensities, and to help understand the influence of crystal growth on the apparent observed symmetry in the electron diffractogram.

In turn, knowledge of the three-dimensional structure of polysaccharides has established a basis for understanding how crystalline arrangements form and predicting the packing relationship between two polymer chains. It is quite straightforward to predict the different ways that a polysaccharide chain is going to interact with other chainlike molecules. Whereas only a few of these arrangements would correspond to chain pairing capable of generating efficient packing, the other ones may represent situations that are likely to occur in the amorphous state or at the surface of polymeric materials. It is also reasonable to extend this predictive methodology to low-symmetry systems, such as gels, where chain–chain interactions may occur to promote the formation of the so-called "junction zones."

Acknowledgments

The author would like to express appreciation to Dr. Anne Imberty for helpful discussions and careful reading of the manuscript. Appreciation is extended to Mrs. Chantal Nicolas for her dedicated help with the artwork.

[28] Molecular Design and Modeling of Protein–Heparin Interactions

By ALAN D. CARDIN, DAVID A. DEMETER, HERSCHEL J. R. WEINTRAUB, and RICHARD L. JACKSON

Introduction

Limited information is available on the specific mechanisms by which heparin and other glycosaminoglycans (GAG) affect the structure and function of proteins; for a review see Jackson *et al.*[1] An understanding of these interactions is complicated by the complexity of mucopolysaccharide structure and the diversity of GAG-binding proteins. It has generally been assumed that the major interactions are electrostatic and involve the negatively charged sulfates and carboxylates on the GAG and positively charged residues on the protein. In previous studies,[2,3] the heparin-binding regions of two protein constituents of human plasma lipoproteins, apolipoprotein E (apoE) and apoB, were isolated and sequenced. ApoE contained two sequence domains and apoB five sequence domains involved in heparin binding. These GAG-binding regions contained clusters of two to three positively charged residues separated by noncharged amino acids. It was hypothesized that this sequence organization of basic residues formed domains of high positive charge density capable of interacting with the negative charge density on heparin. Examination of amino acid sequences

[1] R. L. Jackson, S. J. Busch, and A. D. Cardin, *Physiol. Rev.* **71,** 481 (1991).

[2] A. D. Cardin, N. Hirose, D. T. Blankenship. R. L. Jackson, J. A. K. Harmony, D. A. Sparrow, and J. T. Sparrow, *Biochem. Biophys. Res. Commun.* **134,** 783 (1986).

[3] N. Hirose, D. T. Blankenship, M. A. Krivanek, R. L. Jackson, and A. D. Cardin, *Biochemistry* **26,** 5505 (1987).

of other heparin-binding proteins led to the discovery of regions of clustered basic residues with sequence organizations similar to those of apoE and apoB. Based on these sequence similarities, motifs or consensus sequences for GAG binding were proposed that provided a predictive power to identify primary GAG interaction sites in proteins based on initial sequence inspection.[4] Lending support to this hypothesis, the predicted sequence domains in neural cell adhesion molecule (NCAM),[5] fibronectin,[6] antithrombin III (AT III),[4,7] heparin cofactor II,[4,8] and leuserpin[9] have now been shown to bind heparin.

One approach to investigate the detailed molecular basis of GAG–protein interactions is by the simulated computational modeling of the binding of heparin to peptide domains.[4,10] In this chapter we will review the methods and approaches to the computer-aided molecular design and analysis of the two heparin-binding regions of apoE and their interactions with heparin. Three predictive algorithms based on Chou–Fasman,[11] Schiffer and Edmunson wheel diagrams,[12] and hydrophobic moment calculations[13] have been utilized and the results compared to solution studies on peptide–heparin interactions by circular dichroism (CD). With these data, the structures of limited peptide domains that bind heparin have been constructed by molecular modeling approaches and their interactions with heparin investigated via molecular dynamics calculations.

Identification of Apolipoprotein E Heparin-Binding Domains

ApoE is a 299-amino acid residue protein that is associated with plasma lipoproteins.[14] The apolipoprotein binds to low-density lipoprotein (LDL) receptors on cells and to heparin. To identify the GAG-binding region(s) of apoE, the lipid-free protein is cleaved between Arg-191 and Ala-192 with

[4] A. D. Cardin and H. J. R. Weintraub, *Arteriosclerosis* **9,** 21 (1989).
[5] G. J. Cole and R. Akeson, *Neuron* **2,** 1157 (1989).
[6] F. J. Bober-Barkalow and J. E. Schwarzbauer, *J. Biol. Chem.* **266,** 7812 (1991).
[7] J. W. Smith and D. J. Knauer, *J. Biol. Chem.* **262,** 11964 (1987).
[8] M. A. Blinder and D. M. Tollefsen, *J. Biol. Chem.* **265,** 286 (1990).
[9] H. Ragg, T. Ulshöfer, and J. Gerewitz, *J. Biol. Chem.* **265,** 5211 (1990).
[10] H. J. R. Weintraub, D. A. Demeter, A. D. Cardin, and R. L. Jackson, "Molecular Description of Biological Membranes by Computer Aided Conformational Analysis" (R. Brasseur, ed.), Vol. 2, Chap. 3.4, p. 161. CRC Press, Boca Raton, Florida, 1990.
[11] P. Y. Chou and G. D. Fasman, *Biochemistry* **13,** 222 (1974).
[12] M. Schiffer and A. B. Edmunson, *Biophys. J.* **7,** 121 (1967).
[13] D. Eisenberg, R. M. Weiss, and T. C. Terwilliger, *Proc. Natl. Acad. Sci. U.S.A.* **81,** 140 (1984).
[14] R. W. Mahley, *Science* **240,** 622 (1988).

thrombin[15] to yield the peptide fragments E(1 – 191) and E(192 – 299). The fragments are then assessed for heparin binding by ligand blotting with ^{125}I-labeled heparin.[16] Heparin was enriched in its apoE-binding affinity by fractionating commercial heparin on a column of LDL attached to Affi-Gel 10 to yield a highly reactive heparin (HRH) fraction. This heparin (2.5 mg uronic acid) was reacted with 1 mg of the *N*-hydroxysuccinimide ester of 3-(*p*-hydroxyphenyl)propionic acid and then radioiodinated with carrier-free Na^{125}I in the presence of ICl as described by McFarlane.[17] E(1 – 191) and E(192 – 299) were separated on a linear 6 – 20% polyacrylamide gel containing 6 *M* urea and 0.1% sodium dodecylsulfate (SDS) and the fragments were transferred to nitrocellulose (0.45 μm) for 4 hr at 0.22 A. The nitrocellulose was then incubated with ^{125}I-labeled heparin at room temperature for 8 hr, washed, and subjected to radioautography. By this method, apoE and the two thrombin fragments were shown to bind heparin, indicating a minimum of two heparin-binding domains. To determine the minimal residues required for GAG recognition in these fragments and their relative affinities for GAG, the regions that were rich in basic amino acids in E(1 – 191) and E(192 – 299) were examined for their ability to bind ^{125}I-labeled heparin. Synthetic peptides with successive deletions from the amino- and carboxyl-terminal ends were then investigated (Table I[18]). To assess heparin binding, a dot-blot assay was developed.[18] The peptides were applied to nitrocellulose and then incubated with ^{125}I-labeled heparin. The amount of heparin bound was determined by gamma counting.

Table I and Fig. 1 summarize the heparin-binding activities of the various peptides. The left-hand column of Fig. 1 (inset) represents E(129 – 169) incubated with ^{125}I-labeled heparin in the presence of increasing concentrations of NaCl; the right column represents E(144 – 169). With the exception of E(139 – 169), which binds heparin as well as E(129 – 169) (Table I), the shorter peptides had decreased heparin-binding affinity. Several conclusions can be drawn from these experiments. Deletion of Arg-134 and Arg-136 does not decrease heparin binding, indicating they are not critical. Deletion of Arg-142 and Lys-143 decreased binding by 40%. However, simultaneous deletion of Arg-145, Arg-147, and Lys-146 abolished binding, showing that one or more of these residues is critical. Deletion of residues 156 – 169 from the carboxyl terminus had limited effect on binding. These findings indicated that the critical residues for

[15] S. H. Gianturco, A. M. Gotto, Jr., S.-L. C. Hasang, J. B. Karlin, A. H. Y. Lin, S. C. Prasad, and W. A. Bradley, *J. Biol. Chem.* **258,** 4526 (1983).

[16] A. D. Cardin, K. R. Witt, and R. L. Jackson, *Anal. Biochem.* **137,** 368 (1984).

[17] A. S. McFarlane, "Mammalian Protein Metabolism" (M. W. Munro and J. B. Allison, eds.), p. 331. Academic Press, New York, 1964.

[18] N. Hirose, M. Krivanek, R. L. Jackson, and A. D. Cardin, *Anal. Biochem.* **156,** 320 (1986).

TABLE I
HEPARIN-BINDING PEPTIDES OF APOLIPOPROTEIN E

Residues	Amino acid sequence[a]	Heparin[b] binding activity
129–169	130 140 150 160 ST EELRVRLASH LRKLRKRLLR DADDLQKRLA VYQAGAREG	+++
139–169	140 150 160 H LRKLRKRLLR DADDLQKRLA VYQAGAREG	+++
144–169	150 160 LRKRLLR DADDLQKRLA VYQAGAREG	++
148–169	150 160 LLR DADDLQKRLA VYQAGAREG	–
141–155	150 LRKLRKRLLR DADDL	++
202–243	210 220 230 240 PLQERAQAW GERLRARMEE MGSRTRDRLD EVKEQVAEVR AKL	++
211–243	220 230 240 GERLRARMEE MGSRTRDRLD EVKEQVAEVR AKL	++
219–243	220 230 240 EE MGSRTRDRLD EVKEQVAEVR AKL	–

[a] One-letter amino acid code: A, Ala; C, Cys; D, Asp; E, Glu; F, Phe; G, Gly; H, His; I, Ile; K, Lys; L, Leu; M, Met; N, Asn; P, Pro; Q, Gln; R, Arg; S, Ser; T, Thr; V, Val; W, Trp; and Y, Tyr.
[b] Determined by dot-blot analysis with ^{125}I-labeled heparin.[18]

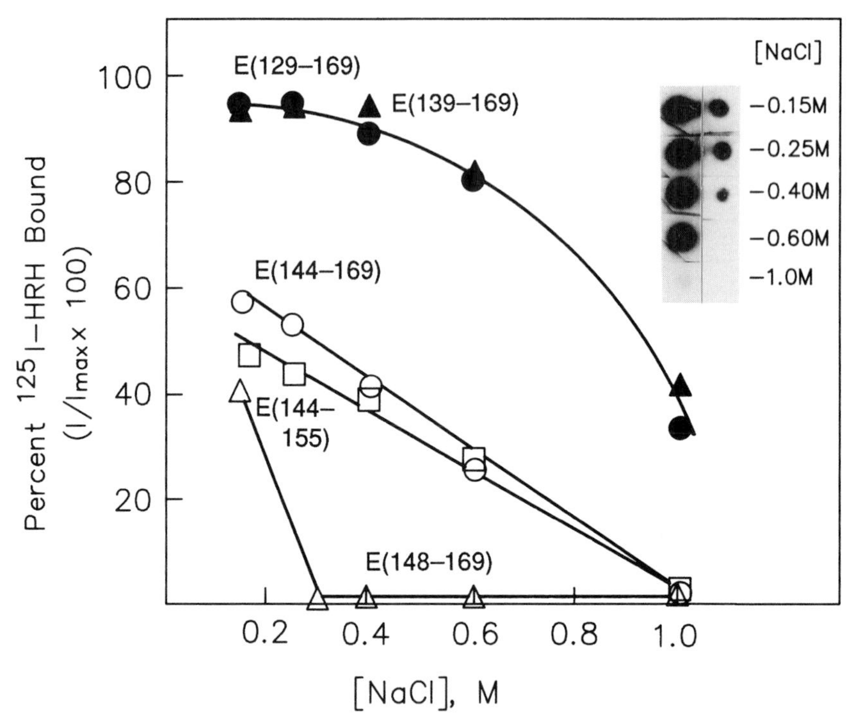

FIG. 1. Effect of ionic strength on the electrostatic binding of an ^{125}I-labeled-highly reactive heparin fraction (HRH) to various synthetic apoE fragments. *Inset:* Dot-blot autoradiographic analysis of ^{125}I-labeled HRH binding to E(129–169) (left column) and E(144–169) (right column). Peptides are E(129–169) (●); E(139–169) (▲); E(144–169) (○); E(141–155) (□); and E(148–169) (△). The amino acid sequences are shown below:

```
129 136                                   158        169
  S-----RLASHLRKLRKRLLRDADDLQKR-----REG
      139
        SHLRKLRKRLLRDADDL        QKR-----REG
          144
            LRKRLLRDADDL         QKR-----REG
             148
              LLLRDADDL          QKR-----REG
        141
          LRKLRKRLLRDADDL
```

heparin binding resided between amino acids 144 and 150 (Leu-Arg-Lys-Arg-Leu-Leu-Arg). In similar experiments, peptides E(202–243), E(211–243), but not E(219–243) bound heparin (Table I), suggesting that the critical residues for GAG recognition in the E(202–243) heparin-binding domain involved residues 213–218 (Arg-Leu-Arg-Ala-Arg-Met).

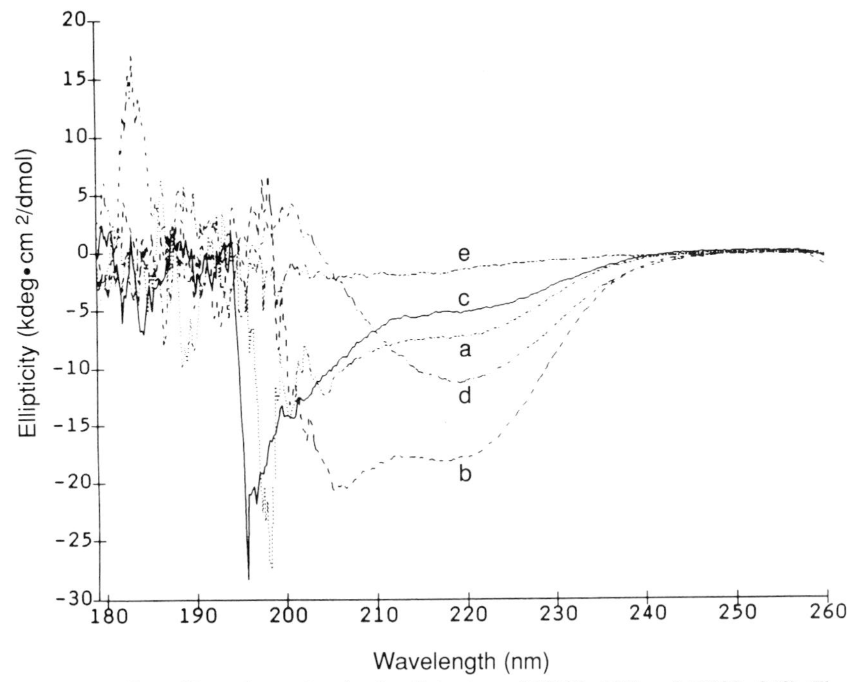

FIG. 2. Effect of heparin on the circular dichroism of E(129–169) and E(202–243). The spectra represent the following: (a) 100 μg/ml E(129–169); (b) 100 μg/ml E(129–169) + 60 μg/ml heparin; (c) 100 μg/ml E(202–243); (d) 100 μg/ml E(202–243) + 60 μg/ml heparin; and (e) 60 μg/ml heparin. The buffer was 20 m*M* HEPES, 0.15 *M* NaCl, pH 7.4.

Solution Conformations of Apolipoprotein E Heparin-Binding Domains

The above experiments suggested that specific amino acid sequences in apoE were involved in heparin binding. To determine whether a GAG could alter peptide structure, the effect of heparin on the solution conformations of E(129–169) and E(202–243) was examined by circular dichroism (CD) (Fig. 2). The CD spectrum of E(129–169) (Fig. 2a) revealed limited ordered structure. However, heparin addition dramatically increased the negative ellipticity of E(129–169) with pronounced minima at 208 and 222 nm (Fig. 2b). The qualitative band shape was characteristic of model α-helical peptides.[19] As summarized in Table II,[20] the $[\theta]_{208}$ increased from -4590 to $-19{,}500$ deg cm^2 dmol^{-1} and the $[\theta]_{222}$ increased

[19] W. C. Johnson, Jr., *Annu. Rev. Biophys. Biophys. Chem.* **17,** 145 (1988).

[20] N. Greenfield and G. D. Fasman, *Biochemistry* **8,** 4108 (1969).

TABLE II
CIRCULAR DICHROISM (CD) PARAMETERS AND SECONDARY STRUCTURES FOR APOE HEPARIN-BINDING PEPTIDES

Peptides	$[\Theta]_{208}$[a]	$[\Theta]_{222}$[a]	Determined by CD[b]		Predicted[c]		Simulated[d]	
			α Helix (%)	β Strand (%)	α Helix (%)	β Strand (%)	α Helix (%)	β Strand (%)
ApoE(129–169)								
− Heparin	−4,590	−2,448	22	5	—	—	25 (40)	
+ Heparin	−19,500	−17,500	49	<1	63	17	56 (65)	—
ApoE(202–243)								
− Heparin		−4,200	11	16	—	—	—	13 (26)
+ Heparin		−12,000	9	67	40–60	—	—	32 (48)

[a] deg cm^2/dmol.
[b] Determined by deconvolution of CD spectra.[20]
[c] Values were determined by the predictive method of Chou and Fasman.[11]
[d] Simulations were determined by molecular dynamics calculations as described in the text for peptides E(129–159) (α helix) and E(211–234) (β strand) complexed to a heparin octasaccharide and hexadecasaccharide, respectively. For the E(211–234) and E(129–159) studies, two (in parentheses) and three residue windows were used to define β-strand and α-helical character, respectively. The values represent the results averaged over the last 20 psec of the 80-psec simulations (60 psec $\leq \tau \leq$ 80 psec). See text for details.

from −2448 to −17,500 deg cm^2 $dmol^{-1}$ as a result of heparin addition. Deconvolution of the CD spectra for E(129–169) by the method of Greenfield and Fasman[20] yielded 22% α helix in the absence of heparin and 49% in the presence. A limited but significant increase in the α-helical content of E(144–169) was observed as a result of heparin addition (not shown). In Fig. 2c, the CD spectrum of E(202–243) in the absence of heparin indicated limited ordered structure. Heparin addition induced a significant increase in the negative ellipticity and a change in the CD band shape characteristic of a β-strand structure (Fig. 2d). The $[\theta]_{222}$ increased from −4200 to −12,000 deg cm^2 $dmol^{-1}$ on addition of the GAG. Heparin alone (60 μg/ml) did not contribute significantly to the spectrum (Fig. 2e). By CD analysis, heparin increased the β-strand structure of E(202–243) by 51%. This finding was not anticipated, as the 203–243 region of apoE had been proposed to form an amphipathic α-helix based on Chou-Fasman predictions.[21] That the CD spectra reflected intrinsic cotton effects attributable to heparin was unlikely. The same heparin added to E(129–

[21] S. C. Rall, Jr., K. H. Weisgraber, and R. W. Mahley, *J. Biol. Chem.* **257,** 4171 (1982).

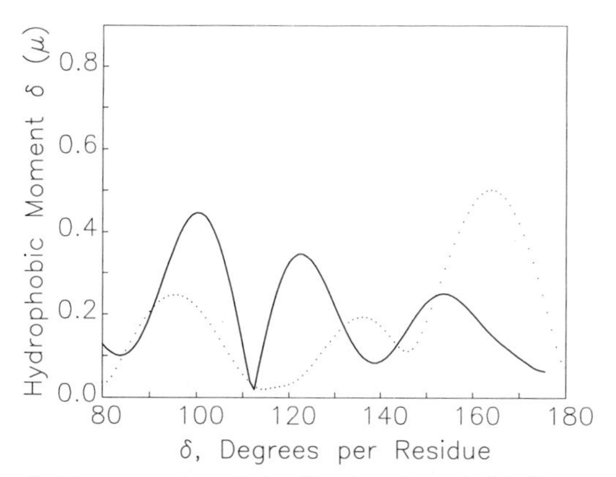

FIG. 3. Hydrophobic moment analysis of various heparin-binding sequences of apoE. (—), ApoE(136–157) (RLASHLRKLRKRLLRDADDLQK); (· · ·), apoE(211–228) (GERLRARMEEMGSRTRDR).

169) and E(202–243) induced different CD spectra classic for known α-helical and β-strand model peptides.[19] Furthermore, E(211–243) yielded a CD spectrum identical to that of E(202–243) in the presence of the GAG (not shown), whereas heparin did not alter the CD spectrum of E(219–243), a peptide that does not bind heparin (see Table I).

Predictive Algorithms of Apolipoprotein E Peptide Conformations

Residues 130–150 of apoE(1–191)[22] and residues 203–221 and 226–243 of apoE(192–299)[21] were previously predicted to form amphipathic α helices involved in lipid binding. For example, the incorporation of E(129–169) into lipid increased its α-helical content.[22] One face of the helix is apolar and interacts with lipid at the surface of the lipoprotein particle, while the other face is highly charged and presumably can bind GAGs.

Analysis of the heparin-binding regions E(136–157) and E(211–228) of apoE by the method of Eisenberg *et al.*[13] reveals a hydrophobic periodicity in peptide secondary structure (Fig. 3). As anticipated from previous predictions of the α-helical forming potential of the E(136–157) domain[23]

[22] J. T. Sparrow, D. A. Sparrow, A. R. Culwell, and A. M. Gotto, Jr., *Biochemistry* **24,** 6984 (1985).

[23] A. D. Cardin and R. L. Jackson, "Eicosanoids, Apolipoproteins, Lipoprotein Particles, and Atherosclerosis" (C. L. Malmendier and P. Alaupovic, eds.), p. 157. Plenum, New York, 1988.

and, consistent with the CD results on the effects of heparin on E(129–169) (Fig. 2b), E(136–157) has a major amphipathic periodicity at 100°/residue. By definition, a model amphipathic α helix has a periodicity in hydrophobic moment at 100°/residue (i.e., 360°/3.6 residues per α-helical turn). Interestingly, E(211–228), which was predicted to be an α helix based on Chou–Fasman analysis,[21] has a prominent amphipathic periodicity at 165°/residue. A model amphipathic β strand yields a periodicity at 180°/residue (360°/2 residues per β strand). This hydrophobic moment prediction of E(211–228) is consistent with the CD results (Fig. 2d). These assignments sum to approximate the global percentages of α helix and β strand determined by CD (see Table II).

The heparin-binding amphipathic α helix of E(135–152) is shown in Fig. 4. One side of the helical face consists of hydrophobic amino acids while the basic residues on the opposite face of the helix segregate into a positive charged cluster (Fig. 4, left). As is shown in the helical net diagram (Fig. 4, right), this cluster aligns along the helical axis. Presumably, a rigid polyanionic structure would maximize the electrostatic interaction by inducing the α-helical structure and stabilizing the conformation by aligning its negative charges along the helical axis. The binding interaction between peptide and heparin (see Fig. 1) had a strong electrostatic component, since at high [NaCl] binding was completely reversed. Moreover, as shown below, NaCl did not significantly affect the peptide conformation in the absence of heparin, indicating that increased salt concentration did not promote self-association of the peptide in solution. Thus, the electrostatic interaction would appear to be a major driving force for the heparin-induced increase in peptide conformation observed by CD.

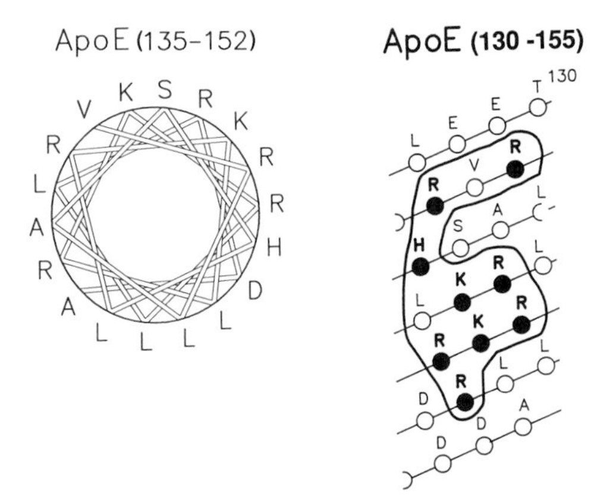

FIG. 4. Helical wheel (left) and helical net (right) diagrams of the heparin-binding amphipathic helix of apoE.

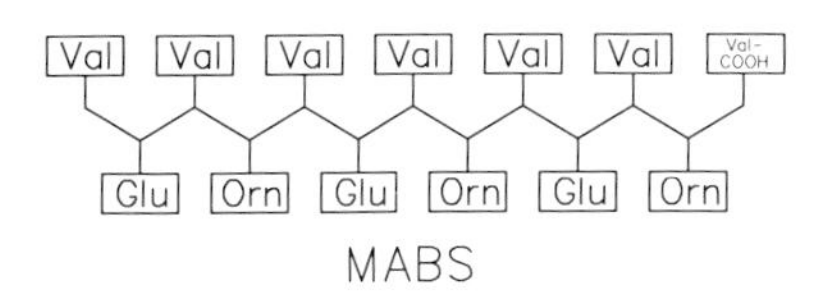

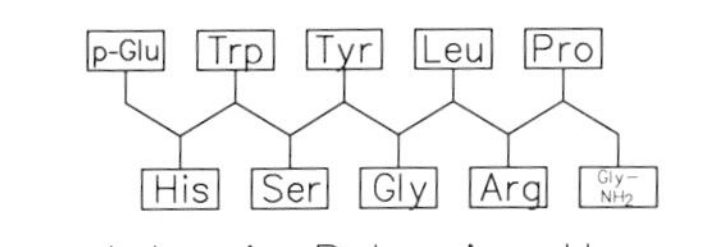

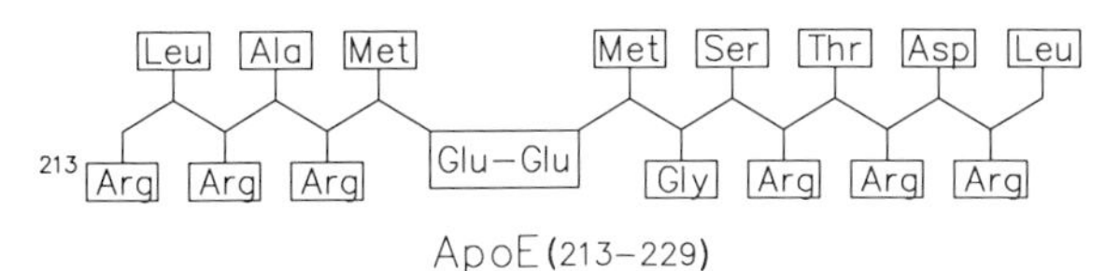

FIG. 5. Proposed amphipathic β-strand structure of the apoE(213–229) region when bound to heparin. Also shown for comparison to this structure are the known secondary structures of the model amphipathic β strand (MABS) and gonadotropin-releasing hormone peptides proposed by Osterman *et al.*[24]

The amphipathic periodicity of E(211–228) is similar to that of the model amphipathic β-strand (MABS) peptide reported by Osterman *et al.*[24] Figure 5 shows several amphipathic β strand-type structures. The feature characteristic of this structure is a hydrophobic face and an opposite face that is either highly charged or hydrophilic, as is shown for MABS and gonadotropin-releasing hormone (GRH),[25] respectively. The apolar face of these structures is stabilized by hydrophobic interactions with neighboring residues on the β strand or, potentially, with other apolar surfaces, such as receptors and membranes. The hydrophilic face of gonadotropin releasing hormone (GRH) is stabilized in part by solvent interactions and with adjacent residues on the β strand; that of MABS is stabilized by the electrostatic interaction between alternating glutamic acid and ornithine side chains. E(213–229) also has a hydrophobic face, whereas the opposite face of the β strand forms a highly positively charged face. However, the electrostatic repulsion between neighboring arginine residues

[24] D. Osterman, R. Mora, F. J. Kézdy, E. T. Kaiser, and S. C. Meredith, *J. Am. Chem. Soc.* **106,** 6845 (1984).
[25] E. T. Kaiser and F. J. Kézdy, *Science* **223,** 249 (1984).

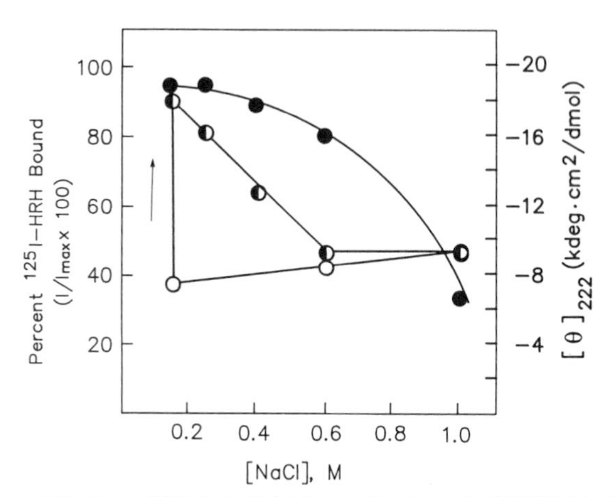

FIG. 6. Effect of NaCl on ^{125}I-labeled highly reactive heparin (HRH) binding (●) and on the solution conformation of E(129–169) in the absence (○) and presence (◐) of heparin. Heparin binding was determined as in Fig. 1. Solution conformations were determined by circular dichroism (CD) as described in Fig. 2.

would mitigate against this as a stable conformation. In fact, this peptide structure is not observed in solution in the absence of heparin. As this proposed structure is ligand induced (Fig. 2d), the acidic groups of heparin would stabilize the conformation via electrostatic attractive forces and dampen the repulsive forces between adjacent arginine side chains.

Solution Studies on Stability of Apolipoprotein E Peptide Conformations in Presence of Glycosaminoglycans

E(129–169) and E(202–243) were investigated by CD in the presence and absence of NaCl to determine the stability of their heparin-induced solution conformations. At physiological salt concentration, E(129–169) had limited ordered structure as determined from its molecular ellipticity value $[\theta]_{222}$ (Fig. 6); increasing the NaCl concentration to 1 *M* had no significant effect on its CD spectrum. The addition of heparin to E(129–169) at 0.15 *M* NaCl caused an increase in the negative ellipticity of the peptide, indicating the formation of an α-helical structure. The $[\theta]_{222}$ decreased with increasing NaCl concentration such that by 0.6 *M* NaCl the heparin-induced CD change was fully reversed; however, 80% of the heparin remained bound to E(129–169). That NaCl reversed the GAG-induced peptide conformation at salt concentrations where heparin remained bound without affecting the conformation of E(129–169) alone indicated that heparin stabilized the peptide structure.

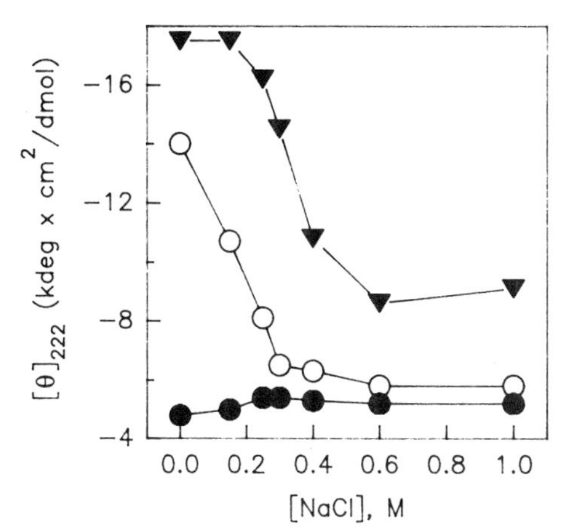

FIG. 7. Effect of NaCl on the solution conformations of E(202–243) in the absence (●) and presence (○) of heparin. The stability profile of the E(129–169) peptide complexed to heparin is shown for comparison (▼).

Figure 7 shows the effect of NaCl on the heparin-induced β-strand conformation of E(202–243). NaCl had no significant effect on the solution conformation of E(202–243) alone. In the absence of NaCl, heparin addition increased the β-strand structure of E(202–243). As the NaCl concentration of the complex was increased, the molecular ellipticity decreased such that by 0.3 *M* NaCl limited β-strand structure remained. Relative to the α-helical structure of E(129–169) induced by GAG, the β-strand conformation of E(202–243) was less stabilized by heparin. These findings are consistent with those summarized in Table I showing that E(129–169) has a higher affinity for heparin than does E(202–243).

Computational Chemistry Methods

Model Assembly. Heparin oligomers are built as four or eight 1,4-linked disaccharide repeats of 2-sulfate α-L-idopyranosyluronic acid and 2-deoxy-2-sulfamino-α-D-glucose 6-sulfate. Their structures are shown in Fig. 8. The repeat is constructed by the addition of carboxylate and sulfate groups to cyclohexane and substitution of one of the cyclic methylene groups with oxygen. Charges are determined by the semiempirical AM1 method[26] provided in the program AMPAC (1.0). The starting structures of the apoE

[26] M. J. S. Dewar, E. G. Zoebish, E. F. Healy, and J. J. P. Stewart, *J. Am. Chem. Soc.* **107,** 3902 (1985).

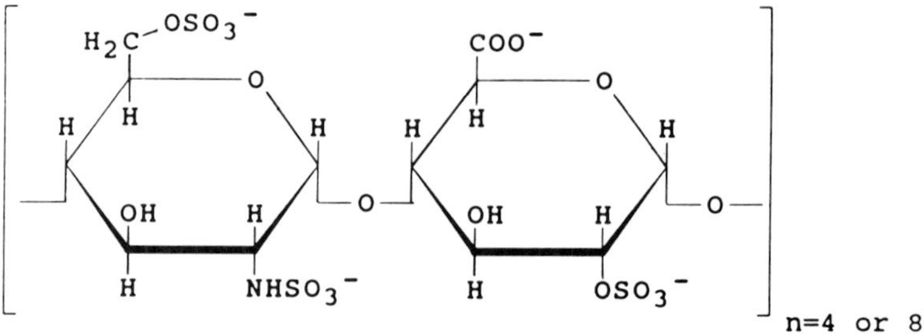

FIG. 8. Disaccharide structure of heparin used in the molecular dynamics simulations: $n = 4$ for apoE-α [α-helix model for E(129–159)]; $n = 8$ for apoE-β [β-strand model for E(211–234)].

peptides are constructed and optimized with the Insight 2.4 program,[27,28] which utilizes the molecular mechanics valence force field calculations. Partial atomic charges are determined with the Insight (version 2.51) look-up tables.

The sequences of the peptides that were modeled are as follows:

ApoE-α ^{129}S-T-E-E-L-R-V-R-L-A-S-H-L-R-K-L-R-K-R-L-L-R-D-A-D-D-L-Q-K-R-L^{159}

ApoE-β ^{211}G-E-R-L-R-A-R-M-E-E-M-G-S-R-T-R-D-R-L-D-E-V-K-E^{234}

The octasaccharide is used in simulations with the E(129–159) α-helix (α) model. A hexadecasaccharide is used with the E(211–234) β-strand (β) model due to the more extended structure of this peptide.

Method 1: Peptide Simulations. Molecular dynamics calculations are performed on each model using Discover 2.51.[28] The peptide models are individually minimized prior to the molecular dynamics calculations to obtain reasonable starting structures for the simulations. The following parameters are set:

CUTOFF =	20.0	(Å)
CUTDIS =	19.0	(Å)
SWTDIS =	1.5	(Å)
FRMS =	5.0	(kcal/Å)
KRMS =	1	(for tethering during dynamics)
KRMS =	0	(for no tethering during dynamics)
DIELECTRIC =	$2.0 \cdot R$	(R = distance between atom pairs)

[27] H. E. Dayringer, A. Tramontano, S. R. Sprang, and R. J. Fletterick, *J. Mol. Graphics* **4,** 82 (1986).

[28] Insight and Discover, Biosym Technologies, Inc., San Diego, California.

In order to reduce the magnitude of the calculations, nonbonded interactions are limited according to the following rules:

1. Certain atoms (switching atoms) defined by Discover determine the amino acids considered in the nonbonded interactions.
2. Atoms of residues whose "switching atoms" are less than 20 Å apart are placed in the atom pairs list. The 20-Å distance is referred to as the CUTOFF distance.
3. Atoms in the atom pairs list which are within 19 Å of one another are considered in the nonbonded interaction calculations. The 19-Å distance is referred to as the CUTDIS distance.
4. Those atom pairs whose interatomic distances fall within the range of 17.5–19 Å ($\Delta = 1.5$ Å, SWTDIS), are computed such that the interaction energy drops to zero at 19 Å.

The distance-dependent dielectric term is used to approximate aqueous solution conditions. For E(129–159) (α), the main-chain atoms of residues Ser-129 and Leu-159 are tethered to their original coordinates with a forcing constant (FRMS) of 5.0 kcal/Å for 20 psec of dynamics simulation at 300 K. The time step increment is 1 fsec. Harmonic bond stretching terms are employed; morse functions and cross-terms are not utilized. The molecular geometries are saved every 0.1 psec. An additional 60 psec of dynamics is then performed with no tethering.

Dynamics are also performed for E(211–234) (β) with the main-chain atoms of residues Gly-211 and Glu-234 tethered as above for the first 20 psec. The purpose of tethering is to maintain conditions similar to those used in the peptide docking experiments with oligosaccharides (Method 2, see below), thus allowing qualitative comparisons to be made. These tethering procedures have been described in earlier simulations on apoE peptides,[10] vitronectin,[4] and the RP-135 region of the envelope glycoprotein gp120 of human immunodeficiency virus type 1 (HIV-1).[1]

Method 2: Peptide–Oligosaccharide Simulations. Docking simulations with oligosaccharides are performed on each peptide. The heparin structures are individually minimized prior to the molecular interaction studies to obtain stable starting structures. Parameters are identical to those of method 1, except that the main-chain atoms of all residues are tethered with a forcing constant (FRMS) of 100.0 kcal/Å for the first 20 psec. Nonbonded interactions are limited according to the rules described in method 1.

For E(129–159) (α), the heparin octamer is placed approximately 16–17 Å from the peptide such that all sulfate and carboxylate groups fall within the 20-Å cutoff distance. A hexadecamer of heparin is placed a similar distance from the E(211–234) (β) structure.

Data Analysis. The ϕ and ψ torsion angles of each of the amino acids in the peptides are tabulated for each saved geometry of the dynamics trajectory. The percentage of α-helix and β-strand conformation in each 0.1-psec frame of the trajectory is determined using both a two- and a three-residue "window." The data are averaged over 20-psec intervals (201 frames). The use of the variable window to define α-helicity or β-strand character of a given amino acid requires either two or three consecutive residues with the appropriate ϕ and ψ dihedral angles.

Figure 9 illustrates the portion of ϕ, ψ torsional space defining the α-helical and β-strand structures. The shaded borders outline the "relaxed" ($\pm 5°$) definitions discussed below (see Table III). The area within these shaded borders corresponds to the dihedral angle definitions that are applied for the α-helical and β-strand conformations:

α Helix: $-118° \leq \phi \leq -19°$ $\quad -92° \leq \psi \leq -28°$

β Strand: $-180° \leq \phi \leq -96°$ $\quad -180° \leq \psi \leq -146°$ or

$60° \leq \psi \leq 180°$

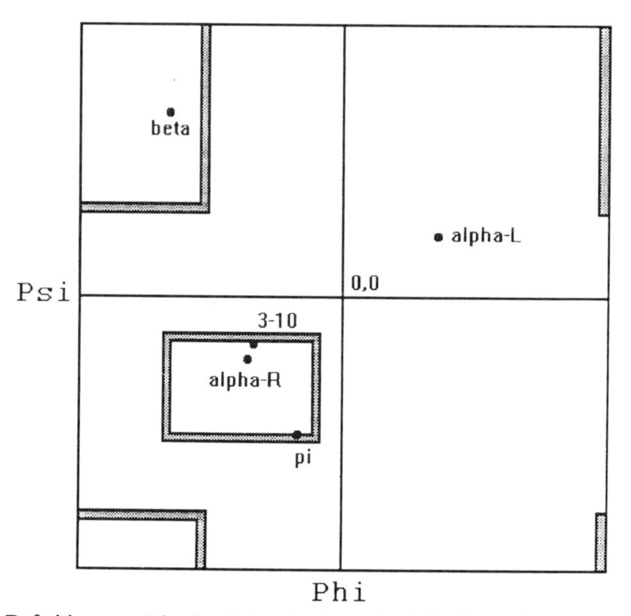

FIG. 9. Definitions used in the determination of α-helical and β-strand conformations: left-handed helix (alpha-L), right-handed helix (alpha-R), 3–10 helix (3-10), and π helix (pi). β-Strand (beta) structures are also indicated on this diagram. The boxed areas show the ϕ and ψ angle limits used in the calculations. The shaded borders outline the "relaxed" $\pm 5°$ definitions (see text).

TABLE III
SECONDARY STRUCTURES OF APOE PEPTIDES[a]

Structure	Simulation	Experimental CD (%)
α-Helix		
E(129–159)	31.1 (46.5)	22
E(129–159)–heparin	60.4 (69.8)	49
Difference	29.3 (23.3)	27
β-Strand		
E(211–234)	16.9 (32.4)	<16
E(211–234)–heparin	41.4 (56.2)	67
Difference	24.5 (23.8)	>51

[a] Average α-helical and β-strand contents of the apoE peptide sequences in the absence and presence of heparin. Secondary structures calculated only for energy-equilibrated structures (60 psec $\leq \tau \leq$ 80 psec). The simulation results are reported for the "relaxed" angle definitions (±5°) for both the three-residue window and the two-residue window calculations (in parentheses).

Computational Considerations

The purpose of the simulations was to determine the conformational characteristics of the E(129–159) (α) and the E(211–234) (β) peptide models in the presence and absence of heparin and to compare these results to the experimental CD measurements. An additional goal was to characterize the geometric properties of the complexes. The calculations were performed *in vacuo,* as the computational demands of a solvated peptide system would exceed the capacity of our VAX computers. To approximate a solution environment, however, both the distance-dependent dielectric term and the tethering of the terminal residues were previously found to be necessary.[10] Tethering prevented excessive folding of the peptides during the initial stages of the simulation, a phenomenon that was not observed in previous solution simulations conducted on a Cray XMP-416 supercomputer.[4,29]

For the simulations with heparin oligosaccharides, the main-chain atoms of E(129–159) and E(211–234) are tethered to their original coordi-

[29] H. J. R. Weintraub, A. D. Cardin, and D. A. Demeter, "Molecular Dynamics Investigation of the Interaction of Heparin with Human Vitronectin" (Proc. Int. Symp. Computational Chemistry on Cray Supercomputers), p. 100. Cray Research, Chicago, Illinois, 1988.

nates for the first 20 psec of dynamics simulation to maintain the initial secondary structure of the peptide during the approach and docking of the GAG. The 100.0-kcal/Å forcing constant is used only during the first 20 psec to prevent peptide distortion during docking as a result of the strong electrostatic attractive forces between the peptide and heparin molecules. An additional 60 psec of dynamics is then performed without tethering.

Study 1: Peptide Simulations. Figure 10A and B are plots of energy

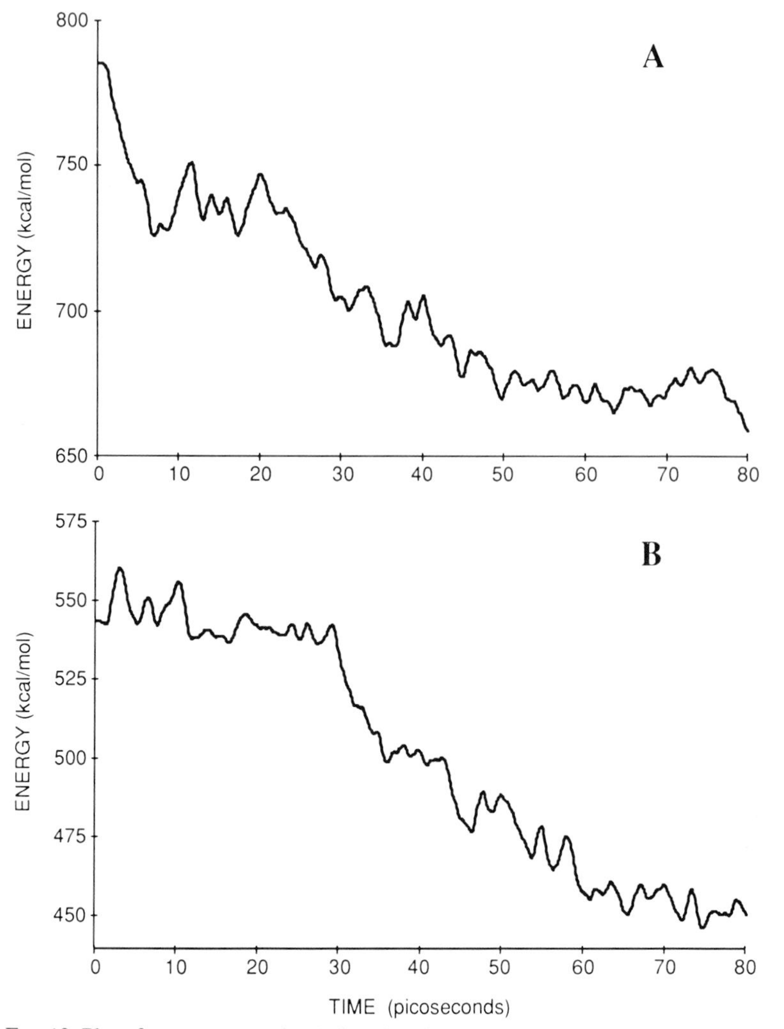

FIG. 10. Plot of energy versus simulation time for (A) the E(129–159) alpha (α) structure and (B) the E(211–234) beta (β) structure.

FIG. 11. Plot of energy versus simulation time for (A) the E(129–169) (α) peptide with an octamer of heparin and (B) E(211–234) (β) peptide with a hexadecasaccharide of heparin.

versus simulation time for the isolated E(129–159) (α) and E(211–234) (β) peptide structures. By 50–60 psec the energy of these systems has equilibrated and remains relatively constant for the remainder of the simulation. In these experiments, equilibration is obtained prior to analyzing the secondary structures of the peptides.

In the docking simulations, the peptides and the heparin oligosaccharides are separated by approximately 16–17 Å and are allowed to "dock" as a result of their electrostatic attractive forces. The molecules are prepositioned in opposing orientations with the long axes of their structures coaligned. Although this procedure greatly reduces the bias associated with manual docking, bias is not totally eliminated due to the assumed parallel starting orientations of the peptides and the oligosaccharides. Figure 11A is a plot of energy versus simulation time for the docking of the E(129–159) (α) sequence with an octamer of heparin. Docking has occurred by 30 psec and the energy of the system has equilibrated by 40 psec. A similar plot is shown in Fig. 11B for E(211–234) with the heparin hexadecasaccharide. In this simulation docking is complete by 30 psec and the system has equilibrated by 60 psec.

Figure 12A illustrates the secondary structure content of E(129–159) (α) in the presence and absence of heparin. Helicity is determined using a 3-residue window and smoothed by averaging over 20 "frames" (or 2-psec intervals) during the last 20-psec time interval of the simulation (60 psec $\leq \tau \leq$ 80 psec). As described above, the peptides are constrained by tethering during the initial 20 psec of simulation and both systems have equilibrated by 50 psec (40 psec for isolated E(129–159); 50 psec for E(129–159)–heparin). The E(129–159) peptide (Fig. 12A, solid curve) exhibits less α-helical content than does the heparin–peptide complex (dashed curve). The results of the three-residue window analysis of secondary structure for E(129–159) in the absence and presence of heparin are in close agreement with the experimental CD measurements. For E(129–159) the simulation yields 25% α helix vs 22% by CD; in the presence of heparin the simulation gives 56 vs 49% determined experimentally (see Table II). The simulation of Fig. 12A further shows that heparin stabilizes the conformation of E(129–159). The isolated peptide retains much of its helical content for approximately 45 psec, then rapidly degrades and stabilizes at about 55–60 psec. In contrast, the heparin-bound complex retains its α-helical content throughout the simulation.

As is shown in Fig. 12B and summarized in Table II, the secondary structure of E(211–234) is higher in the presence of the hexadecasaccharide (dashed curve) than in its absence (solid curve), showing a stabilization of the peptide structure by heparin in the complex. Although the simulations on the equilibrated E(211–234) peptide structures are in qualitative agreement with the CD results in that the heparin stabilizes the peptide structure, the quantitative aspects show a greater variance. On inspection the qualitative appearance of the bound structure suggests local β-strand character with several turns. However, the measured dihedral angles do not support the observed β structure. Relaxing the β-strand ϕ, ψ dihedral angle definitions by 5° slightly improves the consistency with experiment. Table

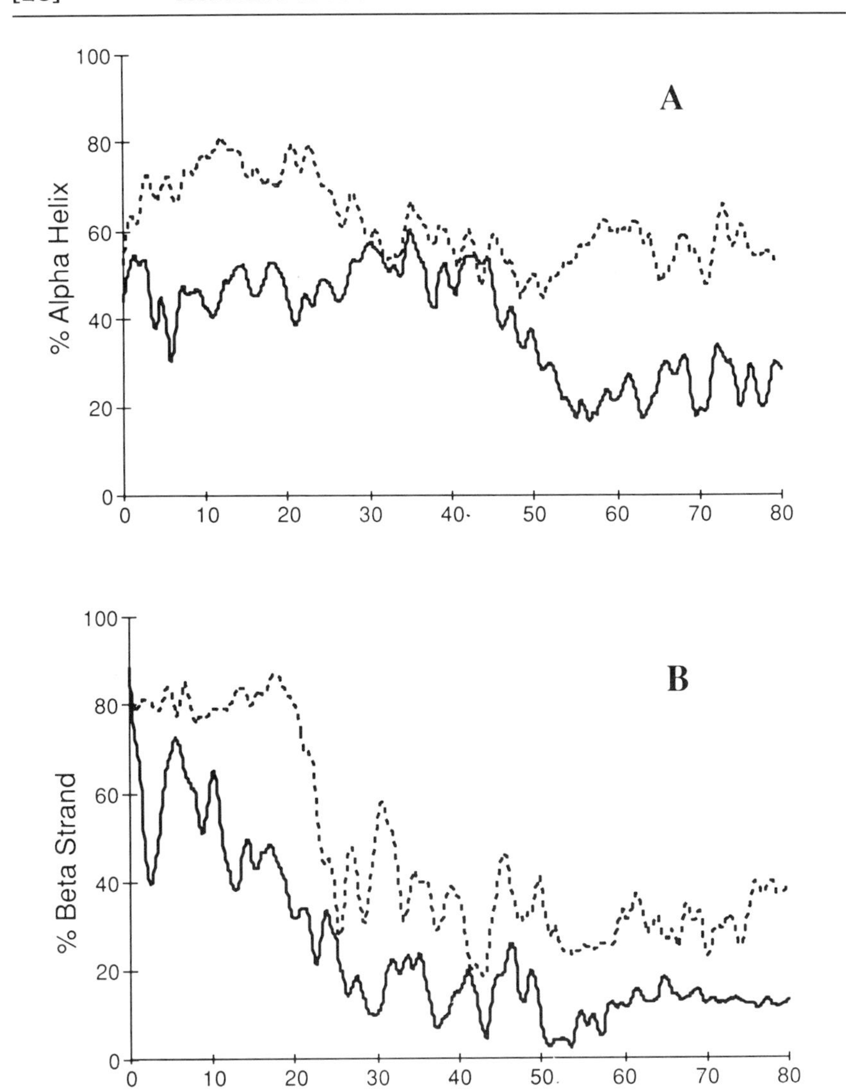

FIG. 12. Stabilization of peptide secondary structures by heparin oligosaccharides. (A) Plot of α-helical secondary structure versus time for E(129–159) (α) peptide in the absence (solid curve) and in the presence of the heparin octamer (dashed curve). (B) Plot of β-strand secondary structure versus time for E(211–234) (β) peptide in the absence (solid curve) and in the presence of the heparin hexadecasaccharide (dashed curve).

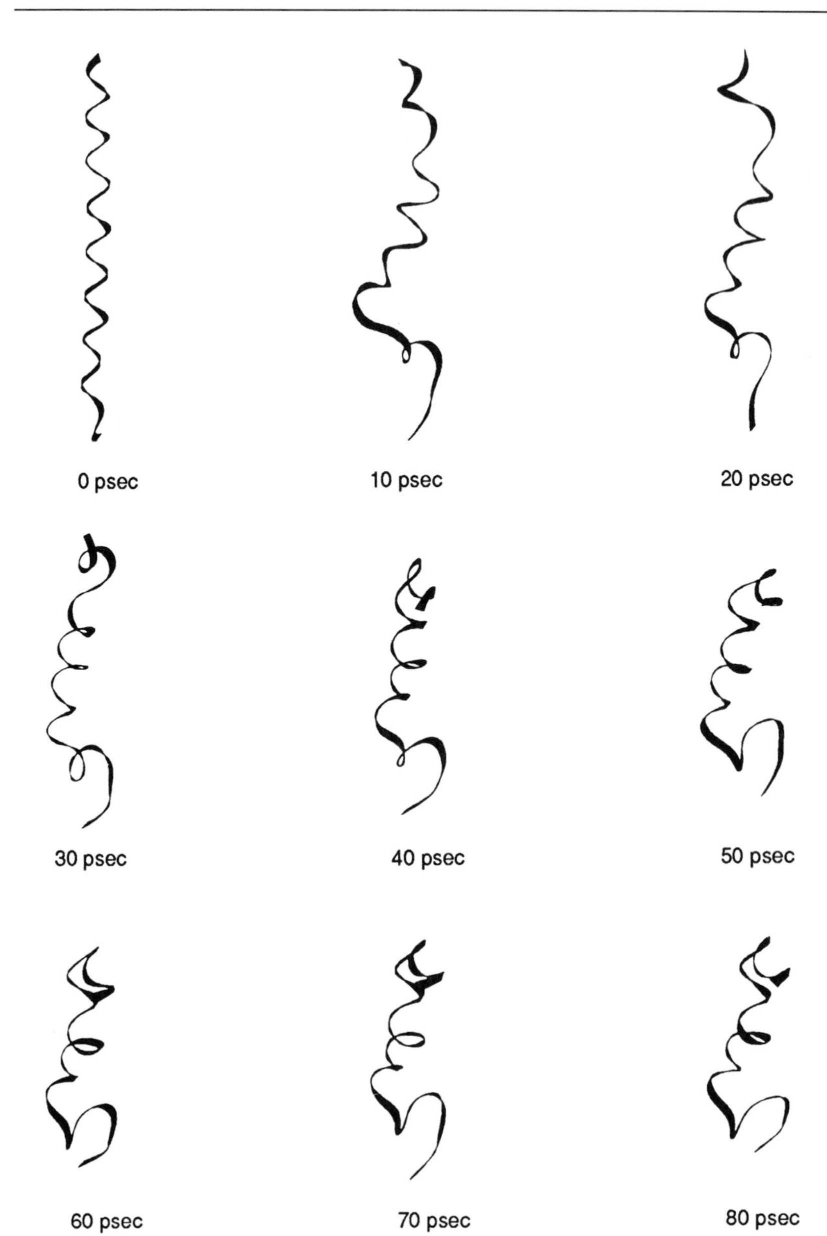

FIG. 13. Structures of E(129–159) (α) taken from the dynamics trajectory.

III summarizes the results for the "relaxed" dihedral angle definitions for both the E(129–159) and E(211–234) sequences.

Representative structures taken from the molecular dynamics trajectories of the isolated E(129–159) peptide and the E(129–159)–oligosaccharide complex are shown in Figs. 13 and 14, respectively. The

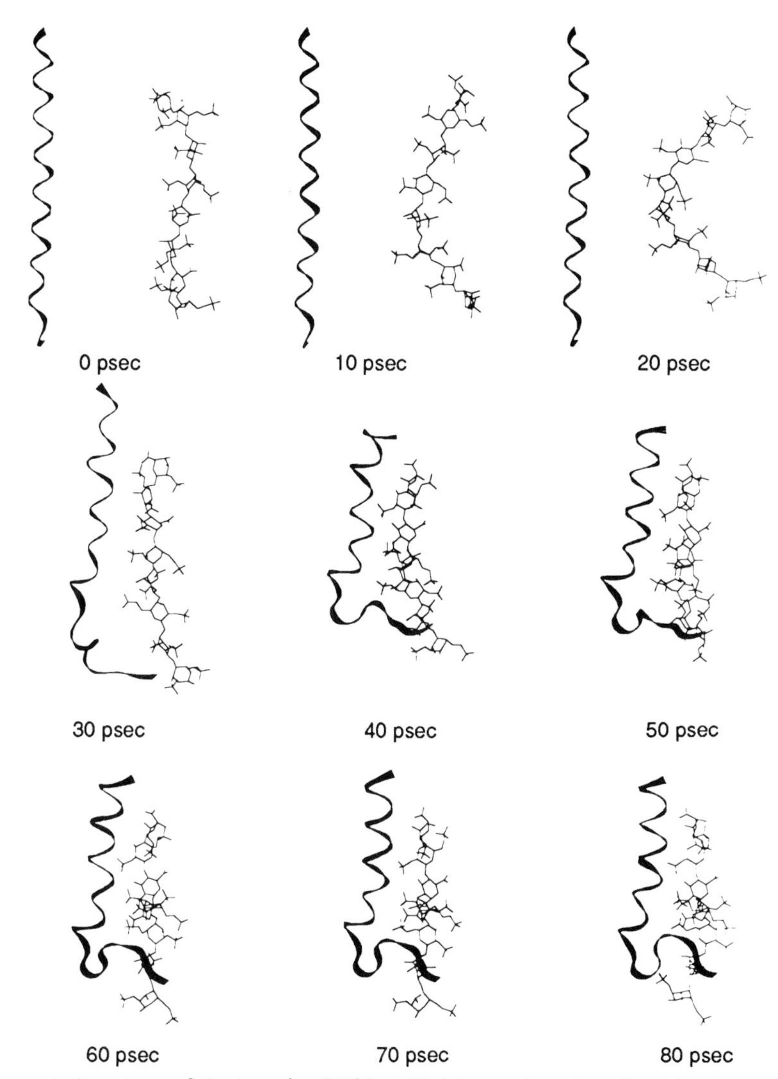

FIG. 14. Structures of the heparin–E(129–159) (α) complex taken from the dynamics trajectory.

ribbon structures provide a visual cue to the peptide helical conformation. In the isolated peptide (Fig. 13), substantial distortion of the helical structure occurs within the first 10 psec of the simulation. Untethering at 20 psec results in a compression that shortens the peptide length relative to the starting tethered structure. Beyond 30 psec, the structure of the isolated peptide further degrades into a coiled structure with substantial loss of helical character. By 80 psec only residues 145–150 retain the α-helical structure. Interestingly, this region represents the consensus sequence of basic residues previously hypothesized to form a nucleation site for the initial electrostatic binding event with the GAG.[4] This region may assume an intrinsic potential to retain helical character in solution in the absence of binding ligands (see Figs. 2–4). Whether the 145–150 region accounts for the α-helical character of the free peptide in solution observed by CD analysis remains to be determined experimentally.

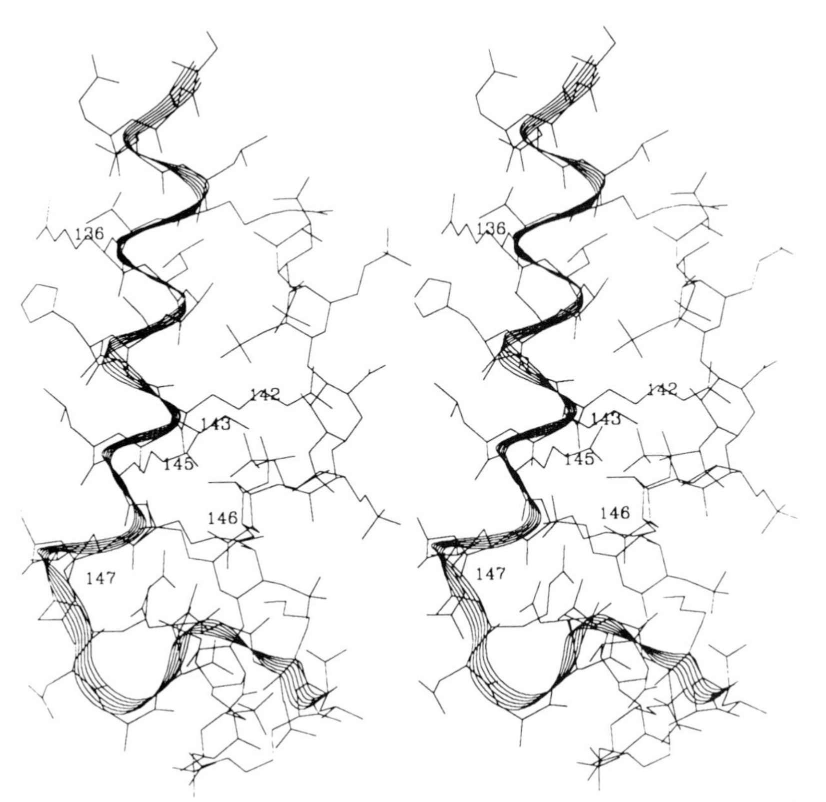

FIG. 15. The lowest energy (nonminimized) structure of the E(129–159) (α)–octasaccharide complex from the dynamics trajectory (52.1 psec).

As is shown in Fig. 14, the heparin and E(129–159) peptide have docked at approximately 30 psec. As discussed above, the peptide conformation is essentially locked in its initial conformation for the first 20 psec of the simulation. During the final 50 psec of simulation, the molecules

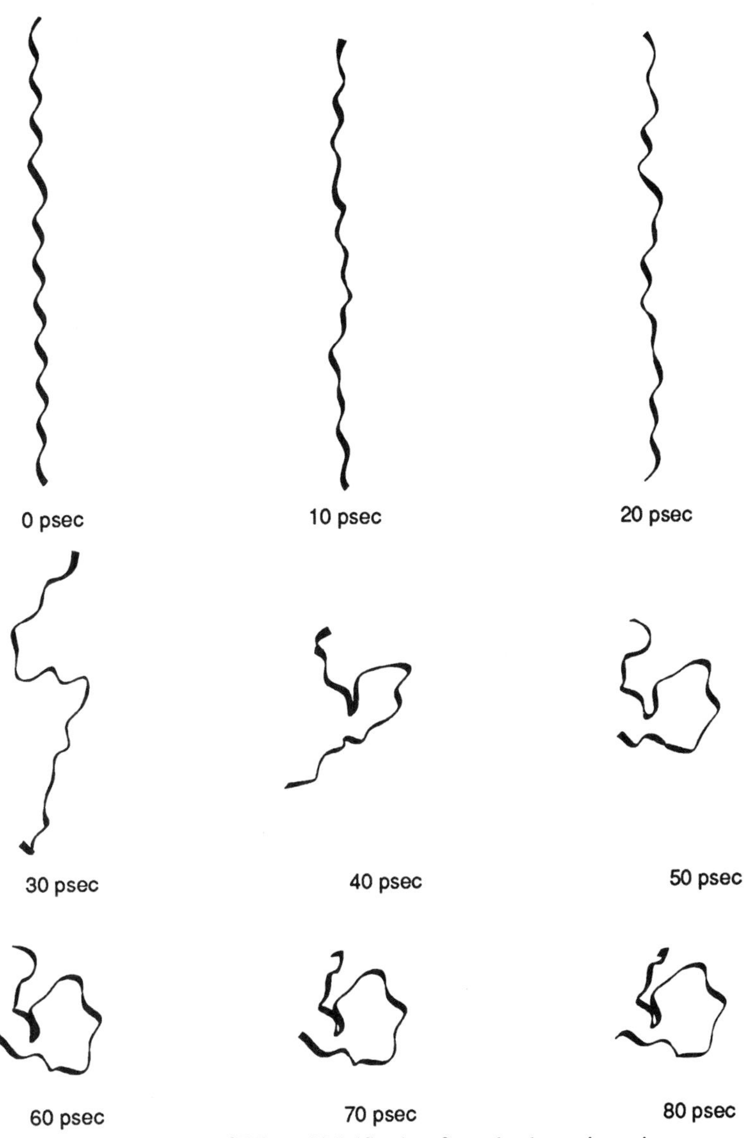

FIG. 16. Structures of E(211–234) (β) taken from the dynamics trajectory.

adjust their conformations to maximize the electrostatic interactions between their respective charged groups. At 80 psec substantial α-helical structure remains in the E(129–159) peptide bound to the octasaccharide. Thus, much of the α helix lost in the simulation of Fig. 13 is retained in the docked structures of Fig. 14.

Figure 15 illustrates a more detailed representation of the 52.1-psec "frame" of the dynamics trajectory. This nonminimized structure has the lowest energy. Note that residues Arg-136 and Arg-147 are not involved in the binding. All residues considered "critical" to binding, with the exception of Arg-147, participate. It has been shown experimentally that deletion of Arg-145, Lys-146, and Arg-147 abrogate heparin binding (Fig. 1). The simulation results would suggest that only Arg-145 and Lys-146 may be critical in this three-residue sequence.

As is shown by the ribbon structures in Fig. 16, the E(211–234) (β) peptide loses much of its secondary structure during the course of the simulation. By 30 psec, the β character had substantially decreased, and by the end of the simulation period (80 psec), the isolated peptide has developed several folds.

Figure 17 illustrates the binding of the heparin hexadecasaccharide to the E(211–234) peptide. As noted earlier, tethering is released at 20 psec. However, the heparin has not yet docked with the peptide. This premature release of the tethering constraints allows the strong electrostatic interactions between the two molecules to distort both structures during docking. One would predict that the β-strand content of the peptide complex would have been higher had the tethering constraints been retained until the initial docking had occurred.

Figure 18 shows the lowest energy (nonminimized) structure located during the dynamics simulation. This structure occurs at 52.7 psec. As is evident from an inspection of this figure, all Arg and Lys residues are involved in the binding with the hexadecasaccharide.

Summary

The methods and approaches taken to investigate heparin–apoE peptide interactions have involved a series of steps, including (1) identification of the heparin-binding domains of apoE, (2) determination of the minimal amino acid sequence regions involved in heparin binding, heparin-induced conformational changes, and stability of apoE peptide structures in solution, (3) modeling of these peptide and oligosaccharide structures, and (4) examination of their behavior during molecular dynamics calculations to determine if the modeled complexes simulate the results of the solution study.

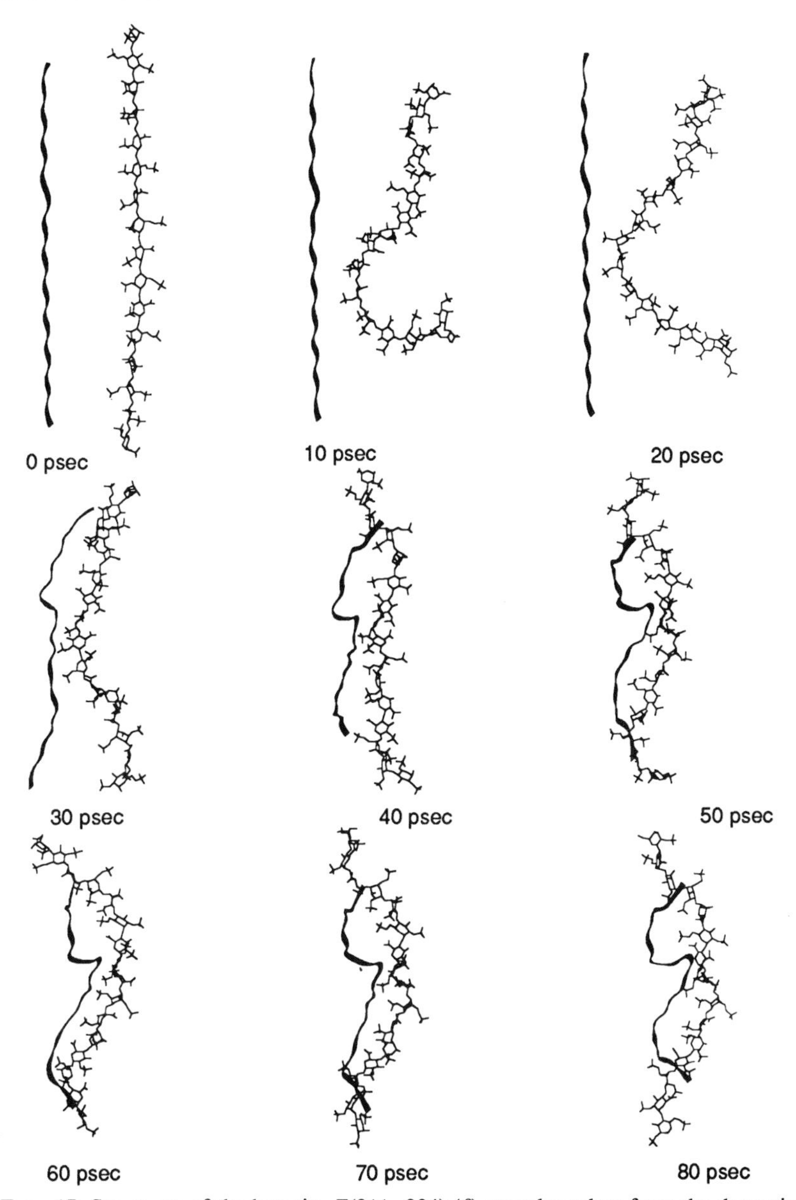

FIG. 17. Structures of the heparin–E(211–234) (β) complex taken from the dynamics trajectory.

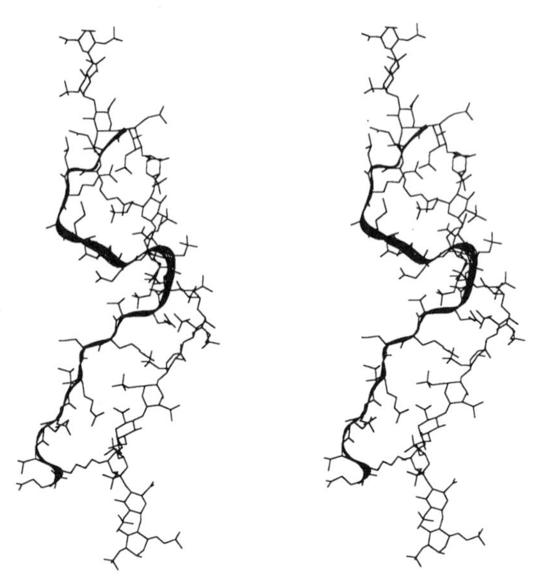

FIG. 18. The lowest energy (nonminimized) structure of E(211–234) (β)-hexadecasaccharide complex from the dynamics trajectory (52.7 psec).

The heparin-binding regions of apoE were determined by fragmentation of the protein and identification of the heparin-binding fragments by ligand-blotting procedures using ^{125}I-labeled heparin. Studies with synthetic peptide fragments of various lengths and dot-blot procedures with ^{125}I-labeled heparin identified the minimal residues critical for heparin-binding and CD studies established the prominent secondary structures of these domains. These studies also showed that heparin binds to the apoE(211–243) and apoE(129–169) regions to induce and stabililze β-strand and α-helical peptide conformations. Secondary structure algorithms were used to identify the specific residues with the highest probabilities of forming α-helix and β-strand structures.

Based on the predictive algorithms, the apoE(211–234) and apoE(129–159) structures were built using the Insight program and their molecular interactions with various heparin oligosaccharide models were investigated by molecular dynamics. In agreement with the solution studies in the presence of salt, the molecular dynamics studies showed that the oligosaccharides stabilized the β-strand and α-helical peptide configurations against simulated thermal denaturations. Further modeling studies are in progress to examine the mechanism of the heparin-induced increase in ordered structure of these peptides.

Acknowledgments

We wish to thank Ms. Mary Lynn Points for the preparation of this manuscript, Mrs. Linda S. Raymond for the figures, John L. Krstenansky, James T. Sparrow and Doris A. Sparrow for synthetic peptides and Cornelius Van Gorpe of Celsus Laboratories for heparin.

Section III

Drugs

[29] Computer-Assisted Rational Drug Design

By YVONNE CONNOLLY MARTIN

Overview

The computer design of novel potent small molecules as potential new drugs is now a reality.[1,2] Molecular modeling is a part of modern chemical and biochemical research and teaching. The aim of this chapter is to show how these tools have been extended and integrated with other concepts to produce computer-based methods for drug design. What strategies are available? What type of experimental information is necessary? What type of results will be produced: novel compounds or analogs of known structures, qualitative or quantitative forecasts of biological activity? What are the limitations of the current methods and what will it take to remove these limitations?

The role of the computer in rational drug design is to integrate the available information, both that specific to the problem and general chemical knowledge. Because the specific information available may not be sufficient, the computer modeler will also help set priorities for collecting new experimental information. Many aspects of drug discovery are empirical; therefore, computer methods are most valuable if they suggest diverse families of molecules.

To design a new bioactive molecule rationally, the scientist uses an experimental or hypothetical picture of the molecular features required. Usually, one wishes to design a new class of molecules that bind to a particular macromolecular target. For this, one would consider the types of required intermolecular interactions between the ligand and the target macromolecule; the geometric relationships between these interacting groups; and the size, shape, and polar vs nonpolar nature of the remainder of the binding site. This information might be available from experiment: in other cases, the information might be merely a list of chemical structures and biological potency. Thus, this chapter will discuss various uses of the computer to formulate hypotheses about the required features if experimental data are not available.

Four computer techniques are used to aid in the design of new molecules. The first requires a three-dimensional (3D) structure of the macromolecule: (1) identification of the location and chemical nature of pre-

[1] J. H. Van Drie, D. Weininger, and Y. C. Martin, *J. Comput.-Aided Mol. Des.* **3**, 225 (1989).

[2] Y. C. Martin, *Tetrahedron Comput. Methodol.* (in press).

ferred interaction sites on the macromolecule.[3] The next two use a known or hypothetical 3D structure of the binding site: (2) molecular graphics design[4-6] and (3) searching 3D structures to identify templates to which to add the required groups.[1,7-14] The final method searches databases of substituent constants to design an analog with the physical properties optimum for bioactivity. This requires the prior establishment of a statistical relationship between potency and physical properties, a quantitative structure activity relationship, or QSAR.[15-19] Of course, these techniques, particularly molecular graphics and QSAR, have uses other than compound design.

Just suggesting a new structure is not enough. One would also like to forecast its potency. There are three methods for this: (1) QSAR,[15-18] (2) QSAR methods based on 3D properties of molecules,[19,20-21] and (3) *de novo* perturbation free energy calculations.[22]

Table I summarizes the information generated by each of the various

[3] P. J. Goodford, *J. Med. Chem.* **28,** 849 (1985).

[4] G. R. Marshall and C. B. Naylor, *in* "Comprehensive Medicinal Chemistry" (C. Hansch, P. G. Sammes, and J. B. Taylor, series eds., C. A. Ramsden, vol. ed.), p. 431. Pergamon, Oxford and New York, 1990.

[5] R. Langridge and T. E. Klein, *in* "Comprehensive Medicinal Chemistry" (C. Hansch, P. G. Sammes, and J. B. Taylor, series eds., C. A. Ramsden, vol. ed.), p. 413. Pergamon, Oxford and New York, 1990.

[6] N. C. Cohen, J. M. Blaney, C. Humblet, P. Gund, and D. C. Barry, *J. Med Chem.* **33,** 883 (1990).

[7] R. A. Lewis and P. M. Dean, *Proc. R. Soc. London, Ser. B* **236,** 125 (1989).

[8] R. A. Lewis and P. M. Dean, *Proc. R. Soc. London, Ser. B* **236,** 141 (1989).

[9] I. D. Kuntz, J. M. Blaney, S. J. Oatley, R. Langridge, and T. E. Ferrin, *J. Mol. Biol.* **161,** 269 (1982).

[10] R. L. DesJarlais, R. P. Sheridan, G. L. Seible, J. S. Dixon, I. D. Kuntz, and R. Venkataraghavan, *J. Med. Chem.* **31,** 722 (1988).

[11] R. P. Sheridan and R. Venkataraghavan, *J. Comput.-Aided Mol. Des.* **1,** 243 (1987).

[12] Y. C. Martin, M. G. Bures, and P. Willet, *in* "Computational Chemistry Reviews" (K. Lipkowitz and D. Boyd, eds.), p. 213. VCH Press, New York, 1990.

[13] P. A. Bartlett, G. T. Shea, S. J. Telfer, and S. Waterman, *in* "Chemical and Biological Problems in Molecular Recognition" (S. M. S. M. Roberts, ed.), *R. Soc. Chem. Spec. Publ.* **78,** 182 (1989).

[14] Y. C. Martin and J. H. Van Drie, in press.

[15] Y. C. Martin, "Quantitative Drug Design." Dekker, New York, 1978.

[16] Y. C. Martin, *J. Med. Chem.* **24,** 229 (1981).

[17] M. S. Tute, *in* "Comprehensive Medicinal Chemistry" (C. Hansch, P. G. Sammes, and J. B. Taylor, series eds., C. A. Ramsden, vol. ed.), p. 1. Pergamon, Oxford and New York, 1990.

[18] T. Fujita, *in* "Comprehensive Medicinal Chemistry" (C. Hansch, P. G. Sammes, and J. B. Taylor, series eds., C. A. Ramsden, vol. ed.), p. 497. Pergamon, Oxford and New York, 1990.

[19] R. D. Cramer III, D. E. Patterson, J. D. Bunce, *J. Am. Chem. Soc.* **110,** 5959 (1988).

TABLE I
INFORMATION GENERATED

Method	Novel structures	Quantitative forecast of		Bioactive conformation	Geometric requirements and superposition rule	Shape of binding site	Computer resources[a]
		Biodata	Affinity				
3D geometric search	Yes	No	No	Confirm or reject	Confirm or reject	Confirm or reject	+-+++
3D steric search	Yes	No	No	No	No	Confirm or reject	+-+++
Graphics	Partial	No	No	Confirm or reject	Confirm or reject	Confirm or reject	+
GRID	Partial	No	No	No	Yes	No	++-+++
QSAR	Analogs	Yes	Yes	No	Implied	Possible	+
CoMFA	No	Yes	Yes	Confirm or reject	Confirm or reject	Yes	++-+++
Perturbation ΔG	No	No	Yes	No	No	No	++++
Active analog receptor mapping	No	No	No	Yes	Yes	Yes	++-+++
Ensemble distance geometry	No	No	No	Yes	Yes	Yes	++-+++
Maximum common 3D substructure searching	No	No	No	Confirm or reject	Yes	No	+-++
Quantum chemistry	No	No	No	Confirm or reject	Yes	No	++-++++
Search Cambridge Structural Database	No	No	No	No	Yes	No	++
Generate 3D protein structure from sequence	No	No	No	No	No	Yes	++-++++
Systematic search	No	No	No	Confirm or reject	Confirm or reject	Confirm or reject	+-+++
Potential energy minimization	No	No	No	Confirm or reject	Confirm or reject	Confirm or reject	+-+++
Molecular dynamics	No	No	No	Confirm or reject	Confirm or reject	Confirm or reject	++-++++
Distance geometry	No	No	No	Confirm or reject	Confirm or reject	Confirm or reject	++-+++
Rapid generation of 3D structures	No	No	No	No	No	No	+

[a] +, pc or <1 min on VAX; ++, minutes on VAX; +++, hours on VAX; ++++, days on VAX.

TABLE II

INFORMATION NEEDED/USEFUL FOR VARIOUS METHODS

Method	Ensemble of structures tested in same assay	Quantitative biodata	Quanitative affinity biodata	Reasonable ligand 3D structure	Bioactive conformation of ligand	Geometric requirements and superposition rule	Shape of binding site	3D structure of protein	3D structure of protein–ligand complex
3D geometric search	No	No	No	"Database" of many	Must be in "database"	Yes	No/yes	No/yes	No/yes
3D steric search	No	No	No	"Database" of many	Depends on the program	No	Yes	No/yes	No/yes
Graphics	No/yes	No	No	Yes	No	No	No	No/yes	No/yes
GRID	No	No	No	No	No	No	No	Yes	No/yes
QSAR	Yes (analogs)	Yes	No/yes	No	No	Implied	No	No	No
CoMFA	Yes (diverse or analogs)	Yes	No/yes	Yes	Yes	Yes	No	No	No
Perturbation ΔG	No	No	Yes on reference	Yes	Yes	No	Yes	Yes	Yes on reference
Active analog receptor mapping	Yes (diverse, 3D constrained)	No	No	Yes	No	Yes	No	No	No
Ensemble distance geometry	Yes (diverse, 3D constrained)	No	No	No	No	Yes	No	No/yes	No/yes

Maximum common 3D substructure	Yes	No	No	Yes	Yes	No	No	No	No
Quantum chemistry	Yes for most uses	No	No	Yes	No	No	No	No	No/yes
Search Cambridge Structural Database	No	No	No	Database of experimental	No	No	No	No	No
Generate 3D protein structure from sequence	No	No	No	No	No	No	No	Yes, of homologous	No/yes of homologous
Systematic search	No	No	No	Yes	No	No	No/yes	No/yes	No/yes
Potential energy minimization	No	No	No	Yes	No	No	No	No/yes	No/yes
Molecular dynamics	No	No	No	Yes	No	No	No	Only if this is studied	Only if this is studied
Distance geometry	No/yes	No	No	No	No	No	No	No/yes	No/yes
Rapid generation of 3D structures	No	No	No	No	No	No	No	No	No

categories of approaches to rational drug design, and Table II, the information needed to use them. The focus of this chapter will be to give a flavor of these various approaches.

Assumptions and Concepts behind Computer Tools

Assumed Characteristics of Macromolecular Binding Sites of Ligands

The three-dimensional structures of protein–ligand complexes form much of the experimental basis for the concepts used in the computer design of bioactive molecules. Clearly, it is attractive to use an experimentally determined 3D structure of a protein or DNA–drug complex as a basis for the design.[23–26]

Beyond this, we use insights gained from the structures of proteins as we imagine the properties we expect of a binding site of unknown structure. Thus, we expect that usually it has a defined three-dimensional structure and chemical properties. We also assume that it will change only slightly with changes in structure of the bound ligand. Hence, the binding site can be probed by making small changes in the structure of the ligand and observing the resulting change in biological properties. Last, we expect that not every atom of the ligand is close to the biomolecule. These concepts form part of the basis of the receptor-mapping techniques discussed below.[4]

However, this static viewpoint sees proteins as rigid objects with only one defined three-dimensional structure. In fact, there is abundant proof that, to function, the three-dimensional structure of a protein may change.

[20] H. Weinstein, R. Osman, and J. P. Green, *in* "Computer-Assisted Drug Design" (E. C. Olson and R. E. Christoffersen, eds.), Symp. 112, p. 161. American Chemical Society, Washington, D.C., 1979.

[21] G. Loew and S. K. Burt, *in* "Comprehensive Medicinal Chemistry" (C. Hansch, P. G. Sammes, and J. B. Taylor, series eds., C. A. Ramsden, vol. ed.), p. 105. Pergamon, Oxford and New York, 1990.

[22] J. A. McCammon, *in* "Comprehensive Medicinal Chemistry" (C. Hansch, P. G. Sammes, and J. B. Taylor, series eds., C. A. Ramsden, vol. ed.), p. 139. Pergamon, Oxford and New York, 1990.

[23] J. M. Blaney and C. Hansch, *in* "Comprehensive Medicinal Chemistry" (C. Hansch, P. G. Sammes, and J. B. Taylor, series eds., C. A. Ramsden, vol. ed.), p. 459. Pergamon, Oxford and New York, 1990.

[24] L. F. Kuyper, B. Roth, D. P. Baccanari, R. Ferone, C. R. Beddell, J. Champness, D. Stammers, J. Dann, F. Norrington, D. Baker, and P. Goodford, *J. Med. Chem.* **28,** 303 (1985).

[25] P. J. Goodford, *J. Med. Chem.* **27,** 557 (1984).

[26] W. G. J. Hol, *Angew. Chem., Int. Ed. Engl.* **25,** 767 (1986).

Furthermore, in solution a protein probably is an ensemble of interconverting structures, some of which may be quite different from others. Last, the 3D structure of a protein may differ when different ligands are bound. Although the static crystallographic view of protein structure is a powerful aid to rational ligand design, one must be aware of its limitations.

Availability of Programs to Generate, Optimize, and Calculate Properties of Three-Dimensional Structures

The speed and reliability of potential energy calculations is a key element in the explosion of interest in computer-aided drug design.[22,27] With them one can quickly calculate the relative energy of various conformations of small molecules and project the relative strengths of various intermolecular interactions. Two of the three methods for the quantitative forecast of the potency of a molecule are based on these calculations.[19,22]

Molecular mechanics optimizes the 3D structure of a molecule input, generally moving to the nearest local minimum energy structure.[27] These methods treat the molecule as though it were a set of balls (atoms) connected by springs (bonds) to the other atoms in the structure. Parameters are assigned for the size and hardness of the atoms, various twisting and bending motions of the framework, and the strength of electrostatic and hydrogen-bonding interactions. The parameters are generally established by fitting experimental or high-quality quantum chemical data on small molecules. These same functions describe the interaction energy between two atoms not in the same molecule.

The parameters of molecular dynamics programs have been fitted to data on molecular motion as well as structure: they are used to simulate the movement of the atoms in a molecule.[22] In a molecular dynamics simulation one starts with a previously minimized molecule that has then been equilibrated at a higher temperature. Then thousands of energy evaluations are made at small time steps. Supercomputers are usually used.

Quantum chemical calculations complement potential energy calculations.[20,21] They can be used to optimize 3D structures and also to calculate chemical properties of molecules. The various procedures differ in that semiempirical methods use experimental data to approximate some of the functions and different *ab initio* basis sets vary in the flexibility given to the final solution by allowing more coefficients to be fit. Semiempirical methods such as CNDO/2, MNDO, and AM1 use at least 10 times more computer time to do an energy evaluation than does molecular mechanics

[27] G. L. Seibel and P. A. Kollman, *in* "Comprehensive Medicinal Chemistry" (C. Hansch, P. G. Sammes, and J. B. Taylor, series eds., C. A. Ramsden, vol. ed.), p. 125. Pergamon, Oxford and New York, 1990.

and 10 times less computer time than *ab initio* calculations. Because of their current parameterization, the semiempirical methods do not treat intramolecular hydrogen bonds well. However, such methods, particularly AM1, are often used for geometry optimization of molecules for which molecular mechanics parameterization is not available. *Ab initio* calculations are used to derive parameters for molecular mechanics and dynamics and to calculate the electrostatic potential around a small molecule to aid in superposition in receptor mapping.

It is not now feasible to do quantum chemical calculations on macromolecules: however, one can study a biochemical reaction by combining a quantum chemical description of that part of the molecule that changes chemical structure and a molecular mechanics description of the rest of the molecular environment.[28]

Quick methods for generating 3D structures are also known. For example, CONCORD generates a low-energy 3D structure from 2D using logical rules and cyclic strain minimization in 1–8 sec on a VAX 11/785.[29] It handles most organic molecules and produces quite good 3D structures. A similar but slower program, COBRA, generates "all" low-energy conformations of a molecule.[30] Workers at several pharmaceutical companies have built 3D databases of compounds synthesized by company chemists.[12]

Role of Structure–Activity Relationships in Drug Design

Structure–activity analysis is a central theme of medicinal chemistry. The challenge is to extract information that is not obvious from the 2D structures of the compounds and to transform this information into the structures of new compounds to synthesize. Implicit in such analyses is the assumption that the features common to the active molecules establish the primary interactions with the target biomolecule and that the variable features are perturbations.

In its simplest form, structure–activity analysis involves comparing the relative potency of one compound with that of a close analog. For example, if substitution of a particular hydrogen atom always abolishes activity, one postulates that in the complex the hydrogen atom is close to the macromolecule. If this hydrogen atom is attached to a polar atom, one postulates a hydrogen bond; but if it is attached to a carbon, one postulates

[28] A. Warshel and R. M. Weiss, *J. Am. Chem. Soc.* **101,** 6218 (1980).

[29] A. Rusinko III, J. M. Skell, R. Balducci, C. M. McGarity, and R. S. Pearlman, "CONCORD, A Program for the Rapid Generation of High Quality Approximate 3-Dimensional Molecular Structures." Tripos Associates, St. Louis, Missouri, 1988.

[30] A. R. Leach, D. P. Dolata, and K. Prout, *J. Chem. Inf. Comput. Sci.* **30,** 316 (1990).

that close van der Waals contacts are made. In either case, the size of the binding site is limited. Ideally the structure–activity analysis is a collaboration with a synthetic chemist and a biologist. It is most useful when the original structure is rather simple in structure, because fewer compounds must be synthesized and analogs are easily synthesized.

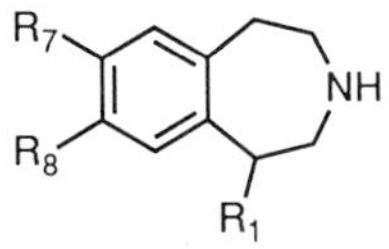

I $R_7 = R_8 = OH$, $R_1 = H$
II $R_7 = R_8 = OH$, $R_1 = C_6H_5$
III $R_7 = OH$, $R_8 = H$, $R_1 = C_6H_5$
IV $R_7 = R_8 = OMe$, $R_1 = C_6H_5$

For example, consider structures **I–IV**. In a test for D1 dopamine agonist activity the relative concentration for half-maximal stimulation of adenylate cyclase is 5.2 and 0.07 μM for **I** and **II** respectively. Compounds **III** and **IV** are inactive. Thus a phenyl at R_1 increases affinity by more than 90-fold and an OH appears to be required at R_8.[31]

Structure–activity data are often voluminous and so are stored in a chemical information database, that is, one that can be searched for compounds that have certain chemical substructural features.[12] We also store the coordinates of various conformations of molecules in such a database for use by our molecular graphics and our 3D searching programs.

Difference between Molecule that Binds to Target Biomolecule and One that Is Also Useful in Treatment of Disease

A ligand, even a tightly bound one, is not necessarily useful as a therapeutic agent.[32] A drug must survive long enough in the body to reach the target biomolecule and it must move through the body to the correct location. Particularly if this location is in the brain, difficulties may be encountered. For very potent compounds, it may be desirable that their effects are over within a few hours of dosing so that a slight overdose would cause problems for a shorter period of time. Sometimes a compound administered is a prodrug that the body converts to the active form. Thus, the computer modeler may examine the structure–activity relationships of

[31] J. Weinstock, J. P. Hieble, and J. W. Wilson III, *Drugs Future* **10,** 646 (1985).

[32] Y. C. Martin, V. Austel, and E. Kutter, eds., "Modern Drug Research: Paths to Better and Safer Drugs." Dekker, New York, 1989.

absorption, metabolism, excretion, and tissue distribution. QSAR is especially attractive for this purpose.

For a molecule to become a drug it must also be selective—that is, it must not affect the function of macromolecules other than the target. Therefore, one may try to use the computer to analyze the unwanted response to build out this feature of the molecules.

If the compound is to be a product, then probably the company will want a patent on it. A new drug must also be chemically stable and have physical properties that allow it to be formulated into a suitable dosage form. Also, one must be able to synthesize and transport it safely and economically. Thus, for many reasons it is important to suggest many structurally novel compounds: emphasis can then be given to synthesis and testing of those that appear most likely to meet these various complex criteria.

Useful Computer Tools for Rational Drug Design

Computer Tools for Design of Novel Molecules

Molecular Graphics. Armed with an experimental or hypothetical structure of a binding site, one expects to use molecular graphics programs to design other active analogs.[4-6,23] Indeed, several examples of this have been reported.[5,23,24] For molecular design the molecular graphics program needs capabilities for structure building, docking, and energy minimization of molecules of diverse chemical nature.

For example, the 3D structure of the antibiotic trimethoprim, **V**, bound to dihydrofolate reductase was established by X-ray crystallography. Molecular graphics inspection of the complex suggested that **VI** would form an additional salt bridge to a basic group in the active site.[24] **VI** indeed gains 50-fold potency. However, it is inactive against intact bacteria, presumably because its added negative charge prevents entry into the microorganism.

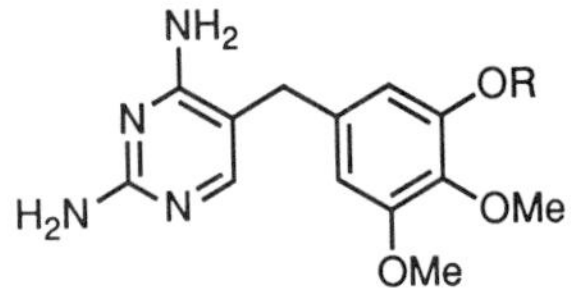

V R = Me
VI R = $(CH_2)_5CO_2H$

If structure–activity relationships form the basis for the design, then one might display as many as 50 molecules simultaneously. Color coding the structures by affinity values and displaying only the heavy atoms helps to improve the relative information content on the screen. Tying the structures to a database helps one organize what is to be displayed.[12] The molecular graphics display often includes derived properties such as surfaces, contours surrounding regions of a particular property, or color denoting regions of particular calculated electrostatic potential on a molecular surface.[4]

Since molecular graphics are just pictures, there is no intrinsic limit on the image that can be displayed. One is not restricted to low-energy conformations, to known molecules, or even to molecules that could exist. Obviously, while this is an advantage it may also be a disadvantage, since it is easy to believe that the images on the screen are the truth.

Molecular graphics has several limitations when used unaided for molecular design. First, it depends only on the creativity and experience of the scientist for suggestions. Second, molecular graphics design is tedious if it takes many tries to find a molecule that has the geometric or steric relationships desired. Third, molecules built on the screen may minimize to a 3D structure that does not fit. Fourth, molecular graphics alone does not provide the energy values to identify the strong interactions that may increase affinity for the macromolecule. Last, since molecular graphics is interactive, one may not have good documentation of the ideas explored. However, it is an essential component of most of the other techniques discussed in this chapter. Clearly the correct graphics display and a convenient and versatile computer program can be a great aid to the scientist.

Tools that Identify Favorable Binding Sites on Protein of Known Three-Dimensional Structure. Proteins have sites especially favored for interaction with a particular ligand atom.[3,33] These sites arise from an especially favorable arrangement of hydrogen bonding or charged groups in a sterically favorable region. The program GRID uses potential energy calculations to identify these favored locations.[3,33] Calculations are done on each proposed type of ligand group; for example, OH, O^-, NH_2^+, CH_3, aromatic carbon, and heterocyclic NH groups. The interaction energy is calculated at each point on a lattice (1-Å spacing, typically) surrounding the target biomolecule.

One output is a contour of favorable regions for the particular ligand atom to display with the protein. This provides a visual tool to supplement molecular graphics for ligand design. The investigator will try to design a

[33] D. N. A. Boobbyer, P. J. Goodford, P. M. McWhinnie, and R. C. Wade, *J. Med. Chem.* **32,** 1083 (1989).

molecule that places appropriate ligand atoms into several points of most favorable interaction energy. If the calculation was done on the protein with a known ligand already bound, then sites are identified for additional binding groups.

Steric Searches of Three-Dimensional Databases. A dominant feature in protein–ligand interactions is the fit of the ligand into the binding site. If the ligand does not fit, it will not bind even if electrostatic and hydrogen-bonding interactions are ideal. The computer program DOCK capitalizes on this knowledge to search a 3D database of small molecules to find those that fit into a binding pocket.[9–11,34]

Fast algorithms have been implemented to identify possible orientations of the potential ligand in the binding site.[9] Orientations with favorable fit are retained for molecular graphics viewing. Flexibility of the ligand may be considered by manually dividing the ligand into overlapping parts and checking that the needed parts overlap.[34]

For example, the method was used to design compounds that fit into a model of the nicotinic agonist binding site.[11] The model had been developed from compounds **VII–XIV**. The shape of the binding site was mod-

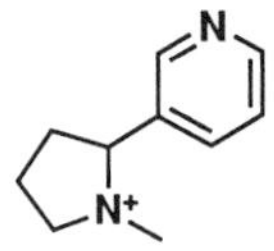

VII 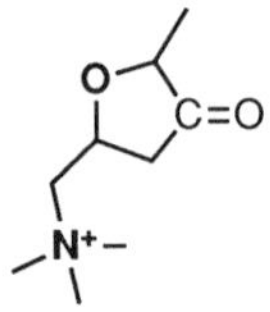 VIII 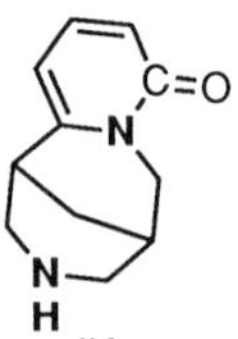 IX 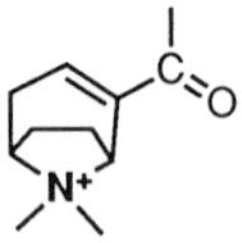 X 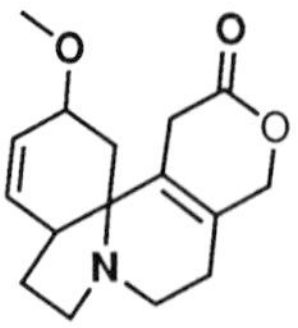 XI

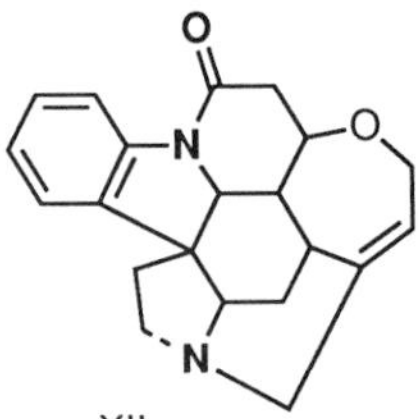

XII 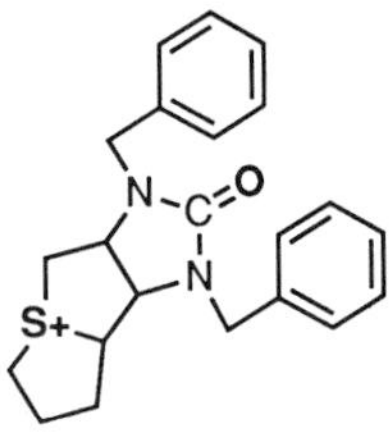 XIII 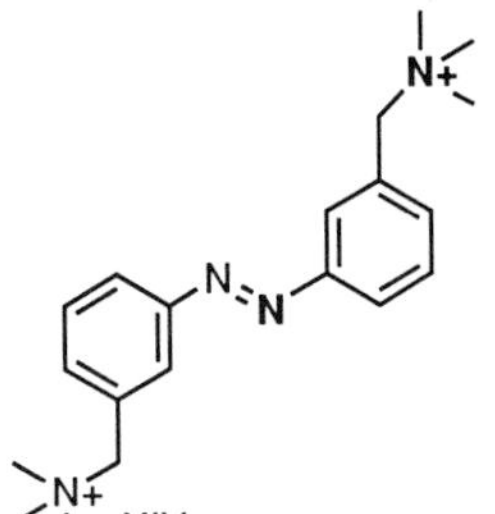 XIV

eled as the union surface of compounds **VII–X** shown in Fig. 1. As well as steric fit, the search specified that in the binding orientation the candidates have atoms within 0.5 Å of each pharmacophore atom (a cationic center and a hydrogen bond acceptor). A search of 7852 3D structures identified

[34] R. L. DesJarlais, R. P. Sheridan, J. S. Dixon, I. D. Kuntz, and R. Venkataraghavan, *J. Med. Chem.* **29,** 2149 (1986).

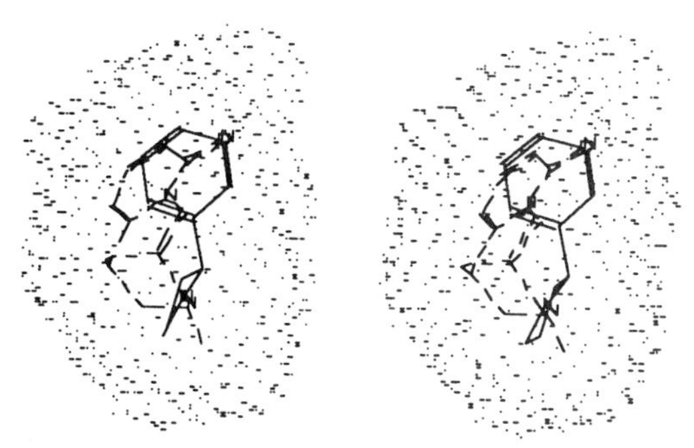

FIG. 1. The superposition of (−)-**VII** (solid line) and (−)-**IX** (dashed line) for modeling the agonist-binding site on the nicotinic receptor. The dots enclose the volume occupied by superimposed (−)-**VII**, (+)-**VII**, (−)-**VIII**, (+)-**VIII**, (−)-**IX**, (−)-**X**, and one-half of **XIV**.

5690 diverse molecules of the correct shape and distance relationships. The 62 unique best fitting structures were further studied with molecular graphics. The found molecules were manually transformed into those expected to have the proper chemical groups to interact with the target biomolecule. This led to four known nicotinic agonists. Compounds **XV–XIX** are five molecules identified by the original authors as especially interesting.[11]

XV XVI XVII XVIII XIX

Geometric Searches of Three-Dimensional Databases. The geometric properties of the interacting groups play a key role in establishing ligand–macromolecule binding: the angles and distances between the groups of the ligand that interact with different sites on the macromolecule must be optimal. Computer programs have been written that search many small molecules to identify those that have backbones or templates, that could hold the interacting groups in a specified geometric orientation.[1,7–8,12–14] Usually the molecules are stored in a chemical information database. Our program can also change the found structure into the 3D structure of the molecule with potential bioactivity.[2,14]

For the design of potential dopaminergic compounds we derived a receptor model based on conformationally defined compounds such as those shown in **XX–XXVIII**.[2] Their superposition is shown in Fig. 2. We

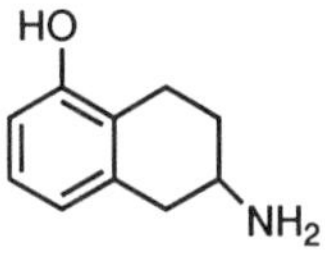

XX 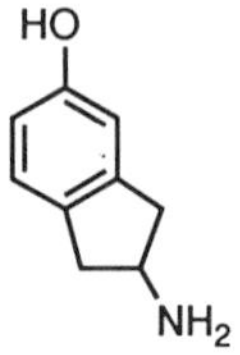 XXI 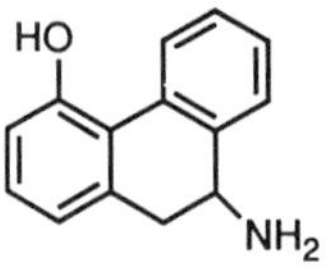 XXII 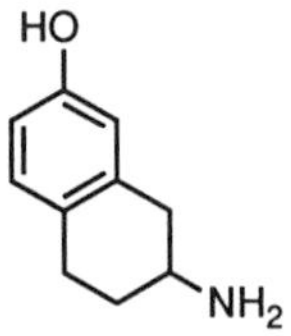 XXIII 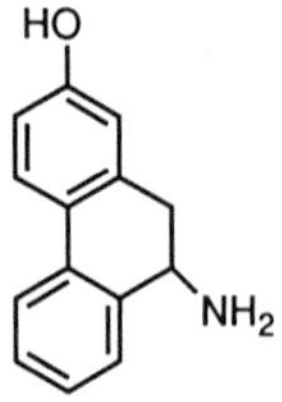 XXIV

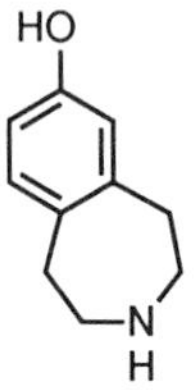

XXV 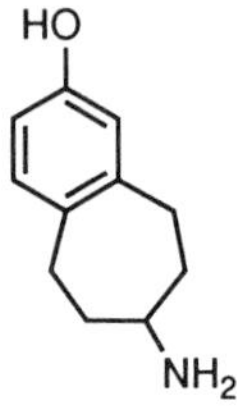 XXVI 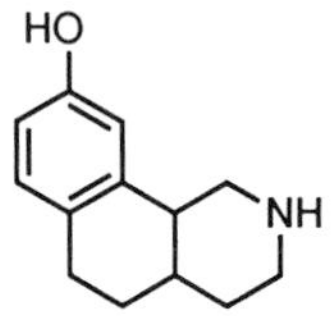 XXVII 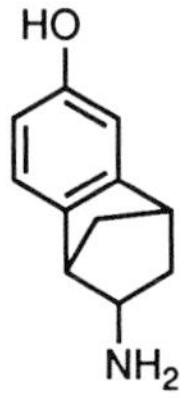 XXVIII

searched for compounds in which there is the proper distance between (1) an aromatic atom that will be changed in the computer into an aromatic carbon bearing an OH group and (2) an aliphatic atom that will be changed into a basic nitrogen atom. We then used the program to generate the 2D structures of the proposed compounds and remove extraneous substituents. Next we generated the 3D structures with CONCORD. Last, we double checked with a second search that the designed compounds do meet the geometric criteria. A search of 54,296 3D structures generated 553 unique molecules that meet the geometric requirements. It identified 8 of 9 classes of known fused ring phenolic dopaminergic compounds (structures **XX–XXVIII**) and 62 other classes of fused ring compounds with

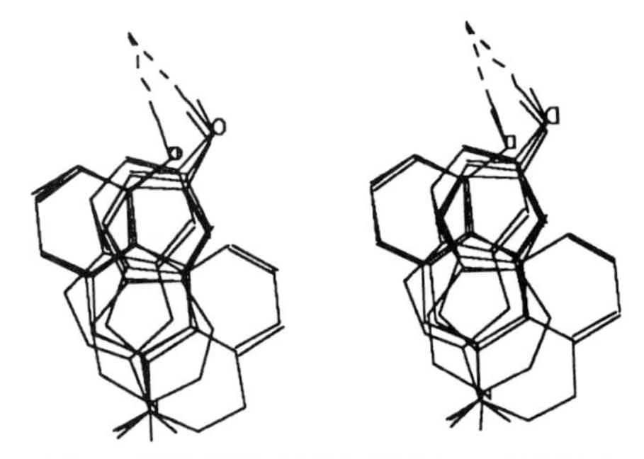

FIG. 2. The superposition of **XX**, **XXI**, **XXII**, **XXIII**, and **XXV** for modeling the agonist binding site on the D2 receptor. The dashed lines show the proposed essential hydrogen bond to the receptor.

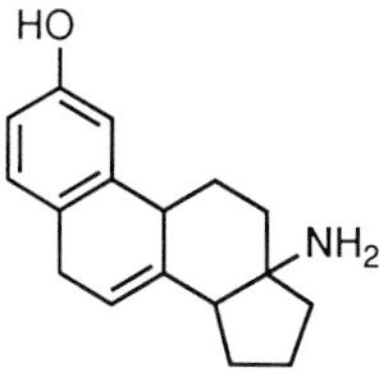

XXIX 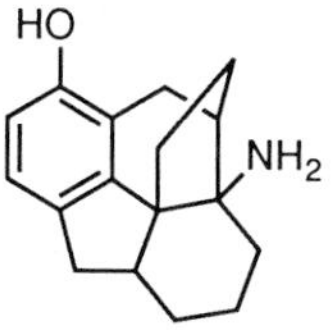 XXX 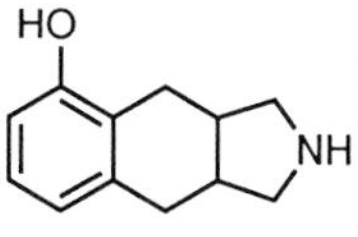 XXXI 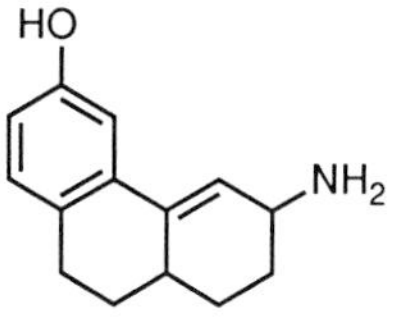 XXXII

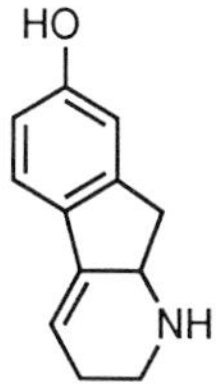

XXXIII 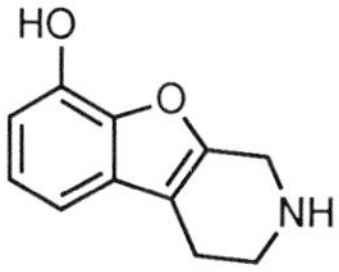 XXXIV 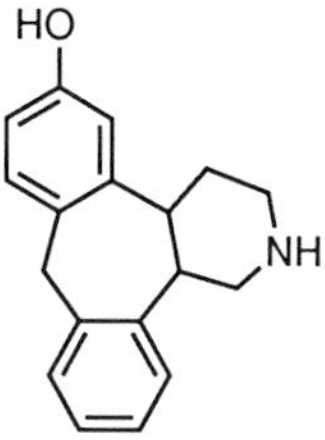 XXXV 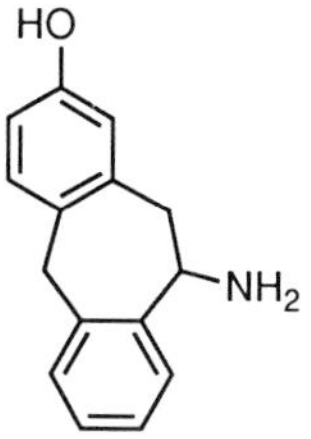 XXXVI

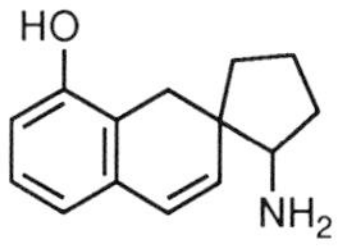

XXXVII 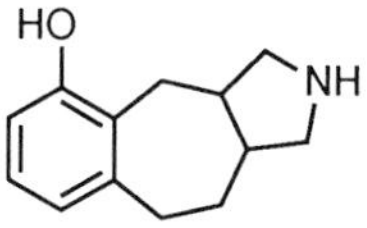 XXXVIII

potential activity. Structures **XXIX–XXXVIII** are examples of some compounds suggested.

These programs can also be used to find known compounds that might have a previously unsuspected biological property.[1,35–40] For example, in a search of the Abbott collection (Abbott Park, IL) structure **XXXIX** was correctly identified as a potential D2 dopaminergic agent.[1]

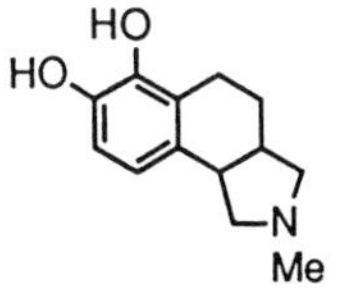

XXXIX

[35] S. E. Jakes and P. Willett, *J. Mol. Graphics* **4,** 12 (1986).

[36] S. E. Jakes, N. Watts, P. Willett, D. Bawden, and J. D. Fisher, *J. Mol. Graphics* **5,** 41 (1987).

The 3D geometric searching programs differ in several details.[12,41] Some use prescreens based on precalculated distance or torsion angle ranges to speed the search.[35-39] They also differ regarding the substructural information that can be included in the definition of the search atoms, in the complexity of geometric properties that can be explored, and in the closeness of the tie of the program to molecular graphics. By analogy with the steric search programs described in the previous section, several also provide a test if the found molecule or its enantiomer can fit into the space of the binding site. For this the user supplies a reference molecule, the superposition rule, and the location of points on the surface of the site. There is also exploration of methods to consider conformational flexibility.[42]

Other workers have taken a somewhat different approach with planar structures.[7,8] They prepared a limited number of polycyclic spacer skeletons that contain rings of different sizes of interest. Geometric searching of these skeletons identifies those that have atoms at the correct spatial relationships to fit the binding site. By first comparing the interatomic distance matrices of the binding site requirements with those of the spacer skeletons, many possible matches are eliminated quickly. The final step is to clip off the atoms of the spacer that do not span the binding site and to add the appropriate ligand atoms. The latter has been automated.

Computer Tools for Quantitative Forecast of Potency of Suggested Molecules

QSAR. A major advance in rational drug design was the development in the 1960s of QSAR.[15-18] In QSAR one correlates, in a statistical sense, the relative biological potency of analogs with the relative values of physical properties, such as the logarithm of octanol–water partition coefficient (log P), pK_a value, or size of the substituent. Earlier work had established substituent constants, σ and E_s, for the change in electronic or steric properties of molecules that result from adding the substituent of interest.

[37] A. T. Brint and P. Willett, *J. Mol. Graphics* **5,** 49 (1987).

[38] R. P. Sheridan, A. Rusinko III, R. Nilakantan, and R. Venkataraghavan, *Proc. Natl. Acad. Sci. U.S.A.* **86,** 8165 (1989).

[39] R. P. Sheridan, R. Nilakantan, A. Rusinko III, N. Bauman, K. S. Haraki, and R. Venkataraghavan, *J. Chem. Inf. Comput. Sci.* **29,** 255 (1989).

[40] M. G. Bures and G. Gardner, *J. Comput.-Aided Mol. Des.* (in press).

[41] In addition to the programs mentioned in Ref. 12 (GEOM from the Cambridge Structural Data Centre, ALADDIN from Daylight Chemical Information Systems, and MACCS-3D from Molecular Design Ltd.) Chemical Design Ltd. has recently started distributing a 3D search program.

[42] N. W. Murrall and E. K. Davies, *J. Chem. Inf. Comput. Sci.* **30,** 312 (1990).

Since hydrophobicity plays a key role in determining the biological properties of molecules, Hansch and Fujita showed how to include this factor in the analysis.[17] They proposed the use of the substituent constant π for the substituent effect on log P and demonstrated that a parabolic relationship in lipophilicity often is present:

$$\log (1/C_i) = a + \rho\sigma_i + \delta E_{s_i} + b\pi_i + c\pi^2 \tag{1}$$

Most important, QSAR has proved to be extremely useful in the design of biologically active molecules: it may be the method for which there are the largest number of successful design efforts.[16,18]

Spinoffs of traditional QSAR include the use of other statistical techniques for analyzing structure–activity data, other mathematical forms of the equation, other descriptors of molecular properties, and strategies for the proper design of series for QSAR analysis.[17]

If the properties for a QSAR are calculated by position, then the result is a map of the chemical properties of the ligand-binding site on the macromolecule. Such QSAR maps of a protein of known 3D structure agree with the crystallographic information as to the nature of the groups that line the binding pocket.[23]

The attractive feature of QSAR is that one need not postulate a 3D structure of the binding site (or even of the ligand). The limitations are the need for experimentally based parameters, the ambiguity as to how to analyze noncongeneric series, and the lack of supplied 3D information.

Three-Dimensional QSAR, Particularly Comparative Molecular Field Analysis. Comparative molecular field analysis (CoMFA) is a combination of receptor mapping, potential energy calculations, and QSAR.[19] In CoMFA one analyzes the quantitative influence on potency of a specific chemical property at a certain region in space; that is, one derives quantitative 3D design criteria from structure–activity relationships.

A CoMFA uses the relative potency of molecules and their superimposed bioactive conformations. The latter might result from the application of one of the methods discussed in the section on receptor mapping below. One calculates for each molecule the steric, electrostatic, and hydrogen-bonding interaction energies at each of hundreds of points on a lattice surrounding the molecule. Statistical analysis explores the relationship between these energy values (descriptors) and biological potency or affinity. Since there are so many more descriptors than observations, the statistical method used (partial least squares) capitalizes on the intrinsic correlations between the descriptors. Cross-validation identifies if a dataset is overfit.

CoMFA equations directly forecast the affinity of proposed molecules. Also, the coefficients of the energy values in a CoMFA can be displayed as

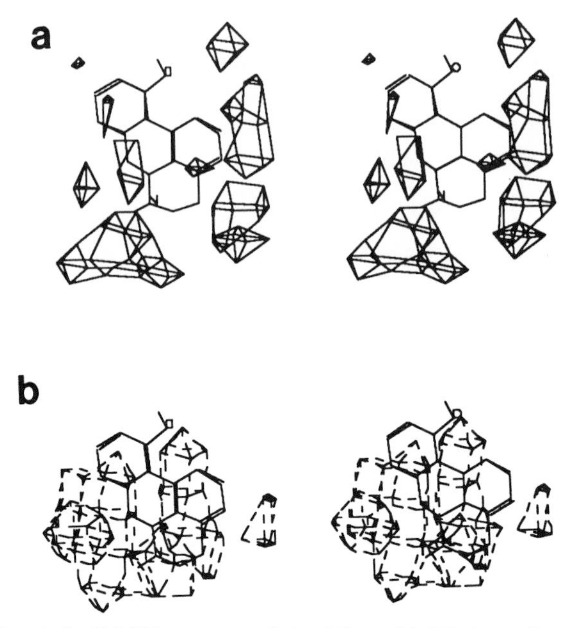

FIG. 3. The steric CoMFA contours derived from 31 D2 dopaminergic agonists. (a) Contours that enclose regions that contribute 0.01 log units to the binding affinity. Note especially the positive contour near the N resulting from the observation that *N*-alkyl groups increase affinity. (b) Contours that enclose regions that contribute −0.01 log units to the binding affinity. These contours are mainly on the back face of the molecules and suggest that the most intimate contact with the receptor is from this side.

contours in a molecular graphics display as an aid to new compound design. Figure 3 shows the CoMFA contours from the analysis of 31 D2 agonists. Here only steric features were statistically important.[43]

CoMFA electrostatic properties reproduce the traditional electronic substituent parameter σ very well and predict the pK_a values of 21 of 23 other compounds to within 0.27 kcal/mol.[44] For the "steric" parameter E_s, although the steric term dominates, an electrostatic term is also significant.[44a] Thus, CoMFA is a 3D analog of QSAR.

There are other quantitative methods that use 3D properties, but have

[43] Y. C. Martin, E. B. Danaher, C. S. May, D. Weininger, and J. H. VanDrie, *in* "QSAR: Quantitative Structure-Activity Relationships in Drug Design" (J. L. Fauchere, ed.), p. 177. Alan R. Liss, New York, 1989.

[44] K. H. Kim and Y. C. Martin, *J. Org. Chem.* **56**, 2723 (1991).

[44a] K. H. Kim and Y. C. Martin, *in* "QSAR: Rational Approaches on the Design of Bioactive Compounds" (C. Silipo and A. Vittoria, eds.), Elsevier, Amsterdam, in press.

so far been used in only one laboratory.[45,46] One can also use properties calculated from 3D structures as descriptors for a QSAR: partial atomic charges, energy of the highest occupied or lowest empty molecular orbital, volume in common, and extra volume compared to the most potent analog, etc.[21,47]

Perturbation Free-Energy Calculations. The direct calculation of the effect of a small change in structure on the affinity of a ligand for a protein, $\Delta\Delta G$, is now possible.[22]

$$\Delta G_1: \quad A + B \rightleftharpoons AB$$
$$\Delta G_2: \quad A' + B \rightleftharpoons A'B$$
$$\Delta\Delta G = \Delta G_2 - \Delta G_1$$

The calculations as written are very difficult to evaluate because they would require accurate functions for the solvation of the ligands and for their interactions with the macromolecule. However, it is possible to calculate the effect of a slight change in structure on the ΔG values of solvation and interaction with the macromolecule, ΔG_3 and ΔG_4:

$$\Delta G_3: \quad A + B \rightleftharpoons A' + B$$
$$\Delta G_4: \quad AB \rightleftharpoons A'B$$

Because all processes are equilibria,

$$\Delta\Delta G = \Delta G_2 - \Delta G_1 = \Delta G_4 - \Delta G_3$$

In these molecular dynamics calculations one slowly perturbs solvated A to A′ and AB to A′B by changing the parameters that describe the ligand. The calculations are done in such a way that ΔG_4 and ΔG_3 are directly calculated. From a knowledge of ΔG_1 one can then calculate $\Delta\Delta G$. These calculations take such a large amount of computer time that they are usually done on a supercomputer. It is fair to quote one of the leaders in the field: "Such applications are far from routine at the present time" and "will be used increasingly in molecular design."[22]

For example, the effect of replacing an -NH- group in **XL** with an -O- (**XLI**) on its inhibitory potency vs thermolysin was forecast to be 4.21 ±

[45] A. K. Ghose and G. M. Crippen, *in* "Comprehensive Medicinal Chemistry" (C. Hansch, P. G. Sammes, and J. B. Taylor, series eds., C. A. Ramsden, vol. ed.), p. 715. Pergamon, Oxford and New York, 1990.

[46] H. D. Holtje and L. B. Kier, *J. Pharm. Sci.* **64,** 418 (1975).

[47] A. J. Hopfinger, *J. Am. Chem. Soc.* **102,** 7196 (1980).

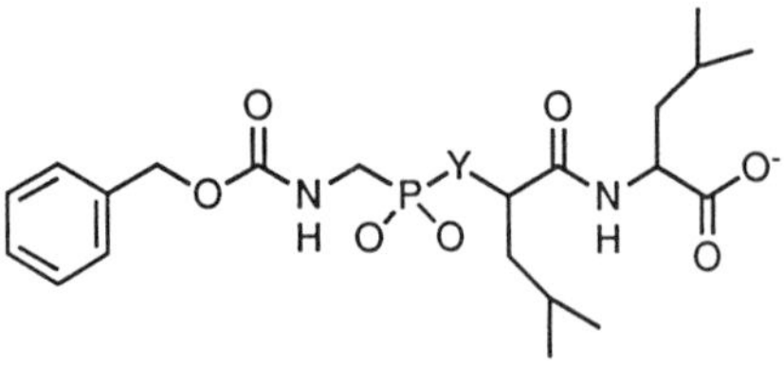

XL Y = NH
XLI Y = O

0.54 kcal/mol, whereas it was observed to be 4.1 kcal/mol.[48] The change in attractive and repulsive interactions amounted to 7.5 kcal/mol but differences in solvation of the two ligands lower the figure to 4.21.

Computer Tools for Proposing Shape and Chemical Properties of Macromolecular Binding Site

Prediction of Three-Dimensional Structure of Protein from Its Amino Acid Sequence. The 3D structure of a protein is a result of its amino acid sequence: proteins that show a high degree of sequence homology show a high degree of similarity in 3D structure. Therefore, there are strategies for modeling the 3D structure of a protein from its sequence and the 3D structure of a homologous protein.[49] Additionally, enzymes that have a similar mechanism of action often are similar in the geometric relationships of the catalytic groups. Thus, the design of enzyme inhibitors can be based on the hypothesis that the structure of the target enzyme is similar to that of a protein with the same mechanism.

For example, consider the design of a potent inhibitor of the human immunodeficiency virus (HIV) protease in the absence of a 3D structure of the enzyme.[50] The sequences of the proteases from HIV and Rous sarcoma viruses are highly homologous. The 3D structure of the Rous sarcoma protease made it clear that these viral proteases are also related in 3D structure to the aspartyl proteases of higher organisms. However, in the viral enzymes two identical subunits form the active site, whereas in the human enzymes the active site is not symmetric. Therefore, it was postulated that a symmetric compound might preferentially inhibit the HIV

[48] P. A. Bash, U. C. Singh, F. K. Brown, R. Langridge, and P. A. Kollman, *Science* **235**, 574 (1987).
[49] J. Greer, *Proteins* **7**, 317 (1990).
[50] J. Erickson, D. J. Neidhart, J. VanDrie, D. J. Kempf, X. C. Wang, D. W. Norbeck, J. J. Plattner, J. W. Rittenhous, M. Turon, N. Wideburg, W. E. Kohlbrenner, R. Simmer, R. Helfrich, D. A. Paul, and M. Knigge, *Science* **249**, 527 (1990).

protease. Such an inhibitor, **XLIII**, was designed from the 3D structure of a reduced peptide inhibitor, **XLII**, as bound to the active site of *Rhizopus* pepsin. To do this **XLII** was docked into the active site of the Rous sarcoma protease and the C-terminal portion of the inhibitor removed.

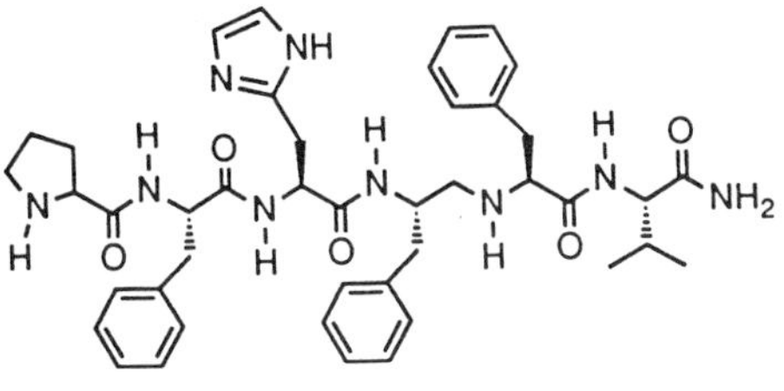

XLII

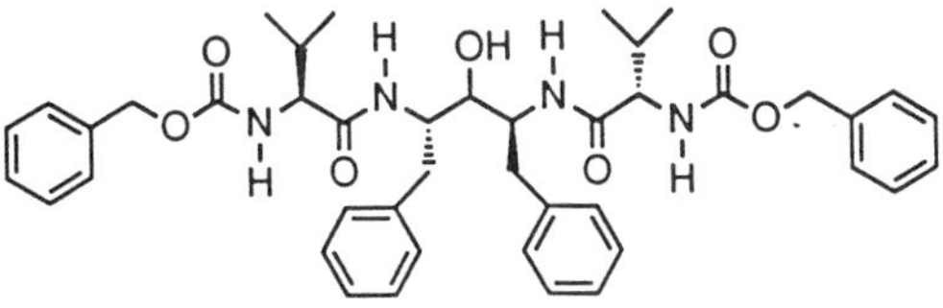

XLIII

The second half of the proposed inhibitor was generated with symmetry operations and the side chains changed to mimic the substrate. The result is a novel, potent inhibitor of HIV protease and the start of a new structural class of such inhibitors. The precision of the design was verified by crystallography of the HIV protease–inhibitor complex.[50]

Mapping Binding Site by Analysis of Structure–Activity Relationships (Receptor Mapping). Probing the shape of a macromolecular binding site by the analysis of the 3D structure–activity relationships of ligands is often called receptor mapping. The objective is to propose the detailed shape and electrostatic properties of the binding site. A receptor mapping analysis requires that one establish both the bioactive conformations and how to superimpose the compounds. The availability of definitive structure–activity information is the key to the success of such a study.

Establishing superposition rule: The atoms to be involved in the superposition are those shown by structure–activity analysis to be required for bioactivity.

How does one superimpose the compounds as they would lie in the binding site? Typically the positions of the protein atoms that interact with the essential atoms of the ligands are overlapped. Proposals regarding the location and chemical nature of these hypothetical atoms may be found from quantum chemical[20,21] or potential energy calculations of the regions

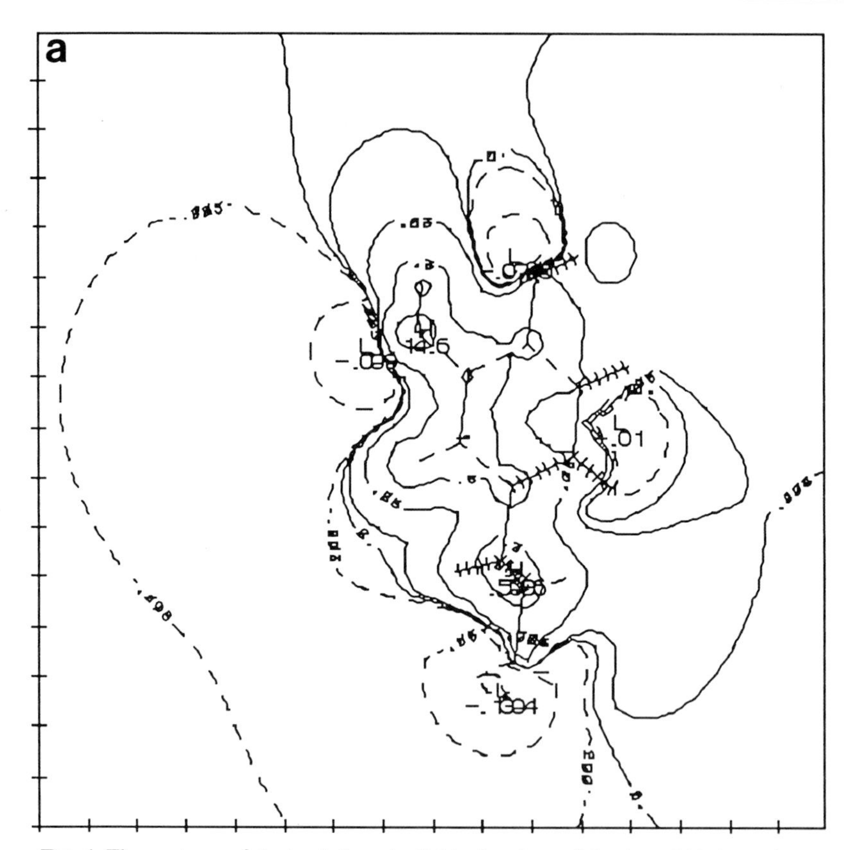

FIG. 4. The contours of electrostatic potential in the plane of the ring of (a) dopamine and (b) EOE, D2 dopaminergic agonists. Note that both have positive potential toward the top center of the figure. This is characteristic of the hydrogen bond proposed to be required for bioactivity.

surrounding the ligands[3,33] or from an investigation of crystallographic packing of related small molecules.[51] For example, Fig. 4 shows the contours of electrostatic potential (energy of interaction with an H^+) calculated in the plane of the aromatic ring for two D2 dopaminergic agonists. The regions of similarity are the positive contours (repulsion to an H^+) at the top of the page. Similar regions were identified by GRID calculations using an O^- probe to show the contours at which an O^- atom in the protein would find a favorable interaction with the dopaminergic ligands. Finally, a search of the Cambridge Structural Database of small-molecule crystal

[51] F. H. Allen, O. Kennard, and R. Taylor, *Acc. Chem. Res.* **16,** 146 (1983).

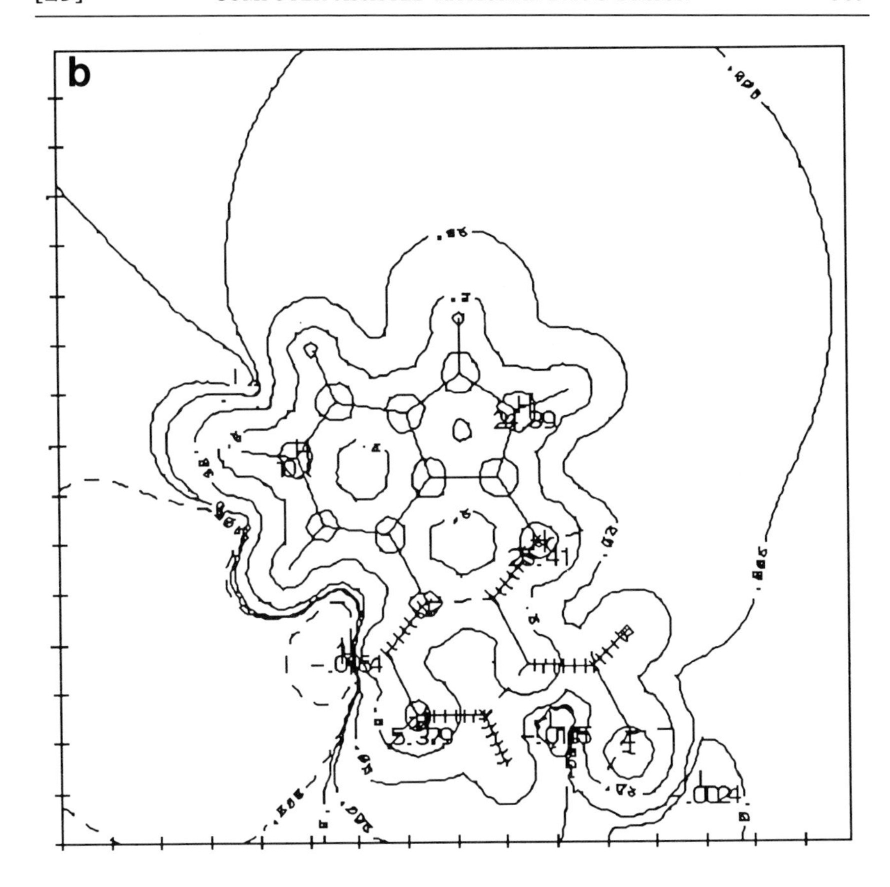

structures showed that catechols interact with neighboring molecules using the OH as a hydrogen bond donor, to a carboxylate if possible. All three considerations led to our dopamine receptor model with a carboxylate as an interaction site with the hydrogen of the catecholic OH group.

Computer programs are available to help superimpose a pair of molecules that contain several alternative groups that might be superimposed with a corresponding group on the other molecule.[12]

Selecting bioactive conformation: The bioactive conformation may not be the minimum energy structure. However, a very high-energy conformation is not a good candidate either, since an energy difference of 1.4 kcal/mol corresponds to a 10-fold difference in binding constant at 38°.

The only ways to establish the bioactive conformation of a ligand are to determine the 3D structure of the ligand–macromolecule complex or to discover a ligand with good bioactivity and little conformational flexibility, the "rigid analog" approach. As noted above, 3D search methods may be used to design conformationally restricted analogs: however, several compounds must be designed to mimic each conformation since the target biomolecule may not tolerate the groups added to constrain the conformation.

Our design of D1 dopaminergic agonists used the rigid analog approach. At the time we started our work the only D1 selective agonist was compound **II**. Conformational searching revealed that it has two low-energy conformations, axial and equatorial. We established its bioactive conformation by the synthesis and testing of compounds in which the pendant phenyl group mimicks either the axial or equatorial conformation, **XLIV–XLVI** or **XLVII–XLIX**, respectively.[52] Each compound was

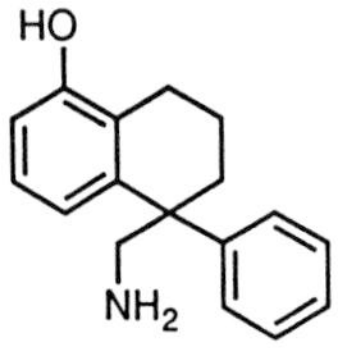

XLIV

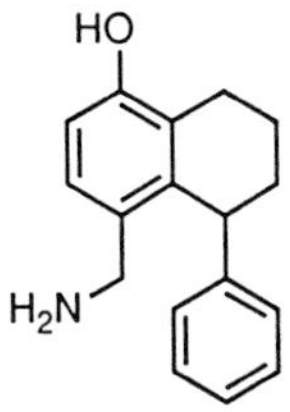

XLV

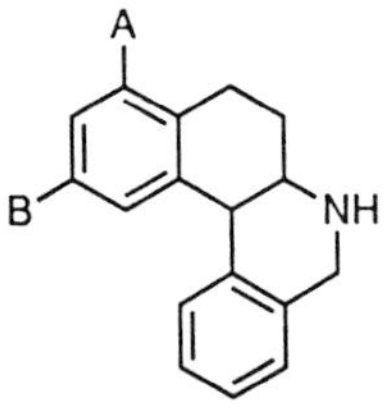

XLVI A = OH
XLVII B = OH

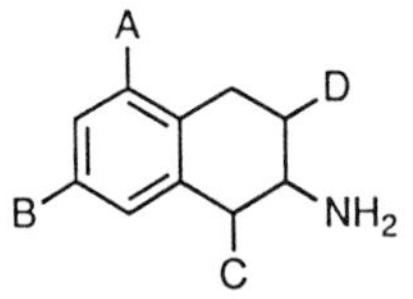

XLVIII A = OH, D = F *trans*
XLIX B = OH, C = F *trans*

compared with the corresponding des-phenyl analog. **XLIV–XLVI** are inactive whereas the phenyl group in **XLVII–XLIX** increases affinity. Thus, the equatorial conformation of **II** is the bioactive conformation.

Previously the project had identified the structurally novel compound **L** as a nonselective dopamine agonist with a D1 pK_i of 5.90. We designed compound **LI** to mimic the bioactive conformation of **II**. Its pK_i for the D1

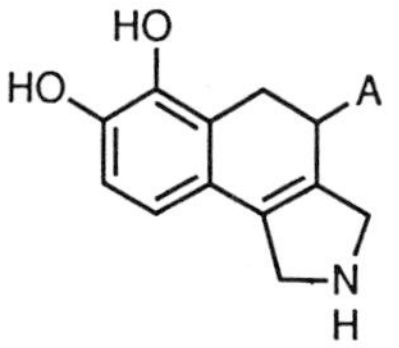

L A = H
LI A = ϕ

[52] Y. C. Martin, J. Kebabian, R. Mackenzie, and R. Schoenleber, *in* "QSAR: Rational Approaches on the Design of Bioactive Compounds" (C. Silipo and A. Vittoria, eds.), Elsevier, Amsterdam, in press.

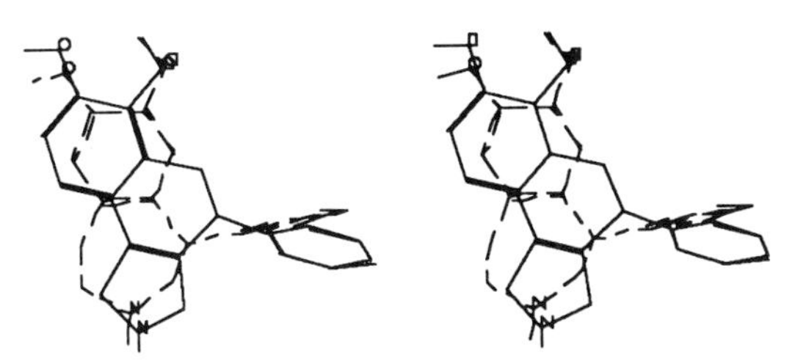

FIG. 5. The superposition of the bioactive conformations of **II** (dashed line) and **LI** (solid line).

receptor is 7.24 and its pEC_{50} for the stimulation of adenylate cyclase at the D1 receptor is an order of magnitude higher than that of **II**. Figure 5 shows the bioactive conformations of **II** and **LI**.

The "active analog"[4] and ensemble distance geometry[53] approaches generate zero, one, or many hypotheses about the bioactive conformations of members of a set of noncogeneric compounds. The user supplies the atoms to be superimposed, the pharmacophore atoms. These methods do not provide the relative energy of the proposed bioactive conformation compared to the global minimum: this must be established by a separate calculation.

The "active analog" approach first searches by systematic bond rotation the low-energy conformations of the most constrained compound and the allowed distances between the pharmacophore atoms is recorded. The search on each subsequent compound is restricted to examine only low-energy conformations that could have the pharmacophore distances common to all molecules so far considered.[4] Thus any conformations that correspond to distances not common to all active compounds are rejected.

Distance geometry programs randomly generate three-dimensional structures consistent with a set of input distance bounds between every atom pair in a molecule or molecular complex.[45] This distance-bounded matrix is automatically generated for a simple conformational search. The program randomly chooses a distance between the upper and lower bound for each atom pair and then projects these distances into three dimensions using eigenvector analysis. The resulting structures are energy minimized to refine the structure and to rank the conformations according to relative energy.

In ensemble distance geometry one simultaneously generates corresponding conformations of each of a set of molecules. These conforma-

[53] R. P. Sheridan, R. Nilakantan, J. S. Dixon, and R. Venkataraghavan, *J. Med. Chem.* **29**, 899 (1986).

tions obey the added constraint that the distances between specified atoms in different molecules is zero, i.e., these atoms should overlap.[52]

Proposing shape of binding site: If the binding site does not change in 3D structure with different structures of the ligand, any region in space occupied by one active molecule will be available to another. Hence, computer graphics programs for receptor mapping display the union of the space occupied by the active compounds.[4]

Often there are inactive compounds that have the correct geometric relationships between the groups needed for binding. The new regions in space that they occupy suggest regions occupied by the macromolecule or "forbidden regions" for the ligand. Thus, a molecular graphics program would display the unique regions in space occupied by inactive analogs.

Unmet Challenges of Computer-Assisted Rational Drug Design

It would be misleading if this chapter ended without a discussion of the current limitations of computer-assisted rational drug design. Of course, one is always restricted by limitations of the experimental data. For example, the design should be based on the known 3D structure and molecular mechanism of the target biomolecule and the biomolecules that lead to the side effects; however, this information is not usually available. The biological affinity may not be measured in a fashion acceptable to a physical chemist. Last, computers are never fast enough nor do they hold enough information. Aside from these problems, there are limitations in the approaches that the computer person can use.

First, we do not know how to treat in potential energy calculations the influence of water specifically and the environment in general; i.e., its effect on 3D structure and intermolecular interactions of macromolecules and their ligands. Therefore, we do not know how to treat electrostatic interactions in potential energy calculations, where to obtain accurate partial atomic charges, how to determine if charges centered at atomic nuclei are realistic enough, or how to treat the medium between these partial atomic charges. Do we have to include explicit water out to a large radius around the molecule? How large is it? How can we tell? We still do not know how to establish the location and number of ionizable sites on the protein that are protonated and the identity, location, and number of counterions in the system. Maybe the availability of huge amounts of computer power will solve these problems.

Second, although there has been a brave and enthusiastic start, we still do not know how to design molecules automatically to meet our criteria. We definitely do not know if we have suggested every possible compound.

Furthermore, we are just now learning how to forecast the bioactivity of the suggested compounds.

Third, someone must design the synthesis, search the literature for related compounds that may make it impossible to patent the suggested compound, and study if known structure–activity relationships on other series or other biological end points make the forecasted potency unreliable. In principle these tasks could be computerized, but presently they are not.

Last, the ability to patent a compound depends on its novelty: its unobviousness to one skilled in the art. If the computer designs a compound, additional questions are raised: Does this mean that it is obvious? Who is the inventor? Can a computer assign patent rights to a company? Will our legal system have to change if mankind is to benefit from these new powers of computers?

[30] Pattern Recognition Methods in Rational Drug Design

By DAVID J. LIVINGSTONE

Introduction

Pattern recognition methods, also referred to as chemometrics or multivariate statistics, are being used increasingly in rational drug design.[1-5] The aim of this chapter is to give an introduction to some of the pattern recognition techniques and to show examples of their use in the investigation of quantitative structure–activity relationships (QSAR). It is not possible here to describe any of the methods in much detail, but there are some excellent texts available that deal with pattern recognition,[6,7] chemomet-

[1] S. Clementi, *in* "Drug Design. Fact or Fantasy?" (G. Jolles and K. R. H. Wooldridge, eds.), p. 73. Academic Press, London, 1984.

[2] R. Franke, "Theoretical Drug Design Methods," pp. 188, 203, and 289. Elsevier, Amsterdam, 1984.

[3] H. Van De Waterbeemd and B. Testa, *in* "Advances in Drug Research" (B. Testa, ed.), Vol. 16, p. 98. Academic Press, London, 1987.

[4] R. M. Hyde and D. J. Livingstone, *J. Comput.-Aided Mol. Des.* **2,** 145 (1988).

[5] D. J. Livingstone, *Pestic. Sci.* **27,** 287 (1989).

[6] K. Varmuza, "Pattern Recognition in Chemistry," Springer-Verlag, New York, 1980.

[7] K. Fukunaga, "Introduction to Statistical Pattern Recognition." Academic Press, New York, 1972.

rics,[8] and multivariate statistics.[9-12] Of particular use to those seeking *practical* help is the book by Wolff and Parsons,[13] which compares and explains the output from three commonly used statistics packages.

The term pattern recognition serves to describe any mathematical or statistical method that may be used to detect or reveal patterns in data. Thus, this description may be applied to such well-known analytical techniques as simple and multiple linear regression (MLR), although these methods would not normally be thought of as pattern recognition. The literature abounds with examples of MLR equations in drug design and it is described in many standard texts (e.g., see Refs. 14 and 15), so it will not be mentioned further here except for comparison with some techniques. A common feature of pattern recognition methods is that they are intended for use with data sets containing many variables, usually considering all the variables simultaneously. This is an important characteristic for an analytical technique used with any kind of data set, but is particularly so for the type of data sets encountered in drug design where complex relationships are the norm, rather than the exception. The increasing use of computational chemistry and molecular modeling systems in drug design has made the generation of large multivariate data sets, containing, say, 50–100 parameters, a matter of routine.[4,5,16-19] For a variety of reasons, such data sets demand the use of analytical methods other than MLR and pattern recognition techniques are well suited to these data.

Some pattern recognition methods have their origins in artificial intelligence research, attempting to devise algorithms that could learn to distin-

[8] M. A. Sharaf, D. L. Illman, and B. R. Kowalski, "Chemometrics." Wiley, New York, 1986.

[9] R. J. Harris, "A Primer of Multivariate Statistics." Academic Press, New York, 1975.

[10] C. Chatfield and A. J. Collins, "Introduction to Multivariate Analysis." Chapman & Hall, London, 1980.

[11] W. R. Dillon and M. Goldstein, "Multivariate Analysis Methods & Applications." Wiley, New York, 1984.

[12] P. E. Green, "Mathematical Tools for Applied Multivariate Analysis." Academic Press, New York, 1976.

[13] D. D. Wolff and M. L. Parsons, "Pattern Recognition Approach to Data Interpretation." Plenum, New York, 1983.

[14] C. Daniel and F. S. Wood, "Fitting Equations to Data." Wiley, New York, 1980.

[15] N. R. Draper and H. Smith, "Applied Regression Analysis." Wiley, New York, 1981.

[16] P. G. De Benedetti, M. C. Menziani, M. Cocchi, and C. Frassinetti, *Quant. Struct.-Act. Relat.* **6,** 51 (1987).

[17] O. Kikuchi, *Quant. Struct.-Act. Relat.* **6,** 179 (1987).

[18] L. M. Nilsson, R. E. Carter, O. Sterner, and T. Liljefors, *Quant. Struct.-Act. Relat.* **7,** 84 (1988).

[19] D. J. Livingstone, M. G. Ford, and D. S. Buckley, *in* "Neurotox '88: Molecular Basis of Drug and Pesticide Action" (G. G. Lunt, ed.), p. 483. Elsevier, Amsterdam, 1988.

guish patterns in a data set, e.g., the linear learning machine.[20] The result of this learning process is the establishment of rules that can be used to classify objects in the data set. In the context of drug design the objects in a data set are compounds and the desired classification is based on their biological properties. Two types of data set are involved in this process: a training set (or learning set) for which the classification is known and a test set for which the classification is unknown. The training set is used to train the method, i.e., generate the classification rules, and having trained the method predictions may be made for the test set. The "success" of a particular technique may be judged in terms of both its training and test set predictions. A third type of set may also be used that consists of compounds whose classification is known, but which are not used in the training process. This is an evaluation set and, as the name implies, it can be used to evaluate a technique by comparison of the predicted and observed classifications. It should be noted here that the term "classification" has been used with reference to biological properties. All pattern recognition methods may be used with classified biological data, e.g., active/inactive, +/++/+++, but some techniques will deal with quantitative data such as the dose to produce a standard response e.g., IC_{50}, ED_{50}. Most of the methods can make use of either quantitative or classified descriptor data or mixtures of both types.[21]

The concept of learning allows pattern recognition techniques to be classified as supervised and unsupervised learning methods. Supervision refers to the use made of the biological data in the analytical process. In unsupervised learning the known categories of the training set compounds are not used; in supervised learning the categories of the training set are used to produce the classification rules. An example of unsupervised learning is shown in Fig. 1, where the parameter values of two physicochemical descriptors, e.g., π and σ, are plotted against one another. This may appear a trivial example but it illustrates a number of useful points both here and later. Figure 1 contains a training set, coded for biological activity and toxicity, and a test set consisting of the three untested compounds X, Y, and Z. The analytical process, i.e., the act of plotting the compounds, was unsupervised but it can be seen that the active compounds are clearly separated from the inactive and intermediate compounds. It can be further seen that the test set compound Y falls quite clearly into an area of active compounds and a prediction of "active" for this compound would be quite

[20] N. J. Nilsson, "Learning Machines: Foundations of Trainable Pattern Classifying Systems." McGraw-Hill, New York, 1965.

[21] Caution should be exercised in the use of too many dichotomous (e.g., 1/0) parameters in a data set (see pp. 5–7 of Ref. 13).

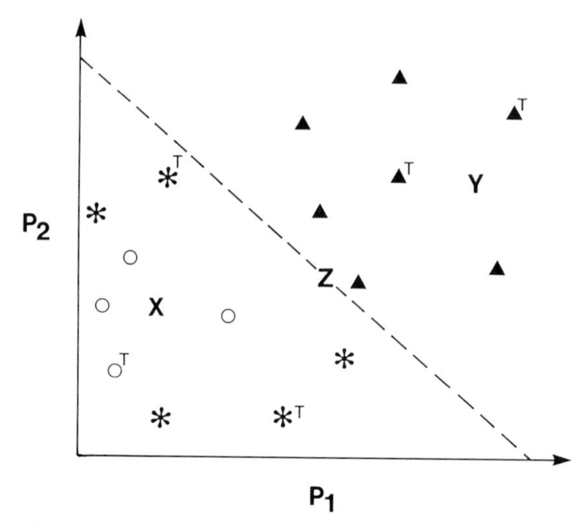

FIG. 1. Plot of two physicochemical parameters for a set of compounds. ▲, Active; ○, intermediate; *, inactive; T, toxic. Dashed line, linear discriminant surface; *X*, *Y*, *Z* are untested compounds.

reasonable. A supervised learning example can also be seen in this figure as the linear decision surface (see section entitled "Classification Methods") between the active compounds and the rest. This surface, or line, in this two-dimensional example, is a classification rule derived from the known categories of the test set.

There are important differences between supervised and unsupervised learning, mostly concerned with the number of compounds in a data set (m) compared with the number of variables (n). When $n \geq m$, some supervised techniques may not work because of failure to invert a matrix, or they may give a false, apparently correct, classification. This is not a problem for unsupervised methods, although the presence of extra variables that do not contain useful information may serve to obscure meaningful patterns. Another problem with supervised learning concerns the matter of chance correlations. Since supervised techniques seek to establish a classification rule, there is a possibility that such a rule may arise by chance. The greater the number of variables that are screened for inclusion in the rule then the higher the probability that a variable will be included by chance. This has been tested by experiments with random numbers for multiple regression[22] and discriminant analysis,[23-26] although, since real

[22] J. G. Topliss and R. P. Edwards, *J. Med. Chem.* **22,** 1238 (1979).

[23] A. J. Stuper and P. C. Jurs, *J. Chem. Inf. Comput. Sci.* **16,** 238 (1976).

[24] E. K. Whalen-Pedersen and P. C. Jurs, *J. Chem. Inf. Comput. Sci.* **19,** 264 (1979).

data contains a correlation structure, results based on random numbers may not be entirely appropriate. Studies have also been carried out on real data sets[27,28] to investigate the use of cross-validation[29] and bootstrapping[30] as measures of chance correlations. These methods test the utility of a model by repeated calculations using subsets of the starting data. Unsupervised learning techniques should not suffer from chance correlations in the same way as the supervised methods, but there is always the possibility that any observed pattern may be due to chance effects.

What do pattern recognition methods do, and how are they best used? In general, these techniques may be thought of as aids for the investigation of n-dimensional spaces (n-dimensional parameter sets). This assistance may be in the form of dimension reduction so as to produce a lower dimensional (usually two- or three-dimensional) display of the data; it may be in the form of a "report" about the n space or it may be an attempt to "fit" or classify objects in the n-dimensional space. Many of the methods are complementary and may be thought of as giving a different view or perspective of the n-dimensional data set. Because of this it is not possible to say, *a priori,* which technique will work best with a particular data set. The only advice that can be given is rather unsatisfactory, and that is to use the method(s) that appear best to describe the data. However, experience has shown that the comparison of results from a number of techniques usually leads to a consistent interpretation of a data set. Because of the different requirements of different data sets, it is also difficult to recommend a procedure by which the various methods should be applied. However, the techniques are described in the following sections in an order that might reasonably be followed in the analysis of a data set. It might be desirable to go back and rerun some of the methods as the data analysis proceeds, e.g., data display following variable deletion.

Data Display Techniques

These methods, which may be divided into linear and nonlinear techniques, are unsupervised learning and thus may be used with any number of parameters and compounds. They are particularly useful at the start of an analysis, where they may serve to confirm that a data set contains some

[25] T. R. Stouch and P. C. Jurs, *J. Chem. Inf. Comput. Sci.* **25,** 45 (1985).

[26] T. R. Stouch and P. C. Jurs, *J. Chem. Inf. Comput. Sci.* **25,** 92 (1985).

[27] H. Mager, *Quant. Struct.-Act. Relat.* **3,** 147 (1984).

[28] R. D. Cramer, J. D. Bunce, D. E. Patterson, and I. E. Frank, *Quant. Struct.-Act. Relat.* **7,** 18 (1988).

[29] S. Geisser, *J. Am. Stat. Assoc.* **70,** 320 (1975).

[30] P. Diaconis and B. Efron, *Sci. Am.* **248,** 116 (1983).

"useful" information, but may also be used at several stages during the analysis. These techniques can be used to examine the relationships between compounds (rows) in a data set but they may also be used to examine the relationships between parameters (columns). This is equivalent to looking at m compounds in an n-dimensional parameter space or n parameters in an m-dimensional compound space.

We have already seen an example of a linear data display method in the two-dimensional plot shown in Fig. 1. The advantage of a plot such as this is that it is very easy to interpret. The active region of this parameter space, for example, consists of high values of both P_2 and P_1. A disadvantage of this type of plot is that for a large data set there are many such possible plots, $n(n-1)/2$ for n variables. A more serious drawback is that the plot contains only information relating to two variables. It can be seen, for example, that these two parameters do not describe the toxicity of these compounds; toxic compounds are seen at all values of both descriptors. A linear technique that considers multiple variables is principal components (PC) analysis.[31] Principal components are new variables created from linear combinations of the starting variables:

$$\begin{aligned} PC_1 &= A_{1,1}V_1 + A_{1,2}V_2 + \cdots + A_{1,n}V_n \\ PC_2 &= A_{2,1}V_1 + A_{2,2}V_2 + \cdots + A_{2,n}V_n \\ PC_k &= A_{k,1}V_1 + A_{k,2}V_2 + \cdots + A_{k,n}V_n \end{aligned} \tag{1}$$

where V_1 to V_n are the n starting parameters and for each principal component, i, there are n weighting coefficients, $A_{i,1}$ to $A_{i,n}$, one for each of the variables. As many principal components may be calculated as the smaller of m compounds or n parameters, and these components have the following properties: (1) the first principal component contains the largest part of the variance of the data set, and subsequent components contain progressively smaller amounts of variance; (2) each principal component is orthogonal to all of the other components.

The weighting coefficients, also referred to as the loadings of the variables on the principal components, are calculated so as to comply with these requirements. Since the first two principal components contain the largest amount of variance in a data set, a plot of the values of these components, called the principal component scores, for a set of compounds should give a "good" two-dimensional picture of the data set. An

[31] H. Seal, "Multivariate Analysis for Biologists," p. 101. Methuen, London, 1968; C. Chatfield and A. J. Collins, "Introduction to Multivariate Analysis," p. 57. Chapman & Hall, London, 1980; R. J. Harris, "A Primer of Multivariate Statistics," p. 57. Academic Press, New York, 1975; W. R. Dillon and M. Goldstein, "Multivariate Analysis Methods & Applications," p. 23. Wiley, New York, 1984.

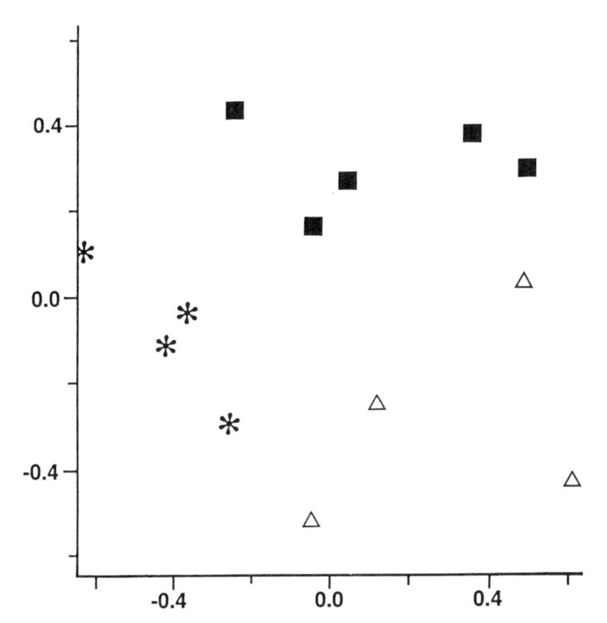

FIG. 2. Principal components plot for a set of γ-aminobutyric acid (GABA) analogs. ■, Potent agonist, △, weak agonist; *, no agonist activity. [Reprinted with permission from *J. Comput. Aid. Mol. Des.* **3,** 55 (1989); copyright 1989 Wellcome Foundation Ltd.]

example of this is shown in Fig. 2 for a set of analogs of γ-aminobutyric acid (GABA). Principal component scores for these compounds were calculated for the first two principal components derived from four physicochemical descriptors and the figure shows the disposition of these compounds in the space described by these components. It can be seen that the three activity classes are clearly distinguished and that an activity prediction may be made for an untested compound by plotting the values of its scores. Although the first two components contain the largest amount of variance, subsequent components may contain useful information in terms of "explaining" the biological data. Plots of other combinations of principal components may reveal significant patterns in the data and in some cases the inclusion of a third component in a three-dimensional plot is required.[32-34] In addition to plotting the scores of the compounds on principal component axes, the loadings of the variables may also be plotted

[32] G. K. Menon and A. Cammarata, *J. Pharm. Sci.* **66,** 304 (1977).
[33] J. U. Clarke, *Chemosphere* **15,** 275 (1986).
[34] B. Hudson, D. J. Livingstone, and E. Rahr, *J. Comput.-Aided Mol. Des.* **3,** 55 (1989).

to give an indication of the relationship between variables.[35–39] When both scores and loadings are plotted the resultant display is known as a biplot.[40] A related technique to the biplot is spectral map analysis (SMA), which operates on contrasts in data, expressed as log ratios.[41] A comparison of SMA with principal components analysis and correspondence factor analysis highlights some advantages of this method.[42]

Principal components have found other uses in drug design, e.g., the construction of new physicochemical descriptors[43] and the analysis of multiple response data,[35–38,44] and they may be used in various correlation techniques as described later. There are, however, some disadvantages to this approach. Principal components are often difficult to interpret, since they are composed of a combination of variables and it is not possible to translate the coordinates of a point on a principal components plot to values of individual variables. Since this method seeks to preserve the overall variance in a data set, variables that contain a small amount of variance, but which nevertheless may be important in the explanation of biological properties, may be obscured in the generation of the components. Techniques exist for the improvement of principal components, by simplification of the correlation structure of the components so as to make them easier to interpret. An example of this is a procedure known as Varimax rotation,[45] which involves a rotation of the principal component axes so as to maximize the high correlations (loadings) of individual variables with the components. Table I shows the first 4 principal components derived from a data set of 18 naphthalene derivatives described by 7 physicochemical parameters[46]; the parameters making a large contribution (high loadings) to each component are shown in boldface type.

The application of Varimax rotation to these principal components resulted in a considerable simplification of their structure, as seen in Table

[35] S. Dove, E. Coats, P. Scharfenberg, and R. Franke, *J. Med. Chem.* **28,** 447 (1985).
[36] Z. Szigeti, T. Cserhati, and B. Bordas, *Gen. Physiol. Biophys.* **4,** 321 (1985).
[37] M. Wiese, J. K. Seydel, H. Pieper, G. Krüger, K. R. Noll, and J. Keck, *Quant. Struct.-Act. Relat.* **6,** 164 (1987).
[38] M. Nendza and J. K. Seydel, *Quant. Struct.-Act. Relat.* **7,** 165 (1988).
[39] H. Van De Waterbeemd, N. El Tayar, P.-A. Carrupt and B. Testa, *J. Comput.-Aided Mol. Des.* **3,** 111 (1989).
[40] H. Hotelling, *J. Educ. Psychol.* **24,** 417 (1933).
[41] P. J. Lewi, *Eur. J. Med. Chem.* **21,** 155 (1986).
[42] P. J. Lewi, *Chemom. Intell. Lab Syst.* **5,** 105 (1989).
[43] C. G. Swain and E. C. Lupton, *J. Am. Chem. Soc.* **90,** 4328 (1968).
[44] R. H. Davies and T. R. Morris, *Int. J. Quant. Chem.* **23,** 1385 (1983).
[45] H. H. Harmon, "Modern Factor Analysis," p. 487. University of Chicago Press, Chicago, Illinois, 1976.
[46] T. W. Schultz and M. P. Moulton, *Bull. Environ. Contam. Toxicol.* **34,** 1 (1985).

TABLE I
PARAMETER LOADINGS FOR FOUR PRINCIPAL COMPONENTS

Parameter[a]	1	2	3	4
π	**0.698**[b]	−0.537	−0.121	−0.258
MR	**0.771**	0.490	−0.302	−0.002
F	0.261	0.389	**0.745**	−0.423
R	0.405	−0.012	0.578	**0.697**
H_a	−0.140	**0.951**	0.071	−0.101
H_d	**−0.733**	0.373	−0.271	0.172
${}^1\chi^v_{sub}$	**0.739**	0.412	−0.404	0.163

[a] See Ref. 46 for details of the parameters.
[b] Boldface numbers indicate parameter making a large contribution to each component.

II. Interestingly, the bulk (MR) and hydrophobicity (π) parameters have been separated in the rotated components and the hydrogen bond acceptor (H_a) and donor (H_d) parameters no longer make a contribution to any of these components.

Principal components analysis imposes a linear structure on the combination of starting variables; if nonlinear combinations of the variables are necessary to explain the biological data, this may be hidden by this technique. A method for multivariate data display that does not suffer from this latter problem is nonlinear mapping (NLM). As the name implies, this is a nonlinear technique for the projection of points from an n-dimensional

TABLE II
PARAMETER LOADINGS AFTER VARIMAX ROTATION

Parameter	1	2	3	4
π	0.200	**0.919**[a]	0.012	0.012
MR	**0.891**	0.195	0.093	0.061
F	0.020	−0.003	**0.975**	0.123
R	0.081	0.018	0.115	**0.982**
H_a	0.272	0.451	0.318	−0.083
H_d	−0.159	−0.285	−0.138	−0.148
${}^1\chi^v_{sub}$	**0.974**	0.037	−0.024	0.064

[a] Boldface numbers indicate parameter making a large contribution to each component.

space to a two- (or three-) dimensional space. NLM operates by consideration of the distances between pairs of points in n space (D^*_{ij}) and in two-dimensional space (d_{ij}) and seeks to minimize the differences between these interpoint distances using an error function such as that shown below:

$$E = \sum_{i>j} (D^*_{ij} - d_{ij})^2/(D^*_{ij})^\rho \tag{2}$$

where ρ is a weighting factor that determines the emphasis that is put on large and small distances. A value of 2 for this parameter gives equal weight to all distances, whereas a value of -2 for ρ gives greater weight to large interpoint distances.[47] This has the effect of making similar objects group together in tighter clusters.[34] Figure 3 shows a nonlinear map of 63 hallucinogenic phenylalkylamine derivatives described by 24 physicochemical parameters.[48] It can be seen from Fig. 3 that the majority of the active compounds are to be found in one region of the map but no clear separation is shown between the intermediate and inactive compounds. At this stage of the analysis the nonlinear map simply suggests that the physicochemical data contain information that will be useful in classification of the compounds. This was confirmed by the generation of a significant regression equation that appeared to have good predictive ability. In much the same way that principal components analysis can be used to show the relationships between variables by plotting their loadings, nonlinear mapping may be used for this purpose[19,49] by using the coefficient of nondetermination ($1 - R^2$) as a distance measure.

Nonlinear mapping has the advantage that it does not impose a linear combination on the variables in order to produce the two-dimensional display, but it does suffer from a number of disadvantages. Since the method relies on the minimization of a set of interpoint distances it is not possible to simply plot a new compound on the map. The whole map must be recalculated and in some cases the inclusion of new compounds may alter the map considerably, although useful patterns are generally preserved. Using a principal components display, a new compound may be plotted by the calculation of its principal component scores from an equation such as Eq. (1). It is not possible using either method to select a point on the two-dimensional display and determine the parameter values required to place a compound there. While principal components may be difficult to interpret, the nonlinear combination of variables in the axes of a nonlinear map is unknown. One report[50] makes use of a technique that is

[47] B. R. Kowalski and C. F. Bender, *J. Am. Chem. Soc.* **95,** 686 (1973).
[48] B. W. Clare, *J. Med. Chem.* **33,** 687 (1990).
[49] T. Takagi, A. Iwata, Y. Sasaki, and H. Kawaki, *Chem. Pharm. Bull.* **30,** 1091 (1982).
[50] R. Benigni, C. Andreoli, and A. Giuliani, *Carcinogenesis* **10,** 55 (1989).

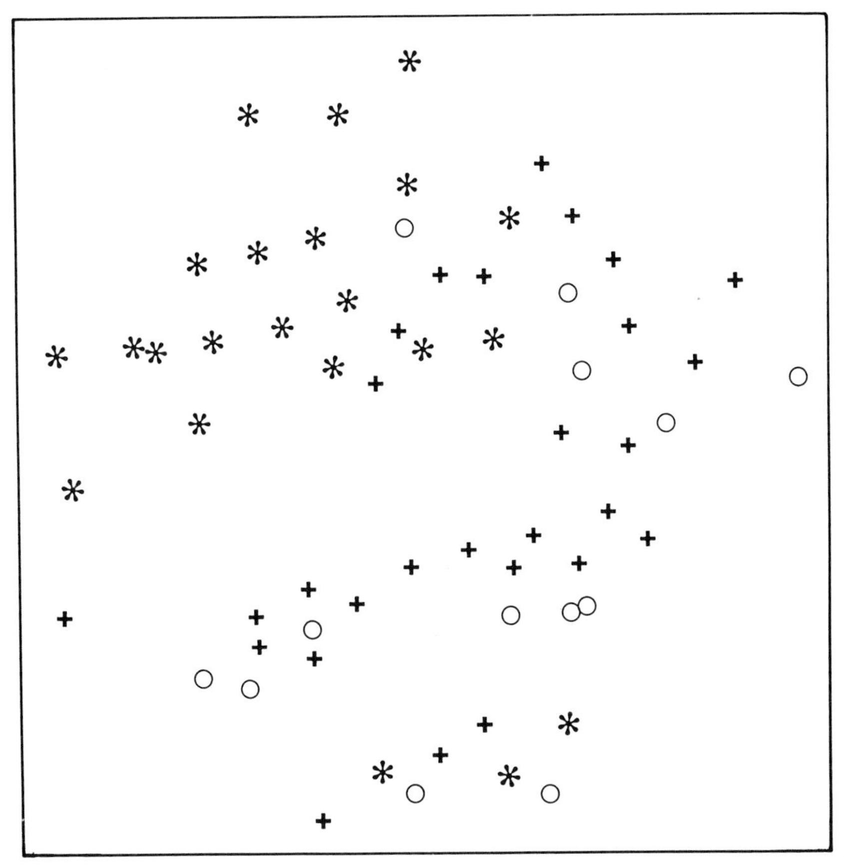

FIG. 3. Nonlinear map for a set of phenylaklylamine derivatives. [Reprinted with permission from *J. Med. Chem.* **33,** 687 (1990); copyright 1990 American Chemical Society.] ○, Inactive; +, low activity; *, high activity.

claimed can take account of both linear and nonlinear relationships between variables.

The final display method that will be discussed here is cluster analysis (CA). This technique, like nonlinear mapping, makes use of interpoint distances to describe a data set. The interpoint distances are used as the basis of a measure of similarity such as that shown below:

$$S_{jk} = 1 - D_{jk}/\mathrm{MAX}(D_{jk}) \tag{3}$$

where D_{jk} is the Euclidean distance between points j and k and $\mathrm{MAX}(D_{jk})$ is the largest interpoint distance in the distance matrix. The similarity index shown here, S_{jk}, can take values between 0 (most dissimilar) and 1 (most similar) and is just one form of similarity index (see Ref. 51 for other similarity measures). Based on these values of similarity, clusters can be constructed in one of two ways, divisive or agglomerative. As the name implies, the divisive methods start out by assigning all the points to one cluster, which is then subdivided in successive steps until each point forms a cluster of one. The agglomerative techniques operate in the opposite fashion and build up clusters from individual points, forming clusters of clusters until one cluster is formed containing all the points. Just as there are several methods for calculating a similarity index, so there are a number of different ways in which clusters may be defined, e.g., the distance between nearest neighbors and the distance between furthest neighbors.[51]

Although cluster analysis is usually used to produce a display of the data, in the form of a dendrogram as shown in Fig. 4, this may also be thought of as a report on the multidimensional data set. The dendrogram has values of similarity marked on one axis and identifiers for the objects that are being clustered on the other axis. The level of similarity of any two (or more) objects in a given cluster may be obtained by reading off the similarity measure at the top of the cluster where the objects join together. An alternative, but not so easily interpreted, presentation of the results of CA is to choose one or more levels of similarity and compile lists of the members of each cluster that fall within that level. This technique may be applied to the rows of a data set; for example, Hansch *et al.* showed the relationship between substituents using CA.[52] These authors used another method for reporting the results of cluster analysis in that they chose different numbers of clusters, 5, 10, 20, and 60, and listed the substituents belonging to each cluster at the different clustering levels. This is useful in the choice of compounds for synthesis since the selection of substituents from different clusters will ensure the widest possible coverage of physicochemical parameter space in the resulting set of compounds. Alternatively, if a particular substituent is desired but is difficult to include for synthetic reasons, a substitute may be found among other members of the same cluster. The division of the substituents into clusters at four different levels facilitates the choice of compounds for different sizes of training set. Cluster analysis may also be applied to the columns (parameters) of a data set, as shown in Fig. 4. This dendrogram shows the relationships between a set of parameters, calculated by computational chemistry methods, used to describe a series of antimalarial compounds. The solid lines indicate the variables retained after a data reduction procedure (described in the next

[51] P. Willett, "Similarity and Clustering in Chemical Information Systems." Research Studies Press, Wiley, Chichester, 1987.

[52] C. Hansch, S. H. Unger, and A. B. Forsythe, *J. Med. Chem.* **16,** 1217 (1973).

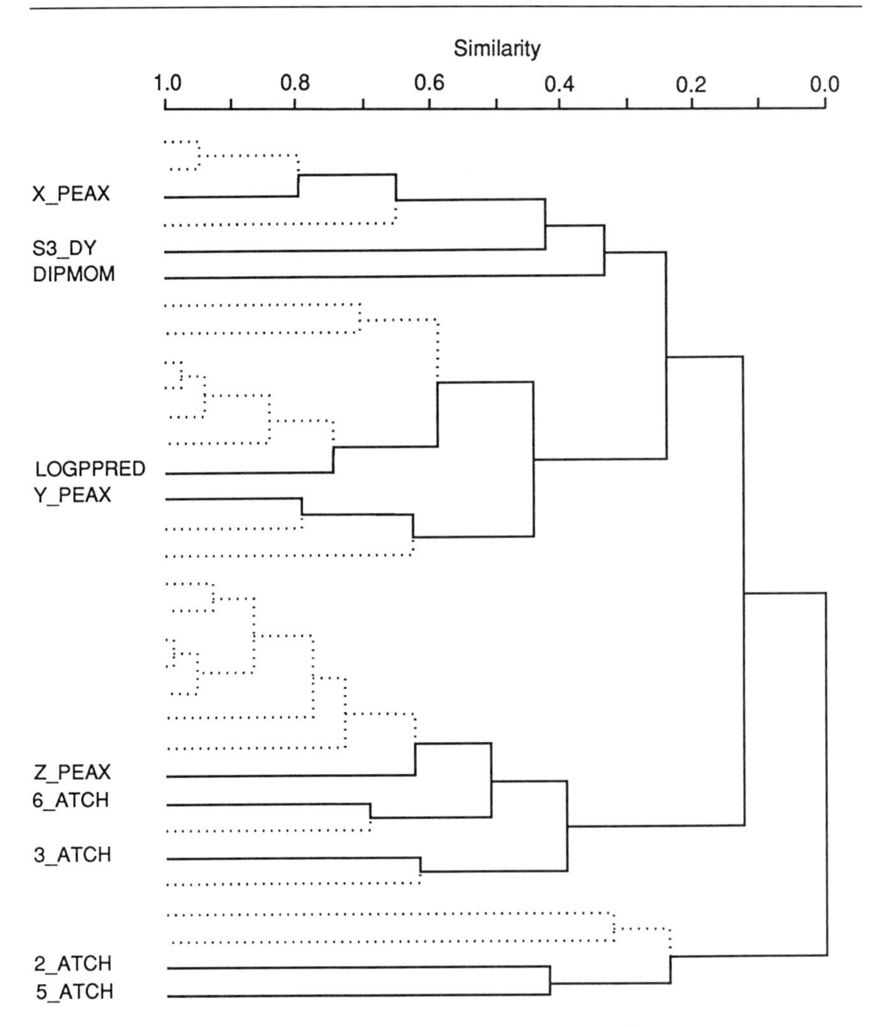

FIG. 4. Dendrogram of parameter associations for a set of antimalarials. For details see text. [Reprinted with permission from *Pestic. Sci.* **27,** 287 (1989); copyright 1989 Society of Chemical Industry.]

section) and the dotted lines show the full clusters of variables that were present in the starting set. Cluster analysis has been applied in drug design to a number of problems, including similarities between substituents[39,52,53] and substituent constants,[49,53] the properties of amino acids,[54] and the

[53] B.-K. Chen, C. Horvath, and J. R. Bertino, *J. Med. Chem.* **22,** 483 (1979).

[54] A. Kidera, Y. Konishi, M. Oka, T. Ooi, and H. Scheraga, *J. Protein Chem.* **4,** 23 (1985).

binding of neuroleptics to a variety of neurotransmitter binding sites.[55] This method has also been used in the matching of dissimilar molecules[56] and studies have been reported of the significance of clusters obtained by CA and other display methods.[57-59]

Data Preprocessing and Data Reduction

Data preprocessing is carried out at a very early stage in an analysis, although a display method may be used to examine the starting data and may reveal unusual data points or parameters. One of the aims of preprocessing is to identify compounds with extreme values of one or more variables, since these will have greatest influence when using many of the techniques. This may be achieved by the consideration of such statistics as the mean, range, and standard deviation, or by the use of various graphical displays. If such compounds are discovered it may point to errors in the calculation or measurement of some of the parameters in the starting set. On the other hand, such data points may be "real" and may indicate a poor choice of compounds for the training set, or a poor choice of descriptor variables, or both.

The question of the choice of compounds for the training set is most important, since this decides the information content of the resulting data set. Various methods have been proposed for this selection,[52,60-62] including a technique that considers synthetic feasibility,[63] and a number of methods have been compared.[64] The concepts of experimental design have also been employed[65-67] and a strategy has been proposed for the toxic hazard ranking of environmentally occurring compounds based on training set choice.[68]

[55] R. Testa, G. Abbiati, R. Ceserani, G. Restelli, A. Vanasia, D. Barone, M. Gobbi, and T. Mennini, *Pharm. Res.* **6,** 571 (1989).

[56] P. M. Dean, P. Callow, and P.-L. Chau, *J. Mol. Graphics* **6,** 28 (1988).

[57] P. Willett, *J. Chem. Inf. Comput. Sci.* **25,** 78 (1985).

[58] J. W. McFarland and D. J. Gans, *J. Med. Chem.* **29,** 505 (1986).

[59] J. W. McFarland and D. J. Gans, *J. Med. Chem.* **30,** 46 (1987).

[60] P. N. Craig, *J. Med. Chem.* **14,** 680 (1971).

[61] R. Wootton, R. Cranfield, G. C. Sheppey, and P. J. Goodford, *J. Med. Chem.* **18,** 607 (1975).

[62] M. L. de Winter, *Eur. J. Med. Chem.* **20,** 175 (1985).

[63] K.-J. Schaper, *Quant. Struct.-Act. Relat.* **2,** 111 (1983).

[64] W. J. Streich, S. Dove, and R. Franke, *J. Med. Chem.* **23,** 1452 (1980).

[65] V. Austel, *Eur. J. Med. Chem.* **17,** 9 (1982).

[66] S. Hellberg, M. Sjöström, B. Skagerberg, and S. Wold, *J. Med. Chem.* **30,** 1126 (1987).

[67] M. Orozco and F. Sanz, *Anal. Quim.* **85,** 34 (1989).

[68] L. Eriksson, J. Jonsson, M. Sjöström, and S. Wold, *Chemom. Intell. Lab. Syst.* **7,** 131 (1989).

Examination of the parameters in a data set is also an important feature of preprocessing. For example, because of the chemical nature of the compounds in a set, it may be found that some physicochemical properties all have more or less the same value. Such descriptors are unlikely to be useful in the explanation of biological properties. Similarly, the nature of the compounds may be such that some physicochemical properties can only take one of two values and so an apparently quantitative descriptor becomes qualitative for that particular set. Another important function of preprocessing is scaling, so as to remove differences in the apparent importance of variables due to differences in the magnitude of their values. Perhaps the most popular form of scaling is autoscaling:

$$X'_{ij} = (X_{ij} - \overline{X}_j)/\sigma_j \tag{4}$$

where X'_{ij} represents the autoscaled value of variable j for row i, X_{ij} is the raw data value, and $\overline{X}_j$ and σ_j are the mean and standard deviation, respectively, for variable j. Autoscaled data have a mean of 0, which makes them less sensitive to outliers, and a standard deviation of 1, which is useful for variance-related methods, such as principal components analysis, since each parameter contributes 1 unit of variance to the total variance.

Data reduction might also be considered a preprocessing step but in fact it is philosophically different from other preprocessing methods. The aim of data reduction is to remove variables from a data set, so as to reduce its complexity, while retaining as much of the information content of the data as possible. One means by which this may be achieved is by examination of the pairwise correlations between variables and the removal of one variable from each pair.[69] Choice of the parameter to remove may be made by counting correlations with other parameters or by examination of the distribution of values for each parameter. The correlation reduction procedure CORCHOP[69] advises the removal of the parameter with the largest number of high correlations (above a set limit) with other parameters. In the case of a tie on correlation count the procedure will select the parameter with the highest kurtosis, in other words the greater deviation, in terms of one aspect of distribution, from a normal distribution. Since many of the pattern recognition methods are nonparametric, i.e., they do not rely on assumptions concerning the statistical distribution of the data, it is not necessary to use variables that conform to a normal distribution. However, the use of kurtosis as a criterion for rejection of correlated variables does help to ensure that the variables retained have well distributed data values.

Figure 4 shows the complete correlation structure for a set of antima-

[69] D. J. Livingstone and E. Rahr, *Quant. Struct.-Act. Relat.* **8**, 103 (1989).

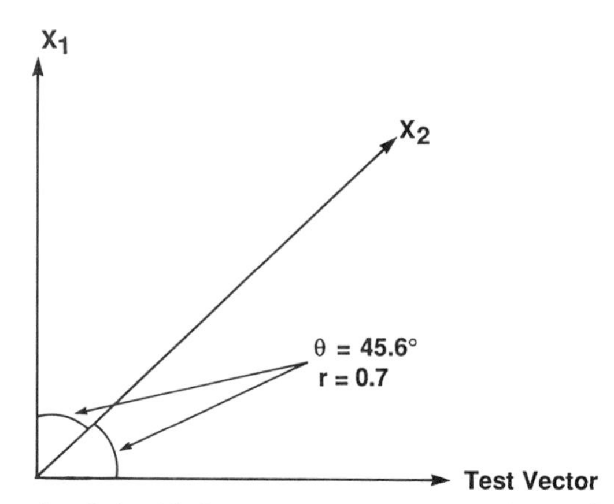

FIG. 5. Geometric relationship between vectors and correlation coefficients. [Reprinted with permission from *Quant. Struct.-Act. Relat.* **8,** 103 (1989); copyright 1989 VCH.]

larials described by 32 physicochemical parameters. The variables retained after processing by CORCHOP are shown as solid lines and it can be seen that a descriptor has been retained to represent most of the clusters. An exception is perhaps the two clusters at the bottom of the diagram where a better choice might have been to select a parameter from the upper cluster rather than both variables from the lower cluster. It is important in data reduction to retain the original variables and a description of their correlation structure, since the parameters that remain are simply representatives of groups of variables in the starting set. If such a nominated descriptor proves to be of use in biological prediction, this may be due to a correlation with one of the original variables in the cluster which it represents. Another reason for the retention of the starting set is that it is possible to remove the "wrong" variable. A reasonable criterion for the removal of correlated variables might be a correlation coefficient of 0.7, which represents just under 50% shared variance between the variables. However, this correlation coefficient also represents an angle between two vectors of just over 45°, as shown in Fig. 5. If the correlation between a biological variable, shown as a test vector in Fig. 5, and variable X_2 is 0.7 then the choice of variable X_1 to represent that pair would result in orthogonality with the biological data.

The removal of variables on the basis of their collinearity not only results in a less complex data set but also improves the properties of the set for use in techniques such as regression.[70] This procedure, however, does

[70] D. W. Salt and M. G. Ford, *in* "Neurotox '88: Molecular Basis of Drug & Pesticide Action" (G. G. Lunt, ed.), p. 469. Elsevier, Amsterdam, 1988.

not address another problem that may occur in data sets, that being multicollinearity. This occurs when a variable is correlated with a linear combination of other variables in the set and may be identified using a technique such as factor analysis (FA). This method is related to principal components but is based on the assumption that the variance in a data set is composed of common variance, which relates to all the variables, and unique variance.[5,9-12,70] The common variance is partitioned into factors and the variables associated with these factors are identified. It has been proposed that FA is a suitable method for the choice of variables and the removal of multicollinearity[71] and its use has been compared with other procedures.[72]

Supervised learning methods may also be used for data reduction by the selection of variables that best "explain" the biological data. An orthogonal feature selection method has been reported[73] and other techniques described.[74,75] The main disadvantage of supervised learning for data reduction is the possibility of chance correlations, as discussed earlier.

Classification Methods

Classification methods are supervised learning techniques, since they use the class membership information in order to produce a classification rule or rules. These methods operate directly in the n-dimensional data space and thus may be more "successful" than data display techniques, or reporting methods, which must necessarily lose some information as they operate. Unsupervised learning may also be used for classification by, for example, including test set compounds in bivariate plots (Fig. 1), principal component plots, and nonlinear maps (Figs. 2 and 3) or in a dendrogram from CA. One classification method, k-nearest neighbors (KNN), makes use of the same matrix of interpoint distances as NLM or CA. Compounds are assigned to membership of a particular class on the basis of the class membership of their k-nearest neighbors in terms of interpoint distances; k is normally taken as an odd number so that a majority vote will decide on the class assignment. This can be illustrated in two dimensions by means of Fig. 1, where X, Y, and Z represent a test set of compounds. On the basis of three nearest neighbors all three compounds will be unambiguously predicted as intermediate, active, and active, respectively. The choice of the

[71] R. Franke, "Theoretical Drug Design Methods," p. 184. Elsevier, Amsterdam, 1984.

[72] M. G. Ford and D. J. Livingstone, *Quant. Struct.-Act. Relat.* **9,** 107 (1990).

[73] B. R. Kowalski and C. F. Bender, *Pattern Recognit.* **8,** 1 (1976).

[74] A. J. Stuper and P. C. Jurs, *J. Am. Chem. Soc.* **97,** 182 (1975).

[75] G. S. Zander, A. J. Stuper, and P. C. Jurs, *Anal. Chem.* **47,** 1085 (1975).

appropriate number for k must be made by examination of the predictions for the training set. A number of studies have been reported of the use of KNN in QSAR.[46,76–80]

A more commonly used classification method is the generation of a linear discriminant function or surface. There are various ways in which such functions can be generated and several names are used for the techniques, such as the linear learning machine,[20] discriminant analysis,[81] adaptive least squares,[82] and others. These methods have some features in common with KNN in that they seek to make classifications of compounds based on their positions in a multidimensional space. This is achieved by the calculation of a surface or hyperplane in the n space so that compounds of one class lie on one side of it and compounds of another class on the other side. This surface is characterized by weighting coefficients for each of the variables in the n-dimensional set thus:

$$W = A_1V_1 + A_2V_2 + \cdots + A_nV_n \tag{5}$$

so that $W < 0$ for compounds on one side of the plane and >0 for compounds on the other side. An example of such a linear decision surface can be seen in Fig. 1, where the surface separates the active compounds from the other two classes. It can also be seen from this figure that it is not possible to generate one such linear surface, which would separate the inactive and intermediate compounds, although a nonlinear surface would. Test set predictions are made by calculation of W for each compound and it can be seen that this surface would predict test compound Y as active and X as intermediate/inactive. Compound Z represents a problem for this type of approach since it lies on the decision surface and thus a prediction would not be possible.

In the use of all classification methods, care must be taken over the ratio of observations (compounds) to dimensions (parameters) since deceptive classifications may occur if this is too small[83–85]; one recommenda-

[76] K. L. H. Ting, R. C. T. Lee, C. L. Chang, and A. M. Guarino, *Comput. Biol. Med.* **4,** 301 (1975).

[77] K. C. Chu, R. J. Feldman, M. B. Shapiro, G. F. Hazard, and R. I. Geran, *J. Med. Chem.* **18,** 539 (1975).

[78] R. Franke, "Theoretical Drug Design Methods" p. 296. Elsevier, Amsterdam, 1984.

[79] D. R. Henry, P. C. Jurs, and W. A. Denny, *J. Med. Chem.* **25,** 899 (1982).

[80] P. C. Jurs, M. N. Hasan, D. R. Henry, T. R. Stouch, and E. K. Whalen-Pedersen, *Fundam. Appl. Toxicol.* **3,** 343 (1983).

[81] Y. C. Martin, J. B. Holland, C. H. Jarboe, and N. Plotnikoff, *J. Med. Chem.* **17,** 409 (1974).

[82] I. Moriguchi, K. Komatsu, and Y. Matsushita, *J. Med. Chem.* **23,** 20 (1980).

[83] N. A. B. Gray, *Anal. Chem.* **48,** 2265 (1976).

[84] C. P. Weisel and J. L. Fasching, *Anal. Chem.* **49,** 2114 (1977).

[85] G. L. Ritter and H. B. Woodruff, *Anal. Chem.* **49,** 2116 (1977).

tion is that this ratio should be ≥ 3.[23] The number of compounds in each class is also an important consideration, since if this is too small, or if there is a large imbalance between the classes, chance effects may also be seen.[24] Unlike regression, where the fit to the data is achieved by least squares, there is not generally a unique solution for a decision surface although for a suitable data set this might lie at 90° to a plane generated by regression.[86] This has a number of implications regarding the stability of the decision surface and the use of discriminant analysis for variable selection.[86] Many applications of these methods have been reported for the analysis of biological activity data,[76–82,86–92] partition coefficients,[93] and the prediction of protein structure from amino acid sequence.[94] The final classification method that will be mentioned here is SIMCA.[95] This technique seeks to describe each activity class separately by generating a principal component model that describes the class. These models can be thought of as "hyperboxes" in the n-dimensional parameter space that enclose all the compounds of a particular class. Test set predictions are made by seeing into which hyperbox or boxes (they may overlap) a test compound falls. This appears a powerful technique and it is claimed that it has a number of advantages over other classification methods.[96,97]

Correlation Techniques

Correlation methods seek to find a relationship in the n-dimensional parameter space that correlates with one or more biological variables. For convenience, the discussion in this section will be limited to applications that involve just a single biological variable. In the same way that the classification techniques aim to achieve their classification directly in the n-dimensional space, correlation methods are designed to find a "fit" in the n space. The most popular correlation method used in drug design is multiple linear regression. This technique has many advantages, including the ability to make quantitative predictions of activity and the possibility of assigning confidence limits to the fit and the predictions. MLR also suffers

[86] T. R. Stouch and P. C. Jurs, *Environ. Health Perspect.* **61,** 329 (1985).
[87] G. Prakash and E. M. Hodnett, *J. Med. Chem.* **21,** 369 (1978).
[88] D. R. Henry and J. H. Block, *J. Med. Chem.* **22,** 465 (1979).
[89] K. Yuta and P. C. Jurs, *J. Med. Chem.* **24,** 241 (1981).
[90] V. K. Gombar, *Arzneim. Forsch.* **35,** 1633 (1985).
[91] S. Nesnow, R. Langenbach, and M. J. Mass, *Environ. Health Perspect.* **61,** 345 (1985).
[92] V. K. Gomber, E. P. Jaeger, and P. C. Jurs, *Quant. Struct.-Act. Relat.* **7,** 225 (1988).
[93] A. J. Harget and D. S. Ellis, *Croat. Chem. Acta* **62,** 449 (1989).
[94] M. Kanehisa, *Protein Eng.* **2,** 87 (1988).
[95] S. Wold, *Pattern Recognit.* **8,** 127 (1976).
[96] W. J. Dunn, S. Wold, and Y. C. Martin, *J. Med. Chem.* **21,** 922 (1978).
[97] W. J. Dunn and S. Wold, *J. Med. Chem.* **23,** 595 (1980).

from drawbacks,[5,70] including the danger of chance correlations, as discussed earlier,[22] and attempts have been made to introduce various "quality" criteria into the use of MLR in drug design.[27,28,98] Other techniques are available for the generation of regression models besides the method of ordinary least squares, although Moriguchi and co-workers concluded that the use of robust methods was not warranted except under extreme circumstances.[99] One particular problem with the MLR approach is seen when there is collinearity or multicollinearity in the data set. When such conditions exist, the order in which variables are considered for inclusion in the regression equation becomes important. Backward stepping regression may give a very different result to forward stepping, for example. The addition or removal of compounds from the set may affect the values of the regression coefficients, and in extreme cases their sign, and attempts at mechanistic interpretation may be thwarted. These problems may be overcome to some extent by the selection of both compounds and parameters in the training set. Another way in which these problems may be averted is by the use of principal components as the independent variables in MLR. The advantage of the use of principal components in this way is that they are orthogonal to one another and thus there are no collinearity effects. For example, Lukovits reexamined[100] the reported pharmacological and quantum chemical data for a set of benzodiazepines.[101] This work showed that log ED_{50} could be "explained" by three principal components, as shown below.

$$\log ED_{50} = -0.25\ (\pm 0.07)X_1 + 0.20\ (\pm 0.08)X_2 + 0.34\ (\pm 0.16)X_3 + 1.34 \tag{6}$$

for $n = 19$, $R^2 = 0.89$, $s = 0.23$, and $F = 39.26^2$;[101a] and where X_1, X_2, and X_3 are the first three principal components derived from the analysis of a set of six quantum chemical descriptors. In another example of the use of MLR and PCA, Kubota and co-workers[102] demonstrated a correlation between the antileukemic activity, $\log(1/C)$ (where C is the optimum concentration for increased life span), of a set of substituted benzoquinones and principal components (Z_n) derived from six physicochemical parameters:

$$\log(1/C) = 0.052Z_1 + 0.072Z_2 + 0.119Z_4 \tag{7}$$

[98] D. W. Osten, *J. Chemom.* **2,** 39 (1988).

[99] I. Moriguchi, K. Komatsu, and Y. Matsushita, *Quant. Struct.-Act. Relat.* **3,** 106 (1984).

[100] I. Lukovits, *J. Med. Chem.* **26,** 1104 (1983).

[101] T. Blair and G. A. Webb, *J. Med. Chem.* **20,** 1206 (1977).

[101a] The statistics quoted here are the number of compounds (n), the square of the multiple correlation coefficient (R^2), the standard error of the estimate (s), and the F statistic (F).

[102] T. Kubota, J. Hanamura, K. Kano, and B. Uno, *Chem. Pharm. Bull.* **33,** 1488 (1985).

for $n = 37$, $r = 0.92$,[102a] $s = 0.25$, and $F = 57.8$. Equation (7) raises an interesting point, that being the question of how to decide how many principal components are "significant." Although there are no "rules," a commonly accepted procedure is to reject principal components that have eigenvalues (which are proportional to the variance they explain) smaller than 1. This seems intuitively reasonable, since it will be remembered that autoscaled data have a variance of 1 and thus an eigenvalue smaller than 1 means that a particular principal component explains less variance than was present in one of the original variables in the data set. The third term in Eq. (7), however, involves a principal component (Z_4, the fourth PC generated by PCA of the descriptor data) with an eigenvalue of 0.499. In this same paper, results are also reported of the analysis of a set of antibacterial quinoline derivatives where two principal components with eigenvalues less than 1 were found to be important in the explanation of the biological data. It is of course possible that a principal component with a small eigenvalue contains just that proportion of the variance of one (or more) of the starting variables that is necessary to explain some facet of the biological data. Work has been reported on the use of cross-validation[103] and procrustes analysis[104] in the choice of dimensionality in PCA.

The combination of MLR and PCA has been shown here as being carried out in two steps. This process can be combined into a single step by the use of a procedure called partial least squares (PLS) regression.[105] This technique generates principal components from the descriptor variables in such a way that their correlation with one or more dependent variables (see the next section) is maximized. The PLS method has been successfully applied to a number of data sets,[18,28,66,68,106–108] although some claims for its utility are perhaps exaggerated. A report of the use of PLS in the analysis of a set of dipeptide inhibitors of angiotensin-converting enzyme (ACE) neatly illustrates the use of this method and also the importance of training set selection.[109] In an initial analysis of the data set, using the results for 58 dipeptides, a 2-component PLS model was found to explain 82% of the

[102a] Quoted as the correlation coefficient.

[103] W. J. Krzanowski, *Biometrics* **43,** 575 (1987).

[104] W. J. Krzanowski, *Appl. Stat.* **36,** 22 (1987).

[105] P. Geladi and B. R. Kowalski, *Anal. Chim. Acta* **185,** 1 (1986).

[106] S. Clementi, G. Coata, C. Ebert, L. Lassiani, P. Linda, S. Hellberg, M. Sjöström, and S. Wold, *in* "QSAR in Drug Design and Toxicology" (D. Hadzi and B. Jerman-Blazic, eds.), p. 19. Elsevier, Amsterdam, 1987.

[107] S. Hellberg, S. Wold, W. J. Dunn, J. Gasteiger, and M. G. Hutchings, *Quant. Struct.-Act. Relat.* **4,** 1 (1985).

[108] J. Johnsson, L. Eriksson, S. Hellberg, M. Sjöström, and S. Wold, *Acta Chem. Scand.* **43,** 286 (1989).

[109] M. L. Tosato, S. Marchini, L. Passerini, A. Pino, L. Eriksson, F. Lindgren, S. Hellberg, J. Jonsson, M. Sjöström, B. Skagerberg, and S. Wold, *Environ. Toxicol. Chem.* **9,** 265 (1990).

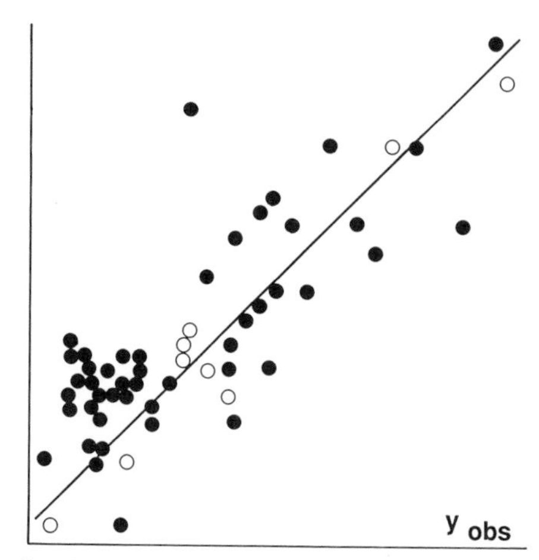

FIG. 6. Plot of predicted vs. observed angiotensin converting enzyme (ACE) inhibition. ○, y_{calc}; ●, y_{pred}. [Reprinted with permission from *Environ. Toxicol. Chem.* **9**, 265 (1990); copyright 1990 Pergamon Press.]

variance in the biological activity. Peptides may be described by a number of physicochemical parameters and it is not clear which properties are of most importance. One approach to this problem is to apply PCA to the descriptor data and to use the resulting principal components as "generalized" peptide parameters.[54,66] Using two such properties as design variables, a fractional factorial design of a training set from the 58 ACE inhibitors resulted in a subset of just 9 dipeptides. A PLS model constructed from this subset had good predictive ability, as shown in Fig. 6, although the fraction of biological variance accounted for by the model was not quoted. In Figure 6 the training set samples are shown as open circles and it can be seen that, in the main, this set is a good choice to span the biological activity space.

Multivariate Dependent Data

There are many situations in which more than one item of biological information is available for each compound, e.g., different strains of bacteria, multiple enzyme assays, multiple pharmacological preparations, and *in vitro* and *in vivo* data. In much the same way that it is often advantageous to treat multiple descriptor data simultaneously, multiple biological data may be handled by pattern recognition techniques. Some fairly obvious applications are the use of PCA or FA to analyze a dependent variable

TABLE III
ROTATED FACTOR LOADINGS FOR 10 BIOLOGICAL RESPONSE VARIABLES

Response variable[a]	Factor 1	Factor 2	Factor 3
KA	−0.93	0.28	0.01
Time to onset	0.84	−0.1	0.15
Time max frequency	0.88	−0.29	−0.1
MTC	0.71	−0.57	0.00
Maximum frequency	0.74	0.39	−0.31
Time to block	0.79	0.04	0.1
Slope	−0.16	0.23	0.93
Intercept	0.21	−0.18	0.94
Maximum burst frequency	0.06	0.90	−0.04
KDA	−0.50	0.72	0.16

[a] KA, Kill activity; KDA, knockdown activity; MTC, minimum threshold concentration for effect; slope and intercept are of the dose response curve.

set[35–38,44] and then to use the resulting components or factors as representative descriptors of the set. An example of this can be seen in the reported factor analysis of *in vivo* insecticidal potencies and *in vitro* neurophysiological responses of a set of pyrethroid analogs.[19] The *in vivo* data consisted of median effective doses for knockdown (KDA) and kill (KA) activity and the *in vitro* responses are as shown in Table III. Three significant factors were identified by this analysis and it can be seen from the factor loadings in Table III that factor 1 is associated with both knockdown and killing activity, whereas factor 2 discriminates KDA and KA.

Reports have demonstrated that the response of different tumor cell lines to platinum complexes may be classified into four groups[110] and that the antitumor activity of 9-aminoacridine derivatives, measured by four biological responses, may be characterized by two principal components.[111]

Multivariate techniques are not restricted to the analysis of one data matrix, either the response or descriptor set, at a time. Methods exist for the *simultaneous* analysis of both biological and physicochemical data. One such technique is canonical correlation analysis (CCA), which calcu-

[110] H. Kuramochi, A. Motegi, S. Maruyama, K. Okamoto, K. Takahashi, O. Kogawa, H. Nowatari, and H. Hayami, *Chem. Pharm. Bull.* **38,** 123 (1990).

[111] Z. Mazerska, J. Mazerski, and A. Ledochowski, *Anti-Cancer Drug Des.* **5,** 169 (1990).

lates a linear combination of q responses:

$$W_1 = a_{11}Y_1 + a_{12}Y_2 + \cdots + a_{1q}Y_q \tag{8}$$

and a linear combination of p descriptors:

$$Z_1 = b_{11}X_1 + b_{12}X_2 + \cdots + b_{1p}X_p \tag{9}$$

So that the pairwise correlation between W_1 and Z_1 is a maximum. Further pairs of canonical variates as described by Eqs. (8) and (9), as many pairs as the smaller of p or q, may be calculated according to the following criteria: (1) they are generated in descending order of correlation coefficient (between W and Z) and (2) successive pairs of variates are orthogonal.

Bordas[112] has compared the use of MLR, PCA, and CCA in the analysis of the inhibitory activity against 5 species of fungi of a set of 26 8-quinolinol derivatives described by 15 physicochemical parameters. Regression analysis of the first principal component derived from the antifungal data against the descriptor data yielded the following equation:

$$\begin{aligned} PC_1 = &-0.17\,(\pm 0.04)MR_x + 1.95\,(\pm 0.40)Es_y \\ &+ 2.68\,(\pm 0.25)(\pi_x + \pi_y) + 1.02 \end{aligned} \tag{10}$$

for $n = 26$, $R = 0.92$, $s = 0.81$, and $F = 40.79$. Regression of the first canonical variate (CV) of the biological set against the descriptor data gave:

$$\begin{aligned} CV_1 = &\ 0.86\,(\pm 0.15)Es_y - 0.66\,(\pm 0.1)B_{4x} \\ &+ 1.10\,(\pm 0.09)(\pi_x + \pi_y) + 1.11 \end{aligned} \tag{11}$$

for $n = 26$, $R = 0.94$, $s = 0.30$, and $F = 56.92$. Both of these equations are better than regression of the average antifungal response and Eq. (11) is a better fit than any of the individual response regressions. It can be seen here that the full canonical correlation was not reported but rather the result of the regression of the response canonical variate against the descriptor variables. This was presumably done in order to aid the interpretation of the correlation and this highlights one of the problems of this type of approach. Successful canonical correlations may be used predictively but the process is not straightforward, since it involves two sets of linear combinations of variables.

As was mentioned in the previous section, the PLS technique is also able to treat simultaneously a multivariate response set and multivariate descriptor set. A report of the analysis of the toxic effects of aromatic hydrocarbons to four aquatic organisms illustrates this.[113] A one-dimen-

[112] B. Bordas, *in* "QSAR Strategies in the Design of Bioactive Compounds" (J. K. Seydel, ed.), p. 389. VCH Press, Weinheim, Germany, 1985.

[113] S. Galassi, M. Mingazzini, L. Vigano, D. Cesareo, and M. L. Tosato, *Ecotoxicol. Environ. Saf.* **16,** 158 (1988).

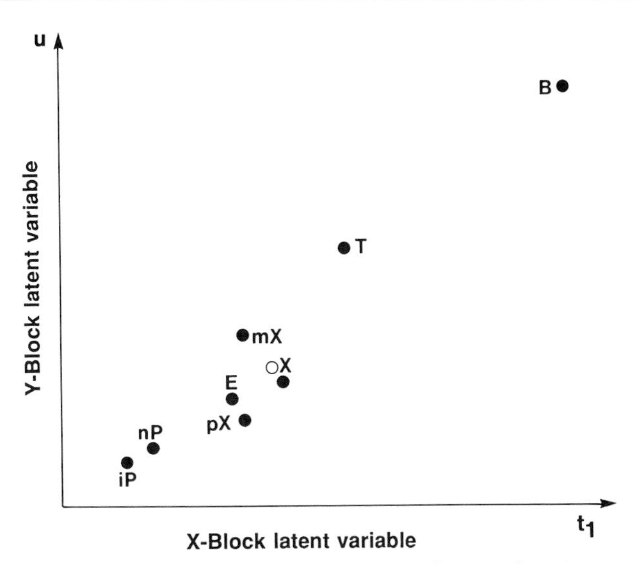

FIG. 7. Plot of the first response latent variable vs. the first descriptor latent variable for a set of aromatic hydrocarbons. [Reprinted with permission from *Ecotoxicol. Environ. Saf.* **16,** 158 (1988); copyright 1988 Academic Press.]

sional PLS model was able to account for 63% of the variance in the biological data set. A plot of the correlation between the first latent Y-block variable (responses) versus the first latent X-block variable (descriptors) is shown in Fig. 7. These latent variables are similar in many respects to canonical variates and, as in CCA, further pairs of latent variables may be generated. Further improvement in the explanation of biological variance was not seen beyond a two-dimensional model and it was shown that the worst prediction of activity for one of the organisms using this model was a mean difference of 0.18 log units. The best prediction gave a mean difference of 0.07 log units.

Summary

Pattern recognition methods have much to offer the drug designer, particularly as the calculation and collation of data, both biological and physicochemical, becomes easier with the widespread use of computer databases, molecular modeling systems, and property prediction packages. Some of the techniques, however, suffer from difficulties in interpretation and the dangers of chance effects have received little attention. The wider use and understanding of these methods is expected to enhance their utility in drug design.

Finally, it should be mentioned here that these methods are becoming applied increasingly in other areas of pharmaceutical research, e.g., the analysis of clinical data,[114-116] and that new techniques for analysis continue to be developed and applied in this field.[117-119]

[114] J. A. Pino, J. E. McMurray, P. C. Jurs, B. K. Lavine, and A. M. Harper, *Anal. Chem.* **57,** 295 (1985).
[115] M. L. Cingolani, L. Re, and L. Rossini, *Pharmacol. Res. Commun.* **17,** 1 (1985).
[116] S. Lanteri, P. Conti, A. Berbellini, G. Centioni, A. Polzonetti, R. Marassi, and S. Zamponi, *J. Chemom.* **3,** 293 (1988).
[117] J.-C. Dore, J. Gilbert, T. Ojasoo, and J.-P. Raynaud, *J. Med. Chem.* **29,** 54 (1986).
[118] E. L. Kinney and D. D. Murphy, *Comput. Biomed. Res.* **20,** 467 (1987).
[119] D. J. Livingstone, G. Hesketh, and D. Clayworth, *J. Mol. Graphics* (in press).

[31] Molecular Electrostatic Potentials for Characterizing Drug–Biosystem Interactions

By PIERRE-ALAIN CARRUPT, NABIL EL TAYAR, ANDERS KARLÉN, and BERNARD TESTA

I. Introduction

No system can exist without a constant flow of information between its constitutive parts to regulate and to integrate their activities. In biological systems, this flow is characterized by a very high intensity and involves two main languages, the electrical and the molecular. In the latter biological language, molecules are the physical support of information, the semantics being the effect on the receptor.[1] Molecular recognition between receptors and molecules such as hormones, pheromones, substrates, and products is thus a *sine qua non* condition for the existence of the molecular language. Drugs, which represent a human attempt to speak the molecular language, are also more or less felicitous supports of information, and more precisely of exogenous information.

Molecular recognition and the converse concept of specificity[2] are explained in mechanistic (and reductionistic?) terms by a stereoelectronic complementarity between the binding molecule and the receptor.[3] In this

[1] E. Schoffeniels, "Anti-Chance," p. 80. Pergamon, Oxford, 1976.
[2] A. Sarai, *J. Theor. Biol.* **140,** 137 (1989).
[3] A. C. T. North, *J. Mol. Graphics* **7,** 67 (1989).

context, it is obvious at the onset that knowledge of the stereoelectronic attributes and properties of binding molecules will contribute significantly to the elucidation of biochemical phenomena,[4] be they elicited by endogenous or exogenous molecules. These are the principle and goal of the present chapter.

A. Binding Recognition and Underlying Intermolecular Forces

The binding of any ligand, be it endogenous or exogenous, to biological macromolecules always involves intermolecular forces whose variety is rather limited and which contribute to both affinity and specificity. Excluding irreversible binding, which results from covalent bonds, the weak intermolecular interactions involved in reversible binding can be categorized into Coulombic and hydrophobic forces.[5] They can also be classified according to their contribution to specificity in that nonspecific, poorly directional interactions such as London forces (instantaneous dipole-induced dipole interactions) and hydrophobic bonds can be contrasted with hydrogen bonds, ionic bonds, and Keesom forces (permanent dipole–permanent dipole interactions), which account for the stereoelectronic fit.[6]

A compilation of weak intermolecular forces contributing to reversible binding is given in Table I. The attractive forces are ranked according to their contribution to specificity, a somewhat arbitrary yet useful scheme. As for the repulsive forces, whose role should neither be forgotten nor underestimated, they may contribute markedly to specificity at the cost of some binding energy. Their role in chiral recognition has, for example, been stressed.[7]

B. Concept of Molecular Structure

A description of molecular structure as informative and complete as possible is an indispensable condition for the successful interpretation of structure–activity relationships. The concept of molecular structure has been analyzed in terms of conceptual levels of structural descriptions to which correspond structural attributes and molecular properties.[8] This approach, as presented in simplified form in Table II, begins with levels that contain comparatively little information and in which molecules are

[4] P. Politzer, P. R. Laurence, and K. Jayasuriya, *Environ. Health Perspect.* **61,** 191 (1985).

[5] G. Naray-Szabo, *J. Mol. Graphics* **7,** 76 (1989).

[6] O. Mekenyan, D. Bonchev, N. Trinajstic, and D. Peitchev, *Arzneim.-Forsch.* **18,** 421 (1986).

[7] B. Testa, *Acta Pharm. Nord.* **2,** 137 (1990).

[8] B. Testa and L. B. Kier, *Med. Res. Rev.* **11,** 35 (1991).

TABLE I
INTERMOLECULAR RECOGNITION FORCES

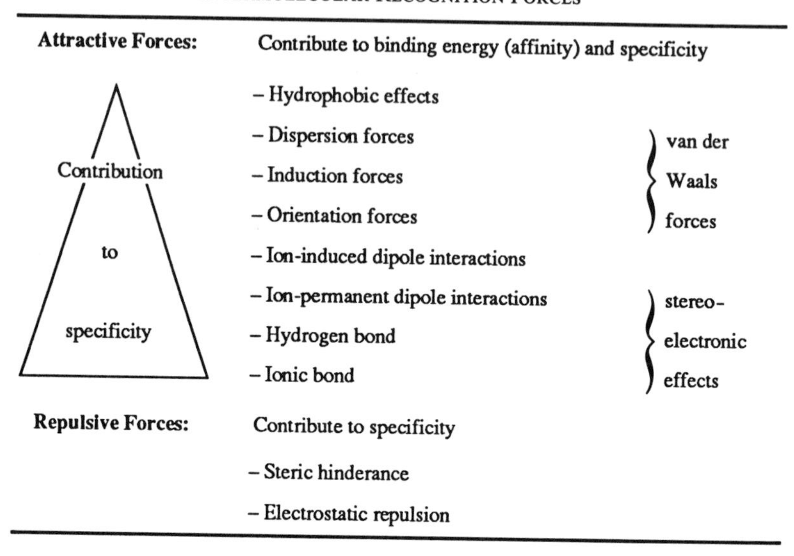

Attractive Forces:	Contribute to binding energy (affinity) and specificity	
Contribution to specificity	– Hydrophobic effects	
	– Dispersion forces	van der Waals forces
	– Induction forces	
	– Orientation forces	
	– Ion-induced dipole interactions	
	– Ion-permanent dipole interactions	stereo-electronic effects
	– Hydrogen bond	
	– Ionic bond	
Repulsive Forces:	Contribute to specificity	
	– Steric hinderance	
	– Electrostatic repulsion	

viewed as abstract, rigid goemetric objects. At the stereoelectronic levels, the description becomes a realistic one, the molecules being seen as dynamic yet lonely entities endowed with electronic properties. It is only at the next level, that of interactions with the molecular environment, that the description becomes complete and able to account for all properties; here, molecules are viewed as "social" entities in that intermolecular interactions are seen as a "social" behavior influencing many properties and permitting the emergence of others.

Two aspects of such a multilevel description must be stressed. The first, as indicated above, is that the information content of the description increases when going from the elementary to the geometric to the stereoelectronic to the interactive levels. Second, each level of description contains in implicit or explicit form the full structural information of the lower levels. For example, lipophilicity contains in implicit form the full geometric and stereoelectronic information of the molecule. In other words lipophilicity is influenced by all geometric and stereoelectronic attributes.

Molecular electrostatic fields (MEFs), and molecular electrostatic potentials (MEPs), which are their visualization, appear in Table II at the stereoelectronic levels of description. This position makes it immediately clear that MEFs are dependent on, and influenced by, such molecular attributes as the geometric, stereodynamic, and stereoelectronic structures.

The geometric, abstract entities of level B cannot generate the intermo-

TABLE II
MULTILEVEL DESCRIPTION OF MOLECULAR STRUCTURE AND PROPERTIES

Conceptual levels of structural description	Structural attributes and molecular properties[a]
A. Elementary level	**Molecular weight**
B. Geometric levels	
2D structure	**Atom connectivity**
	***Z/E* configuration**
3D structure	**Relative configuration**
	Absolute configuration
C. Stereoelectronic levels	Attributes and properties of isolated molecules
Bulk	**Molar volume, surface area**
Stereodynamic structure	**Flexibility, conformation**
	Prototropic equilibria
Stereoelectronic structure	Ionization, **electron distribution,** polarizability, **Molecular electrostatic field, molecular electrostatic potential**
D. Level of intermolecular interactions	Medium-influenced properties of level C
	Emergent properties
	Melting point, boiling point, solvation and hydration, chromatographic properties, (lipophilicity), colligative properties
E. Interactions with biological environment	Biological properties

[a] Structural attributes are set in boldface type, properties are underlined, and attributes/properties are set in boldface type and underlined.

lecular forces presented in Table I. This capacity appears only with the realistic molecules described in level C, and provides the basis for level D. But what is the correspondence between intermolecular forces of Table I and attributes and properties of levels C and D? This question is answered in a schematic manner in Fig. 1,[9] showing that to hydrophobicity, electronic properties, and geometric properties correspond, respectively, hydrophobic bonds plus van der Waals interactions, electrostatic interactions, and the geometric fit.[9] It thus becomes clear that MEPs offer an informative description of the capacity of molecules to generate stereoelectrostatic forces. Yet despite the richness of the structural information they convey, MEPs fail to describe the capacity to generate hydrophobic bonds and dispersion interactions. As such, and this is an important point that must be stressed here, MEPs will reveal only the stereoelectronic compo-

[9] H. van de Waterbeemd and B. Testa, *in* "Advances in Drug Research" (B. Testa, ed.), Vol. 16, p. 85. Academic Press, London, 1987.

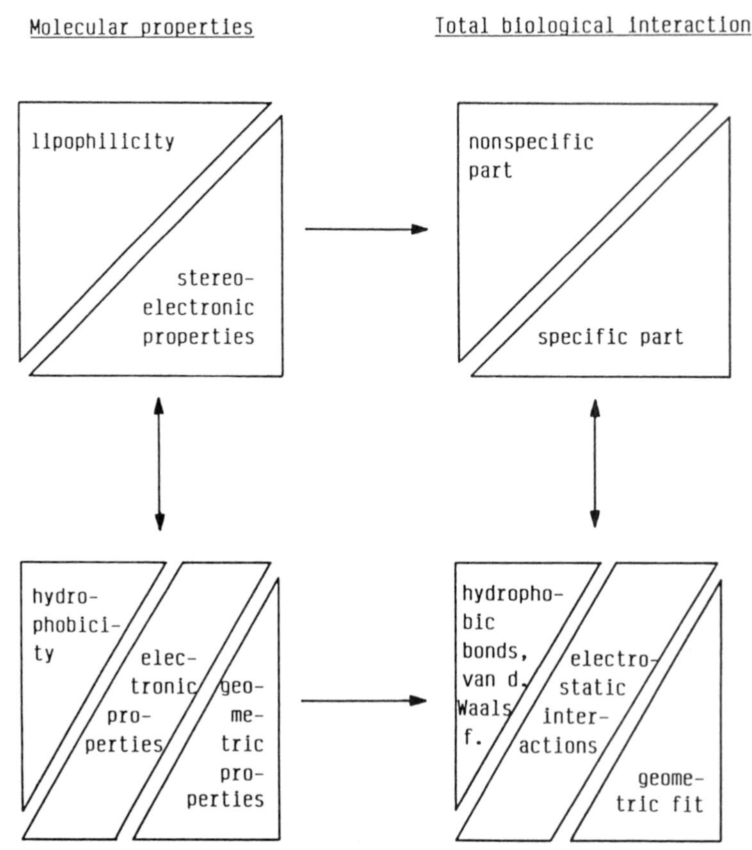

FIG. 1. Schematic illustration of the fundamental analogy between molecular properties and the total interaction between chemical compounds and biological systems.[9]

nents of intermolecular recognition forces. Structure–affinity relationship studies based on MEPs only are thus bound to be limited and partial. This point will be taken up again later.

II. Molecular Electrostatic Potentials as Descriptors of Electronic Structure

A. Calculation of Electronic Structure

To unravel the structure of a molecule and to identify the attributes responsible for its properties have long since been a major challenge to chemistry in its objective to create new compounds endowed with original chemical, physical, or pharmacological properties. Traditionally, to meet this challenge was the exclusive task of spectroscopy. Since about 1975, however, the explosive development of informatics has provided us with numerous tools capable of predicting molecular structure with vastly improved reliability due to a much more rigorous application of the principles of quantum mechanics.[10]

The Born–Oppenheimer approximation, which neglects the motion of nuclei, reduces the full quantum problem to one of electronic structure. In other words, the shape of the surface of potential energy originates in the wave functions and electronic energies of specific geometric structures. The electronic structure at various locations of the surface of potential energy allows the calculation not only of equilibrium geometries but also of nuclear motions (vibrational properties of molecules).

Calculating the electronic structure of molecules allows such molecular properties as dipolar and quadrupolar electric moments, magnetic susceptibility, polarizability, vibration frequencies, and electric field gradients to be approximated. These properties, which are a direct function of the position of atomic nuclei, in turn affect molecular structure.

To obtain the electronic structure of a molecule calls for the resolution of Schrödinger's equation. Thanks to the ever-growing power of computers, techniques of increasing sophistication have been elaborated to approach the exact solution of this equation.[11,12] Although a description of these techniques lies outside the scope of the present chapter, it must be stressed that they allow the computation of wave functions and energies from which the electronic density function $\rho(r)$ is obtained. Molecular electrostatic potential is one of the means used to express the calculated electronic density.

Indeed, the molecular electrostatic potential (MEP) is a genuine molecular property rigorously defined[13] as the energy of Coulombic interaction between a distribution of static (i.e., unperturbated) molecular charge and a unitary positive charge located at point r_1:

$$V(r_1) = \sum_{A} Z_A/|r_1 - R_A| - \int \rho(r)/|r_1 - r|$$

where Z_A is the electric charge of nucleus A located at point R_A. The two

[10] C. E. Dykstra, "*Ab Initio* Calculation of the Structures and Properties of Molecules." Elsevier, New York, 1988.

[11] D. B. Cook, "*Ab Initio* Valence Calculations in Chemistry." Butterworth, London, 1974.

[12] A. Szabo and N. S. Ostlund, "Modern Quantum Chemistry: Introduction to Advanced Electronic Structure Theory." MacMillan, New York, 1982.

[13] P. Politzer and D. G. Truhlar, *in* "Chemical Applications of Atomic and Molecular Electrostatic Potentials" (P. Politzer and D. G. Truhlar, eds.), p. 1. Plenum, New York, 1981.

terms on the right-hand side of the equation represent the nuclear and electronic contributions to the MEP, respectively.

The genuine physical nature of the MEP allows its determination and validation by experimental methods such as X-ray diffraction[14-16] and electronic diffraction.[17-19]

The accurate computation of the MEP requires that the molecular electronic distribution be known beforehand, and the more accurate the latter the closer the MEP will be to physical reality. While the calculation of MEPs is a relatively rapid one, wave functions require important resources in terms of computer capacity and time, especially for the more precise computations at *ab initio* levels. Despite the power of current supercomputers and the development of highly effective methods of computation,[20] obtaining even an approximate electronic structure of large molecules of pharmacological or biological interest by *ab initio* methods is not possible at present within a reasonable time. As a result, a number of alternative methods of calculation have been proposed that, while being comparatively fast, reproduce *ab initio*-computed MEPs with remarkable precision. A few such approaches are presented below.

MEPs have been calculated from wave functions obtained by semiempirical methods. In the CNDO and INDO approximations, interesting results were reported,[21-23] provided that the zero differential overlap (ZDO) approximation was canceled by deorthogonalization of Slater-type orbitals and by inclusion of all integrals in the MEP calculation.[24] A study[25] has shown that MNDO wave functions allow a correct reproduction of relative variations in MEPs generated by even a sophisticated *ab initio* basis set (6-31 G*). While it always overestimated minima, the MNDO

[14] M. A. Spackman and R. F. Stewart, *in* "Chemical Applications of Atomic and Molecular Electrostatic Potentials" (P. Politzer and D. G. Truhlar, eds.), p. 407. Plenum, New York, 1981.

[15] G. Moss and D. Feil, *Acta Crystallogr. Sect. A* **37,** 414 (1981).

[16] G. Moss and P. Coppens, *in* "Chemical Applications of Atomic and Molecular Electrostatic Potentials" (P. Politzer and D. G. Truhlar, eds.), p. 427. Plenum, New York, 1981.

[17] P. Politzer, *J. Chem. Phys.* **72,** 3027 (1980).

[18] M. Fink and R. A. Bonham, *in* "Chemical Applications of Atomic and Molecular Electrostatic Potentials" (P. Politzer and D. G. Truhlar, eds.), p. 93. Plenum, New York, 1981.

[19] S. R. Gadre and R. D. Bendale, *Chem. Phys. Lett.* **98,** 86 (1983).

[20] J. T. Egan and R. D. MacElroy, *J. Comput. Chem.* **5,** 72 (1984).

[21] C. Giessner-Prettre and A. Pullman, *Theor. Chim. Acta* **25,** 83 (1972).

[22] A. J. Duben, *Theor. Chim. Acta* **59,** 81 (1981).

[23] J. C. Culberson and M. C. Zerner, *Chem. Phys. Lett.* **122,** 436 (1985).

[24] A. Chung-Phillips, *J. Comput. Chem.* **10,** 17 (1989).

[25] F. J. Luque, F. Illas, and M. Orozco, *J. Comput. Chem.* **11,** 416 (1990).

results did not display the deficiencies observed with the INDO and CNDO parameterizations.[26,27]

Discrete atomic charges, which are usually derived from a Mulliken population analysis,[28] can also be used to calculate MEPs. This method is extremely fast, especially if based on atomic charges generated by semiempirical methods, but it displays a number of well-known and severe shortcomings. Major approaches used to remove or compensate these deficiencies include the use of atomic charges calibrated for *ab initio* MEPs,[29-34] the addition of fractional charges to Mulliken charges,[35] the incorporation of high-order multipolar moments,[36-40] and the use of diffuse Gaussians in multipolar expansion.[41,42] Yet despite these refinements, MEPs generated from discrete atomic charges are not very reliable near the nuclei.

Other approaches to rapidly calculate the MEP of large molecules are based on the transferability of bond fragments[43-45]; agreement with *ab initio* MEPs is semiquantitative only.[46]

B. Frozen Molecules: Conformational Problem

The MEP, as calculated from the electron density of a molecular system, strongly depends on the relative spatial position of atoms, itself a

[26] G. Naray-Szabo and P. R. Surjan, *in* "Theoretical Chemistry of Biological Systems" (G. Naray-Szabo, ed.), p. 1. Elsevier, Amsterdam, 1986.

[27] J. J. Kaufman, P. C. Hariharan, and F. L. Tobin, *in* "Chemical Applications of Atomic and Molecular Electrostatic Potentials" (P. Politzer and D. G. Truhlar, eds.), p. 335. Plenum, New York, 1981.

[28] R. S. Mulliken, *J. Chem. Phys.* **23,** 1833 (1955).

[29] F. A. Momany, *J. Phys. Chem.* **82,** 592 (1978).

[30] S. R. Cox and D. E. Williams, *J. Comput. Chem.* **2,** 304 (1981).

[31] U. C. Singh and P. A. Kollman, *J. Comput, Chem.* **5,** 129 (1984).

[32] R. Lavery, K. Zakrzewska, and A. Pullman, *J. Comput. Chem.* **5,** 363 (1984).

[33] K. Zakrzewska and A. Pullman, *J. Comput. Chem.* **6,** 265 (1985).

[34] L. E. Chirlian and M. E. Franckl, *J. Comput. Chem.* **8,** 894 (1987).

[35] H. Kubodera, S. Nakagawa, and H. Umeyama, *Chem. Pharm. Bull.* **35,** 1673 (1987).

[36] R. Lavery, C. Etchebest, and A. Pullman, *Chem. Phys. Lett.* **85,** 266 (1982).

[37] C. Etchebest, R. Lavery, and A. Pullman, *Theor. Chim. Acta* **62,** 17 (1982).

[38] J. R. Rabinowitz and S. B. Little, *Int. J. Quant. Chem. Quant. Biol. Symp.* **13,** 9 (1986).

[39] D. E. Williams, *J. Comput. Chem.* **9,** 745 (1988).

[40] M. Berrondo, S. W. Eggleston, and E. G. Larson, *Int. J. Quant. Chem.* **36,** 749 (1989).

[41] G. G. Hall and K. Tsujinaga, *Theor. Chim. Acta* **69,** 425 (1986).

[42] J. R. Rabinowitz and S. B. Little, *Int. J. Quant. Chem. Quant. Biol. Symp.* **22,** 721 (1988).

[43] R. Bonaccorsi, E. Scrocco, and J. Tomasi, *J. Am. Chem. Soc.* **98,** 4049 (1976).

[44] R. Bonaccorsi, E. Scrocco, and J. Tomasi, *J. Am. Chem. Soc.* **99,** 4546 (1977).

[45] G. Naray-Szabo, *Croat. Chem. Acta* **57,** 901 (1984).

[46] G. Naray-Szabo, A. Grofcsik, K. Kosa, M. Kubinyi, and A. Martin, *J. Comput. Chem.* **2,** 58 (1981).

function of the conformation of the molecule. Indeed, each three-dimensional arrangement of atoms in a molecule corresponds to a different electronic distribution and hence a different MEP. This makes it imperative, before computing the electronic properties of a molecule, to define its three-dimensional geometry exactly in terms of absolute configuration and relevant conformation. By relevant conformation, we mean, for example, a low-energy one, or one that is topographically congruent with other active compounds in a series.

While rigid or relatively rigid compounds can exist only in a small number of three-dimensional geometries, flexible compounds can adopt a variety of dynamically interconverting conformations. In many cases, it has been possible to ascribe the pharmacological or biological properties of a compound to a given conformation, termed the active conformation; the latter, however, may not correspond to the global energy minimum in the conformational space of the molecule. Exploring the conformational energy hypersurface is thus necessary to locate spatial arrangements of low energy (local minima and global minimum). Analyzing the electronic structure of these conformers allows one to refine the three-dimensional description of a molecule and to better assess common stereoelectronic features in a series of compounds of comparable pharmacological activity.

Theoretical conformational analysis has witnessed significant progress; no longer restricted to the systematic search of conformers generated by rotation around acyclic bonds,[47] it has been enriched by elegant tools allowing analysis of cyclic compounds and restriction of the conformational space when the latter is too complex for systematic exploration. Modern theoretical tools include systematic search strategies optimized for cyclic compounds,[48-51] artificial intelligence,[52] Monte Carlo methods,[53] distance geometry calculations,[54-56] and the ellipsoid algorithm.[57-59]

[47] B. Pullman, "Quantum Mechanics of Molecular Conformations" (B. Pullman, ed.), Wiley, New York, 1976.

[48] N. C. Cohen, P. Colin, and G. Lemoine, *Tetrahedron* **37,** 1711 (1981).

[49] G. M. Smith and D. F. Veber, *Biochem. Biophys. Res. Commun.* **134,** 907 (1986).

[50] P. R. Gerber, K. Gubernator, and K. Müller, *Helv. Chim. Acta* **71,** 1429 (1988).

[51] R. A. Dammkoehler, S. F. Darasek, and E. F. Berkely Shands, *J. Comput.-Aided Mol. Des.* **3,** 3 (1989).

[52] D. P. Dolata, A. R. Leach, and K. Prout, *J. Comput.-Aided Mol. Des.* **1,** 73 (1987).

[53] G. Chang, W. C. Still, and W. C. Guida, *J. Am. Chem. Soc.* **111,** 4379 (1989).

[54] J. C. Wenger and D. H. Smith, *J. Chem. Inf. Comput. Sci.* **22,** 29 (1982).

[55] P. K. Weiner, S. Profeta, Jr., G. Wipff, T. Havel, I. D. Kuntz, R. Langridge, and P. A. Kollman, *Tetrahedron* **39,** 1113 (1983).

[56] M. G. Crippen and T. F. Havel, "Distance Geometry and Molecular Conformation" (D. Bawden, ed.), Research Studies Press, Wiley, New York, 1988.

[57] M. Billeter, T. F. Havel, and K. Wutrich, *J. Comput. Chem.* **8,** 132 (1987).

[58] M. Billeter, T. F. Havel, and I. D. Kuntz, *Biopoplymers* **26,** 777 (1987).

C. Lonely Molecules: Solvent Problem

In a first approximation, the conformational analysis of an isolated molecule takes only intramolecular forces into account. However, molecules of pharmacological or biological interest never act in the ideal gaseous state: the many solvent molecules that surround them influence not only their conformation but also their electronic distribution. For a realistic description of molecules, stereoelectronic calculations increasingly take into explicit account intermolecular interactions such as electrostatic, dispersive, and hydrophobic forces. Simulating the influence of solvent on potential energy surfaces[60,61] and complex biochemical phenomena[62] is currently a privileged task of molecular dynamics.

D. Molecular Electrostatic Potentials for Describing Chemical Reactivity

MEPs have been a major tool in the stereoelectronic study of even complex molecules, as documented in the remainder of this chapter. In contrast, little use has been made of MEPs to investigate even simple chemical reactions. Examples of studies include regioselectivity of protonation reactions,[63–65] basicity of amines,[66] selectivity of electrophilic additions to allylic double bonds,[67,68] nucleophilic attack on aromatic compounds,[69] and nucleophilic addition.[70] The limited success in applying the MEP approach to simple chemical reactions is mainly due to the partiality

[59] M. Billeter, A. E. Howard, I. D. Kuntz, and P. A. Kollman, *J. Am. Chem. Soc.* **110,** 8385 (1988).

[60] J. L. Rivail, *in* "New Theoretical Concepts for Understanding Organic Reactions" (J. Bertan and I. G. Csizmadia, eds.), p. 219. Kluwer Academic Publishers, Dordrecht, The Netherlands, 1989.

[61] J. Bertran, *in* "New Theoretical Concepts for Understanding Organic Reactions" (J. Bertan and I. G. Csizmadia, eds.), p. 231. Kluwer Academic Publishers, Dordrecht, The Netherlands, 1989.

[62] W. F. van Gunsteren and P. K. Weiner, "Computer Simulations of Biomolecular Systems" (W. F. van Gunsteren and P. K. Weiner, eds.), ESCOM, Leiden, The Netherlands 1989.

[63] G. Naray-Szabo, *Acta Phys. Acad. Sci. Hung.* **51,** 65 (1981).

[64] K. Osapay and G. Naray-Szabo, *J. Mol. Struc. (THEOCHEM)* **92,** 57 (1983).

[65] P. Politzer, I. M. Elminyawi, P. Lane, and L. Robert, Jr., *J. Mol. Struc. (THEOCHEM)* **60,** 117 (1989).

[66] P. Nagy, K. Novak, and G. Szasz, *J. Mol. Struc. (THEOCHEM)* **60,** 257 (1989).

[67] S. D. Kahn, C. F. Pau, A. R. Chamberlin, and W. J. Hehre, *J. Am. Chem. Soc.* **109,** 650 (1987).

[68] S. D. Kahn and W. J. Hehre, *J. Am. Chem. Soc.* **109,** 666 (1987).

[69] P. Politzer and R. Bar-Adon, *J. Phys. Chem.* **91,** 2069 (1987).

[70] P. Sjoberg and P. Politzer, *J. Phys. Chem.* **94,** 3959 (1990).

of the model: The electrostatic term alone cannot describe chemical reactions in their dynamic complexity, the influence of polarization and charge transfer being even more important.

III. Biochemical Applications of Molecular Electrostatic Potential

The MEP approach has provided invaluable information on the stereoelectronic character of drugs, as discussed in Sections IV–VI. In addition, a number of other bioactive molecules and macromolecules have been the objects of such studies, sometimes leading to insightful results. A number of papers selected for their representativeness, variety, and pharmacological relevance are briefly discussed below; the reader is referred to the original publications for full details.

A. Some Biological Molecules

Endogenous compounds such as neurotransmitters and hormones have been investigated by the MEP approach in an effort to better understand the structural factors accounting for their biological activities. Thus, the MEP maps of acetylcholine and analogs,[71,72] dopamine and analogs,[72,73] noradrenaline and analogs,[74,75] and serotonin and analogs[76] have allowed a better view of their mechanism of action, as developed in Section VI.

Kothekar and colleagues[77–79] have compared the MEP maps of several types of prostaglandins, i.e., $PGF_{2\alpha}$, $PGF_{1\beta}$, PGA_1, PGE_1, and PGE_2. It was found that $PGF_{2\alpha}$, the most potent analog, had the lowest MEP minimum on the ring fragment. The MEPs of vasopressin analogs have been reported,[80] revealing relationships with the biological activity of these compounds. The MEPs of guanosine 3′,5′-cyclic monophosphate (cGMP), a second messenger in various receptor-mediated events, and its C-6-substituted analogs have provided evidence that electrostatic interactions are

[71] B. Odell, *J. Comp.-Aided Mol. Des.* **2,** 191 (1988).
[72] H. Kubodera and H. Umeyama, *Chem. Pharm. Bull.* **35,** 3087 (1987).
[73] H. van de Waterbeemd, P. A. Carrupt, and B. Testa, *Helv. Chim. Acta* **68,** 715 (1985).
[74] D. Hadzi, D. Kocjan, and T. Solmajer, *Period. Biol.* **83,** 13 (1981).
[75] N. El Tayar, P.-A. Carrupt, H. van de Waterbeemd, and B. Testa, *J. Med. Chem.* **31,** 2072 (1988).
[76] L. E. Arvidsson, A. Karlén, U. Norinder, L. Kenne, S. Sundell, and U. Hacksell, *J. Med. Chem.* **31,** 212 (1988).
[77] V. Kothekar and A. K. Srivastava, *J. Theor. Biol.* **93,** 7 (1981).
[78] V. Kothekar and S. Kailash, *J. Theor. Biol.* **93,** 25 (1981).
[79] V. Kothekar and S. Dutta, *Indian J. Biochem. Biophys.* **19,** 266 (1982).
[80] A. Liwo, A. Tempczyk, and Z. Grzonka, *J. Comput.-Aided Mol. Des.* **3,** 261 (1989).

responsible for the relative stability of the syn-conformers in this series of compounds.[81]

B. Peptides and Proteins

A number of studies have shown that electrostatic interactions contribute largely to the three-dimensional structure and reactivity of peptides and proteins.[82,83] A telling study showing the importance of electrostatic potentials in protein–protein interactions is that of Geysen and co-workers.[84] In this work, the process of antigen–antibody recognition was examined using electrostatic potential calculations of the antigen protein myohemerythrin. The results have clearly demonstrated that the sites most frequently recognized in the antigen molecule correspond to regions of strong negative potential and that the first, molecular steps in the immune response are sensitive to electrostatic forces.

C. Nucleic Acids

Nucleic acids as highly charged polyanions interact selectively with positively charged molecules, a reaction believed to account for their interaction with some carcinogenic and mutagenic chemicals. This evidence suggests that the electrostatic properties of nucleic acids should play an important role in their reactivity. Although computing and describing the electrostatic potential of nucleic acids was considered an all but impossible task, B. Pullman, A. Pullman, and collaborators[85,86] have developed a strategy to overcome this problem. What they did was to expand the computations progressively from the "simple" monomeric units to the addition of their MEPs to the macromolecules themselves. Figure 2 illustrates the MEP maps of the fundamental purine and pyrimidine bases of nucleic acids.

Although pharmacological applications will be discussed in Section IV, we note here the usefulness of the MEP technique in examining biological properties of nucleic acids. In another study,[87] the MEPs of 8-hydroxyguanine and guanine were compared to understand the molecular mecha-

[81] S. Topiol, T. K. Morgan, Jr., M. Sabio, and W. C. Lumma, Jr., *J. Am. Chem. Soc.* **112,** 1452 (1990).

[82] D. G. Wallace, *Biopolymers* **29,** 1015 (1990).

[83] A. Pullman and C. Etchebest, *FEBS Lett.* **163,** 199 (1983).

[84] H. M. Geysen, J. A. Tainer, S. J. Rodda, T. J. Mason, H. Alexander, E. D. Getzoff, and R. A. Lerner, *Science* **235,** 1184 (1987).

[85] B. Pullman and A. Pullman, *Stud. Biophys.* **86,** 95 (1981).

[86] A. Pullman and B. Pullman, *Q. Rev. Biophys.* **14,** 289 (1981).

[87] M. Aida and S. Nishimura, *Mutat. Res.* **192,** 83 (1987).

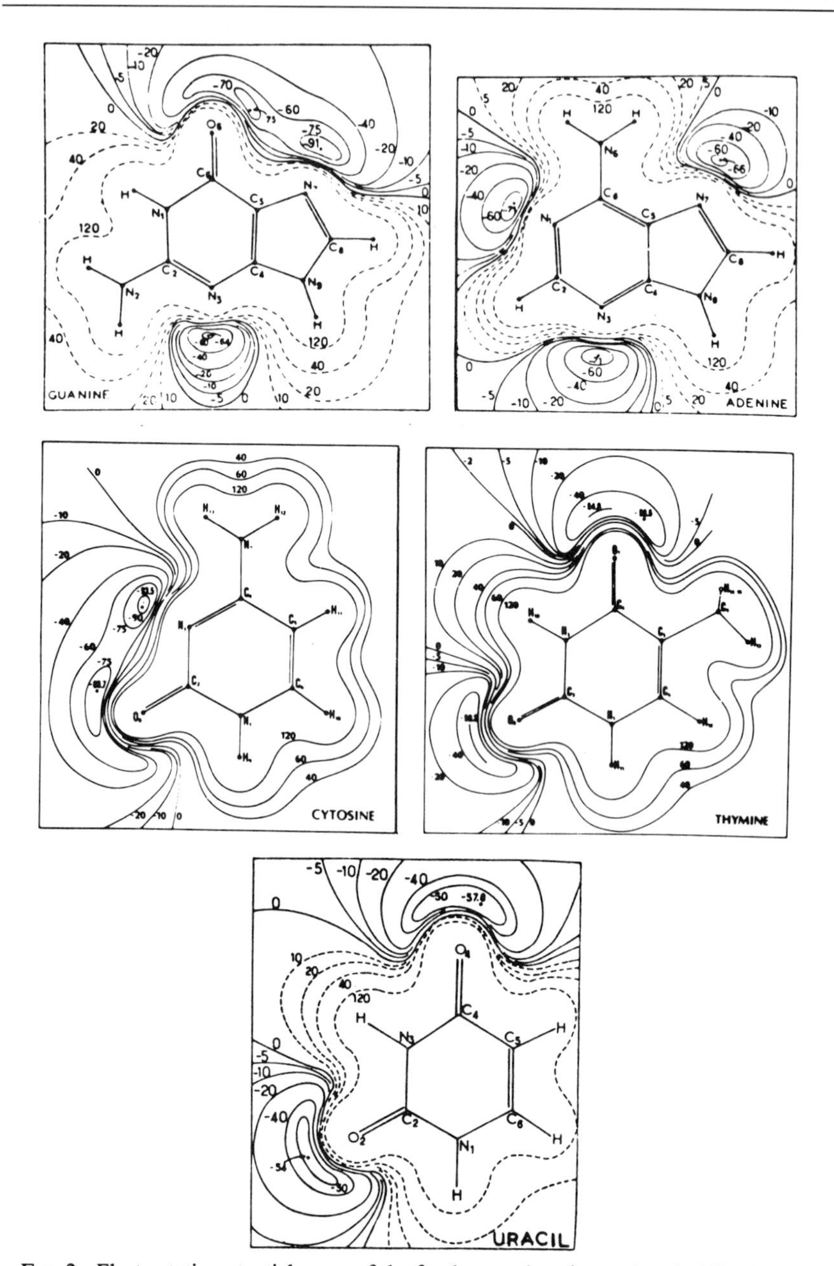

FIG. 2. Electrostatic potential maps of the fundamental purine and pyrimidine bases of nucleic acids in the plane of the bases. Isopotential curves in kilocalories per mole. Full lines, negative potentials; dotted lines, positive potentials. Reproduced from Ref. 86 with the permission of Cambridge University Press.

nisms by which an 8-hydroxyguanine residue could affect the fidelity of DNA replication. The results showed that the addition of an oxygen atom in the 8-position of guanine changes the electrostatic potential of the entire molecule and markedly perturbates the local structure of 8-hydroxyguanine-containing DNA, preventing normal recognition and leading to erroneous DNA replication. Based on MEP calculations, it was also shown that the stacking interactions between nucleic acid bases, a major structural determinant of DNA, are determined by electrostatic forces.[88] Of course, other forces and factors also influence the conformation and reactivity of nucleic acids, e.g., charge transfer and dispersion energies, steric factors, and counterions and solvent.[89]

D. Enzymatic Mechanisms

Several studies have established that complementary electrostatic interactions play an important role in facilitating electron-transfer reactions between enzymes. Theoretical studies of electron-transfer reactions between flavodoxin and *c*-type cytochromes[90,91] have suggested that the location and magnitude of electrostatic potential maxima and minima on protein surfaces control the specificity of the interaction between the reaction partners. In addition, it was shown that optimum reactivity is achieved when the reactant proteins have regions of opposite electrostatic potential adjacent to the sites of electron transfer.

The interaction between cytochrome *c* and a modified cytochrome b_5 is another attractive example demonstrating the important role of electrostatic forces in facilitating the formation of a productive protein–protein electron-transfer complex.[92] The three-dimensional MEPs of both oxidized and reduced forms of cytochrome *c*-551 indicate that the apoprotein modifies the electrostatic potential of the heme edge, resulting in a rearrangement of surrounding water molecules in the reduced form of the enzyme.[93]

The NADH dinucleotide is known to be transferred directly between dehydrogenases of opposite stereospecificity, i.e., A and B dehydrogenases.[94] A calculation of the MEP surface of the NADH-binding site of three

[88] C. Nagata and M. Aida, *J. Mol. Struct. (THEOCHEM)* **179,** 451 (1988).

[89] B. Pullman, R. Lavery, and A. Pullman, *Eur. J. Biochem.* **124,** 229 (1982).

[90] P. C. Weber and G. Tollin, *J. Biol. Chem.* **260,** 5568 (1985).

[91] G. Cheddar, T. E. Meyer, M. A. Cusanovich, C. D. Stout, and G. Tollin, *Biochemistry* **25,** 6502 (1986).

[92] M. R. Mauk, A. G. Mauk, P. C. Weber, and J. B. Matthew, *Biochemistry* **25,** 7085 (1986).

[93] G. Pèpe, B. Serres, D. Laporte, G. Del Re, and C. Minichino, *J. Theor. Biol.* **115,** 571 (1985).

[94] D. K. Srivastava and S. A. Bernhard, *Biochemistry* **23,** 4538 (1984).

dehydrogenases of known structure has revealed large areas of negative potential in the two A dehydrogenases, and large areas of positive potential in the B dehydrogenase.[95] These calculations explain the stereospecificity of NADH transfer between A and B dehydrogenases.

IV. Drug–DNA Recognition

The understanding of the structural factors governing molecular recognition by nucleic acids, and the elucidation of molecular mechanisms of action, are among the most important themes of research in the field of drug–DNA interactions. Much attention has been devoted to a large series of nonintercalating groove-binding antitumor agents that exhibit a remarkable specificity to a given base sequence in one of the grooves of B-DNA.[86,96] Netropsin (Fig. 3), an antibiotic with antitumor and antiviral activities, binds in a highly specific manner to the minor groove of (A-T)-rich sequences of B-DNA.[97] Until recently, it was generally accepted that the ability of the charged ends of netropsin and analogs to form hydrogen bonds was essential to the above-described specificity.[98,99] This has now proved to be incorrect, since SN 18071 (Fig. 3), a bisquaternary ammonium heterocycle that cannot form hydrogen bonds, also binds to B-DNA and shows a similar A-T minor groove specificity.[100] Based on comparative computations of the binding energy of netropsin and SN 18071 to B-DNA (taking into account the electrostatic and the steric components and allowing for conformational adaptability), Zakrzeska and co-workers[101] demonstrated that the formation of hydrogen bonds is not a condition for binding to B-DNA or for the A-T minor groove specificity. The same calculations also showed that these interactions are governed to a large extent by steric fit and the highly negative electrostatic potential generated in the minor groove by A-T sequences.

The significance of electrostatic potentials was also evidenced in the base sequence and groove specificity of 9-aminoacridine-4-carboxamide, a new intercalating antitumor drug (Fig. 3). Theoretical computations of the intercalative interactions of the dicationic form of 9-aminoacridine-4-car-

[95] D. K. Srivastava, S. A. Bernhard, R. Langridge, and J. A. McClarin, *Biochemistry* **24,** 629 (1985).

[96] B. Pullman, *in* "Electrostatic and Specificity in Nucleic Acid Reactions" (C. Chagas and B. Pullman, eds.), Vol. 55, p. 1. Pontificae Academiae Scientiarum Scripta Varia, Rome, 1984.

[97] C. Zimmer, G. Luck, and G. Burckhardt, *Stud. Biophys.* **104,** 247 (1984).

[98] P. K. Weiner, R. Langridge, J. M. Blaney, R. Schaefer, and P. A. Kollman, *Proc. Natl. Acad. Sci. U.S.A.* **79,** 3754 (1982).

[99] D. J. Patel, *Proc. Natl. Acad. Sci. U.S.A.* **79,** 6424 (1982).

[100] B. C. Baguley, *Mol. Cell. Biochem.* **43,** 167 (1982).

[101] K. Zakrzewska, R. Lavery, and B. Pullman, *Nucleic Acids Res.* **11,** 8825 (1983).

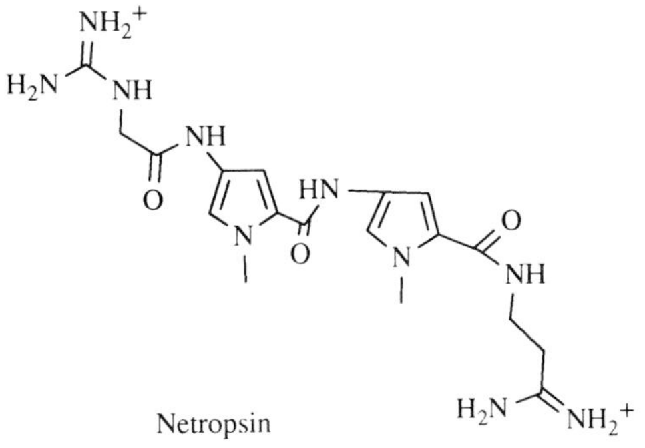

Netropsin

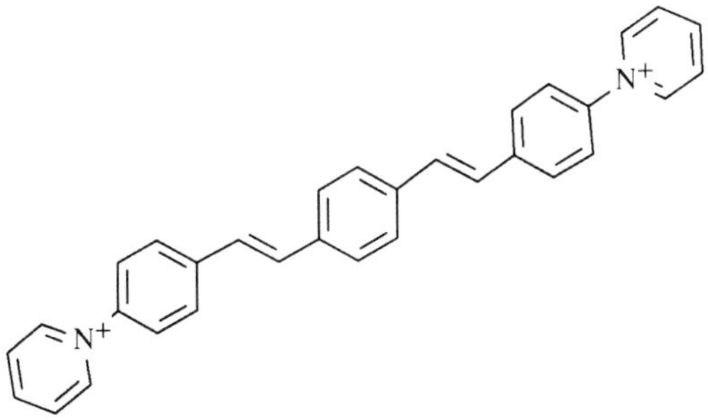

SN 18071

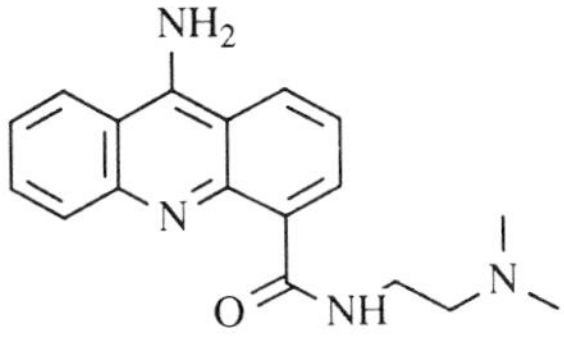

9-Aminoacridine-4-carboxamide

Fig. 3. Chemical structure of netropsin, SN 18071, and 9-aminoacridine-4-carboxamide.

boxamide with AT and GC sequences in B-DNA clearly demonstrated the GC specificity of the intercalating drug; in addition, the side chain is preferentially located in the major groove of these sequences, a selectivity determined essentially by the electrostatic term of the interaction energy. The interested reader is referred to an excellent chapter by Pullman,[102] in which DNA–antitumor drug interactions are comprehensively reviewed.

[102] B. Pullman, *in* "Advances in Drug Research" (B. Testa, ed.), Vol. 18, p. 1. Academic Press, London, 1989.

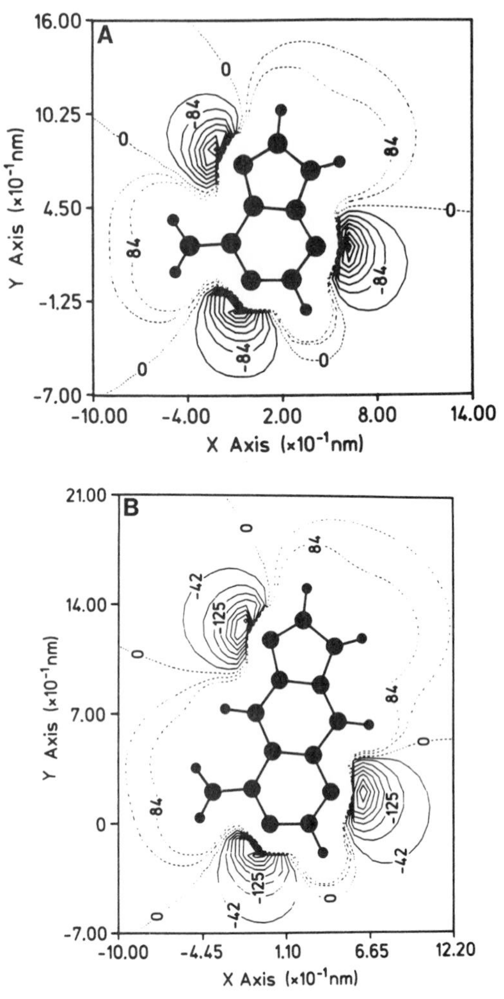

FIG. 4. *Ab initio* molecular electrostatic potentials (MEPs) (in the molecular plane) of (A) adenine, (B) *lin*-benzoadenine, and (C) *prox*-benzoadenine. Reproduced from Ref. 105 with the permission of Springer-Verlag, New York.

V. Drug–Enzyme Interactions

Several studies have suggested that electrostatic properties of both enzyme and ligand determine long-range interactions in diffusion-controlled reactions,[103,104] and here again the MEP approach has proved its useful-

[103] S. Allison, G. Ganti, and J. McCammon, *J. Phys. Chem.* **89,** 3899 (1985).
[104] T. Head-Gordon and C. L. Brooks III, *J. Phys. Chem.* **91,** 3342 (1987).

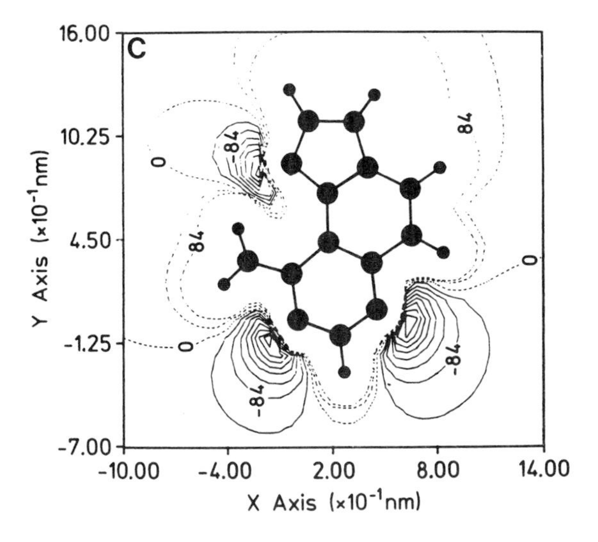

FIG. 4C.

ness. *Ab initio* MEP calculations of adenine and benzoadenine derivatives as substrates of adenosine deaminase have demonstrated the close stereoelectronic analogies between these molecules (Fig. 4[105]) and confirmed the existence of nonlinear hydrogen bonds inside the active site of the enzyme.[105] A coherent explanation of the substrate properties of benzoadenine derivatives has thus been offered.

Another study[106] reports the MEPs of a series of substrates, products, and inhibitors of dihydrofolate reductase (DHFR). These data were used in conjunction with the energy calculations and the crystallographically determined enzyme structure to elucidate the structural basis for the different modes of binding the DHFR substrates versus inhibitors. The results showed that the differences in electrostatic potentials between inhibitors and substrates provide a simple explanation for their inverted binding modes.

In order to design new inhibitors of papain, a protease containing a catalytic cysteinyl residue in the active site, Akahane and Umeyama[107] calculated the MEPs of papain (whose tertiary structure is known) and of its irreversible peptide inhibitor benzyloxycarbonyl-L-phenylalanyl-L-alanine chloromethyl ketone. The results illustrated the high electrostatic complementarity between enzyme and inhibitor.

[105] M. Orozco, E. I. Canela, and R. Franco, *Eur. J. Biochem.* **188,** 155 (1990).

[106] P. R. Andrews, M. Sadek, M. J. Spark, and D. A. Winkler, *J. Med. Chem.* **29,** 698 (1986).

[107] K. Akahane and H. Umeyama, *Enzyme* **36,** 141 (1986).

Several researchers have attempted to understand the nature of enzymatic reaction mechanisms by designing and preparing artifical enzymes that mimic the behavior of a particular type of enzyme.[108-110] Using cyclic ureas as mimics of the serine-195 residue in α-chymotrypsin, a reliable computational protocol was developed to examine the enzymatic features accounting for ligand recognition.[111] The relative contribution of electrostatic interactions to molecular recognition was thus assessed. For example, comparing the MEP patterns of the active site of α-chymotrypsin with that of its cyclic urea mimics showed how electrostatic interactions determine substrate binding and influence the hydrolytic mechanism of the enzyme. Molecular models were also computed to represent the main functional properties of the active site of carboxypeptidase[112,113]; quantum mechanical calculations of the interaction between the enzyme model and substrates or inhibitors again indicated a close relationship between affinity/reactivity and MEPs. In addition, the MEPs also accounted for the relative orientation of enzymatic reactive sites and substrate or inhibitor target sites in the ligand–enzyme complex. These examples eloquently document the significance of electrostatic interactions in the molecular recognition between an enzyme and its ligands, be they substrates or inhibitors. The comprehensive review of Venanzi[111] affords further examples and in-depth discussions.

VI. Drug–Receptor Interactions

MEP computations of agonists and antagonists have been successfully applied to the topographical analysis of a number of pharmacological receptors. Representative examples are compiled in Table III[114-124]; the latter is far from being complete but simply aims at guiding the reader to some other receptors in addition to the ones discussed below.

[108] D. J. Cram, I. B. Dicker, C. B. Knobler, and K. N. Trueblood, *J. Am. Chem. Soc.* **106,** 7150 (1984).
[109] G. L. Trainor and R. Breslow, *J. Am. Chem. Soc.* **103,** 158 (1981).
[110] C. A. Venanzi and J. D. Bunce, *Int. J. Quant. Chem.* **12,** 69 (1986).
[111] C. A. Venanzi, *in* "Environmental Influences and Recognition in Enzyme Chemistry" (J. F. Liebman and A. Greenberg, eds.), p. 251. VCH Publisher, New York, 1988.
[112] H. Weinstein, R. Osman, S. Topiol, and C. A. Venanzi, *in* "Quantative Approaches to Drug Design" (J. C. Dearden, ed.), p. 81. Elsevier, Amsterdam, 1983.
[113] M. N. Liebman, C. A. Venanzi, and H. Weinstein, *Biopolymers* **24,** 1721 (1985).
[114] P. J. M. van Galen, H. W. T. van Vlijmen, A. P. I. Jzerman, and W. Soudijn, *J. Med. Chem.* **33,** 1708 (1990).
[115] G. Loew, J. R. Nienow, and M. Poulsen, *Mol. Pharmacol.* **26,** 19 (1984).
[116] S. Guha and D. Majumdar, *Int. J. Quant. Chem. Quant. Biol. Symp.* **13,** 19 (1986).

TABLE III
EXAMPLES OF A FEW ADDITIONAL RECEPTORS WHOSE LIGANDS HAVE BEEN STUDIED BY MEP APPROACH

Receptors	Ref.
Adenosine A_1	114
Benzodiazepine	115
γ-Aminobutyric acid (GABA)	116
Histamine H_2	117–119
Opiate	120–124

A. Ligands of Dopamine D_2 Receptors

Central and peripheral dopamine (DA) receptors can be divided into D_1 and D_2 types based on their ability to stimulate (D_1)[125] or inhibit (D_2)[126] the DA-sensitive adenylate cyclase.

The D_2-receptor antagonists belonging to the class of substituted benzamides (orthopramides) display a wide variety of interesting pharmacological properties, such as antipsychotic and antiemetic effects.[127] These drugs are characterized by marked structural and pharmacological differences as compared to "classical" neuroleptics (e.g., phenothiazine and butyrophenone derivatives).[127] To better understand how orthopramides interact with D_2 receptors, the conformational, electrostatic, and physicochemical properties of these agents have been studied by several research groups.

It was shown by X-ray crystallography,[128,129] conformational calcula-

[117] A. Donetti, G. Trummlitz, G. Bietti, E. Cereda, C. Bazzano, and H. U. Wagner, *Arzneim.-Forsch.* **35,** 306 (1985).

[118] A. P. Mazurek, R. Osman, and H. Weinstein, *Mol. Pharmacol.* **31,** 345 (1987).

[119] F. J. Luque, F. Sanz, F. Illas, R. Pouplana, and Y. G. Smeyers, *Eur. J. Med. Chem.* **23,** 7 (1988).

[120] G. Loew, C. Keys, B. Luke, W. Pollgar, and L. Toll, *Mol. Pharmacol.* **29,** 546 (1986).

[121] T. P. Lybrand, P. A. Kollman, V. C. Yu, and W. Sadée, *Pharm. Res.* **3,** 218 (1986).

[122] B. V. Cheney, *J. Med. Chem.* **31,** 521 (1988).

[123] D. T. Manallack, M. G. Wong, M. Costa, P. A. Andrews, and P. M. Beart, *Mol. Pharmacol.* **34,** 863 (1988).

[124] A. Grassi, G. C. Pappalardo, and A. Marletta, *J. Theor. Biol.* **140,** 551 (1989).

[125] J. W. Kebabian and D. B. Calne, *Nature (London)* **277,** 93 (1979).

[126] P. Onali, M. C. Olianas, and G. L. Gessa, *Eur. J. Pharmacol.* **99,** 127 (1984).

[127] P. Jenner and C. D. Marsden, *Life Sci.* **25,** 479 (1979).

[128] M. Cesario, C. Pascard, M. E. Moukhtari, and L. Jung, *Eur. J. Med. Chem.* **16,** 13 (1981).

[129] C. Houttemane, J. C. Boivin, G. Nowogrocki, D. J. Thomas, and J. P. Boute, *Acta Crystallogr. Sect. B* **37,** 981 (1981).

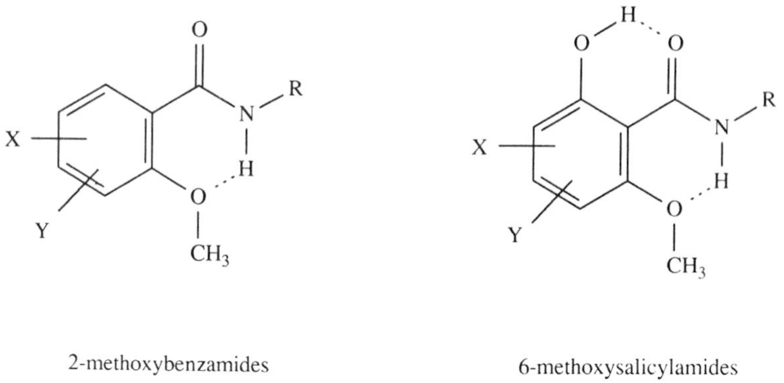

2-methoxybenzamides 6-methoxysalicylamides

FIG. 5. Intramolecular hydrogen bond in 2-methoxybenzamides and 6-methoxysalicylamides.

tions,[130] and nuclear magnetic resonance (NMR) spectroscopy[131] that orthopramides form an intramolecular hydrogen bond between their amide hydrogen and their *o*-methoxy oxygen atom. This hydrogen bond, which exists predominantly in media of low polarity, forms a pseudoring (virtual cycle) and stabilizes the planar conformation of the benzamide moiety (Fig. 5). In 6-methoxysalicylamides the pseudoring is stabilized by an additional hydrogen bond as shown by X-ray crystallography[132–135] (Fig. 5). As for the side chain of orthopramides, it may assume a folded (gauche) or extended (anti) form, depending on its nature (R = aminoethyl, 2-pyrrolidinylmethyl, 3-pyrrolidinyl, or 4-piperidinyl).[130,133,136,137] However, the pharmacologically active conformation is not known.

[130] A. Pannatier, L. Anker, B. Testa, and P. A. Carrupt, *J. Pharm. Pharmacol.* **33,** 145 (1981).
[131] L. Anker, J. Lauterwein, H. van de Waterbeemd, and B. Testa, *Helv. Chim. Acta* **67,** 706 (1984).
[132] T. Högberg, S. Rämsby, T. de Paulis, B. Stensland, I. Csöregh, and A. Wägner, *Mol. Pharmacol.* **30,** 345 (1986).
[133] T. Högberg, U. Norinder, S. Rämsby, and B. Stensland, *J. Pharm. Pharmacol.* **39,** 787 (1987).
[134] T. de Paulis, H. Hall, S. O. Ögren, A. Wägner, B. Stensland, and I. Csöregh, *Eur. J. Med. Chem.* **20,** 273 (1985).
[135] A. Wägner, B. Stensland, I. Csöregh, and T. de Paulis, *Acta Pharm. Suec.* **22,** 101 (1985).
[136] H. van de Waterbeemd and B. Testa, *Helv. Chim. Acta* **64,** 2183 (1981).

FIG. 6. MEPs (calculated by the STO-3G *ab initio* method) of five model compounds with typical substitution patterns of orthopramides. The isoenergy contours (in kilocalories per mole) are in a plane 1.75 Å above the plane of the aromatic ring. Reproduced form Ref. 138 with the permission of the American Chemical Society.

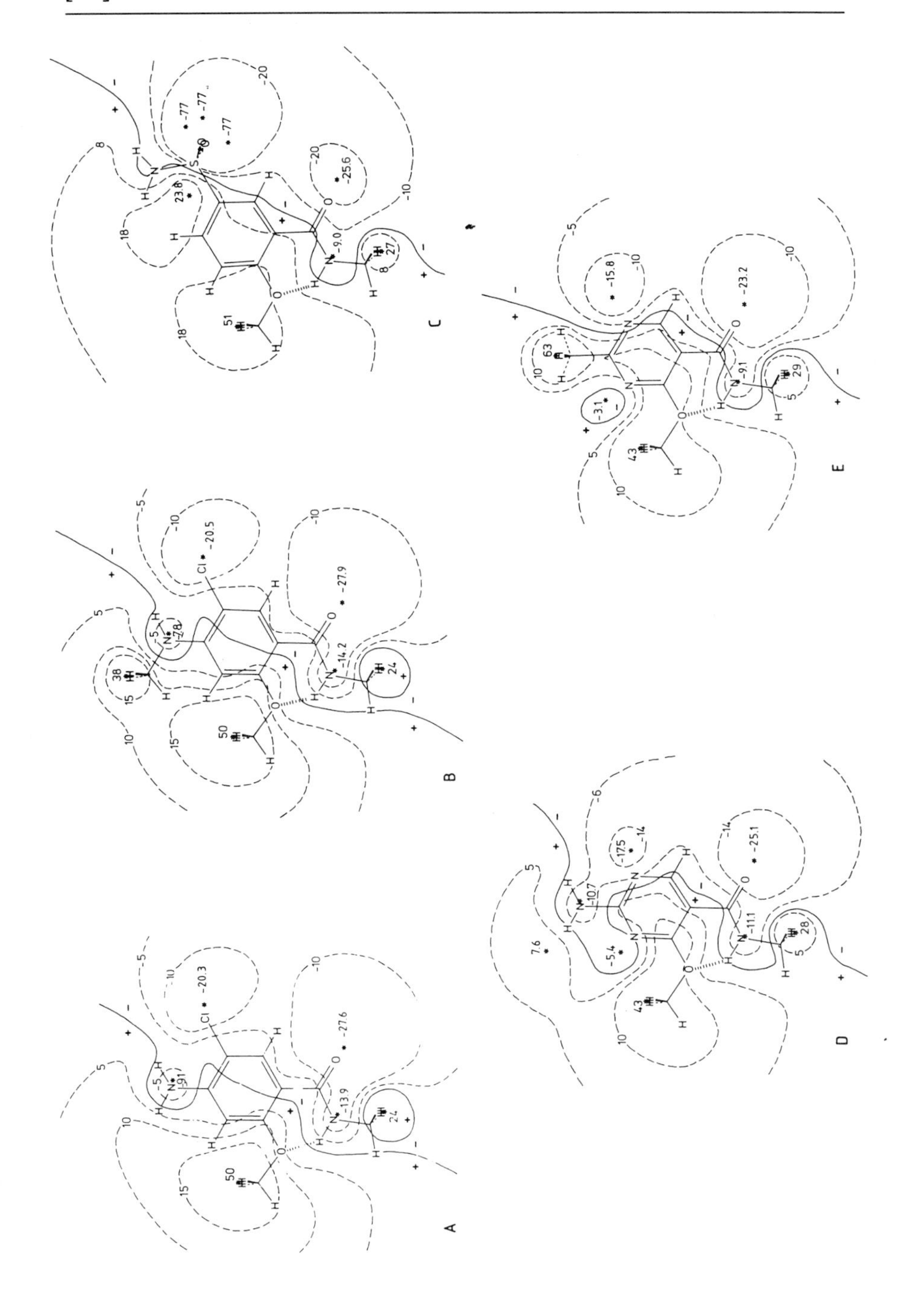
A
B
C
D
E

A topographical comparison between orthopramides and dopamine has shown that the key pharmacophoric groups, namely the aromatic ring and the basic nitrogen atom, are not superimposable.[130] In the fully extended conformation of DA the distance between the nitrogen atom and the center of the aromatic ring (N-ArC) measures about 5 Å as compared to orthopramides, where the same distance is 6 Å in the folded and 7 Å in the extended form. This has led to the suggestion that the aromatic ring of DA is topographically equivalent to the pseudoring of orthopramides,[137] and to the implication that the electrostatic potential of the aromatic ring in DA and of the pseudoring in orthopramides should be comparable.[73] To test this hypothesis and to define stereoelectronic features of importance for DA antagonistic activity, the MEPs of some model 2-methoxybenzamides with the aromatic substitution patterns of typical orthopramides were calculated.[138] The results indicated that the MEPs of the nonprotonated orthopramides could be divided into a positive and negative region by a boundary surface resembling a curtain (Fig. 6). In addition, common features were seen in the MEPs of the five model compounds, suggesting a distance pharmacophore and a contact pharmacophore. The distance pharmacophore, which was postulated to be responsible for the recognition and alignment of the ligands, extends a few angstroms away from the molecules. It consists of the field generated by the basic nitrogen, a positive maximum near the methoxy group, and two negative minima around the carbonyl group and the 5-substituent, respectively. At the shorter distances (ca. 2 Å) corresponding to the actual binding, the pharmacophoric pattern becomes more complex. Support for this model was obtained[139] from a QSAR study using a large series of orthopramides. Supplementing these results with those of MEP studies of analogs of different structure but similar activity (e.g., piquindone and zetidoline, Fig. 7) led investigators to hypothesize the superpositions of orthopramides and analogs[139,140] shown in Fig. 8 and to the pharmacophoric model of the dopamine D_2 receptor[141] shown in Fig. 9. It is worth noting in Fig. 8 that there is a good overlap between salient electrostatic features (minima and maxima), but not between some corresponding moieties such as the aromatic rings. This is not surprising, since receptors are expected to recognize stereoelectronic features and not atoms per se.

[137] H. van de Waterbeemd and B. Testa, *J. Med. Chem.* **26,** 203 (1983).

[138] H. van de Waterbeemd, P. A. Carrupt, and B. Testa, *J. Med. Chem.* **29,** 600 (1986).

[139] N. El Tayar, G. J. Kilpatrick, H. van de Waterbeemd, B. Testa, P. Jenner, and C. D. Marsden, *Eur. J. Med. Chem.* **23,** 173 (1988).

[140] H. van de Waterbeemd, P. A. Carrupt, and B. Testa, *J. Mol. Graphics* **4,** 51 (1986).

[141] S. Collin, G. Evrard, D. P. Vercauteren, F. Durant, P. A. Carrupt, H. van de Waterbeemd, and B. Testa, *J. Med. Chem.* **32,** 38 (1989).

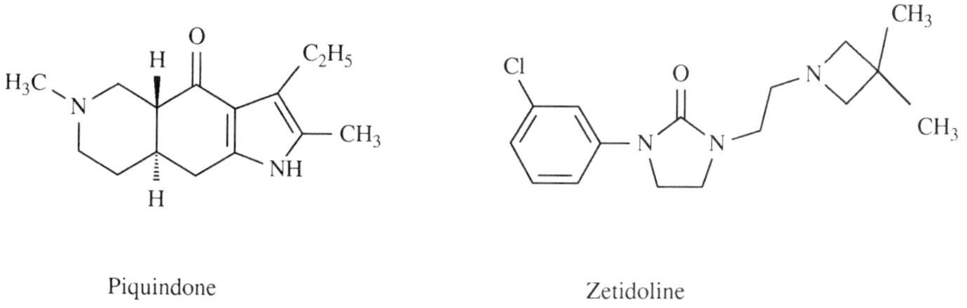

Piquindone Zetidoline

FIG. 7. The chemical structure of piquindone and zetidoline.

The development of some 6-methoxysalicylamides (Fig. 5) that show promising antipsychotic activity has been reviewed.[142] MEP calculations on some of these compounds were performed in order to study the electronic characteristics in their aromatic region.[143] This work demonstrated that small modifications of the benzamide geometry could change the two-dimensional MEP surface significantly, especially at a distance of 1.75 Å. When comparing the MEPs of the salicylamides with those of the orthopramides, some similarities and differences were noted. The basic nitrogen, the positive area between the methoxy and amide groups, and the negative minimum close to the carbonyl oxygen are congruent. However, an out-of-plane methoxy group can project a negative potential in the region above or below the plane of the aromatic ring. Furthermore, the aromatic ring in the salicylamides can be positive, partly positive, partly

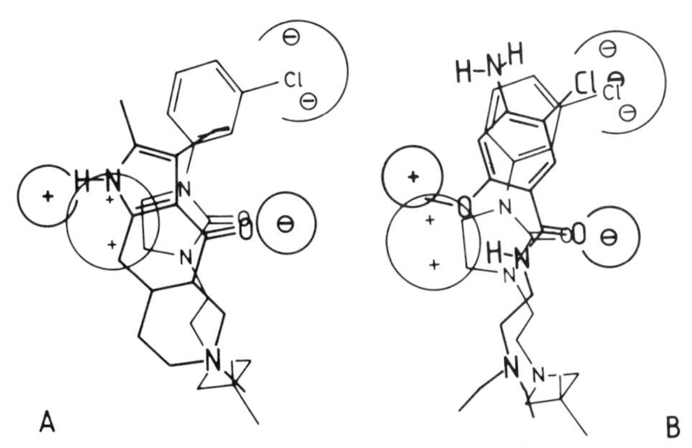

FIG. 8. Superposition of the pharmacophoric points of (A) zetidoline (thin lines) and piquindone (bold lines), and (B) zetidoline (thin lines) and metoclopramide (bold lines). Reproduced from Ref. 141 with the permission of the American Chemical Society.

[142] T. Högberg, S. Rämsby, S. O. Ögren, and U. Norinder, *Acta Pharm. Suec.* **24,** 289 (1987).
[143] U. Norinder and T. Högberg, *Acta Pharm. Nord.* **1,** 75 (1989).

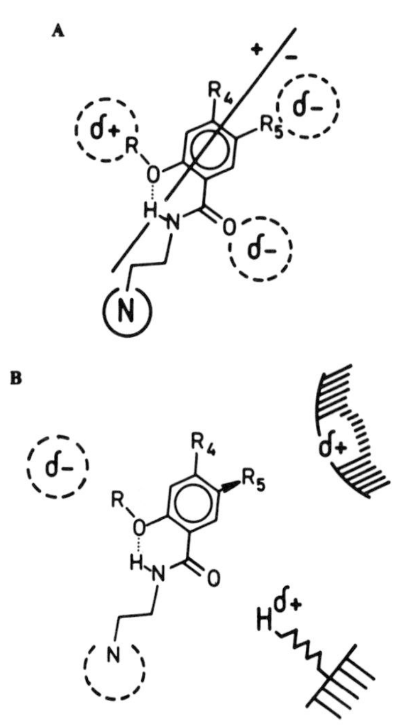

FIG. 9. (A) Pharmacophoric model of orthopropamides and analogs as deduced from their MEPs.[138,140] The pharmacophoric elements in this model are drawn with bold lines; δ^+ and δ^- indicate regions in which a positive or negative potential, respectively, is generated by the molecule. (B) Postulated binding sites in the dopamine D_2 receptor (bold lines) interacting with complementary groups in orthopramides, as deduced from Fig. 9A. Reproduced from Ref. 139 with the permission of Elsevier Sci. Publ.

negative, or negative, in contrast to the longitudinal line that separates the positive and negative zones above and below the aromatic ring in orthopramides. The difference in the MEP pattern between the two classes may indicate that they bind in slightly different ways to the receptor, as also suggested by QSAR studies.[142] A conclusion of a similar nature was reached when comparing the MEPs of orthopramides with that of dopamine itself.[73]

The usual way to fit molecules is to select three or more atoms from each molecule and superimpose them in the best possible manner. However, Kocjan and co-workers, in their work on ergolines and analogs, used another approach and based their fit on the MEPs.[144] Many ergolines and analogs with the tricyclic pyrrole (**I**) and pyrazole (**II**) structures shown in

[144] D. Kocjan, M. Hodoscek, and D. Hadzi, *J. Med. Chem.* **29,** 1418 (1986).

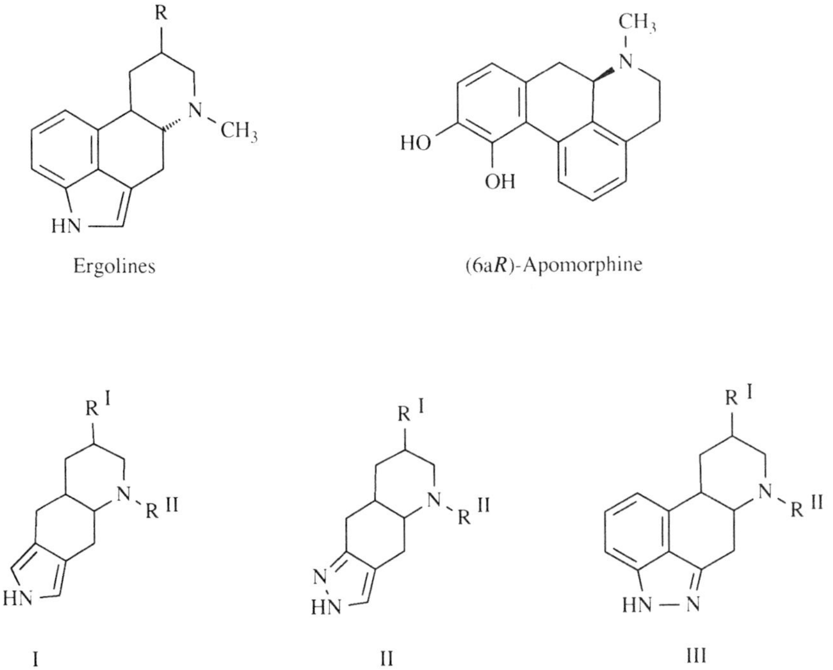

Ergolines

(6a*R*)-Apomorphine

I II III

FIG. 10. Chemical structure of ergoline, (6a*R*)-apomorphine, octahydropyrrolo[3,4-*g*]quinoline (**I**), octahydropyrazolo[3,4-*g*]quinoline (**II**), and 2-azaergoline (**III**).

Fig. 10 are potent DA receptor agonists[145]; in contrast, the 2-azaergolines (Fig. 10, structure **III**) are devoid of dopaminergic activity.[146] MEP maps of model compounds of these molecules were calculated. Using the MEP of the potent and stereoselective D_2-receptor agonist (6a*R*)-apomorphine (Fig. 10) as template and superimposing the nitrogens and the most negative MEP regions of ergoline and compounds **I–III** yielded the fits shown in Fig. 11. From Fig. II it can be seen that ergoline and compounds **I** and **II** display a good steric fit to (6a*R*)-apomorphine, while the inactivity of compound **III** may be explained by the poor steric fit.

B. Ligands of Serotonin Receptors

Research into the area of 5-hydroxytryptamine (5-HT) receptors has shown an enormous growth. The following receptor subtypes are now

[145] N. J. Bach, E. C. Kornfeld, N. D. Jones, M. O. Chaney, D. E. Dorman, J. W. Paschal, J. A. Clemens, and E. B. Smalstig, *J. Med. Chem.* **23,** 481 (1980).

[146] N. J. Bach, E. C. Kornfeld, J. A. Clemens, E. B. Smalstig, and R. C. A. Frederickson, *J. Med. Chem.* **23,** 492 (1980).

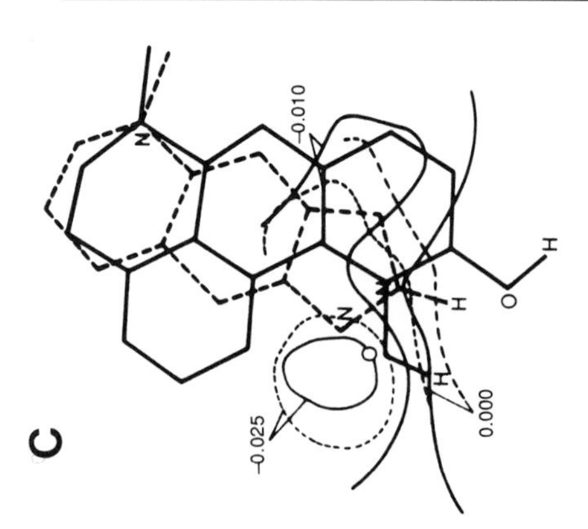
C
N
−0.010
−0.025
0.000
N
O
H
H
H
O

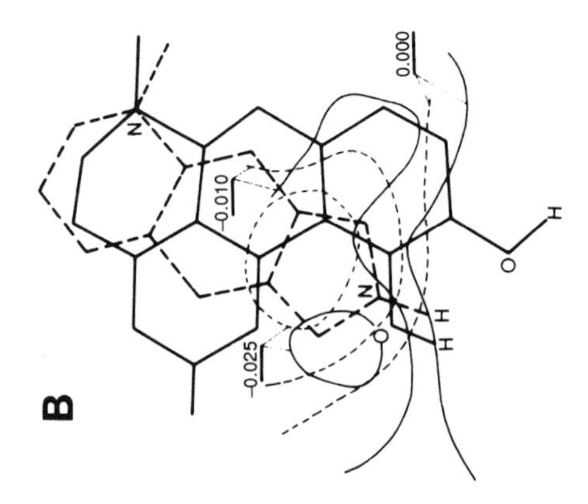
B
N
−0.010
0.000
−0.025
N
O
H
H
H
O

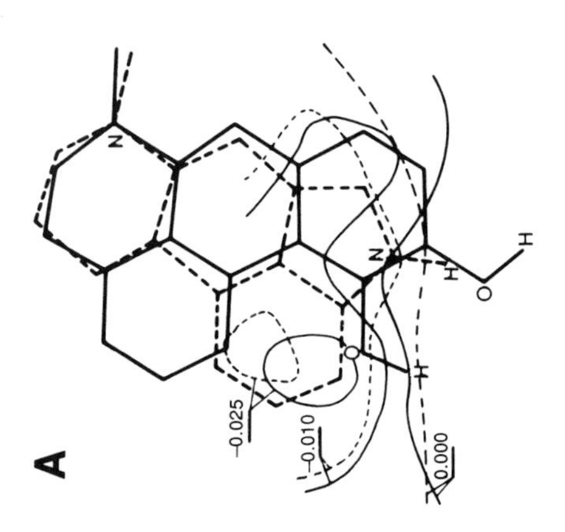
A
N
−0.025
−0.010
0.000
N
O
H
H
H
O

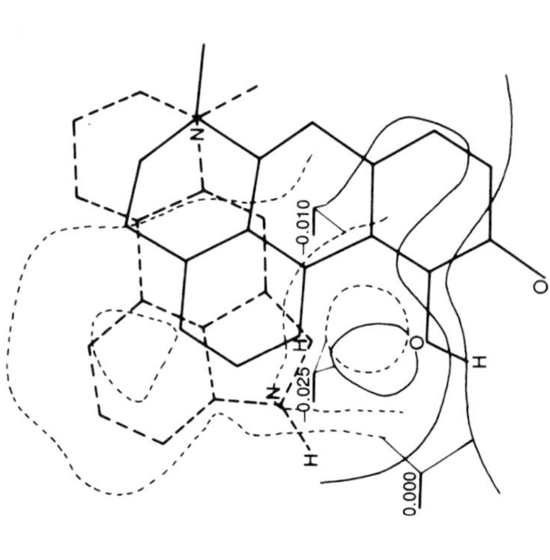
D
N
−0.010
−0.025
0.000
N
H
H
O
H
O

recognized[147,148]: 5-HT$_{1A}$, 5-HT$_{1B}$, 5-HT$_{1C}$, 5-HT$_{1D}$, 5-HT$_2$, and 5-HT$_3$. A 5-HT$_4$ site has also been proposed.[149]

Much work has been conducted on modeling the recognition to and activation of a class of 5-HT receptors with high affinity for 5-HT and lysergic acid diethylamide (LSD). This work has been initiated long before the current 5-HT receptor classification and involved a small series of tryptamine derivatives.[150–154] The resulting model should be able to account for agonistic as well as antagonistic interactions at this receptor.[153] Basic to these studies was the assumption that the interaction of different tryptamine analogs with the same binding site as 5-HT depends on the ability of these molecules to mimic the MEP characteristics of 5-HT. More specifically, an electrostatic orientation vector was obtained by connecting the MEP minima above the substituted indole ring; its direction depends on the substitution pattern in the indole ring and may serve as a guide to rank the potencies of the different congeners.

To describe recognition by and binding to the receptor, the imidazolium cation was chosen as a receptor model on which the substituted tryptamines were aligned so as to maximize the electrostatic interaction. Optimal alignment to the model was assumed for 5-HT. To model the activation step the imidazolium cation, acting as a proton donor, was coupled to an ammonia molecule acting as a proton acceptor. This proton transfer model (PTM) provided a mechanistic description of the receptor events that follow ligand binding. When 5-HT (in the aligned geometry) was brought to interact with this model it served to generate a driving force for the proton transfer by stabilizing the products over the reactants and

[147] S. J. Peroutka, *Trends Neurosci.* **11,** 496 (1988).

[148] P. R. Hartig, Trends Pharmacol. Sci. **10,** 65 (1989).

[149] D. E. Clarke, D. A. Craig, and J. R. Fozard, *Trends Pharmocol. Sci.* **10,** 385 (1989).

[150] H. Weinstein, D. Chou, S. Kang, C. L. Johnson, and J. P. Green, *Int. J. Quant. Chem. Quant. Biol. Symp.* **3,** 135 (1976).

[151] H. Weinstein, R. Osman, W. D. Edwards, and J. P. Green, *Int. J. Quant. Chem. Quant. Biol. Symp.* **5,** 449 (1978).

[152] H. Weinstein, R. Osman, S. Topiol, and J. P. Green, *Ann. N. Y. Acad. Sci.* **367,** 434 (1981).

[153] R. Osman, S. Topiol, L. Rubenstein, and H. Weinstein, *Mol. Pharmacol.* **32,** 699 (1987).

[154] P. H. Reggio, H. Weinstein, R. Osman, and S. Topiol, *Int. J. Quant. Chem. Quant. Biol. Symp.* **8,** 373 (1981).

FIG. 11. Superposition of the MEPs (*ab initio*, STO-3G, 1.6 Å above the plane of the aromatic ring) of apomorphine (full lines) with those (in dotted lines) of (A) ergoline, (B) compound **I**, (C) compound **II**, and (D) compound **III** in Fig. 10. Reproduced from Ref. 144 with the permission of the American Chemical Society.

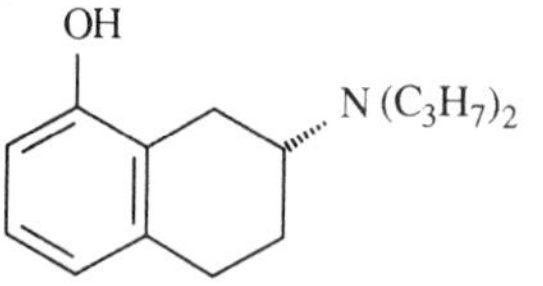

(2*R*)-IV 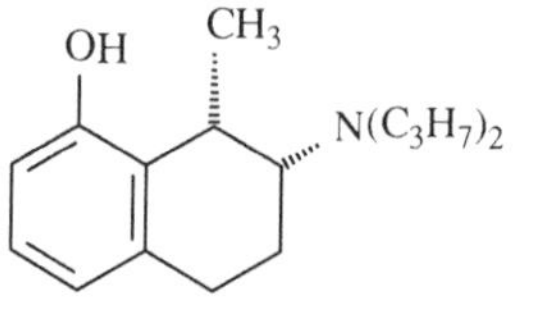 (1*S*,2*R*)-V 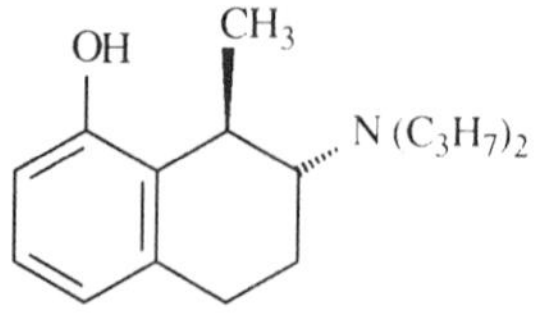 (1*R*,2*R*)-VI

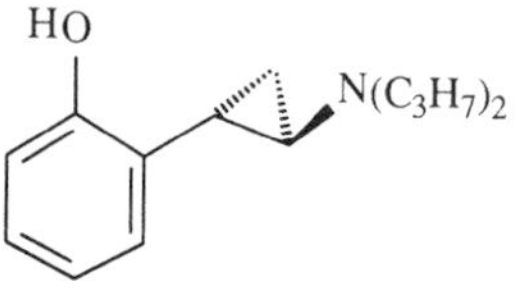

(1*R*,2*S*)-VII 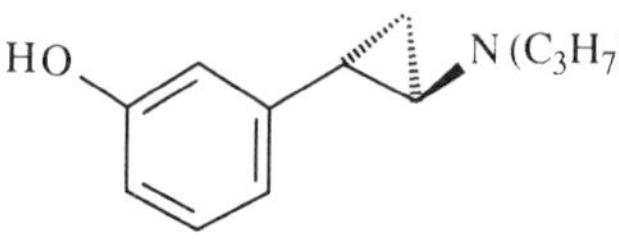 (1*R*,2*S*)-VIII

FIG. 12. Chemical structures **(IV–VIII)** of 2-aminotetralin and *trans*-2-phenylcyclopropylamine derivatives.

lowering the energy barrier. In the PTM an antagonist may be oriented over the imidazolium cation in the same way as an agonist but will not be able to trigger the proton transfer.[153]

Other investigators studied the conformational and electrostatic characteristics of a series of 2-aminotetralin and *trans*-2-phenylcyclopropylamine derivatives (Fig. 12) in order to rationalize their activities and stereoselectivities at the 5-HT receptor.[76] 8-Hydroxy-2-(di-*n*-propylamino) tetralin (IV; 8-OH-DPAT) is a highly potent 5-HT receptor agonist with selectivity for 5-HT_{1A}-binding sites,[155] the 2*R*-enantiomer being twice as potent as its antipode. Introduction of a *cis*-C-1-methyl group in (2*R*)-**IV** led to the 5-HT receptor agonist (1*S*, 2*R*)-**V**, which was enantioselective and approximately equipotent with (2*R*)-**IV**; (1*R*, 2*S*)-**V** and the trans isomer (±)-**VI** were inactive.[156] The derivatives (1*R*, 2*S*)-**VII** and (1*R*, 2*S*)-**VIII** were enantioselective but 5–10 times less active than (2*R*)-**IV**.[157] These results could not be rationalized without MEP calculations and conformational analyses.

[155] M. Hamon, S. Bourgoin, H. Gozlan, M. D. Hall, C. Goetz, F. Artaud, and A. S. Horn, *Eur. J. Pharmacol.* **100,** 263 (1984).

[156] L. E. Arvidsson, A. M. Johansson, U. Hacksell, J. L. G. Nilsson, K. Svensson, S. Hjorth, T. Magnusson, A. Carlsson, B. Andersson, and H. Wikström, *J. Med. Chem.* **30,** 2105 (1987).

[157] L. E. Arvidsson, A. M. Johansson, U. Hacksell, J. L. G. Nilsson, K. Svensson, S. Hjorth, T. Magnusson, A. Carlsson, P. Lindberg, B. Andersson, D. Sanchez, H. Wikström, and S. Sundell, *J. Med. Chem.* **31,** 92 (1988).

To study the electrostatic characteristics of these compounds MEP maps were calculated (*ab initio,* STO-3G basis set) in planes parallel to that of the indole ring and 1.6 Å distant. The MEP maps thus obtained were all quite similar and could not be correlated with differences in activity. However, conformational analysis indicated that (±)-**VI** preferentially adopts half-chair conformations with a pseudoaxial nitrogen substituent. In contrast, the pharmacophoric conformation was deduced to be a half-chair conformation with a pseudoequatorial nitrogen substituent, thus explaining the inactivity of (±)-**VI**. Compounds (1*S*, 2*R*)-**V**, (1*R*, 2*S*)-**VII**, and (1*R*, 2*S*)-**VIII** easily assumed pharmacophoric conformations, their stereoselectivities being rationalized in terms of steric factors. To summarize, compounds **IV**–**VIII** (Fig. 12) represent an example where, the MEP features being similar, conformational and steric factors apparently govern the pharmacological differences.

C. *Ligands of β-Adrenergic Receptors*

The majority of β-adrenergic ligands belong to one of the two chemical classes, arylethanolamines (AEAs, agonists and antagonists) or aryloxypropanolamines (AOPAs, antagonists only),[158] the only structural difference being the OCH_2 group. The β-adrenoceptor antagonists sotalol and propranolol are clinically useful representatives of these classes (Fig. 13).

It is generally believed that β-adrenoceptor agonists and antagonists from these classes bind to the same receptor site and that the same pharmacophoric groups are involved (aromatic ring, side-chain hydroxyl group, and amino group). This poses a dilemma as far as structure–activity relationships are concerned, since the pharmacophoric groups of these different classes cannot be superposed when the compounds are considered in their low-energy conformations. One hypothesis that was put forward to explain how AEAs and AOPAs can bind to the same receptor site was deduced from theoretical and experimental results. Based on results from X-ray diffraction studies it was shown that the OCH_2 group is coplanar with the aromatic ring in AOPA derivatives.[159–161] This Aryl-OCH_2 plane was hypothesized to simulate the steric and electrostatic pattern of the aryl group in AEAs owing to the conjugation of the oxygen atom with the

[158] B. G. Main and H. Tucker, *in* "Progress in Medicinal Chemistry" (G. P. Ellis and G. B. West, eds.), Vol. 22, p. 122. Elsevier, Amsterdam, 1985.

[159] Y. Barrans, M. Cotrait, and J. Dangoumau, *Acta Crystallogr. Sect. B* **29,** 1264 (1973).

[160] M. Gadret, M. Goursolle, J. M. Léger, and J. C. Colleter, *Acta. Crystallogr. Sect. B* **31,** 2780 (1975).

[161] H. L. Ammon, D. B. Howe, W. D. Erhardt, A. Balsamo, B. Macchia, F. Macchia, and W. E. Keefe, *Acta Crystallogr. Sect. B* **33,** 21 (1977).

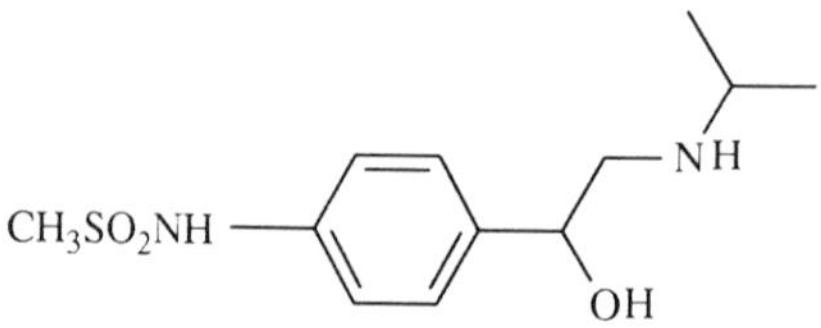

Sotalol (AEA)

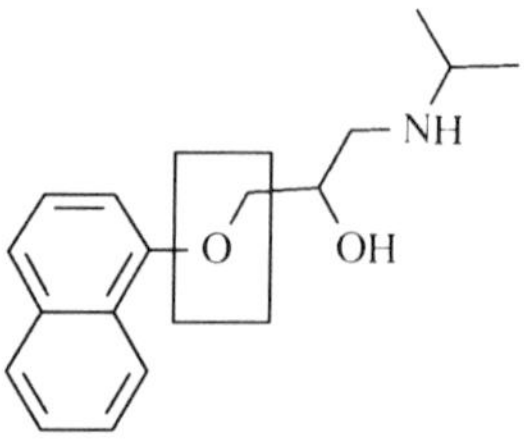

Propranolol (AOPA)

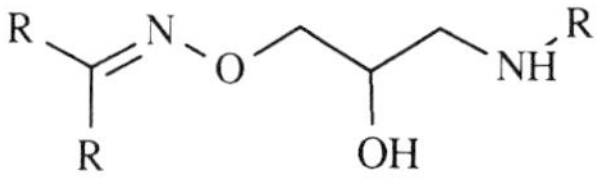

Oxime ethers

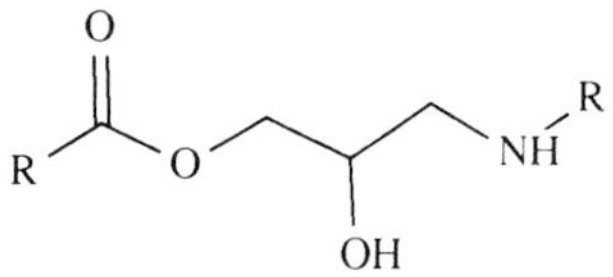

3-(Acyloxy)propanolamines

FIG. 13. The chemical structure of sotalol and propranolol [as representatives of arylethanolamines (AEAs) and aryloxypropanolamines (AOPAs), respectively], and that of oxime ethers and 3-(acyloxy)propanolamines.

aromatic ring.[162] This hypothesis was later strengthened by MEP calculations performed on some AEA and AOPA derivatives with comparable pharmacological properties, showing that there is some spatial correspondence between the sign of the MEP in the aromatic region and in the region above the planar zone of the Aryl-OCH_2 moiety.[163-165] To further test this hypothesis, aliphatic oxime ethers and aliphatic 3-(acyloxy)propanolamines (Fig. 13) were prepared and tested for pharmacological activity. These compounds proved to be β-adrenoceptor antagonists; while lacking an aromatic group, they generate an electrostatic distribution similar to that of AEAs and AOPAs.[166,167] Taken together, these results confirm that

[162] H. L. Ammon, A. Balsamo, B. Macchia, F. Macchia, D. B. Howe, and W. E. Keefe, *Experientia* **31,** 644 (1975).

[163] C. Petrongolo, B. Macchia, F. Macchia, and A. Martinelli, *J. Med. Chem.* **20,** 1645 (1977).

[164] B. Macchia, F. Macchia, and A. Martinelli, *Eur. J. Med. Chem.* **15,** 515 (1980).

[165] B. Macchia, F. Macchia, and A. Martinelli, *Eur. J. Med. Chem.* **18,** 85 (1983).

[166] B. Macchia, A. Balsamo, A. Lapucci, A. Martinelli, F. Macchia, M. C. Breschi, B. Fantoni, and E. Martinotti, *J. Med. Chem.* **28,** 153 (1985).

[167] B. Macchia, A. Balsamo, A. Lapucci, F. Macchia, A. Martinelli, H. L. Ammon, S. M. Prasad, M. C. Breschi, M. Ducci, and E. Martinotti, *J. Med. Chem.* **30,** 616 (1987).

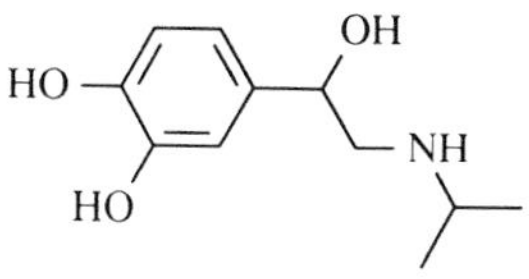

Isoproterenol

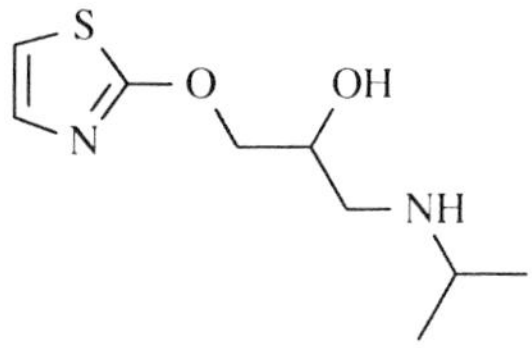

Tazolol

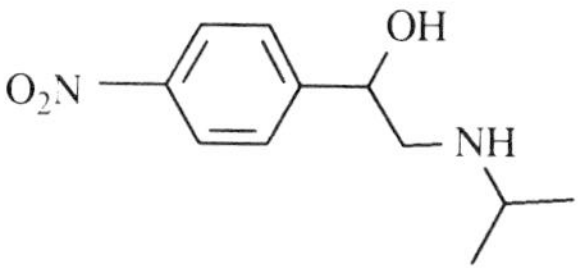

INPEA

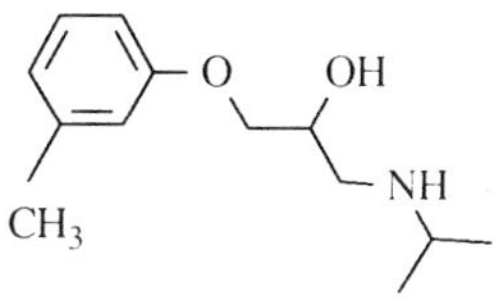

Doberol

FIG. 14. Chemical structure of the β-adrenoceptors isoproterenol and tazolol and of the antagonists *N*-isopropyl(4-nitrophenyl)ethanolanine (INPEA) and doberol.

the OCH_2 group in AOPAs may simulate a portion of the aromatic ring in AEAs.

The molecular electrostatic characteristics of the two β-agonists isoproterenol and tazolol and the two β-antagonists INPEA and doberol (Fig. 14) were investigated,[163,164] showing that the agonists have an essentially negative MEP around the aromatic ring or the corresponding OCH_2 region, and the antagonists an essentially positive one. Successful attempts at a more quantitative description were also performed,[164,165] suggesting that the above pattern might be used to predict the agonistic or antagonistic activity of other β-adrenoceptor ligands.[165] One weakness of this model, as also pointed out by the authors, lies in the limited number of compounds on which it was based. It is also difficult to consistently interpret the MEP around the Aryl-OCH_2 group. Other authors also noted that this model is unsuccessful in predicting the agonist–antagonist activity of sotalol (Fig. 13) and some other β-adrenergic ligands.[75,168]

In a comprehensive study, the MEPs of a series of β-adrenoceptor agonists and antagonists were calculated (*ab initio,* STO-3G basis set).[75] The compounds investigated were 4 phenylethanolanines (PEAs), 23 AOPA derivatives, 1 oxime ether, and 4 compounds with miscellaneous structures, the aim being to explore the possible role of electrostatic interactions in determining β_1/β_2 selectivity. The molecules were set in con-

[168] J. Rouvinen, E. Pohjala, J. Vepsäläinen, and P. Mälkönen, *Drug Des. Delivery* **5,** 281 (1990).

gruent extended conformations, which are often found to be the preferred ones. Salient electrostatic (maxima and minima) and topographic features are schematized in Fig. 15.

1. In the four PEA derivatives a negative minimum (M1) located in the vicinity of the meta substituent is found.
2. In all AOPAs a negative minimum is generated above and below the aromatic ring close to the ortho carbon. This minimum is designated M2 and corresponds to M1 in AEAs.
3. In the AOPAs, a second negative zone located beyond the meta position and designated M3 is found in all β_1-selective antagonists and in some nonselective and β_2-selective antagonists.
4. The β_1-selective antagonists display an additional zone in the para position that is positive (P4) in the full antagonists and negative (M4) in the partial antagonists (i.e., with intrinsic sympathomimetic activity). It was suggested that β_1-selectivity results from the simultaneous presence of M3 and either M4 or P4.

The limitations of these stereoelectronic rationalizations have been stressed,[75] since the model takes into consideration neither the lipophilic zone(s) believed to increase β_2-selectivity,[169] nor the influence of N-substituents. In addition, the problem of the receptor-bound conformer(s), i.e., the active conformation(s), is still unresolved. The fact that extended conformations lead to a coherent picture is a mere indication that these *may* be the receptor-bound ones, as already concluded by other workers.[170]

D. Ligands of Cholinergic Receptors

Two main types of cholinergic receptors have long been known to exist, namely the muscarinic and the nicotinic receptor. Beers and Reich in 1970 proposed a pharmacophoric model for nicotinic receptor ligands[171] that served successfully as a template for the synthesis of some nicotinic agonists, e.g., the potent and semirigid nicotinic agonist isoarecolone methiodide[172–174] (Fig. 16).

[169] N. El Tayar, B. Testa, H. van de Waterbeemd, P. A. Carrupt, and A. J. Kaumann, *J. Pharm. Pharmacol.* **40,** 609 (1988).
[170] J. M. Léger, M. Gadret, and A. Carpy, *Mol. Pharmacol.* **17,** 339 (1980).
[171] W. H. Beers and E. Reich, *Nature (London)* **228,** 917 (1970).
[172] C. E. Spivak, T. M. Gund, R. F. Liang, and J. A. Waters, *Eur. J. Pharmacol.* **120,** 127 (1986).
[173] J. A. Waters, C. E. Spivak, M. Hermsmeier, J. S. Yadav, and R. F. Liang, *J. Med. Chem.* **31,** 545 (1988).
[174] C. E. Spivak, J. S. Yadav, W. C. Shang, M. Hermsmeier, and T. M. Gund, *J. Med. Chem.* **32,** 305 (1989).

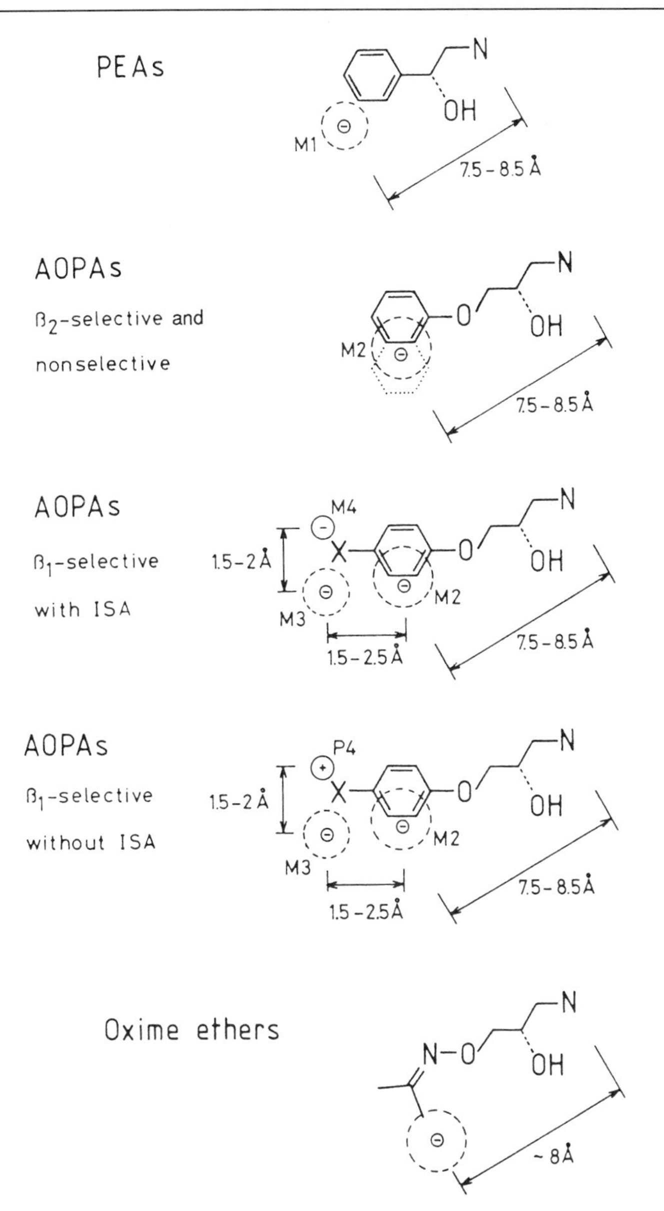

FIG. 15. MEP-based pharmacophoric elements of various chemical and pharmacological classes of β-adrenoceptor ligands; for explanations and discussion see text. Reproduced from Ref. 75 with the permission of the American Chemical Society.

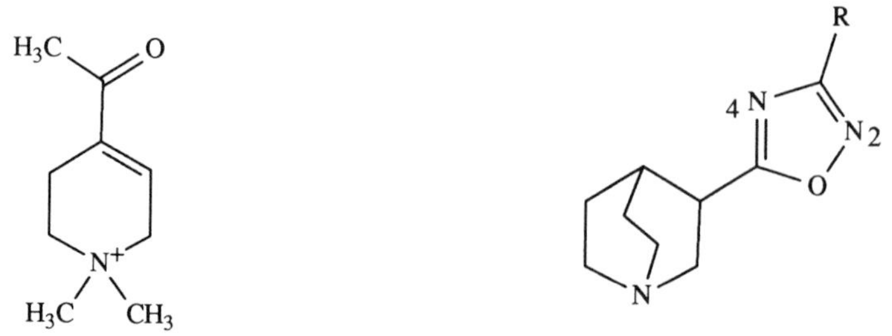

Isoarecolone methiodide

Quinuclidine derivatives

FIG. 16. Chemical structure of isoarecolone methiodide and quinuclidine derivatives.

Using isoarecolone methiodide as a lead compound resulted in derivatives whose activity varied over a 10,000-fold range; their conformational and electrostatic properties were examined in an attempt to explain this activity range.[173,174] Based on the results of conformational analysis a "bioactive" conformation was postulated for each compound, the criteria of selection being that the conformation should fit the pharmacophoric model of Beers and Reich[171] and be within 5 kcal/mol of the global energy minimum. Identified conformers were superimposed on the template molecule isoarecolone methiodide in different ways to obtain an optimal fit. Three-dimensional MEPs were generated at the van der Waals surface of the bioactive conformers and color coded according to values of the electrostatic potential. The areas of the different ranges in electrostatic potential were calculated in an attempt to relate the observed areas with potency. Potency differences could be rationalized by combining (1) the energy penalty needed to assume pharmacophoric conformations, (2) the presence of a methyl group adjoining the carbon atom bearing the hydrogen-bond acceptor (carbonyl oxygen), (3) superimposability of the pharmacophoric groups onto the template, (4) the strength of the electrostatic potential at the cationic head, and (5) that at the hydrogen-bond acceptor site. Solvation factors were also speculated to be of importance.

Saunders and co-workers sought to develop nonquaternary muscarinic agonists in order to study cholinergic mechanisms in memory and cognition.[175] More specifically, they wanted to develop agonists with high efficacy at cortical receptors that would readily cross the blood–brain barrier, and for this purpose prepared quinuclidine-based derivatives with the general structure shown in Fig. 16. Structure–activity relationship studies

[175] J. Saunders, M. Cassidy, S. B. Freedman, E. A. Harley, L. I. Iversen, C. Kneen, A. M. Macleod, K. J. Merchant, R. J. Snow, and R. Baker, *J. Med. Chem.* **33**, 1128 (1990).

demonstrated that by gradually removing the heteroatoms in the oxadiazole ring resulted in a loss of efficacy. In order to study this effect in more detail, MEPs were calculated. A good correlation was found between the efficacy and the negative electrostatic potentials near atoms 2 and 4 in the oxadiazole ring, suggesting that hydrogen-bond acceptance at these sites is necessary for high-affinity binding. In addition, it was observed that agonistic activity required a small 3-substituent, and that increasing the size and lipophilicity of the 3-substituent resulted in compounds with an antagonistic profile.

VII. Molecular Electrostatic Potential in Molecular Toxicology

A vast array of xenobiotics that may elicit toxic effects do not act directly, but only following toxification reactions occurring in the organism. These xenobiotics may be natural compounds, environmental chemicals, or drugs. Their toxification reactions, which are catalyzed by xenobiotic-metabolizing enzymes, can be investigated by quantum chemical methods like any enzymatic reaction (see Sections III,D and V). Only a few examples of MEP applications to molecular toxicology will be given here; in fact, the number of meaningful studies reported to date is a relatively limited one.

A number of nitroheterocyclic compounds are known to be effective radiosensitizers, their effects being unwanted (toxicity) or wanted (radiotherapy) ones depending on circumstances. Structure–activity relationships of radiosensitizers and prediction of their efficiency are thus doubly useful, rendering all the more interesting the study by Zhu and coworkers[176] of 11 nitroimidazolyl and other nitroheterocyclic derivatives. It was found that efficient radiosensitizers are characterized by a wide and deep negative potential in their MEP, while this potential is narrow for compounds devoid of significant activity. This phenomenon was suggested to be related to an interaction with certain biomacromolecules.

Mechanisms of chemical carcinogenesis have also been investigated by the MEP approach.[4] For example, polycyclic aromatic hydrocarbons (PAHs) such as benzo[a]pyrene are activated to highly reactive intermediates via a sequence of toxification steps that are competitive with detoxification reactions. Toxification is initiated by cytochrome *P*-450-mediated epoxidation at certain key positions; the same reaction occurs in halogenated olefins. Quantum mechanical studies indicated that the product selectivity of PAH epoxidation, and the substrate selectivity of olefin epoxidation, are largely influenced by electronic distribution and MEPs. The same

[176] A. Zhu, S. Xu, J. Huang, and Z. Luo, *Int. J. Radiat. Biol.* **56,** 893 (1989).

applies to the next reaction, namely epoxide hydration, which may be a toxification or a detoxification step. For olefin epoxides, the magnitude of the potential minimum associated with the epoxide oxygen was found to be correlated with substrate selectivity toward epoxide hydrolase.[4]

MEP calculations have also shed light on the covalent binding of reactive metabolic intermediates and other electrophiles to DNA (see also Section V). For example, these calculations correctly predicted the reactivity of the amino groups in DNA to decrease in the order NH_2 (guanine) > NH_2 (adenine) > NH_2 (cytosine), although this was not the ordering of their potential minimum in the isolated bases.[177] This implies that correlations based on MEP calculations of isolated bases[178] must be viewed with caution.

Clearly MEP calculations are a valuable tool in helping to rationalize and predict toxification reactions, but the greatest care is needed to avoid misleading results.

VIII. Concluding Remarks

This chapter has given attention mainly to drugs and their interactions with binding sites such as receptors and enzymes. As discussed in Section I, the intermolecular forces that contribute to both affinity and specificity can be schematically classified as hydrophobic and electrostatic ones. In this context, MEPs are of particular values in that they permit visualization and assessement of the capacity of a molecule to interact electrostatically with a binding site. This is the fundamental argument on which the present chapter is based. As illustrated, MEPs can be interpreted in terms of a *stereoelectronic pharmacophore* condensing all available information on the electrostatic forces that underlie affinity and specificity.

The first limitation of this approach is neglect of hydrophobic forces, which also contribute to affinity. *Molecular lipophilicity potentials* (MLPs) appear as an appealing and promising approach to calculate and display the capacity of a molecule to interact by hydrophobic forces. A few methods exist to calculate MLPs,[179–181] but significant improvements will have to be made before these methods can become really useful. In particular, reliable and sound distance functions for interacting polar groups are needed.

Combining MEPs and MLPs in order to obtain a visualization and

[177] A. Pullman and B. Pullman, *Int. J. Quant. Chem. Quant. Biol. Symp.* **7,** 245 (1980).
[178] D. F. V. Lewis and V. S. Griffiths, *Xenobiotica* **17,** 769 (1987).
[179] E. Audry, J. P. Dubost, J. C. Colleter, and P. Dallet, *Eur. J. Med. Chem.* **21,** 71 (1986).
[180] P. Furet, A. Sele, and N. C. Cohen, *J. Mol. Graphics* **6,** 182 (1988).
[181] J. L. Fauchère, P. Quarendon, and L. Kaetterer, *J. Mol. Graphics* **6,** 203 (1988).

assessment of the full interaction capacity of a ligand is an ambitious but necessary task that should lead to highly informative *hydrophobic/stereoelectronic pharmacophores.* In this context, the most sophisticated and powerful tool now available to medicinal chemists is the comparative molecular field analysis (CoMFA).[182] Here, steric and electrostatic fields are calculated and sampled at many points in space. By congruently superimposing a number of active compounds and sampling their fields, a data table is obtained that can be completed with biological activities and subjected to multivariate correlation analysis. The results generated take the form of a three-dimensional model where steric and electrostatic interactions, be they favorable or unfavorable, are quantitatively related to activity. We have successfully applied this method to, e.g., serotoninergic ligands (manuscript in preparation) and the toxification of MPTP analogs by monoamine oxidase.[183]

But whatever the method used and the type of pharmacophore obtained, it must be borne in mind that only *short-range recognition* is taken into account in the form of the ordinary lock-and-key model. This is, of course, a very partial and limited view of ligand binding, a view that fails to fully account for the exceedingly low physiological concentrations of many enzyme substrates and receptor agonists.[184] Clearly, *long-distance recognition* must be involved and can no longer be ignored despite considerable difficulties, in particular the calculation of electrostatic potentials and interactions in solution.[185,186] It is of singular significance in this context that the electrostatic field around a protein should reach so far (10–20 Å and more).[186] That such a field should play a major role in long-distance recognition and enzyme turnover acceleration is proven by computer simulations of the diffusion of the superoxide anion to superoxide dismutase.[187] It was indeed shown that the electric field of the enzyme enhances the association rate constant of the superoxide anion by a factor of 30 or more.

But even such long-distance recognition may not be enough to account for the efficiency of many enzymes and receptors despite the minute concentrations of their physiological substrates and agonists. Another mechanism can be hypothesized to be operative in enhancing efficiency, namely binding to a random point on the macromolecule, followed by

[182] R. D. Cramer III, D. E. Patterson, and J. D. Bunce, *J. Am. Chem. Soc.* **110,** 5959 (1988).
[183] G. Maret, N. El Tayar, P. A. Carrupt, B. Testa, P. Jenner, and M. Baird, *Biochem. Pharmacol.* **40,** 783 (1990).
[184] Z. Blum, S. Lidin, and S. Andersson, *Angew. Chem., Int. Ed. Engl.* **27,** 953 (1988).
[185] A. Warshel and S. T. Russell, *Q. Rev. Biophys.* **17,** 283 (1984).
[186] J. B. Matthew, *Annu. Rev. Biophys. Biophys. Chem.* **14,** 387 (1985).
[187] K. Sharp, R. Fine, and B. Honig, *Science* **236,** 1460 (1987).

diffusion to the receptor/catalytic site along lines of force on the surface of the macromolecule. Eigen[188] has, for example, shown that the *lac* repressor acts on its DNA operator site at a rate larger by one or two orders of magnitude than the rate predicted by simple diffusion. Eigen was then able to prove that the repressor is able to bind to any DNA site via electrostatic interaction, after which it is guided through one-dimensional diffusion along the helix axis into the operator site. This is like picturing the substrate or agonist being caught in a *spider's web.* In such a model, electrostatic and hydrophobic forces should control both the nonspecific binding and the guided diffusion, opening up a vast field of investigation in which MEPs and MLPs would be major tools.

The final remark we wish to make concerns *postbinding events.* Most of the chapter, excepting parts of Sections III,D, IV, and VII, has dealt with binding recognition. Such a vision, however, is a limited one since three conceptual steps can be seen to underlie the biological effects of xenobiotics in general, and the pharmacological effects of drugs in particular. Recognition, as expressed by selectivity, can occur at all three steps (Fig. 17). First, the active compound must penetrate into the compartment of action, e.g., tissue, cells, organelles; selectivity in the penetration step is a well-documented phenomenon, although it is seldomly investigated per se. Binding to the site of action (receptor, enzyme active site) follows this penetration, and is in turn followed by the *activation step*, i.e., molecular events that trigger the response seen at the macroscopic level. In the case of enzymes, the activation step can be equated with the catalytic step, in other words formation of the transition state and release of the product(s). In the case of receptors, the activation step involves the biochemical reactions triggered by agonist binding, often a rearrangement of the receptor followed by, e.g., enzyme activation, release of second messenger, or ion channel opening. How much MEPs and other quantum mechanical approaches can contribute to our understanding of postbinding events remains to be seen. As far as enzymatic catalysis is concerned, the few relevant examples discussed in Sections III,D, V, and VII suggest that valuable insights can be expected from quantum chemistry and molecular graphics provided that sound use is made of supercomputers and X-ray crystallography of proteins. Comparable advances in receptor activation lie farther in the future and await unraveling of the three-dimensional structure of receptors.

In brief, this chapter has illustrated a number of biological and pharmacological applications of MEPs, discussing their interests and limitations. If a single conclusion must emerge, it is that the MEP approach, despite its

[188] M. Eigen, *Curr. Contents/Life Sci.* **33**(6), 16 (1990).

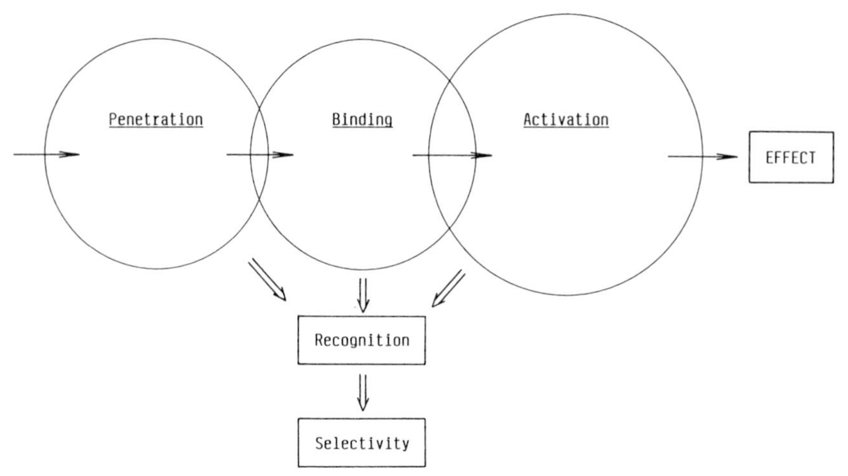

FIG. 17. The three conceptual steps underlying the biological action of xenobiotics, e.g., the pharmacological action of drugs.

power and sophistication, is but one tool among the many available to medicinal chemists. Only in combination with other disciplines and approaches (e.g., synthetic, physical, and structural chemistry, biochemistry, molecular pharmacology) does it yield results that are meaningful rather than fanciful. As always in science, context gives import.

Acknowledgments

B.T., P.A.C., and N.E.T. are indebted to the Swiss National Science Foundation for research Grants 31-8859.86 and 31-27531.89. The postdoctoral studies of A.K. in Lausanne were supported by The Swedish Institute, the Stiftelsen Bengt Lundqvist Minne, The I.F. Foundation for Pharmaceutical Research, The Swedish Academy of Pharmaceutical Sciences, and the Fred Anderson Stipendium.

[32] Pharmacophore for Nicotinic Agonists

By T. M. GUND and C. E. SPIVAK

Our objectives in this chapter are to provide a concise review of the salient elements of the pharmacophore for nicotinic agonists, to describe the methods and insights obtained through studies employing computer-assisted molecular modeling, and to point out some of the unsolved problems. The earlier pharmacology of the agonists has been reviewed.[1-4]

Activities of Nicotinic Agonists

Before one can evolve a pharmacophore, one must have consistent, reliable data on the activity of the agonists. Here one encounters a number of complications.

First, nicotinic activities have been measured in a variety of ways, such as contracture of a suitable muscle, e.g., the frog rectus abdominis muscle or the chick biventer cervicis, depolarization, measured by intracellular recording of frog muscle or eel electroplax, or by rise in a mammal's blood pressure (after atropine). The last reflects action at the *ganglionic* receptor, which requires a somewhat different pharmacophore. Even when determined at one type of tissue, such as by the contracture of the frog rectus abdominis muscle, one must, before comparing data, consider the variables that can influence the result, such as the species of frog, whether a cholinesterase inhibitor was used (when acetylcholine or other substrates for acetylcholinesterase were tested), and whether the contracture was isotonic or isometric (and the resting tension).

Second, the output of most test systems is a consequence of several steps leading from receptor occupation to muscle contraction. Even at the most elementary level, that of single receptors, the rate constants of the steps leading to the opening and shutting of the ion channel are influenced by the particular agonist occupying the recognition site of the receptor.[5,6] Thus, recognition can be viewed as a *sequential* process, and each step may require different elements of the pharmacophore. Some progress in discovering these influences of agonist structure on channel kinetics has begun,[6,7] but these studies are highly labor intensive.

Third, it is probably safe to say that no modification of the structure of a drug introduces a single, isolated, discrete change. For example, replacing a hydrogen by a halogen may be intended to demonstrate the influence of

[1] R. B. Barlow, "Introduction to Chemical Pharmacology." Wiley, New York, 1955.

[2] W. F. Riker, *Pharmacol. Rev.* **5,** 1 (1953).

[3] D. J. Triggle, "Neurotransmitter–Receptor Interactions." Academic Press, New York, 1971.

[4] C. E. Spivak and E. X. Albuquerque, *in* "Progress in Cholinergic Biology: Model Cholinergic Synapses" (I. Hanin and A. M. Goldberg, eds.), p. 323. Raven, New York, 1982.

[5] D. Colquhoun and B. Sakmann, *J. Physiol. (London)* **369,** 501 (1985).

[6] R. L. Papke, G. Millhauser, Z. Lieberman, and R. E. Oswald, *Biophys. J.* **53,** 1 (1988).

[7] C. E. Spivak and J. A. Waters, *Abstr. Neurosci.* **13,** 146 (1987).

inductive effects, but steric bulk at that position, solvation, ground state conformation, polarizability, etc., are all altered concomitantly.

Fourth, one is tempted to account quantitatively for variations in potencies among agonists in terms of the energies of the weak and (necessarily) transient attachments of the agonist to the recognition site. The pharmacophore suggests the sorts of bonds at work between the agonist and its recognition site. The estimated energies of these bonds can be summed for a given agonist to yield total binding energy. The difference in total binding energies for two agonists then predicts a ratio of potencies, which can be compared to measured values. Usually, however, the *uncertainties* in the estimated binding energies overwhelm the potency ratio one attempts to rationalize. For example, a difference in bond energy of only 2 kcal/mol will alter the dissociation constant by 31-fold.

With these caveats in mind, the activities of some agonists spanning a 20,000-fold range of potency are presented in Table I. We have confined most of these entries to agonists whose conformations are constrained by rings. All the activities shown were determined by contracture of the frog (mostly *Rana pipiens*) rectus abdominis muscle. They are ranked in order of potency, which is estimated in reference to carbamylcholine (an easily available, potent agonist that resists hydrolysis by cholinesterase).

Elements of Nicotinic Cholinergic Pharmacophore

In the subsections below, we refer to salient aspects of the pharmacophore for the nicotinic acetylcholine receptor (nAChR) and note problems for which intuition and mechanical modeling [e.g., Corey-Pauling-Koltun (CPK) models] are inadequate.

Cationic Head

The tetramethylammonium ion (agonist **29** in Table I) is the simplest nicotinic agonist. A cationic head, be it ammonium, sulfonium, or phosphonium, is essential for activity[1-4]; all potent agonists have it. Quaternary amines are most common, but secondary (agonists **1, 2,** and **18,** Table I) and tertiary (agonists **22, 23, 25,** and **26**) amines may be active in cyclic compounds. The active nonquaternary amines are expected to be overwhelmingly protonated at physiological pH. Among the simple, alkylammonium compounds, an R-N(Me)$_3$ ion seems necessary, but increasing steric bulk on the N diminishes activity (e.g., compare agonists **29, 34,** and **38**). Among bi- and tricyclic agonists, however, activity as a function of successive methylation shows no consistent trend. The secondary amine cytisine **(18)** loses activity when methylated **(26)** and becomes still weaker

TABLE I
RELATIVE POTENCIES OF NICOTINIC AGONISTS ASSAYED BY CONTRACTURE OF FROG MUSCLE

Number	Agonist	Relative potency[a]
1	(+)-Anatoxin *a*	110[b]
2	Pyrido[3,4-*b*]homotropane	~100[c]
3	Isoarecolone methiodide	50[d]
4	3-Hydroxyphenylpropyltrimethylammonium	28[e]
5	4-Aminophenylethyltrimethylammonium	15[e]
6	3-Hydroxyphenylethyltrimethylammonium	9.8[e]
7	Dihydroisoarecolone methiodide	9.1[f]
8	Arecolone methiodide	8.6[g]
9	Coryneine (i.e., dopamine methiodide)	5.3[h]
10	1-Methyl-4-(trifluoroacetyl)piperazine methiodide	4.6[f]
11	(−)-Ferruginine methiodide	3.3[g]
12	3-Aminophenylethyltrimethylammonium	3.0[e]
13	Sulfoisoarecoline methiodide	2.9[i]
14	Isoarecoline methiodide	2.7[i]
15	1-Methyl-4-acetylpiperazine methiodide	2.6[d]
16	(±)-Isoarecolol methiodide	2.0[f]
17	Arecoline methiodide	1.3[g]
18	Cytisine	1.1[g]
19	(−)-Nicotine methiodide	1.05[e]
20	1-Methyl-4-carbamyl-1,2,3,6-tetrahydropyridine methiodide	0.77[j]
21	(±)-Muscarone	0.77[g]
22	(−)-Nicotine	0.56[e]
23	Isoarecolone	0.48[k]
24	1-Methyl-1,2,3,6-tetrahydropyridine-4-methanol methiodide	0.35[f]
25	(−)-Anabasine	0.29[e]
26	Caulophylline (*N*-methylcytisine)	0.26[l]
27	Dihydroisoarecolol methiodide	0.25[f]
28	4-Hydroxyphenylpropyltrimethylammonium	0.23[e]
29	Tetramethylammonium	0.20[g]
30	1-Methyl-4-carbamylpiperazine methiodide	0.15[j]
31	(−)-Norferruginine	0.14[m]
32	Pyrrolidinedimethylammonium	0.096[n]
33	Isoarecoline	0.09[i]
34	Ethyltrimethylammonium	0.073[n]
35	Nornicotine	0.069[o]
36	1-Methyl-4-carbamylpiperidine methiodide	0.052[j]
37	(−)-Ferruginine	0.04[g]
38	Diethyldimethylammonium	0.019[n]
39	Piperidinedimethylammonium	0.018[n]
40	Cyclopentanetrimethylammonium	0.018[m]
41	(±)-Octahydro-2-methyl-*trans*-(1*H*)-isoquinolone methiodide	0.015[m]
42	Cyclohexanetrimethylammonium	0.012[m]
43	Arecoline	0.011[p]
44	1-Methyl-4-piperidone oxime methiodide	0.0055[f]

TABLE I *(continued)*

[a] Potencies are reciprocals of the equipotent molar rations determined at frog rectus abdominis muscles, estimated in reference to carbamylcholine.

[b] K. L. Swanson, C. N. Allen, R. S. Aronstam, H. Rapoport, and E. X. Albuquerque, *Mol. Pharmacol.* **29,** 250 (1986).

[c] D. B. Kanne and L. G. Abood, *J. Med. Chem.* **31,** 506 (1988). These authors found that this agonist is equipotent with anatoxin *a* in binding assays using *Torpedo* electric organ and rat brain. Other work, comparing the potency of agonists at frog rectus abdominis muscle to the inhibition of binding of α-bungarotoxin in *Torpedo* electric organ, show good correlation [C. E. Spivak, J. A. Waters, and R. S. Aronstam, *Mol. Pharmacol.* **36,** 177 (1989)]. Although this agonist has not been tested directly on the isolated rectus, its outstanding potency merits inclusion.

[d] C. E. Spivak, T. M. Gund, R. F. Liang, and J. A. Waters, *Eur. J. Pharmcol.* **120,** 127 (1986).

[e] R. B. Barlow, G. M. Thompson, and N. C. Scott, *Br. J. Pharmacol.* **37,** 555 (1969). Potencies given here were reevaluated to be in reference to carbachol by comparing original potencies to the potency of cytisine as determined by these authors.

[f] J. A. Waters, C. E. Spivak, M. Hermsmeier, J. S. Yadav, R. F. Liang, and T. M. Gund, *J. Med. Chem.* **31,** 545 (1988).

[g] C. E. Spivak, J. Waters, B. Witkop, and E. X. Albuquerque, *Mol. Pharmacol.* **23,** 337 (1983).

[h] R. B. Barlow, *Br. J. Pharmacol.* **57,** 517 (1976).

[i] H.-D. Holtje, G. Lambrecht, U. Moser, and E. Mutschler, *Arzneim.-Forsch.* **33,** 190 (1983). Original data reported in reference to acetylcholine + 10 μM physostigmine. Conversion to a carbachol standard was estimated by the observation that acetylcholine is 14 times as potent as carbachol after inhibition of cholinesterase (footnote *b* of this table).

[j] C. E. Spivak, J. S. Yadav, W.-C. Shang, M. Hermsmeier, and T. M. Gund, *J. Med. Chem.* **32,** 305 (1989).

[k] C. Reavill, C. E. Spivak, I. P. Stolerman, and J. A. Waters, *Neuropharmacology* **26,** 789 (1987).

[l] R. B. Barlow and L. J. McLoed, *Br. J. Pharmacol.* **35,** 161 (1969). They report that cytisine is 4.2 times as potent as caulophyllin.

[m] C. E. Spivak, unpublished observations, 1987.

[n] R. B. Barlow, N. C. Scott, and R. P. Stephenson, *Br. J. Pharmacol.* **31,** 188 (1967). Potencies given here were reevaluated to be in reference to carbachol by comparing original potencies to the potency of tetramethylammonium as determined by these authors.

[o] M. J. Mattila, L. Ahtee, and A. Vartiainen, *in* "Tobacco Alkaloids and Related Compounds. Proceedings of the Fourth International Symposium" (U. S. von Euler, ed.), p. 321. MacMillan, New York, 1965. These authors present relative potencies of nornicotine to nicotine.

[p] A. Burgen, *J. Pharm. Pharmacol.* **16,** 638 (1964).

when quaternized[8,9] (the observation that the optical rotatory dispersion of this quaternary agonist was similar to the secondary parent indicated no major change in the conformation of the agonist[8]). Conversely, nicotine gains potency through the progressive methylation of nornicotine **(35)** to nicotine **(22)** to nicotine methiodide **(19)**. The ferruginines defy simple order; the potencies are quaternary **(11)** $\gg$ secondary **(31)** $>$ tertiary **(37).**

[8] R. B. Barlow and L. J. McLoed, *Br. J. Pharmacol.* **35,** 161 (1969).

[9] C. E. Spivak, J. A. Waters, and R. S. Aronstam, *Mol. Pharmacol.* **36,** 177 (1989).

Hydrophobic Region

In the series of alkyltrimethylammonium agonists, potency generally increases with the alkyl chain length up to about butyl or pentyl.[2,3] Pentyltrimethylammonium, for example, may be among the most potent agonists known (it is about 520 times as potent as tetramethylammonium at the frog rectus abdominis muscle[3]). Among these one can include the bisonium compounds, especially decamethonium.[10,11] However, quaternary amines attached to saturated, cyclic hydrocarbons of similar chain lengths are weak (Table I, agonists **32, 39, 40, and 42**). This observation suggests that the alkyltrimethylammoniums lie in a roughly extended conformation in a hydrophobic crevice of the nAChR to anchor the quaternary head. Without the constraints of rings, the conformation required is unknown. The extent by which this site overlaps the site that binds agonists with hydrogen bond acceptors is also unknown.

Hydrogen Bond Acceptor

The suggestion by Barlow, in 1955,[1] that the carbonyl oxygen, present in many agonists, may function as a hydrogen bond acceptor was incorporated as a critical element of the pharmacophore for nicotinic agents proposed by Beers and Reich in 1970.[12] The hydrogen bond acceptor may be a carbonyl oxygen (e.g., agonists **1, 3, 7, 8, 10, 11, 13, 14, 15, 17, 18, 20,** and **21** among others in Table I), a phenolic OH (agonists **4, 6, 9,** and **28**), an alcoholic OH (agonists **16** and **24**), or an amine group (agonists **2, 5, 12, 19, 22, 25,** and **35**). Although a hydrogen bond has not been proved, the variety of auxiliary groups noted above indicates that other factors, such as a permanent dipole moment, are not critical.

Three Dimensional Structure

The over 100-fold ratio of potencies for positional isomers **4** and **28** (Table I) illustrates steric selectivity by the receptor. Less dramatic comparisons of positional isomers exist, such as for **3** and **8** (5.8-fold ratio of potencies), **5** and **12** (5.0-fold), **14** and **17** (2.1-fold), and **33** and **43** (8-fold).

The Beers–Reich pharmacophore[12] still serves as a common template for most agonists. Although inadequate to predict potency, as we discuss below, it is a criterion that is useful in screening synthetic candidates. The pharmacophore is constructed as follows: A plane is defined by the center of positive charge and a line through the carbonyl bond (or its equivalent).

[10] R. B. Barlow and H. R. Ing, *Nature (London)* **161,** 718 (1948).
[11] W. D. M. Paton and E. J. Zaimis, *Nature (London)* **161,** 718 (1948).
[12] W. H. Beers and E. Reich, *Nature (London)* **228,** 917 (1970).

The distance between the center of charge and the van der Waals surface of the hydrogen bond acceptor that intersects this line must be 5.9 Å. We refer to this as the Beers–Reich distance. In addition, the angle between the charge center, the center of the hydrogen bond acceptor, and the line through the carbonyl bond (or its equivalent) directed toward the donor is about 120°. Although not explicitly discussed by Beers and Reich, this angle appears in all their figures. Beers and Reich derived their pharmacophore from physical models of semirigid agonists and antagonists (e.g., nicotine, cytisine, strychnine, and *β*-erythroidine). More recently, Sheridan *et al.*[13] have derived practically the identical pharmacophore using a distance geometry method based on four semirigid agonists. The Beers–Reich distance they derived was identical, but the angle was a little larger, 137°. Critical review of the Beers–Reich pharmacophore points out inadequacies. Whereas most of the agonists in Table I fit the pharmacophore, some, such as agonists **36, 41,** and **44,** fit but are exceedingly weak; the pharmacophore may be necessary but insufficient. Anatoxin *a* (agonist **1**) and the ferruginines (agonists **11, 31,** and **37**), in the *s-cis* conformation, fit the Beers–Reich distance but not the angle. Conversely, in the *s-trans* conformations, these agonists fit the angle fairly well but not the distance. Arecolone methiodide (agonist **8**) is similar. The 4-substituted phenyl derivatives, **5** and **28,** cannot be contorted to a conformation close to this angle. The Beers–Reich distance may be the more critical of the two. Good support for this postulate comes from agonist **2,** which is locked into a conformation corresponding to the *s-cis* conformation of anatoxin *a*. In addition, two theoretical objections to the pharmacophore may be raised. First, although Beers and Reich emphasize the charge *center* (usually a quaternary nitrogen) as a focus of Coulombic attraction, *ab initio* molecular orbital calculations show that the positive charge is dispersed among the attached methyl groups[14,15]; it is not concentrated on the nitrogen atom. Second, Beers and Reich assume that the hydrogen bond forms along the line defined by the carbonyl bond. One may guess, however, that the bond should be directed toward an orbital containing an unshared pair of electrons (i.e., at ±60° from this line and in the plane of the carbonyl carbon and its substituents for the sp^2-hybridized orbitals of a carbonyl oxygen). Evidence supporting this tendency has been reported.[16,17]

Physical models of proposed agonists can be compared to the Beers–

[13] R. P. Sheridan, R. Nilakantan, J. S. Dixon, and R. Venkataraghavan, *J. Med. Chem.* **29,** 899 (1986).

[14] A. Pullman and G. N. J. Port, *Theor. Chim. Acta* **32,** 77 (1973).

[15] A. N. Barrett, G. C. K. Roberts, A. S. V. Burgen, and G. M. Clore, *Mol. Pharmacol.* **24,** 443 (1983).

[16] P. Murray-Rust and J. P. Glusker, *J. Am. Chem. Soc.* **106,** 1018 (1984).

[17] R. Taylor and V. Versichel, *J. Am. Chem. Soc.* **105,** 5761 (1983).

Reich pharmacophore as a first screen for candidates, but the unsatisfying inadequacy of the pharmacophore to predict potency led us to seek explanations obtainable through computer-assisted modeling. The remainder of this chapter describes the methods and findings we have obtained using modeling studies on analogs of isoarecolone methiodide (**3** in Table I). The structures of these analogs (**3**, **7**, **10**, **15**, **16**, **20**, **24**, **27**, **30**, **36**, **41**, and **44**) are shown in Fig. 1.

Computer-Aided Molecular Modeling of Nicotinic Agonists

Computer-aided molecular modeling offers certain advantages over mechanical models (e.g., Dreiding). For instance, most modeling programs generate and superimpose conformations and structures. From these representations, they can extract quantitative data, such as interatomic distances, relative conformation energies, and electrostatic contours.[18-22] Molecular modeling software and functions have been reviewed.[18] Nevertheless it must be remembered that these are still models that are being generated and manipulated, and the value of these models should be judged by their ability to rationalize and predict experimental results.

Relevance of Molecular Modeling to Receptor Agonist Activities

Some consideration of the complexity of the receptor–agonist interactions will be useful for interpreting the results of the more simplistic molecular modeling results. Consider an agonist, A, interacting with a receptor, R, in solution:

$$\langle A \rangle + \langle R \rangle \xrightarrow[\text{recognition}]{k_1} \langle A + R \rangle \xrightarrow[\substack{\text{activation +}\\ \text{biological}\\ \text{effect}}]{k_2} \langle A'\text{–}R' \rangle \xrightarrow[\text{recovery}]{k_3} \langle A \rangle + \langle R \rangle$$

There is an ensemble of conformations of agonist ⟨A⟩ and receptor ⟨R⟩ in solution. Step one consists of one or more conformations of A, the "bioactive" conformation(s), "recognizing" one or more conformations of receptor R. Step two consists of the actual binding of A and R, possibly with one or both changing conformation (to A′ and R′), to elicit the biological effect. Step three enables A and R to separate and relax to their initial, ground state conformations.

In the absence of a structural model of the receptor or the detailed

[18] N. C. Cohen, J. M. Blaney, C. Humblet, P. Gund, and D. C. Barry, *J. Med. Chem.* **33**, 883 (1990).

[19] M. Sadek and S. Munro, *J. Comput. Aided Mol. Des.* **2**, 81 (1988).

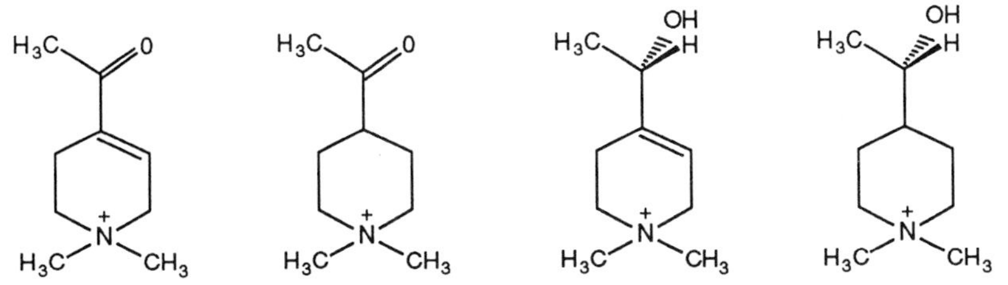

3 7 16 27

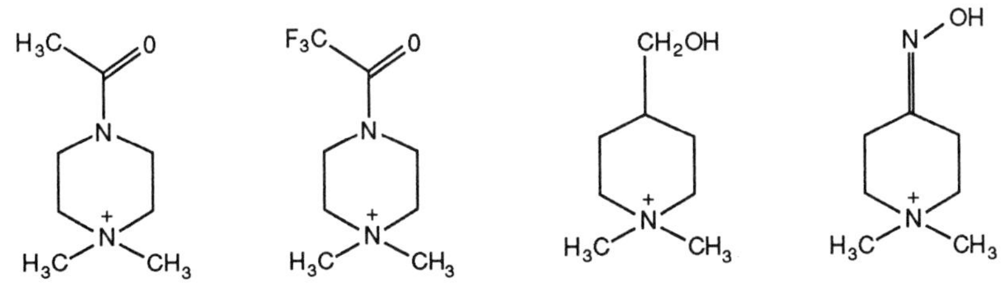

15 10 24 44

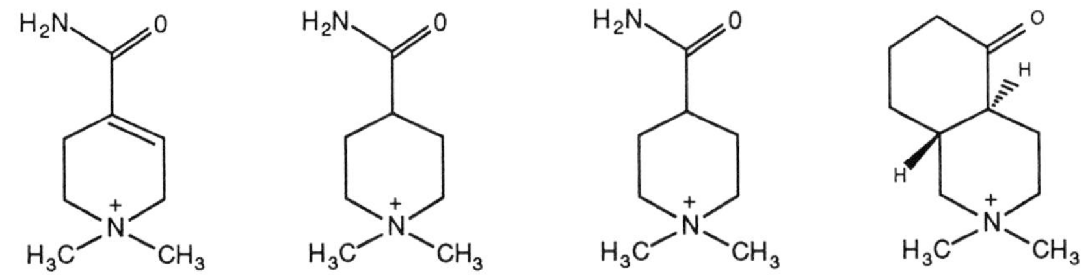

20 36 30 41

FIG. 1. Modeled nicotinic agonists: isoarecolone methiodide (**3**), dihydroisoarecolone methiodide (**7**), isoarecolol methiodide (**16**), dihydroisoarecolol methiodide (**27**), 1-methyl-4-acetylpiperazine methiodide (**15**), 1-methyl-4-(trifluoroacetyl) piperazine methiodide (**10**), 1-methyl-1,2,3,6-tetrahydropyridine-4-methanol methiodide (**24**), 1-methyl-4-piperidone oxime methiodide (**44**), 1-methyl-4-carbamyl-1,2,3,6-tetrahydropyridine methiodide (**20**), 1-methyl-4-carbamylpiperidine methiodide (**36**), 1-methyl-4-carbamylpiperazine methiodide (**30**), (+)-octahydro-2-methyl-*trans*-(1*H*)-isoquinolone methiodide (**41**).

kinetics of its activation, molecular modeling must focus on the first, recognition step. Effects on k_1 include the following: (1) the closer conformation A′ is to the ground state conformation, the less energy is required to activate A for the recognition step; (2) the greater the electrostatic complementarity between receptor and agonist, the stronger the "recognition"; and (3) the greater the steric complementarity of R′ and A′, the stronger the recognition. Effects on the activation and recovery step, while difficult to calculate, can be inferred: (4) energy gained by binding between R and A in step two is available to drive conformational changes or other processes necessary to elicit the biological response; and (5) if binding is too tight between R and A, the recovery step is disfavored and the agonist may become an antagonist or partial agonist.

Molecular Modeling Methods

Most of our modeling studies were performed on an Evans and Sutherland PS330 color vector/VAX 11/785 computer system. Some of the more extensive conformation searches were done on the CYBER 205 supercomputer at the former John von Neuman Center (Princeton, NJ). Display, manipulation, and superposition of molecules were done with programs Sybyl[23] and ChemX.[24]

Generation of Molecular Structures and Conformations

Because no X-ray diffraction studies have been done on these agonists, initial structures were built from the fragment libraries in ChemX or Sybyl. Ring conformations were explored with the popular molecular mechanics program MM2[25,26] or with RCG5[27] by breaking a bond, systematically rotating the remaining bonds, and reforming the broken bond. Standard parameters were used for MM2, augmented to account for formally

[20] T. M. Gund, *IEEE Eng. Med. Biol.* **7,** 21 (1988).

[21] P. Gund, T. A. Halgren, and G. M. Smith, *Annu. Rep. Med. Chem.* **22,** 269 (1987).

[22] T. M. Gund and P. H. Gund, *in* "Molecular Structures and Energetics" (J. Liebman and A. Greenberg, eds.), Vol. 4, p. 319. VCH Publishers, New York, 1987.

[23] SYBYL: Tripos Associates, St. Louis, Missouri.

[24] Chemical Design, Oxford, England.

[25] N. L. Allinger and Y. H. Yuh, *Quantum Chem. Prog. Exch.* Prog. No. 395 (1980).

[26] U. Burkert and N. L. Allinger, "Molecular Mechanics" ACS Monograph 177. American Chemical Society, Washington, D.C., 1982.

[27] G. M. Smith and D. F. Veber, *Biophys. Chem. Commun.* **34,** 907 (1986).

charged molecules,[28] ammonium salts,[29] and amides.[30] MM2 was used to evaluate the respective energies of these conformations, and the lowest energy ring conformations were studied further. Next, the energies of rotatable bonds were evaluated (in 5° steps) using the dihedral driver of MM2. The variable dihedral angle is that angle between the two planes, defined as follows: (1) the carbonyl oxygen, carbonyl carbon, and 4-position of the ring, and (2) the carbonyl carbon, the 4-position, and the 3-position of the ring.

The energies calculated in this work strictly pertain to a molecule in a vacuum, far from the real constraints of solvation and the powerful electrostatic fields that likely prevail within the recognition site of a receptor. Consequently we take our calculated ground state, and conformations within about 6 kcal of that minimum energy structure, as candidates for the bioactive form.

Choice of a "bioactive conformation" is critical because all other computations, comparisons, and conclusions depend on it. Since isoarecolone methiodide (**3**) is the most potent of the homologous series of agonists we examined, it served as a model. Energy calculations indicate that the acetyl group resides in two possible energy wells, *s-cis* or *s-trans* to the double bond, to maximize π orbital overlap in the conjugated enone system. The *s-cis* was selected as the bioactive conformation in analogy with anatoxin *a*[31,32] and pyridohomotropane,[33] despite its being 0.8 kcal/mol above the ground state. Choice of bioactive conformation of the other agonists was governed by (1) energy above the ground state (only conformations within 6 kcal/mol were considered), (2) the Beers–Reich distance (only conformations with a Beers–Reich distance between 5.5 and 6.4 Å were allowed) and (3) analogy with isoarecolone methiodide. Agonist **7** has energy minima at dihedral angles of about 0 and 240°. The ranges of angles that conform to the Beers–Reich criteria are 345 to 40° and 200 to 255°. Adopting an angle of 15° costs 1.3 kcal/mol, but it is well within the Beers–Reich constraint, and it places the acetyl group in conformity with the acetyl group in **3** (i.e., the oxygen is very nearly coplanar with ring position-2, -3, and -6). The bioactive conformations of the other agonists were adopted in similar fashion.

[28] T. Halgren, Merck Sharpe and Dohme, unpublished, 1982.

[29] J. Snyder, Searle Pharmaceutical Corp. (Skokie, IL) and T. Gund, NJIT.

[30] G. Marshall, Washington University, St. Louis, Missouri, private communication, 1983.

[31] C. E. Spivak, B. Witkop, and E. X. Albuquerque, *Mol. Pharmacol.* **18**, 384 (1980); C. S. Huber, *Acta Crystallogr., Sect. B* **28**, 2577 (1972).

[32] A. M. P. Koskinen and J. Rapoport, *J. Med. Chem.* **28**, 1301 (1985).

[33] D. B. Kanne and L. G. Abood, *J. Med. Chem.* **31**, 506 (1988).

Molecular Superpositions

The pharmacophore of a drug is the minimum collection of atoms spatially disposed in a manner that elicits a biological response.[34,35] The search for active drug molecules can begin with a search for the pharmacophoric pattern, although a drug may have the proper pharmacophoric pattern and still not be active. The pharmacophoric pattern may be topologic (graphtheoretic or connectivity-based structural fragments) or topographic[36] (geometric, usually three-dimensional). Topographic patterns are dependent on molecular conformations and should be suitable for describing drug activity, even for chemically dissimilar compounds having the same type of bioactivity. There are steric and electrostatic components to such pharmacophores.[37]

A common method of determining a pharmacophoric pattern is to derive a small number of possible bioactive conformations for similarly acting molecules and to superimpose these structures to find a common arrangement of functional groups in all active compounds. This is the method used in the present work.

Receptor Mapping

Receptor maps may be generated using the excluded volume technique of Marshall.[38-40] Briefly stated, one computes the union of volumes of active drugs and, separately, the union of volumes of inactive drugs. Inactive drugs are assumed to be so because they sterically impose on the receptor, so the difference between these two composite volumes represents space occupied by the receptor, the receptor map. A good receptor map will appear as a cavity that accommodates active ligands. It serves to rationalize the activities of known ligands and to predict structures of new ones.

Sybyl, ChemX, and ARCHEM all generate receptor maps. We use Sybyl predominantly. None of these packages, however, has the capability

[34] P. Ehrlich, *Chem. Ber.* **42,** 17 (1909); E. J. Ariens, *Prog. Drug Res.* **10,** 429 (1966).

[35] P. Gund, *Annu. Rep. Med. Chem.* **14,** 299 (1979).

[36] P. Gund *in* "Progress in Molecular and Subcellular Biology" (F. E. Hahn, ed.), Vol. 5, p. 117. Springer-Verlag, New York, 1977.

[37] C. Humblet and G. Marshall, *Annu. Rep. Med. Chem.* **15,** 267 (1980).

[38] C. Humblet and G. R. Marshall, *Drug Dev. Res.* **1,** 409 (1981).

[39] G. R. Marshall and N. A. Simkin, *in* "Medicinal Chemistry VI," p. 225. Proc. 6th Int. Symp. Med. Chem., Brighton, U.K., September 4–7, 1978.

[40] G. R. Marshall, C. D. Barry, H. E. Bosshard, R. A. Dammkoehler, and A. Dunn, *in* "Computer Aided Drug Design" (E. C. Olson and R. E. Christoffersen, eds.), ACS Symp. Ser., Vol. 112, p. 205. American Chemical Society, Washington, D.C., 1979.

of incorporating electrostatic information into their maps or evaluating steric interactions of new ligands as they are fit into an existing map.

Electrostatic Potential Surfaces

After selecting the bioactive conformation, partial charges were calculated using Dewar's MNDO program from MOPAC.[41] For agonists **3** and **15,** Mulliken or potential derived charges were calculated by a Gaussian, *ab initio* program,[42,43] using the STO-3G basis set. Although the partial charges calculated by these two methods differed somewhat, the electrostatic potentials projected onto surfaces in space were similar.[44] Electrostatic potentials[45] were calculated with reference to an incoming, unitary, positive charge. Since all the agonists are ammonium cations, the energies are repulsive. ChemX was used to project isopotential surfaces into space with a dielectric constant of unity.

We prefer to project the electrostatic potential onto the van der Waals surface to avoid the huge uncertainties in the dielectric constant of the medium beyond the van der Waals surface. The ARCHEM[46] program accomplishes this projection and provides additional options, including vector representation of the electric field, stereoscopic pairs (relaxed or crossed eye), alternate pattern selection according to atom type or potential range, the embedding of a stick representation of the molecule within the van der Waals surface, scaling of atomic radii or constant extension of the radii, full flexibility to reorient the molecule and select a window on the plotting surface, and depiction of the outside or inside of the molecular surface with full hidden line removal. ARCHEM can produce high-resolution renditions of the electrostatic properties of molecules without expensive graphics terminals. Although visual comparison of the electrostatic potentials on the van der Waals surface is usually useful, a quantitative comparison should be advantageous. ARCHEM can provide sums of surface areas whose potentials are within defined ranges. These areas depend critically on atomic radii, which are as follow[46] (Å): C (sp^3), 1.70; C (sp^2), 1.74; O, 1.50; N, 1.60; H, 1.30; H (hydroxyl), 1.00; F, 1.45.

[41] M. J. S. Dewar, University of Texas, Houston, Texas. MOPAC program available from the Quantum Chemistry Program Exchange.

[42] J. S. Binkely, R. A. Whiteside, R. Kirshnan, R. Seeger, D. J. Defrees, H. B. Schlegel, S. Topiol, L. R. Kahn, and J. A. Pople, Gaussian 80 QCPE, 1980. Bull. 2, p. 17, 1982.

[43] U. C. Singh and P. Kollman, Gaussian 80, UCSF, QCPE Program.

[44] J. S. Yadav, M. Hermsmeier, and T. Gund, *Int. J. Quant. Chem. Quant. Biol. Symp.* **16,** 101 (1989).

[45] P. Politzer, *in* "Chemical Applications of Atomic and Molecular Electrostatic Potentials" (P. Politzer and D. G. Truhbar, eds.), p. 1. Plenum, New York, 1981.

[46] M. Hermsmeier and T. M. Gund, *J. Mol. Graphics* **7,** 150 (1989).

Results

Superpositions and Bioactive Conformations

The bioactive conformations of homologous agonists **3, 7, 10, 15, 16,** and **24** were superimposed at sets of three analogous points. Two of these superpositions were meaningful in the same way. The Beers–Reich atoms (N, carbonyl C, and O; see Fig. 2a) were superimposed.[47–49] The positions of the acetyl methyl groups align with that of **3** in a rank order (**3** > **7** > **10** ≈ **15** > **27** > **16**) that nearly parallels potency.[48] Similarly, if one superimposes the N, the carbonyl C, and the acetyl C (Fig. 2b), the oxygen atoms align with the oxygen of **3** in the same rank order. We interpret this observation as follows: First, the position of the *center* of the cationic head is important (as Beers and Reich implied), but not the precise steric placement of the methyl groups. In other words, the receptor requires a cationic knob of limited size that may function in one of several possible ways. For example, it may (1) oppose a fixed anion and displace water and the mobile counterion of the resting state, (2) sterically trip the mechanism into the activated state, or (3) act as a pivot, like a ball and socket, to permit optimal docking of the acetyl end. Thus, bulky bi- or tricyclics, such as anatoxin *a* **(1)** and cytisine **(18),** both of which are secondary unmethylated amines, work. Second, the fit seems to depend on the cationic center and the whole of the acetyl group. One may imagine a diffuse, Coulombic fit of the cationic head and a fairly precise cavity for the whole acetyl group, such that if placement of this group is suboptimal, a strained, disfavored hydrogen bond forms.

[47] C. E. Spivak, T. M. Gund, R. F. Liang, and J. A. Waters, *Eur. J. Pharmacol.* **120,** 127 (1986).

[48] J. A. Waters, C. E. Spivak, M. Hermsmeier, J. S. Yadav, R. F. Liang, and T. M. Gund, *J. Med. Chem.* **31,** 545 (1988).

[49] C. E. Spivak, J. S. Yadav, W. C. Shang, M. Hermsmeier, and T. M. Gund, *J. Med. Chem.* **32,** 305 (1989).

FIG. 2. Stereo representations of the superposition of agonists **3, 7, 10, 15, 16,** and **27.** Color code: green **(3),** red **(7),** magenta **(10),** orange **(15),** yellow **(16),** blue **(27).** (a) agonists superimposed at N, carbonyl C, and O (Beers–Reich fit); (b) agonists superimposed at N, carbonyl C, and acetyl C.

FIG. 3. Stereo representation of the superposition of agonists **41, 3,** and **7** at N, carbonyl C, and O (Beers–Reich fit). Color code: red **(3),** blue **(7),** green **(41).**

FIG. 4. Stereo drawings of the electrostatic potentials on the van der Waals surface. Color code: purple (0–40 kcal), blue (40–60 kcal), turquoise (60–80 kcal), green (80–100 kcal), orange (100–120 kcal), brown (120–140 kcal), red (140–160 kcal).

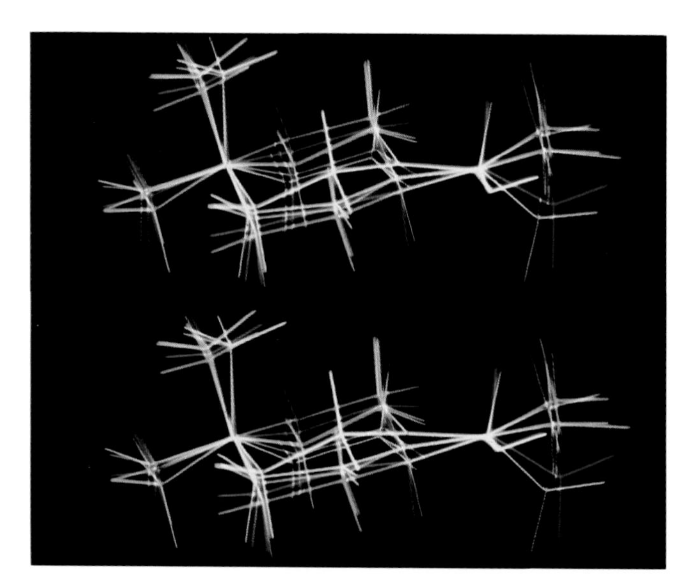

FIG. 2A.

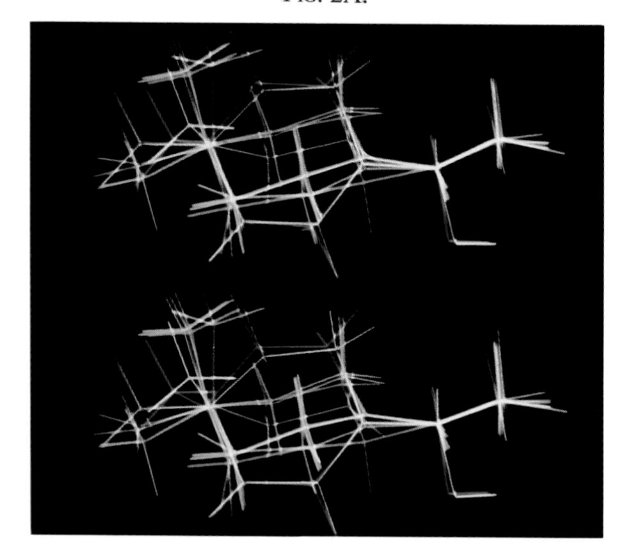

FIG. 2B.

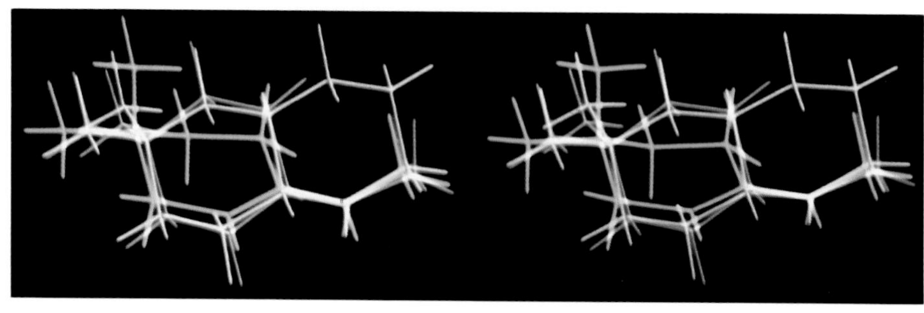

FIG. 3.

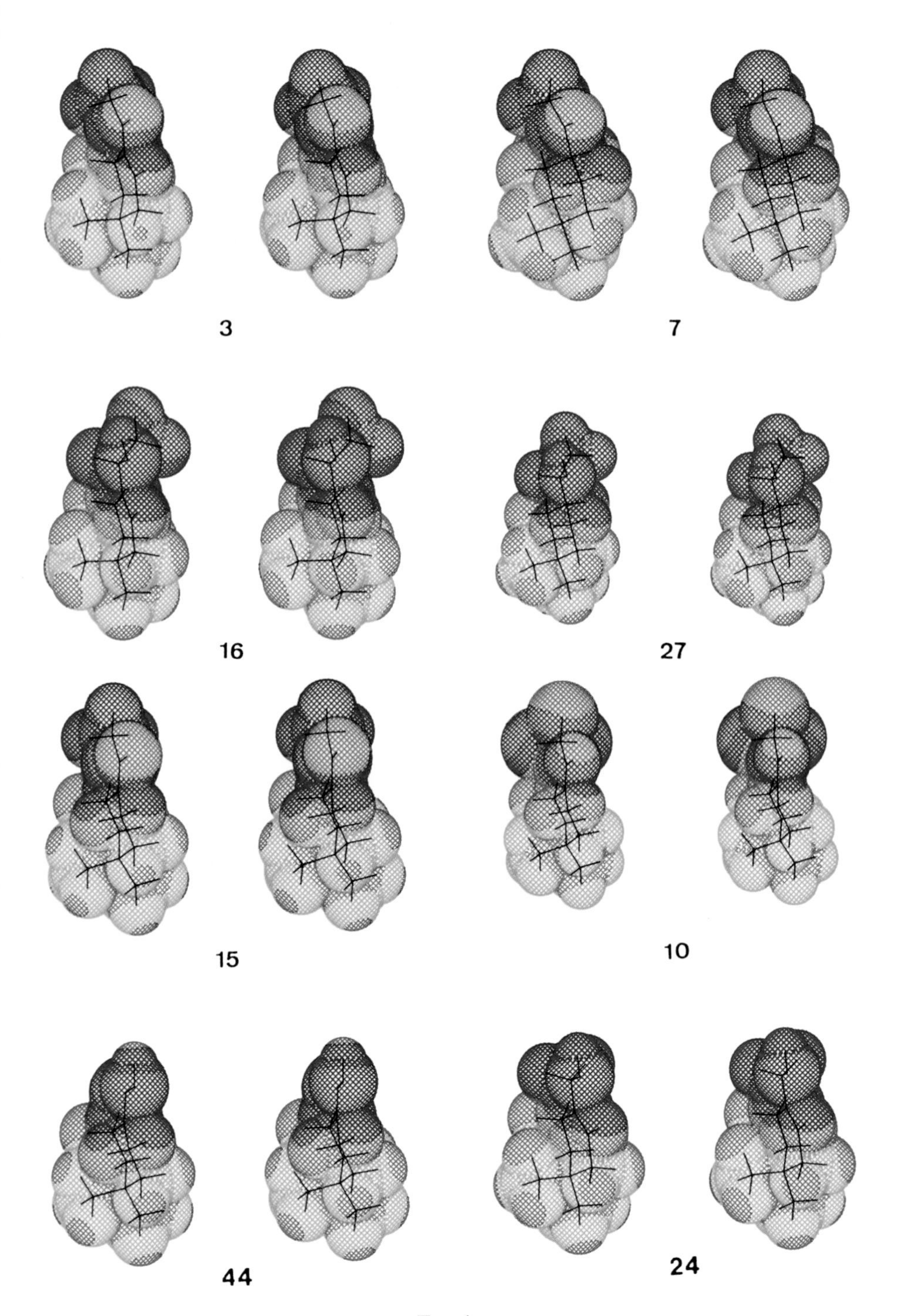

FIG. 4.

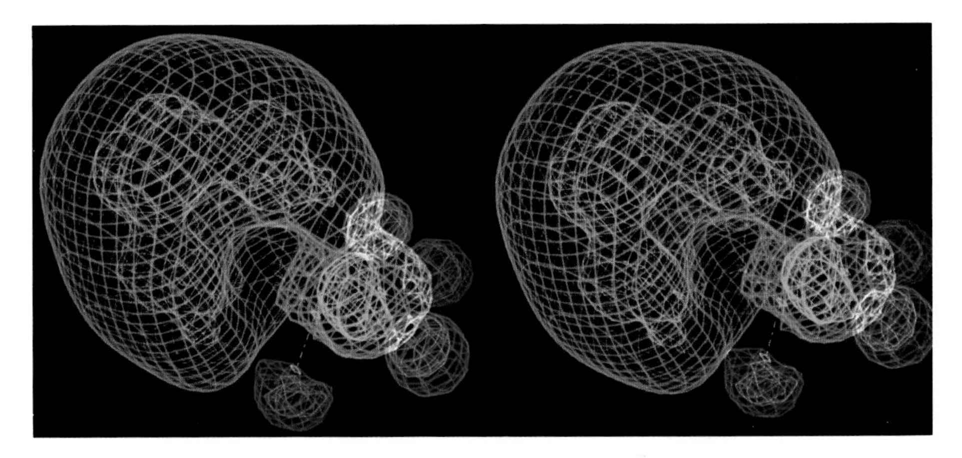

FIG. 5A.

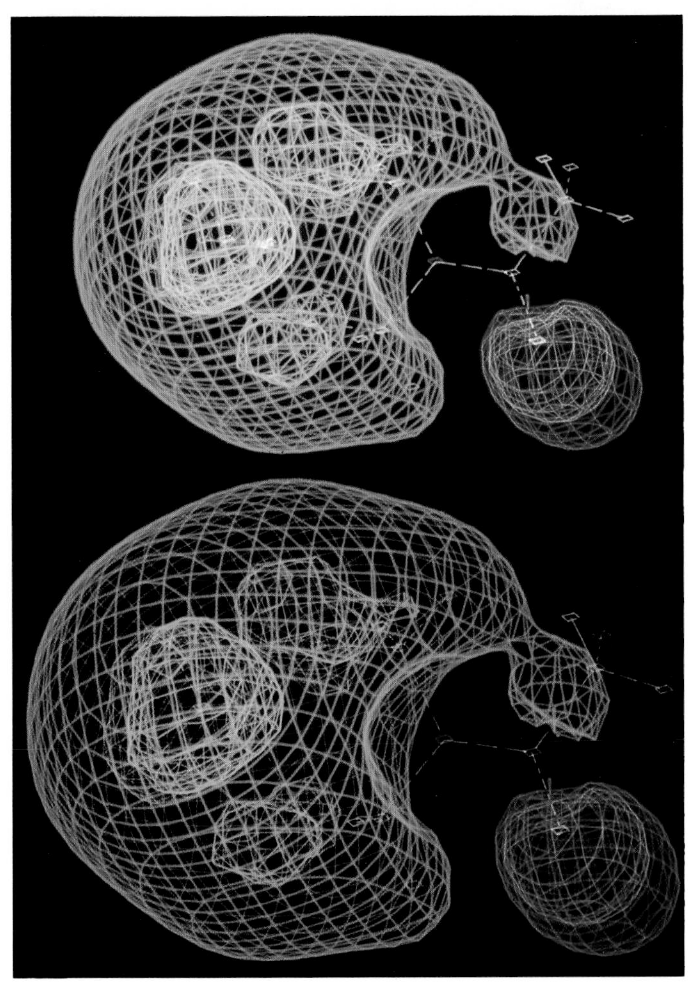

FIG. 5B.

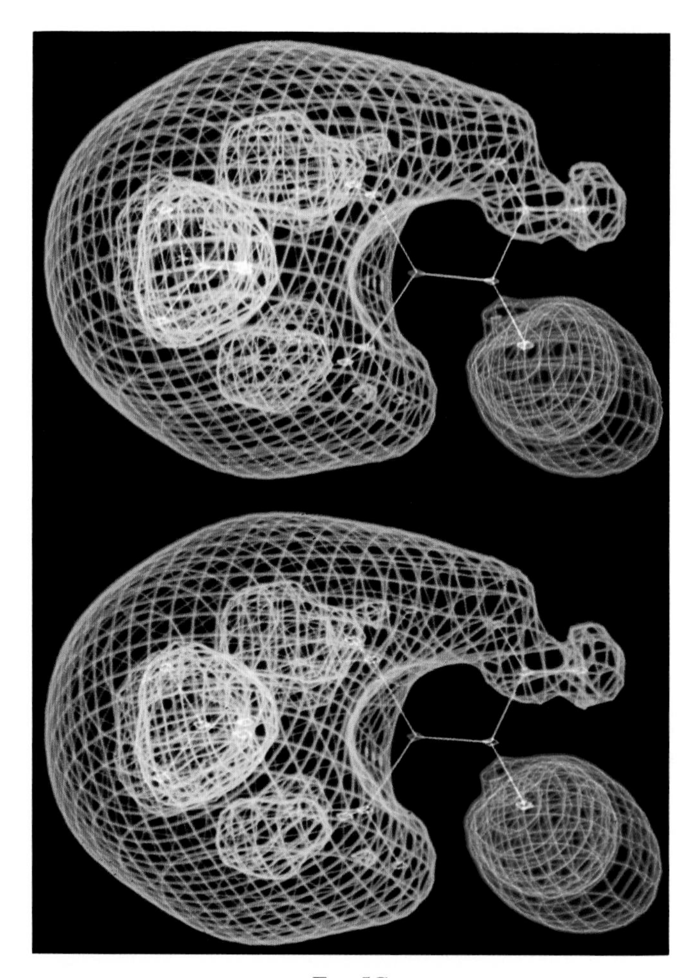

FIG. 5C.

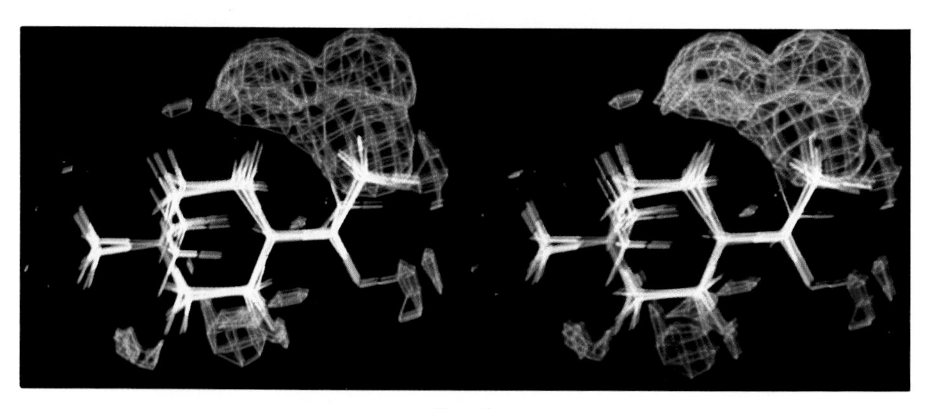

FIG. 6.

The overall shape of the agonist remains nearly unchanged if the acetyl methyl group is replaced by an NH_2; the steric requirements of the recognition site would still be met, and the Coulombic effect at the cationic head should be small, since the modification is distant. This was the rationale for synthesizing the carbamyl analogs to the potent agonists **3**, **7**, and **15**. These carbamyl analogs (**20**, **36**, and **30**, respectively) were compared pairwise to their acetyl counterparts. It was expected that the increased solvation, resulting from the amide NH_2 donating a hydrogen bond to water, would diminish the partitioning of the agonist into the recognition site, thereby diminishing activity of all the carbamyl analogs by a certain, constant factor. The results[49] can be summarized as follows: *All three* pairs, in their bioactive conformations, superimposed almost perfectly. However, to achieve this conformation, agonist **36** required 2.8 kcal/mol due to steric crowding of the NH_2 group with ring methylenes. Consequently, the potency ratio for **7** and **36** was the highest, 175. Agonists pairs **3**–**20** and **7**–**36** also deviated most in their electrostatic potential distributions; the potency ratio for **3** vs. **20** was 65. The pair **15**–**30** was closest electrostatically, and we attribute this potency ratio, 17, mostly to the greater solvation of the carbamyl agonists.

The Beers–Reich pharmacophore and the template (agonist **3**) both require the acetyl group to be nearly coplanar with the ring. We expected, therefore, that the isoquinolone (**41**), a trans-fused bicyclic analog of **7**, would be highly potent. Indeed, as shown in Fig. 3, it superimposes well on **7**, but its activity is only 0.0016 as great.[50] Because the electrostatic picture of **41** was similar to **7**, we returned to steric considerations. The explanation cannot lie in steric bulk or ring rigidity because cytisine (**18**) and pyridohomotropane (**2**) are at least as large and rigid as **41**, and both are far more potent. However, the rings of **2** and **18** corresponding to the acetyl group are unsaturated and planar, whereas the corresponding, *saturated*, ring of **41** puckers a little. This is the only major difference we can find, and we conclude that this puckering has encountered a steric obstacle in the recognition site. Therefore **41** finds use in constructing the receptor map (see Receptor Mapping, below).

[50] C. E. Spivak, J. A. Waters, J. S. Yadav, W. C. Shang, M. Hermsmeier, R. F. Liang, and T. Gund, *J. Mol. Graphics* **9**, 105 (1991).

FIG. 5. Stereo drawings of the electrostatic potentials as isopotential surfaces of agonists (a) **10**, (b) **15**, and (c) **30**. Color code: light blue (20 kcal), blue (40 kcal), red (100 kcal), yellow (160 kcal).

FIG. 6. Stereo receptor map fitted with active ligands.

Electrostatic Potentials

Figures 4 and 5 illustrate the electrostatic potential calculations on the van der Waals surfaces.[48] The energies are color coded according to ranges in electrostatic potentials. The most positive areas, associated with the highest electrostatic potential energy (140–160 kcal/mol) correspond to the vicinities of the cationic head. The least positive areas, associated with the lowest electrostatic potential energy (0–40 kcal/mol), correspond to the hydrogen-bonding areas. In general one observes that for the acetyl derivatives, activity diminishes with the electron density at the carbonyl region (Fig. 4). For the derivatives in which the acetyl group was substituted by electron-donating and electron-withdrawing groups, one observes that the electrostatic potential of the acetyl group correlates with potency.[48,49] Electron-withdrawing groups enhance activity[48] and electron-donating groups decrease it.[49] In electrostatic maps, such as shown in Fig. 5, one notices that electron-withdrawing groups induce a negative cap around the acetyl region and electron-donating groups induce a corresponding positive region.

Receptor Mapping

The active volume of the recognition site was calculated from the set of active agonists, which included all the isoarecolone methiodide analogs but the isoquinolone **(41)**, the amide of acetyl piperazine **(30)**, and the oxime **(44)**. The inactive volume was generated from the inactive derivatives, and the difference between the inactive and active volumes produced the receptor map depicted in Fig. 6 (shown with the active analogs fitted). The periphery indicates the repulsion with the receptor of the inactive analogs, of which **41** constitutes the main contributor. This map was used to fit cytisine and to calculate the amount of steric hindrance produced by this moderately weak agonist.[50] The map can be refined by adding more agonists into the calculation.

Shape Group Studies of Molecular Similarity

A mathematical method of comparing steric or electrostatic surfaces could be of value to quantify the visual differences apparent in receptor maps. The shape group method has been applied to finding the similarity of nicotinic agonist surfaces by a model based on group theory.

For the four agonists isoarecolone methiodide, dihydroisoarecolone methiodide, isoarecolol methiodide, and dihydroisoarecolol methiodide, the method suggests that hydrogenating the carbonyl double bond affects surface shape differently than does hydrogenating the carbon–carbon double bond.[51]

Conclusions

The binding of nicotinic agonists to the nicotinic receptor has been shown to depend on the ammonium head, the hydrophobic groups, and the hydrogen-bonding groups of a ligand. For the semirigid nicotinic analogs, computer modeling has offered ways of viewing the chemical structures that inspire new hypotheses, explanations, and predictions. Previously the notion that only the Coulombic and hydrogen-bonding sites determined potency is now augmented by acknowledging the importance of the acetyl methyl group. Whether there is another binding site that complements that region, or whether the acetyl methyl group is exerting its influence on the hydrogen bonding of the carbonyl or alcohol oxygen, remains to be explored. Nevertheless the factors that seem important in controlling activity are (1) ground state conformation, (2) closeness of fit between the ground state (or local minimum energy) conformation and the template, **3**, (3) high positive potential around the cationic head, (4) electron density around the acetyl group, where high electron density increases activity and low density decreases it, and (5) steric environment around the acetyl region.

More work is needed to assess the effect of solvent on conformational and electrostatic properties of the ligands. Once the receptor structure becomes known, modeling techniques will be useful in docking experiments to further elucidate the binding mechanism.

[51] G. A. Arteca, V. B. Jammed, P. G. Mezey, J. S. Yadav, M. A. Hermsmeier, and T. M. Gund, *J. Mol. Graphics* **6,** 45 (1988).

Section IV

Cross-Index to Prior Volumes

[33] Related Chapters Published in Previous Volumes of *Methods in Enzymology*

Related to Section I,A

Vol. 70 [1]. Basic Principles of Antigen–Antibody Reactions. E. A. Kabat.
Vol. 115 [28]. Domains in Proteins: Definitions, Location, and Structural Principles. J. Janin and C. Chothia.
Vol. 153 [32]. Engineering for Protein Secretion in Gram-Positive Bacteria. S. Chang.
Vol. 178 [32]. Production and Properties of Chimeric Antibody Molecules. S.-U. Shin and S. L. Morrison.
Vol. 178 [33]. Expression of Engineered Antibodies and Antibody Fragments in Microorganisms. M. Better and A. H. Horowitz.
Vol. 178 [34]. Expression of Functional Antibody Fv and Fab Fragments in *Escherichia coli.* A. Plückthun and A. Skerra.
Vol. 178 [35]. Recombinant Antibodies Possessing Novel Effector Functions. T. W. Love, M. S. Runge, E. Haber, and T. Quertermous.

Related to Section I,B

Vol. 70 [1]. Basic Principles of Antigen–Antibody Reactions. E. A. Kabat.
Vol. 121 [8]. Preparation of Monoclonal Antibodies to Preselected Protein Regions. M. Z. Atassi.
Vol. 121 [9]. The Auto-Anti-idiotypic Strategy for Preparing Monoclonal Antibodies to Receptor Combining Sites. W. L. Cleveland and B. F. Erlanger.
Vol. 176 [2]. Two-Dimensional Nuclear Magnetic Resonance Spectroscopy of Protein: An Overview. J. L. Markley.
Vol. 176 [5]. Heteronuclear Nuclear Magnetic Resonance Experiments for Studies of Protein Conformation. G. Wagner.
Vol. 177 [6]. Determination of Three-Dimensional Protein Structures in Solution by Nuclear Magnetic Resonance: an Overview. K. Wüthrich.
Vol. 177 [7]. Proton Nuclear Magnetic Resonance Assignments. V. J. Basus.
Vol. 177 [8]. Computer-Assisted Resonance Assignments. M. Billeter.
Vol. 177 [9]. Distance Geometry. I. D. Kuntz, J. F. Thomason, and C. M. Oshiro.
Vol. 177 [10]. Molecular Dynamics Simulation Techniques for Determination of Molecular Structures from Nuclear Magnetic Resonance Data. R. M. Scheek, W. F. van Gunsteren, and R. Kaptein.
Vol. 177 [11]. Heuristic Refinement Method for Determination of Solution Structure of Proteins from Nuclear Magnetic Resonance Data. R. B. Altman and O. Jardetzky.
Vol. 177 [16]. Ligand–Protein Interactions via Nuclear Magnetic Resonance of Quadrupolar Nuclei. C. R. Sanders II and M.-D. Tsai.
Vol. 177. Appendix: Computer Programs Related to Nuclear Magnetic Resonance: Availability, Summaries, and Critiques.
Vol. 178 [1]. Idiotypic Networks and Nature of Molecular Mimicry: An Overview. H. Köhler, S. Kaveri, T. Kieber-Emmons, W. J. W. Morrow, S. Müller, and S. Raychaudhuri.

Vol. 178 [38]. Use of Hydrophilicity Plotting Procedures to Identify Protein Antigenic Segments and Other Interaction Sites. T. P. Hopp.

Vol. 178 [39]. Computer Prediction of B-Cell Determinants from Protein Amino Acid Sequences Based on Incidence of β Turns. V. Krchňák, O. Mach, and A. Malý.

Vol. 178 [47]. Protein Footprinting Method for Studying Antigen–Antibody Interactions and Epitope Mapping. H. Sheshberadaran and L. G. Payne.

Related to Section I,C

Vol. 178 [36]. Production of Antibodies That Mimic Enzyme Catalytic Activity. A. Tramontano and D. Schloeder.

Vol. 178 [37]. Design of Catalytic Antibodies. S. J. Pollack, G. R. Nakayama, and P. G. Schultz.

Vol. 178 [40]. Identification of T-Cell Epitopes and Use in Construction of Synthetic Vaccines. J. L. Cornette, H. Margalit, C. DeLisi, and J. A. Berzofsky.

Vol. 178 [41]. Use of Synthetic T-Cell Epitopes as Immunogens to Induce Antibodies to Hepatitis B Components. D. R. Milich and G. B. Thornton.

Vol. 178 [42]. Peptide Vaccines Based on Enhanced Immunogenicity of Peptide Epitopes Presented with T-Cell Determinants or Hepatitis B Core Protein. M. J. Francis and B. E. Clarke.

Vol. 178 [43]. Production and Properties of Site-Specific Antibodies to Synthetic Peptide Antigens Related to Potential Cell Surface Receptor Sites for Rhinovirus. J. McCray and G. Werner.

Vol. 178 [44]. Custom-Designed Synthetic Peptide Immunoassays for Distinguishing HIV Type 1 and Type 2 Infections. J. W. Gnann, Jr., L. L. Smith, and M. B. A. Oldstone.

Vol. 178 [45]. Production and Use of Synthetic Peptide Antibodies to Map Region Associated with Sodium Channel Inactivation. H. Meiri, M. Sammar, and A. Schwartz.

Vol. 178 [46]. Multiple Antigenic Peptide Method for Producing Antipeptide Site-Specific Antibodies. D. N. Posnett and J. P. Tam.

Vol. 178 [22]. Development and Use of Antireceptor Antibodies to Study Interaction of Mammalian Reovirus Type 3 with Its Cell Surface Receptor. W. V. Williams, D. B. Weiner, and M. I. Greene.

Vol. 178 [23]. The Use of Anti-idiotypic Antibodies to Treat Lymphoid Tumors. G. T. Stevenson.

Vol. 178 [24]. Preparation and Use of Anti-idiotypic Antibodies Armed with Holotoxins or Hemitoxins in Treatment of B-Cell Neoplasms. S. Bridges, D. L. Longo, and R. J. Youle.

Vol. 178 [25]. Use of Anti-idiotypic Antibody–Drug Conjugates to Treat Experimental B Cell Tumors. E. Hurwitz and J. Haimovich.

Vol. 178 [26]. Monoclonal Anti-idiotypic Antibody Vaccines against Poliovirus, Canine Parvovirus, and Rabies Virus. G. F. Rimmelzwaan, E. J. Bunschoten, F. G. C. M. UytdeHaag, and A. D. M. E. Osterhaus.

Vol. 178 [27]. Use of Anti-idiotypic Antibodies in Approaches to the Immunoprophylaxis of Schistosomiasis. J.-M. Grzych, F. Roussel-Velge, and A. Capron.

Vol. 178 [28]. Selective Immunotoxins Prepared with Mutant Diphtheria Toxins Coupled to Monoclonal Antibodies. M. Colombatti, L. Dell'Archiprete, R. Rappuoli, and G. Tridente.

Vol. 178 [29]. Immune Suppression of Anti-DNA Antibody Production Using Anti-idiotypic Antibody–Neocarzinostatin Conjugates. T. Sasaki, Y. Koide, and K. Yoshigaga.

Vol. 178 [31]. Inhibition of Autoimmune Reactivity against the Acetylcholine Receptor with Idiotype-Specific Immunotoxins. K. A. Krolick.

Related to Section II

Vol. 138 [4]. Oligosaccharide Structure by Two-Dimensional Nuclear Magnetic Resonance Spectroscopy. T. A. W. Koerner, J. H. Prestegard, and R. K. Yu.

Vol. 152 [74]. *In Vivo* Footprinting of Specific Protein–DNA Interactions. P. D. Jackson and G. Felsenfeld.

Vol. 155 [33]. Hydroxyl Radical Footprinting: A High Resolution Method for Mapping Protein–DNA Contacts. T. D. Tullius, B. A. Dombroski, M. E. A. Churchill, and L. Kam.

Vol. 155 [36]. Computer Programs for Analyzing DNA and Protein Sequences. F. I. Lewitter and W. P. Rindone.

Vol. 170 [6]. Electrophoretic Analyses of Nucleosomes and Other Protein–DNA Complexes. S. Y. Huang and W. T. Garrard.

Vol. 180 [16]. Enzymatic Approaches to Probing of RNA Secondary and Tertiary Structure. G. Knapp.

Vol. 180 [17]. A Guide for Probing Native Small Nuclear RNA and Ribonucleoprotein Structures. A. Krol and P. Carbon.

Vol. 180 [18]. Phylogenetic Comparative Analysis of RNA Secondary Structure. B. D. James, G. J. Olsen, and N. R. Pace.

Vol. 180 [20]. Computer Prediction of RNA Structure. M. Zuker.

Vol. 180 [21]. RNA Pseudoknots: Structure, Detection, and Prediction. C. W. A. Pleij and L. Bosch.

Vol. 180 [29]. Analysis of Ultraviolet-Induced RNA–RNA Cross-Links: A Means for Probing RNA Structure–Function Relationships. A. D. Branch, B. J. Benenfeld, C. P. Paul, and H. D. Robertson.

Vol. 180 [34]. Genetic Methods for Identification and Characterization of RNA–RNA and RNA–Protein Interactions. R. Parker.

Vol. 183 [1]. GenBank: Current Status and Future Directions. C. Burks, D. Benton, *et al.*

Vol. 183 [2]. EMBL Data Library. P. Kahn and G. Cameron.

Vol. 183 [3]. Protein Sequence Database. W. C. Barker, D. G. George, and L. T. Hunt.

Vol. 183 [5]. Rapid and Sensitive Sequence Comparison with FASTP and FASTA. W. R. Pearson.

Vol. 183 [6]. Searching through Sequence Databases. R. F. Doolittle.

Vol. 183 [7]. Finding Protein Similarities with Nucleotide Sequence Databases. S. Henikoff, J. C. Wallace, and J. P. Brown.

Vol. 183 [8]. Use of Homology Domains in Sequence Similarity Detection. C. B. Lawrence.

Vol. 183 [9]. Profile Analysis. M. Gribskov, R. Luthy, and D. Eisenberg.

Vol. 183 [17]. Predicting Optimal and Suboptimal Secondary Structure for RNA. J. A. Jaeger, D. H. Turner, and M. Zuker.

Vol. 183 [18]. Detecting Pseudoknots and Other Local Base-Pairing Structures in RNA Sequences. H. M. Martinez.

Vol. 183 [19]. Computer Modeling and Display of RNA Secondary and Tertiary Structures. D. Gautheret, F. Major, and R. Cedergren.

Vol. 183 [12]. Searching for Patterns in Protein and Nucleic Acid Sequences. R. Staden.

Vol. 183 [13]. Consensus Patterns in DNA. G. D. Stormo.

Vol. 183 [14]. Consensus Sequences for DNA and Protein Sequence Alignment. M. S. Waterman and R. Jones.

Vol. 186 [56]. Footprinting Protein–DNA Complexes with γ-Rays. J. J. Hayes, L. Kam, and T. D. Tullius.

Vol. 195 [19]. Synthetic Peptide Antisera with Determined Specificity for G Protein α or β Subunits. S. M. Mumby and A. G. Gilman.

Vol. 198 [19]. Assessment of Biological Activity of Synthetic Fragments of Transforming Growth Factor α. G. Schultz and D. Twardzik.

Related to Section III

Vol. 124 [1]. Computationally Directed Biorational Drug Design of Peptides. F. A. Momany and H. Chuman.

Vol. 178 [16]. Production and Characterization of Antimorphine Anti-idiotypic Antibodies and Their Antiopiate Receptor Activity. J. A. Glasel.

Vol. 178 [17]. Production and Characterization of Anti-idiotypic Antiopioid Receptor Antibodies. C. Gramsch, R. Schultz, S. Kosin, A. H. S. Hassan, and A. Herz.

C

D

E

F

G

H

I

J

K

L

M

N

O

P

Q

R

S

W

X

Y

Z

Subject Index

A

B

C

E

F

G

H

I

J

K

L

M

O

P

Q

R

T

U

V

W

X

Z